Activated Charcoal in Medical Applications

DAVID O. COONEY
University of Wyoming
Laramie, Wyoming

CRC Press
Taylor & Francis Group
Boca Raton   London   New York

CRC Press is an imprint of the
Taylor & Francis Group, an **informa** business

CRC Press
Taylor & Francis Group
6000 Broken Sound Parkway NW, Suite 300
Boca Raton, FL 33487-2742

First issued in paperback 2019

© 2009 by Taylor & Francis Group, LLC
CRC Press is an imprint of Taylor & Francis Group, an Informa business

No claim to original U.S. Government works

ISBN-13: 978-0-8247-9300-5 (hbk)
ISBN-13: 978-0-367-40191-7 (pbk)

This book contains information obtained from authentic and highly regarded sources. While all reasonable efforts have been made to publish reliable data and information, neither the author[s] nor the publisher can accept any legal responsibility or liability for any errors or omissions that may be made. The publishers wish to make clear that any views or opinions expressed in this book by individual editors, authors or contributors are personal to them and do not necessarily reflect the views/opinions of the publishers. The information or guidance contained in this book is intended for use by medical, scientific or health-care professionals and is provided strictly as a supplement to the medical or other professional's own judgement, their knowledge of the patient's medical history, relevant manufacturer's instructions and the appropriate best practice guidelines. Because of the rapid advances in medical science, any information or advice on dosages, procedures or diagnoses should be independently verified. The reader is strongly urged to consult the relevant national drug formulary and the drug companies' and device or material manufacturers' printed instructions, and their websites, before administering or utilizing any of the drugs, devices or materials mentioned in this book. This book does not indicate whether a particular treatment is appropriate or suitable for a particular individual. Ultimately it is the sole responsibility of the medical professional to make his or her own professional judgements, so as to advise and treat patients appropriately. The authors and publishers have also attempted to trace the copyright holders of all material reproduced in this publication and apologize to copyright holders if permission to publish in this form has not been obtained. If any copyright material has not been acknowledged please write and let us know so we may rectify in any future reprint.

Except as permitted under U.S. Copyright Law, no part of this book may be reprinted, reproduced, transmitted, or utilized in any form by any electronic, mechanical, or other means, now known or hereafter invented, including photocopying, microfilming, and recording, or in any information storage or retrieval system, without written permission from the publishers.

For permission to photocopy or use material electronically from this work, please access www.copyright.com (http://www.copyright.com/) or contact the Copyright Clearance Center, Inc. (CCC), 222 Rosewood Drive, Danvers, MA 01923, 978-750-8400. CCC is a not-for-profit organization that provides licenses and registration for a variety of users. For organizations that have been granted a photocopy license by the CCC, a separate system of payment has been arranged.

Trademark Notice: Product or corporate names may be trademarks or registered trademarks, and are used only for identification and explanation without intent to infringe.

Library of Congress Cataloging-in-Publication Data

Cooney, David O.
    Activated charcoal in medical applications / David O. Cooney.
        p. ; cm.
    Includes bibliographical references and index.
    ISBN-13: 978-0-8247-9300-5 (hardcover : alk. paper)
    ISBN-10: 0-8247-9300-5 (hardcover : alk. paper)
    1. Carbon, Activated--Therapeutic use. 2. Poisoning--Treatment.
    I. Title. [DNLM: 1. Charcoal--therapeutic use. 2. Charcoal--pharmacology.
3. Poisoning--drug therapy. QV 601 C775a 1995]

RM666.C34C663 1995
615.9'08--dc20   94-40860

**Visit the Taylor & Francis Web site at
http://www.taylorandfrancis.com**

**and the CRC Press Web site at
http://www.crcpress.com**

To Eric and Jon

FOREWORD

Because there are few specific antagonists for ingested toxic substances, medical treatment consists primarily of supportive measures and attempts to decrease systemic absorption by a variety of gastrointestinal decontamination measures. Over the past several years it has become increasingly clear that, at least for drug ingestions treated in emergency departments, stomach-emptying techniques such as induced emesis and gastric aspiration/lavage result in little benefit in terms of improved clinical outcome. These measures also result in delayed administration of activated charcoal, whose clinical efficacy has been well demonstrated. Current clinical practice for most cases of drug ingestion treated in the hospital is to forego gastric emptying in favor of early administration of activated charcoal. Repeated doses of activated charcoal constitute an active therapy for removing a number of systemically absorbed substances.

In this extensive and comprehensive review, Professor Cooney explores the history of activated charcoal in medical practice, describes its chemistry and pharmacology, details its past and current clinical applications for poisoning treatment, and elucidates the need for studies of its potential clinical application in gastroenterology and other medical fields. Health care professionals who treat poisoned patients, as well as pharmacology and toxicology researchers, will find a wealth of information about activated charcoal in this volume, which lays a firm foundation for the rational use of activated charcoal in medical therapy.

Barry H. Rumack, M.D.
Director Emeritus
Rocky Mountain Poison and Drug Center
Clinical Professor of Pediatrics
University of Colorado Health Sciences Center
Denver, Colorado

Alan H. Hall, M.D., F.A.C.E.P.
Senior Consultant
Rocky Mountain Poison and Drug Center
Clinical Assistant Professor of Preventive Medicine
   and Biometrics
University of Colorado Health Sciences Center
Denver, Colorado

PREFACE

Activated charcoal, long known to the ancients as a substance of therapeutic value in a variety of maladies, has been relatively recently "rediscovered" to be of great general use as an oral antidote for drug overdoses and poisonings. Over the past several decades, orally administered activated charcoal has been proven to be highly effective in reducing the systemic absorption of analgesics and antipyretics, sedatives and hypnotics, alkaloids, tricyclic antidepressants, cardiac glycosides, and a wide variety of other kinds of drugs and chemicals. The efficacy of activated charcoal in binding drugs and poisons, especially nonpolar or lipophilic ones, has been demonstrated emphatically by a host of in vitro studies and in vivo tests with humans, dogs, rabbits, pigs, cattle, sheep, rats, and mice.

It is rather surprising that prior to about the mid-1960s activated charcoal was essentially neglected, considering that its therapeutic effects had already been extensively reported in the medical literature. This neglect perhaps resulted from the fact that many medical personnel were never really aware of how and why activated charcoal really works (i.e., the physical chemistry of adsorption by charcoal in various biological media). Such a lack of knowledge would explain the origin of the absurd notion, prevalent in the 1950s, that burnt toast be used in place of activated charcoal. "Universal antidote" is another example of an antidote that, on a chemical basis, can be shown to be decidedly inferior to plain activated charcoal as a general antidote. The promotion and acceptance of such ill-founded notions undoubtedly did considerable harm to the reputation of activated charcoal and deterred people from using it.

Fortunately, the virtues of activated charcoal have finally been recognized by large numbers of people in the medical community. Nevertheless, very few private homes today have activated charcoal on hand as a ready antidote. One reason may be that charcoal in water makes a rather gritty, tasteless mixture that is difficult to swallow. With the relatively recent development of ways in which to add lubricants and flavors to charcoal slurries without interfering with their efficacy, palatable formulations are now possible. Even "flavors" such as chocolate syrup and ice cream, which do decrease the adsorption capacity of charcoal to some extent, are useful, since one can overcome this drawback by simply giving a larger charcoal dose. However, as of this writing, no palatable formulations are being commercially marketed to the general public. (Formulations containing the sweetener/cathartic sorbitol *are* marketed commercially, but only to hospitals.) Again we must speculate: perhaps the sale of such formulations to the general public is not a financially attractive venture, considering that most people might never need to use such an antidote. (Thus, repeat sales would be small.) And so there continues to be a wide gap between the appreciation of charcoal by medical researchers and the acceptance and utilization of charcoal by the general public.

This book represents an attempt to gather in one volume most of what has been reported to date on the use of activated charcoal for medical purposes. A successor to *Activated Charcoal: Antidotal and Other Medical Uses* (Cooney, D. O., New York: Marcel Dekker, 1980), it incorporates nearly all significant studies on the use of activated charcoal for antidotal purposes that have appeared between about 1979 and early 1994. In addition to reviewing most of the research on the in vitro and in vivo adsorption of a wide variety of drugs and poisons over the past several decades, I review what was known about charcoal by our ancestors, beginning at about 1500 B.C.

The coverage has been expanded to include a wide variety of medical uses of activated charcoal other than as an antidote for drug overdoses and poisonings, e.g., the use of activated charcoal for treating pruritus, neonatal jaundice, porphyria, hyperlipidemia, and wounds. A wide range of studies involving activated charcoal as an adsorbent for viruses, bacteria, vitamins, hormones, endogenous and exogenous toxins, etc., are also reviewed.

Some rather intriguing uses of activated charcoal in other biological contexts, e.g., to promote plant growth, in assays for hormones, and even in the removal of "congeners" from whiskeys, have been included to demonstrate the breadth of activated charcoal applications.

New chapters have been added that will provide a basis for understanding the large numbers of clinical studies discussed. These include chapters on: general methods for treating overdoses and poisonings, basic gastrointestinal tract physiology, the design of clinical studies and the statistical analysis of data, and basic pharmacokinetic modeling.

New chapters also have been added on various hazards of using antidotal charcoal, on currently available charcoal formulations, and on interactions between ipecac/cathartics and charcoal. There is an expanded discussion of multiple-dose charcoal therapy. Detailed comparisons are given of the effects of induced emesis alone, gastric lavage alone, and other types of treatments versus activated charcoal alone.

In the writing of this book, several policies have been established that should be mentioned and explained:

1. Activated charcoal is referred to as an "antidote" throughout this volume. To many readers, the word *antidote* connotes a specific chemical agent that has the ability to inhibit or block the mechanisms of toxicity of drugs, poisons, or toxins (e.g., diphenhydramine is an antidote for bee-sting toxin). Activated charcoal is thus not an antidote in this sense but a simple nonspecific adsorbent. However, dictionaries generally define *antidote* as any remedy or agent used to counteract the effects of a poison. In this sense, activated charcoal is an antidote. Additionally, reference to activated charcoal as an antidote by the medical community is so common and so established that there is more than sufficient justification for referring to it as such.

2. Reference is made thousands of times to *charcoal*. Including the adjective *activated* in every instance seems unnecessary, and furthermore would increase the text by roughly 2%! Thus, except in a brief part of Chapter 2 involving a discussion of

early charcoals (those made prior to about 1900), the term *charcoal*, if used without *activated*, should be taken to mean "activated charcoal."

3. Many commercially marketed brands of charcoals (e.g., Norit), charcoal formulations (e.g., InstaChar), and drugs (e.g., Theo-Dur) are mentioned. It seems unnecessary to superscript these names with the trademark symbol (™) or the registered trademark symbol (®) in every instance, or even just at the first reference to each product (this latter policy would lead to sections of text where some products are "tagged" and others are not, which would appear awkward). Thus, these symbols are omitted. The reader will have no difficulty discerning product names, as they always carry initial capital letters and are quite obviously brand names.

4. Researchers often use many different units for concentrations, such as mg/L, µg/mL, ng/L, mg%, mg/100 mL, mg/dL, and g/L. Some of these are equivalent (e.g., mg/L and µg/mL, mg/100 mL and mg/dL). To provide for a degree of consistency in this volume, most concentrations are expressed on a "per liter" basis. Thus, the reader will typically find mg/L, µg/L, or ng/L for concentrations. Exceptions are cases in which amounts per mL or per 100 mL (or "per dL") seem more appropriate.

5. It is assumed that the reader is reasonably familiar with medical terminology common to the field involved (e.g., anatomical terms, names for various medical conditions, and parameters such as $t_{1/2}$, AUC, LD_{50}, etc.). However, because this volume may be read by persons having significantly different backgrounds (e.g., clinical MDs, emergency medical technicians, pharmacists), the text often explains medical terms and concepts in language approaching that of a pure layman, particularly the first time these terms or concepts are encountered. Readers who are familiar with more formal scientific terminology are implored to be tolerant of this, and not take offense. Such writing is not intended to insult the reader. One very important specific point regarding terminology that must continually be kept in mind is the great difference in the two similar nouns *adsorption* and *absorption*. They are used very frequently in this work.

6. There is no consistent pattern among researchers when reference is made to the adsorption of a substance *by*, *to*, *on*, or *onto* charcoal. In this volume, all four prepositions are used, with no consistent pattern, and should be construed to mean the same thing.

7. The medical community's practice of using the adjectives *maximal* and *minimal* rather than *maximum* and *minimum* has generally been followed. With respect to the practice of referring to certain drugs in the sodium salt form as *phenobarbital sodium* rather than *sodium phenobarbital*, usage in the medical community is not consistent (chemists generally use the latter form). Thus, both forms are used in this volume. The choice of form often follows that used in the journal paper being discussed, and at other times is simply arbitrary.

8. While this volume is concerned primarily with activated charcoal, it should be recognized that other sorbents are sometimes used for similar purposes (*sorbent* is a general term for substances that bind molecules by physical adsorption, chemical adsorption, ion exchange, absorption, etc.). Specifically, anion- and cation-exchange resins and clays have a substantial history of use in treating intoxications. Many of the in vitro and in vivo studies discussed in this book have employed one or more of these alternative sorbents in addition to activated charcoal. In such cases, the results obtained with these other sorbents are generally presented and compared to

those obtained with charcoal, since there are situations in which these other sorbents are more effective than charcoal, and readers may wish to learn about them. Also, one chapter (Chapter 22) is devoted to studies in which these alternative adsorbents alone have been used.

Thanks are due to all those researchers who have cited the first edition of this work. It was those citations that provided the inspiration for this new volume. As with any work, the coverage and interpretations offered here reflect the personal decisions of the writer. The field being addressed has now grown to the point where it is clearly impossible to cover everything, or to cover many excellent research studies in the depth they deserve.

David O. Cooney

CONTENTS

Contents

Contents

Contents

1

Introduction

To indicate the potential impact that activated charcoal could have in treating drug overdoses and poisonings, just in the USA alone, we begin this chapter with some statistical data which demonstrate the tremendous number of incidents of intoxication by ingestion of drugs and poisons in the USA in only one selected year. We then present some data which indicate how infrequently charcoal is actually used as an initial treatment measure. On a more hopeful note, we end the chapter with a discussion of the growing awareness of the benefits of charcoal, as evidenced by a large number of recent review articles in the medical literature.

I. DATA ON POISONING INCIDENTS IN THE USA

Acute incidents involving the accidental or intentional ingestion of drugs, poisons, and various household chemicals continue to be a serious problem. In 1991 more than 1.84 million cases of exposure to toxic substances were reported to the American Association of Poison Control Centers. As Krenzelok and Dunmire (1992) have mentioned, estimates for the USA indicate that over 18,000 cases of poisoning per year per million people occur, and this would translate to a total of about 4.5 million cases per year (thus, only about 40% of poisoning cases are reported to poison control centers).

The nature of poisoning instances in the USA is indicated, in part, by data collected from the American Association of Poison Control Centers (AAPCC). The 1991 annual report (see Litovitz et al., 1992) reports data compiled from 73 participating centers, serving a population of 200.7 million persons (about 80% of the US population).

It should be pointed out that these data reflect only telephone calls made to Poison Control Centers, and do not reflect cases where persons were found with some sort of intoxication and were transported to emergency departments of hospitals (EDs) or other medical facilities without any call being made to a Poison Control Center. In that sense, the data reflect primarily exposures of moderately low severity (the fatality rate in the data set is only 0.042%) in home environments.

Table 1.1 Reason for Exposure to Toxic Substances

	Reason	Percent
Accidental	General	78.5
	Misuse[a]	5.4
	Occupational	2.0
	Environmental	1.4
	Unknown	0.1
	Total	87.4
Intentional	Suicidal	7.2
	Misuse[b]	1.5
	Abuse[c]	1.0
	Unknown	0.9
	Total	10.6
Adverse reaction	Drug	1.0
	Food	0.4
	Other	0.2
	Total	1.6
Unknown		0.4
Total		100.0

[a] Accidental overdose or misreading of the label, for example.
[b] Intentional excessive overdose to obtain a faster or stronger "therapeutic" effect.
[c] Improper use of a substance in which the person was seeking a psychotropic effect.
Adapted from Litovitz et al. (1992). Reproduced by permission of the W. B. Saunders Company and the American Association of Poison Control Centers.

In 1991, there were 1,838,000 exposures to toxic substances reported to the 73 AAPCC centers, or roughly 9.2 per 1000 of population. Children younger than 3 years of age were involved in 45.6% of the cases, while 59.9% of the cases occurred in children younger than 6 years. Overall, the data show an equal breakdown in terms of male/female (49.5% each, with 1% unknown). However, males predominate among victims younger than 13 years, and females predominate in the teenage and adult populations.

Almost all exposures (92.0%) occurred in the home. At the time of contact with the AAPCC center, 63.1% of the victims were asymptomatic. In 90.7% of the cases, only a single substance was involved. The primary routes of exposure were ingestion (76.3%), dermal (7.4%), ophthalmic (6.2%), inhalation (5.6%), and bites/stings (3.7%). All other routes were 0.3% or less. The reasons for exposure are given in Table 1.1. One can see that the vast majority of cases (87.4%) were accidental in nature.

Of the 1,838,000 exposures in the 1991 data base, only 764 fatalities were reported (405 males, 358 females, 1 unknown), or a fatality rate of only 0.042%. For the fatalities, the primary routes were ingestion (75.7%) and inhalation (14.1%). The age distribution data for the fatalities are summarized in Table 1.2. Since many fatalities involved adults, the suggestion is that many of these were due to intentional suicides.

Table 1.2 Age Distribution of Fatalities

Age interval	Percent
< 6 years old	5.8
6–12	0.5
13–19	8.4
20–29	15.6
30–39	23.4
40–49	14.9
50–59	8.8
60–69	7.2
70–79	6.9
80–89	4.3
90–99	1.4
Unknown	2.8
Total	100.0

Adapted from Litovitz et al. (1992). Reproduced by permission of the W. B. Saunders Company and the American Association of Poison Control Centers.

II. INITIAL APPROACHES TO THE TREATMENT OF POISONING

Several initial approaches to the emergency treatment of drug overdoses and poisonings, in addition to supportive measures, are commonly used: ipecac-induced emesis, gastric lavage, administration of cathartics, single- and multiple-dose activated charcoal, combinations of therapies, etc.

In the Litovitz 1991 survey, data are given on initial therapies which were employed in over 1,440,000 patients. Table 1.3 is a summary of some of the data. The present tabulation does not include measures to enhance elimination (diuresis, dialysis, hemo-

Table 1.3 Initial Therapies Employed

Therapy	Percent
Dilution	45.3
Irrigation/washing	25.4
Activated charcoal	9.0
Cathartic	7.4
Ipecac syrup	6.6
Gastric lavage	4.1
Other emetics	0.3
Total	98.1

Adapted from Litovitz et al. (1992). Reproduced by permission of the W. B. Saunders Company and the American Association of Poison Control Centers.

perfusion, exchange transfusion, urine alkalinization/acidification), which accounted for a total of 0.6% of the cases, nor specific antidote administration (which accounted for a total of 1.3% of the cases) in the Litovitz summary.

The surprising thing about this table is the high incidence of dilution, which is contraindicated except in cases of mineral acid or base ingestion (see Chapter 7 for a more detailed discussion of this point) and the small percentages for charcoal and ipecac.

The agents comprising the "other emetic" category are unknown. They could be liquid dishwashing detergents and/or various types of salt solutions. Because syrup of ipecac is relatively safe and can be administered easily in the home, many people have recommended it as the only emetic of choice. (Although apomorphine is faster, it is difficult to prepare and has to be given parenterally. Also, it can cause significant central nervous system and respiratory depression. Thus, apomorphine has not been in common clinical use in human medicine for 10–20 years.)

Ipecac syrup has a reasonably long latency period. In one study (Robertson, 1962) it was found that in 250 patients given 20 mL ipecac syrup the average delay before emesis was 19 min. Although this may seem to be a long time, even more time is usually needed to carry out gastric lavage. For quickly absorbed drugs like barbiturates, neither emesis nor gastric lavage may be rapid enough to be of significant value.

Other emetics that have been recommended in the past are now considered to be dangerous. For example, several deaths have occurred with table salt solutions. Other emetics (e.g., copper sulfate and potassium antimony tartrate) are relatively toxic; therefore, if emesis fails to occur, poisoning from such agents can occur.

It is common practice in many emergency departments to perform gastric lavage on patients, particularly if charcoal is to be given afterwards (if one employs induced emesis as the initial therapy, there is considerable likelihood that charcoal given afterwards will be vomited and thus useless). However, because some of the drug or poison is sequestered in the recesses of the stomach and is thus not accessible to the lavage tube, lavage is often not very efficient. The efficiencies of both emesis and gastric lavage have been reported to be on the order of 40% or less. One careful study (Corby et al., 1968) has shown, for example, an average recovery of stomach contents of only 27% (range 0–77%) from ipecac-induced emesis. Additionally, induced emesis and gastric lavage involve definite risks and potential complications.

Oral dilution with large volumes of water has also been recommended as a first aid treatment for overdoses. It was believed that such dilution would retard the rate of absorption of the toxic material. However, Henderson et al. (1966) and Borowitz et al. (1971) have clearly shown that oral dilution often greatly increases the rate of absorption, the reason for this being that dilution promotes gastric emptying and hence exposes the drug or poison to a larger absorptive surface. Thus, as Ferguson (1964) has observed, oral dilution increases the lethality of drugs (lowers their LD_{50} values). Thus, except for ingested mineral acids and bases, for which dilution with a reasonable amount (200–300 mL) of plain water is of benefit, oral dilution should not be used.

Many other methods of treating drug overdose and poison victims exist, and they will be discussed and evaluated in later chapters. They include: whole-bowel irrigation, forced diuresis, urine acidification or alkalinization, hemoperfusion, hemodialysis, and exchange transfusion. In general, such techniques are either not very effective or are complicated, or both. They are useful only in very limited and specific cases.

In summary, it appears that the time delay required to promote emesis or carry out lavage, combined with the unpredictable and low efficiencies of these two methods, and attendant potential complications, makes them unsatisfactory in most cases, as are the other specialized techniques just enumerated. A better approach, which has been proved over and over to be quite effective, involves the use of orally administered activated charcoal.

Activated charcoal has the ability to adsorb a wide variety of drugs or poisons. It is nontoxic, indefinitely potent if kept in a closed container, and can be administered without undue difficulty. Studies have shown that slurries of powdered activated charcoal in water are fairly well accepted by children; moreover, if palatable formulations containing lubricants (such as carboxymethylcellulose) along with some flavoring agent (e.g., sorbitol, sucrose, saccharin, etc.) were to be made widely available to the general public, a reasonably high level of charcoal use in the home could probably be achieved in the future.

Activated charcoal has been proved effective in vitro and in vivo with all kinds of subjects (humans, dogs, rats, mice, rabbits, cattle, sheep, pigs) for many different intoxications. Charcoal quickly adsorbs most organic chemicals and many inorganic ones in significant amounts. If enough charcoal is used, the extent of desorption due to dissociation of the drug/charcoal complex further down in the digestive tract is inconsequential. Charcoal is a poor adsorbent only for inorganic substances that are normally in dissociated form in aqueous solution, such as mineral acids, alkalis, and salts (e.g., $NaCl$, $FeSO_4$, etc.).

Some of the chemical compounds effectively adsorbed by activated charcoal are listed in Table 1.4. It should be mentioned that cyanide falls into the "poorly-adsorbed" class because the maximal amount that can be bound by charcoal is on the order of 35 mg/g (Andersen, 1946), as compared to values on the order of 300–500 mg/g (or higher) for "well-adsorbed" drugs. However, quite often the amount of cyanide ingested is small (50 mg is toxic, and 200–300 mg is fatal) and thus the use of charcoal doses in the range of 20–100 g can be reasonably effective in cyanide intoxications, but only if the charcoal is given very quickly.

III. THE GROWING USE OF ACTIVATED CHARCOAL

Several editorials and short commentary papers have spoken of "activated charcoal rediscovered" (Anonymous, 1972), the "ascendency of the black bottle" (Greensher et al., 1987), of "activated charcoal reborn" (Spyker, 1985), and of "new roles for activated charcoal" (Spector and Park, 1986). The consensus now emerging among clinical physicians is that the best way of handling overdoses consists of the immediate administration of relatively large (50–100 g) doses of powdered activated charcoal, sometimes followed every few hours by additional smaller (25–50 g) doses. The charcoal, as an aqueous slurry, can be taken in the home. However, few homes have such mixtures on hand. Thus, it is usually given in a hospital environment. If the patient is not conscious or cooperative, the activated charcoal slurry must be administered using a nasogastric or orogastric tube.

Many recent studies have concerned whether the concomitant use of osmotic cathartics (sorbitol) or saline cathartics with oral charcoal provides any added benefit. In subsequent chapters, we will review these studies. The general conclusions from these investigations are that sorbitol, in limited amounts, can sometimes enhance the in vivo efficacy of charcoal but that saline cathartics are rarely of significant benefit. Of course, cathartics can be hazardous if used in excess.

Table 1.4 Adsorption of Drugs and Other Substances in Vitro

Well adsorbed	Moderately well adsorbed	Poorly adsorbed
Aflatoxins	Aspirin	Cyanide
Amphetamines	Other salicylates	Ethanol
Antidepressants	DDT	Ethylene glycol
Antiepileptics	Disopyramide	Iron
Antihistamines	Kerosene, benzene, and	Lithium
Atropine	dichloroethane	Methanol
Barbiturates	Malathion	Mineral acids
Benzodiazepines	Mexiletine	and alkalis
Beta-blocking agents	Nonsteroidal anti-inflammatory	
Chloroquine and primaquine	drugs (e.g., tolfenamic acid)	
Cimetidine	Paracetamol (acetaminophen)	
Dapsone	Polychlorinated biphenyls	
Dextropropoxyphene and	Phenol	
other opioids	Syrup of ipecac constituents	
Digitalis glycosides	Tolbutamide, chlorpropamide,	
Ergot alkaloids	carbutamide, tolazamide	
Furosemide		
Glibenclamide and glipizide		
Glutethimide		
Indomethacin		
Meprobamate		
Nefopam		
Phenothiazines		
Phenylbutazone		
Phenylpropanolamine		
Piroxicam		
Quinidine and quinine		
Strychnine		
Tetracyclines		
Theophylline		

From Neuvonen and Olkkola (1988). Reproduced by permission of ADIS International.

An impressive testimonial to the benefits of activated charcoal has been stated by Hayden and Comstock (1975). They recount their own clinical experience with treating poisoned and overdosed patients in intensive care wards. From 500 patients not treated with charcoal and 1000 who were, Hayden and Comstock found that administering activated charcoal after gastric lavage effectively reduced further drug absorption, as indicated by a leveling off of central nervous system (CNS) depression within about 2 hr. When charcoal was not used, progressive CNS depression occurred for up to 48 hr or longer. They estimate that the use of charcoal could reduce the average stay in an intensive care unit of 3 or 4 days by at least 1 day. This alone could save hundreds of millions of dollars in the value of care avoided, not to mention the great benefits of reductions in morbidity and mortality which can also be expected.

IV. GENERAL REVIEW PAPERS

In the following chapters, we will explore in detail the nature and properties of activated charcoal, and the myriad of past and present studies relating to the in vitro and in vivo adsorption powers of this truly extraordinary material. Some of the topics we shall consider have been discussed to some extent in the review papers of Gosselin and Smith (1966); Corby et al. (1970); Holt and Holz (1963); Corby and Decker (1974); Hayden and Comstock (1975); Lawrence and McGrew (1975); Picchioni (1967, 1970, 1971, 1974); Greensher et al. (1979); Ilo (1980); Neuvonen (1982); Cooney (1984); Boehnert et al. (1985); Moores and Spector (1986); Pond (1986a, 1986b); Park et al. (1986); Derlet and Albertson (1986); Neuvonen and Olkkola (1988); Watson (1987); Katona et al. (1987); Olkkola and Neuvonen (1989); Howland (1990); Palatnick and Tenenbein (1992); Lovejoy et al. (1992); and Vale and Proudfoot (1993). However, we will be covering these topics in greater detail and shall discuss many additional topics in depth.

V. REFERENCES

Andersen, A. H. (1946). Experimental studies on the pharmacology of activated charcoal. I. Adsorption power of charcoal in aqueous solutions, Acta Pharmacol. Toxicol. 2, 69.

Anonymous (1972). Activated charcoal rediscovered (editorial), Br. Med. J., August 26, p. 487.

Boehnert, M. T., Lewander, W. J., Gaudreault, P., and Lovejoy, F. H., Jr. (1985). Advances in clinical toxicology, Pediatr. Clin. North Am. 32, 193.

Borowitz, J. L., Moore, P. F., Yim, G. K. W., and Miya, T. S. (1971). Mechanism of enhanced drug effects produced by dilution of the oral dose, Toxicol. Appl. Pharmacol. 19, 164.

Corby, D. G., Decker, W. J., Moran, M. J., and Payne, C. E. (1968). Clinical comparison of pharmacologic emetics in children, Pediatrics 42, 361.

Cooney, D. O. (1984). Medicinal applications of adsorbents, in Proceedings of the 1983 Engineering Foundation Conference on Fundamentals of Adsorption, Schloss Elmau, Bavaria, 1983, A. L. Myers and G. Belfort, Eds., Engineering Foundation Press, New York, pp. 153-162.

Corby, D. G., Fiser, R. H., and Decker, W. J. (1970). Re-evaluation of the use of activated charcoal in the treatment of acute poisoning, Pediatr. Clin. North Am. 17, 545.

Corby, D. G. and Decker, W. J. (1974). Management of acute poisoning with activated charcoal, Pediatrics 54, 324.

Derlet, R. W. and Albertson, T. E. (1986). Activated charcoal—Past, present and future, West. J. Med. 145, 493.

Ferguson, H. C. (1962). Dilution of dose and acute oral toxicity, Toxicol. Appl. Pharmacol. 4, 759.

Gosselin, R. E. and Smith, R. P. (1966). Trends in the therapy of acute poisonings, Clin. Pharmacol. Ther. 7, 279.

Greensher, J., Mofenson, H. C., Picchioni, A. L., and Fallon, P. (1979). Activated charcoal updated, JACEP 8, 261.

Greensher, J., Mofenson, H. C., and Caraccio, T. R. (1987). Ascendency of the black bottle (activated charcoal), Pediatrics 80, 949.

Hayden, J. W. and Comstock, E. G. (1975). Use of activated charcoal in acute poisoning, Clin. Toxicol. 8, 515.

Henderson, M. L., Picchioni, A. L., and Chin, L. (1966). Evaluation of oral dilution as a first aid measure in poisoning, J. Pharm. Sci. 55, 1311.

Holt, L. E., Jr. and Holz, P. H. (1963). The black bottle—a consideration of the role of charcoal in the treatment of poisoning in children, J. Pediatrics *63*, 306.

Howland, M. A. (1990). Antidotes in depth: Activated charcoal, in *Toxicologic Emergencies*, 4th edition, L. R. Goldfrank, Ed., Appleton & Lange, Norwalk, CT, pp. 129-133.

Ilo, E. (1980). Charcoal: Update on an old drug, J. Emerg. Nurs. *6*, 45.

Katona, B. G., Siegel, E. G., and Cluxton, R. J., Jr. (1987). The new black magic: Activated charcoal and new therapeutic uses, J. Emerg. Med. *5*, 9.

Krenzelok, E. P. and Dunmire, S. M. (1992). Acute poisoning emergencies: Resolving the gastric decontamination controversy, Postgrad. Med. *91*, 179.

Lawrence, F. H. and McGrew, W. R. (1975). Activated charcoal: A forgotten antidote, J. Maine Med. Assoc. *66*, 311.

Litovitz, T., Holm, K. C., Bailey, K. M., and Schmitz, B. F. (1992). 1991 Annual Report of the American Association of Poison Control Centers National Data Collection System, Am. J. Emerg. Med. *10*, 452.

Lovejoy, F. H., Jr., Shannon, M., and Woolf, A. D. (1992). Recent advances in clinical toxicology, Curr. Probl. Pediatr. *22*, 119.

Moores, K. and Spector, R. (1986). Activated charcoal for gastrointestinal decontamination of the poisoned patient, Iowa Med. *76*, 231.

Neuvonen, P. J. (1982). Clinical pharmacokinetics of oral activated charcoal in acute intoxications, Clin. Pharmacokin. *7*, 465.

Neuvonen, P. J. and Olkkola, K. T. (1988). Oral activated charcoal in the treatment of intoxications: Role of single and repeated doses, Med. Toxicol *3*, 33.

Olkkola, K. T. and Neuvonen, P. J. (1989). Treatment of intoxications using single and repeated doses of oral activated charcoal, J. Toxicol. Clin. Exp. *9*, 265.

Palatnick, W. and Tenenbein, M. (1992). Activated charcoal in the treatment of drug overdose: An update, Drug Safety *7*, 3.

Park, G. D., Spector, R., Goldberg, M. J., and Johnson, G. F. (1986). Expanded role of charcoal therapy in the poisoned and overdosed patient, Arch. Intern. Med. *146*, 969.

Picchioni, A. L. (1967). Activated charcoal as an antidote for poisons, Am. J. Hosp. Pharm. *24*, 38.

Picchioni, A. L. (1970). Activated charcoal—a neglected antidote, Pediatr. Clin. North Am. *17*, 535.

Picchioni, A. L. (1971). Management of acute poisonings with activated charcoal, Am. J. Hosp. Pharm. *28*, 62.

Picchioni, A. L. (1974). Research in the treatment of poisoning, in *Toxicology Annual—1974*, pp. 27-51.

Pond, S. M. (1986a). A review of the pharmacokinetics and efficacy of emesis, gastric lavage, and single and repeated doses of charcoal in overdose patients, in *New Concepts and Developments in Toxicology*, P. L. Chambers, P. Gehring, and F. Sakai, Eds., Elsevier, Amsterdam.

Pond, S. M. (1986b). Role of repeated oral doses of activated charcoal in clinical toxicology, Med. Toxicol. *1*, 3.

Robertson, W. O. (1962). Syrup of ipecac: A fast or slow emetic?, Am. J. Dis. Child. *103*, 58.

Spector, R. and Park, G. D. (1986). New roles for activated charcoal (editorial), West. J. Med. *145*, 511.

Spyker, D. A. (1985). Activated charcoal reborn. Progress in poison management (letter), Arch. Intern. Med. *145*, 43.

Vale, J. A. and Proudfoot, A.T. (1993). How useful is activated charcoal?, Br. Med. J. *306*, 78, January 9.

Watson, W. A. (1987). Factors influencing the clinical efficacy of activated charcoal, Drug Intell. Clin. Pharm. *21*, 160.

2

Historical Background of Activated Charcoal

This chapter will review the early history of charcoals, which originally were not "activated" but which—from about 1900 onwards—were "activated." A wide variety of methods for testing the adsorptive "powers" of activated charcoals evolved during the mid 1910s until the late 1930s; we review them in some detail. First, however, we must dispense with the matter of exactly what to call the remarkable material we shall be discussing throughout this volume.

I. TERMINOLOGY

The terms "activated charcoal," "activated carbon," and "active carbon" all occur in the literature and are generally used interchangeably. Although some scientists prefer the adjective "active" to "activated," almost all charcoals in modern use have been purposely activated by taking the charcoal resulting from the controlled pyrolysis of the starting material and subjecting it to the action of an oxidizing gas such as steam, carbon dioxide, or air at elevated temperatures. This enhances the adsorptive power of the charcoal by developing an extensive internal network of fine pores in the material. Thus, the adjective "activated" (which is preferred by almost all industrial and medical personnel) is more correct than the adjective "active" and will be used in this work (when an adjective is used).

Concerning the second half of the term, engineers and most manufacturers seem to prefer the term "carbon," whereas the *U. S. Pharmacopeia* (USP), most of the medical community and a few manufacturers prefer "charcoal." To many, the designation "charcoal" implies vegetable or animal origin (e.g., wood or bone charcoal) and it is argued that "carbon" is a better term, as it includes all source materials. However, regardless of the source, no charcoal is purely carbon but is rather a combination of carbon plus many impurities (e.g., inorganic materials). Thus, the term "carbon" is not strictly correct. For this reason and because of its overwhelming traditional usage in the medical literature, we prefer the term "charcoal" over the term "carbon."

Thus, the preferred adjective will be "activated" and the preferred noun will be "charcoal" in this work. However, as stated in the Preface, we will frequently just use the word "charcoal" in the interest of saving some space, in which case the term should be construed to denote "activated charcoal" (AC). But, we should also point out that, prior to about 1900, charcoals were not activated (the activation process had not been invented). Hence, the proper term for such materials is simply "charcoal." Since we will be considering unactivated charcoals only briefly in this chapter and not thereafter, there should be no confusion over the use of the word "charcoal" alone.

II. EARLY HISTORY

The use of charcoal for medicinal purposes is ancient. In an Egyptian papyrus of 1550 B.C., various kinds of charcoals are specified for medicinal use. Over succeeding centuries, those who practiced as physicians believed greatly in the healing properties and therapeutic values of wood charcoal. In the times of Hippocrates (400 B.C.) and Pliny (50 A.D.), wood charcoal was used to treat epilepsy, vertigo, chlorosis, and anthrax. These practices gradually fell into disuse but were still mentioned, often even into the nineteenth century. D. M. Kehls (1793) wrote of the external application of charcoal to gangrenous ulcers to remove bad odors. Charcoal was also recommended for internal use in the treatment of *fievre putride* at a dosage of 1/16 oz charcoal six times daily. Kehls also recommended that charcoal suspended in water be used as a mouthwash and, additionally, at the first indications of any bilious condition.

The discovery of how charcoal really works (i.e., of the phenomenon of adsorption as we presently understand it) is generally attributed to Scheele (Deitz, 1944), who in 1773 described some experiments on gases exposed to charcoal. In 1777, Fontanna wrote about some experiments in which glowing charcoal was introduced into an inverted tube which contained a gas and whose lower end was submerged in mercury; most of the gas disappeared, as evidenced by the rise of the mercury into the tube. Relative to liquid phase systems, the earliest notice of adsorption seems to have been in 1785, when Lowitz observed that charcoal would decolorize many liquids. Soon after, wood charcoal was used to clarify cane sugar in a sugar refinery. By 1808, its use had been extended to the growing beet sugar industry in France. Wood charcoal, as prepared in those days, decolorized sugar solutions imperfectly.

In 1811, Figuier discovered that animal charcoal made from the decomposition of bones ("bone char") was superior for sugar decolorizing. Actually, bone char is quite different in nature from charcoals prepared from most other starting materials. Bone char contains roughly 78% calcium phosphate, 9% calcium carbonate, and only 11% carbon, plus various other minor constituents. By contrast, charcoals made from vegetable types of starting materials (e.g., wood, coal) generally contain about 90% carbon. Considering the low carbon content of bone chars, it may seem surprising that they are quite effective for decolorizing various solutions. The explanation for this is that bone chars have been found to consist of an intricate skeleton of calcium phosphates and calcium carbonates which are entirely coated with carbon in a state of very fine subdivision (Mantell, 1951, p. 107).

During the nineteenth century many attempts were made to produce decolorizing charcoals from other sources. In 1822, Bussy found that by heating blood with potash, a charcoal vastly superior to bone char resulted. Hunter, in 1865, reported on the great capacity of a charcoal derived from coconut shells for adsorbing gases. Other charcoals were made by Lee, in 1863, from peat, and by Winser and Swindells, in 1860, from paper mill wastes (Hassler, 1963).

The charcoals made in the 1800s and before were not activated in the sense we are familiar with. That is, they were made by pyrolysis alone, without any subsequent treatment using oxidizing gases. Thus, their adsorption abilities, while decent, were still far lower than that of modern activated charcoals. Ostrejko, a Russian, is credited with introducing the concept of activation. In 1900 and 1901 he patented several processes involving the treatment of pyrolyzed matter with superheated steam or carbon dioxide and showed that these greatly enhanced the adsorbing powers of the charcoals.

Gradually, over the period from 1870 to about 1920, more and more processes and source materials were tried. A complete list of source materials that have been tried would include such diverse materials as blood, cereals, fish, fruit pits, kelp, corn cobs, rice hulls, and distillery waste. However, the superiority of charcoals made from certain sources and the costs of purchasing and processing different materials have by now reduced the number of practical choices. Today, charcoals are essentially made only from petroleum coke, coals (bituminous, lignite), peat, sawdust and wood char, paper mill waste (lignin), and coconut shells.

According to Holt and Holz (1963), the first systematic studies of charcoal as an antidote were performed in France in the early 1800s. A chemist named Bertrand studied arsenic poisoning in animals around 1811 and observed that charcoal was efficacious in preventing toxicity. It is claimed that in 1813 he gave a public demonstration of its effectiveness by swallowing 5 g arsenic trioxide mixed with charcoal. Touéry, a French pharmacist, also did studies using animals in the years 1820–1840 and was reported to have swallowed a mixture of 15 g charcoal and strychnine (in the amount of 10 times a lethal dose) as a demonstration for the French Academy of Medicine in 1831. According to Andersen (1946), an American physician named Hort succeeded in 1834 in saving a patient from bichloride of mercury poisoning by having the patient ingest large amounts of powdered charcoal.

Garrod (1846) reported some extensive early studies he performed in England using strychnine and other poisons administered to dogs, cats, rabbits, and guinea pigs. He carefully studied the effectiveness of charcoal as affected by: (1) the poison dose, (2) the charcoal dose, and (3) the time interval between the ingestion of the poison and the charcoal. Garrod found charcoal to be effective, not only against strychnine but also against opium, morphine, aconite, ipecac, veratrum, elaterium, stramonium, cantharides, delphinium, hemlock, and mineral poisons (e.g., bichloride of mercury, silver nitrate, and lead salts).

Rand (1848), an American physician, extended Garrod's types of studies to humans. He reported on observations made using various drugs, including digitalis, morphine, strychnine, arsenic, camphor, iodine, and bichloride of mercury. He, like Garrod, determined what ratio of charcoal to drug was required to reduce clinical symptoms of toxicity to a barely detectable level.

An interesting application of charcoal was reported by Graham and Hofmann (1853), when it was alleged that strychnine was being added to certain English pale ales. To show that strychnine could be detected, these authors shook 2 oz animal charcoal with 1/2 gallon ale to which 1/2 grain strychnine had been added. The charcoal was then filtered off, the ale was found drinkable, and the strychnine was recovered quantitatively from the charcoal by extraction with alcohol. During the late 1800s and early 1900s there continued to be many reports on the efficacy of charcoal as an antidote, mainly in the European literature. In America, interest grew in charcoal as an aid in curing intestinal disorders. For example, the 1908 catalog of Sears, Roebuck and Co. (reprinted in 1969) carried the following advertisement:

Willow Charcoal Tablets
Every person is well acquainted with the great benefit derived from willow charcoal in gastric and intestinal disorder, indigestion, dyspepsia, heartburn, sour or acid stomach, gas upon the stomach, constant belching, fetid breath, all gaseous complications and for the removal of the offensive odor from the breath after smoking.

A similar advertisement of the same period touts claims of antibacterial and antiparasitic activity:

Bragg's Vegetable Charcoal and Charcoal Biscuits
Absorb all impurities in the stomach and bowels. Give a healthy tone to the whole system, effectually warding off cholera, smallpox, typhoid, and all malignant fevers. Invaluable for indigestion, flatulence, etc. Eradicate worms in children. Sweeten the breath.

It was not until much later that scientific research demonstrated that most of the claims made in such advertisements are indeed valid. It is now known that activated charcoal can adsorb poisons, bacterial toxins, and such, in the gut.

III. EVOLUTION OF METHODS FOR TESTING MEDICINAL CHARCOALS

With the realization by 1900 that charcoals have the capacity to adsorb toxins, drugs, and other chemicals, a growing variety of proposed methods evolved soon after 1910 for determining the adsorption powers of the wide variety of charcoals that were beginning to appear on the market. Table 2.1 shows the molecular weights of most of the substances which have been used, both by early researchers and by present day researchers, to test the adsorption powers of charcoals. Note that many compounds are listed as hydrates, since the hydrates are the usual forms of those compounds. Because molecular weight is a rough index of molecular size and since molecular size determines the sizes of pores in a charcoal which a molecule can enter, then molecular weight is an important characteristic of a test substance. Note also that the compounds in Table 2.1 are listed in order of increasing molecular weight. Not included in Table 2.1 are various toxins and antitoxins (tetanus, diphtheria) which have been used.

Table 2.2 summarizes most of the methods used by the early researchers in terms of the various kinds of test substances involved. One of the first tests proposed was that by Helch (1914) for "animal" (i.e., bone char) charcoals. He suggested that 0.1 g finely divided dry charcoal should be able to completely decolorize 20 mL of a 1% methylene blue-HCl solution within 1 min, when shaken with it in a closed flask. As an in vivo test, Helch recommended that if 2.3 g charcoal is shaken with 65 mL of this same solution, on drinking the mixture no color should appear in the urine for 24 hr.

Table 2.1 Molecular Weights of Substances Used to Test Charcoals

Substance	Formula	Molecular weight (daltons)
Oxalic acid	$C_2H_2O_4$	90.0
Phenol	C_6H_6O	94.1
Neurine	$C_5H_{13}NO$	103.2
Nicotine	$C_{10}H_{14}N_2$	162.2
Antipyrine (phenazone)	$C_{11}H_{12}N_2O$	188.2
Iodine	I_2	253.8
Mercuric chloride	$HgCl_2$	271.5
Morphine	$C_{17}H_{19}NO_3$	285.4
Strychnine	$C_{21}H_{22}N_2O_2$	334.4
Methylene blue	$C_{16}H_{18}ClN_3S \cdot 3H_2O$	373.9
Morphine HCl	$C_{17}H_{19}NO_3 \cdot HCl \cdot 3H_2O$	375.9
Strychnine nitrate	$C_{21}H_{22}N_2O_2 \cdot HNO_3$	397.4
Strychnine sulfate	$(C_{21}H_{22}N_2O_2)_2 \cdot H_2SO_4 \cdot 5H_2O$	857.0

Joachimoglu (1916) reported soon after on the use of I_2 as a test substance. Charcoal (0.2 g) was shaken with 50 mL of a 0.1 N I_2 solution for 30 min, then centrifuged. The clear supernate was then titrated with 0.1 N $Na_2S_2O_3$ to indicate how much residual I_2 remained. Joachimoglu stated that charcoals exhibiting large I_2 capacity likewise exhibited large affinities for methylene blue and for tetanus antitoxin. In a later paper, Joachimoglu (1920) reported on in vivo studies in which charcoal/strychnine mixtures were given to dogs. He found that the charcoal that gave the best in vivo protective power was not the one which showed the highest in vitro I_2 capacity. Keeser (1924) later reported similar findings, i.e., that there was "no parallelism between the adsorption capacity of charcoal for alkaloids and iodine in vitro and its detoxifying effect in the intestinal tract of the dog." However, another study by Joachimoglu (1923), on ten charcoals, revealed "a fair degree of parallelism" between the in vitro I_2 adsorption power and the protective action against strychnine nitrate administered to dogs. Joachimoglu concluded that on the average 10 parts of charcoal will detoxify 1 part of strychnine nitrate in vivo.

A likely reason for inconsistencies noted in these studies between in vitro adsorbing powers and in vivo detoxifying powers is that the charcoals had different pore-size distributions (see Chapter 3 for a detailed discussion of pore-size distributions). Thus, a charcoal with small pores might adsorb iodine (mol. wt. 253.8) well in vitro and not adsorb larger molecules (such as toxins) in vivo.

Hörst (1921) reported that in vitro studies on five kinds of charcoals showed that chemically well-defined compounds (e.g., methylene blue, I_2, strychnine) give consistent indications of the relative adsorption capacities of the charcoals, but with the use of true toxins (e.g., tetanus, diphtheria) inconsistent results are obtained. Such incongruities could again be easily explained by the fact that the pore-size distributions of the charcoals were probably quite different.

Koenig (1923) tested six charcoals in vitro with methylene blue, I_2, and cyanide compounds, and recommended these compounds as test substances "for the new *German*

Table 2.2 Charcoal Testing Methods (1910–1940)

Investigators	Number of charcoals tested	I_2	Methyl-ene blue	Alkaloid[a]	$HgCl_2$	Others
Helch (1914)	?		X			
Joachimoglu (1916)	Several	X	X			Tetanus antitoxin
Joachimoglu (1920)	Several			S		
Hörst (1921)	5	X	X	S		Tetanus and diph-theria toxins, neurine
Joachimoglu (1923)	10	X		SN		
Koenig (1923)	6	X	X			Cyanides
Keeser (1924)	4	X		SN Mor-HCl		
Merck (1924)	?	X	X		X	
Dingemanse and Laqueur (1925)	Several		X	SS, Mor		Oxalic acid
Koenig (1925)	Several		X		X	
Laqueur and Sluyters (1925)	3	X	X	SN, MorS		Oxalic acid, $K_2C_2O_4$
Amstel (1926)	3	X			X	Phenol
Dingemanse and Laqueur (1926)	2			SN	X	
Irgang (1927)	14		X			Phenol
Brindle (1928)	18				X	
Sabalitschka and Oehlke (1928)	Several	X	X	Nic	X	Phenol
Langecker (1930)	Several		X	Mor	X	
Rohmann and Gericke (1932)	10		X			Antipyrine
Bari (1933a,b)	31	X	X		X	Antipyrine
Rausch (1935)	13	X	X			Benzene, H_2S, peptone
Sjögren and Wallden (1935)	20		X		X	
Suchy and Rice (1935)	Various			SS		
Kunzova (1937)	Various			S		

[a] S = strychnine, SN = strychnine nitrate, SS = strychnine sulfate, Mor = morphine, MorS = morphine sulfate, Nic = nicotine.

Pharmacopeia." Merck (1924) proposed the following standards in 1924: (1) on shaking 0.1 g charcoal with 35 mL 0.15% methylene blue solution, decolorization should occur in 5 min, and (2) 1 g charcoal should adsorb at least 800 mg $HgCl_2$. Koenig (1925) suggested that the number of milligrams of $HgCl_2$ adsorbed by 1 g charcoal from a 0.02 N $HgCl_2$ solution be used (along with the methylene blue test) as a quantitative index of adsorption power.

Amstel (1926), Dingemanse and Laqueur (1925), and Laqueur and Sluyters (1925) reported on the in vitro adsorptive power of several charcoals (Supra-Norit, Carbo animalis Merck, medicinal Norit, etc.) for $HgCl_2$, I_2, phenol, oxalic acid, potassium oxalate, strychnine sulfate, strychnine nitrate, morphine sulfate, and methylene blue. A great variation was found among the charcoals, with Supra-Norit being the best (probably because, like present-day Norit Supra charcoals, it had a large internal surface area for adsorption).

Dingemanse and Laqueur (1926) further showed that, in vivo, in the pig stomach and intestine, after 55 min, 47% of an $HgCl_2$ dose was adsorbed from the stomach and nearly all was adsorbed from the intestine. Similar results were found with strychnine nitrate.

Brindle (1928), Irgang (1927), Langecker (1930), and Sabalitschka and Oehlke (1928) all did extensive in vitro tests of medicinal charcoals using I_2, methylene blue, $HgCl_2$, phenol, and alkaloids such as morphine and nicotine. Brindle claimed that the $HgCl_2$ results were erratic (although a statement of his indicates that by this date, 1928, the $HgCl_2$ test was standard in the *German Pharmacopeia*). The erratic nature of Brindle's $HgCl_2$ results could be due to pH variations in his tests (the state of $HgCl_2$ in solution is quite sensitive to pH). Sabalitschka and Oehlke, in contrast, claimed the $HgCl_2$ test to be valid, as well as the methylene blue test, but stated that a charcoal should pass both tests to be considered satisfactory. Langecker found that of methylene blue, morphine, and $HgCl_2$, results with the latter two substances correlated well and occasionally disagreed with results on methylene blue (e.g., strong methylene blue adsorption and weak morphine and $HgCl_2$ adsorption). He also concluded that animal charcoals were superior to plant charcoals.

Rohmann and Gericke (1932) used antipyrine as a test substance and found results with it correlated well with methylene blue adsorption. Bari (1933a,b) also tested charcoals using antipyrine, I_2, $HgCl_2$, and methylene blue, and he recommended use of the former two substances as better tests of adsorption power. Rausch (1935) suggested that the I_2 and $HgCl_2$ tests are not valid as a measure of in vivo detoxifying power, since "in the products of food metabolism, substances of such a strong chemical reactivity are absent." He recommended more attention be given to gases, such as H_2S, and to decomposition products such as peptones. A new peptone test was proposed.

Sjögren and Wallden (1935) went so far as to state that "the reduction of a lethal dose of $HgCl_2$ or any other poison by a charcoal cannot be judged by the adsorption of methylene blue or any other substance," but the exact basis for this opinion is unclear.

Charonnat and Leclerc (1949) suggested that the most accurate assessments of the adsorbing powers of charcoals using the I_2 method are achieved when the I_2 solution concentration is adjusted so that 50% of the I_2 is adsorbed (it would seem easier to adjust the charcoal amount, however).

Kunzova (1937) reported some interesting in vivo work with frogs. Strychnine was dissolved in Ringer's solution, treated with various charcoals, filtered, and then injected into the dorsal lymph sacs of frogs. When the ratio of charcoal to strychnine was 65:1, convulsions appeared in 25 min; when the ratio was larger, convulsions failed to appear.

Only small variations between the different charcoals were observed. It appeared that
the amount of charcoal needed to adsorb 1 mg strychnine ranged from 65 to 76 mg.
Other tests involving frogs have been described by Hofmann and Neubauer (1948).
Various charcoals were given along with strychnine to frogs, and their in vivo detoxifying
powers were related to their in vitro iodine and methylene blue adsorption powers.
Animal charcoals were found to be superior to vegetable charcoals.

Other in vivo studies were reported by Saunders et al. (1931). They shook activated
charcoal with solutions of various drugs, filtered them, and injected the filtrates into
dogs. Strychnine, brucine, adrenaline, histamine, and tyramine were completely inactivated; acetylcholine and ephedrine solutions were partly inactivated.

The use of methylene blue and strychnine sulfate to test the adsorptive power of
medicinal charcoals has carried over to the present day. The latest *U. S. Pharmacopeia*
(1989) includes methods involving both of these substances. Apparently, methylene blue
and strychnine sulfate must be considered to be representative of the molecular sizes
and organic characters of most drugs and poisons, but whether this is an accurate
assumption is perhaps debatable. The *British Pharmacopoeia* specifies phenazone (antipyrine) as the test substance for determining adsorption power.

It is interesting to note that the phenol and iodine tests have also carried over to
the present day, although not for medicinal charcoals (however, an exception is *The
International Pharmacopeia*, which specifies iodine and strychnine sulfate for testing
medicinal charcoals). Additional tests which are currently used for evaluating commercial
activated charcoals involve the decolorization of molasses and the adsorption of potassium permanganate (an intensely purple compound).

IV. REFERENCES

Andersen, A. H. (1946). Experimental studies on the pharmacology of activated charcoal. I.
    Adsorption power of charcoal in aqueous solutions, Acta Pharmacol. Toxicol. 2, 69.
Amstel, P. J. (1926). Medicinal charcoal, Deutsch. Med. Wochenschr. 52, 1986.
Bari, Z. (1933a). Evaluation of Carbo medicinalis and its preparations, Ber. Ungar. Pharm. Ges.
    9, 52.
Bari, Z. (1933b). Evaluation of medicinal charcoals and certain charcoal preparations, Pharm. Ztg.
    78, 284.
Brindle, H. (1928). A new test for the activity of medicinal charcoal, Pharm. J. 121, 84.
British Pharmacopoeia, Her Majesty's Stationery Office, London, 1988.
Charonnat, R. and Leclerc, J. (1949). [Determination of the adsorbing power of activated charcoal],
    Ann. Pharm. Franc. 7, 625.
Deitz, V. R. (1944). Bibliography of Solid Adsorbents, U.S. Cane Sugar Refiners and Bone Char
    Manufacturers and the National Bureau of Standards, Washington, D.C.
Dingemanse, E. and Laqueur, E. (1925). The adsorption of poisons on charcoals. II. Biochem. Z.
    160, 407.
Dingemanse, E. and Laqueur, E. (1926). Adsorption of poisons on charcoal. III. The distribution
    of poisons between stomach and intestine wall and charcoal, Biochem. Z. 169, 235.
Garrod, A. B. (1846). On purified animal charcoal as an antidote to all vegetable and some mineral
    poisons, Trans. Med. Soc. London 1, 195.
Graham, T. and Hofmann, A. W. (1853). Report on the alleged adulteration of pale ales by
    strychnine, J. Chem. Soc. 5, 173.
Hassler, J. W. (1963). *Activated Carbon*, Chemical Publishing Co., New York.

Helch, H. (1914). Animal charcoal, Pharm. Post *47*, 949.

Hofmann, H. and Neubauer, M. (1948). Medicinal charcoal and the differences in adsorption capacity of various charcoal preparations, Pharmazie *3*, 529.

Holt, L. E., Jr. and Holz, P. H. (1963). The black bottle - a consideration of the role of charcoal in the treatment of poisoning in children, J. Pediatrics *63*, 306.

Hörst, F. (1921). The adsorption capacity of various charcoals, Biochem. Z. *113*, 99.

Irgang, I. (1927). Some comparative investigations of medicinal carbon preparations, Dansk. Tids. Farm. *1*, 167.

Joachimoglu, G. (1916). Adsorption capacity of animal charcoal and its determination, Biochem. Z. *77*, 1.

Joachimoglu, G. (1920). The theoretical principles of charcoal therapy, Chem.-Ztg. *44*, 780.

Joachimoglu, G. (1923). Adsorption and detoxification power of certain charcoals, Biochem. Z. *134*, 493.

Keeser, E. (1924). Adsorption and the distribution of drugs in the organism. III, Biochem. Z. *144*, 536.

Kehls, D. M. (1793). Memoire sur le charbon végétal, observations et journal sur la physique, de chemie et l'histoire naturelle et des arts, Paris, Tome XLII, 250.

Koenig, F. (1923). Evaluation of Carbo medicinalis, Pharm. Zentralhalle *64*, 205.

Koenig, F. (1925). Adsorption value of Carbo medicinalis, Pharm. Zentralhalle *66*, 645.

Kunzova, H. (1937). The evaluation and the use of animal charcoals, Prakticky Lekar *17*, 337.

Langecker, H. (1930). The measurement of the adsorption capacity of medicinal charcoals, Klin. Wochenschr. *9*, 2298.

Laqueur, E. and Sluyters, A. (1925). Adsorption of poisons on a new plant charcoal "Supra-Norit," Biochem. Z. *156*, 303.

Mantell, C. L. (1951). *Adsorption*, McGraw-Hill, New York.

Merck, E. (1924). Estimation of the adsorptive power of Carbo medicinalis, Pharm. Ztg. *69*, 523.

Rand, B. H. (1848). On animal charcoal as an antidote, Med. Examiner *4*, 528.

Rausch, A. (1935). Comparative investigations of the adsorptive power of charcoals for medicinal purposes, Arch. Chem. Farm. *2*, 182.

Rohmann, C. and Gericke, P. (1932). Evaluation of medicinal carbons, Pharm. Ztg. *77*, 653.

Sabalitschka, T. and Oehlke, K. (1928). Adsorption from solutions and testing the adsorptive capacity of medicinal charcoal, Pharm. Zentralhalle *69*, 629, 646.

Saunders, F., Lackner, J. E., and Schochet, S. S. (1931). Studies in adsorption. I. The adsorption of physiologically active substances by activated charcoal, J. Pharmacol. *42*, 169.

Sears, Roebuck, and Company (1969). 1908 Catalog No. 117 (reprint), Digest Books, Northfield, Illinois, p. 790.

Sjögren, B. and Wallden, E. (1935). The adsorbing power of medicinal charcoal, Svensk. Farm. Tid. *39*, 617, 632, 673, 697, 720.

Suchy, J. F. and Rice, R. V. (1935). Adsorption of strychnine sulfate by various charcoals and by Lloyd's reagent, J. Am. Pharm. Assoc. *24*, 120.

The International Pharmacopeia, 3rd edition, World Health Organization, 1981.

United States Pharmacopeia, Twenty-Second Revision, United States Pharmacopeial Convention, Rockville, Maryland, 1989.

3

Fundamentals of Activated Charcoal and the Adsorption Process

Although those persons who use activated charcoal for medical purposes may not require detailed information on how activated charcoals are manufactured, a general knowledge of such can be useful in understanding how different adsorption properties arise and why various activated charcoals often behave differently as antidotes. Thus, we present a brief discussion of how activated charcoals are made. We then proceed to describe the various basic properties of activated charcoals (e.g., internal surface area, pore-size distribution) and then discuss in some detail the nature of "adsorption" and the wide variety of chemical and physical factors which affect the adsorption process. Finally, we explain how adsorption "isotherms" are determined experimentally and discuss the significance of such isotherms.

I.   THE MANUFACTURE OF ACTIVATED CHARCOAL

While most carbonaceous substances can be converted into activated charcoal, the final properties of the charcoal will nevertheless reflect to a considerable extent the nature of the source material. This is particularly true with respect to the hardness of the final product; for example, coconut shells yield a strong, dense charcoal which resists mechanical abrasion well.

A great many methods of manufacture of activated charcoals exist, and indeed hundreds of patents have been issued covering specific procedures. However, most processes consist of the pyrolysis (or carbonization) of the starting material, followed by a stage of controlled oxidation. This latter stage is what is meant by "activation" of the charcoal.

A. Carbonization

The pyrolysis step is usually carried out by heating the source material to temperatures ranging between 600 and 900°C in the absence of air. Development of a porous material during this first step has been found to be aided greatly by incorporating metallic chlorides in the starting mixture. A typical example of this is the European process in which pulverized peat or sawdust is mixed with a concentrated zinc chloride solution, dried, and carbonized in a kiln at 600–700°C. The zinc salt is removed from the product by washing it with dilute acid and water. If only large pores are desired in the charcoal (e.g., for decolorizing sugars, where the color bodies to be adsorbed are fairly large molecules which can penetrate into only large pores) then this first step may be all that is needed. However, most charcoals are subjected to a second, oxidation, step.

B. Activation with Oxidizing Gases

The basic character of a charcoal is determined in the first (pyrolysis) step and the subsequent oxidation step must be tailored to fit in with the first stage. Oxidation is usually carried out using steam, although air (and more rarely, CO_2), is sometimes employed. Temperatures are normally in the range of 600-900°C. Under the proper conditions, the oxidizing gas selectively erodes the internal surfaces of the charcoal, develops a greater and finer network of pores in the charcoal, and converts the atoms lying on the surfaces to specific chemical forms (e.g., oxides) which may have selective powers of adsorption.

The reactions which occur during activation are of the type:

$$C + H_2O \rightarrow H_2 + CO$$

$$C + CO_2 \rightarrow 2CO$$

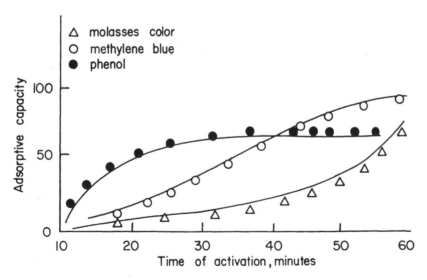

Figure 3.1 Rate of development of specific adsorptive capacities during steam activation of a charcoal. From Hassler (1963). Reproduced by permission of the Chemical Publishing Company.

.

$$C + \tfrac{1}{2}O_2 \rightarrow CO$$

One can see that these reactions cause solid carbon, C, to be converted to gaseous species, thereby creating pores in the charcoal.

The various changes which occur do not all happen at the same rate. Some types of changes develop early and some develop late in the process. Depending on the properties desired, the total time allowed for activation is a very important variable. As shown in Figure 3.1, the ability of a charcoal to adsorb different materials can depend strongly on the activation time. The primary effect of time is on the sizes of the pores which develop. As time goes on, a succeedingly greater number of pores are generated and the internal surface area gradually increases. However, as the process continues, solid material separating adjacent pores may be eaten away; the net result is the generation of larger pores and a reduction in the total internal surface area. Figure 3.1 shows that the ability to adsorb the smallest molecule [phenol, molecular weight (mw) 94 daltons] develops first, the ability to adsorb an intermediate-sized molecule (methylene blue, mw 374 daltons) comes next, and the affinity for molasses color bodies (mw variable but on the order of 2000 daltons or more) develops last. This behavior clearly relates to the creation of increasingly larger pores by oxidation.

II. THE PROPERTIES OF ACTIVATED CHARCOAL

Activated charcoals contain inorganic constituents derived from the source material or from ingredients (such as metallic chlorides) added during manufacture. Typical elemental compositions of activated charcoals are shown in Table 3.1.

The "ash" consists of the residue remaining when a sample of the charcoal, placed in a porcelain crucible, is heated in air in a furnace at 600°C until the carbon has been entirely burned. Ash contents range typically from about 1 to 5% of the original charcoal weight. A detailed listing of the properties of some common charcoals which were available at the time the first edition of this volume appeared (Cooney, 1980) is given in Table 3.2.

A similar compilation of the properties of charcoals available as of 1993 is presented in Table 3.3. The 1993 data sheets do not give any detailed pore-size distribution figures. In addition to some items explained in the "Notes" following Table 3.3, there are some other points of information which round out the information: (1) Norit SX 3 and Norit SX 4 charcoals, similar to Norit SX 2, with surface areas of 750 and 650 m²/g, respectively, are also available, (2) Norit SX 2, Norit SX 3, and Norit SX 4 were formerly called

Table 3.1 Elemental Composition of Activated Charcoals (wt%)

Charcoal	Ash	Carbon	Hydrogen	Sulfur	Nitrogen
1	4.3	94.4	1.1	0.04	0.62
2	3.2	91.7	1.7	0.07	0.38
3	1.2	95.3	0.6	0.62	0.54
4	2.0	87.5	2.2	0.16	0.39

Adapted from Hassler (1963). Reproduced by permission of the Chemical Publishing Company.

Table 3.2 Some Properties of Powdered Charcoals (1980)

	Darco G-60	Darco S-51	Darco KB	Norit A	Norit SG Extra
Source material	Lignite	Lignite	Wood	Peat	Peat
Surface area (m^2/g)	600	650	1500	720	810
Bulk density (g/mL)	0.40	0.51	0.45	0.34	0.30
Pore volume (mL/g)	1.0	1.0	1.8	—	—
Ash (%)	3.5	—	3.0	5–9	6.0
Acid solubles (%)	1.0	—	—	—	0.5
H_2O solubles (%)	0.3	1.0	—	0.5	0.2
H_2O (%)	8	8	25	10	10
Size (%)					
< 100 mesh	95	98	99	98–100	98–100
< 325 mesh	70	70	70	60–65	60–65
Pore volume distribution (%)					
< 20 Å	10	25	20	—	—
20–50 Å	10	15	27	—	—
50–100 Å	30	10	13	—	—
100–500 Å	35	20	30	—	—
> 500 Å	15	30	10	—	—
Mean pore radius (Å)	25	30	23	—	—

An Angstrom (Å) is 1×10^{-8} or 0.00000001 centimeters.

Table 3.3 Properties of Charcoals from 1993 Data Sheets

	Norit XXII	Norit B Supra	Norit A Supra	Norit SX 2	Darco KB	Darco G-60	Darco S-51
Source material	Peat	Coconut shells	Coconut shells	Peat	Wood	Lignite	Lignite
Surface area (m^2/g)	900	1400	2000	800	1500	—	650
Bulk density (g/mL)	0.35	0.30	0.32	0.41	0.45	0.40	0.51
Pore volume (mL/g)	—	—	—	—	1.8	—	1.0
Ash (%)	4	4	3	5	3	5.5	—
H_2O (%)	8	8	8	10	33	12	10
Acid solubles (%)	0.5	0.5	0.5	1.0	—	1.0	—
Water solubles (%)	—	—	—	0.3	1.5	0.5	1.5
pH	7	7	7	7	5	7	5
MB (%)	18	30	40	14	(185)	[15]	(95)

1. The manufacturer's data sheet for Darco KB-B is identical to that for Darco KB. Thus, data for the Darco KB-B charcoal are not shown. The difference between the KB and KB-B types is believed to be that the KB-B type is more highly acid-washed.
2. The MB (methylene blue) decolorizing values are in percent MB color adsorbed, except for values in ( ) parentheses, which are in terms of "decolorizing efficiency", and the value in the [ ] brackets, which is in terms of g MB adsorbed/g charcoal.
3. Particle sizes: In the order listed, from left to right, the percentages of particles less than 150 μm (100 mesh) are 97, 97, 97, 97, 99, 90, and 98.
4. Phenazone (antipyrine) adsorption values are given for the Norit B Supra and Norit A Supra as 41 and 50%, respectively (phenazone adsorption criteria appear in the pharmacopeias for some countries).

Norit SG Extra, Norit SG, and Norit 211, respectively, and (3) the values given for moisture content, acid-solubles, and water-solubles are generally maximum expected values.

It is common to designate the amount of ash which is water-soluble and the amount of ash which is acid-soluble. As Tables 3.2 and 3.3 suggest, the acid-soluble part is on the order of 1% or less of the original charcoal weight and the water-soluble portion is about 1.5% or less. Some of the inorganic constituents of the ash that are commonly found are iron, calcium, sodium, copper, sulfates, chlorides, and phosphates.

The 22nd edition of the *U. S. Pharmacopeia* (1989) specifically states that USP activated charcoals should be fine, black, and odorless powders which possess the following additional characteristics: less than 15% weight loss on drying, less than 4% residue after ignition (i.e., ash), less than 3.5% acid-soluble substances, less than 0.2% alcohol-soluble substances, less than 0.15% sulfate, less than 0.02% chloride, less than 0.005% heavy metals content, and essentially zero sulfides, cyanogens, and uncarbonized constituents. To meet these standards, USP-grade activated charcoals are normally washed with acid (usually HCl) at elevated temperatures to remove the major part of the inorganic constituents. Any components of the ash not extracted by acid-washing would not normally be solubilized when the charcoal is subsequently used.

A.   Densities

There are several densities that one can speak of when referring to particulate materials. "Bulk density" is the weight per unit volume in a packed bed. The volume on which it is based includes that of the interstices between the particles, as well as that of the particles themselves. Typical values are 0.3–0.5 g/mL.

"Particle density" is the density of a single particle. The volume on which it is based includes the volume of the pores as well as the volume of the solid carbonaceous material. Typical values are 0.7–1.0 g/mL. The "real density" is that of the solid carbonaceous material alone. Values are typically about 2.2 g/mL.

One can now easily estimate the approximate value of the external area per gram of powdered activated charcoal for comparison to typical internal area values. Let us assume that the charcoal powder particles are approximately 400 mesh (a large fraction of most powdered charcoals passes through a 325 mesh sieve, as indicated in Table 3.2, so assuming a smaller sieve size such as 400 mesh as characteristic of the "average" particle size is a reasonable guess). A 400 mesh sieve has openings of 0.0038 cm. Now, if we take the charcoal particle density to be 0.7–1.0 g/cm^3, the volume of 1 g of charcoal is 1.43 to 1.0 cm^3. If we divide by the volume per particle ($\pi D^3/6$), where D is the particle diameter, we get the number of particles per gram of charcoal. If we then multiply by the surface area of a particle (πD^2), assuming the particle is a smooth sphere, we get the total external surface area. Thus the surface area is a number in the range of 1.0–1.43 times ($6\pi D^2/\pi D^3$) or 1.0–1.43 times 6/D, which gives 1.0–1.43 times 6/0.0038, or 1580–2260 cm^2, or 0.158–0.226 m^2. If we allow for the surface to be rough, so that the actual surface area is, say, five times greater than that of a smooth sphere, the area is about 1.0 m^2. By comparison to internal surface areas of 900–2000 m^2/g for currently available USP-grade charcoals, one can see that the external area is indeed negligible.

B.   Pore Volume and Pore-Size Distribution

Pore volume represents the total volume of the pores in a charcoal particle per unit weight of the charcoal. Values are usually on the order of 1.0–1.8 mL/g.

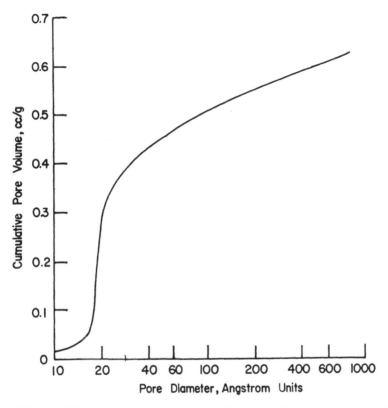

Figure 3.2 Cumulative pore volume distribution for a typical activated charcoal.

Activated charcoals contain a complex network of pores of various shapes and sizes. The pores have an irregular geometry, are branched, and are interconnected by passages which may or may not be constricting. Pore sizes range from less than 10 Å to more than 100,000 Å (0.001 cm). The so-called pore-size distribution depends on the source materials used and on the method and extent of activation. Pores are often classified as macropores (those greater than 100 Å in diameter), micropores (those less than 10–20 Å in diameter), and transitional pores (those with diameters between the micropore and macropore values). Through well-established methods (e.g., determining the volume of mercury that can be forced into the pores, as a function of pressure), it is possible to determine the relative numbers of pores of a given size.

This information is usually expressed as a plot of "cumulative pore volume" versus pore diameter. The cumulative pore volume corresponding to any pore diameter value represents the total volume contained in pores of that diameter or less. Figure 3.2 shows a typical distribution. The steepest part of the curve represents the pore diameter range with which most of the pore volume is associated.

Pore-size distributions are useful in selecting charcoals with high adsorptive capacities for particular types of molecules. For the removal of "color bodies" from liquids, a charcoal having larger pores (> 20 Å) is needed, since color bodies are of relatively large molecular size. For gas adsorption, small pores (< 10 Å) are best. For adsorption of drugs and poisons, which are of moderate molecular weight (say 100–800 daltons), a charcoal having many pores on the order of 10–20 Å would be best.

C. Surface Area

Internal surface areas of porous materials such as activated charcoals are generally determined by use of the so-called BET equation, derived by Brunauer, Emmett, and Teller (1938). This equation describes the formation of multilayers of condensed gas which occurs when an adsorbable gas is contacted with the porous solid. In such a test, one normally uses nitrogen at a temperature corresponding to its atmospheric boiling point (−196°C or 77K). That is, the temperature employed is that at which nitrogen would change from a liquid to a gas if the pressure, P, were equal to exactly 1 atmosphere. However, the actual pressure in the testing system is varied from essentially zero (very high vacuum conditions) to about $P = 0.3$ atm. As the pressure rises, more and more nitrogen adsorbs (the attractive forces of the surface cause the nitrogen to essentially "condense" on the surface). One keeps track of the weight gain of the solid (due to the increase in adsorbed nitrogen) versus pressure. Data are usually taken over the range $P = 0.05$ atm to $P = 0.3$ atm.

The BET equation was derived assuming that: (1) in the first layer, the rate of condensation of molecules on bare sites is equal to the rate of evaporation of molecules from sites covered by only one molecule, and (2) the heat of adsorption beyond the first layer is constant and is equal to the heat of liquefaction. These assumptions are only approximately correct. The BET equation is

$$\frac{Q}{Q_m} = \frac{b\,(P/P_s)}{(1 - P/P_s)(1 - P/P_s + bP/P_s)}$$

where P_s is the saturation vapor pressure at the temperature used in the test (again, one usually uses a temperature such that P_s is 1 atmosphere), Q is the amount of substance adsorbed at any given pressure P, and Q_m is the amount of substance adsorbed when one complete monomolecular layer of surface coverage is attained. This equation may be rearranged to the form

$$\frac{P}{Q(P_s - P)} = \frac{1}{bQ_m} + \left(\frac{1}{bQ_m}\right)(b - 1)\left(\frac{P}{P_s}\right)$$

One then takes the data on the amounts adsorbed (Q) at the different pressures (P) used, and computes values of $P/Q(P_s - P)$. These values are then plotted versus the corresponding P/P_s values, and the whole set of data points is fitted with the best straight line (using the linear least-squares procedure discussed below in Section IV.C). From the last equation, one can see that the intercept on the vertical axis at $P/P_s = 0$ on such a plot is equal to $1/bQ_m$ and the slope of the line is equal to $(1/bQ_m)(b - 1)$. From the values of the slope and intercept, one can easily determine b and Q_m. As just mentioned, Q_m represents the amount (e.g. moles) of nitrogen corresponding to exactly one complete monomolecular layer of coverage of the surface. If one then uses a value of 16.2 square Angstroms as the area covered by a single N_2 molecule (a standard, accepted value), then the surface area of the solid can be readily calculated. Because of the approximate nature of the assumptions in the BET model, the computed area may not be quite correct,

Table 3.4 BET Areas of a NH₃ Synthesis Catalyst
Determined Using Different Gases as Sorbates

Sorbate gas	Temp (K)	Area (m^2/g)
Nitrogen	77	580
Bromine	352	470
Carbon dioxide	195	460
Carbon monoxide	90	550

From Ruthven (1984). Reprinted by permission of John Wiley &
Sons, Inc.

but as long as all researchers agree to use the BET method as their "standard", then at least values for the relative areas of different porous solids will be essentially correct.

Ruthven (1984) has shown how the BET area of a certain ammonia synthesis catalyst varies with the type of gas used in the test. The results are shown in Table 3.4. These values suggest that the absolute area may be determined to within about 20–25% by the BET procedure. Typical surface areas for activated charcoals are in the range of 600–2000 m^2/g, with 800–1200 m^2/g being about average.

Figure 3.3 shows a typical relationship between pore size and total surface area. The cumulative surface area is the area contributed by all pores of a given diameter or larger. It is clear that most of the internal surface area is associated with the smaller pores.

It should be pointed out in passing that even if a charcoal is very finely powdered, its *external* surface area will still be very small compared to its *internal* surface area. There have been misconceptions about this in the literature in which writers have implied that the great surface area for drug adsorption by charcoal is due to its being finely divided. This implies that it is the external area which contributes most to the total surface area for adsorption. Such is not the case. External area is very important but for a different reason. The external area defines the amount of contact area between the charcoal and a surrounding solution. The *rate* of transfer of adsorbable molecules from the fluid phase to the charcoal phase will be proportional to the external area; in contrast, the internal surface area determines the *amount* of adsorption that can occur at equilibrium. To reiterate, external surface area determines the rate of adsorption and internal surface area determines the amount of adsorption.

D. Nature of the Charcoal Surface

A very extensive discussion of the nature of the charcoal surface, how it depends on the conditions used to activate the charcoal, and how these surface properties affect the adsorption behavior of charcoals has been given in a book by Mattson and Mark (1971) entitled *Activated Carbon: Surface Chemistry and Adsorption from Solution* and will not be repeated here in any detail. Space allows for only a brief discussion of these topics.

Depending on the nature of the partial oxidation reactions which take place during the activation of a charcoal, two forms of charcoal can be produced. So-called H-type charcoals are obtained under normal activation conditions, in which high temperatures

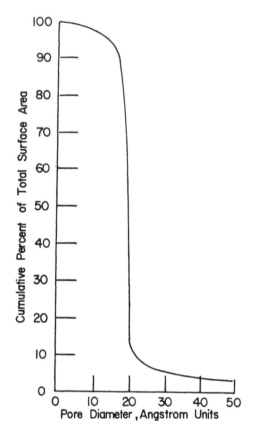

Figure 3.3 Cumulative surface area distribution for a typical activated charcoal.

(> 750°C) are used and exposure to steam or CO_2 is employed, followed by exposure to air at room temperature. Such charcoals are hydrophobic and take on a positive charge when immersed in water, by adsorbing H^+ ions, thus making the water alkaline. Frumkin (1930), for example, stripped all surface oxides from an H-type charcoal by exposing it to a high vacuum at 1000°C and found that acid adsorption by this charcoal was critically dependent on how much oxygen was subsequently exposed to the charcoal. The charcoals normally produced by most manufacturers are of the H-type. So-called L-type charcoals can be made by activating the charcoals at low temperatures (200–400°C) in air. Such L-type charcoals are hydrophilic and take on a negative charge when immersed in water, by adsorbing OH^- ions, thus making the water acidic.

Data obtained by Steenberg (1944) on the HCl and NaOH adsorption capacities of charcoals derived from sugar are shown in Figure 3.4. This figure shows clearly that charcoals activated at low temperatures (L-type charcoals) adsorb hydroxyl ions, and charcoals activated at high temperatures (H-type charcoals) adsorb hydrogen ions.

The differences between H-type and L-type charcoals can be explained in terms of the oxidation processes which occur during activation and the resultant types of surface functional groups (carbon/oxygen groups) which are formed. With H-type charcoals, basic surface oxides predominate, whereas with L-type charcoals, acidic oxides predominate. Among the various types of oxide groups which can be formed on L-type

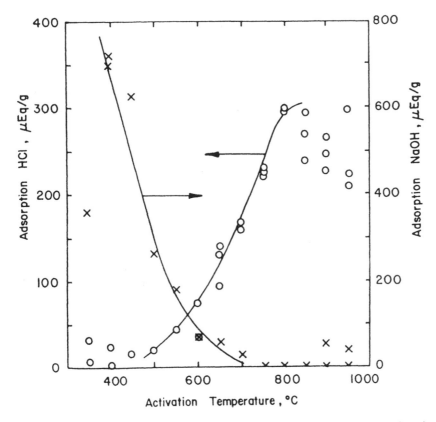

Figure 3.4 Adsorption of acid and base by activated sugar charcoals as a function of activation temperature. From Mattson and Mark (1971).

charcoals are carboxyl, phenolic hydroxyl, carbonyl (quinone type), carboxylic acid, anhydride, lactone, and cyclic peroxide groups (see Corapcioglu and Huang, 1987). Ishizaki and Marti (1981) have determined that for an H-type charcoal, the predominant surface oxides are lactones, quinones, phenolic hydroxyls, and carboxylates.

Neglecting metallic impurities for the moment, it can be said (see Snoeyink and Weber, 1967) that two major types of surfaces occur in charcoals: the planar surfaces of the carbon microcrystallites, and the edges of the carbon planes which make up the sides of the microcrystallites. The planar surfaces are quite uniform and have few functional groups, since the electrons of the carbon atoms are covalently bonded to neighboring carbon atoms. However, the sides of the microcrystallites are not uniform and contain various functional groups. It is known that the adsorption of organic compounds can be strongly affected by the nature and amount of surface oxide groups. For example, phenol forms strong donor-acceptor complexes with oxygen groups and thus its adsorption occurs by a donor-acceptor complex mechanism involving carbonyl oxygens of the carbon surface acting as the electron donor and the aromatic ring of the phenol acting as the acceptor. After the carbonyl oxygen sites are fully utilized, further adsorption can occur by complexation of the phenol with the graphitic rings of the carbon basal planes (Mattson et al., 1969).

Most of the extensive literature that exists on the topic of surface oxides and how they affect adsorption behavior has been generated by chemists and chemical engineers in their professional journals. However, there is a recent interesting paper in the pharmaceutical literature by Burke et al. (1992). They did an x-ray photoelectron spectroscopic analysis of a charcoal surface, and determined that carbon-carbon (C–C), carbon-hydrogen (C–H), hydroxyl or ether groups (C–O), carbonyl groups (C=O), and carboxylic acid or ester groups (O–C=O) all exist on the surface. The C–O functional state comprised 60-70% of the oxygen-containing groups. They showed that the amount of phenobarbital adsorbed on the surface was proportional to the percentage of the surface carbon/oxygen groups which were of the C–O form, and thus concluded that phenobarbital adsorption occurs on those sites. Further details may be found in Burke (1991).

III. THE NATURE OF THE ADSORPTION PROCESS

The chemical nature of the internal surfaces created when charcoal is activated is such that the surface has an attraction for certain molecules if they are present in the liquid phase which fills the pores. The intermolecular forces are of a relatively weak type and, therefore, desorption of the adsorbed solute can occur. That is, if one has adsorbed a drug onto charcoal by exposing the charcoal to an aqueous solution having a significant concentration of the drug, one can desorb the drug to some extent by exposing the charcoal to pure water (a more substantive discussion of desorption is given in Chapter 9, Section III).

Various factors affect the extent to which a given compound will adsorb to charcoal. We will now briefly discuss the major factors.

A. Effect of Temperature

In general, raising the temperature decreases adsorption somewhat because the adsorbed molecules have greater vibrational energies and are therefore more likely to desorb from the surface. However, since all applications of interest to us are for a single temperature (37°C), we need not explore temperature effects any further.

B. Nature of the Solvent

The solvent has an important effect, since it competes with the charcoal surface in attracting the solute. Thus, the adsorption of an organic solute out of an organic solvent is much less than its adsorption out of an aqueous solution. However, since in all cases of interest to us the solvent will be water, we need not consider solvent effects further.

C. Surface Area of the Charcoal

At first thought, the amount of substance which a charcoal can adsorb would seem to be directly proportional to the internal surface area. Yet such is not always true. One must be aware that the surface area is usually inferred from the amount of N_2 (a small molecule) which can be adsorbed at −196°C (N_2 boiling point). If one wishes to adsorb a large molecule, much of the internal surface area may not be accessible.

Table 3.5 gives an example of this. The adsorption power for aniline blue dye increases greatly in the later stages of activation, the reason being that many pores which

Table 3.5 Relation of Surface Area to Adsorptive Power

Time of activation (min)	Surface area (m^2/g)	Phenol adsorption[a]	Aniline blue adsorption[a]
10	402	0.09	0.05
20	600	0.15	0.11
30	815	0.15	0.14
40	990	0.15	0.20
50	1,065	0.15	0.28

[a] Grams adsorbed per gram charcoal at $C = 0.10$ g/liter. From Hassler (1963). Reprinted by permission of the Chemical Publishing Company.

were initially too small to admit the large molecules of aniline blue finally become large enough to admit them. Note that increases in the surface area above 600 m^2/g have no effect on phenol adsorption. It is not clear why this happens.

D. Pore Structure of the Charcoal

As discussed earlier, the pore structure is important because pore diameters, which range from less than 10 Å to over 100,000 Å, control which sizes of molecules are accessible to them. Table 3.6 shows typical minimum pore diameters for ensuring accessibility of pores to selected solutes.

E. Nature of the Solute

Inorganic compounds display a wide range of adsorbability. On one hand, strongly dissociated salts like sodium chloride and potassium nitrate are essentially not at all adsorbed by activated charcoal. On the other hand, nondissociating solutes like iodine and mercuric chloride are adsorbed very well. The key factor seems to be whether the solute exists in neutral or in ionized form.

As for organic substances, which are of most interest to us, several generalizations can be made. The more "organic" the solute is (i.e., nonpolar, of lower solubility in water), the better it is adsorbed. Also, an increase in size of the solute molecule usually enhances adsorption, especially for compounds that are similar. There is, in fact, a generalization known as Traube's rule which states: "The adsorption of organic substances from aqueous solutions increases strongly and regularly as we ascend the homol-

Table 3.6 Minimum Pore Diameter for Selected Solutes

Solute	Minimum pore diameter (Å)
Iodine	10
Potassium permanganate	10
Methylene blue	15
Erythrosine red	19
Molasses color bodies	About 28

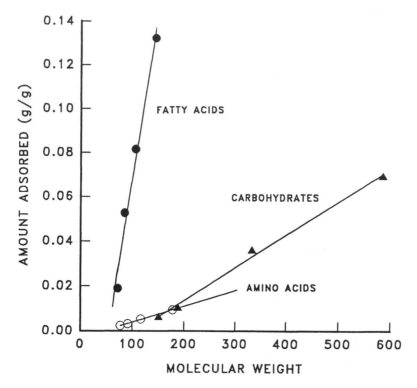

Figure 3.5 Adsorption capacity of an activated charcoal for three homologous series of compounds. Adapted from Tebbutt and Bahiah (1977).

ogous series." Figure 3.5, from a paper by Tebbutt and Bahiah (1977), illustrates this principle for three homologous series of compounds. Of course, an obvious limit to this rule is reached when the molecular sizes become large enough so that a significant number of pores become inaccessible.

With dissimilar compounds, the size factor is often outweighed by constitutive effects. For example, we may indicate the effects of the presence of certain substituent groups as follows:

Hydroxyl groups	Generally reduce adsorption
Amino groups	Like hydroxyl, but greater effect
Sulfonic groups	Usually decrease adsorption
Nitro groups	Often increase adsorption

Other groups (halogens) and bond types (double bonds, carbonyl bonds) have variable effects, depending on the nature of the host molecule.

Molecular structure is also an important factor. For example: (1) aromatic compounds are usually more adsorbable than aliphatic compounds of a similar molecular size, (2) branched-chain molecules are generally more adsorbable than straight-chain molecules, and (3) the effect of a substituent group depends much on the position (e.g., ortho, meta, para, other) where it is introduced.

The solubility of a compound in aqueous solution is often regarded as a good indicator of whether the compound is likely to be well adsorbed. The reason is that solubility reflects the degree of attraction of the solute by the solvent. For example, water, being a very polar solvent, has a great affinity (and therefore solubility) for polar solutes. This means that water will tend to hold onto a polar solute and prevent its being bound by the charcoal surface.

F. pH of the Solution

The effect of the solution pH is extremely important when the adsorbing species is capable of ionizing in response to the prevailing pH (we speak of ionization rather than dissociation, because with basic substances a low pH causes a group such as $-NH_2$ to become protonated and exist in the form $-NH_3^+$; thus, the molecule is *ionized* but it is not *dissociated*). It is well known that substances adsorb poorly when they are ionized. When the pH is such that an adsorbable compound exists in ionized form, adjacent molecules of the adsorbed species on the charcoal surface will repel each other to a significant degree because they carry the same electrical charge (forces of repulsion/attraction between actual ions are strong, as compared to weak forces such as van der Waal's forces). Thus, the adsorbing species can not pack together very densely on the surface and the equilibrium amount of adsorbed solute is only modest. In contrast, when the adsorbing species is not ionized, no such electrical repulsion exists and thus the packing density on the surface can be much higher. This explains the common observation that nonionized forms of acidic and basic compounds adsorb much better than their ionized counterparts. Acidic species (e.g., barbiturates) thus adsorb better at low pH, and basic species (e.g., alkaloids) adsorb better at high pH.

Figure 3.6 shows data obtained by Cooney (1992) on the adsorption of sodium salicylate to activated charcoal in buffer solutions of different pH values. In these tests, 0.057 g charcoal (this weight was selected to give 50% adsorption at pH 7.4) was added to 20 mL of various solutions of 1 g/L sodium salicylate. The trend of the percentage salicylate adsorbed at equilibrium can be compared to the curve shown in Figure 3.7, which shows the charge on the salicylate molecule as a function of pH. It is clear that the percent salicylate adsorbed essentially varies inversely with the charge on the salicylate molecule. Cooney and Wijaya (1987) have presented similar information on the percent adsorption of various aromatic compounds to charcoal as a function of pH. Again, the amount of charcoal, the solution volume, and the initial concentration of the solute in solution were held constant in each test series. Figures 3.8, 3.9, and 3.10 show the results for three aromatic carboxylic acids, for two basic aromatic compounds, and for an aromatic amphoteric compound (i.e., one that has both basic and acidic characteristics). It is clear that all of the compounds adsorb most strongly under conditions where they are uncharged (i.e., nonionized).

Of course, there have been many studies, particularly in the chemistry, chemical engineering, and environmental engineering literature, on the effect of pH on the adsorption of organic substances to charcoal. We will mention here only four such studies. Ward and Getzen (1970) have studied the adsorption of three aromatic acidic herbicides as a function of pH, Martin and Iwugo (1982) have examined the pH dependence of the adsorption of organic compounds from sewage effluents, and Wang et al. (1975) have looked at the variation of the adsorption of organics from industrial waste waters with pH. Müller et al. (1980) have provided a theoretical model for the variation of the adsorb-

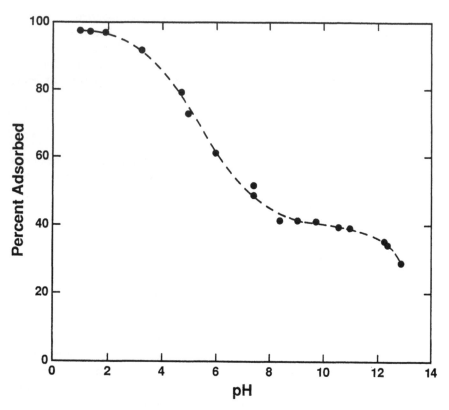

Figure 3.6   Effect of pH on adsorption of salicylate by activated charcoal. From Cooney, 1992.

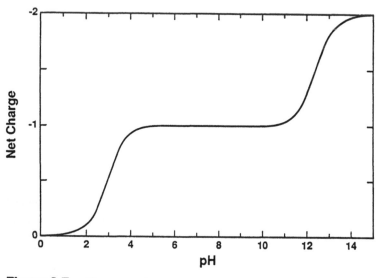

Figure 3.7   Charge on the salicylate molecule as a function of pH. From Cooney, 1992.

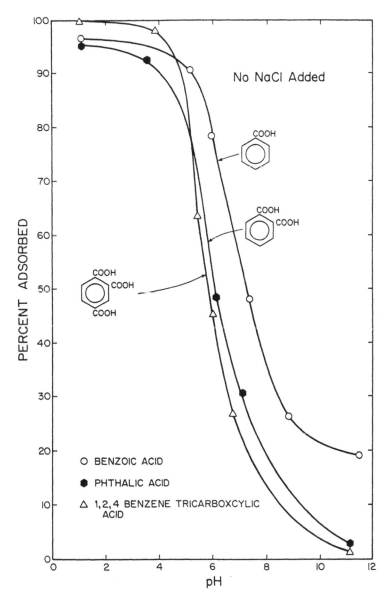

Figure 3.8 Adsorption of three aromatic carboxylic acids to activated charcoal as a function of pH. From Cooney and Wijaya (1987). Reproduced by permission of the Engineering Foundation.

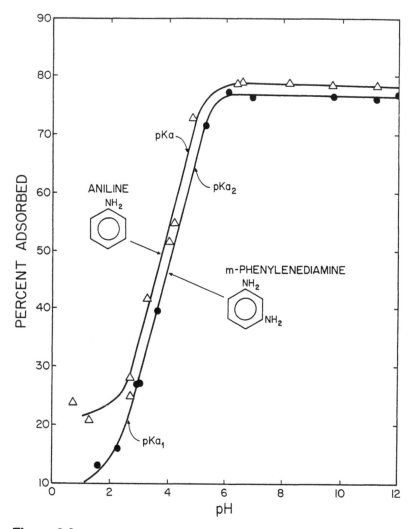

Figure 3.9 Adsorption of two basic aromatic compounds to activated charcoal as a function of pH. From Cooney and Wijaya (1987). Reproduced by permission of the Engineering Foundation.

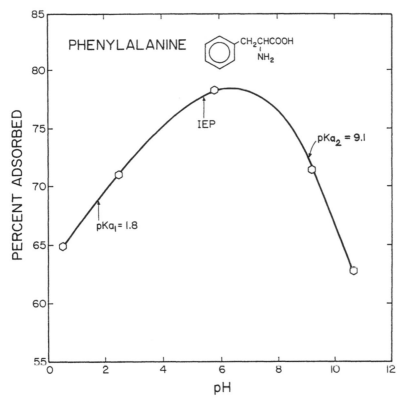

Figure 3.10 Adsorption of an amphoteric compound to activated charcoal as a function of pH. IEP = the isoelectric point. From Cooney and Wijaya (1987). Reproduced by permission of the Engineering Foundation.

ability of weak organic electrolytes with pH, and have applied it to data of theirs on *p*-nitrophenol and benzoic acid adsorption. All of these studies confirm the general principle that adsorption varies inversely with the amount of charge on the adsorbing species.

G. Presence of Inorganic Salts

Several investigators have shown that the presence of inorganic salts (e.g., NaCl) can enhance the adsorption of organic species to charcoal. The effect of adding a simple salt like NaCl to a system in which adsorption of an organic ion to charcoal is occurring is fairly simple: those salt ions which carry a charge opposite to that of the adsorbed organic ions are attracted to the spaces between adjacently adsorbed organic ions, and reduce the strength of repulsion of the adjacently adsorbed organic ions. Thus, a greater packing density of the adsorbed organic ions on the charcoal surface can be achieved.

Figure 3.11 shows data of Sorby et al. (1966) on the adsorption of promazine to charcoal from distilled water, and with the addition of 0.01 N NaCl or 0.10 N NaCl. The beneficial effect of adding sodium chloride is clear.

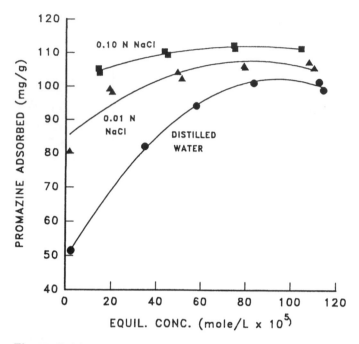

Figure 3.11   Effect of added sodium chloride on the adsorption of promazine to activated charcoal. From Sorby et al. (1966). Reproduced by permission of the American Pharmaceutical Association.

    Snoeyink et al. (1969) found that adding 1 M NaCl to solutions of *p*-nitrophenol (pK_a = 7.15) at pH 10 (at which the *p*-nitrophenol is essentially fully ionized) dramatically increased *p*-nitrophenol adsorption to charcoal, whereas at pH 2 (at which the *p*-nitrophenol is totally nonionized) there was no effect. Similarly, Coughlin and Tan (1968) observed a doubling in the adsorption of sodium benzene sulfonate (which strongly ionizes in solution) to different charcoals when 0.002–0.004 M $CaCl_2$ was added to their test solutions. Cooney and Wijaya (1987) have shown that NaCl significantly increased the adsorption of various organic compounds (benzoic acid, aniline, *m*-phenylenediamine, and anthranilic acid) when the pH was such that these compounds were ionized. An excellent and extensive review of salt effects on adsorption has been given by Randtke and Jepsen (1982); thus we refer the reader to that article for further information.

H.   Competing Solutes

Since any activated charcoal has a given surface area, the presence of other adsorbable solutes invariably implies competition for available adsorption sites. It is well known that the adsorption of a drug by activated charcoal in vivo is much less (generally about one-half) than that measured in vitro from simulated gastric fluid. The main reason for this is that in the gastrointestinal tract there exist many other solutes (enzymes, breakdown products of food such as amino acids and fatty acids) which are also capable of adsorbing to charcoal. This competition reduces the adsorption of drugs. For example, the competitive effect of tannic acid on the adsorption of the sodium salicylate is well illustrated by Figure 9.6 (see Chapter 9).

IV. DETERMINING ADSORPTION ISOTHERMS

In batch adsorption tests, a selected weight, W, of activated charcoal is mixed into a certain volume, V, of a solution of a drug (or other test substance) at an original (i.e., initial) concentration C_o. The charcoal/solution mixture is well stirred. After adsorption equilibrium is achieved (which, for typical powdered charcoal takes only around 20 min), the final concentration of drug in the solution is C_f and the amount of drug adsorbed onto the charcoal, per unit weight of charcoal, is Q. A mass balance states the obvious fact that the amount of drug adsorbed onto the solid must equal the amount of drug removed from the solution, or, in mathematical terms:

$$WQ = V(C_o - C_f)$$

Any consistent set of units for the quantities in this mass balance is acceptable. One such set might be: W = grams of charcoal, V = liters of solution, C_o and C_f = g drug/liter of solution, and Q = grams drug adsorbed/gram of charcoal. Thus, both sides of the mass balance represent amounts of drug (grams, in this case).

One can generate different C_f values in different batch samples by simply using different amounts of charcoal, W, in each sample container (in principle, one could also do this by varying C_o or V, but doing so is less convenient). One generally separates the charcoal from the solution after equilibrium is achieved using a membrane filter or by centrifugation, and analyzes the clear solution to determine the residual drug concentration (C_f). One can do the analysis by gas chromatography or liquid chromatography, ultraviolet spectrophotometry, or other suitable means. Since one then knows C_o, V, W, and C_f for each sample, then one can compute Q values for each sample from the equation

$$Q = \frac{V(C_o - C_f)}{W}$$

A plot of the pairs of Q and C_f values (Q on the ordinate, or vertical axis, and C_f on the abscissa, or horizontal axis) gives the adsorption isotherm plot. The word "isotherm" refers to the fact that the tests are conducted at a constant temperature (i.e., under isothermal conditions). One can, of course, repeat the batch adsorption tests at a different constant temperature and thereby generate another set of Q versus C_f data for that new temperature.

Figure 3.12 shows a plot of what happens during the batch adsorption process. The starting point is at $Q = 0$ and the initial fluid-phase concentration C_o. As adsorption proceeds, the system follows a path on the straight line shown; when equilibrium is achieved the system is represented by the point of intersection of the straight-line adsorption path and the curved adsorption isotherm. The slope of the adsorption path is equal to $-V/W$ [the equation of the path line is $Q = (V/W)(C_o - C)$, where Q and C are the coordinates of any point on the path line and thus the derivative, or slope, dQ/dC, is equal to $-V/W$]. Thus, if one were to decrease V and/or increase W, the slope would be lower and the point of intersection with the adsorption isotherm would be further to the left—that is, at a lower C_f value, as one would expect.

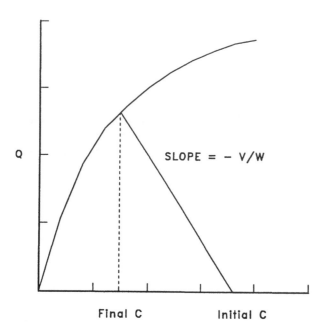

Figure 3.12   Locus of the Q and C path in a batch adsorption process.

Many isotherm equations have been proposed to fit data on Q versus C_f. We have already discussed the BET isotherm equation (the BET equation is useful for determining charcoal surface areas from the multilayer adsorption of a condensible gas like N_2 but it does not apply to the adsorption of drugs from liquid solution). We will restrict further discussion to only two other isotherm equations: the Langmuir equation and the Freundlich equation. Both of these are "two parameter" equations. That is, other than the variables Q and C_f, there appear only two constants (or parameters) in the equations. The two constants are evaluated by "fitting" the data to the equations, by methods which will be discussed shortly. These two equations are used more than 99% of the time to describe the equilibrium adsorption of drugs/poisons/toxins to activated charcoal. The reason is simple: in almost every case, one of these two equations fits the data quite well. Thus, there is no need for more elaborate isotherm equations, particularly those involving three or more parameters.

A.   The Langmuir Isotherm Equation

An adsorption "isotherm" equation is an expression of the relation between the amount of solute adsorbed and the concentration of the solute in the fluid phase, at a given constant temperature. One particular mathematical form of an isotherm which is often found to fit experimental data is the so-called Langmuir isotherm equation. The mathematical form of this equation results from the following development.

When a solution is contacted with charcoal and the system is allowed to attain equilibrium, in that equilibrium state the rate at which molecules are adsorbing to the surface is equal to the rate at which molecules are leaving the surface. This is what the concept of "equilibrium" implies. That is, equilibrium does not mean that adsorption and desorption cease to occur, but rather that their rates are equal and therefore no further net adsorption takes place.

Whether or not it is obvious, the Langmuir equation development has three important underlying assumptions: (1) it is assumed that adsorption occurs at definite localized "sites" on the surface; (2) it is assumed that each site can bind only one molecule of the adsorbing species; (3) it is assumed that the energy of adsorption (i.e., the strength of the bond created between the surface and the adsorbing species) is the same for all sites and that there are no forces of interaction between adjacently adsorbed molecules. Because the number of "sites" per unit weight of the adsorbent is fixed, adsorption can only take place until the condition is reached where every site is occupied. This generally corresponds to the condition of one complete monomolecular layer of coverage of the adsorbing species on the surface. Thus, while so-called multilayer adsorption is physically possible for certain systems, such as those involving "condensible" gases, it is assumed not to occur in liquids (the BET isotherm equation discussed above is an example of an approach which specifically does allow for multilayer adsorption).

If the rate of adsorption is presumed to be proportional to the solute concentration in the fluid (C_f) and to the fraction of the charcoal's surface area which is vacant $(1 - \theta)$, where θ = the fraction of the surface covered, we can write

Rate of adsorption $= kC_f(1 - \theta)$

where k is some constant. The assumed homogeneity of the surface sites (energy-wise) means that k has the same value for all sites. The fact that only a monomolecular layer of coverage is possible means that the adsorption rate is proportional to $(1 - \theta)$, namely, adsorption must cease when $\theta = 1$.

The rate of desorption is presumed to be proportional to the amount of solute on the surface, therefore

Rate of desorption $= k' \theta$

where k' is some other constant. Equating the two rates gives

$kC_f(1 - \theta) = k'\theta$

or

$$\theta = \frac{kC_f}{k' + kC_f}$$

$$= \frac{KC_f}{1 + KC_f}$$

where $K = k/k'$.

It is usually preferable to work in terms of the quantity Q, the weight of solute adsorbed per unit weight of charcoal, rather than θ. Since Q and θ are obviously proportional to each other, we can write

$$Q = \frac{KC_f Q_m}{1 + KC_f}$$

where Q_m is another constant (mathematically, Q_m represents the maximum value that Q tends toward as C_f becomes large). Physically, Q_m clearly represents the concentration

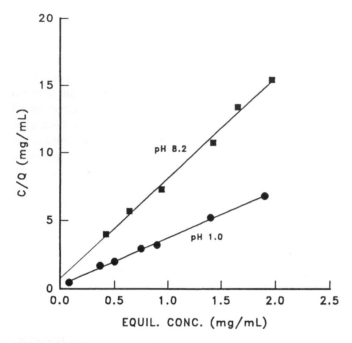

Figure 3.13 Plot for determining Langmuir isotherm parameter values, for aspirin adsorption at 37°C to activated charcoal at pH 1.0 and pH 8.2. From Tsuchiya and Levy (1972). Reproduced by permission of the American Pharmaceutical Association.

of the adsorbed species on the surface when one complete monomolecular layer of coverage is achieved.

The constants K and Q_m can be determined from experimental data on Q versus C_f by noting that the last equation can be written as

$$\frac{1}{Q} = \frac{1}{Q_m} + \frac{1}{KC_fQ_m}$$

and therefore a plot of $1/Q$ versus $1/C_f$ will yield a straight line of slope $1/(KQ_m)$ and intercept $1/Q_m$. Knowing values for the slope and the intercept allows one to easily calculate values of the two parameters K and Q_m. The "best" straight line through the data is usually obtained by a linear least-squares fitting procedure.

Alternatively, it can be noted that the previous equation may be multiplied through by C_f to give

$$\frac{C_f}{Q} = \frac{C_f}{Q_m} + \frac{1}{KQ_m}$$

and so a plot of C_f/Q versus C_f will yield a "best fit" line having a slope of $1/Q_m$ and an intercept of $1/(KQ_m)$. Such a plot is given in Figure 3.13. This figure shows data of Tsuchiya and Levy (1972) on the adsorption of aspirin at 37°C to charcoal at pH 1.0 and at pH 8.2. To use the pH 8.2 data as an example, the intercept $1/(KQ_m)$ is equal to 0.671 mg charcoal/mL, and the slope $1/Q_m$ of the line through the data is 7.524 mg charcoal/mg aspirin. From this, one can easily compute: $Q_m = 0.133$ mg aspirin/mg charcoal, and $K = 11.21$ mL/mg aspirin. Thus, the Langmuir equation for pH 8.2 is

$$Q = \frac{1.49C_f}{1 + 11.21C_f}$$

where Q has units of mg aspirin/mg charcoal and C_f has units of mg aspirin/mL.

B. The Freundlich Isotherm Equation

Data on adsorption from a liquid phase frequently are fitted better by the so-called Freundlich isotherm equation

$$Q = KC_f^{1/n}$$

where n is a constant which is usually greater than 1 (the power on C_f is written as $(1/n)$ rather than just n for traditional reasons, perhaps because this makes n have a usual value greater than 1). If one takes the logarithm of each side of this equation one obtains

$$\log Q = \log K + \left(\frac{1}{n}\right)\log C_f$$

Either logarithms to the base 10 (usually denoted by "log") or logarithms to the base e, i.e., "natural" logarithms ("ln") may be used. We have arbitrarily used "log" here. Thus, a plot of $\log Q$ versus $\log C_f$ on rectilinear paper will yield a straight line with a slope of $(1/n)$ and an "intercept" of $\log K$. From the slope and intercept values, one can readily determine the values of the two parameters K and n (equivalently, one can regard $1/n$ itself as the parameter). The "best" straight line is again usually determined using a linear least-squares fitting procedure, taking the variables as $\log Q$ and $\log C_f$ rather than as Q and C_f.

Alternately, one can plot the Q versus C_f data on log-log graph paper and obtain a straight line with the same slope and intercept. One note concerning the "intercept" is that, because $\log Q$ and $\log C_f$ do not have zero values, except when Q and C_f equal minus infinity, there is no 0,0 point on the graph one is dealing with here (this is true whether one uses rectilinear paper or log-log paper). Hence, the intercept in this case refers to the intercept of the best fit straight line through the data and a vertical line drawn upwards from the abscissa at a value corresponding to $C_f = 1$, at which point $\log C_f = 0$ and the Freundlich equation reduces to $\log Q = \log K$. This intercept gives one the value for K. An equivalent and simpler way to get values for $(1/n)$ and K, especially if one's graph does not have an abscissa which includes a $C_f = 1$ point, is to just pick two pairs of Q,C_f values near the ends of the best fit straight line, insert these into the last equation, thereby generating two equations having numerical values for Q and C_f, and solving these two equations simultaneously for $(1/n)$ and K.

The Freundlich isotherm is derivable on a theoretical basis, but the procedure is involved and is not important for our present purposes. We will, however, mention two things about the assumptions and derivation of the Freundlich isotherm equation. First, the Freundlich model does not impose any requirement that the coverage must approach a constant value corresponding to one complete monomolecular layer as C_f gets large. Indeed, the form of the equation shows that Q can continue to increase without bound

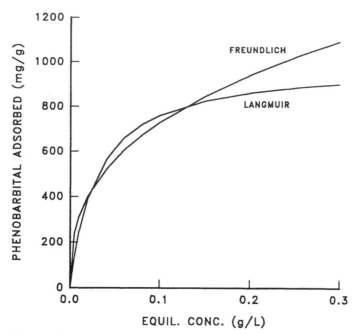

Figure 3.14 Comparison of fits of the Langmuir and Freundlich isotherm equations to data on phenobarbital adsorption from simulated gastric fluid to SuperChar activated charcoal. From Wurster et al. (1988). Reproduced by permission of the Plenum Publishing Corp.

as C_f increases, which—being physically impossible—means that the Freundlich equation should fail to fit experimental data at very high C_f values. However, most systems that one is concerned with in real adsorption processes (including those of interest in this book) are sufficiently "dilute" that one never encounters the region where the Freundlich equation breaks down for this reason.

The Freundlich isotherm equation also implies that the energy distribution for the adsorption sites is of essentially an exponential type rather than of the uniform type assumed in the Langmuir development. There is much experimental evidence that real energy distributions, while perhaps not being strictly exponential, are indeed roughly of this sort. Thus, some sites are highly energetic and bind the adsorbed solute strongly, whereas some are much less energetic and bind the adsorbed solute weakly. The rates of adsorption/desorption vary with the strength or energy of the sites. This leads to the possibility of more than just one monomolecular layer of coverage, and to a different shape of the isotherm equation.

Figure 3.14 compares the shapes of the two types of isotherm equations fitted by Wurster et al. (1988) to their data on the adsorption of phenobarbital from simulated gastric fluid to SuperChar charcoal. One can see that the Freundlich isotherm has a sharper rise at low C_f values (caused by the high energy sites, which bind well even at low C_f) and lacks a distinct plateau at high C_f values. In general, the Freundlich equation tends to fit data on adsorption from liquid solutions better, whereas the Langmuir equation sometimes fits data on the adsorption of gases to solids a bit better.

C. Least-Squares Fitting of Data to Determine Isotherm Constants

The linear least-squares fitting procedure used to obtain the two parameters in either the Langmuir or Freundlich isotherm equations can be explained rather simply. We will explain it in terms of fitting a general linear equation of the form $y = mx + b$ to some data, where y is one variable (the "dependent" variable), x is a second variable (the "independent" variable), and m and b are two constants. The constant m is the slope of a straight line on a y versus x plot and b is the intercept on the vertical axis, i.e., the value of y when $x = 0$. The aim of linear least-squares fitting is to give one the "best" values for m and b, in a statistical sense.

The statistically "best" values of m and b are determined by a mathematical procedure which minimizes the sum of the squares of the differences between the observed y values and the y values predicted from the equation $y = mx + b$ by substituting the corresponding x values. Any elementary statistics book may be consulted by the reader for the mathematical details. We will give here only the formulas for determining the best m and b values from a set of N pairs of y_i, x_i values. They are

$$m = \frac{N \sum x_i y_i - \sum x_i \sum y_i}{N \sum x_i^2 - \sum x_i \sum x_i}$$

$$b = \frac{1}{N}\left(\sum y_i - m \sum x_i\right)$$

A specific example will illustrate the determination of m and b. Let us assume we have the ten pairs of observed y and x values shown in Table 3.7. From these values, we readily compute:

$$\sum x_i = 34.5$$

$$\sum y_i = 5460$$

Table 3.7 Values of y and x for Linear Fit Example

y value	x value
425	0.9
420	1.3
480	2.0
495	2.7
540	3.4
530	3.4
590	4.1
610	5.2
690	5.5
680	6.0

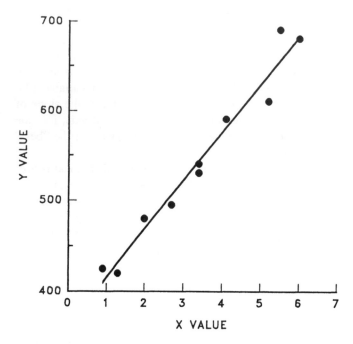

Figure 3.15 Data of the linear fit example, shown with the least-squares best fit straight line.

$$\sum x_i y_i = 20329$$

$$\sum x_i^2 = 147.01$$

Thus,

$$m = \frac{10\,(20329) - (34.5)\,(5460)}{10\,(147.01) - (34.5)\,(34.5)}$$

$$= 53.31$$

$$b = \frac{1}{10}\,[5460 - 53.31\,(34.5)]$$

$$= 362.07$$

Hence, the equation of the best fit straight line is $y = 53.31\,x + 362.07$. Figure 3.15 shows a plot of the y,x data from Table 3.7 along with the best fit straight line just determined.

A measure of how well two random variables are linearly related is given by the correlation coefficient, r. More specifically, the value of r^2 indicates what proportion of the variation of y (the dependent variable) can be attributed to a linear relationship to x (the independent variable). The quantity r is given by

$$r = \frac{N \sum x_i y_i - \sum x_i \sum y_i}{\left(N \sum x_i^2 - \sum x_i \sum x_i\right)^{0.5} \left(N \sum y_i^2 - \sum y_i \sum y_i\right)^{0.5}}$$

which, in the present example, has a numerator of

$$[10(20329) - (34.5)(5460)] = 14920$$

and a denominator of

$$[10(147.01) - (34.5)(34.5)]^{0.5} [10(3063650) - (5460)(5460)]^{0.5} = 15193.7.$$

Hence, $r = 14920/15193.7 = 0.982$ and $r^2 = 0.964$. This means that 96.4% of the variation of y can be explained by the variation of x. The correlation coefficient r is thus an index of how linear is the relationship between y and x. If y is perfectly linear in x with a positive slope, $r = 1$, whereas if y is perfectly linear in x with a negative slope, $r = -1$. If y scatters in such a way that the best fit line relating y and x is exactly horizontal, then $r = 0$. If the y values trend upward with increasing x, but scatter considerably above and below the best fit straight line, a value of r between 0 and 1 is indicated.

D. Theoretical Aspects of Isotherm Equations

Gessner and Hasan (1987a) commented on the fact that many research papers present plots of the fraction of drug adsorbed from solution in an in vitro test versus the charcoal:drug ratio. They showed mathematically that if the adsorption process is governed by either a Freundlich or a Langmuir isotherm equation, the fraction adsorbed is not a unique function of the charcoal:drug ratio, unless the charcoal amount or the initial concentration of the solution is kept constant throughout the study (a third constant parameter which would keep the function unique is the final, or equilibrium, concentration of the drug in the solution, but in practical terms there is no easy way in which to do this). Thus, such a plot giving only information on the fraction adsorbed versus the charcoal:drug ratio would not allow one to determine the isotherm equation for the system, nor to predict what fraction would be adsorbed if a different initial concentration or different charcoal amount were used.

This point can be clearly seen by an analysis which is much simpler than they present. Since we have shown that

$$QW = V(C_o - C_f)$$

then it follows that

$$QW = VC_o - VC_o \left(\frac{C_f}{C_o} \right)$$

$$= VC_o \left[1 - \left(\frac{C_f}{C_o} \right) \right]$$

If we recognize that the original dose of drug, D, in the solution is VC_o and that the fraction, F, of drug adsorbed from solution is given by $[1 - (C_f/C_o)]$, then this last equation may be written as $QW = DF$. Rearranging this gives $Q = DF/W$. If we let R stand for the ratio of charcoal used to the original amount of drug in the solution, i.e., $R = W/D$, then $Q = F/R$. Thus, if F were a unique function of R, Q would have to be constant. But we know from Figure 3.12 (the plot of the adsorption path and adsorption isotherm) as well as from the equation $WQ = V(C_o - C_f)$ that the equilibrium Q value is not constant but rather depends on C_o, W, and V. Thus, it is clear that F is not a unique function of R. Because $F/R = Q$, the relationship between F and R depends on V, W, and C_o. This is true not just for the Langmuir and Freundlich isotherm equations but for any isotherm equation.

Gessner and Hasan (1987b) have also considered the question of which type of isotherm equation tends to fit available in vitro adsorption data better: the Langmuir equation or the Freundlich equation. In their study, they generated adsorption data for paraldehyde and metaldehyde adsorbing to three different charcoals and showed that the Freundlich equation fit the data better. They then considered the fit of data presented by Sorby and Plein (1961), Sorby (1965), Sorby et al. (1966), and Andersen (1946, 1947, 1948) to these two equations, and reached the same conclusion. Gessner and Hasan indicate that even when the "linearized" Langmuir plots ($1/Q$ versus $1/C_f$) produce high correlation coefficient values (e.g., 0.99), even higher correlation coefficient values (e.g., 0.999) can occur on Freundlich plots of the same data (i.e., Q versus C_f on log-log paper or a plot of log Q versus log C_f on rectilinear paper).

Kleeman and Bailey (1988a) produced two highly theoretical papers dealing with determining the intrinsic strength of adsorption of solutes on activated charcoal. In the first paper, they derived an equation for determining the so-called Gibbs free energy of adsorption from high pressure liquid chromatography (HPLC) data. The HPLC method involved passing a solution of the adsorbable drug in a liquid consisting of 35 volume percent acetonitrile and 65 volume percent water through a chromatographic column packed with 10–20 μm sized Norit F charcoal. The Gibbs free energy value is an index of the strength of adsorption of the solute (i.e., the drug) on the charcoal and theoretically can be used to give a general prediction of how well a given type of drug could be expected to adsorb to a given type of charcoal. The Gibbs free energy is determined experimentally primarily by the retention volume of the solute in the HPLC column, i.e., how much fluid volume is required to carry an injected "slug" of the solute from the column inlet to the column outlet. The larger this volume, the larger is the free energy, and thus the greater is the strength of adsorption for that solute/charcoal pairing.

In their second paper, Kleeman and Bailey (1988b) used their Gibbs free energy equation along with experimental HPLC data to calculate values for the Gibbs free energy for 16 different drug compounds. Values ranged from −227.27 kJ/mole for antipyrine to −0.62 kJ/mole for morphine (the way free energy is defined, values are negative for the adsorption process, so what really matters is the magnitude of the value, and not its sign). The only problem with this method, in addition to its relative complexity, is that chromatographic methods inherently determine solute/solid interactions only for low solute concentrations and do not give information for higher solute concentrations. Thus, such a method is equivalent to giving information on the initial slope of an adsorption isotherm (near the 0,0 origin of a Q versus C_f plot) but not information at higher C_f values where the isotherm typically bends over sharply. What happens in real systems is often determined more by the magnitude and shape of the isotherm at higher C_f values than what the isotherm is like at very low C_f values. Thus, the Kleeman/Bailey method is limited in usefulness to "dilute" systems.

V. SUMMARY

Perhaps the most important property of an activated charcoal which is created by the pyrolysis/activation manufacturing process is its internal surface area, since its adsorption capacity depends directly on its magnitude. Present-day charcoals have quite large surface areas, on the order of 900–2000 m^2/g. The second most important property is the charcoal's pore-size distribution, since pore size determines which molecules can enter the pores. Actually, surface area and pore size are linked, since the attainment of large surface areas requires that the pores be fairly small. However, the pores should not be too small, otherwise molecules having sizes typical of drugs and poisons could not enter. Fortunately, available medicinal grade charcoals have the proper range of pore sizes. Other properties such as particle density and total pore volume are of secondary importance. However, the surface character (e.g., the types of oxides developed on the surface) is of significance and present-day charcoals are purposely activated at high temperatures to give the "H-type" characteristics which have been shown to be desirable.

With respect to the adsorption process itself, the key factors (other than the already-mentioned important properties of the charcoal—surface area, pore-size distribution) are, in decreasing order of importance, the fundamental nature of the solute (e.g., aromaticity, aliphaticity, substituent groups, size), the solution pH (which affects the ionization state of the solute), the presence of competing solutes, and the influence of inorganic salts in the solution.

Determining adsorption isotherms by preparing batch samples, each with different amounts of charcoal (or different volumes or concentrations of solution) has been reviewed. We have also discussed the use of least-squares fitting procedures to determine whether the data fit classic equations such as those of Langmuir or Freundlich. With this background, the reader should be ready to better understand subsequent discussions of the adsorption of specific types of drugs and poisons onto charcoal, and especially why the extent of adsorption varies under different conditions.

VI. REFERENCES

Andersen, A. H. (1946). Experimental studies on the pharmacology of activated charcoal. I. Adsorption power of charcoal in aqueous solutions, Acta Pharmacol. Toxicol. 2, 69.

Andersen, A. H. (1947). Experimental studies on the pharmacology of activated charcoal. II. The effect of pH on the adsorption by charcoal from aqueous solutions, Acta Pharmacol. Toxicol. 3, 199.

Andersen, A. H. (1948). Experimental studies on the pharmacology of activated charcoal. III. Adsorption from gastro-intestinal contents, Acta Pharmacol. Toxicol. 4, 275.

Brunauer, S., Emmett, P. H., and Teller, E. (1938). Adsorption of gases in multimolecular layers, J. Am. Chem. Soc. 60, 309.

Burke, G. M. (1991). Adsorptivity and Surface Characterization of Activated Charcoals, Ph.D. thesis, University of Iowa, Iowa City.

Burke, G. M., Wurster, D. E., Berg, M. J., Veng-Pedersen, P., and Schottelius, D. D. (1992). Surface characterization of activated charcoal by x-ray photoelectron spectroscopy (XPS): Correlation with phenobarbital adsorption data, Pharm. Res. 9, 126.

Cooney, D. O. (1980). Activated Charcoal: Antidotal and Other Medical Uses, Marcel Dekker, New York.

Cooney, D. O. and Wijaya, J. (1987). Effect of pH and added salts on the adsorption of ionizable organic species onto activated carbon from aqueous solution, in Fundamentals of Adsorption, A. I. Liapis, Ed. (Proceedings of the Second Engineering Foundation Conference on Fundamentals of Adsorption), Engineering Foundation, New York, pp. 185-194.

Cooney, D. O. (1992). Desorption of sodium salicylate from powdered activated charcoal, unpublished paper.

Corapcioglu, M. O. and Huang, C. P. (1987). The surface acidity and characterization of some commercial activated carbons, Carbon *25*, 569.

Coughlin, R. W. and Tan, R. N. (1968). Role of functional groups in adsorption of organic pollutants on carbon, Chem. Eng. Progr. Symp. Ser. *90*, (No. 64), 207.

Frumkin, A. (1930). The adsorption of electrolytes by activated carbon, Kolloid-Z. *51*, 123.

Garten, V. A. and Weiss, D. E. (1957). Ion and electron exchange properties of activated carbon in relation to its behavior as a catalyst and adsorbent, Rev. Pure Appl. Chem. *7*, 69.

Gessner, P. K. and Hasan, M. M. (1987a). Toxicant adsorption on activated charcoal: Is the fraction adsorbed a unique function of the charcoal:adsorbate ratio?, J. Pharm. Sci. *76*, 707.

Gessner, P. K. and Hasan, M. M. (1987b). Freundlich and Langmuir isotherms as models for the adsorption of toxicants on activated charcoal, J. Pharm. Sci. *76*, 319.

Hassler, J. W. (1963). *Activated Carbon*, Chemical Publishing Co., New York.

Ishizaki, C. and Marti, I. (1981). Surface oxide structures on a commercial activated carbon, Carbon *19*, 409.

Kleeman, W. P. and Bailey, L. C. (1988a). Thermodynamic evaluation of activated charcoal as a poison antidote by high-performance liquid chromatography I: Derivation and validation of an equation for Gibbs free energy of liquid-solid adsorption, J. Pharm. Sci. *77*, 500.

Kleeman, W. P. and Bailey, L. C. (1988b). Thermodynamic evaluation of activated charcoal as a poison antidote by high-performance liquid chromatography II: In vitro method for the evaluation of activated charcoal as a poison antidote, J. Pharm. Sci. *77*, 506.

Martin, R. J. and Iwugo, K. O. (1982). The effects of pH and suspended solids in the removal of organics from waters and wastewaters by the activated carbon adsorption process, Water Res. *16*, 73.

Mattson, J. S., Mark, H. B., Jr., Malbin, M. D., Weber, W. J., Jr., and Crittenden, J. C. (1969). Surface chemistry of active carbon: Specific adsorption of phenols, J. Colloid Interface Sci. *31*, 116.

Mattson, J. S. and Mark, H. B., Jr. (1971). *Activated Carbon: Surface Chemistry and Adsorption from Solution*, Marcel Dekker, Inc., New York.

Müller, G., Radke, C. J., and Prausnitz, J. M. (1980). Adsorption of weak organic electrolytes from aqueous solution on activated carbon: Effect of pH, J. Phys. Chem. *84*, 369.

Randtke, S. J. and Jepsen, C. P. (1982). Effects of salts on activated carbon adsorption of fulvic acids, J. Am. Water Works Assoc. *74*, 84.

Ruthven, D. M. (1984). *Principles of Adsorption and Adsorption Processes*, Wiley and Sons, New York, pp. 52-55.

Snoeyink, V. L. and Weber, W. J., Jr. (1967). The surface chemistry of active carbon: A discussion of structure and surface functional groups, Environ. Sci. Technol. *1*, 228.

Snoeyink, V. L., Weber, W. J., Jr., and Mark, H. B., Jr. (1969). Sorption of phenol and nitrophenol by active carbon, Environ. Sci. Technol. *3*, 918.

Sorby, D. L. and Plein, E. M. (1961). Adsorption of phenothiazine derivatives by kaolin, talc, and Norit, J. Pharm. Sci. *50*, 355.

Sorby, D. L. (1965). Effect of adsorbents on drug absorption: I. Modification of promazine absorption by activated attapulgite and activated charcoal, J. Pharm. Sci. *54*, 677.

Sorby, D. L., Plein, E. M., and Benmaman, J. D. (1966). Adsorption of phenothiazine derivatives by solid adsorbents, J. Pharm. Sci. *55*, 785.

Steenberg, B. (1944). *Adsorption and Exchange of Ions on Activated Charcoal*, Almqvist and Wiksells, Uppsala, Sweden.

Tebbutt, T. H. Y. and Bahiah, S. J. (1977). Studies on adsorption with activated carbon, Effluent Water Treat. J. *17*, 123.

Tsuchiya, T. and Levy, G. (1972). Relationship between effect of activated charcoal on drug absorption in man and its drug adsorption characteristics in vitro, J. Pharm. Sci. *61*, 586.

Wang, L. K., Leonard, R. P., Wang, M. H., and Goupil, D. W. (1975). Adsorption of dissolved organics from industrial effluents onto activated carbon, J. Appl. Chem. Biotechnol. *25*, 491.

Ward, T. M. and Getzen, F. W. (1970). Influence of pH on the adsorption of aromatic acids on activated carbon, Environ. Sci. Technol. *4*, 64.

Wurster, D. E., Burke, G. M., Berg, M. J., Veng-Pedersen, P., and Schottelius, D. D. (1988). Phenobarbital adsorption from Simulated Intestinal Fluid, U.S.P., and Simulated Gastric Fluid, U.S.P., by two activated charcoals, Pharm. Res. *5*, 183.

United States Pharmacopeia, Twenty-Second Revision, United States Pharmacopeial Convention, Rockville, Maryland, 1989.

4

Properties of Antidotal Charcoal

In this chapter, we review the criteria that medicinal-grade charcoals must satisfy according to standards set in the official Pharmacopeia of the United States, as well as standards of other pharmacopeias (e.g., that of Britain). We then list and discuss the wide variety of charcoals and charcoal formulations currently available in the USA and elsewhere in the world. As a side topic, we describe the history of "superactive" charcoal, which unfortunately is no longer available. Then, we discuss issues relating to whether certain ready-mixed charcoal formulations "deliver" what their labels imply. Finally, we review several studies in which the in vitro and in vivo performance of different brands and types of charcoals have been compared.

I. REQUIREMENTS OF THE *U. S. PHARMACOPEIA*

Activated charcoal for antidotal purposes must, in the USA, conform to the specifications of the twenty-second edition of the *United States Pharmacopeia* (USP) (1989), which states that USP charcoal must have the following main attributes:

1. Less than 15 wt% water.
2. No more than 4 wt% residue upon ignition (i.e., ash).
3. No more than 3.5 wt% acid-soluble substances.
4. No more than 0.2 wt% chloride, no more than 0.2 wt% sulfate, no more than 0.005 wt% heavy metals, and essentially zero sulfides and cyanogen compounds.
5. Water boiled with the charcoal shows a neutral pH, after cooling.
6. The charcoal shows an absence of *Salmonella* and *E. coli* organisms when tested using standard microbial limit tests.

These reflect the specifications on page 269 of the "Official Monograph" portion of the USP. However, on page 1734 under the "Reagents" section, the specifications for activated charcoal are in many cases different and the test methods also vary. For example, under the "Reagents" specifications there is no mention of microbial limits, packaging/storage methods, loss on drying, cyanogens, heavy metals, and the methylene blue adsorption test. Also, the sulfate requirement is stricter (0.15 wt% versus the 0.2 wt% on page 269) and there is an alcohol-soluble substance limit of 0.2 wt% which does not appear on page 269. The amounts of charcoal to be used in the pH test and the volumes of solutions used in the pH, acid-solubles, chloride, and sulfate tests are different on the two pages.

Some of these differences are logical, e.g., there is no reason why reagent charcoals used primarily for decolorizing or otherwise purifying solutions which are not intended for human consumption should have to meet microbial, packaging/storage, loss on drying, cyanogen, and heavy metals requirements. But, differences in the amounts of charcoal and solutions used in the various other tests seem strange. One would think that the tests for things like chlorides, sulfates, and reaction (i.e., pH) should be the same.

The requirement for an absolutely neutral pH seems unnecessary for charcoals which are to be used as antidotes (however, for other medical applications, such as the purification of pharmaceuticals, there is justification for such a requirement). Most charcoals are acid-washed with HCl solution in order to meet the acid-soluble and ash requirements. They are then washed repeatedly with pure water to remove essentially all of the HCl (an expensive process) so that the "neutral pH" criterion can be met. However, in most cases, these charcoals are then introduced into the stomach—a medium which is essentially a pH 1–3 HCl solution. The other advantage of allowing some residual HCl in the charcoal is that, if the charcoal is used in some type of ready-mixed aqueous suspension, a certain degree of acidity would exist which would kill (or prevent the growth of) microorganisms.

The adsorptive power of the charcoal must meet certain requirements for the adsorption of an alkaloid (strychnine sulfate) and for a dye (methylene blue). In the strychnine sulfate test, 1 g of charcoal is shaken for 5 minutes with 50 mL of a 2 g/L strychnine sulfate solution. The residual amount of strychnine sulfate in the solution should be essentially zero, as determined qualitatively by a turbidity test using added reagents. A simple calculation shows that zero turbidity corresponds to a Q value (g adsorbed/g charcoal) of 0.10, which should be achievable for any "typical" activated charcoal.

In the dye test, 50 mL of a 1 g/L methylene blue (molecular weight 374 daltons) solution is shaken for 5 minutes with 250 mg of the charcoal. In another flask, no charcoal is added. Then, in a long series of filtrations, additions of reagents, shaking at 10 minute intervals for 50 minutes, and titrations, the residual methylene blue in the two cases is determined. The difference (expressed in terms of the volume of the final titrating solution) should be at least 0.7 mL or greater.

Two comments can be made concerning these adsorptive power tests. First, the shaking time of 5 minutes is too brief. As Figure 4.1 shows, even with the vigorous shaking which was involved in the tests shown in the figure, it takes on the order of 25 minutes for adsorption equilibrium to be attained for solutes of medium molecular weight (sodium barbital). The 5 minutes specified for the adsorptive power tests is too short for equilibrium to be reached. The question of how the 5 minute period was selected, especially with respect to what may be typical times of contact between drugs and charcoal in actual poisoning situations, has never been explained or defended. In actual fact, however, by the time one sets up the filters which are used in the post-contacting steps of the tests and/or performs the filtrations, the effective time of contact is probably much larger than 5 minutes and thus adsorption equilibrium may indeed be attained anyway.

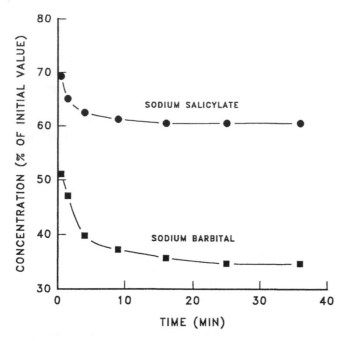

Figure 4.1 Rate of adsorption of two selected drugs to powdered activated charcoal as a function of time. From Cooney and Kane, 1980.

Secondly, the methylene blue test, in particular, is unnecessarily complicated. If one filters a methylene blue/charcoal aqueous suspension, adds an appropriate volume of water to dilute the filtrate, and measures the color in an ordinary colorimeter (i.e., a spectrophotometer employing visible light), a rapid assessment of the residual methylene blue concentration can be obtained. Treating the "control" sample (i.e., the one where no charcoal was added) similarly tells one quickly what percent of the methylene blue was adsorbed by the charcoal. This value can then be compared to some minimum acceptable standard value. Such a procedure would avoid the addition of the many reagents, the further shaking and waiting, the performance of a second filtration, and the performance of an exacting titration which the current test requires.

A final comment concerns the rationale for selecting strychnine sulfate and methylene blue as the two test substances. The question is: why these and not others? The one clear advantage of using a solute such as methylene blue—namely, the ability to quickly determine its concentration by measuring the intensity of color—is not taken advantage of in the current test; thus, there must be another reason for its use (such as its molecular weight making it representative of a variety of common drugs/poisons). These adsorptive power tests need to be to reviewed with respect to whether some other solutes might be better choices.

The *British Pharmacopoeia* (1988) has some criteria for activated charcoal which are very similar to those of the USP. For example, the ash content is set at a maximum of 5 wt% (versus 4 wt% in the USP), acid-soluble substances are set at 3 wt% (versus 3.5 wt% in the USP), the maximum water content is 15 wt% (as in the USP), alcohol-soluble substances are 0.5 wt% (versus 0.2 wt% in the USP), and there must be a virtual absence of sulfides (as in the USP). However, the *British Pharmacopoeia* sets limits of 10 ppm for lead, 25 ppm for zinc, and 25 ppm for copper, whereas the USP merely sets a limit of 0.005 wt% for all "heavy metals." The *British Pharmacopoeia* has a limit on alkali-soluble substances, whereas the USP makes no mention of these. On the other hand, the USP has limits for cyanides, chloride, and sulfate, while the *British Pharmacopoeia* has no such specifications. Regarding adsorbing power, the *British Pharmacopoeia* uses "not less than 40% of its own weight of phenazone", whereas the USP adsorbing power tests are in terms of strychnine sulfate and methylene blue.

The International Pharmacopeia (1981), published by the World Health Organization, has specifications which are both similar to and different from those of the USP and the *British Pharmacopoeia*. For adsorbing power, it uses a strychnine sulfate test (like the USP) and has an iodine test (which used to be a common test, as explained in Chapter 2, but which no longer appears in either the USP or the *British Pharmacopoeia*). Many of the criteria are similar to those of the *British Pharmacopoeia*, and some (e.g., for heavy metals and cyanides) are similar to those of the USP. The variations among these three pharmacopeias alone (many others exist in other specific countries) are enough to suggest that a consistent and universal set of criteria might be worthwhile.

It would be interesting to compare the specifications for activated charcoal in various other pharmacopeias, such as those of France, Germany, Spain, Sweden, Australia, etc. However, the writer has not been able to access these, and so a comparison of this sort can not be presented here.

II. "SUPERACTIVE" CHARCOAL

In early 1976, Amoco Research Corporation (Naperville, Illinois), a subsidiary of the Standard Oil Company of Indiana, announced the development of experimental powdered activated charcoals having very high surface areas. These charcoals were then being produced in a 1000 ton/week pilot plant and Amoco stated that their immediate plans were to evaluate the market potential of these charcoals. A prime market that was envisioned was for powdered activated charcoal waste water treatment processes (similar to DuPont's PACT process) and demonstration facilities for such were established.

Two of these new charcoals, the PX-21 and PX-23 grades, had internal surface areas of 2800–3500 m^2/g and 3000–3300 m^2/g, respectively. A third powdered charcoal, PX-24, had a surface area of 1300–1400 m^2/g, while a granular type of charcoal (GX-22) made from powdered charcoal mixed with a binder was also produced.

Assuming that a charcoal's adsorption capacity is proportional to its internal surface area (this implies that all pores are large enough for the drug to enter), one would expect the PX-21 and PX-23 charcoals to adsorb roughly three times as much drug as conventional charcoals of that time having about 1000 m^2/g area.

This writer obtained samples of all of these new charcoals from Amoco in late 1976 and proceeded to conduct some simple in vitro experiments (Cooney, 1977) on the adsorption of sodium salicylate from Simulated Gastric Fluid (pepsin omitted) with the PX-21 charcoal. For comparison, the same tests were run on Norit A charcoal. Figure 4.2 shows the results. The Amoco PX-21 charcoal had an adsorption capacity 2.5–2.8 times that of the Norit A charcoal over the concentration range of 0.1 to 1.0 g/L sodium salicylate in the liquid phase, respectively. This brief report brought the existence of "superactive" charcoal to the attention of researchers in the antidotal charcoal and related

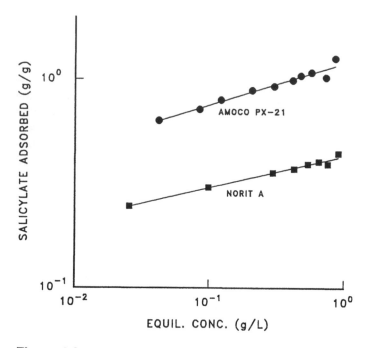

Figure 4.2 Salicylate adsorption from simulated gastric fluid by two types of activated charcoals. Originally published in Cooney (1977), American Society of Hospital Pharmacists, Inc. All rights reserved. Reprinted with permission.

fields, and led to its use over the next dozen years in a large variety of in vitro and in vivo studies.

Not long after Cooney's report, Medema (1979) questioned the advisability of using Amoco PX-21 charcoal, arguing that its very large surface area could only mean that its internal pore structure must be exceedingly fine; hence, adsorption to such a charcoal might be very slow. However, a subsequent article by Cooney and Kane (1980) demonstrated that the Amoco PX-21 charcoal adsorbed test drugs (sodium barbital, sodium salicylate) even somewhat faster, in vitro, than Norit A charcoal. Thus the Amoco PX-21 was shown to be to be superior to Norit A with respect to kinetics as well as with respect to equilibrium adsorption capacity.

The story of the Amoco PX-21 "superactive" charcoal appears throughout this volume, in the very many studies which made use of it. This charcoal appeared in commercially marketed antidotal products in 1984 known as SuperChar Liquid (with sorbitol) and SuperChar Powder (to which water was to be added prior to use), both marketed by Med-Corp, Inc. (Dallas), as listed in the 1984 edition of the *Annual Pharmacists' Reference* (Drug Topics Redbook). In 1984, Med-Corp also marketed two products with "ordinary" charcoal, ActaChar Liquid (with sorbitol) and ActaChar Powder. In the 1985–1990 editions of the Redbook, Med-Corp is still listed as the marketer of the ActaChar products, but in 1985 (and until 1989) the SuperChar products were marketed by Gulf Bio-Systems (Dallas). Gulf Bio-Systems also marketed a veterinary charcoal product containing SuperChar. Curiously, both Med-Corp and Gulf Bio-Systems had the same street address in Dallas: 5310 Harvest Hill Road. The exact connection between these two companies is unknown to the writer, but it appears that the companies had certain persons associated with both concerns, and other persons associated with just one or the other.

The complete Amoco PX-21 story is rather involved. In the mid 1980s (1984 would be a rough guess), Amoco apparently decided not to commercialize their superactive charcoal products. The manufacturing process was expensive, thus requiring a much higher than normal price, and not enough customers were willing to pay a higher price per pound, even though fewer pounds of the PX-21 would be required (due to its higher adsorption capacity). Thus, plans for a reputed $30 million full-scale production plant were cancelled. By this time, the antidotal products community was enamored with "superactive" charcoal. This market was too small to interest Amoco, but they were willing to sell off their remaining stock of PX-21 and a 100-year license to produce it to a company named Anderson Development Company (Adrian, MI). The account obtained by the writer is that Mr. Andrew Anderson, with the financial involvement of other investors, spent on the order of $1,000,000 to build a production facility and make SuperChar over the next 3–5 years. The charcoal was marketed by the new Gulf Bio-Systems company mentioned above. However, there apparently were persistent problems with the charcoal involving excessive heavy metals contents and bacterial contamination. When Mr. Anderson, who reportedly owned 51% of Gulf Bio-Systems, received an offer from Japanese investors to buy the company, Mr. Anderson sold it. The Japanese buyers soon after shut down the company, perhaps because they simply wished to buy the technology and bring it back to Japan. However, the writer is not aware of any Japanese company producing superactive charcoal, although such might exist. In any event, after 1989, no listing of SuperChar products or any mention of Gulf Bio-Systems can be found in the Redbook. For that matter, the Med-Corp company and its products disappeared after the 1990 Redbook.

After Amoco decided not to commercialize their "superactive" charcoals, two of their researchers published a scientific paper which describes the unique method of preparing these charcoals. The authors, O'Grady and Wennerberg (1986), mention that these charcoals were "evaluated over several years by nearly a hundred industrial, government and academic organizations worldwide in several hundred tests for numerous applications." In most of these tests, these charcoals outperformed other available charcoals by factors "in the 2 to 4 range." They further state that, "of considerable interest has been the use of this carbon for poison and drug overdose control for both humans and animals." By virtue of this paper by O'Grady and Wennerberg, we now know how to prepare these charcoals, even if we can not buy them.

III. CURRENTLY AVAILABLE USP CHARCOALS IN THE USA

At the time of this writing (late 1993), there is apparently only one manufacturer of USP grade activated charcoals in the USA. This is the American Norit Company (Atlanta), a subsidiary of the well-known European company Norit N. V. (Amersfoort, The Netherlands). Norit markets three USP charcoals in the USA: Norit USP XXII, Norit B Supra, and Norit A Supra, with surface areas of 900, 1400, and 2000 m^2/g, respectively. The properties of these three charcoals were given in Table 3.3. The designation "XXII" on the first one means that it conforms to the requirements of the XXII edition of the *U. S. Pharmacopeia* (1989). The two Supra types also meet these criteria. American Norit also markets the line of Darco charcoals, previously made by ICI United States, Inc. (Wilmington, Delaware). American Norit bought out ICI in 1985. However, none of the Darco charcoals are USP grade, as none are packaged in a sterile environment, and thus do not meet the USP sterility requirement (despite this, manufacturers have been known to claim that some Darco types meet all USP criteria).

Other Norit charcoals marketed by the American Norit Company are Norit SX2 and Norit SX Plus. The Darco charcoals marketed by the American Norit Company include Darco S-51, Darco S-51A, Premium Darco, Darco KB, Darco KBB, Darco FM-1, Darco GFP, Darco G-60, Darco TRS, Darco FGD, and HydroDarco B. American Norit also markets charcoals with the names Gro-Safe and HDC/HDR. None of these other charcoals meet USP specifications, however. The variations in these charcoals relate to their surface area and pore-size distributions. Many are tailored to specific applications, such as uses in the fine chemicals, industrial chemicals, electroplating, dry cleaning, cane sugar, corn syrup, candy, lactose, vegetable fats/oils, fruit juice, wine, distilled spirits, and agricultural industries. Others find uses in desulfurization, as catalyst supports, and for potable water or waste water treatment. For the general purification of pharmaceuticals and other medicinal products, the three Norit USP charcoals, and the Norit SX2, Darco KB, Darco KBB, and Darco G-60 charcoals are used.

IV.   CURRENTLY AVAILABLE CHARCOAL PRODUCTS IN THE USA

Many companies purchase USP charcoals from American Norit and then repackage them into smaller quantities, make tablets or capsules with them, or prepare liquid suspensions (with or without sorbitol) from them. Despite any claims to the contrary, it is believed that all medicinal charcoals in the USA originate from the American Norit Company.

Using the *Annual Pharmacists' Reference* (Drug Topics Redbook), which has been published in the USA for more than 90 years, and looking back over the past ten years or so of these, several things are apparent with regard to products listed under the heading of "Charcoal." First, charcoal products fall into four categories: (1) tablets, (2) capsules, (3) powdered charcoal sold in large quantities (such as 500 g) by chemical companies, or in smaller packaged quantities (e.g., 10, 30, 50 g) by pharmaceutical type companies, and (4) ready-mixed aqueous suspensions of powdered charcoal, with or without sorbitol. Second, one notes that several manufacturers suddenly "appear" in the listings, only to equally suddenly "disappear" a few years later (commonly only 2–3 years later).

As of the 1992 Redbook, only two makers of capsules were listed, and only three makers of pure charcoal tablets were listed, plus one company which makes tablets of charcoal and simethicone combined. Both the capsules and tablets appear to be aimed at "historical" remedy-type uses of charcoal (e.g., reduction of intestinal gas). Powdered USP charcoal in bulk (lots on the order of 1 lb) is offered by six companies. However, as noted above, these bulk suppliers must obtain their charcoals from the American Norit Company. Bottles containing charcoal doses on the order of 15–50 g, to which water is intended to be added (with subsequent shaking) just prior to administration, first appeared in about 1982; several such formulations are currently marketed. Formulations with charcoal already mixed with sterile water or water/sorbitol solutions seem to have appeared beginning in about 1984. The use of sorbitol in such formulations is currently a topic of considerable debate (as discussed in Chapters 7 and 15), and whether sorbitol should be permitted is being examined by the FDA.

Table 4.1 Antidotal Charcoal Suspensions Available in the USA Having No Sorbitol[a]

Product name	Charcoal amount (g)	Volume (mL)	Charcoal type	Surface area (m^2/g)
LiquiChar[b]	12.5	60	Norit USP XXII	900
	25	120	"	"
	30	120	"	"
	50	240	"	"
InstaChar[c]	15	120	Norit USP XXII	900
	50	240	"	"
Actidose Aqua[d]	15	72	(undisclosed)	
	25	120	"	
	50	240	"	
CharcoAid 2000[e]	50	240	Norit A Supra	2000
Activated charcoal USP[f]	25	120	Norit USP XXII	900

[a] In addition to the water/charcoal formulations shown above, Lex Pharmaceutical (Medley, FL) markets 15 and 30 g packages of *dry* Norit USP XXII activated charcoal under the brand name Charcolex.
[b] Jones Medical, Canton, OH.
[c] Kerr Chemical Co., Novi, MI.
[d] Paddock Laboratories, Minneapolis, MN.
[e] Requa, Greenwich, CT.
[f] UDL Laboratories, Rockford, IL.
Adapted from McFarland and Chyka (1993). Reprinted by permission of the Harvey Whitney Books Company.

Since this book is not primarily concerned with the use of charcoal capsules, tablets, or bulk charcoal, we will mention here only those currently available products which consist of: (1) bottles containing charcoal in amounts of roughly 10–50 g, to which water is to be added, (2) ready-mixed formulations of charcoal and water, and (3) ready-mixed formulations with charcoal, water, and sorbitol.

Tables 4.1 and 4.2 show summaries of the charcoal/water and charcoal/water/sorbitol formulations currently available in the USA. Similar tabulated information has been compiled and presented by McFarland and Chyka (1993), along with price data. Some of the "volumes" are often quoted on the labels and in the manufacturers' literature in terms of fluid ounces. Since a fluid ounce is 30 mL (actually, 29.57 mL, but most pharmacopeias accept 30 mL as the equivalent), then 60 mL = 2 oz, 120 mL = 4 oz, and 240 mL = 8 oz.

One can see that there are only six marketers of prepackaged antidotal formulations at present in the USA. Concerning the formulations which contain sorbitol, the total amounts of sorbitol in each of the formulations are: LiquiChar with Sorbitol; 27 and 54 g in the 25 and 50 g charcoal mixtures; Actidose with Sorbitol, 48 and 96 g in the 25 and 50 g charcoal mixtures; Charcoaid, 110 g in the 30 g charcoal mixture; and Activated Charcoal USP with Sorbitol, 27 g in the 25 g charcoal mixture.

The label on the Charcoaid bottle states that the sorbitol content is "110 g". The product is made by mixing 30 g charcoal with approximately 120 mL of Sorbitol Solution USP (this gives a final volume after mixing with the charcoal of 150 mL). Since Sorbitol Solution USP is 70% w/w (not 70% w/v) and has a density of 1.285 g/mL, the sorbitol

Table 4.2   Antidotal Charcoal Suspensions Available in the USA Having Sorbitol

Product name	Charcoal amount (g)	Volume (mL)	Sorbitol (mg/mL)	Charcoal type	Surface area (m^2/g)
LiquiChar with Sorbitol[a]	25	120	225	Norit USP XXII	900
	50	240	225	"	"
Actidose with Sorbitol[b]	25	120	400	(undisclosed)	
	50	240	400	"	
Charcoaid[c]	30	150	720	Norit B Supra	1400
Activated charcoal USP with Sorbitol[d]	25	120	225	Norit USP XXII	900

[a] Jones Medical, Canton, OH
[b] Paddock Laboratories, Minneapolis, MN
[c] Requa, Greenwich, CT
[d] UDL Laboratories, Rockford, IL
Adapted from McFarland and Chyka (1993). Reprinted by permission of the Harvey Whitney Books Company.

content is 120 mL × 1.285 g/mL × 0.70 g/g = 108 g. Rounding off gives the "110 g" figure. The actual sorbitol content of the product in mg/mL is thus 108,000/150 = 720 mg/mL.

From Table 4.2, one may note that the sorbitol contents of the Jones Medical Liqui-Char with Sorbitol products and of the UDL Activated Charcoal USP with Sorbitol product are only 56.25% as much as those of the comparable Paddock Actidose with Sorbitol products, and that the Requa Charcoaid has the highest sorbitol content of all. On a mg/mL basis, Requa Charcoaid has 80% more sorbitol than the Paddock Actidose with Sorbitol products.

The CharcoAid 2000 product was released for sale by Requa in late 1993, and is the only charcoal formulation to employ the 2000 m^2/g surface area Norit A Supra charcoal (Requa has an exclusive license with the American Norit Company for its use). Requa also markets Charcocaps, containing 260 mg charcoal, in capsule form and also in caplet form, for "relief from intestinal gas" and "Activated Charcoal Tablets", with 250 mg charcoal and other unstated ingredients, for the same purpose.

While McFarland and Chyka list the surface areas of Norit USP XXII and Norit B Supra as 950 and 1500 m^2/g, respectively, recent data sheets from the American Norit Company give the figures of 900 and 1400 m^2/g (and the figure of 2000 m^2/g for Norit A Supra) quoted earlier in Chapter 3, and cited in Tables 4.1 and 4.2. The InstaChar formulation is made with Norit USP XXII charcoal (Saulson, 1993) (McFarland and Chyka listed the charcoal identity as "undisclosed" in their compilation). The type of charcoal used in the Paddock formulations has not been disclosed by Paddock. In response to an inquiry, Mr. Bruce G. Paddock, President, stated that: "We choose not to disclose the source of the activated charcoal, nor the specifics of the formulations, since they are proprietary information and trade secrets." (Paddock, 1994).

It should be mentioned that all charcoal suspensions must contain some type of preservative to inhibit microbial growth. It is believed that at least one of the currently marketed formulations contains propylene glycol as a preservative. One advantage of this chemical is that it adsorbs poorly to charcoal itself because of its simple non-ring structure. This keeps most of the chemical in solution, which prevents microbial growth in the solution. Also, because it adsorbs poorly to charcoal, it does not interfere to any significant extent with drug adsorption. However, propylene glycol imparts a slightly acrid taste to a formulation (which, however, can be masked to some extent by a sweetener).

Many formulations have paraben compounds or benzoate compounds as preservatives (e.g., CharcoAid 2000 contains sodium methyl paraben, sodium propyl paraben, and sodium benzoate in the amount of about 0.1–0.2% of each). Since paraben and benzoate compounds contain benzene ring moieties (paraben compounds are esters of simple alcohols and *p*-hydroxybenzoic acid), they adsorb strongly to charcoal . Because the amounts of such preservatives used are small, this does not cause any significant competition for adsorption by drug molecules. However, it does mean that only small amounts are left unbound in solution. The writer filtered samples of LiquiChar, InstaChar, Actidose Aqua, and CharcoAid 2000, and ran ultraviolet absorption scans on the filtrates using a Beckman DU-6 spectrophotometer. Only the filtrate from the Actidose Aqua showed a significant UV absorption, implying that the amounts of unbound preservatives in the other three products were very small, while the levels of unbound ingredients (preservatives and/or sweeteners) in the Actidose Aqua product were much higher. Apparently, however, the amounts of unbound preservatives left in the LiquiChar, Insta-Char, and CharcoAid 2000 solutions must be sufficient to inhibit microbial growth.

The writer recently performed some simple tests using charcoal obtained from Paddock's Actidose Aqua. A sample of the Actidose Aqua slurry was diluted with distilled water and placed in a 50 mL syringe. It was then filtered through a 0.45 µm pore-size Millipore cellulose filter, leaving the charcoal on the filter pad. The charcoal was rinsed several times by putting distilled water into the syringe and forcing it through the filter unit. In this way, the water-soluble constituents of the Actidose Aqua were presumably removed. The charcoal was then placed into a glass beaker and dried for several hours at 400°F. Samples of Norit B Supra and Norit USP XXII charcoals (obtained from the American Norit Company) were dried in the same way. Then, 0.0300 g amounts of each charcoal were contacted with 20 mL of a solution of 1 g/L sodium salicylate in 0.1 N HCl overnight in new, capped glass vials, with shaking. After filtering off the supernatants and analyzing for residual salicylate using Trinder's Reagent (a colorimetric analysis at 540 nm), the percentages of salicylate adsorbed by each charcoal were found to be: Norit B Supra, 78.7%; Norit USP XXII, 56.2% (average of two trials), and the Actidose Aqua charcoal, 54.3% (average of two trials). Thus, it was concluded that the type of charcoal in Actidose Aqua is Norit USP XXII (the possibility exists that Paddock has chosen to stay with the charcoal they used to use, Darco KB-B, but this charcoal does not meet USP requirements). The difference between 56.2% and 54.3% could be experimental error or a normal variation in the charcoal from lot to lot, or some small interference caused by an ingredient of the Actidose Aqua formulation.

With respect to Actidose Aqua, it should be pointed out that the vehicle is not just water. The suspension has a distinct sweet taste, similar to that of Actidose with Sorbitol, but less intense. When the writer filtered some water-diluted Actidose Aqua through a membrane filter to obtain a clear filtrate and placed this filtrate on a watch glass for several days to let the water evaporate, white crystals formed. The substance looked like ordinary sugar, and tasted like it. However, there was a smell of a volatile substance in the suspension when it was poured out of the bottle, and a slight sharp taste in the dry residue, suggesting that more than just some type of sugar was present. This is very likely the preservative.

The writer has learned (J. W. Ford, Vice President of Sales and Marketing, UDL Laboratories, 1993) that UDL Laboratories does not prepare their formulations but relies on other suppliers. Moreover, in 1993 they changed from Jones Medical to Paddock as their supplier, and they now market aqueous suspensions of Activated Charcoal USP in 2.5, 4, and 8 oz bottles, plus Activated Charcoal USP with Sorbitol in 4 and 8 oz bottles. These are the same formulation sizes offered by Paddock. Thus, except for the labeling, the Paddock and UDL formulations are presumably now very similar.

In a study of several ready-mixed formulations published by Krenzelok and Lush (1991), mention was made of two other major suppliers: Med-Corp. (Dallas), which sold the product line called ActaChar (with and without sorbitol), and Gulf Bio-Systems (Dallas), which sold the product line called SuperChar (with and without sorbitol). These companies were discussed earlier, and it was mentioned that they no longer exist.

Although our discussion of charcoal products available in the USA has mentioned a fair number of items, ready-mixed powdered charcoal formulations are not, to this writer's knowledge, available to the general public in regular pharmacies. It appears that the only powdered charcoal product being sold in regular pharmacies is the Requa product called Charcocaps, mentioned above, and this product is labeled as being for the relief of intestinal gas. Thus, almost nobody would have the insight to purchase Charcocaps for antidotal use.

The present situation in the USA is perhaps indicated by a recent article by Lee (1992). He describes a USA survey he made in which he called 50 community pharmacies at random and asked whether they carried any activated charcoal suspension or powder. Only two of the 50 pharmacies carried charcoal in a dosage form suitable for use in poisoning (the survey report does not say what these were). When the other 48 pharmacies were asked if they knew where such were available, only one had any such knowledge. It is clear from this survey that community pharmacists are not familiar with activated charcoal as a poison antidote.

Before leaving this subject, we will briefly mention a few antidotal products which were available as of 1983 (see the compilation of Picchioni, 1983) but which are no longer marketed. We present this information for its historical interest. The predecessor of Jones Medical in Canton, OH (same physical location, apparently), Bowman Pharmaceuticals, used to produce "Activated Charcoal USP in Liquid Base," which was an aqueous suspension in three different sizes (12.5 g in 60 mL, 25 g in 120 mL, and 50 g in 240 mL). Propylene glycol (amount unspecified) was included. The reason for this is unclear. One person to whom the writer has spoken indicated that he had once tasted such a suspension, and that the propylene glycol made it "taste awful." Bowman also marketed a Poison Antidote Kit containing four of their 12.5 g bottles of the charcoal suspension and a 30 mL bottle of syrup of ipecac.

Another product available in 1983 was CharcolantiDote (U. S. Products, Miami Lakes, FL), which was simply dry Norit USP powdered charcoal in 15 g and 30 g bottles. This same company also sold Liquid-Antidose containing 40 g charcoal in 200 mL of an aqueous solution having some carboxymethylcellulose and sodium benzoate (preservative) in it. As we shall see in Chapter 16, many researchers have used carboxymethylcellulose as a lubricating agent in charcoal suspensions, generally with success. Thus, it is interesting that one company made use of this additive and it is also of note that the product has since disappeared.

V.   OTHER CURRENTLY AVAILABLE CHARCOAL PRODUCTS

The *European Drug Index* (1992) is a compilation of all drug products sold in Europe (17 countries are included). Products are listed alphabetically by brand name, not by product type (e.g., charcoals); thus, it is difficult to pick out charcoal products from this massive listing (over 1000 pages). However, an attempt to do this has produced the listing of European charcoal products shown in Table 4.3. The entries are listed exactly as they appear in the source, and so readers will have to interpret various abbreviations, the different names for charcoal (e.g., carbo activ, carbo activatus, charbon activ, carvao activado, kohle, kool), and the names of other ingredients. Martindale's *The Extra Pharmacopeia* (Twenty-Ninth Edition, 1989) also lists some of the charcoal products available in Europe (primarily in the UK) as well as in Canada.

The vast majority of charcoal products listed in Table 4.3 are marketed for purposes other than antidotal use (e.g., intestinal problems). This can be seen from the inclusion of ingredients such as salicylates, bismuth compounds, antacids (aluminum hydroxide), and baker's yeast (*Saccharomyces cerevisiae*). The use of charcoal as a remedy for intestinal ailments is quite popular in Europe, whereas such use in the USA is very small.

We now discuss some of the specific European charcoal formulations which *are* intended for antidotal applications, including for the most part those listed in Martindale's *The Extra Pharmacopeia*. One formulation is called Carbomix (Penn Pharmaceuticals, UK; Medica, Sweden; S.C.A.T., France) and consists of 50 g charcoal in a bottle, to which water is to be added at the time of use (Table 4.3 indicates that Carbomix is also available in a 5 g charcoal dose, and is also marketed as a pre-mixed suspension of 50 g charcoal in water).

Another European charcoal product, which has been used in many studies discussed in this volume, is Medicoal (Lundbeck, Ltd., UK), which consists of 5 g charcoal mixed with an effervescent medium, in a sachet containing a total of 10 g of contents. It is to be mixed with water prior to administration. According to a remark made in a paper by Neuvonen and Olkkola (1988), this formulation has 1.5 g $NaHCO_3$ per each 5 g charcoal. It is presumed that effervescence is attained by incorporation of a powdered acid such as citric acid in the mixture. When mixed with water, the bicarbonate and acid react, forming tiny bubbles of CO_2. Assuming that citric acid is the acid and that a 1:1 mole ratio of $NaHCO_3$:acid is used, then 3.43 g citric acid would be needed. Thus, a packet would contain 5.0 + 1.5 + 3.43 = 9.93 g of material. Since this number is exceedingly close to the 10.0 g of contents stated by the manufacturer, the guess that citric acid is an ingredient is probably reasonable. In any event, an inquiry by the writer to Lundbeck elicited a reply by Mr. Ian Bruce, Marketing Manager, dated June 7, 1993, which states "we no longer market this product." The reason for discontinuance of this product and whether any other firm has taken up its manufacture are unknown. Thus, we must assume for the present that this product is no longer available.

A tableted charcoal product called Carbellon (Torbet Laboratories, UK) made from charcoal, belladonna extract, magnesium hydroxide, and peppermint oil is also available. It is intended for treating intestinal problems. Other charcoal products mentioned in the Martindale source include: Kolsuspension (AC, Sweden), Kullsuspensjon (Norway), Carbomucil (Norgine, UK), Kaltocarb (BritCair, UK), Intosan (Sauter, Switzerland), Norit Medicinaal (The Netherlands), Charcosorb (Seton, UK), and Charcodote and Aqueous

Table 4.3 Some Products Containing Charcoal Listed in the 1992 Edition of the *European Drug Index*

Acticarbine cpr drg: papaverine chlorhydrate 14 mg, 280 mg charbon activ 280 mg (France)

Activadone 1% 10 ml colirio: chromocarbo 100 mg (Spain)

Activadone 200 mg 45 capsulas: chromocarbo 200 mg (Spain)

Carbo Medicinalis "Chepharin"—Kapseln: Carbo adsorbens 175 mg, Saccharosum 140 mg (Austria)

Carbo Medicinalis "Chepharin"—Tabletten: Carbo adsorbens 0.25 g, Saccharosum 0.2 g (Austria)

Carbo medicinalis 250 mg tabl: carbo activ 250 mg (Finland)

Carbo medicinalis tab: carbo activatus 300 mg (Czechoslovakia)

Carbocit sevac tab: bismuthi subgallus 25 mg, carbo activatus 250 mg, acidium citricum monohydricum 3 mg (Czechoslovakia)

Carbofagyl gelule: *Saccharomyces cerevisiae* 108.5 mg, charbon activ 109 mg (France)

Carboguan "Aesca"—Tabletten: Carbo medicinalis 0.3 g, Sulfaguanidin 0.25 g (Austria)

Carbolactanose: adsorberende kool 50 mg *Streptococcus lactis* caps (Belgium)

Carbolevure adulte gelule: *Saccharomyces cerevisiae* 108.5 g, charbon active 109 mg (France)

Carbolevure enfant gelule: *Saccharomyces cerevisiae* 0.0477 g, charbon active 0.0480 g (France)

Carbolevure Kapseln Erwachsene: *Saccharomyces cerevisiae* vivant 108.5 mg, Carbo activatus 109 mg (Switzerland)

Carbolevure Kapseln Kinder: *Saccharomyces cerevisiae* vivant 48 mg, Carbo activatus 48 mg (Switzerland)

Carbomix 5.0 g/annos: carbo activ 5.0 g (Finland)

Carbomix 50.0 g/annos: carbo activ 50.0 g (Finland)

Carbomix 50 g granules pr susp buv: charbon active 50 g (France)

Carbomucil granules enr: aluminium hydroxide et carbonate 0.7 g/5 g, gomme sterculia 1.65g/5 g, charbon active 1.5 g/5 g (France)

Carbon ausonia 20 comprimidos: carbon adsorbente 0.13 g, ftalilsulfatiazol 0.13 g, pectina 0.13 g (Spain)

Carbonaphtine forminee granules: benzonapthol 5 g/100 g, salicylate aluminium 1 g/100 g, methenamine 1.5 g/100 g, charbon vegetal 15 g/100 g (France)

Carbonaphtine pectinee granules: benzonaphthol 1 g/100 g, salicaire ext 3 g/100 g, salicylate aluminium 5 g/100 g, charbon vegetal 15 g/100 g, kaolin 10 g/100 g, pectine 2 g/100 g (France)

Carbonesia*30 cpr mast., carbo activatus + magnesii oxidum leve + calcii carbonas + magnesii peroxidum (Italy)

Carbonesia*32 ciald., carbo activatus + magnesii oxidum leve + calcii carbonas + magnesii peroxidum (Italy)

Carbonesia*os granul. 120 G, carbo activatus + magnesii peroxidum leve + calcii carbonas + magnesii peroxidum (Italy)

Carbophos cpr: bromure sodium 0.10 g, reglisse ext 0.20 g, calcium carbonate 0.20 g, charbon vegetal 0.4 g, phosphate tricalcique 0.10 g (France)

Carbosorb (ostacol 85) plv: carbo activatus 25 g (Czechoslovakia)

Carbosorb tab: carbo activatus 250 mg, massa tabularum a 800 mg (Czechoslovakia)

Carbosylane caps: carvao activado, dimeticona (Portugal)

Carbosylane gelule: dimeticona 45 mg, charbon active 140 mg (France)

Carboticon capsules: Carbo activatus 140 mg, Dimeticonum 45 mg (Switzerland)

Carbotox tab: carbo activatus 250 mg, massa tabularum a 800 mg (Czechoslovakia)

Charbon de belloc 0.90 g pastille: charbon vegetal 900 mg (France)

(Continued)

Table 4.3 (Continued)

Charbon de belloc 75 g pdr: charbon vegetal 75 g/100 g (France)

Charbon tissot granules: anis 145 mg/10 g, cassia angustifolia 145 mg/10 g, badiane 290 mg/10 g, bourdaine 145 mg/10 g, rhubarbe 70 mg/10 g, cascara 50 mg/10 g, reglisse 340 mg/10 g, charbon vegetal 6.75 g/10 g, gluten 50 mg/10 g (France)

Charbon vegetal cooper 500 mg cpr: charbon vegetal 500 mg (France)

Charbon euguanidine: adsorberende kool 127 mg, bismutsubnitraat 42 mg, pectine 42 mg, sulfaguanidine 127 mg compr. (Belgium)

Kohle—Granulat "Merck": Carbo adsorbens 0.75 g (Austria)

Kohle-Compretten, Medizinische Kohle 0.25 g (former East Germany)

Kohle-Compretten, Tabletten; 1 Tbl.: Med. Kohle 250 mg (Former West Germany)

Kohle-compretten tab 250 mg medicinal charcoal (Netherlands)

Kohle-Granulat, 100 g: Medizinische Kohle 80 g (former East Germany)

Kohle-Hevert, Tabletten; 1 Tbl.: Carbo medicinalis 250 mg (former West Germany)

Kohle-Pulvis, Pulver; 1 Schraubdose: Medizinische Kohle (Eur. Pharm III) 10 g (former West Germany)

Medicoal; sachets 5 g: charcoal 5 g (UK)

Norit capsule 200 mg: kool geactiveerd 200 mg (Netherlands)

Norit Kapseln: Carbo activatus 200 mg (Switzerland)

Norit medicinal: adsorberende kool caps. 200 mg compr. 125 mg (Belgium)

Norit poeder: kool geactiveerd 1000 mg (Netherlands)

Norit tablet 125 mg: kool, geactiveerd 125 mg (Netherlands)

Charcodote (Pharmascience, Canada). The exact nature of these products (e.g., what other ingredients may be present) is not described in Martindale's.

McLuckie et al. (1990) have mentioned that the following ready-made charcoal formulations are available in Australia: Carbosorb (Delta West Ltd.), containing 50 g BPC activated charcoal in 300 mL sterile water, and Carbosorb S (Delta West Ltd) with 50 g of the same BPC charcoal in 150 mL 70% sorbitol plus enough sterile water to bring the volume to 300 mL. Both are packaged in a polyvinylchloride bag with a twist-off nozzle, which can be connected directly into an orogastric tube.

The availability of charcoal in powdered form in local pharmacies must be better in Europe than in the USA. A study by Lamminpää et al. (1993) in Finland showed that of 174 families who contacted the Finnish Poison Information Center for cases of poisoning of a child under 5 years of age, 63 (36.2%) had charcoal at home. Of these 63, 31 had charcoal tablets, 24 had charcoal powder, and 8 had charcoal powder along with syrup of ipecac in an "emergency kit." Of 103 families who had no charcoal at home when they contacted the Poison Center, 78 were able to buy it readily from a local pharmacy, usually as a powder. Such is not the case in the USA, where (to the writer's knowledge) local pharmacies do not sell powdered charcoal intended for antidotal use.

However, in other areas of the world (e.g., parts of Africa) appreciation of activated charcoal as an antidote is virtually nil. Orisakwe (1992) has described a survey of five hospitals in Nigeria which showed that only 18% of doctors and pharmacists had ever prescribed activated charcoal for poisoning. Orisakwe also mentions a remark made by the editors of the East African Medical Journal concerning an article on charcoal which she published in that journal. The remark was "that activated charcoal was not used in their hospital or any hospital known to them for the past ten years."

We might finally mention that there are several cloth-type charcoal dressings available in the UK (Actisorb, Actisorb Plus, Carbonet, Lyofoam C) which will be described in Chapter 23 under the "Effect of Charcoal on Surface Wounds."

VI. THE "CONTAINER RESIDUE" ISSUE

Krenzelok and Lush (1991) examined five different commercially available aqueous charcoal products (none with any sorbitol) in an effort to determine how much charcoal typically remained in their containers after use. The concern was that, after storage for long periods, the charcoal would settle to the bottom of these mixture and be difficult to resuspend by normal agitation. Thus, bottles of ActaChar, Actidose Aqua, InstaChar, LiquiChar, and SuperChar were stored for periods of 3 and 12 months, then were agitated (40 shakes over 15 seconds) by the same person and the contents poured into a beaker. The ActaChar, Actidose Aqua, and LiquiChar products were labeled as containing 50 g charcoal, and the SuperChar was labeled as having 30 g. The InstaChar was used "as two 25 g containers." The actual total amounts of charcoal in each product were determined by subsequently removing all of the residual charcoal, determining the residue weights, and adding them to the weights of charcoal delivered upon the initial pouring.

The total amounts (averages of the 3-month and 12-month samples) were found to be: 58.23 g for LiquiChar, 50.50 g for ActaChar, and 47.46 g for Actidose, compared to the "50 g" labeled amounts. The SuperChar had 34.22 g charcoal versus the "30 g" cited on the label. The InstaChar had 31.15 g charcoal versus the nominal 50 g promised by the "two 25 g containers". Thus, three of the products delivered somewhat more than the "promised" amounts of charcoal, one a bit less, and one (InstaChar) had "approximately 40% less than labeled" (we will shortly explain the error in this last statement).

The results on the amount of charcoal delivered when the mixtures were poured out after the shaking showed that the Actidose delivered nearly all (94–95%) of the charcoal that was in the container, while the LiquiChar, InstaChar, and the SuperChar delivered only 7.9, 13.8, and 12.0% of the charcoal, respectively (these percentages are the averages for the 3-month and 12-month samples). The ActaChar was in the middle (28 and 57% delivered). It was concluded that, despite container instructions to "Shake Well", a vigorous manual shaking for 15 seconds was clearly inadequate to resuspend the charcoal.

In response to this study (see Saulson, 1992), the Vice President of the company which markets InstaChar (the Frank W. Kerr Chemical Company) wrote a letter to the journal which published the Krenzelok/Lush study, pointing out that Kerr had never made a 25 g charcoal product. Saulson stated that Kerr has made only 15 g and 50 g products. Krenzelok's reply [see Krenzelok (1992a)] to this mentions that, since the total charcoal amounts in the 3 month and 12 month samples were 31.55 and 30.74 g, respectively, "this suggests that two 15-g containers ... were used." Krenzelok simultaneously acknowledged the possibility of error on his part and the possibility that the containers were mislabeled; thus, he could neither "refute nor support Mr. Saulson's claim ..."

An apparent resolution of the scientific aspects of the original study seems to have occurred. Krenzelok (1992b) recently has published an abstract of a second study in which the shaking time was prolonged beyond the 15 seconds used before. In the new study, 30, 60, and 120 seconds of manual shaking of products which had been stored for 12 months was employed. All products were nominal "50 g" charcoal aqueous suspensions. Additionally, mechanical shaking for 60 seconds (at 150 cycles/minute) was

Table 4.4 Grams Activated Charcoal Residue in Containers After Shaking

Product	Duration or Type of Shaking			
	30 sec	60 sec	120 sec	Mechanical
Actidose	6.2	4.4	3.9	26.7
InstaChar	4.8	4.9	4.5	37.7
LiquiChar	22.2	9.0	7.8	58.2

Adapted from Krenzelok (1992b). Reprinted with permission of *Veterinary and Human Toxicology*.

performed. The ActaChar and SuperChar products were not tested (they had disappeared from the market, as mentioned above). The results are shown in Table 4.4.

These new results show that, after 30 sec shaking, the InstaChar left less residue than the Actidose product. Other striking results of the study are the relatively large residue at 30 sec for the LiquiChar and the inferiority of mechanical shaking. It was concluded that vigorous manual shaking for at least 60 sec should be carried out.

Although the essence of this story is more or less complete, we mention two other references, both by Harchelroad, which fall in the time period between the Krenzelok/Lush study and the second Krenzelok study, in order to provide all of the literature references dealing with this issue.

In response to the original Krenzelok/Lush study, Harchelroad (1991) reported in a letter that he had measured the amount of residue in 82 used containers of Actidose with Sorbitol collected from an ED. He found the average residue was 0.55 g, or 2.2% of the 25 g dose listed on the container. An analysis of unused containers showed they contained an average of 25.89 g charcoal. Thus, an average delivered dose of 25.89 − 0.55 = 25.34 g, or 101.4% of the 25 g figure on the label, could be expected. Hence, this product appears to deliver the expected amount. Unfortunately, no other charcoal products were tested. These results agree well with those of Krenzelok and Lush, which gave 94–95% delivery of the charcoal in Actidose with their manual shake protocol.

The second reference (Harchelroad, 1992) is simply the full-length journal paper which gives the details of what was summarized in Harchelroad's 1991 letter. This paper does offer some additional insights: (1) the nature of the shaking used by the ED nurses is not stated, (2) the length of time that the containers had been immobile prior to use was "likely" to have been less than 3 months, and (3) there was no strict control over whether the ED nurses may have rinsed the containers after use. With respect to shaking, the ED nurses "were advised to deliver the dosage as they normally do," whatever that might mean.

With respect to the ability of commercial charcoal preparations to "flow" easily, not just out of the original container but also through a tube inserted into a patient, Schneider and Michelson (1993) have done a very illuminating study. Samples of both Actidose Aqua and Actidose with Sorbitol were taken with a 60 mL syringe from continuously-stirred 360 mL stock suspensions; they were then pumped with a syringe pump, at a constant inlet pressure of 250 mm Hg, through a 16 French Salem sump tube, a 16

French Cantor tube, and an 8 French Duo-tube. Flow rates for the Actidose Aqua were 9.5, 1.8, and 1.5 mL/min, respectively. For Actidose with Sorbitol, the flow rates were 8.5, 1.2, and 1.2 mL/min, respectively. Thus the sorbitol, by increasing the suspension viscosity, reduced the flow rates by 10–20%. Precoating the tubes with mineral oil gave Actidose Aqua flow rates of 13, 3, and 1.2 mL/min, respectively (increases of 37, 67, and −20%). Thus, the mineral oil pretreatment gave inconsistent effects. Since flowing liquids never "slip" relative to a solid surface, the use of a mineral oil coating on the tube wall would not be expected to enhance the flow rate, unless the mineral oil has a lower viscosity that the charcoal/suspension (however, no viscosity values are given). Studies with diluted charcoal suspensions were then done for the Duo-tube only, using only the sorbitol-free suspension (Actidose Aqua). Dilutions of 10, 20, and 30% with tap water increased the flow rate from 1.5 mL/min to 5.25, 16.2, and 26.5 mL/min, respectively. Thus, huge increases were obtained by moderate dilution. To express these results in a different way, the times required to deliver 240 mL (8 oz) of a suspension containing 25 g charcoal per 120 mL through the Duo-tube would be 160 minutes with no dilution, and 50.3, 17.8, and 11.8 minutes, respectively, for 10, 20, and 30% dilutions (note that, upon dilution, the volume of the 240 mL suspension becomes 264, 288, and 312 mL, respectively, in these three cases, and the times cited are based on these increased volumes). Thus, it is clear that to deliver 240 mL of a charcoal suspension to a patient through an 8 French Duo-tube in a reasonably short time span would require a dilution on the order of 20% or more. While manufacturers cannot market diluted charcoal/sorbitol suspensions because of bacterial concerns (standard 70% w/w sorbitol solutions are bacteriostatic), there seems to be no reason why formulations without sorbitol could not be marketed in a more dilute form. Clearly, there is no reason why dilution of any formulation with tap water just prior to administration would pose any problems. Thus, a modest amount of dilution of commercial charcoal suspensions prior to delivery should be strongly considered, especially if the delivery tube to be used is of relatively small bore.

An earlier study of the effect of dilution on the ability of charcoal formulations to flow through tubes was done by Holmes et al. (1984). They hung nasogastric tubes of different diameters vertically, with syringes connected to the tops. Five charcoal preparations (Charcoaid, Actidose, InstaChar, Armachar, and a generic charcoal/sorbitol) were tested for flow rate. Various dilutions of these were tested. All products showed increased flow rate with increased dilution. Armachar flowed most readily and Charcoaid was the slowest, at all dilutions. None of the five formulations would flow undiluted through a 3 mm tube.

VII. COMPARATIVE STUDIES OF ANTIDOTAL CHARCOALS

Despite the fact that the charcoals which are available in the world at present are different from those which were available when many comparative studies of different charcoals were conducted, and despite the fact that some (or many) of such charcoals did not meet some pharmacopeial requirements, we will review a few of those comparative studies here. There is some historical value, perhaps, in doing so.

Picchioni (1970) compared Darco G-60 and Merck charcoals with respect to their abilities to inhibit aspirin and chlorpheniramine uptake into the tissues of rats given these drugs, alone or in combination with either of the two charcoals. In the aspirin tests, the

Table 4.5   Percentage of Drug Adsorption by Four Activated Charcoals[a]

Drug	Merck	Darco G-60	Nuchar C	Norit A
Strychnine	39.5	26.5	39.0	38.0
Pentobarbital	17.5	10.7	17.8	17.2

[a] Each value is the mean of eight determinations.
From Picchioni et al. (1974).

tissue concentrations in mg/100 mL were: controls, 22.3; Darco G-60, 8.8; and Merck, 4.2. For chlorpheniramine the values were: controls, 9.1; Darco G-60, 3.8; and Merck, 1.3. Both charcoals reduced tissue drug levels but the Merck brand was clearly much better than the Darco G-60 brand.

Dozzi et al. (1974) carried out a study in Australia in which the in vitro adsorption of aspirin from simulated gastric fluid was determined for Norit powder, Sigma powder, Merck's Medicinal Charcoal Powder, Hill and Sons powder, and Langley's Char-bons (both as whole and crushed tablets). The Merck and Norit powders were far superior to the other charcoals.

Boehm and Oppenheim (1977) evaluated the abilities of four activated charcoals for adsorbing six drugs (three acidic, three basic) from simulated gastric and intestinal fluids in vitro. The best charcoal was Medicinal Norit DAB VI. Medicinal Norit A adsorbed about 5% less strongly, Merck charcoal adsorbed about 10% less strongly, and BDH charcoal adsorbed about 50% less strongly than the Norit DAB type, respectively.

Picchioni et al. (1974) evaluated the efficacy of 300 mg each of Merck, Darco G-60, Nuchar C, and Norit A charcoals for adsorbing strychnine and pentobarbital from aqueous solutions of 40 mg drug in 40 mL water. Their results are shown in Table 4.5. Clearly the Darco G-60 was inferior to the other three charcoals (which were essentially equal). In vivo studies were also carried out in which aspirin, pentobarbital, or chlorpheniramine given to rats were followed by the administration of either the Merck or Darco G-60 charcoal. The Darco G-60 was again shown to be decidedly less effective than the Merck charcoal. The probable reason for the poorer performance of the Darco G-60 is simply that its internal surface area is only about 650 m^2/g, whereas the surface areas of the other charcoals are probably on the order of 1000 m^2/g.

The activated charcoal preparation called Medicoal (Lundbeck, Ltd., UK) has been discussed earlier. It consists of 10 g of mixture vacuum-sealed in packets. About half of the mixture, by weight, is activated charcoal powder. Upon mixing the packet contents with water an effervescent slurry is produced which "is not unpleasant to drink," according to Dawling et al. (1978). These investigators report that Medicoal adsorbs "about four times" as much nortriptyline in vitro, per unit weight of charcoal, as Norit A charcoal. Braithwaite et al. (1978) reported the specific data on which this conclusion was based: 1 g Medicoal adsorbed 318 mg nortriptyline, whereas Norit A adsorbed 130 mg nortriptyline, in certain in vitro tests. For amitriptyline, the amounts adsorbed were 282 mg (Medicoal) and 125 mg (Norit A). The ratios adsorbed (Medicoal/Norit A) are thus 2.45 for nortriptyline and 2.26 for amitriptyline; these would be twice as high if one expressed the Medicoal data on a "per gram of charcoal" basis. In the writer's view, the Norit A results seem lower than one would expect. Indeed, a personal communication from Dr. Crome has revealed that the Norit A had been left exposed to room air for a substantial time and had not been heated to high temperature prior to use so as to drive off any gases that undoubtedly had adsorbed to it. Hence, the comparison between Medicoal and Norit A in this study was flawed.

Dawling et al. (1983) studied the adsorptive capacities of two forms of powdered charcoal, both in vitro and in vivo, using aspirin as the test compound. Adsorption in vitro from 2.5 g/L aqueous aspirin solutions, pH 1.0, showed maximal adsorption capacities of 359 mg/g for Norit A charcoal, 482 mg/g for Norit A charcoal which had been dried for 4 hr at 100°C, and 477 mg/g for the Medicoal formulation. The implication is that the Norit A charcoal, as received, contained about 25 wt% moisture, which is higher than usual but conceivable (powdered charcoals often have about 10–15 wt% moisture if left in contact with air). There was thus no difference between dried Norit A and the charcoal in Medicoal. In vivo studies with the heated Norit A (12 g) and Medicoal (12 g charcoal), both given orally 30 min after 1.2 g aspirin had been given to six healthy subjects, showed 0–8 hr AUC (area under the curve of drug concentrations versus time) reductions of 38% for Norit A and 35% for Medicoal, as compared to controls. Peak plasma drug concentrations were reduced by 37 and 34%, respectively. Thus, these two charcoals also showed similar capacities in vivo.

Krenzelok and Heller (1987) studied five ready-mixed charcoal formulations which were available at that time for their effectiveness in treating human subjects who had been given 2,592 mg aspirin (eight 324 mg tablets) and 240 mL water. They were then given 25 g aliquots of five available formulations: ActaChar, Actidose Aqua, InstaChar, LiquiChar, and SuperChar (these formulations were described above; however, note that ActaChar and SuperChar are no longer available). The formulations did not come in 25 g packages, so 25 g aliquots had to be measured out. AUC values were determined based on aspirin blood levels determined up to 8 hr. The study was a double-blinded one and subjects served as their own controls. Reductions in aspirin absorption relative to controls were: SuperChar, 57.8%; Actidose Aqua, 50.4%; InstaChar, 39.6%; Liqui-Char, 33.4%; and ActaChar 27.5%. The higher surface area products, SuperChar (3,150 m^2/g) and Actidose Aqua (1,500 m^2/g in 1987) performed better than the other three products, whose charcoal (same type for all) had a surface area of 950 m^2/g.

Weaver (1988) wrote a letter on this study in which he stated that giving the charcoal products immediately after the aspirin "brings into question the validity of the results relative to an actual clinical setting," since one study (Robertson, 1962) showed an average elapsed time of 68 minutes between ingestion of a toxic substance and arrival at an ED. Weaver then mentioned a study reported in abstract form by Dillon et al. (1987) in which 25 g each of SuperChar and Actidose Aqua were given to 12 volunteers 1 hr after they had been dosed with 20 mg/kg aspirin (for the complete paper, see Dillon et al., 1989). Urine was collected from 12 hr prior to the aspirin until 3 days afterwards and the samples were analyzed for salicylates. The results showed that SuperChar reduced the aspirin absorption by 48%, while the Actidose Aqua reduced it by only 19%. Thus, in this study, the SuperChar was much more effective than the Actidose Aqua. Krenzelok and Heller (1988) responded by stating that "delays in the administration of activated charcoal for an hour or more may not truly represent the charcoal's ability to adsorb large amounts of toxin, but instead the ability to adsorb drug passively secreted into the gastrointestinal tract in significantly lower concentrations."

Harou-Kouka et al. (1989) in France have compared Norit A Supra, Norit B Supra, a charcoal called Picactif Medicinal, Cooper charcoal, a charcoal called L40S Acticarbone, and the suspension called Carbomix for their abilities to adsorb amitriptyline, clomipramine, chlorpromazine, chloroquine, and paracetamol in vitro in artificial gastric fluid. After 15 min of contact for adsorption, the test solutions were filtered using microporous membrane filters and the entire charcoal recovered was then put into a drug-free artificial duodenal fluid medium for 15 min for desorption to occur. Adsorption and desorption isotherms were determined for all drug/charcoal combinations by varying the amount of charcoal used in each test. In general, the Norit A Supra adsorbed the test drugs the strongest and allowed the smallest amount of desorption to occur. Since Norit A Supra has the highest surface area (2,000 m^2/g) of the charcoals tested, this is not surprising. The Norit B Supra (surface area 1,400 m^2/g) and the Carbomix also showed good but lesser adsorption capacities. The degrees of desorption were generally small and were rarely more than a few percent of the amounts adsorbed.

M'Bisi et al. (1989) evaluated 21 charcoals which met the specifications of the *French Pharmacopeia* valid at that time. Tests were done with phenazone (antipyrine) and with methylene blue, after drying the charcoals at 150°C for 5 hr. The five best charcoals and the percentages of phenazone adsorbed by each were: Norit A Supra, 62.74%; Norit B Supra, 49.40%; Picactif Médicinal, 48.60%; L40S Acticarbone 47.80%; and Cooper charbon activé, 47.60% (the identities of the other 16 charcoals were not stated). Freundlich isotherm plots for these five charcoals adsorbing methylene blue were constructed and the efficacy of each charcoal relative to the best charcoal (Norit A Supra) was quantified by an Efficacité Relative (ER) index defined as the Q value for that charcoal at $C_f = 0.01$ g/L divided by the Q value for Norit A Supra at $C_f = 0.01$ g/L, with the result multiplied by 100 to give a percentage. The ER values for the charcoals were: Norit B Supra, 62.06%; Picactif médicinal, 60.35%; L40S Acticarbone, 51.72%; and Cooper charbon activé, 51.72%.

M'Bisi and coworkers also determined the effect of pH on the degree of adsorption of phenazone (a base with a pK_a of 1.4) using simulated gastric fluid (pH 1.2), an unbuffered aqueous solution (pH 6.9), simulated intestinal fluid (pH 7.5), a phosphate buffer (pH 8.0), and a borate buffer (pH 9.0). As expected, the Norit A Supra charcoal was the best and the Norit B Supra charcoal second best, at all pH levels, and the degree of phenazone adsorption rose with pH as the extent of phenazone ionization decreased. For Norit A Supra, the percentages of phenazone adsorbed at the five pH values mentioned were 55.05, 62.74, 65.60, 72.31, and 73.10, respectively.

Ilkhanipour et al. (1991), in a report presented as an abstract, compared Norit A Supra charcoal (2,000 m^2/g) versus Norit USP XXII charcoal (950 m^2/g) in enhancing the elimination of theophylline in five healthy adult men. Aminophylline (8 mg/kg) was given IV, followed by 50 g doses of charcoal at 0, 4 and 8 hr (aminophylline is a compound of theophylline and 1,2-ethanediamine; it dissociates upon dissolution). Blood samples were taken over 0.5-12 hr. The 0–∞ AUC values in mg-hr/L were: control, 223; Norit USP XXII, 112; and Norit A Supra, 111. Thus, while multiple-dose charcoal therapy did enhance theophylline elimination, there was no effect of surface area with the two charcoals tested. This is somewhat surprising as one would expect a more than two-fold difference in surface area to have had a noticeable effect.

What is even more surprising about this work is that exactly the same study was published two years later in a journal (see Ilkhanipour et al., 1993) except that all references to Norit USP XXII charcoal are replaced with "Actidose Aqua" (Paddock Laboratories), which is stated to be "1,500 m^2/g" in surface area. And yet, Paddock has consistently refused to divulge the type of charcoal it uses in its products and tests by the writer

Table 4.6 Pharmacokinetic Parameters for Healthy Volunteers Given 200 mg Oral Phenobarbital Plus 30 g of Three Charcoal Formulations

Parameter	No charcoal	Actidose	SuperChar	Charcoaid
C_{max} (mg/L)	5.40	4.93	4.97	5.16
t_{max} (hr)	0.42	0.42	0.36	0.38
$AUC_{0-504\ hr}$ (mg-hr/L)	610	435	354	438

From Berg et al. (1993).

strongly suggest that the charcoal in Actidose Aqua is Norit USP XXII. Thus, the statement in the journal article of Ilkhanipour and coworkers that the Actidose Aqua charcoal has 1,500 m^2/g area (which is the surface area of Norit B Supra charcoal) seems incorrect.

Berg et al. (1993) evaluated three commercial charcoal preparations—Actidose, Charcoaid, and SuperChar—with respect to their effects on phenobarbital absorption in human subjects. Healthy volunteers were given 200 mg phenobarbital orally, then a single 30 g dose of each preparation was given 30 min later. In IV tests, phenobarbital sodium equivalent to 200 mg of the free acid was infused over 1 hr. Immediately thereafter, single 30 g doses of each charcoal preparation were given orally. Blood samples were taken for up to 504 hr in all tests. Table 4.6 shows the results for the orally administered phenobarbital. Clearly, all of the charcoal preparations were effective in reducing the 0–504 hr AUC, with the high surface area SuperChar being the most effective. For the IV phenobarbital tests, the 0–504 AUC values in mg-hr/L were: control, 506; Actidose, 423; SuperChar, 387; and Charcoaid, 439. Thus, again the same conclusions apply. However, the reductions in the AUC caused by charcoal in the oral drug tests are greater than in the IV drug tests, as one would expect. Other pharmacokinetic parameters (e.g., C_{max}, t_{max}, AUC values for shorter time intervals) generally showed no differences with or without charcoal, but the 0-504 AUC values are the important ones in terms of overall drug absorption.

VIII. SUMMARY

The focus of this chapter has been to review the requirements of the *U. S. Pharmacopeia* and to discuss those charcoals and charcoal-containing products in the USA which utilize USP-grade charcoals. Along the way, mention was made of the large variety of charcoal-containing products marketed in Europe but no attempt to evaluate them was offered.

We described all of the ready-mixed charcoal formulations being marketed in the USA, the list being divided into those which incorporate sorbitol and those which do not. Sorbitol contents vary widely—from 225 to 700 mg/mL.

All USP-grade charcoals originate from the American Norit Company and consist of three choices: Norit USP XXII, Norit A Supra, and Norit B Supra. The Requa company (Greenwich, CT) appears to be the only company using Norit A Supra (in CharcoAid 2000) and Norit B Supra (in Charcoaid). All other companies appear to be using Norit USP XXII.

The stories of "superactive" charcoal and of the "container residue" issue were related. The latter story points to a specific conclusion: ready-mixed charcoal formulations should be manually shaken vigorously for at least 60 seconds before use.

Many comparative studies of charcoals were reviewed. However, many of the charcoals involved are either no longer available, are not widely available, or are inferior (in terms of surface area) to present-day charcoals. The performance of charcoals has usually, but not always, been found to be proportional to their surface areas. Thus, the Norit Supra charcoals should be more effective than Norit USP XXII charcoal. Other charcoals sold in Europe and other areas of the world which have comparable surface areas should be roughly equivalent.

IX. REFERENCES

Annual Pharmacists' Reference: Drug Topics Redbook, Medical Economics Company, Oradell, New Jersey, published annually.

Berg, M. J., et al. (1993). Effect of three charcoal preparations on oral and intravenous phenobarbital, Clin. Res. Reg. Affairs *10*, 81.

Boehm, J. J. and Oppenheim, R. C. (1977). An in vitro study of the adsorption of various drugs by activated charcoal, Aust. J. Pharm. Sci. *6*, 107.

Braithwaite, R. A., Crome, P., and Dawling, S. (1978). The in vitro and in vivo evaluation of activated charcoal as an adsorbent for tricyclic antidepressants, Br. J. Clin. Pharmacol. *5*, 369.

British Pharmacopoeia (1988), Her Majesty's Stationery Office, London.

Cooney, D. O. (1977). A "superactive" charcoal for antidotal use in poisonings, Clin. Toxicol. *11*, 387.

Cooney, D. O. and Kane, R. P. (1980). "Superactive" charcoal adsorbs drugs as fast as standard antidotal charcoal (letter), Clin. Toxicol. *16*, 123.

Dawling, S., Crome, P., and Braithwaite, R. (1978). Effect of delayed administration of activated charcoal on nortriptyline absorption, Eur. J. Clin. Pharmacol. *14*, 445.

Dawling, S., Chand, S., Braithwaite, R. A., and Crome, P. (1983). In vitro and in vivo evaluation of two preparations of activated charcoal as adsorbents of aspirin, Hum. Toxicol. *2*, 211.

Dillon, E. C., Jr., Wilton, J. H., Barlow, J., and Watson, W. A. (1987). The effect of two activated charcoal preparations on aspirin absorption in man (abstract), Vet. Hum. Toxicol. *29*, 491.

Dillon, E., Wilton, J. H., Barlow, J. C., and Watson, W. A. (1989). Large surface area activated charcoal and the inhibition of aspirin absorption, Ann. Emerg. Med. *18*, 547.

Dozzi, A. M., Leversha, A., and Stewart, N. F. (1974). Comparison of activated charcoals, Aust. J. Hosp. Pharm. *4*, 40.

European Drug Index, Second Edition, N. F. Muller and R. P. Dessing, Eds., Elsevier, Amsterdam, 1992.

Ford, J. W. (1993). Private communication (letter).

Harchelroad, F., Jr. (1991). Container residue of activated charcoal products (letter), Am. J. Emerg. Med. *9*, 520.

Harchelroad, F. (1992). Container residue after activated charcoal administration in the emergency department, Vet. Hum. Toxicol. *34*, 13.

Harou-Kouka, M., et al. (1989). Efficacité des charbons activés dans le traitement des intoxications médicamenteuses, J. Toxicol. Clin. Exp. *9*, 255.

Holmes, W., Banner, W., and Tong, T. G. (1984). Effect of charcoal dilution on nasogastric transit time (abstract), Vet. Hum. Toxicol. *26* (Suppl. 2), 59.

Ilkhanipour, K., Yealy, D. M., and Krenzelok, E. P. (1991). Activated charcoal surface area and its role in multiple-dose charcoal therapy (abstract), Ann. Emerg. Med. *20*, 474.

Ilkhanipour, K., Yealy, D. M., and Krenzelok, E. P. (1993). Activated charcoal surface area and its role in multiple-dose charcoal therapy, Am. J. Emerg. Med. *11*, 583.

Krenzelok, E. P. and Heller, M. B. (1987). Effectiveness of commercially available aqueous activated charcoal products, Ann. Emerg. Med. *16*, 1340.

Krenzelok, E. P. and Heller, M. B. (1988). Actual clinical settings and the use of activated charcoal (reply to letter), Ann. Emerg. Med. *17*, 663.

Krenzelok, E. P. and Lush, R. M. (1991). Container residue after the administration of aqueous activated charcoal products, Am. J. Emerg. Med. *9*, 144.

Krenzelok, E. P. (1992a). Kerr InstaChar activated charcoal product (reply to letter), Am. J. Emerg. Med. *10*, 265.

Krenzelok, E. P. (1992b). Importance of prolonged agitation of aqueous activated charcoal slurries (abstract), Vet. Hum. Toxicol. *34*, 336.

Lamminpää, A., Vilska, J., and Hoppu, K. (1993). Medical charcoal for a child's poisoning at home: Availability and success of administration in Finland, Hum. Exp. Toxicol. *12*, 29.

Lee, R. J. (1992). Ancient antidote ignored, Am. Pharmacy *NS32*, 34.

M'Bisi, R., Gayot, A. T., Traisnel, M., Harou-Kouka, M., Erb, F., and Haguenoer, J. M. (1989). Etude sélective in vitro des charbons activés pour leur utilisation dans les intoxications aiguës, J. Toxicol. Clin. Exp. *9*, 249.

McFarland, A. K., III and Chyka, P. A. (1993). Selection of activated charcoal products for the treatment of poisonings, DICP -Ann. Pharmacother. *27*, 358.

McLuckie, A., Forbes, A. M., and Ilett, K. F. (1990). Role of repeated doses of oral activated charcoal in the treatment of acute intoxications, Anaesth. Intens. Care *18*, 375.

Martindale: The Extra Pharmacopeia, Twenty-ninth edition, J. E. F. Reynolds, Ed., The Pharmaceutical Press, London, 1989.

Medema, J. (1979). Comments on "A superactive charcoal for antidotal use in poisonings" (letter), Clin. Toxicol. *14*, 205.

Neuvonen, P. J. and Olkkola, K. T. (1988). Oral activated charcoal in the treatment of intoxications: Role of single and repeated doses, Med. Toxicol *3*, 33.

O'Grady, T. M. and Wennerberg, A. N. (1986). High-surface-area active carbon, in *Petroleum-Derived Carbons*, ACS Symposium Series No. 303, J. D. Bacha, J. W. Newman, and J. L. White, Eds., Washington, D. C., pp. 301-309.

Orisakwe, O. E. (1992). Negligence of activated charcoal due to lack of current information, Israel J. Med. Sci. *28*, 751.

Paddock, B. G. (1994). Personal communication to the writer.

Picchioni, A. L. (1970). Activated charcoal - a neglected antidote, Pediatr. Clin. North Am. *17*, 535.

Picchioni, A. L., Chin, L., and Laird, H. E. (1974). Activated charcoal preparations—Relative antidotal efficacy, Clin. Toxicol. *7*, 97.

Picchioni, A. L. (1983). Activated charcoal products for medicinal (antidote) use, Vet. Hum. Toxicol. *25*, 293.

Robertson, W. O. (1962). Syrup of ipecac—a fast or slow emetic?, Am. J. Dis. Child. *103*, 58.

Saulson, S. S. (1992). Kerr InstaChar activated charcoal product (letter), Am. J. Emerg. Med. *10*, 97.

Saulson, S. S. (1993). Private communication (letter).

Schneider, S. M. and Michelson, E. A. (1993). Enhanced activated charcoal delivery through small-bore tubing, Vet. Hum. Toxicol. *35*, 503.

The International Pharmacopeia, 3rd edition, World Health Organization, 1981.

United States Pharmacopeia, Twenty-Second Revision, United States Pharmacopeial Convention, Rockville, Maryland, 1989.

Weaver, W. R. (1988). Actual clinical settings and the use of activated charcoal (letter), Ann. Emerg. Med. *17*, 662.

5

The Nature of Drug Absorption, Distribution, and Elimination

In order to understand how activated charcoal can affect the absorption and elimination of a drug or poison in humans (or in animals), one must have a certain basic knowledge of the physiology of the gastrointestinal tract, including a knowledge of both its mechanical and chemical characteristics. And, in order to understand how charcoal can interrupt the enteroenteric or enterohepatic circulation of a drug or poison, one must be familiar with how drugs or poisons distribute and circulate within the human body. Figure 5.1 (from Donovan, 1987) shows schematically all of the routes by which a drug or poison may be transported into and out of the gastrointestinal tract. This chapter discusses the mechanical and chemical characteristics of gastrointestinal tract physiology, with particular emphasis on the various drug/poison transport processes shown in Figure 5.1. We begin, however, with a brief discussion of the influence of drug dosage form and route of administration.

I. THE EFFECT OF DRUG DOSAGE FORM

Drugs are "packaged" in a variety of forms: in solutions, in suspensions, in emulsions, in capsules, and in tablets. Tablet forms additionally may be coated, sometimes with enteric (i.e., acid-resistant) coatings. Additionally, there are a range of dosage forms of the sustained-release or prolonged-release variety. The dosage form of the drug affects its rate of absorption because when the drug is not already in solution, some dissolution (preceded by disintegration, if the drug is in capsule or tablet form) must occur prior to absorption (a drug must exist as an individual molecule in solution in order for absorption to occur). It is not enough for a drug to simply disperse into finely powdered form, since if it has a low solubility in the gastrointestinal (GI) tract fluid (at the prevailing pH), it will not undergo complete dissolution.

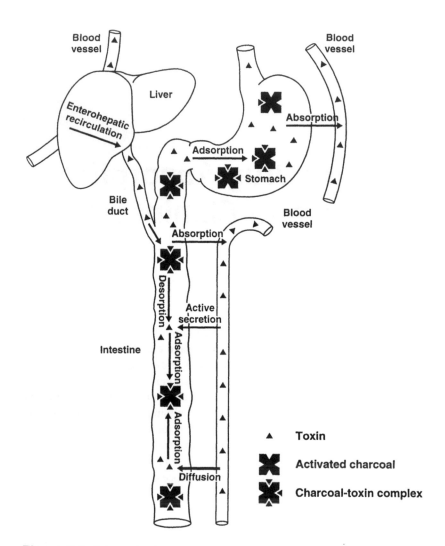

Figure 5.1 Schematic diagram of the gastrointestinal system, showing all drug transport processes. From Donovan (1987). Reproduced by permission of McGraw-Hill Healthcare Publications.

Standard tests for disintegration of tablets and for subsequent dissolution exist. Our purpose here is not to discuss any details of the dosage form effect, except to say that it exists, and that if the processes of disintegration/dissolution are slow, they could be rate-limiting in the overall drug absorption process. Clearly, drugs dissolved in solution give the most rapid absorption, since they already exist as discrete molecules. However, for ease of administration, solid dosage forms (capsules, tablets) are more common.

The case of sustained-release drug forms (implants, ocular devices, transdermal devices, intrauterine devices, solutions of microcapsules containing the drug which can be injected intramuscularly, etc.) is clearly outside of the scope of this volume, since these forms—if properly made and used—do not lead to toxicity.

II. ROUTES OF DRUG ADMINISTRATION

The various routes of drug administration include: oral, buccal, sublingual, rectal, topical (e.g., in the eye, on the skin, etc.), subcutaneous, inhalation, intranasal, intravaginal, intrathecal, intramuscular, intra-arterial, intraperitoneal, and intravenous. It is obvious that some of these routes, such as the intravenous route, bypass the need for drug absorption from the GI tract and thus give faster delivery to the body as a whole. Thus, the kinetics of drug distribution and elimination are strongly influenced by the route of administration. However, space will not permit us to elaborate on the route of administration as a factor in the use of antidotal charcoal. We will only point out that this volume will be confined almost entirely to drugs which are administered orally or intravenously.

III. THE ROLE OF pH IN DRUG ABSORPTION

Some knowledge of gastrointestinal tract physiology is of value in understanding what happens to a drug after it is taken orally. An important aspect of the GI tract is the pH variation along its length. Normal values of pH along the GI tract are: in the stomach, 1–3.5; in the duodenum, 5–6; and in the ileum, about 8. The variation of pH is exceedingly important because most drugs absorb well only in nonionized form. Since acidic drugs are nonionized in a low pH environment, and basic drugs are highly ionized in an acidic environment, the potential for absorption of acidic drugs from the stomach is high and for basic drugs it is very low (however, for reasons we will soon discuss, even the absorption of acidic drugs from the stomach is generally quite small).

In general, once a drug is in solution in the gastrointestinal fluids (i.e, after any dosage form disintegration/dissolution processes have occurred), drug absorption depends on the drug's solubility in the fluids and its lipid solubility. A drug's solubility in aqueous media can be greatly enhanced by using a salt form of the drug, for example sodium salts of acidic drugs (e.g., phenobarbital sodium) or HCl salts of basic drugs (e.g, chlordiazepoxide HCl). Such salts are far more soluble than the parent compound. If the drug is not in the salt form, some idea of its solubility can be judged from its pK_a value. Tables 5.1, 5.2, and 5.3 give pK_a values for some acidic, basic, amphoteric, and dibasic compounds. Such values can be found in standard pharmaceutical handbooks, such as *The Merck Index* (1989), *Remington's Pharmaceutical Sciences* (1990), *AHFS Drug Information '92* (1992), or the *Physicians' Desk Reference* (1994).

The meaning of pK_a for an acid is illustrated by considering the dissociation formula

$$HA \rightleftarrows H^+ + A^-$$

for which the acidic dissociation constant K_a is given by

$$K_a = \frac{[H^+][A^-]}{[HA]}$$

where the brackets [ ] denote the molar concentrations of the undissociated acid HA, the anion A^-, and the proton H^+. When the acid HA is half dissociated, the concentrations [HA] and $[A^-]$ will be equal, and K_a will be equal to $[H^+]$. Thus, pK_a will equal pH (pH is equal to minus the logarithm of $[H^+]$ to the base 10). Hence, the pK_a is the pH

Table 5.1 Approximate pK_a Values for Selected Acidic Compounds

p-Aminosalicylic acid	3.2	Ibuprofen	5.2
Acetylsalicylic acid	3.5	Chloramphenicol	5.5
Furosemide	3.9	Tolbutamide	5.5
Mefenamic acid	4.2	Acetazolamide	7.2
Naproxen	4.2	Barbitals	7.3–8.0
Indomethacin	4.5	Haloperidol	8.3
Valproic acid	4.8	Phenytoin	8.3
Warfarin	4.8	Theophylline	9.0
Piroxicam	5.1	Glutethimide	9.3

Table 5.2 Approximate pK_a Values for Selected Basic Compounds

Dapsone	1.1	Doxepin HCl	8.0
Diazepam	3.4	Tetracaine HCl	8.4
Phenylbutazone	4.5	Promethazine HCl	9.1
Minoxidil	4.6	Procainamide HCl	9.2
Chlordiazepoxide	4.8	Flecainide acetate	9.3
Glipizide	5.9	Amitriptyline HCl	9.4
Propoxyphene	6.3	Imipramine HCl	9.5
Amiodarone HCl	6.6	Nortriptyline HCl	9.7
Trazadone HCl	6.7	Amphetamine	10.0
Cimetidine HCl	7.1	Maprotiline HCl	10.2
Lidocaine HCl	7.9	Disopyramide phosphate	10.4

Table 5.3 Approximate pK_a Values for Selected Dibasic or Amphoteric Compounds

	pK_{a_1}	pK_{a_2}
Metoclopramide HCl	0.6	9.3
Caffeine[a]	0.9	13.9
Desipramine HCl	1.5	10.2
Lorazepam[a]	1.3	11.5
Clonazepam	1.5	10.5
Oxazepam[a]	1.7	11.6
Ranitidine HCl	2.7	8.2
Captopril[a]	3.7	9.8
Quinine/quinidine	4.0	8.6
Ciprofloxacin[a]	6.0	8.8
Chlorothiazide	6.7	9.5
Pilocarpine	7.2	12.6

[a] Amphoteric compound (all others are dibasic).

value at which the acid is 50% dissociated. For a basic compound, the acid form is, for example, RNH_3^+, and the acidic dissociation is represented by

$$RNH_3^+ \rightleftarrows RNH_2 + H^+$$

where R stands for an arbitrary molecular moiety.

If the pH of the GI tract fluid is significantly greater than the pK_a of an acidic drug, the drug will be highly ionized and therefore much more soluble. Likewise, if the pH of the GI tract fluid is significantly less than the pK_a of a basic drug, the drug will be highly ionized and therefore much more soluble. However, conditions which promote drug solubility in the GI tract fluids work against drug absorption. The ionized forms of drugs have much lower lipid solubilities than their nonionized forms, and thus pass through the mucosal membrane (which has a central lipid core) extremely slowly. The most rapidly absorbed drugs are ones which exist in nonionized form in the GI tract environment and which, despite being nonionized, are nevertheless reasonably soluble in the GI tract fluids.

IV. MECHANICAL ASPECTS OF GASTROINTESTINAL PHYSIOLOGY

We now review in some detail what happens to a drug (or poison) after it is swallowed by mouth. After a voluntary swallowing stage, the drug is propelled by reflexive action through the pharynx (this occurs in 1–2 seconds) and into the esophagus. Here, additional involuntary movements (peristalsis) occur. In a time of about 4–8 seconds, the drug is delivered to the lower end of the esophagus, through the gastroesophageal constrictor, and into the stomach. This is an easy process, in general, since the constrictor muscle and the stomach both relax as the peristaltic wave coming down the esophagus approaches them. The constrictor muscle acts primarily to prevent the reflux of the highly acidic stomach contents back into the esophagus (the lower esophageal mucosa, except for the last few centimeters, are not capable of resisting stomach acids).

The stomach mixes the drug in with the stomach contents, by virtue of contractile waves (tonus waves) which begin in the upper part of the stomach wall and move downward along the stomach wall about every 20 seconds or so, and which become stronger in force as the antral portion of the stomach (i.e., the part nearest the outlet sphincter known as the pylorus) is approached. Thus, mixing of the drug with the digestive juices (that are secreted by the gastric glands which cover nearly the entire inner wall of the stomach) becomes more and more complete as the stomach contents come nearer to the pylorus. The greater mixing in the antral portion of the stomach is due to intense peristaltic movements in this region, which also occur about every 20 seconds and are superimposed on the weaker tonus waves. With each peristaltic wave, only a small fraction of the stomach contents are expelled through the pylorus into the duodenum, with the majority of the antral contents being squeezed backwards through the peristaltic ring, thereby strongly mixing them.

The gastric secretions typically amount to 2 liters/day total and normally consist of a solution of hydrochloric acid, pepsin, and other digestive enzymes having a pH of about 1–3 (upon dilution and mixing with the contents of the stomach, the resulting pH of the stomach contents is usually about 3, although foods initially can raise this to as high as 5). The walls of the stomach are such that poor solute absorption occurs across them; only a few highly lipid-soluble substances such as alcohols and lipid-soluble drugs pass through the walls, and then only in small quantity. Weakly acidic substances are nonionized at the pH of the stomach. They, and nonionizable drugs (and even some weakly basic drugs) can absorb somewhat from the stomach. However, if gastric emptying is not impeded, not much drug absorption takes place from the stomach. The time spent in the stomach is simply too brief and the surface area for absorption in the stomach is not large. For example, typically only about 10% of an aspirin dose is absorbed from the stomach, and even for a fairly rapidly absorbed substance such as ethanol, only about 30% is absorbed from the stomach (Gibaldi, 1984).

The pylorus normally remains almost but not completely closed due to tonic contractive forces, but opens somewhat each time an antral peristaltic wave occurs. The rate of passage of stomach contents through the pylorus depends almost entirely on the activity (strength, frequency) of the antral peristaltic waves. Such activity is regulated by the enterogastric reflex (this results from signals in the upper small intestine which reduce the gastric peristalsis whenever the upper small intestine is reasonably full) and by the enterogastrone mechanism (this results from the release of the hormone enterogastrone by the mucosa of the upper small intestine whenever fatty foods are present; this hormone is absorbed into the blood and quickly reduces the gastric motility dramatically). Excessive acidity, hypotonicity, or hypertonicity in the upper small intestine also elicit the enterogastric reflex.

Normally, gastric emptying is a first order process (i.e., the amount of material passed out of the stomach falls off exponentially with time). The normal half-life of emptying is about 20–60 min in healthy adults. However, the rate of emptying is decreased by the presence of fats and fatty acids (as mentioned above), certain diseases, lying on one's left side, and by a high viscosity or high bulk of the stomach contents. Many drugs (atropine, propantheline, narcotic analgesics, various tricyclic antidepressants, chlorpromazine, etc.) also decrease the rate of gastric emptying. For example, in one study (Hurwitz et al., 1977), 30 mg propantheline bromide given to 13 normal subjects 90 min before taking a liquid test meal doubled the half-life for emptying from 68 to 135 min. Certain factors can work in the opposite manner, increasing the rate of gastric emptying. Fasting, hunger, alkaline buffer solutions, anxiety, and drugs such as metoclopramide (a potent antiemetic) all increase the rate.

Clearly, the type of food eaten has a large effect on gastric emptying. As one might guess, liquids empty much faster from the stomach than do solids. Fats can delay the emptying of a fatty meal for as long as three to six hours; proteins have an intermediate effect, and carbohydrates have only a mild effect. Even a "standard" breakfast has been shown to take 4–8 hr to empty completely from the stomach. Sometimes food slows the rate of absorption but not the overall extent of absorption of the drug from the GI tract as a whole. Examples include digoxin, acetaminophen, phenobarbital sodium, and various sulfonamides. Then, with some drugs (e.g., many antibiotics), both the rate of absorption and the total amount absorbed from the GI tract are decreased by food. For a very few compounds (riboflavin is one example), food decreases the rate of absorption but actually increases the total amount absorbed. Food also seems to reduce the metabolism (and hence loss) of certain drugs (beta-blockers such as propranolol and metoprolol) by the liver (all of the mesenteric, or GI tract, blood goes to the liver before proceeding to other parts of the body).

Once the mixed material (chyme) from the stomach enters the small intestine, small ringlike contractions move the material along the entire small intestine. These are often irregular but generally are reasonably regular at a frequency of 8–9 per minute in the upper small intestine and at progressively lower frequencies further down. In the upper small intestine, pancreatic fluid containing various digestive enzymes and bile (which is secreted by the liver and stored in the gallbladder) enter and mix with the chyme. The function of the bile is to break down and emulsify fats, and the rate at which it enters the GI tract is controlled by the fat content of the chyme. The normal daily production of pancreatic fluid is about 1200 mL/day and that of the bile is about 700 mL/day.

About 3 to 10 hours are needed for chyme to pass through the small intestine down to the ileocecal valve, which separates the small and large intestines. Upon reaching this valve, the chyme is usually blocked (sometimes for several hours) until another meal is eaten. The main function of the ileocecal valve is to prevent any backflow of fecal contents from the large intestine into the small intestine.

The functions of the colon (large intestine) are to remove water and electrolytes from the chyme, forming it into a semi-solid mass (feces), and (in the lower part) storing the feces until they can be expelled. Movement in the colon is sluggish but still has the usual characteristics of mixing/propulsion.

The absorption of broken-down food items, drugs, etc., takes place in the small intestine, by contact with the absorptive surfaces of the intestinal mucosa, which consist of folds of tissue covered by millions of small projections called villi. The villi are so numerous that the effective surface area for absorption in the small intestine is about 10 times that of the stomach. This surface area, however, is not uniformly distributed—in the upper small intestine it is about four times as much per unit length as in the lower part of the intestine. The total surface area of the small intestine is about 550 m^2 (the millions of small villi in the small intestine serve the same purpose as the millions of tiny pores in activated charcoal, i.e., an extremely fine degree of subdivision creates large surface areas in both cases).

The total quantity of fluid which is absorbed per day through the GI tract walls is equal to whatever is ingested plus the roughly 8.5 liters of the various GI tract secretions (saliva, gastric secretions, small intestine secretions, bile, and pancreatic fluid). Ninety-five percent of a normal total of about 10 liters is absorbed in the small intestine; the remainder is absorbed in the colon.

V. ABSORPTION IN THE GASTROINTESTINAL TRACT

Any drug must be absorbed before it can have any systemic effect; thus, it is important to describe the factors which govern the rate of drug absorption from the small intestines. The small intestine region is the most important site of drug absorption, due to its very large surface area, as compared to that of the stomach. Thus, many drugs— whether they are weak acids, weak bases, or neutral compounds—are absorbed rapidly from the small intestine. If a drug dissolves readily in the gastric contents, then the rate of gastric emptying becomes the rate-limiting factor for drug absorption. Thus, charcoal is much more effective for drugs which empty quickly from the stomach. For drugs which are emptied slowly from the stomach or which are absorbed slowly from the small intestine and/or which are excreted in the bile or secreted from the GI tract walls into the GI tract lumen to a significant extent, charcoal given even after substantial delay can still be effective.

To be absorbed, a drug must be in solution (i.e., dissolved), be in a nonionized state, and have sufficient lipid solubility to penetrate the membrane surfaces. Drugs vary widely in their lipid solubilities and thus in the rates with which they are absorbed. Additionally, many drugs influence gastrointestinal motility (e.g., drugs with anticholinergic properties). This has been clearly demonstrated by Hurwitz et al. (1982), who studied gastric empty-ing using a gamma camera in volunteers who had been given the anticholinergic drugs propantheline bromide and clidinium bromide followed by a liquid test meal containing a nonabsorbable radioactive indium tracer. They found, for example, that the percent of a swallowed meal which emptied from the stomach in 1 hr decreased from about 48% with no propantheline to about 24% when 30 mg of propantheline had been given. Besides a delay in emptying of the stomach, there is also a decrease in the rate of propulsion of chyme through the intestines. And, by virtue of moving down the intestines more slowly, such a drug will be in contact with a smaller total surface area of the intestines at any given time. Hence, the rate of absorption is greatly decreased.

Absorption through the gastrointestinal mucosa proceeds by regular (passive) diffu-sion, carrier-mediated transport, and by active transport, as it does in the case of transport through all cell membranes. The mucosal membrane contains lipids, proteins, lipo-proteins, and polysaccharides. The membrane structure, like that of most biological membranes, is believed to consist of an unbroken central lipid core, layers of protein/ lipoprotein having pores in them on both sides of the lipid core, and layers of polysac-charides on top of both protein layers. The pores are filled with aqueous fluid, but these pores are only about 7.5 Å in diameter in the upper small intestine (jejunum) and 3.5 Å in diameter in the lower small intestine (ileum); thus, only small hydrophilic substances (e.g., urea) and ions can pass through the pores. All other substances, including most drugs, must traverse the mucosal membrane by passive, carrier-mediated, or active transport through the main part of the membrane (i.e., the part not containing pores). Adjacent to the mucosal membrane itself is a stagnant layer of aqueous fluid. Thus, either this watery layer or the membrane itself can affect the absorption of various substances. Usually, the membrane is more controlling.

Regarding food absorption, carbohydrates are broken down into simple sugars and adsorbed as such, proteins are broken down into amino acids and absorbed as such, and fats (as fatty acids) are broken down into simple glycerides and absorbed as such. All of the breakdown processes proceed by enzymatic hydrolysis which is carried out inside the GI tract lumen. In general, amino acids, sugars, fatty acids, water, monovalent ions, and other simple molecules pass through the GI tract walls fairly easily. Lipid-soluble substances and small hydrophilic molecules are transported by passive diffusion (the small hydrophilic molecules go through the pores), larger polar molecules such as sugars and amino acids are transported by carriers (this is often called "facilitated diffusion") since they are too large to fit in the pores, and some ions, amino acids, bile and bile salts are transported by "active transport" mechanisms. Since drugs are transported mainly by passive diffusion and since their concentrations on the blood-side of the membrane are small, they absorb in a first-order fashion.

Many vitamins (riboflavin, thiamine) are transported by facilitated diffusion. Since facilitated diffusion depends on a carrier and since the amount of the carrier involved is limited, the rate is governed by a Michaelis-Menten type of rate equation (see below) which gives first order kinetics at low concentration values and zero-order kinetics at high concentration values. Electrolytes are absorbed mainly by active transport and water passes through the mucosal walls by osmosis.

The upper small intestine is where most carrier-mediated transport of substances occurs. The side of the mucosal wall opposite to the lumen of the GI tract is richly endow ed with blood capillaries. Thus, solutes transported through the mucosal membrane are readily picked up into the blood stream. The circulatory system is such that the blood all flows into the portal vein which carries it to the liver. That is, the mesenteric blood coming from the intestines proceeds directly to the liver prior to going in succession to the right side of the heart, to the lungs, to the left side of the heart, and then to the various organs or other anatomical regions of the body.

Drugs that pass readily through the GI tract walls have rates which could potentially be affected by the mesenteric blood flow rate, since if the blood flow rate is low and the drug level builds up in the blood, a certain amount of "back pressure" (i.e., reduction in the concentration difference across the wall) would exist which would lower the rate. For drugs which absorb more slowly, no such dependence on the mesenteric blood flow rate would occur. In practice, few drugs pass quickly enough through the GI tract walls to show a significant dependence on the mesenteric blood flow rate.

The liver is able to process all of the blood coming from the intestinal system before that blood is distributed to the rest of the body. This allows the liver, the body's "chemical factory," to have a chance to metabolize, use for synthesis, or otherwise chemically convert the various solutes in the blood stream to forms which are, in one sense or another, more desirable (e.g., ammonia picked up from the large intestines can be converted to less harmful substances such as urea, and drugs may be converted to metabolites which are more easily excreted in the urine). The loss of a drug which is brought about by its metabolism when it is absorbed and carried to the liver is called the "first pass hepatic effect."

A certain amount of drug can be "lost" by other mechanisms. For example, enzyme hydrolysis can occur (e.g., that of penicillin G or digoxin in the stomach) which converts some of the drug to other forms. Or, complexation of the drug (e.g., binding of antibiotics such as streptomycin with polysaccharides which exist in the GI tract mucus) can occur, leading to large molecules that cannot cross the GI tract walls.

One may characterize the efficiency of transport of a drug across the GI tract walls and into the systemic circulation in terms of the drug's "bioavailability." The bioavailability is the fraction of the ingested drug which actually reaches the systemic circulation unaltered, and it can be estimated by comparing the area under the blood drug concentration versus time curve (from time zero to infinity, in principle) to the same area which is observed when the drug is given intravenously (and thus is delivered directly into the systemic circulation). Factors which can reduce bioavailability are:

1. Low permeability of the drug in the GI mucosal membrane, which thus leads to very slow absorption (e.g., neomycin).
2. Poor aqueous solubility, which leads to incomplete dissolution of the drug in the GI tract fluids (e.g., phenytoin).
3. Degradation by enzymatic hydrolysis (e.g., penicillin G and digoxin).
4. Preadsorptive metabolism by enzymes in the upper small intestine (e.g., aspirin) or by bacteria further down in the small intestine (e.g., digoxin).
5. Metabolism during absorption, either in the gut wall itself (e.g., proterenol) or by the liver (e.g., propranolol). This latter phenomenon is the previously described "first pass hepatic effect."

Some interesting results obtained by Neuvonen and Elonen (1980) demonstrate that the rates of drug absorption in human subjects can vary widely. They gave five healthy

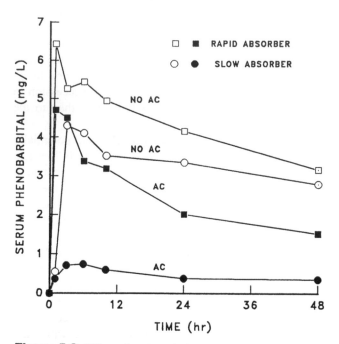

Figure 5.2 Effect of activated charcoal given after 1 hr on the absorption of phenobarbital in subjects having slow and rapid drug absorption. From Neuvonen and Elonen (1980). Figure 9. Copyright Springer-Verlag GmbH & Co. Used with Permission.

volunteers different drugs, followed by doses of oral charcoal within 5 min, or at 1 hr, or at several different times (10, 14, 24, 36, and 48 hr). Figure 5.2 shows results for two of the subjects who were given 200 mg phenobarbital followed by 50 g of Norit A charcoal 1 hr later. For one subject having a slow rate of drug absorption, the charcoal was effective in reducing drug absorption even when administered after 1 hr; for another subject showing rapid drug absorption, the delayed charcoal was much less effective. The authors imply that the differences in the rates of drug absorption were due primarily to differences in the rates of gastric emptying.

This work highlights an important point which can not be overemphasized: the sooner charcoal is administered following the ingestion of drugs which undergo rapid gastric emptying, the more effective the charcoal is likely to be.

VI.   DRUG FATE AFTER ABSORPTION

Figure 5.3 shows all of the distribution and elimination processes which can occur after a drug is absorbed from the GI tract. Once a drug is absorbed through the GI tract walls and enters the blood stream, several things may happen. The first is that the drug may bind to plasma proteins (albumin, fibrinogen, globulins). Secondly, the drug will start to distribute to all of the extravascular fluids (extracellular, intracellular) in the body. This distribution is usually fairly rapid and is reversible. Transfer into various tissues, such as fat tissue and muscle tissue, also takes place.

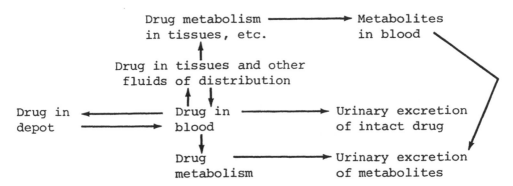

Figure 5.3 A simple model for drug uptake, distribution, and elimination.

With drugs which are highly lipid-soluble, considerable transfer into fatty tissues can occur, and the preference of the drug for the fatty medium makes the drugs exhibit a strong preference for staying in the fatty tissue (the oil/water partition coefficient of a drug is a good indicator of its lipid solubility). Thus, while transfer of a lipid soluble drug into fatty tissue is reversible, the equilibrium relationship characterizing its relative distribution between the systemic circulation and the fatty tissue favors its having a much higher concentration in the fatty tissues. Thirdly, the drug will transfer from the blood into the urine, bile, saliva, and similar body fluids. This type of transfer is usually irreversible and the net result of all of these transfer processes, plus metabolism (which will be discussed shortly) can be collectively labeled "elimination." Renal excretion of both the parent drug and/or its metabolites is the primary route of elimination of most drugs.

Drug distribution and elimination processes occur simultaneously; however, the rate constants for distribution are generally larger than those for elimination. Thus, the distribution processes in the body will essentially be in equilibrium before the elimination processes proceed very far and the body will act pharmacokinetically as if it were a single compartment. (Although the drug concentrations in the various "compartments" of the body may be different, the concentration of the drug in the blood will be proportional to the total amount of drug in the body, at any given time.) In cases where the drug enters the body by intravenous administration, however, the distribution processes do not proceed ahead of the elimination processes and significant elimination can occur before distribution equilibrium is attained.

With respect to elimination of the drug by metabolism in the liver, we can state that after absorption the drug or drug/protein complexes quickly go to the liver, where biotransformation (metabolism) may occur, transforming the drug to some extent into metabolites, which may or may not themselves be pharmacologically active. Transformation is usually due to hepatic metabolism by microsomal enzymes which convert the drug into more water-soluble forms for excretion in the urine. Although many drugs may be converted at rates proportional to their concentrations in the blood (i.e., according to first-order kinetics), saturation of hepatic or other enzyme systems may occur and result in a constant rate of metabolic conversion (zero-order kinetics).

Such a first-order/zero-order response is often modeled by the Michaelis-Menten approach, in which the reaction sequence is considered to be:

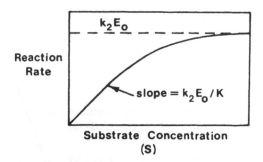

Figure 5.4 Reaction rate dependence on substrate concentration according to the Michaelis-Menten model.

$$E + S \overset{k_1}{\underset{k_{-1}}{\rightleftarrows}} ES \overset{k_2}{\rightarrow} E + P$$

where the enzyme E is regarded as reacting with a substrate (e.g., the drug) S to form a complex ES, which then decomposes to give back the enzyme and yield a product (e.g., a metabolite) P. If one assumes that the decomposition step is slow relative to the forward and backward parts of the $E + S \rightleftarrows ES$ step, then one can consider this first step to be essentially in equilibrium, and therefore governed by an equilibrium constant K which is

$$K = \frac{k_{-1}}{k_1} = \frac{[E][S]}{[ES]}$$

Here, the brackets [ ] around a species designate the molar concentration of that species (e.g., in millimoles/liter or such). Since the total molar enzyme concentration (as either E or as ES) is always constant at a value which we may call E_o, that is

$$E_o = [E] + [ES]$$

then we can easily show that

$$[ES] = \frac{E_o[S]}{[S] + K}$$

Thus, since the overall reaction rate is just that of the second step, we can write

$$\text{Rate} = k_2[ES] = \frac{k_2 E_o[S]}{[S] + K}$$

This is the famous Michaelis-Menten equation. If one plots the reaction rate versus the substrate (drug) concentration, one gets the kind of relationship shown in Figure 5.4. It is clear that at low [S] the reaction is essentially first-order, since, when $[S] \ll K$, the Michaelis-Menten equation reduces to

$$\text{Rate} = \left(\frac{k_2 E_o}{K}\right)[S]$$

Conversely, at high [S] the reaction rate is constant (or of "zero order") since, when [S] $\gg K$, the Michaelis-Menten equation reduces to

$$\text{Rate} = k_2 E_o$$

VII. GASTROINTESTINAL DIALYSIS AND INTERRUPTION OF THE ENTEROHEPATIC CYCLE

In the case of secretion of the drug into the GI tract, reabsorption of the drug from the GI tract can occur by the usual mechanisms which led to its absorption in the first place. Therefore, if activated charcoal can somehow be introduced to the small intestines after the initial absorption of the drug, the net rate of transfer of the drug across the GI tract walls can be enhanced and the drug which passes through the walls can be effectively "trapped" by adsorption to the charcoal. Such a process has been called "gastrointestinal dialysis" by Levy (1982) because he considered the GI tract membrane, with blood vessels on one side and charcoal on the other side, to act in much the same manner as the membrane in a conventional hemodialysis machine (in fact, however, what occurs with the use of charcoal is more like hemoperfusion than hemodialysis).

Additionally, drugs and/or metabolites which are picked up from the blood by the liver and are excreted into the bile can also be trapped by charcoal, when the bile enters the small intestine. Thus, charcoal can effectively interrupt the cycle of enterohepatic circulation. However, although the concentrations of drugs in the bile may be higher than their concentrations in the fluid secreted from the GI tract walls, the amounts of drugs involved in enteroenteric recirculation are usually significantly larger than the amounts involved in enterohepatic recirculation.

However, as Spector and Park (1986) have pointed out, drugs which have a large volume of distribution, and hence a low concentration in the blood, will be secreted/excreted into the GI tract in small amounts. Thus, charcoal in the GI tract will have a small effect on elimination. An example of this is imipramine. Indeed, Goldberg et al. (1985) have shown that even multiple-dose charcoal therapy has no significant effect on imipramine elimination. However, drugs with small volumes of distribution (say, 1 L/kg or less) such as theophylline, phenobarbital, and digitoxin would be expected to be well removed by gastrointestinal dialysis.

We now review a series of studies by Arimori and coworkers on the effectiveness of charcoal in adsorbing drugs secreted into the GI tract lumen directly or via the bile. The main pharmacokinetic parameters involved, the area-under-the-curve (AUC) and the elimination half-life ($t_{1/2}$), are discussed in the next chapter. Readers unfamiliar with how values for these parameters are determined may wish to refer to that chapter first before proceeding here. In all studies dealing with theophylline, the drug was used in the form of aminophylline (a compound of theophylline and 1,2-ethanediamine which dissociates upon dissolution).

Arimori and Nakano (1985) noted from literature reports that the elimination of theophylline administered intravenously is enhanced by charcoal given orally, thereby suggesting that theophylline is conveyed from the blood into the GI tract lumen by some mechanism(s). They decided to investigate theophylline "exsorption" into the small intestine (by "exsorption" they mean the transport mechanisms, such as secretion, by which the drug passes from the blood into the GI tract lumen) and bile excretion into the biliary tracts using Wistar male rats. Their technique involved the IV infusion of aminophylline (10 mg/kg) given as an injection of a 25 mg/mL solution over a 2 min period into the right femoral vein while carrying out single-pass perfusion of the small intestines with isotonic phosphate buffers, and collecting bile via cannulation of the common bile ducts. The drug levels in the perfusate and bile were found to obey first-order kinetics. Over a two-hour period, the amounts of the drug excreted into the perfusate and into the bile juice were 12.1 and 0.2%, respectively, when a pH 6.0 buffer was used, and 13.8 and 0.2%, respectively, when a pH 8.0 buffer was used. Thus, a significant amount of theophylline was exsorbed into the intestinal lumen (where charcoal would be able to bind it, if charcoal were present), whereas a negligible amount was excreted into the bile.

Arimori and Nakano (1986) next studied the effect of oral charcoal on the clearance of theophylline (10 mg/kg) and phenobarbital sodium (10 mg/kg) following their IV administration to rats. Single doses of charcoal (300 mg) at time zero, and multiple doses of charcoal (300 mg at time zero, and 150 mg at 1, 2, 3, and 4 hr) were used. The single and multiple charcoal dose regimens reduced the theophylline $0-\infty$ AUC value from a control level of 138 mg-hr/L to 106 and 88 mg-hr/L, respectively. Half-lives fell from 4.6 hr (control) to 3.4 and 2.8 hr, respectively. For phenobarbital, the $0-\infty$ AUC values were 184 mg-hr/L (control) and 119 mg-hr/L (multiple-dose charcoal); the half-life values were 8.5 hr (control) and 5.7 hr (multiple-dose charcoal). Single-dose charcoal studies were not done with phenobarbital. The results indicate that significant "intestinal dialysis" of both drugs can occur when charcoal is used.

Arimori and Nakano (1987) did another study in which rats were given phenytoin IV with single-pass perfusion of the small intestines. At a dose of 10 mg/kg, 1.1% of the drug appeared in the perfusate over 2 hr, while at a dose of 50 mg/kg 2.5% of the drug appeared in the perfusate over 2 hr. Excretion of phenytoin into the bile over 2 hr varied from 0.10–0.21%, depending on the size of the dose injected, whereas excretion of 5-(p-hydroxyphenyl)-5-phenylhydantoin (p-HPPH), its main metabolite, varied between 7.9–16.3% of the dose injected. When multiple doses of charcoal were given, no significant effects occurred at a drug dose level of 10 mg/kg; however, at a drug dose of 50 mg/kg, the drug serum half-life decreased to 77% of the control value and the AUC value was 75% of the control level. Thus, it appears that charcoal has some benefit at higher phenytoin doses, mainly by adsorbing phenytoin which enters the GI tract in the bile.

Arimori et al. (1987) then studied the effects of oral charcoal on the intestinal dialysis of M79175, an aldose reductase inhibitor, in rats. The M79175 was injected IV. Again, single-pass perfusion and bile duct cannulation were performed. Over 2 hr, the percentages of the drug appearing in the intestinal lumen and in the bile were 3.2 and 2.0%, respectively. Oral administration of multiple doses of charcoal significantly reduced serum drug levels (AUC fell by 46%) and the serum drug half-life ($t_{1/2}$ fell 47%). Total fecal plus urinary excretion was unaffected with charcoal (fecal excretion rose and urinary excretion fell, but the total was essentially unchanged). Thus, intestinal dialysis of this drug by charcoal was substantial.

Arimori and Nakano (1988a) then studied the transport of furosemide into the intestinal lumens of normal rats and rats having acute renal failure (ARF). Again, single-pass perfusion and bile duct cannulation methods were used. Whereas, over 2 hr, only 0.83% of the IV furosemide dose given to normal rats appeared in the intestinal lumen, with ARF rats the value was 1.83%. The amounts excreted into the bile were 1.53% in normal rats and 2.64% in ARF rats. The increased exsorption of furosemide in ARF rats was hypothesized to be due to decreased protein binding of the drug in ARF rats, since only

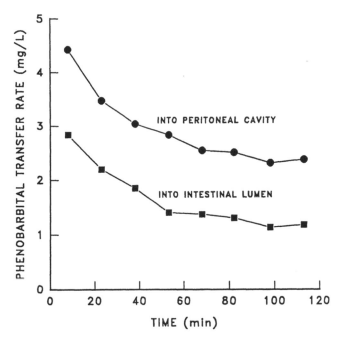

Figure 5.5 Transfer rate of phenobarbital from the blood into the intestinal lumen and into the peritoneal cavity after IV administration (10 mg/kg) to rats. Each point is the mean of the values for four rats. From Arimori et al. (1990). Reprinted with permission of the Royal Pharmaceutical Society of Great Britain.

the unbound drug can permeate the intestinal membrane. Oral charcoal, in multiple doses, had little effect on serum drug levels in ARF rats given a 10 mg/kg drug dose. The lack of effect probably relates to the small amount of drug excreted into the GI tract.

Arimori and Nakano (1988b) then studied the dose-dependency of theophylline exsorption in rats. Aminophylline was given IV at dose levels of 10–50 mg/kg. The amount of drug exsorbed over 2 hr into the intestinal lumen as a function of drug dose varied linearly, from about 450 μg at a dose of 10 mg/kg to about 2300 μg at a dose of 50 mg/kg. The percent of the drug exsorbed over 2 hr was in the range of 12–15% (by comparison, the amount of drug which was excreted into the bile in the same period was only 0.17–0.30%). Multiple oral doses of charcoal were given to rats receiving a drug dose of 50 mg/kg and caused the AUC and $t_{1/2}$ values to decrease by 52 and 50%, respectively, as compared to controls. Total body clearance rose to 188% of controls.

Arimori and Nakano (1989) next studied the drug disopyramide with the same techniques. They found that this drug was appreciably secreted into the bile, where its levels were roughly 10-fold higher than those in the serum. The average amounts of disopyramide appearing in the perfusate over 2 hr were 17.0 and 18.4% at drug doses of 10 and 30 mg/kg, respectively (by contrast, only 1.1 and 1.2% of the drug was excreted into the bile at these dose levels). Again, multiple doses of activated charcoal were highly effective, decreasing the $t_{1/2}$ by 89% and the AUC by 82%, as compared to controls. Thus, charcoal provides highly effective intestinal dialysis for this drug because of its excellent exsorption into the intestinal lumen.

Arimori et al. (1989a) then studied the transport of theophylline into the intestinal lumen of rats having hepatic cirrhosis (HC). Using the same methods as before, with a 10 mg/kg drug infusion, they found that the HC rats had higher serum drug levels and greater exsorption into the GI lumen than did normal rats. Over 2 hr, HC rats exsorbed 13.4% of the drug dose, compared to 8.8% of the drug dose for the normal rats. Oral multiple-dose charcoal reduced AUC and $t_{1/2}$ values in the HC rats by 34 and 48%, respectively, compared to HC rats not receiving charcoal. Thus, charcoal may be a useful method for treating theophylline overdose in patients with hepatic failure.

In sharp contrast to the studies just discussed, Arimori et al. (1989b) turned to using human subjects in one of their next studies. They gave disopyramide (200 mg) orally to six healthy subjects followed by 40 g of charcoal in a suspension containing 250 mL of a magnesium citrate solution (34 g magnesium citrate content) at 4 hr, and charcoal doses of 20 g with 150–200 mL water at 6, 8, and 12 hr. The charcoal treatment decreased serum half life and AUC (0–∞) values by 33 and 19%, respectively, as compared to control values. Total body clearance increased to 122% of the control value. Serum drug level data showed that the peak drug levels occurred between 1–4 hr in the subjects; hence, even though the first charcoal dose was not given until 4 hr, there was still a significant charcoal effect.

Arimori et al. (1990) then studied the intestinal and peritoneal dialysis of theophylline and phenobarbital in rats. A new feature of this study was the addition of peritoneal dialysis. This was carried out by exchanging dialysate fluid in the peritoneal cavity every 15 min. With theophylline (dose = 10 mg/kg IV), peritoneal dialysis alone exsorbed 16.5% of the drug dose in 2 hr. With intestinal perfusion alone, the value was 15.7%, or essentially the same. With phenobarbital, peritoneal dialysis exsorbed 12.5% of the drug in 2 hr, compared to 7.8% exsorbed into the intestinal lumen over 2 hr. Figure 5.5 shows the rates of phenobarbital transfer into the intestinal lumen and into the peritoneal cavity in graphical form. The net water flux showed that secretion was dominant for transport across the peritoneum, while absorption was the dominant mechanism for transport through the intestinal membrane. These results suggest that the use of charcoal, which would adsorb these drugs in the GI lumen, would be a simpler treatment for overdose than peritoneal dialysis, since the transport rates into the GI lumen and into the peritoneal cavity are not widely different.

VIII. SUMMARY

We have reviewed the basic physiology of the gastrointestinal tract, as a preamble to understanding how orally-administered charcoal can affect the absorption and overall bioavailability of drugs. We implicitly assumed that disintegration the drug dosage form was not a significant limiting factor and that drug administration was either oral or IV.

Mechanical aspects of the GI tract were discussed and the physicochemical aspects of absorption processes (passive transport, carrier-mediated transport, active transport) were described. The important role of pH in affecting the state of ionization of a drug (as determined by the drug's pK_a relative to the prevailing pH) and the effect of pH on drug absorption (in general, a drug must not be in ionized form in order to be absorbed) were considered in some detail.

The distribution, metabolism, and elimination of drugs by various routes subsequent to their absorption were discussed, with particular emphasis on the processes of enteroenteric and enterohepatic recycling of drugs back into the GI tract (exsorption). The pharmacokinetic modeling of drug absorption, distribution, and elimination processes will be considered in the next chapter.

To illustrate the importance of exsorption processes in allowing orally-administered charcoal to capture drugs recycled back into the GI tract, a series of studies by Arimori and coworkers on the gastrointestinal dialysis of drugs given intravenously to animals and human subjects were reviewed. These studies demonstrated that already-absorbed drugs (drugs given IV are, of course, already in essence absorbed, since the absorption step is bypassed) can be effectively bound by charcoal in the GI tract, if the drug undergoes significant recycling to the GI tract by secretion through the GI tract membranes or secretion into the bile.

IX. REFERENCES

AHFS Drug Information '92, American Hospital Formulary Service, American Society of Hospital Pharmacists, Bethesda, Maryland, 1992.

Arimori, K. and Nakano, M. (1985). Transport of theophylline from blood to the intestinal lumen following i.v. administration to rats, J. Pharmacobiodyn. *8*, 324.

Arimori, A. and Nakano, M. (1986). Accelerated clearance of intravenously administered theophylline and phenobarbital by oral doses of activated charcoal in rats: A possibility of the intestinal dialysis, J. Pharmacobiodyn. *9*, 437.

Arimori, A. and Nakano, M. (1987). The intestinal dialysis of intravenously administered phenytoin by oral activated charcoal in rats, J. Pharmacobiodyn. *10*, 157.

Arimori, K., Mishima, M., Iwaoku, R., and Nakano, M. (1987). Evaluation of orally administered activated charcoal on intestinal dialysis of intravenously administered M79175, an aldose reductase inhibitor, in rats, J. Pharmacobiodyn. *10*, 243.

Arimori, K. and Nakano, M. (1988a). Transport of furosemide into the intestinal lumen and the lack of effect of gastrointestinal dialysis by charcoal in rats with acute renal failure, J. Pharmacobiodyn. *11*, 1.

Arimori, A. and Nakano, M. (1988b). Dose-dependency in the exsorption of theophylline and the intestinal dialysis of theophylline by oral activated charcoal in rats, J. Pharm. Pharmacol. *40*, 101.

Arimori, K. and Nakano, M. (1989). Study on transport of disopyramide into the intestinal lumen aimed at gastrointestinal dialysis by activated charcoal in rats, J. Pharm. Pharmacol. *41*, 445.

Arimori, K., Wakayama, K., and Nakano, M. (1989a). Increased transport of theophylline into gastrointestinal lumen and gastrointestinal dialysis by activated charcoal in rats with hepatic cirrhosis, Chem. Pharm. Bull. *37*, 3148.

Arimori, K., Kawano, H., and Nakano, M. (1989b). Gastrointestinal dialysis of disopyramide in healthy subjects, Int. J. Clin. Pharmacol. Ther. Toxicol. *27*, 280.

Arimori, K., Hashimoto, Y., and Nakano, M. (1990). Comparison of intestinal and peritoneal dialysis of theophylline and phenobarbitone in rats, J. Pharm. Pharmacol. *42*, 726.

Donovan, J. W. (1987). Activated charcoal in management of poisoning: A revitalized antidote, Postgrad. Med. *82*, 52.

Gibaldi, M. (1984). *Biopharmaceutics and Clinical Pharmacokinetics*, 3rd edition, Lea & Febiger, Philadelphia.

Goldberg, M. J., Park, G. D., Spector, R., Fischer, L. J., and Feldman, R. D. (1985). Lack of effect of oral activated charcoal on imipramine clearance, Clin. Pharmacol. Ther. *38*, 350.

Hurwitz, A., Robinson, R. G., and Herrin, W. F. (1977). Prolongation of gastric emptying by oral propantheline, Clin. Pharmacol. Ther. *22*, 206.

Hurwitz, A., Robinson, R. G., Herrin, W. F., and Christie, J. (1982). Oral anticholinergics and gastric emptying, Clin. Pharmacol. Ther. *31*, 168.

Levy, G. (1982). Gastrointestinal clearance of drugs with activated charcoal, New Engl. J. Med. *307*, 676.

Neuvonen, P. J. and Elonen, E. (1980). Effect of activated charcoal on absorption and elimination of phenobarbitone, carbamazepine, and phenylbutazone in man, Eur. J. Clin. Pharmacol. *17*, 51.

Physicians' Desk Reference, 48th edition, Medical Economics Data Production Company, Montvale, NJ, 1994.

Remington's Pharmaceutical Sciences, 18th edition, Mack Publishing Company, Easton, PA, 1990.

Spector, R. and Park, G. D. (1986). New roles for activated charcoal, West J. Med. *145*, 511.

The Merck Index, 11th edition, Merck and Company, Rahway, NJ, 1989.

6

Basic Details of Pharmacokinetic Modeling

A very large number of in vivo studies involving humans or animals, in which drugs or poisons were ingested in toxic amounts, or administered in subtoxic amounts, will be discussed in later chapters of this volume. In many of these studies, the "pharmaco-kinetics" of the drug or poison were monitored and subsequently modeled, using standard "pharmacokinetic models." The purpose of this type of analysis is to quantify the kinetics (i.e., rate behaviors) of the various processes which the drug or poison undergoes in the living organism. If one quantifies the kinetics of these processes when different treatment methods (such as activated charcoal) are used, as compared to the situation where only simple supportive methods are employed, then one obtains a quantitative basis for evaluating the effectiveness of whatever active therapeutic measure is used. This chapter is aimed at allowing the reader to understand and appreciate the analysis of drug dynamics using such pharmacokinetic modeling procedures.

I. INTRODUCTION

The subject area of pharmacokinetics is concerned with studying and characterizing the time course of drug absorption, distribution, metabolization, and excretion by the human body, and is also concerned with the relationship of these processes to the time course of therapeutic and toxicological effects of drugs.

The general physiological aspects of drug absorption, distribution, and elimination were discussed in the previous chapter. What we wish to discuss now are: (1) actually characterizing these processes quantitatively by making appropriate scientific measurements (such as drug concentrations in the blood, urine, and other body fluids at various times), and (2) how these processes can be modeled mathematically so as to extract from such data values of fundamental parameters such as drug bioavailabilities (as indicated by areas under blood drug concentration versus time curves), rate constants and/or half-lives for absorption and elimination, volumes of distribution, and clearance rates.

As Rosenberg et al. (1981), Sangster and De Kort (1986), and Sue and Shannon (1992) have emphasized in their excellent reviews, the pharmacokinetics of drugs in overdose amounts are often different than the pharmacokinetics observed with therapeutic doses. Several factors are involved: (1) the drug itself may slow down or accelerate absorption, (2) the extent of protein binding may be different, (3) the volume of distribution may be different, (4) saturation of hepatic enzyme systems in overdose may delay metabolism of the drug, (5) overdose may induce alternate metabolic pathways and actually increase drug metabolism, and (6) the difference in the ratio of drug/metabolites and/or saturation effects may alter renal elimination. Such changes in the pharmacokinetics have implications for treatment approaches. For example, delayed absorption would make oral charcoal more effective, while alkalinization or acidification of the urine can enhance the renal elimination of unmetabolized drugs which would normally undergo hepatic transformation.

Two basic types of models have evolved in the pharmacokinetic literature. The type of model used in classical "compartmental analysis" defines the body as consisting of one or more compartments, without making any particular statements regarding the specific contents of the compartments (i.e., the fluids or tissues contained therein). Such compartments usually (but not always) represent fairly large regions, such as the entire body itself, the entire GI tract, etc. Moreover, the contents of each compartment are assumed to be homogenous, although clearly this is often not an accurate assumption. Finally, the inputs and outputs associated with these compartments are not identified as actual fluid flows or diffusive fluxes but are left in general terms as simply "inputs of drug" or "outputs of drug." For many purposes, the use of models with a few large compartments having no fine details is acceptable. Certainly, such models are by far the most common.

A second fundamental group of models has evolved in which the body is modeled as a group of quite distinct compartments which are defined as specific anatomical regions or organs (e.g., the head, the lungs, the upper trunk, the extremities, the liver), as specific types of tissues (muscle tissue, fatty tissue, etc.), and as specific types of fluids (blood, CSF fluid, interstitial fluid, intracellular fluid, etc.). The exact types of specific compartments selected depend much on the drug involved and which types of subdivisions are most appropriate for that drug. While these more detailed models have the potential for characterizing drug transport processes more accurately than simpler "large compartment" models, much more information must be known in order to create such models. In general, the models which have been used in the studies that we discuss in this volume are of the "large compartment" type; thus we restrict our discussion henceforth to them.

Regardless of which type of pharmacokinetic model is constructed, one usually proceeds to write equations for each region which, being "rate" equations, are normally differential equations. These equations characterize the rate of transfer of the drug into and out of each region by transfer to/from adjacent regions, and the rate of elimination of the drug from that region by one or more mechanisms. These equations must be integrated to give one the actual equations which will be "fitted" to the experimental data. This fitting process allows one to extract parameter values (areas-under-curves, rate constants, volumes of distribution, clearances, etc.) from the data. Comparison of the parameter values for data obtained using different treatment modalities (e.g., charcoal, lavage, etc.) gives one a quantitative basis for comparing these modalities against each other and against control data obtained when no intervention was attempted.

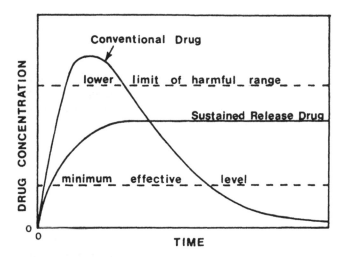

Figure 6.1 Typical behavior of the concentration of a drug in the blood as a function of time, for conventional and sustained-release dosage forms.

Pharmacokinetics could be, and has been, the subject of numerous books; those by Gibaldi (1984), Notari (1987), and Rowland and Tozar (1989) are excellent treatises. In this volume, we will restrict ourselves to those basic principles and concepts which are most relevant to the antidotal charcoal studies discussed in this volume. Thus, for the most part, we will be concerned with models which utilize data on blood concentrations (usually in the serum or in the plasma) versus time or data on the urinary excretion of drugs/toxins versus time. We will generally not be concerned with data on tissue drug concentrations versus time for various organs (liver, lungs, etc.) or for various types of tissues (muscle, fat).

A typical course of blood concentration versus time for an orally administered drug shows the behavior displayed in Figure 6.1. There is first a rise during the time in which absorption from the GI tract is the dominant process, the attainment of a peak at which time the input rate (absorption) equals the sum of the output rates (metabolization, elimination, and perhaps further distribution to other fluids/tissues), and then a gradual fall during the period when the output processes become more and more dominant. Shown on this figure are lines delineating the minimum effective drug concentration level and the level at which undesired symptoms of toxicity occur (for drugs given in proper dosage, this toxic level would lie well above the peak concentration). The range between these two horizontal lines is called the "therapeutic range"; drugs for which the lines are relatively closely spaced are said to have a "narrow therapeutic index."

Distribution processes (i.e., transfer of the drug from the blood to the various other fluids and tissues of the body and, later on, the reverse process) begin as soon as the drug enters the blood stream and continue as long as drug is in the body. At early times, as distribution of the drug from the blood stream to other fluids and tissues occurs, the distribution processes dampen the rate and extent of the rise of the drug concentration in the blood stream. At late times, as elimination processes cause the drug level in the blood to fall, the other fluids and tissues transfer drug back into the blood and ameliorate the rate of fall of the blood drug level. If the distribution kinetics were infinitely rapid, all of the drug concentrations in these other fluids and tissues would always be in equilibrium with the blood drug concentration and the times at which the peak drug levels are attained in all of the fluids and tissues would be identical.

In reality, the distribution kinetics for some fluid/tissue spaces in the body are not fast and their drug level versus time responses lag behind those in the blood (e.g., the peak times occur later than the peak time in the blood). At sufficiently "late" times, drug absorption is over and distribution equilibrium is essentially attained. Then, the fall of the drug level in the blood is due entirely to elimination processes. It is the data collected during this true "elimination phase" which are used to determine the half-life of the drug disappearance processes and the drug's apparent volume of distribution.

Since charcoal in the gut (and, most importantly, the stomach) has its main effect on the drug absorption process and not on the drug disappearance processes, the use of charcoal will have its greatest impact on the area under the drug concentration versus time curve (AUC) and the peak drug level. If charcoal is effective in binding the drug in the gut, the concentration versus time curve will be dramatically lower, especially at "early" times (at quite late times, the drug levels with charcoal use may actually be a bit higher than those without charcoal use, due to some degree of desorption of the drug from the charcoal as the drug/charcoal complex travels down the GI tract). Thus, the AUC will be much lower and the peak concentration will be much lower. Also, charcoal often increases the time to the attainment of the peak level, since charcoal "delays" as well as reduces the absorption process.

Charcoal generally has a much lower impact on the half-life of drug elimination. However, there are many drugs which undergo enterohepatic circulation, so that after absorption from the GI tract the drug is taken up from the blood by the liver and excreted into the bile, which then delivers the drug back to the GI tract. There, the charcoal can bind the drug and prevent its reabsorption. Also, there are drugs which can diffuse directly through the walls of the GI tract (so-called enteroenteric circulation) into the intestinal lumen. In the case of such drugs, the elimination of the drug from the blood can indeed be affected long after the initial absorption process and thus charcoal will indeed affect the rate of decline of the falling concentration part of the blood drug concentration, C_B, versus t curve. Thus, for drugs which undergo enteroenteric and/or enterohepatic circulation, the half-life which characterizes the "tailing" part of the C_B versus t curve can be dramatically decreased by charcoal. As mentioned in Chapter 5, the concentrations of drugs in the bile may be larger than their concentrations in the luminal fluid, but the amounts of drugs involved in enteroenteric circulation are usually significantly larger than the amounts involved in enterohepatic circulation. Thus, orally administered charcoal is often more important in interrupting the enteroenteric circulation of drugs.

As just intimated, the key parameters derived from blood concentration data are area-under-the-curve (AUC) values and half-life ($t_{1/2}$) values. The AUC is just the numerical value of the area under the concentration versus time curve, and this of course depends on the limits of the integration, that is, the time chosen as the starting time (not always "zero", i.e., the time at which the drug was ingested) and the time chosen as the final time. The final time is usually not the time when the blood concentration has fallen to zero (or near-zero) because this time may be impractically large. In practice, one can follow the drug concentration versus time for only a limited period of time, on the order of a day or two. Thus the AUC values obtained will be less than those one would obtain if one could follow the drug concentration until it had decayed to zero. However, for purposes of comparing different intervention strategies, the AUC values thus obtained are usually close enough to the "infinite time" values to be valid for comparing the strategies. In any event, as will be shown later, simple methods exist for estimating how much AUC needs to be added to a "time zero to time t" AUC value in order to arrive at a good estimate of the "time zero to infinite time" AUC value.

Half-life values are obtained by assuming that drug elimination during the "true" or "terminal" elimination phase conforms to first-order kinetics, that is, that the decay of the drug concentration in the blood (or the drug amount in the body) has a monoexponential character. Thus it is assumed that the concentration of drug in the body, or more specifically in the blood, follows the equation

$$C_B = C_{Bo} \exp(-k_e t)$$

which characterizes a first-order decay process, where C_{Bo} is the value of C_B at time zero, and k_e is the first-order elimination rate constant (the writing of the exponential term as $\exp(-k_e t)$ is of course identical to writing it as e raised to the power $-k_e t$, and in this volume the former method will be used exclusively). Hence a plot of the natural logarithm of C_B versus time (i.e., $\ln C_B$ versus t), using data obtained during the true elimination phase, should yield a straight-line plot with a slope equal to $-k_e$. Equivalently, one can plot the C_B versus t data on semilogarithmic graph paper and avoid the need for computing logarithms of the C_B values. In all that follows, references to logarithm are meant to denote the logarithm to the base e, unless otherwise stated. Also, it will be assumed that any graphical analyses will be done by plotting C_B versus time on semilogarithmic graph paper, with a logarithmic ordinate and a linear abscissa, rather than by plotting $\ln C_B$ versus t on graph paper which has both axes linear.

The half-life is the time required for a halving of the concentration. Thus, if one sets

$$\frac{C_B}{C_{Bo}} = 0.5 = \exp(-k_e t_{1/2})$$

one can show that

$$t_{1/2} = \frac{0.693}{k_e}$$

Therefore, $t_{1/2}$ is computed as 0.693 divided by the rate constant determined as described above. The units of $t_{1/2}$ are time. A value of $t_{1/2}$ of 6.4 hr, for example, says that the drug concentration falls by a factor of two (i.e., is cut in half) during each 6.4 hour period. The $t_{1/2}$ value for a true first-order process is independent of the value of C_{Bo}. In contrast, if the drug concentration process were zero-order, and therefore characterized by the equation

$$C_B = C_{Bo} - k_e t$$

the half-life would be given by

$$t_{1/2} = \frac{0.5 C_{Bo}}{k_e}$$

Table 6.1 Half-lives (hr) for Selected Drugs in Normal Adults

Acetaminophen	2–3	Indomethacin	1–2
Amobarbital	24	Meprobamate	8–14
Aspirin	0.25	Penicillins	0.5–1
Chlordiazepoxide	6–24	Phenobarbital	2–4
Diazepam	24–48	Phenytoin	20–30
Digitoxin	96–192	Salicylamide	1
Digoxin	32–45	Tetracyclines	10–20
Griseofulvin	13–24	Theophylline	3–20

Adapted from Notari (1987).

and thus the value of $t_{1/2}$ would depend on C_{Bo}. Table 6.1 gives some typical first-order $t_{1/2}$ values for various drugs.

Some other common first-order processes, all described by the general equation

$$C = C_o \exp{(-kt)}$$

are: absorption of a drug from the GI tract, intercompartmental transfer of a drug, metabolic conversion of a drug (at low concentrations where the amount of needed enzymes exceeds the amount of drug), and the appearance of a drug in the urine. Zero-order processes are generally uncommon but can occur where a carrier or enzyme is needed in a process (e.g., transfer across the GI mucosa by a carrier mechanism, or enzyme-mediated metabolic reactions) where the carrier or enzyme is "saturated" (i.e., all fully utilized). The Michaelis-Menten equation (see Chapter 5) is generally used to characterize such cases where one has first-order kinetics at low drug concentrations and a transition to zero-order kinetics at high drug concentrations.

II. THE PHARMACOKINETICS OF INTRAVENOUSLY ADMINISTERED DRUGS

Because the first part of the C_B versus t curve is dominated by the process of drug absorption from the GI tract, many pharmacokinetic researchers have simplified the situation by administering the chosen drug intravenously. This of course bypasses the absorption process and gets the drug directly into the blood. In this case, the C_B versus t curve starts almost immediately at a high value and decays from the start. The only processes involved are the distribution and elimination processes. If the drug distributes reasonably rapidly, one can assume that the distribution processes reach equilibrium before any substantial amount of drug elimination occurs. Then, the body will in fact behave as a single compartment. Although this single compartment is actually not at all homogenous (e.g., a highly lipid-soluble drug will have a much higher concentration in the fatty tissues than in other tissues, or in the blood), if true distribution equilibrium exists, and if the drug distributions between the blood and the various other fluids and tissues are characterized by constant values of the partition coefficients (i.e., the partition coefficients do not vary with concentration), then the amount of drug in the body will

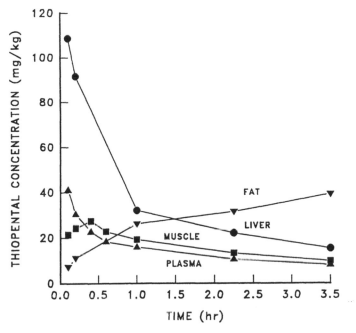

Figure 6.2 Concentration of thiopental in various tissues following an IV bolus dose to a dog. From Brodie et al. (1952). Copyright 1952 Williams & Wilkins. Reproduced with permission of the Williams & Wilkins Company.

nevertheless be proportional to the concentration of the drug in the blood. Thus, one can use the blood drug concentration as an indicator of the amount of drug in the body, and characterize its elimination by following the blood drug concentration versus time. Thus, for a drug which is given intravenously and which distributes fairly rapidly, one can use essentially all of the data obtained (discarding only the very earliest data points, perhaps) to evaluate the elimination rate constant. Many of the studies that we shall consider in later chapters have employed intravenous drug administration.

The fluids and tissues into which different drugs distribute, and the extent of such distribution, is highly variable. Blood flow rates to different tissues, per 100 mL of tissue, range from roughly 500 mL/min to less than 2 mL/min. Various tracers and indicator substances have been used to determine the volumes of the various fluid compartments and the speed with which they tend to equilibrate with the circulating blood. Scholer and Code (1954) have shown that deuterium oxide ("heavy water") given IV initially distributes rapidly into a volume only about 25% as large as the final volume of distribution. About 2 hr are required for nearly complete equilibration. Other investigators suggest that between 1/2 and 2 hr are needed for good equilibration, depending on the person and the drug involved. Data of Brodie et al. (1952) on the distribution of thiopental in a dog following the administration of a single IV dose of 25 mg/kg are shown in Figure 6.2. Clearly, there is an early rise of the drug levels in the well-perfused tissues, such as the liver and the muscles, but the rise in the adipose tissues is much slower. At the end of 3 hr, much of the drug remaining in the body is "hung up in" the fatty tissues. Thus, the assumption of rapid distribution equilibrium is often tenuous.

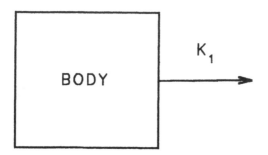

Figure 6.3 One compartment open model for IV drug injection.

A. The One Compartment Open Model

The case of intravenous drug administration is frequently modeled using a "one compartment open model." The word "open", when used in conjunction with compartment models, just means that the value of the drug concentration in the compartment goes to zero as time goes to infinity. This simple model is depicted in Figure 6.3. There are no input streams (the drug is assumed to be introduced into the compartment instantly by IV administration, and the compartment is assumed to be homogenous at all times, in terms of its behavior).

Assuming instantaneous injection/infusion and distribution of a drug in one compartment, then if elimination is by one mode (e.g., excretion in the urine) and is first-order in nature, the concentration of the drug in the blood, C_B, is characterized by the previously cited monoexponential equation

$$C_B = C_{Bo} \exp(-k_1 t)$$

where C_{Bo} is the concentration of drug in the blood at "time zero." (We now designate the elimination rate constant as k_1 rather than the k_e we used earlier, so that we may subsequently allow for elimination by other routes, with rate constants k_2, k_3, etc.) If, as is often the case, obtaining values of C_B very close to $t = 0$ is impractical, one may obtain a good estimate of C_{Bo} by plotting C_B versus t on semilogarithmic graph paper and extrapolating a "best fit" straight line back to $t = 0$. The intercept on the vertical axis will then be C_{Bo}. Since the amount of drug in the body, A, and the concentration in the blood, C_B, are proportional to each other, then it is also true that

$$A = A_o \exp(-k_1 t)$$

where A_o is the amount of drug in the body at "time zero." If elimination occurs by two first-order routes (e.g., first-order metabolism as well as excretion into the urine, the proper equation is the biexponential one

$$C_B = C_1 \exp(-k_1 t) + C_2 \exp(-k_2 t)$$

where k_1 and k_2 are the rate constants characterizing the two elimination processes, and

$$C_1 + C_2 = C_{Bo}$$

A similar biexponential equation describes the time dependence of the amount of drug, A, in the body. Since one of the two rate constants will generally be larger than the

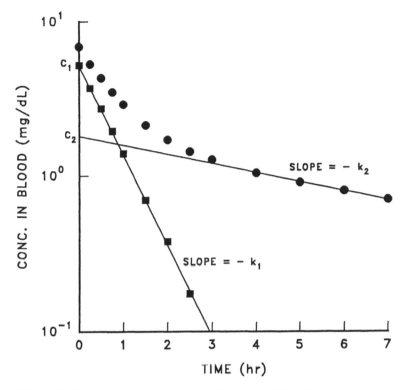

Figure 6.4 Feathering of data to find biexponential model parameters.

other, it is customary to always let k_1 be the larger of the two. Then, as time gets large, the first term in the equation, $C_1 \exp(-k_1 t)$ will decay to insignificance while the second term is still important. For these "late" times, C_B is given to a good approximation by

$$C_B = C_2 \exp(-k_2 t)$$

and thus a plot of C_B versus t on semilogarithmic graph paper can be fitted with a straight line whose slope will be $-k_2$. Thus, as shown in Figure 6.4, one can use a graphical method called "feathering" to obtain the rate constants k_1 and k_2 as follows: (1) one fits the late time data on a plot of C_B versus time on semilogarithmic graph paper with a straight line, determines $-k_2$ from its slope, and determines C_2 from the intercept one gets when the straight line is extrapolated back to the vertical axis at $t = 0$, (2) one then subtracts the values given by $C_2 \exp(-k_2 t)$, that is, the values represented by the straight line fit to the late time data, from the actual data points, and (3) one then plots the resultant values, which are equivalent to $C_B - C_2 \exp(-k_2 t)$. From the equation for the biexponential decay it is clear that

$$C_B - C_2 \exp(-k_2 t) = C_1 \exp(-k_1 t)$$

and so a straight line fit to the "feathered" data will have a slope of $-k_1$ and an intercept on the ordinate of C_1. In this way, all four parameters ($C_1, C_2, k_1,$ and k_2) are found.

In the event that this method does not yield a good straight-line fit to the feathered data, it may be that a triexponential equation (corresponding to three simultaneous modes of elimination) would describe the data better. The triexponential equation is

$$C_B = C_1 \exp(-k_1 t) + C_2 \exp(-k_2 t) + C_3 \exp(-k_3 t)$$

where $k_1 > k_2 > k_3$. All six constants can be determined by following the same method, i.e., fitting the late time data to get k_3 and C_3, subtracting off $C_3 \exp(-k_3 t)$ from the actual data, fitting the resultant values with a straight line over the late time part to get k_2 and C_2, subtracting another $C_2 \exp(-k_2 t)$ from the previously feathered data to get a second feathered set of data, and then fitting those data with a straight line to get k_1 and C_1. Even higher-order (4th, 5th, etc.) exponential equations could be tried, but in practice the data analysis becomes too difficult when one feathers the data so many times.

For the case of intravenous injection with a monoexponential decay of C_B or A, the area under the curve of C_B versus t is given by the integral of $C_B dt$ from $t = 0$ to $t = \infty$. Simple mathematics shows this to result in

$$\text{AUC} = \frac{C_{Bo}}{k_1}$$

and thus AUC is given by the C_{Bo} intercept divided by the rate constant k_1. If the time interval for the AUC is from $t = t_1$ to $t = t_2$, then the general formula for AUC is

$$\text{AUC}(t_1 - t_2) = \left(\frac{C_{Bo}}{k_1}\right)[\exp(-k_1 t_1) - \exp(-k_1 t_2)]$$

For the biexponential case, the AUC over $t = 0$ to $t = \infty$ is given by

$$\text{AUC} = \frac{C_1}{k_1} + \frac{C_2}{k_2}$$

Also, for the IV injection case, the apparent volume of distribution, V_d, is given by the equation

$$V_d = \frac{A}{C_B}$$

where A is the amount of drug in the body at any given time and C_B is the concentration of the drug in the blood at the same time. The volume of distribution really is only "apparent" and is simply the volume of fluid in the body which would be required to contain all of the drug in the body at a given time if the concentration of the drug in the total compartment were the same as its concentration in the blood at the same time. However, quite clearly, drug concentrations do vary significantly in different regions of the body. The V_d, in fact, even includes fatty tissues, in which lipid-soluble drugs are often much more concentrated than in the blood itself. Thus, V_d actually reflects the character of the drug rather than of the biological system; V_d also reflects the extent to which a drug is highly bound in high concentrations in places such as fatty tissues. Drugs with large V_d values, then, are often difficult to eliminate from the body for this very reason. Thus, V_d is a useful index of the difficulty of elimination of a particular drug from the body.

Table 6.2 Apparent Volumes of Distribution per Unit
Body Weight (L/kg) for Selected Drugs

Quinacrine	600	Quinidine	2.6
Chloroquine	220	Procainamide	2.2
Desmethylimipramine	40	Phenobarbital	0.75
Nortriptyline	15	Phenytoin	0.6
Digoxin	8	Digitoxin	0.5
Ethchlorvynol	5	Theophylline	0.4
Amphetamine	4	Salicylic acid	0.15
Propranolol	3.5	Tolbutamide	0.1

The average accessible body water (there is some water in the bones which is inaccessible) for a 70 kg adult male is about 42 liters (i.e., about 0.60 L/kg). About 23 liters consists of intracellular fluid and the remaining 19 liters of extracellular fluid consists of about 3 liters of blood plasma and 16 liters of other extracellular fluids (8 liters of interstitial fluid, and 8 liters of fluids such as cerebrospinal, GI tract, and intraocular fluids. At times the volume of fluid in the GI tract can be as much as 1 liter but it usually is much less.

Table 6.2 shows approximate volumes of distribution per unit body weight for some common drugs. Since the plasma volume is 0.04 liters/kg, the total extracellular fluid plus plasma volume is 0.27 liters/kg, and the total accessible body water volume is 0.60 liters/kg, one can see that most of the drugs listed have volumes of distribution which are much greater than the total body water volume. This means that such drugs preferentially distribute into many of the tissues of the body. Additionally, it should be mentioned that drugs can exist in the blood plasma, in other body fluids, and in tissues, in both unbound and bound forms (drug binding to plasma proteins is particularly common). Binding to proteins outside of the plasma is clearly possible when one realizes that 60% of the body's albumin content, for example, lies outside of the plasma. Thus, the total volume of distribution of a drug can be subdivided into its unbound (or "free") volume of distribution and its bound volume of distribution.

A letter by Huang and Tzou (1986) addresses the subject of the effect of activated charcoal on the volume of distribution of drugs. They mention that charcoal can be considered a "compartment" of the body (a "sink" in which drug can reside, just as in the case of organs such as the liver, or tissues such as fat). If adsorption to charcoal is reversible, the charcoal will have the effect of increasing the unbound, or free, volume of distribution, V_{df} of the drug. If, however, adsorption to the charcoal is irreversible or the rate of desorption is very slow compared to the rate of drug elimination, then the charcoal compartment can be regarded as an elimination compartment. In that case, there would be no increase in the unbound volume of distribution. The authors' attempts to desorb theophylline that had been adsorbed on charcoal, in vitro, by washing, showed that the theophylline was essentially irreversibly adsorbed. This suggests that charcoal should have no effect in vivo on the unbound volume of distribution for theophylline.

Data of other researchers were then cited by Huang and Tzou which indeed show V_{df} values per unit body weight of 0.50 L/kg (no charcoal) versus 0.46 L/kg (with charcoal) in one study with humans, and values of 0.85 L/kg (no charcoal) and 0.79 L/kg (with charcoal) for a study with rabbits. This suggests that, at least for theophylline, that charcoal can be treated, pharmacokinetically, as if it were an elimination compartment.

The calculation of V_d assumes that distribution equilibrium has been achieved, and relies on the use of the terminal phase elimination rate constant. It can be shown that, if an IV drug dose is D_{iv}, then

$$V_d = \frac{D_{iv}}{k_1 \text{AUC}}$$

This is easily proved, for

$$D_{iv} = V_d C_{Bo}$$

and from what was given above

$$\text{AUC} = \frac{C_{Bo}}{k_1}$$

Thus

$$V_d = \frac{V_d C_{Bo}}{k_1 C_{Bo}/k_1} = V_d$$

This V_d equation is written in terms of the elimination constant k_1, assuming that elimination occurs by one first-order mode. More generally, if several simultaneous first-order modes exist,

$$V_d = \frac{D_{iv}}{k_T \text{AUC}}$$

where k_T is the overall terminal elimination rate constant, that is, the smallest rate constant, which is the one which characterizes the "late time" behavior.

The clearance of a drug by a discrete organ (e.g., the kidneys) is defined as the product of the volumetric flow rate of blood going to that organ, Q, and the extraction ratio. The extraction ratio is just the fraction of the drug removed from the blood in one pass through the organ, or

$$\frac{C_{Bin} - C_{Bout}}{C_{Bin}}$$

The clearance has units of flow rate (e.g., mL/min) and represents the equivalent flow of blood from which the drug would be totally removed. An example will illustrate this: if $Q = 200$ mL/min, and the extraction ratio is 0.60, then the clearance is $200(0.60) = 120$ mL/min. Thus, the removal of 60% of the drug from 200 mL/min of blood is equivalent to 100% removal of the drug from 120 mL/min, in terms of the amount of drug removed.

The total body clearance of a drug is the sum of all rates of elimination of the drug from the body, by all routes of elimination (urinary excretion, metabolism, etc.) divided by the concentration of the drug in the blood. One can prove that the total body clearance, CL, is equal to ratio of the total amount of drug ultimately eliminated to the total area under the drug concentration versus time curve. For the case of intravenous administration, the total amount of drug ultimately eliminated is equal to the dose, and so

$$CL = \frac{D_{iv}}{\text{AUC}}$$

where AUC is the total value over $t = 0$ to ∞. For a drug given orally, with only a bioavailability fraction f, the same formula applies, but with a numerator of fD_{iv}. The clearance can also be shown to be given by

$$CL = k_T V_d$$

where k_T is the terminal phase elimination rate constant. This is easily proved, for we have shown that

$$V_d = \frac{D_{iv}}{k_T \text{AUC}}$$

and a simple rearrangement of that result gives

$$\frac{D_{iv}}{\text{AUC}} = k_T V_d$$

The rate of elimination of a drug from the body is just

Rate of elimination $= CL\ C_B$

Thus, if a total body clearance value of, say, 0.5 L/min were determined, and the prevailing C_B value of the drug was 1 mg/L, then the prevailing rate of drug elimination from the body would be 0.5 L/min times 1 mg/L, or 0.5 mg/min.

The renal clearance, CL_R, is given by the equation

$$CL_R = \frac{A_e}{\text{AUC}_{0-t}}$$

where A_e is the cumulative drug excreted unchanged into the urine over the time interval $0-t$. If the end point of the time interval is ∞, then the formula is

$$CL_R = \frac{A_{e\infty}}{\text{AUC}_{0-\infty}}$$

where $A_{e\infty}$ is the total drug excreted unchanged into the urine over $t = 0$ to ∞. $A_{e\infty}$ is equal to $f_e fD$, where f_e is the fraction of the drug absorbed into the blood stream (fD) which is excreted into the urine.

B.   The Two Compartment Open Model

A simple two compartment open model for the case of an intravenous drug injection is shown in Figure 6.5. The injection is assumed to occur into the blood compartment and elimination is assumed to occur only from the blood compartment, while the tissue compartment acts only as a space for drug distribution to and from the blood compartment.

This model can be developed to give the following equation for the concentration of drug in the blood compartment:

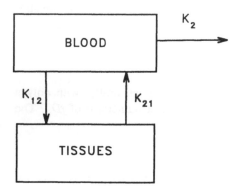

Figure 6.5 Two compartment open model for IV drug injection.

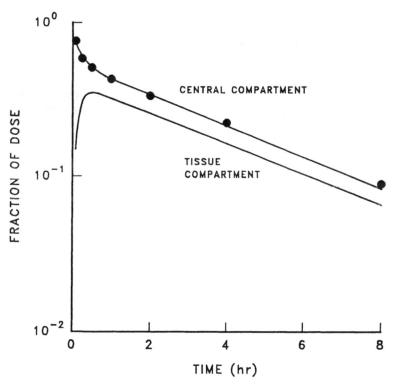

Figure 6.6 LSD data compared to the two compartment open model for IV drug injection. Circles are experimental data, and the lines are predictions from the model. From Levy et al. (1969). Reproduced by permission of the American Pharmaceutical Association.

$$C_B = C_1 \exp(-\alpha t) + C_2 \exp(-\beta t)$$

where

$$C_1 = C_{Bo} \frac{\alpha - k_{21}}{\alpha - \beta}$$

and

$$C_2 = C_{Bo} \frac{k_{21} - \beta}{\alpha - \beta}$$

and C_{Bo} is the initial concentration of the drug in the blood compartment. Data for this biexponential solution can be analyzed by the feathering method discussed earlier. Once one determines the constants C_1, C_2, α, and β using the feathering method, one can determine the individual rate constants from the following relationships:

$$C_{Bo} = C_1 + C_2$$

$$k_{21} = \frac{C_1}{C_{Bo}} \beta + \frac{C_2}{C_{Bo}} \alpha$$

$$k_2 = \frac{\alpha\beta}{k_{21}}$$

$$k_{12} = \alpha + \beta - k_{21} - k_2$$

For describing the concentration of drug in the tissue compartment versus time, the solution is

$$C_T = \left[\frac{k_{12} C_{Bo}(V_1/V_2)}{\alpha - \beta} \right] [\exp(-\beta t) - \exp(-\alpha t)]$$

Levy et al. (1969) used this model to study LSD (lysergic acid diethylamide) pharmacokinetics in humans. The IV drug dose was 2 µg/kg. Figure 6.6 shows the data, the model fit to the blood compartment data, and the predicted tissue concentrations. The blood concentration in µg/L with time in hours was found to be given by:

$$C_B = 5.47 \exp(-7.62t) + 6.92\exp(-0.23t)$$

The kinetic parameter values were $k_2 = 0.41$ hr^{-1}, $k_{12} = 3.08$ hr^{-1}, and $k_{21} = 4.36$ hr^{-1}.

III. MODELS WITH FIRST-ORDER DRUG ABSORPTION

A. One Compartment Open Model

Since most drugs are not given intravenously but rather by mouth, a model which takes into account the drug absorption process is very useful. Figure 6.7 shows a simple one compartment open model with first-order drug absorption and one first-order mode of elimination. The equation describing the concentration of the drug in the compartment is:

$$C_B = \left[\frac{k_a fD}{V_d(k_a - k_e)} \right] [\exp(-k_e t) - \exp(-k_a t)]$$

Figure 6.7 One compartment open model with first-order drug absorption and elimination.

where f is the fraction of the drug dose D that is absorbed and reaches the blood stream. Again, the method of feathering can be used to determine the constants in this biexponential equation. The kind of time course predicted by this equation is just that shown earlier in Figure 6.1. This reflects the usual fairly rapid rise to a peak, followed by a much more gradual decline.

B. Two Compartment Open Model

A more elaborate model, with both blood and tissue compartments and first-order drug absorption, is shown in Figure 6.8. The qualitative nature of the blood and tissue compartment drug concentrations (C_B and C_T) versus time, as a function of the relative values of k_{12} and k_{21} (which determine how the drug partitions between the blood and tissue compartments), has been shown and discussed by Cooney (1976). The mathematical solutions for C_B and C_T as functions of time and the model parameters have also been given by Cooney (1976) and other authors, and will not be repeated here.

IV. EXAMPLES OF PHARMACOKINETIC CALCULATIONS

Many of the equations given suggest clearly how the various pharmacokinetic parameters can be determined. However, we will give some specific examples here. The discussion is not meant to be all-inclusive by any means.

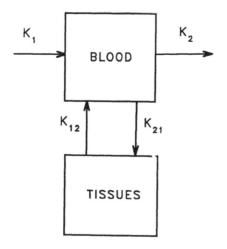

Figure 6.8 Two compartment open model with first-order drug absorption and elimination.

A. Area Under the Curve

The trapezoidal rule is usually used to determine AUC values. For any two successive values of C_B, say C_{B1} and C_{B2}, determined at times t_1 and t_2, respectively, where t_2 is later than t_1, the area under a straight line drawn between the two concentrations on a plot of C_B versus t is just

$$(\tfrac{1}{2})(C_{B1} + C_{B2})(t_2 - t_1)$$

If one adds all areas created by all pairs of adjacent C_B points, one obtains a good estimate of the total area. In the case where the time interval between successive C_B values is the same (represented by Δt) and there are a total of n values of C_B, the area is given by

$$\text{AUC} = 0.5C_{B1} + C_{B2} + C_{B3} + \ldots + C_{Bn-2} + C_{Bn-1} + 0.5C_{Bn}$$

that is, one adds half of the first and last values to the sum of all of the other values.

B. Elimination Half-life

Besides obtaining the elimination half-life by finding $t_{1/2} = 0.693/k_e$, where k_e is the terminal phase elimination rate constant found from the late time part of a plot of C_B versus time on semilogarithmic graph paper, one can find $t_{1/2}$ from urinary excretion data after an intravenous drug dose. One measures the amount of drug in the urine at various times and determines the excretion rate at different times from these data. Then, since

$$\text{Urinary excretion rate} = \left(\frac{CL_R}{V_d}\right)A = \left(\frac{CL_R}{V_d}\right)D_{iv}\exp(-k_e t)$$

then a plot of the excretion rate versus time on semilogarithmic graph paper will have a slope of $-k_e$, since

$$\ln(\text{excretion rate}) = \ln\left[\left(\frac{CL_R}{V_d}\right)D_{iv}\right] - k_e t$$

Then, of course,

$$t_{1/2} = \frac{0.693}{k_e}$$

C. Estimating $t_{1/2}$ for Drug Absorption

The biexponential equation for C_B versus time given earlier for first-order drug absorption and first-order elimination is feathered to yield a value for k_a, the rate constant for absorption. Then the $t_{1/2}$ for absorption is computed as $t_{1/2} = 0.693/k_a$.

D. Mean Residence Time

Although not discussed before, the mean residence time (MRT) of a drug molecule in the body can be determined from urinary excretion data or from C_B versus time data. If A_e is the cumulative amount of drug excreted in the urine at any arbitrary time, and $A_{e\infty}$ is the corresponding value at $t = \infty$, then it can be shown that

$$MRT = \int_{0}^{\infty} \frac{(A_{e\infty} - A_e)dt}{A_{e\infty}}$$

Using drug blood concentration data instead, the correct formula is

$$MRT = \frac{\int_{0}^{\infty} t\, C_B\, dt}{\int_{0}^{\infty} C_B\, dt}$$

V. SUMMARY

This chapter has shown that basic pharmacokinetic modeling makes use of the concepts of zero-order and first-order drug transfer processes (absorption, distribution, elimination) and their representation using compartment models. Simple one-compartment and two-compartment models of the "open" type, for IV drug administration and for oral drug administration with first-order absorption, have been discussed. Only simple models such as these are generally usable for the interpretation of experimental data.

The data one usually has for application of these models are blood drug concentrations (C_B) versus time, or drug amounts excreted in the urine versus time. Thus, we have presented equations for C_B versus time for the models mentioned, with a discussion of how "feathering" of the data can be performed to determine the various parameters in the C_B equations.

Equations have also been presented to allow computation of the area-under-the-curve (AUC) and the drug elimination half-life ($t_{1/2}$), perhaps the two most important pharmacokinetic parameters. Formulas for the volume of distribution (V_d) of a drug and for the total body clearance (CL) have also been given. Finally, examples of pharmacokinetic calculations were presented. With this background, the reader should be able to understand the meanings and significance of pharmacokinetic parameters determined in the many in vivo studies to be discussed in subsequent chapters.

VI. REFERENCES

Brodie, B. B., Bernstein, E., and Mark, L. (1952). The role of body fat in limiting the duration of action of thiopental, J. Pharmacol. Exp. Ther. *105*, 421.

Cooney, D. O. (1976). *Biomedical Engineering Principles: An Introduction to Fluid, Heat, and Mass Transport Processes*, Marcel Dekker, New York.

Gibaldi, M. (1984). *Biopharmaceutics and Clinical Pharmacokinetics*, 3rd edition, Lea & Febiger, Philadelphia.

Huang, J. D. and Tzou, M. C. (1986). The effect of activated charcoal on the volume of distribution of drugs, J. Pharm. Sci. *75*, 923.

Levy, G., Gibaldi, M., and Jusko, W. J. (1969). Multicompartment pharmacokinetic models and pharmacologic effects, J. Pharm. Sci. *58*, 422.

Notari, R. E. (1987). *Biopharmaceutics and Clinical Pharmacokinetics: An Introduction*, 4th edition, Marcel Dekker, Inc., New York.

Rosenberg, J., Benowitz, N. L., and Pond, S. (1981). Pharmacokinetics of drug overdose, Clin. Pharmacokin. *6*, 161.

Rowland, M. and Tozer, T. N. (1989). *Clinical Pharmacokinetics: Concepts and Applications*, 2nd edition, Lea & Febiger, Philadelphia.

Sangster, B. and De Kort, L. A. M. (1986). Toxicokinetics versus pharmacokinetics, in *New Concepts and Developments in Toxicology*, P. L. Chambers, P. Gehring, and F. Sakai, Eds., Elsevier, Amsterdam, pp. 303-313.

Scholer, J. F. and Code, C. F. (1954). Rate of absorption of water from stomach and small bowel of human beings, Gastroenterology *27*, 565.

Sue, Y. J. and Shannon, M. (1992). Pharmacokinetics of drugs in overdose, Clin. Pharmacokinet. *23*, 93.

7

Methods for Treating Poisoning and Drug Overdose

In order to place the use of charcoal for treating victims of drug overdoses and poisonings into perspective and to show later why the use of charcoal is generally superior to other forms of treatment, it is both useful and necessary to review the wide range of treatment methods which are in common use. In a subsequent chapter (Chapter 15) we will present specific and quantitative comparisons of the efficacy of charcoal versus syrup of ipecac, charcoal versus gastric lavage, and charcoal versus whole-bowel irrigation. To understand these comparisons, one needs to have a feeling for what these other methods entail. Also in that later chapter, we will discuss how the use of ipecac and cathartics (saline compounds, sorbitol) affect the in vivo action of charcoal. Thus, some understanding of emetic and cathartic therapies will be of use to the reader. We begin, however, with a brief overview of the general characteristics of intoxicated patients.

I. GENERAL CHARACTERISTICS OF POISONED PATIENTS

Pond (1986) has provided an excellent summary of the general characteristics of poisoned patients, drawing a sharp distinction between children and adults. Her summary (with minor differences) is given in Table 7.1. These differences need to be kept in mind in deciding what treatment approaches might be most easily carried out and which might be most effective.

II. GENERAL OVERDOSE MANAGEMENT STRATEGIES

When poisoning occurs in the home, a call is often made to a Poison Control Center. There are few data on the number of homes in the USA which have powdered charcoal on hand and even less information on whether it is successfully consumed if it is available. However, a study by Dockstader et al. (1986) reported on the feasibility of

Table 7.1 Characteristics of Poisoned Children and Adults

Children	Adults
Present early	Present late
Accidental	Intentional
One substance	Mixtures of drugs
Substance known	Substance(s) often unknown
Low toxicity	Frequently highly toxic
Conscious	In a stupor or coma
Will take syrup of ipecac	Will take syrup of ipecac
Resist taking charcoal	Will take charcoal
Charcoal often causes emesis	Less emesis due to charcoal

Adapted from Pond (1986). Reprinted by permission of Elsevier Science Publishers BV.

home administration of SuperChar charcoal/sorbitol. Fifty selected callers to a Poison Control Center were instructed to obtain SuperChar charcoal at a convenient site in the community (the researchers were responsible for placing the charcoal at these locations). Follow-up calls revealed that there was difficulty in administering the mixture to 35 patients (70%). However, 30 of the 50 (60%) did take the total dose specified. Eleven (22%) vomited the mixture within 30 min. Water given after the mixture appeared to cause the vomiting. Passage of charcoal in the stool was observed in 30 children within 4 hr, with 42 (84%) having normal stool color within 72 hr. In spite of the difficulties encountered, charcoal/sorbitol administration in the home is possible (reducing additional fluids is recommended). We also described a similar study in Finland by Lamminpää et al. (1993) in Chapter 4, in which the home use of charcoal was somewhat more successful. We turn now from charcoal use in the home to treatment methods carried out in the hospital environment.

Summaries of methods for treating overdose victims may be found in several general books on poisoning, such as the treatises by Arena and Drew (1986), Goldfrank (1986), Dreisbach and Robertson (1987), Kaye (1988), Ellenhorn and Barceloux (1988), Noji and Kelen (1989), Haddad and Winchester (1990), and Amdur et al. (1991). A specific review of methods for treating phenobarbital intoxication was presented by Lindberg et al. (1992). Recommendations for ipecac and charcoal use in children have been given in a short article (Anonymous, 1979). General review articles by Easom and Lovejoy (1979), Kunkel (1985), Goldberg et al. (1986), Wheeler-Usher et al. (1986), Haddad and Roberts (1990), Flomenbaum et al. (1990), Hall (1991), Rumack and Lovejoy (1991), Krenzelok (1992), Krenzelok and Dunmire (1992), Kulig (1992), Harris and Kingston (1992), Fine and Goldfrank (1992), Jawary et al. (1992), Vale (1992), Phillips et al. (1993), and Perrone et al. reviewed initial measures and therapies used in treating poisoning victims, and have reviewed the large variety of gastrointestinal decontamination methods available, noting their advantages and disadvantages. We will present such a review here.

The management of poisoning in the emergency room setting is aimed at maintaining life and trying to reduce the further absorption of the drug (or poison), so that the patient's system can safely detoxify the drug which has already been absorbed. If possible, the type of drug or poison involved should be identified at once, as a specific antagonist may be available (however, the number of specific antagonists relative to the 6–12 million or more known toxic substances is exceedingly small). It goes without saying that symptomatic and supportive procedures should be used throughout the treatment of the patient, as indicated. It is especially important to manage the airway so that obstruction or aspiration does not occur, to provide mechanical oxygen delivery if respiration is depressed, and to control the patient's hemodynamics and electrolyte balances. Other specific procedures will be mentioned below.

The most commonly recommended next step (recent thinking on this has changed, however) is to administer syrup of ipecac, if the patient is awake and is likely to remain awake. Ipecac is generally repeated in 15–30 minutes if emesis does not occur. However, if the patient has a decreased or deteriorating level of consciousness, if convulsions are present, if the gag reflex is absent, or if the ingested substance is an acid, an alkali, or a hydrocarbon distillate, emesis is contraindicated. Also, if any delay in waiting for emesis is judged to be unwise (in the case of drugs which are rapidly absorbed or which have seizure-causing potential, for example) then the use of syrup of ipecac should be foregone in favor of lavage or charcoal. However, as we will discuss shortly, there is considerable debate over whether ipecac should be contraindicated in cases involving hydrocarbons and seizure-causing substances.

As we shall soon see, the relative efficacy of induced emesis and gastric lavage is still a topic of controversy. The efficacy of using either one as compared to proceeding directly to oral charcoal therapy is also still open to considerable debate. In this chapter, we shall consider these issues in detail.

Saincher et al. (1993) showed the effect of a delay in giving syrup of ipecac. They gave subtoxic doses of acetaminophen (3.9 g) to ten human volunteers along with 30 mL syrup of ipecac at either 5, 30, or 60 min in a four-limbed randomized cross-over study (see Chapter 8 for a description of randomized cross-over studies). In the control arm, the subjects received no ipecac. AUC values based on 0-8 hr blood samples were: control, 206; ipecac at 5 min, 67; ipecac at 30 min, 183; and ipecac at 60 min, 162 mg-hr/L, respectively. Thus, ipecac at 5, 30, and 60 min reduced acetaminophen absorption by 67, 11, and 21%, respectively. The primary conclusion is that for ipecac to be effective it should be administered as soon as possible.

In addition to induced emesis, gastric lavage, the newer technique of whole-bowel irrigation, and charcoal therapy, the elimination of some drugs may be increased by forced diuresis, alterations in urine pH, the use of hemodialysis, or the use of hemoperfusion. These measures are generally both more complicated and more risky than supportive therapy and are not often used. Some discussion of these methods will be given in this and other chapters. A brief review of forced diuresis, dialysis, and hemoperfusion has been presented by Heath (1986).

In certain cases, specific chemical antidotes are of some use. Examples are: the use of chemicals which form inert complexes with the poison, substances which compete with the toxic material for essential receptor sites, substances which block receptor sites through which the toxic effects are mediated, and substances which accelerate the conversion or elimination of the poison to less toxic forms. Vale and Meredith (1986) have reviewed the pharmacokinetics of some of these chemical antidotes. We will not consider them further here as we wish to focus on more general methods of treatment.

Various problems which arise after serious drug overdoses or poisonings, requiring a variety of specific measures, are too detailed for the present book to cover, and are outside of its focus. Therefore, they will not be discussed. These problems include the treatment of:

- Convulsions
- Coma
- Hyperactivity, delirium, or mania
- Shock
- Congestive heart failure
- Cardiac arrest
- Acute renal failure
- Urine retention
- Liver damage
- Methemoglobinemia
- Intractable pain
- Fluid imbalance
- Electrolyte imbalances
- Acidosis
- Hypoglycemia
- Pulmonary edema
- Prolonged vomiting
- Prolonged diarrhea

The last two problems (vomiting and diarrhea), which can accompany poisoning, may actually aid in the removal of poisons from the body. What we refer to here are the problems created by excessive or prolonged vomiting or diarrhea, which can lead to a serious loss of fluid. The deliberate promotion of vomiting with emetic agents, and the deliberate promotion of diarrhea with cathartics, will be discussed later.

Specific measures which are also outside of the scope of this book, but which are commonly involved in overdose/poisoning cases, are:

- Maintenance of body temperature
- Provision of adequate nutrition to the patient
- Maintenance of unobstructed airways
- Treatment of hypoxia or depressed respiration with artificial respiration and mechanical ventilators

III. ORAL DILUTION

Dilution with water was at one time widely recommended as a first aid measure for the treatment of ingested poisons. Dilution is logical when the ingested substance is a mineral acid or base, because damage by these agents is primarily due to the severe irritation ("burning") of mucosal tissues; lowering the strength of the acid or base reduces this effect. However, excessive dilution may have detrimental effects, especially in the case of caustics where large volumes of fluid may induce emesis and expose the esophagus again to the caustic.

Also, there has also been much fear that neutralizing an acid with a base (or a base with an acid) would cause thermal injury to mucosal tissues due to the large "heat of neutralization" released. Such fears may be overstated, as Kingston et al. (1993) have

Table 7.2 Effect of Dilution on Lethality of Different Drugs

	Volume of water given with drug (% body weight)		
	5%	2.5%	1.25%
Death % with mice	83.3	53.8	21.3
Death % with rats	87.1	55.8	25.8

Adapted from Ferguson (1962). Reprinted by permission of Academic Press, Inc.

shown. They added 45 mL milk of magnesia to 15, 30, or 50 g of an acid toilet bowl cleaner which contained 9.5% HCl. The temperature rises noted were 14, 16, and 19°C, respectively, which would be insufficient to cause thermal injury. In another test, 480 mL water was added along with 45 mL milk of magnesia to 50 g of the cleaner and the temperature rise was 25°C (the extra 6°C probably came from a "heat of dilution" effect). Still, this temperature rise of 25°C (45°F, which would give a final stomach temperature of about 143°F) would probably not be harmful. However, if a product had more than the 9.5% HCl which was in the one product tested, neutralization might indeed be harmful (a similar statement holds for products containing bases such as sodium hydroxide).

In any event, modern handbooks on the treatment of poisoning generally do recommend dilution for mineral acids or bases with modest amounts (200–300 mL) of water, not acids or bases. However, in the case of drugs in general, oral dilution is counterproductive and, fortunately, most poison "handbooks" now recommend against it. A review of the potential harm of dilution has been presented by Chin (1971).

Ferguson (1962) gave doses of 12 drugs of various classes by stomach tube to rats and mice in volumes of water corresponding to 1.25, 2.5, and 5% of their body weights. The number of animals which died over 24 hr was recorded for each group of 20 animals. The drugs involved were: amphetamine sulfate, benactyzine HCl, diphenhydramine HCl, mephenesin, pentobarbital sodium, pentylenetetrazol, phenylbutazone, potassium cyanide, procaine HCl, sodium fluoride, sodium thiocyanate, and yohimbine HCl. The average percentages of the animals which died from all of the drugs are shown in Table 7.2 (see Ferguson's paper for data on each specific drug).

Very clearly, the death rates increased sharply as the volume of water added to dilute the drugs was increased. Ferguson goes on to show that LD_{50} values (dose amounts which cause 50% lethality) determined by him for four drugs (pentobarbital sodium, potassium cyanide, procaine HCl, and sodium fluoride) were decreased by 54.3, 46.3, 46.3, and 46.8%, respectively, when the "water of dilution" was increased from 1% to 5% of body weight. Thus, LD_{50} values reported in the literature need to state how much, if any, water was given with the drug involved.

A second study which shows the effects of oral dilution is that of Henderson et al. (1966). They gave rats pentobarbital sodium, quinine, or aspirin by oral intubation in a volume of 2 mL/kg. Then, 1 min later, the "control" animals were given 1 mL/kg water, while the "test" animals were given 20 mL/kg water. Blood samples were collected over the next several hours. For pentobarbital, the test rats had drug levels 33, 90, and 78% higher at 10, 20, and 40 min, respectively, than the control rats. At 80 min and beyond the drug levels were similar. For quinine, the 15, 60, and 120 min drug levels were 138, 72, and 63% higher, respectively, than for the control animals. For aspirin, the drug levels at 5, 15, and 60 min were not significantly different in the two groups.

Borowitz et al. (1971) have clearly shown that giving drugs diluted with water can greatly enhance their absorption. This is especially true of drugs which are weak acids or weak bases, and drugs having limited water solubility. They found that the onset of sleep in rats given sodium pentobarbital was faster when the drug was given in dilute solution (5 mL/100 g) than in a more concentrated solution (0.5 mL/100 g). The onset of sleep occurred at 4.9 and 7.1 min, respectively. Also, sodium salicylate levels were found to be significantly higher when the drug was given in dilute form (0.8% solution) as opposed to concentrated form (25% solution). When large doses of atropine sulfate, aminopyrine, or sodium salicylate were given in dilute form to rats (1.5, 2, and 1.6% solutions, respectively), deaths were 9/9, 15/15, and 9/10, respectively. However, when the same drugs were given in concentrated form (30, 10, and 64% solutions, respectively), deaths were only 0/9, 7/20, and 1/10, respectively.

The effects of oral dilution are at least three-fold. First, oral dilution reduces the drug concentration and thus reduces the "driving force" for absorption, whether the drug is in the stomach or in the small intestines. Second, oral dilution increases the surface area for absorption in the stomach, by distending the stomach and opening up the recesses (or folds) in the stomach walls. Third, oral dilution greatly increases the rate of gastric emptying, which brings the drug into the small intestine, where the surface area for absorption is far greater than in the stomach. For a drug like quinine, which is poorly absorbed from the stomach, the effect of greater gastric emptying is the dominant one; oral dilution therefore greatly increases absorption. For aspirin, which is well absorbed from the stomach, the dilution effect in lowering the "driving force" for absorption and the increase in the stomach's surface area more or less counterbalance and thus oral dilution has no large effect.

However, the Ferguson study on 12 drugs of a wide variety show without exception that increasing the dilution led to a consistent increase in the absorption of the drug. Thus, as a general rule, it would appear that oral dilution of ingested substances (other than mineral acids and bases) is counterproductive.

IV. INDUCED EMESIS WITH SYRUP OF IPECAC

As summarized by Decker (1971), there have been many emetics used throughout history. Ancient seamen found that sea water could produce emesis. This led to the idea of using common salt (NaCl) in warm water. Mustard in water was used during Roman times; copper sulfate (blue vitriol) was introduced in Iran in the 9th century; tartar emetics (antimony sodium tartrate and antimony potassium tartrate) were used in medieval times, and zinc sulfate (white vitriol) came into use in the 17th century. However, it has long since been learned that tartar salts, copper sulfate, and zinc sulfate are highly toxic. As pointed out in one review (Federal Register, 1985) the emetic dose of zinc sulfate is 2 g, whereas the lethal dose is estimated to be anywhere from 3–15 g. Thus, the margin of safety between its effective dose (2 g) and its lowest reported lethal dose (3 g) makes its use quite dangerous. Mustard is relatively safe but is ineffective. NaCl in water is also ineffective and, in large amounts, can cause fluid shifts in the central nervous system which may cause fatal abnormalities of cellular metabolism. Meester (1980) has summarized five deaths due to the use of salt water as an emetic. He points out that one tablespoon of salt contains about 250 mEq of sodium, enough to raise the serum sodium level by 25 mEq/L in a three-year-old child with an estimated total body water of 10 L.

Chin (1972) also described the hazards of using sodium chloride and other salts, in a review of induced emesis. Several deaths due to table salt administration for non-emetic as well as emetic purposes are detailed.

Lobeline sulfate given subcutaneously (10 mg/kg) has been used as an emetic (Ab-dallah and Tye, 1967) in dogs, and produced emesis in an average of 12.5 min. However, when the dose was increased to 12 mg/kg, the dogs developed hyperactivity and convulsions. Thus, the margin of safety of lobeline is too narrow to recommend its use.

Ipecac, reportedly first used in Brazil in the 16th century, and apomorphine, introduced around 1900, remain as the only two viable emetics at this time, and currently apomorphine is seldom used. Ipecac is the dried rhizome and root of the *Cephaelis acuminata* or *Cephaelis ipecacuanha* plant. Syrup of ipecac is made from powdered ipecac which contains the alkaloids emetine and cephaeline. Emetine comprises more than 50% of the total alkaloids in ipecac, and is more cardiotoxic than cephaeline. On the other hand, cephaeline is more potent as an emetic but is present in less quantity than emetine. Apomorphine (actually apomorphine HCl) is prepared by heating morphine with an excess of HCl in a glass vessel of some sort for 2–3 hr at a temperature of 140–150°C.

The effects of emetic agents (and cathartic agents) on the treatment of toxic ingestion has been discussed in great detail in an extensive review article by Stewart (1983). He points out that the principal emetic agents used, apomorphine (given IV or SC) and syrup of ipecac, both act on areas of the brain, inducing vomiting by activating the chemoreceptor trigger zone in the medulla. Ipecac additionally induces vomiting by action on peripheral areas in the GI tract itself. It is believed that initial episodes of vomiting induced by ipecac are caused by the activation of the peripheral sensors (emetine and cephaeline are the major alkaloids in syrup of ipecac which are responsible). Late episodes of vomiting after ipecac administration are thought to be due to absorption of the emetic alkaloids which then activate the chemoreceptor trigger zones in the brain.

Guidelines for the labeling and use of syrup of ipecac were reviewed by the US Federal Government (Federal Register, 1982) and included the following: "Vomiting should not be induced if the patient: (a) is semiconscious or unconscious; (b) is having convulsions; (c) has swallowed strychnine [which can cause convulsions] unless advised otherwise by a Poison Control Center, emergency medical facility, or physician; (d) has swallowed petroleum distillates such as kerosene, gasoline, paint thinner, or cleaning fluids, unless advised otherwise by a Poison Control Center, emergency medical facility, or physician; (e) has swallowed a corrosive poison such as alkali (lye) or strong acid, unless advised otherwise by a physician or emergency treatment facility."

It was advised that children less than 1 year old be given 5–10 mL of ipecac, followed by 1/2 to 1 glass of water. Patients older than 1 year were to be given 15 mL ipecac, followed by 1–2 glasses of water. Milk or carbonated beverages were not to be used instead. The Federal Register review stated that milk coats the gastrointestinal tract, and "inactivates the ipecac alkaloids through its protein binding action", and that carbonated beverages "may cause overdistention of the stomach." If vomiting does not occur in 20 min, a second dose was recommended.

These guidelines were revisited 3 years later in a more extensive study (Federal Register, 1985) which presented and discussed a wide variety of comments made by members of the medical community in response to the agency's (i.e., the Food and Drug Administration, or FDA) request for comments. Reference was made to a comment that the general notion of giving large volumes of fluids after ipecac "may have detrimental effects, especially in the case of caustics where large volumes of fluid may induce emesis and expose the esophagus again to the caustic." The comment stated "that it would be more prudent to suggest dilution of caustics with one glassful or less of water or milk." Comments were made that the ingestion of petroleum distillates is not an absolute contraindication to the use of ipecac and that an alert patient who has swallowed considerable petroleum distillate, especially one containing fairly toxic components, could safely undergo emesis. This controversy [see Meester (1980) for an excellent discussion of the "hydrocarbon controversy"] was not resolved in the document. Other comments challenged the notion that convulsant substances (strychnine, camphor, amphetamines, tricyclic antidepressants, isoniazid) contraindicate ipecac use. It was stated: "the risk of administering ipecac to a person who has ingested a toxic dose of a rapidly acting convulsant is considerably less than the risk of allowing these toxic compounds to be absorbed into the bloodstream. Therefore, the agency is proposing not to include strychnine or any other convulsant, such as camphor, in the warning."

The document then reviewed data (Uden et al., 1981) which show that giving carbonated beverages instead of water, after ipecac administration, causes no adverse effects. However, other data (Varipapa and Oderda, 1977) have shown that milk does interfere with the ability of ipecac to induce emesis.

In the milk study, it was found that the mean time to emesis in a group of adults was 23.9 min with water and 34.5 min with milk (15 mL ipecac followed by 8 oz fluid). However, in a study by Grbcich et al. (1987), clear fluids or milk were given with ipecac to 120 children aged 5 years-old or less. The mean times to emesis were 23.3 and 24 min for the clear fluid and milk groups, respectively. The mean number of vomiting episodes were 2.7 and 2.95, respectively [a report by Grbcich et al. (1986a) a year earlier describes what appears to be the same study but with slightly different numbers: 23.4, 23.7, 2.8, and 2.96]. Clearly, there was no difference between clear fluids and milk.

An even more recent study with 250 children, by Klein-Schwartz et al. (1991) also showed no difference between clear fluids and milk given with ipecac. In that study, the mean times to emesis were 23.4 and 23.3 min, the mean number of vomiting episodes were 3.5 and 3.4, and the mean durations of vomiting were 45 and 39 min, for the milk and clear fluids groups, respectively. The clear fluids used were: water ($N = 60$), juice ($N = 61$), carbonated beverages ($N = 34$), and mixtures of these or other types ($N = 95$).

In the carbonated beverage study of Uden and coworkers, 24 children were given Sprite or water (6 oz) and then 15 mL ipecac syrup. Mean time to emesis was 21.9 min for the carbonated beverage and 20.4 min for water. Mean volumes of emesis were 181.5 mL for the carbonated beverage group and 185.0 mL for the water group. These are clearly insignificant differences. The carbonated beverage group did show a slight increase in abdominal girth 10 min post-ipecac (53.6 cm versus 50.8 cm before ipecac) but this was not judged to be of significance relative to the efficacy of ipecac.

Bukis et al. (1978) studied whether, if fluids are given with syrup of ipecac, it makes a difference whether the fluids are given before or after the ipecac. Two groups of children were handled in these two manners and no difference in the mean times to emesis (32.8 and 33.0 min, respectively) was found. The volumes of fluids consumed by the "fluids after ipecac" group were 25% larger and the volumes of emesis produced were 53% larger. The authors state that these differences are not significantly different, as measured by the t-test, but the p level is not stated. Thus, these differences are certainly significant at some reasonable p level, suggesting that giving fluids after ipecac does result in larger emesis volumes.

Eisenga and Meester (1978) did a study with healthy volunteers who were given 30 mL syrup of ipecac and then were assigned at random to moving on a treadmill at 2 miles per hour or confined to bed. Motion made no statistical difference ($p < 0.05$) in the time to emesis in the two groups.

The FDA reviewed the question of appropriate ipecac doses and revised its earlier guidelines to state that adults should be given 30 mL of ipecac syrup (the prior guidelines called for 15 mL), followed by 1–2 glasses of water. Also, a dosage specification for children less than 6 months old was deleted. The FDA also concluded that, since emesis from a first dose often occurs near the 20 min mark, the guideline for giving a second dose should state that one wait 30 min rather than 20 min. Data supporting this include those of Rauber (1978), who found a mean time to emesis of 26 min, those of Robertson (1962), who found that 15.4% of his patients vomited in the 20–30 min time interval, and those of Manoguerra and Krenzelok (1978), who found that 18.9% of their patients vomited in the 20–30 min time interval.

Finally, in other matters related to charcoal rather than ipecac, the FDA revised the recommended dosage of charcoal to provide a range of 20–30 g and recommended that it be mixed with no less than 8 oz of liquid. This latter recommendation resulted from a case of charcoal aspiration (Pollack et al., 1981) in which a slurry of 9 g charcoal in 35 mL water gave a mixture which was judged to be too thick and which may have contributed to vomiting and subsequent aspiration.

It should be mentioned that most physicians recommend syrup of ipecac for children 6 months to one year of age only when it can be given in a health care facility, and recommend it not be given to children less than 6 months old because of inability to protect the airway during vomiting.

A.   Toxicity of Ipecac

We briefly review information relative to the toxicity of ipecac before proceeding further. Severe adverse reactions have occurred with ipecac. However, most cases of significant toxicity have occurred with the fluid extract form of ipecac (which has not been available commercially since the mid-1960s), which is 14 times more concentrated than the syrup form. As little as 10 mL of the extract can be fatal, whereas with syrup of ipecac as much as 105 mL has been taken by a child with only minor EKG changes occurring (King, 1980). The important clinical signs of ipecac toxicity include diarrhea, abdominal cramping, skeletal muscle weakness, aching stiffness, mild tremor, edema, and convulsions. Cardiac effects include mild tachycardia, hypotension, dyspnea, precordial pain, and electrocardiographic abnormalities.

The first report of severe ipecac toxicity seems to be that described by Harrison (1908) in 1908, in which an adult ingested an unknown quantity of *Vinum ipecacuanhae* (a preparation which was still official in the 1953 *British Pharmacopoeia* in the form of tincture of ipecacuanha). He was seized with violent and uncontrollable vomiting, had cold and damp extremities, had a rapid pulse, and complained of abdominal pain. He died 2–3 hr after the ipecac ingestion. Autopsy revealed a subacute congestion of the stomach and the first two feet of the intestine.

A second case, reported in 1959 by Allport (1959), involved a 2 1/2-year-old boy who had been given 15 mL of ipecac fluid extract (not syrup of ipecac) after he ingested six 4-mg tablets of chlorpheniramine maleate. He vomited violently for eight hours. After 17 days, he was discharged but began to suffer increased difficulty in swallowing. He was readmitted and was found to have marked inflammation and constriction of the lower third of the esophagus. This condition was corrected over several months' time by bougienage (passage of a tapered instrument into the lumen in order to increase its diameter).

Smith and Smith (1961) described a fatality due to ipecac, involving a 4-year-old boy who had been given two 30 mL doses of an ipecac fluid extract (again, not syrup of ipecac) in an ED as treatment for an overdose of antacid tablets. The boy vomited after the second dose. After discharge to his home, the boy vomited "rather continuously", developed diarrhea and a fever, experienced severe abdominal pain, and then began to vomit blood-like material. Upon readmission, the boy was found to be in profound shock. He died 8 hr later. Autopsy revealed dehydration, moderate esophagitis, acute moderate colitis, etc.

Another report dealing with a fatality due to extract of ipecac is that of Bates and Grunwaldt (1962). A 4-year-old female who was treated for aspirin overdose with extract of ipecac died on the 7th day. Only 10 mL of the extract was given, yet vomiting continued for 6 days. Another case involving a 34-month-old male who was given 5 mL of ipecac extract and survived, after intermittent vomiting episodes covering 5 days, was also described by Bates and Grunwaldt. Other fatalities due to the taking of the fluid extract form of ipecac have been described by Speer et al. (1963) and by Rose (1970).

Robertson (1979) described a fatality associated with syrup of ipecac. A 14-month-old female was given syrup of ipecac per the standard advice of a Poison Control Center after a call was received concerning the child's having chewed leaves of an amaryllis plant. A detailed report of everything which transpired until the child's death at about 50 hr post-ingestion is given. Autopsy revealed that all of the child's stomach down to the pylorus was located in the left pleural space. Since there was no evidence of a congenital abnormality, it was hypothesized that the herniation occurred in response to the emetic action of the amaryllis leaf or the syrup of ipecac or both. The herniation, with its associated complications, was what led to the cardiorespiratory arrest which was the immediate cause of death. Later, a pediatric pathologist and a pediatrician-teratologist both judged that an underlying congenital defect of the diaphragm was, in fact, highly likely.

Brotman et al. (1981) reported a case in which an 18-year-old female with anorexia nervosa used syrup of ipecac as an emetic to further reduce her weight (she had already lost 35% of her original body weight over the preceding 2.5 yr). She consumed about 300 mL per week in 85 mL doses during two 3-month periods and a smaller weekly amount over a 9-month intermediate period. The girl developed severe physical debility and was found to have significant cardiomyopathy as a result of her chronic ipecac use. The patient was treated with a special diet and psychotherapy, and returned to normal after five months.

Adler et al. (1980) described a woman who died of ventricular tachycardia after ingesting syrup of ipecac to lose weight. She had been drinking 3–4 bottles of syrup of ipecac per day over a three-month period. It is reported that "frequently, she would not vomit after taking the ipecac." In a similar case, Schiff et al. (1986) relate the story of a death due to chronic syrup of ipecac use in a patient with bulimia. The patient, a 17-year-old girl, presented with malaise, weakness, palpitations, dysphagia, and myalgias. She developed intractable hypotension, congestive heart failure, arryhthmias, and died.

Rauber (1978) showed that, in therapeutic doses, syrup of ipecac causes no appreciable cardiotoxic effects, and "adds no hazard not inherent in vomiting itself." In 27 children (mean age 2.8 years, range 9 mo to 13 yr) studied after being given 15 mL syrup of ipecac, the mean heart rate increase was only 6%. There was a tendency towards increased QRS times, but these exceeded normal limits only once. Mild drowsiness and diarrhea can occur with ipecac but not to any serious extent.

Additional discussion of the toxicity of ipecac may be found in a brief review by King (1980) and in a very comprehensive review by Manno and Manno (1977). A short review of 14 reports in the literature, from 1908 to 1978, of ipecac poisoning has also been given by Miser and Robertson (1978). This review mentions many of the reports we have discussed above. Of the 14 cases, 11 involved ingestion of the fluid extract of ipecac. Since ipecac extract is no longer available (it was removed from the *U. S. Pharmacopeia* in the mid-1960s), the chances of accidental fatalities from using ipecac formulations essentially no longer exists.

B.   Other Adverse Effects of Ipecac

A variety of gastrointestinal tract problems have been reported to be caused by syrup of ipecac. These include stenosis of the esophagus lumen, bloody vomitus, protracted vomiting, abdominal pain, mucosal ulceration, and tears of the esophagus/stomach junction.

In 1929, Mallory and Weiss described a syndrome of massive upper gastrointestinal hemorrhage due to mucosal lacerations at the esophagogastric junction after violent retching and vomiting. Several reports of Mallory-Weiss syndrome caused by syrup of ipecac have been reported in the literature (Tandberg et al., 1981; Timberlake, 1984). The death of a 30-month-old child due to gastric rupture after more than 24 hr of vomiting subsequent to the administration of 15 mL of syrup of ipecac has been described by Knight and Doucet (1987).

Wolowodiuk et al. (1984) described the case of an 18-year-old woman who developed pneumomediastinum (air collection in the mediastinum) and retropneumoperitoneum (air collection behind the peritoneum) following the induction of emesis with 30 mL syrup of ipecac. It was postulated that the mechanism for the development of these conditions was a transient perforation of the esophagus with spontaneous closure.

Klein-Schwartz et al. (1984) reported the case of an 84-year-old woman who was found to have intracerebral bleeding in the temporal lobe (suggestive of a ruptured aneurysm) subsequent to administration of 30 mL syrup of ipecac. While it could not be proved that ipecac caused the bleeding, this episode suggests that the use of ipecac in elderly patients should be approached with caution.

C.   Success of Ipecac in Producing Emesis

We now review some studies in which the success of ipecac in causing emesis and the time ipecac takes to produce emesis have been determined.

It might be of interest to first mention some data of Dabbous et al. (1965) in which the success of mechanically-induced vomiting was compared to that of syrup of ipecac. They mention that, of 15 children who were gagged at home by their parents, only 2 vomited, and of 30 children gagged in the hospital by a physician, only 2 vomited. By comparison, of 30 children given syrup of ipecac in the hospital, all 30 vomited. Thus, the success of mechanically-induced vomiting was very low (although, when successful, it has the advantage of being immediate) and much inferior to that of ipecac.

Robertson (1962) examined data relating to the onset of vomiting in 214 children treated with 20 mL syrup of ipecac (occasionally, a second 5–20 mL dose was given). Fifty-six percent of the children vomited within 15 min, another 17% vomited between 15–20 min, 8% vomited between 20–25 min, 8% vomited between 25–30 min, and the remaining 12% vomited after 30 min. The mean time to vomiting was 18.7 min. Since 88% of the children vomited in 30 min or less, Robertson concluded that syrup of ipecac is not a "slow" emetic.

Thoman and Verhulst (1966) showed that syrup of ipecac is an effective emetic even in cases where drugs with antiemetic properties have been ingested. They studied the success of ipecac in producing emesis when antiemetic drugs of the phenothiazine class and the antihistamine class were involved. Percent emesis in the phenothiazine group was 92.2%, and 97.5% in the antihistamine group. For a large data base involving a wide range of substances called "all drugs", the percent emesis was 96.2%. Thus, success with antiemetic drugs was no different from that with all drugs in general. The reason for this perhaps unexpected success is that ipecac acts both by stimulation of the chemo- receptor trigger zone in the medulla and by gastric irritation. Antiemetic drugs generally block vomiting by their action on the medulla trigger zone but leave the initiation of emesis by gastric irritation unimpaired.

Manoguerra and Krenzelok (1978) studied patients treated for poisoning or overdose with syrup of ipecac in a major medical center. Adults or children over age 5 were given 30 mL syrup of ipecac, followed by 360 mL water. Children 1–5 years old were given 15 mL syrup of ipecac, followed by 240 mL water. A second dose was administered if emesis failed to occur in 30 min. Of 232 patients, 188 (81%) vomited with the first dose, 34 (15%) required a second dose, and 7 (3%) did not vomit (the remaining 3 patients became lethargic or convulsed after receiving ipecac and before vomiting began, and were treated with lavage). Of 63 patients who had ingested substances with anti-emetic properties (these were part of the 232 patient group), 51 (81%) vomited with one dose of ipecac, 9 (14%) vomited with two doses, and 3 (5%) did not vomit. Average time to emesis was 24.2 min in the whole group. It was concluded that syrup of ipecac was an effective emetic.

A much later study by Krenzelok and Dean (1985a) summarized the results of a study in which 30 mL syrup of ipecac was administered in the home to children 5 years of age or less. Of 2,401 children given a single dose of 30 mL, only 2 (0.08%) failed to vomit. In another study lasting one year, 1,905 children less than 5 years old were treated in the home with 15 mL syrup of ipecac. The number who failed to vomit in this group was 168 (8.85%, or 100 times higher). However, repeating the 15 mL dose caused emesis in 163 of these patients. Thus, only 5 (or 3%) of the 168 patients who did not vomit with one 15 mL dose failed to vomit with a second dose. The overall failure rate of 5/1,905 was only 0.25%. Krenzelok and Dean thus concluded that the failure of ipecac is highly dose-related and recommended increasing the standard pediatric dose from 15 mL to 30 mL.

In contrast to this is a report by Smolinske et al. (1987). One poison center recommended 15 mL syrup of ipecac in 284 cases (with 100% compliance) and another center recommended 30 mL syrup of ipecac in 243 cases (with 72.4% compliance). All cases involved children aged 1–5 years. Emesis in the 15 and 30 mL groups was achieved 97.5 and 99.4% of the time with one dose, respectively (the difference was not statistically significant). Mean times to emesis were 19.9 min (15 mL) and 17.7 min (30 mL). The mean number of episodes of emesis were 3.4 (15 mL) and 3.8 (30 mL). Thus, the higher dose gave somewhat faster emesis and slightly more episodes, but there was no significant dose-related difference in whether emesis was achieved with one dose.

In support of there being no difference between 15 mL and 30 mL in terms of time to emesis, there is a brief report by Boeck et al. (1985) which also states that there is no statistically significant difference in time to emesis. They do not describe the patient population, nor give any information on "success" rates.

A study by Gaudreault et al. (1984) also supported the lack of a dose-response effect with syrup of ipecac. Forty children were given 30 mL doses and ten were given 60 mL doses. The mean times to vomit and the number of vomiting episodes were similar in the two groups. Mean times to vomit as a function of dose in mL/kg were: for 0.6–1.0 mL/kg, 14 min; for 1.1–1.5 mL/kg, 24 min; for 1.6–2.0 mL/kg, 20 min; for 2.1–2.5 mL/kg, 20 min; and for > 2.5 mL/kg, 20 min.

A study which compared 15 mL and 30 mL doses of syrup of ipecac in terms of time to emesis is one reported by Dean and Krenzelok (1985). There were 100 cases in each group. With 15 mL, the mean time to emesis was 25.8 min, whereas for 30 mL the mean time to emesis was 15.8 min. None of the subjects receiving 30 mL required a second dose but 15 of those given 15 mL did. The 10 minute difference in mean times to emesis was statistically significant at $p < 0.001$.

A study of 105 poisoned infants 6–11 months old who were given syrup of ipecac by Litovitz et al. (1985) showed a vomiting success rate of 101/105 or 96.2%. There were no serious side effects. Thus, the authors recommended the use of syrup of ipecac in the home as being safe and effective (when ipecac was not available at home, there was an additional mean delay of 21.8 min when the ipecac was obtained from a pharmacy and an additional mean delay of 38.4 min if it was obtained from an ED).

Krenzelok and Dean (1985b) also studied the use of syrup of ipecac in children less than one year old ($N = 38$), and found that a 10 mL dose caused all of the children to vomit, in a mean time of 26.1 min. No adverse sequelae were noted.

Rauber and Maroncelli (1982) focussed on determining how long the emetic effects of syrup of ipecac last. They determined the mean duration of the emetic effect from standard 15 mL doses in 65 children to be 60.9 min with a standard deviation of 55.0 min (the mean time to first emesis was 21.4 min and the mean number of episodes was 3.0). In a second set of patients, the subjects were asked to try to provoke further vomiting 2 hr after the original ipecac dose by refilling their stomachs with clear liquids. Only 9/88 or 11.2% of the patients were able to vomit by doing this. The authors concluded that the emetic effects of a standard dose of ipecac rarely last more than 2 hr and that any vomiting after this time must be due to other causes.

Bond et al. (1989) studied the relationship between the time of emesis and serum drug levels at 4 hr for children (ages 1–5 yr) who were treated for acetaminophen overdose. The mean drug level at 4 hr in children not treated at all was 33.4 mg/L. For ipecac-treated children, the 4 hr drug levels for different times to emesis were as follows: 0–30 min, 16.6; 30–60 min, 15.7; 60–90 min, 19.9; 90–120 min, 26; and > 120 min, 31.4 mg/L. The p levels for statistically significant differences compared to the control value were 0.003, 0.0001, 0.035, 0.28, and 0.74. Thus, a reasonable conclusion is that emesis within 90 min post-ingestion of acetaminophen is effective in reducing acetaminophen absorption.

Amitai et al. (1987) also performed a similar study of 50 children accidentally overdosed with acetaminophen who were given syrup of ipecac. Ipecac reduced the 4 hr plasma acetaminophen level, for the group as a whole, from a mean predicted value of

97 mg/L (the predicted values were based on the ingested drug doses and the use of a pharmacokinetic model) to a mean actual value of 34 mg/L. More specifically, they found that the ratio of the measured 4 hr acetaminophen plasma levels to predicted 4 hr levels increased with time to vomiting (with prompt ipecac administration, this ratio was often on the order of 0.1 or less, i.e., ipecac lowered the 4 hr plasma drug level by 90% or more).

It should be noted that in the studies of the Bond and Amitai research groups, the acetaminophen blood values were not at hepatotoxic levels, so it is unclear what the effects of ipecac would be if hepatotoxic doses were involved. Also, it is to be noted that no comparison of ipecac versus early charcoal administration was made; thus, there is no proof that ipecac would be better than charcoal.

Another issue related to the giving of syrup of ipecac deals with the fact that fluid in large quantities has been recommended as a way of decreasing the time required for emesis. However, giving large volumes of fluids expands the stomach and increases drug absorption from the stomach. It also increases gastric emptying and thus absorp- tion of the drug from the small intestines. We have already discussed above (under the section on "Oral Dilution") that dilution generally leads to a much greater rate of drug absorption.

Friday et al. (1980) performed a study in which dogs were injected with 0.1 mg/kg triacetoxy-N-n-propyldesmethylnormorphothebaine, which is 33 times more potent than apomorphine, after administration of one of four marker substances in 6 mL/kg saline. In one case, no additional fluid was given; in a second case, 27 mL/kg saline was given 5 min before the emetic drug; and in a third case, 27 mL/kg saline was given 25 minutes before the emetic drug. There were no differences among the three treatments in the recovery of markers or the chronology of vomiting. Thus, it was concluded that the efficacy of induced emesis was not dependent on ingested fluid volume. It might be mentioned that the marker recoveries were: for phenol red, 50%; for tobramycin, 61%; for radioactive 15 μm spheres given 25 min before the emetic drug, 52%; and for radioactive spheres given 5 min before the emetic drug, 66%.

Grande and Ling (1987) likewise have found no statistical correlation between the volume of fluid given with syrup of ipecac and time to emesis (the regression coefficients were the same at the $p = 0.094$ level).

Grbcich et al. (1986b) did a study in which different volumes of clear fluids were given to children during the 5 min after ipecac administration (no fluids were given prior to the ipecac). Although their data showed no statistical correlation between fluid volume and time to emesis at the $p < 0.05$ level, a distinct trend existed, as shown in Table 7.3. Based on these data, it appears that giving fluid after ipecac can only lengthen the time to emesis, and runs the risk of increasing drug absorption.

However, as King (1980) states, "giving syrup of ipecac on an empty stomach is like trying to squeeze an empty balloon." Therefore, it is clear that some amount of fluid needs to be given. The FDA (Federal Register, 1985) has recommended the amount of 1–2 glasses of water after an adult dose of 30 mL syrup of ipecac. Presumably 1 glass of water would be an appropriate maximal amount after a pediatric dose of 15 mL syrup of ipecac.

D. Comparison of Ipecac and Apomorphine in Producing Emesis

There have been several studies which have directly compared the effectiveness of syrup of ipecac and apomorphine in terms of their success in producing emesis, the times to emesis, and side effects (such as CNS depression).

Table 7.3 Fluid Volume After Ipecac
and Mean Times to Emesis

Volume	Mean time to emesis (min)
0–2 oz	15.0
> 2–4 oz	17.0
> 4–6 oz	21.5
> 6–8 oz	20.2
> 8–10 oz	24.4
> 10 oz	27.8

From Grbcich et al. (1986b). Reprinted with permission of *Veterinary and Human Toxicology.*

MacLean (1973) administered syrup of ipecac (15–30 mL, with a second dose if emesis failed to occur in 20–30 min) or apomorphine (0.07 mg/kg SC) to 86 children who presented to an ED for acute poisoning. The mean time for emesis was 14 min (range 3–30 min) with ipecac and 4 min (range 2–15 min) with apomorphine. The mean number of times vomiting occurred with ipecac and apomorphine were 3 and 6, respectively. The percentages of each group that vomited were: ipecac, 90.9%; apomorphine, 87.8%. There was significantly more CNS depression with apomorphine. Since the failure rates of the two emetics were about the same, MacLean recommended ipecac over apomorphine based on CNS depression considerations. Indeed, it is the significant incidence of CNS and respiratory depression which has caused apomorphine to be essentially abandoned in human medicine in the last 10–20 years (occasional use in animal poisoning apparently still exists). The fact that apomorphine must be given parenterally is another factor which has contributed to its demise.

Schofferman (1976) compared apomorphine and syrup of ipecac in 28 adults. Fifteen received 30 mL syrup of ipecac and 13 received 0.1 mg/kg apomorphine subcutaneously. Emesis was successful in 13 of the ipecac cases (87%) and in 10 of the apomorphine cases (77%). The mean latency periods before vomiting were 11.6 min in the ipecac group (range 4–26 min) and 5.3 min in the apomorphine group (range 2–13 min). Even though apomorphine was faster, 8 (62%) of the apomorphine treated patients developed significant CNS depression, five (38%) had hypotension, and one patient had respiratory depression. In the ipecac group, one (7%) patient experienced moderate CNS depression. Thus, Schofferman recommended ipecac over apomorphine.

Table 7.4 gives a summary of most of the studies in which syrup of ipecac and apomorphine have been compared.

E. Effectiveness of Ipecac and Apomorphine in Recovering Stomach Contents

There have been a number of studies aimed at determining the extent of recovery of stomach contents obtained with syrup of ipecac and apomorphine. Abdallah and Tye (1967) gave barium sulfate to dogs, followed by treatment with apomorphine (SC) or syrup of ipecac. They first tried to determine the minimal doses that would produce

Table 7.4 Emesis Success with Syrup of Ipecac and Apomorphine

Emetic	Subjects	Dose[a]	Percent emesis	Mean onset time (min)	Range (min)	Ref
Ipecac	Adults	30	87	11.6	4–26	4
	Children	20	78	15	5–40	2
	Children	20	—	18.7	< 15 to > 30	1
	Children	15–30	91	14	3–30	3
	Puppies	15–30	33	29	8–37	5
Apomorphine (SC)	Adults	0.1	77	5.3	2–13	4
	Children	0.07	88	4	2–15	3
	Children	0.07	100	6	2–17	2
	Puppies	0.07	100	2.9	2.5–3.2	5

[a] Syrup of ipecac dose in mL; apomorphine dose in mg/kg.
References: (1) Robertson, 1962; (2) Corby et al., 1968; (3) MacLean, 1973; (4) Schofferman, 1976; (5) Corby et al., 1967. From Stewart (1983).

emesis in 100% of the dogs, and then tried to establish an optimal dose in terms of barium sulfate recovery. The minimal doses for 100% successful emesis were found to be 0.15 mg/kg apomorphine SC and 1.5 mL/kg for syrup of ipecac. Apomorphine doses of 0.15, 0.30, and 0.60 mg/kg gave barium sulfate recoveries of 54, 78, and 60%, respectively (the lower recovery at 0.60 mg/kg than at 0.30 mg/kg was not explained).

Syrup of ipecac doses of 1.5 and 3.0 mL/kg gave barium sulfate recoveries of 62 and 39%, respectively (the lower recovery at the higher dose was explained by the possibility that ipecac can cause diarrhea and increased peristalsis, and hence less barium sulfate was in the stomach when emesis occurred). Thus, the optimal doses of these two agents were determined to be: apomorphine, 0.30 mg/kg, and ipecac, 1.5 mL/kg.

The optimal apomorphine dose of 0.30 mg/kg removed 78, 64, and 25% of the barium sulfate when given at 0, 30, and 60 min, respectively, and the optimal syrup of ipecac dose of 1.5 mL/kg gave 62, 44, and 28% recoveries for the same delay times (the 28% value is indicated by the graphical presentation of Abdallah and Tye; their tabular data indicate that the 60 min value for ipecac was 31%). Comparison was also made to lavage performed with a rubber tube (size not stated) connected to a 50 mL syringe. Clearly, all three treatments recovered less barium sulfate as the time delay increased, as one would expect. Apomorphine was significantly better than ipecac at 0 and 30 min but about the same at 60 min, and lavage was consistently less effective than either emetic (but note that the rubber tube used for lavage may have been of relatively small diameter).

Corby et al. (1967) gave dogs 1 g barium sulfate orally and then after a 20 min delay performed gastric lavage, or gave 15–30 mL syrup of ipecac, or administered 0.03 mg/lb apomorphine. The type of tube used for the lavage is not stated but was most likely a nasogastric tube, as such a tube was used to instill the marker substance. Lavage required an average of 23 min until returns were clear, ipecac required an average of 29 min for emesis (range 8–37 min) for those dogs which vomited, and apomorphine required an average of 2.9 min to produce emesis (range 2.5–3.2 min). In the ipecac group, of the six dogs, only two vomited with a single 15 mL dose. Of the remaining four, only one vomited when more ipecac was given. The mean percent recoveries of the barium sulfate were: lavage, 29% (range 10–62%); ipecac, 19% (range 2–31%); and apomorphine, 74% (range 54–87%). Thus, in this study, apomorphine was faster and gave better recoveries of stomach contents. Ipecac gave poor recoveries and often failed to produce emesis (this, and other studies, have shown that dogs respond much less well to ipecac than do humans). Lavage was intermediate between these emetic agents both with respect to the time required and with respect to recovery of the marker substance.

Corby et al. (1968) did a follow-up study with children who had ingested potentially poisonous amounts of drugs, using magnesium hydroxide (1 g) in 200 mL of a cherry flavored suspension as the marker. Ipecac (20 mL) and apomorphine (0.03 mg/lb) were employed as emetics. The mean times to onset of emesis were: 6 min (range 2–17 min) for apomorphine and 15 min (range 5–40 min) for ipecac. The percent recovery of the marker was: 31% (range 3–92%) for apomorphine, and 28% (range 0–78%) for ipecac. A double dose of ipecac was needed to produce vomiting in 3 of 14 children, whereas apomorphine produced emesis in all 22 patients in that group. In no case did either emetic produce significant CNS depression, although transient drowsiness did occur with apomorphine. Thus, it was concluded that apomorphine is safe, and faster than ipecac. However, both emetics gave low recoveries of stomach contents.

Tandberg and Murphy (1989) found that even when patients were given syrup of ipecac within 10 min after ingesting a tracer (cyanocobalamin) the mean recovery of the tracer was only 47.1% (range 40.1–54.0%). Their study also showed that a knee-chest position gave the same recovery (47.2% average) as a horizontal position (46.9% average).

F.   General Comments on the Use of Ipecac

Here we review some general articles in which various authors have summarized their overall feelings about whether syrup of ipecac should be used.

Rumack (1976) has pointed out that, while Schofferman's study (1976) with adults showed that apomorphine is faster than ipecac (5.3 versus 11.6 min), the mean delay time between ingestion of the poison/drug and arrival at an ED is 68.7 min, according to Robertson (1962). Thus, the mean time to emesis for apomorphine and ipecac, assuming immediate administration of the emetic upon presentation at the ED, is 74.0 versus 80.3 min, for apomorphine and ipecac, respectively. This is a difference of only about 8%. Rumack (1976) also mentions that, while apomorphine and ipecac give similar recoveries of gastric contents (31 and 28%, respectively, in the 1968 study of Corby et al.), the Schofferman study showed that apomorphine caused much more CNS depression than did ipecac. Thus, because of the risk of severe respiratory depression in already seriously ill patients, he recommended against apomorphine. He further stated that, because of the poor recovery (circa 30%) of stomach contents, charcoal should be used.

Vale et al. (1986) questioned the use of ipecac as a routine treatment method. They state that, in fact, few children admitted to EDs after ingestion of toxic substances really have severe symptoms (they give values of 4–12% in Britain and in West Germany as the percentage of such patients showing severe symptoms, and quote a mortality figure of less than 0.1% in a large Australian study). They go on to cite some of the studies discussed here, showing poor recovery of stomach contents with ipecac and the poorer performance of ipecac as compared to charcoal in reducing drug absorption.

Rumack and Rosen (1981), in an editorial, indicate that the effectiveness of ipecac is still uncertain and state the need for studies which would evaluate "ipecac" versus "no ipecac" in terms of outcome (e.g., duration of coma, time in hospital, etc.) and balance that against the dangers of aspiration, perforation, hemorrhage, etc.

MacGregor (1988) argued in a letter that ipecac should not be abandoned yet since, in remote areas, access to EDs is not available. He cites the high success rate of emesis found by Krenzelok (1985) and the relative safety of ipecac. In reply, Greensher et al. (1988) argued that, if charcoal preparations were stocked at isolated stations or health centers, they would be preferred over syrup of ipecac.

Overall, the studies we have reviewed indicate that syrup of ipecac is reasonably safe, but that the percent removal of stomach contents is not large (usually less than half). As far as apomorphine is concerned, it is faster than ipecac but also produces low recovery of stomach contents. It has the disadvantage of sometimes causing significant CNS depression. Since apomorphine is given only in a hospital environment, then—since the patient is already in such an environment—it would seem that more effective measures such as oral charcoal would be preferable.

In summary, syrup of ipecac is of value for accidental pediatric ingestions managed immediately in the home. It also may be of value for some children who present to EDs soon after an ingestion. Perrone et al. (1994) point out that, for adults, syrup of ipecac should only be used for those few persons who present very early after a large and potentially severe ingestion of agents for which activated charcoal is not effective (e.g., iron, lead, and lithium compounds), and agents with delayed toxicity (e.g., sustained-release preparations). However, in general, the use of syrup of ipecac in a hospital environment has not been shown to be of clinical benefit and only delays the use of activated charcoal. The delay can be considerable if nausea is prolonged.

V. GASTRIC LAVAGE

Gastric lavage is usually considered if the patient's gag reflex is absent and if rapid deterioration of mental status or hemodynamics is anticipated. In performing gastric lavage, the patient is usually placed in the left lateral decubitus head-down position. An orogastric tube (a large-bore tube should be used: 16 to 28 Fr in children and 36 to 40 Fr in adults) is passed gently into the stomach. It is no longer recommended that endotracheal intubation be routinely performed, as the airway in conscious patients can be adequately protected by placing the patient in the position described and having suction readily available. Pulmonary aspiration has been known to occur even when the trachea had been intubated, due to leakage of gastric contents around the balloon of the tube. Endotracheal intubation should be performed only in patients who require it for airway maintenance or controlled ventilation due to respiratory or CNS depression.

Although tap water can be used for lavage, isotonic or half-isotonic solutions are safer. Severe or even fatal electrolyte disturbances can occur in young children if tap water is employed instead of saline solutions. Small amounts of solution (100–200 mL typically) are passed into the stomach and then aspirated. This cycle is repeated 10–12 times or until the returns are clear. A double-lumen tube which delivers fluid continuously and aspirates the stomach contents simultaneously allows the procedure to be done more quickly (e.g., in as little as 5 min). An excellent review of the history and practice of lavage (as well as emesis) in adults and children has been written by Burke (1972). This article also describes a clinical study of gastric lavage in ten adults.

McDougal and Maclean (1981) suggested modifications of the usual lavage technique, based on studies of theirs which showed that "it was nearly impossible to remove pills from an artificial stomach using room temperature water and no mechanical agitation." In their artificial stomach studies, the use of warm tap water and repetitive compressions removed pills easily. Thus, they recommended that traditional lavage be followed by a second step involving warm lavage fluid, larger aliquots, and massage of the epigastrium. Bartecchi (1974), several years earlier, also recommended that massage be used. He states that, after conventional lavage, he has the patient lie flat on his back, with his knees flexed, so that maximal abdominal wall relaxation is achieved. Then, 150–200 mL of saline is instilled into the stomach and the upper left quadrant is massaged with a steady firm stroke, to loosen or break up any concretions. Lavage is then continued further. These recommendations appear not to have been adopted by most practitioners of lavage.

Gastric lavage is not without some hazard, as shown in a study by Scalzo et al. (1989) in which the positions of large-bore (24–40 French) lavage tubes were determined radiographically after insertion. For 14 patients, 7 had the proper tube position. The remaining 7 had improper tube tip placement, with 6 of these being low placement by grossly distending the stomach to the pelvic inlet. There were 2 patients with malpositioning in the esophagus, one at the gastroesophageal junction and one tube whose tip was directed backwards towards the head. The authors concluded that clinicians may underestimate the incidence of improper tube placement, that large bore tubes can easily distend the stomach with a resultant risk of rupture, that anatomical data be used to help determine the appropriate length of tube insertion, and that proper placement be confirmed. This can be done by radiography (most lavage tubes have a stripe or other marking which shows up on x-rays) prior to starting lavage, but radiography is often impractical and certainly is expensive. Tube position is more easily determined by aspirating gastric contents or by injecting air while auscultating the epigastrium. It is important to observe the patient's respiratory status and chest movement before lavage, since inadvertent tracheal intubation usually will produce abnormalities in these.

Askenazi et al. (1984), in fact, reported that a lavage tube caused an esophageal perforation in a case involving a 44-year-old woman who was intubated with a large "monoject" tube (34 French). Charcoal was administered via the same tube after lavage and it entered the right pleural space, further complicating matters. This report does not describe the eventual outcome.

Thompson et al. (1987) found that gastric lavage can produce significant cardiorespiratory effects. It usually increased the pulse rate, decreased the P_{O2} from 95 to 80 mm Hg, and frequently caused potentially serious electrocardiogram changes.

An interesting study on the use of gastric lavage as a "gastric dialysis" procedure was published more than two decades ago by Hart et al. (1969). They state that: "Reports that basic drugs appear in gastric juice in high concentration after systemic administration suggested that such compounds could be removed from the body by gastric lavage using the gastric mucosa as a dialyzing membrane." Obviously, the lavage would have to be carried out for an extended period of time. In experiments with rats in which the renal pedicles were ligated to prevent renal excretion of the test drugs, 10–25% of IV doses of basic drugs (pK_a 4.6–5.1) could be removed over 2 hr by lavage with 0.1 N HCl. Basic drugs with higher pK_a values were removed only to the extent of 2–3% over 2 hr, and acidic drugs (pK_a 3.0–10.6) were removed only to the extent of 0.7–1.6% over 2 hr by lavage with a 0.3 M pH 10 tris buffer. Thus, this technique would have potential benefit only in the case of basic drugs of certain pK_a values and, in general, would be inferior to the use of activated charcoal.

One of the first systematic studies of the efficacy of gastric lavage was performed in Copenhagen by Harstad et al. (1942). The major findings reported in this article (in German) were reported in English in an editorial in *JAMA* (Anonymous, 1947). The Harstad study showed that lavage was generally of quite low and highly variable effectiveness in removing a variety of drugs, particularly barbiturates. No more than 3% of any drug, with the exception of quinine (12%), was recovered by lavage. However, Matthew (1970) has pointed out that "Harstad's methods of qualitative and quantitative analysis of barbiturate ... were by determination of melting point and by weighing. The former is certainly valuable for identification but the latter is an extremely crude method of quantification." Matthew then went on to discuss some relatively successful experiences of his with gastric lavage in barbiturate overdose.

Matthew et al. (1966) studied the effectiveness of gastric lavage in 259 patients, particularly with respect to barbiturate and salicylate overdoses. They found that in 24 of 65 barbiturate cases (37%) more than 200 mg of the drug was recovered by lavage, and in salicylate cases as much as 20 g of drug was recovered nine hours after ingestion. Thus, they recommended lavage for barbiturate overdoses if it can be done within 4 hr of ingestion (unless it can be established that fewer than 10 tablets or capsules were taken) and in all cases of salicylate overdose. For other drugs, the guidelines for barbiturates were recommended as being reasonable.

Fane et al. (1971) studied whether a large bore lavage tube and a large volume of lavage fluid are more effective than a smaller tube and smaller volumes. Twelve dogs given phenobarbital (750 mg) or meprobamate (4 g) in 20 mL tap water via a 16 French tube were then lavaged immediately with one of the following protocols: (1) 200 mL of fluid in 50 mL increments via a 16 French tube, (2) 2000 mL of fluid in 200 mL increments via a 16 French tube, (3) 200 mL of fluid in 50 mL increments via a 32 French tube, or (4) 2000 mL of fluid in 200 mL increments via a 32 French tube. Mean recoveries of phenobarbital obtained were 36, 40, 21, and 49%, respectively; for meprobamate, mean recoveries were 21, 43, 41, and 37%, respectively. Because of the significant variation in the recovery values, no clear effect of tube size or fluid volume was possible. Note, however, that the drug recoveries averaged only 36%, overall.

Comstock et al. (1981) reported on the efficacy of gastric lavage alone as practiced in a large metropolitan hospital. In 76 patients treated for sedative/hypnotic overdose by gastric lavage, two or more therapeutic doses were recovered from 15.8% of the patients and 10 or more therapeutic doses were recovered from 6.6% of the patients. Drugs such as diazepam and amitriptyline were found to be more effectively recovered than sedative/hypnotic drugs. Poor recoveries were obtained when lavage was delayed more than 2 hr after drug ingestion, except in cases of amitriptyline overdose or massive sedative/hypnotic overdose. Thus, lavage can be effective if done as early as possible, while much of the drug is still in the stomach. For drugs which produce delayed gastric emptying (like tricyclics), lavage even up to 5 hr post-ingestion can still be effective.

An obvious but neglected consideration in gastric lavage is the simple question of whether pharmaceutical tablets or pills can fit through the holes in the lavage tube. Agocha et al. (1986) collected all solid dosage forms from a large hospital pharmacy, 293 in all, and determined if they would fit through the holes of size 18, 28, and 40 French tubes. Zero percent of the 293 fit through the holes in the size 18 tube, 4.4% fit through the holes in the size 28 tube, and 45.3% fit through the holes of the size 40 tube. Thus, 54.7% of the tablets/pills could not pass through the holes of the largest tube. Of course, as tablets/pills dissolve in the stomach, they decrease in size and often fragment into smaller pieces. Thus, as time goes on, a larger proportion of the tablets/pills can indeed pass into the lavage tube, but for freshly ingested tablets/pills the problem of the tablets/pills simply being too big for the holes in the lavage tube is real. One ED technician that this writer talked to about this problem stated that he cuts the end off the lavage tube, "so that there will at least be one big hole." However, this could leave sharp edges which could traumatize the esophagus or stomach and a better method (which is apparently used by many clinicians) is to enlargen the holes in the side of the lavage tube.

VI. COMPARISON OF IPECAC AND LAVAGE

In trying to compare the relative efficacies of ipecac-induced emesis versus gastric lavage, it is important that the size of the lavage tube be known, as many older studies employed small-bore nasogastric tubes rather than large-bore orogastric tubes.

Meester (1980) provided an excellent review of emesis and lavage (besides ipecac as an emetic agent, this review also discusses mechanically-induced vomiting, salt water, and copper sulfate). Voldeng (1976) has discussed gastric lavage versus induced emesis briefly, pointing out that the two methods often give similar results, with two important differences: (1) vomiting induced by ipecac often reclaims material from beyond the pyloric sphincter, and (2) when a volume of fluid is given with the ipecac, 84–100% of the fluid volume is usually returned. He states that gastric lavage should be considered as an alternative to induced emesis when the patient is not comatose or convulsing and: (1) vomiting has not been intense, or (2) the patient's gag reflex is absent.

Arnold et al. (1959) evaluated the relative efficacy of lavage and induced emesis in salicylate poisoning, using dogs. After giving the dogs 0.5 g/kg sodium salicylate orally, one of several treatments was instituted. One was lavage done immediately, another was lavage delayed for 1 hr, a third was syrup of ipecac (25 mL) given at 15–20 min time, and a fourth was ipecac given at 1 hr. The lavage tube was a number 16 French orogastric tube. They found that with immediate lavage, salicylate recovery ranged from 2–69% of the dose (average 38%) and with ipecac at 15–20 min the salicylate recovery ranged from 7–75% (average 45%). Thus, both "early" treatments had a very large range and on average removed less than half of the drug. For the "delayed" lavage, the drug recovery ranged from 0–40% (average 13%) and for the "delayed" emesis the drug recovery ranged from 5–74% (average 39%). Thus, when treatment was delayed for 1 hr, emesis was more effective than lavage.

Boxer et al. (1969) compared the efficacy of ipecac and gastric lavage (using a nasogastric tube) for the treatment of acute salicylate ingestion in 20 children (aged 12–20 months) presenting to an ED. Immediate lavage was done with one group, followed by 15 mL of syrup of ipecac given afterwards (if no emesis occurred within 15 min, a second 15 mL dose was given). This was compared to the use of 15 mL syrup of ipecac (again, a second 15 mL was given if emesis had not occurred by 15 min), followed by lavage started at 30 min (by which time emesis had occurred in all patients). In the group in which lavage was followed by emesis, the amount of salicylate recovered by emesis averaged 2.02 times that recovered by lavage, despite the lavage being performed first. In the group in which emesis was followed by lavage, the same ratio was about 6.3. Thus, emesis recovered more salicylate than lavage, even when lavage was done first. The authors concluded that ipecac-induced emesis alone is better than any treatment involving lavage. However, it must be noted that the lavage was done with a small-bore tube (nasogastric tube).

Tenenbein (1985a) described three cases in which ipecac and gastric lavage were performed on poisoning victims. In the first, a 17-month-old boy was treated for ingestion of an estimated 10-16 tablets of ferrous sulfate. An x-ray upon admission showed 9 tablets in the stomach and 1 in the small intestine. After ipecac administration, with three episodes of vomiting, a repeat x-ray showed the same ten tablets, but in a different arrangement. After subsequent lavage (using a 24 French orogastric tube with enlarged side holes), one tablet was vomited. A third x-ray showed 8 tablets (the one in the small intestines had presumably dissolved). In a second case, a 16-year-old girl was treated for ingestion of approximately 100 ferrous sulfate tablets. Ipecac administration, with four episodes of vomiting, brought up 15 tablets. An x-ray showed more than 50 tablets still in the stomach. Gastric lavage was performed using a 34 French orogastric tube, and no tablets or particles were recovered. In a third case, another 16-year-old girl was treated, this time for ingestion of 80–90 TheoDur tablets. Spontaneous vomiting produced an estimated 20 tablets prior to arrival of the ambulance and further spontaneous vomiting produced 5 more tablets. Large-bore gastric lavage (using a 34 French orogastric tube with enlarged side holes) gave 5 more tablets. Charcoal (100 g) plus 15 g magnesium sulfate was administered, and shortly thereafter the patient vomited, producing 51 tablets. Ipecac and lavage were dramatically ineffective in the first two cases. In case three, ipecac was not given (spontaneous vomiting occurred) and lavage itself was of little help (the spontaneous ejection of 51 tablets after the lavage is proof of this). The ineffectiveness of ipecac and lavage, combined with the uncomfortable nature of these procedures for the patient, make them open to serious question.

Kulig et al. (1985) studied the effect of gastric emptying (by gastric lavage or by use of syrup of ipecac) versus no gastric emptying in a large group of patients presenting to an ED. Alert patients were treated randomly with either syrup of ipecac, followed by oral charcoal, or with oral charcoal only. Obtunded patients were randomly treated either with gastric lavage (using size 30 to 40 French orogastric tubes) followed by charcoal, or with activated charcoal only (given by nasogastric tube). Syrup of ipecac did not significantly affect the clinical outcome of alert patients, probably because the patients presented well after the drug ingestion and by such time the drug had been mostly absorbed. Gastric lavage in obtunded patients led to a better clinical outcome only if performed within one hour of drug ingestion, probably for the same reason (i.e., after more than an hour, most of the drug would usually have been absorbed or would have passed into the intestines, and thus would no longer be in the stomach). Thus, gastric emptying—whether by syrup of ipecac or lavage—is effective only if done relatively soon after drug ingestion.

Everson (1986) remarked in a letter concerning the Kulig study that, since there was not control group one can not conclude whether charcoal itself improved either the alert group or the obtunded group. He further stated that until gastric emptying is shown to be ineffective or unsafe, it should be used in conjunction with charcoal. The reply of Kulig (1986) to this was an agreement with the lack of any demonstration of a charcoal effect and the point of view that until gastric emptying can be shown to be effective and safe, it should not be used in conjunction with charcoal.

Tandberg et al. (1986) carried out a controlled study in normal adults. Eighteen volunteers each ingested twenty-five 100-μg cyanocobalamin tablets. On one occasion, ipecac-induced emesis was used, while on another occasion gastric lavage with a modified 32 French orogastric tube was performed. The collected aspirates, lavage returns, and vomitus were homogenized and assayed for cobalt. Emesis returned a mean of 28% of the tracer, versus 45% in the gastric aspirate and lavage fluid. Thus, this study showed lavage to be more effective than emesis. Litovitz (1986), however, pointed out in an editorial that the investigators gave 1,000 mL tap water after the ipecac, much more than is commonly recommended. This large water volume would enhance gastric emptying into the small intestine and reduce tablet returns from the stomach during emesis. Litovitz also pointed out that subjects in the lavage treated group were placed in the left lateral decubitus position with the head lowered 10 degrees immediately after tablet ingestion, in order to minimize passage of gastric contents into the small intestine. Subjects in the ipecac group were not placed in this position. Finally, Litovitz mentions the 10 minute delay in the ipecac group (a delay which would not necessarily occur in practice). The lavage group had preparations for lavage begun immediately, although the lavage itself was not started until the 10 min mark. In summary, the results of Tandberg and coworkers are inconclusive to a significant degree.

Auerbach et al. (1986) compared the efficacy of gastric emptying by lavage and ipecac-induced emesis. Thiamine was used as the marker substance. Eighty-eight patients presenting to an ED were assigned to two types of treatments. One involved instillation of 100 mg thiamine by gastric tube along with 100 mL of irrigation fluid; then, lavage was begun 5 min later. The lavage tube employed was a 24 French Harris Flush Tube having additional holes cut by the physician and passed either by the nasal or oral route. The other group was given a mixture of 100 mg thiamine in 30 mL syrup of ipecac, followed by 200–400 mL water. The ipecac was repeated if no emesis occurred in 20 min. Recovery of thiamine by lavage averaged 90%, whereas ipecac gave an average recovery of 50%. However, the ranges of recoveries were extremely large for both treatments. With lavage, thiamine recoveries ranged from about 28% to 191%, and with ipecac the range was 0% to 161% (based on a reading of a figure in their paper). The authors offered several hypotheses as to why recoveries greatly in excess of 100% were indicated. Because of the wide ranges of the recoveries, and the number of values exceeding 100%, this study must be viewed with some caution. The major conclusion that lavage is more effective than ipecac-induced emesis must be viewed in the context of the extremely wide range of recoveries observed with both treatments.

Vasquez et al. (1988) gave radiolabeled sucralfate (RSC) to human volunteers, followed by 30 mL syrup of ipecac and 240 mL water at 5, 30, or 60 min. Gastrointestinal tract images done at the time of RSC ingestion and 1 hr after syrup of ipecac administration showed that the percentages of the RSC removed in 1 hr by emesis for the three groups were 83, 59, and 44%, respectively. Thus, syrup of ipecac was quite efficient when given early (i.e., at 5 min).

Pond (1986) reviewed studies dealing with the effectiveness of emesis induced by syrup of ipecac and of lavage. She summarized her results in the tabular form shown in Table 7.5. (Pond cites recovery ranges of 0–100% for both ipecac and lavage in the Auerbach study, although the Auerbach paper indicates the ranges stated in the two paragraphs above.) These data show that both emesis and lavage are not very efficient if delayed. After a one-hour delay, the percent recovery is no better than about one-third, at best. When ipecac or lavage is given immediately, percent recoveries are about

Table 7.5　Comparison of Induced Emesis and Lavage in Removing Marker Substances

Subjects	Marker	Time after marker given (min)	% Recovery (range)		Ref
			Ipecac	Lavage	
Adult dogs	Salicylate	15	45 (7–75)	38 (2–69)	1
		15–60	39 (5–74)	13 (0–40)	
Young dogs	Barium sulfate	0	62	54	2
		30	44	26	
		60	31	9	
Puppies	Barium sulfate	30	19 (2–31)	29 (10–62)	3
Overdose patients	Thiamine	0	50 (0–100)	90 (0–100)	4

References: (1) Arnold et al., 1959; (2) Abdallah and Tye, 1967; (3) Corby et al., 1967; (4) Auerbach et al., 1986. From Pond (1986). Reprinted by permission of Elsevier Science Publishers BV.

50–60% for ipecac and 50–90% for lavage. What is disturbing, however, in cases where the average recoveries are reasonably high, is the range of results encountered. In actual overdose patients, both treatments had ranges covering essentially 100% and in dogs the ranges (albeit with some delay) were very large also (5–75% for ipecac, and 0–69% for lavage). Thus, it is probably the inconsistency of reasonable recovery more than the absolute amounts which make these treatments far less dependable than oral charcoal.

Pond has also summarized the side-effects which have been reported in conjunction with the use of syrup of ipecac and gastric lavage. Among those cited are:

Ipecac: Intractable vomiting, diarrhea, lethargy, muscle weakness or stiffness, arrhythmias, aspiration, esophagitis, dehydration, and shock.
Lavage: Emesis, mechanical injury, epistaxis, aspiration, laryngospasm, hypernatremia, gastric hemorrhage, and cardiac arrhythmias.

The prolonged vomiting and/or diarrhea associated with excessive ipecac doses can cause severe fluid and electrolyte imbalances, with a variety of associated adverse symptoms and hazards.

Adverse effects due to the use of charcoal are discussed in detail in Chapter 17, and include aspiration, constipation, and obstructions of the intestinal tract. However, if one avoids the use of repeated doses of fairly large amounts of charcoal, gives a reasonable dose of laxative, and uses care in administering the charcoal, these problems can usually be avoided. Thus, in general, charcoal is better-liked and better-tolerated, and has fewer adverse side effects than do ipecac and lavage.

Rodgers and Matyunas (1986) reviewed gastrointestinal decontamination for acute poisoning in children, and have made the following recommendations: (1) ipecac syrup is to be preferred (but not for children less than 6 months old), if contraindications to its use are not present, (2) gastric lavage should be considered to be of limited use in pediatric patients and, if done, should be performed with large-bore orogastric tubes, not with nasogastric tubes, (3) administration of a charcoal slurry prior to lavage should be considered, if lavage is contemplated, and (4) patients with sufficiently significant toxic symptoms should be given repeated doses of charcoal and cathartics until symptoms resolve (the use of multiple doses of cathartics is presently discouraged, due to the risk of severe fluid and electrolyte imbalances).

Danel et al. (1988) gave a brief report on the use of charcoal, emesis, and gastric lavage (performed with a 30 French orogastric tube) in aspirin overdose. Twelve healthy subjects were given 1.5 g aspirin, and then were treated with either syrup of ipecac (30 mL, repeated 30 min later if emesis did not occur), or 50 g charcoal, or by gastric lavage. Urine was collected for 24 hr and analyzed for salicylates. In this study, the three protocols were about equally effective. The mean recovery of salicylate in the urine was 60.3% (control), 55.6% for the ipecac group, 55.6% for the lavage group, and 52.5% for the charcoal group. Lavage was time-consuming and ipecac produced fairly long-lasting nausea and fatigue. Thus, charcoal, being easier to administer and producing no side effects, was recommended over lavage or ipecac.

Underhill et al. (1990) also compared gastric lavage, ipecac, and charcoal. Patients 16 years of age or over who had ingested 5 g or more of paracetamol (acetaminophen) within 4 hr of admission to an ED ($N = 60$) were treated with lavage (using a 36 French orogastric tube), ipecac (30 mL, repeated after 30 min if no response occurred), or activated charcoal (Carbomix, in a charcoal:drug ratio of 10:1). Blood samples taken prior to treatment and 60, 90, and 150 min later were analyzed for paracetamol. The mean percentage decreases in the drug level between the first and last blood samples were: with lavage, 39.3%; with ipecac, 40.7%; and with charcoal, 52.2%. The difference between lavage and ipecac was not significant and charcoal was somewhat better than both.

An editorial by Olson (1990) summarized much of the available data on induced emesis and gastric lavage. He mentions: the study of Kulig et al. (1985) discussed above, which showed that lavage was effective only if performed within an hour of drug ingestion; work of Watson et al. (1989) which showed that only 8.7% (range 0.4–21.7%) of the estimated ingested dose was removed by lavage in seven patients treated for tricyclic antidepressant overdose (the orogastric tube sizes used were 18 French for one patient, 34 French for two patients, and 36 French for the other four patients); and a study by Albertson et al. (1989) which confirmed Kulig's finding concerning ipecac versus charcoal in an ED setting and found increased risk of aspiration pneumonia in patients who had been given ipecac. In the same issue of the journal which contained the editorial, a report by Merigian et al. (1990) showed that in over 800 self-poisoned patients, there was no benefit from gastric emptying prior to charcoal administration. Merigian et al. treated 357 patients who were symptomatic either with gastric emptying (gastric lavage with a large-bore Ewald tube for obtunded patients or ipecac-induced emesis for alert patients) plus charcoal, or with charcoal alone. Gastric emptying produced no significant differences in the length of stay in the ED, the mean length of time intubated, or mean length of time of stay in the intensive care unit. The gastric lavage group had a higher incidence of admission to the intensive care unit and of aspiration pneumonia. However, it should be noted that the gastric lavage patients were all obtunded, so that the likelihood of more serious sequelae is perhaps not surprising. Dr. Merigian has been quoted (Anonymous, 1991) as stating: "I now rarely empty anyone's stomach, although, when I think it's appropriate, I will put down a nasogastric tube to deliver charcoal if the patient isn't alert enough to drink it."

Merigian et al. also gave charcoal or no charcoal to 451 asymptomatic patients, and found no effect due to charcoal. Not a single patient in this group showed clinical deterioration. It is not surprising that in patients with such a low level of toxicity, that no effect of charcoal, as assessed by somewhat subjective measures, was noted. Olson

(1990) also remarked on design flaws in many studies, for example, the lack of quantitative measurements of outcome (such as drug levels, amount of drug removed by gut emptying, and depth or duration of coma), and the mixing of patients with varying severities of drug overdose (patients with mild overdoses are likely to have satisfactory outcomes, regardless of whatever gastric decontamination procedure is employed). Olson states his views that ipecac, while useful in the home for immediate treatment, is of little value in an ED and that gastric lavage is of questionable value considering the risk of pulmonary aspiration. In any event, any gastric emptying procedure done after 1–2 hr post-ingestion is probably of little value, except perhaps for cases of very serious overdose. The use of charcoal is recommended as early as possible.

Saetta and Quinton (1991) used flexible endoscopy to assess the intragastric residue after either ipecac-induced emesis or gastric lavage (using a size 33 Faucher tube) in 30 self-poisoned patients. The group of 13 patients treated with 30 mL syrup of ipecac all vomited (mean time 23 min; range 11–25 min). Of this group, 5 out of 13 (38.5%) had significant residual solid in their stomachs. In the lavage group of 17 patients, 15 (88.2%) had significant residual intragastric solids. Thus, both techniques, particularly lavage, failed to clear the stomachs.

Saetta et al. (1991) gave pellets containing barium to self-poisoned patients before either ipecac-induced emesis or gastric lavage (the lavage tube size is not stated) and then x-rayed the patients to determine if the gastric contents had been forced into the small intestine by either technique. In the ipecac group, 234 pellets (58.5% of those given) were retained in the GI tract after emesis, 92 of which (39.3%) were noted in the small intestine. In the lavage group, 207 pellets (51.8% of those given) were retained in the GI tract after lavage, 69 of which (33.3%) were in the small intestine. These results show unequivocally that: (1) neither ipecac nor lavage removed even half of the pellets from the GI tract, and (2) both procedures resulted in a significant (33–39%) emptying and/or forcing of the pellets from the stomach into the small intestine. Presumably, if the pellets were tableted drugs, the same sorts of results would occur.

VII. USE OF SALINE CATHARTICS OR SORBITOL

Shannon et al. (1986) reviewed the role of cathartics in the management of poisoned patients. Among other things, they mention some arguments against the efficacy of cathartics: (1) cathartics act primarily in the colon and yet most drug absorption occurs much further up, in the small intestine, (2) they can delay gastric emptying due to their hyperosmolarity, thus allowing both more time for drug absorption from the stomach and slowing down the passage of charcoal into the small intestine, and (3) the hyperperistalsis which they induce could lead to greater tablet/pill breakup and dissolution, thereby increasing drug absorption. On the other hand, decreased GI transit time offers the possibility of decreased drug absorption.

Pietrusko (1977) also provided an excellent review of cathartics. We mention here some facts cited by Pietrusko which relate to saline cathartics and to sorbitol, the cathartics most commonly employed in conjunction with overdose management and the use of charcoal. Table 7.6 summarizes the major items. Note that sorbitol acts much more quickly than do the saline cathartics, in general. Saline cathartics, by virtue of creating hypertonic conditions, induce fluid transfer into the GI tract and its subsequent osmotic retention, thereby increasing the intraluminal bulk and stimulating the peristaltic

Table 7.6 Comparison of Saline and Sorbitol Cathartics

Cathartic	Onset of action	Site of action	Probable mechanism of action	Systemic absorption?
Saline types (MgSO$_4$, NaSO$_4$, Mg citrate)	0.5–3 hr	Small and large intestine	Release of cholecystokinin; Osmotic action	Yes
Hyperosmotic types (sorbitol)	0.5 hr	Colon	Hygroscopic	Poor

motility reflex in the intestines. This reduces transit time and causes a watery stool. The small intestine is unable to absorb the extra water well enough to counteract the diarrheal condition. Research has shown that, in addition to this osmotic effect, saline cathartics stimulate the release of cholecystokinin which increases small bowel motility and inhibits the absorption of fluid from the jejunum and ileum.

The idea of using saline cathartics or sorbitol alone for drug or poison ingestion is that an increase in the rate of propulsion of the drug or poison through the GI tract would reduce its absorption, either because of there being less time for absorption, and/or the drug being more dilute (reducing the concentration gradient across the GI tract walls and thus, presumably, its rate of transfer through the walls). However, as Stewart states: "There are no experimental data in animals or man that demonstrate unequivocally that saline catharsis results in reduced absorption of drugs or poisons from the gastrointestinal tract." Riegel and Becker (1981) similarly state: "Despite the appealing theoretical concept that catharsis of the overdosed patient will decrease gastrointestinal transit time ... and thus prevent some absorption of the toxic substance, proof of such action is lacking." They go on to enumerate contraindications to catharsis (e.g., for very old or very young patients, in patients with absent bowel sounds, etc.). Specific contraindications for sodium and magnesium cathartics are given also.

It is well known that substances generally empty from the stomach at a rate dependent on their physical and chemical characteristics. Thus, liquids empty faster than solids; however, for liquids, the volume, osmolarity, acidity, and chemical composition all have an effect. The rate of gastric emptying of liquids becomes slower as the solution osmotic pressure increases. This explains why saline cathartics have been found to inhibit the rate of gastric emptying in rats. Saline cathartics are almost always given as hypertonic solutions, in order to induce water retention and/or secretion into the GI tract. The slowing of gastric emptying by saline cathartics should have important implications for their use in combination with charcoal, for the combination of a saline cathartic and charcoal should provide the charcoal with greater time in which to adsorb toxic substances in the stomach. However, as stated earlier, there are no experimental studies which definitely prove that saline cathartics enhance the antidotal effectiveness of activated charcoal.

It should be mentioned, in connection with the slower gastric emptying caused by saline cathartics, that if one looks at a plasma or serum concentration versus time curve, the lower initial values might be taken as a sign that the cathartic has reduced the amount of drug absorption, whereas in fact only the rate of drug absorption has been changed, as an examination of the complete drug concentration versus time curve would show.

There has been some concern that saline cathartics, while perhaps not enhancing the effect of charcoal, might actually interfere with charcoal by adsorbing on it and taking up some of the adsorption "sites" on the charcoal. However, in vitro studies by Ryan et al. (1980) and by LaPierre et al. (1981) have shown that magnesium citrate has no significant effect. Cooney and Wijaya (1986), however, have shown that at low pH citrate ions transform to undissociated citric acid, which adsorbs to charcoal and has a slight deleterious effect. But, cathartics with inorganic anions such as the sulfate ion would not be expected to produce any interference, as inorganic ions generally do not adsorb to charcoal appreciably.

Sue et al. (1991) carried out a study in children to determine the efficacy of magnesium citrate alone as a treatment method. They gave children 6 years old or less, who presented to an ED with poisonings, the following randomized treatments: (1) charcoal in water, (2) charcoal in 4 mL/kg magnesium citrate solution, or (3) charcoal in 6 mL/kg magnesium citrate solution. The onset and frequency of stools was monitored. The mean times to the first stool were 17, 16, and 15 hr, respectively. The mean numbers of stools over 48 hr were 2.1, 2.8, and 2.4. It was concluded that, in the doses used (which reflected dose levels currently recommended as safe), magnesium citrate did not hasten the onset of stools nor increase their numbers.

As mentioned by Krenzelok (1987), sorbitol is known to produce strong effects in children when dose levels of 0.3–0.5 g/kg or 20–30 g are attained or exceeded. Cathartic effects in adults occur when sorbitol doses of 1–1.5 g/kg are employed. Sorbitol is capable of producing catharsis in less than 60 min. Charney and Bodurtha (1981) have described a case of intractable diarrhea (1–2 L/day of watery stool for several weeks) in a 16-year-old boy due to sorbitol. The physicians were unaware that one of the medications given to the boy (hydralazine HCl) was in a 70% sorbitol vehicle and that the patient was thereby receiving 60 g/day of sorbitol. Upon recognition of the sorbitol and its discontinuance, diarrhea stopped in 24 hr.

There are a few studies in which the effects of sorbitol alone on drug absorption have been assessed. Most of these studies have involved several treatment regimens: no therapy (control), charcoal alone, sorbitol alone, and charcoal/sorbitol. Thus the "sorbitol alone" treatment provides information on the effects of just using sorbitol. Here we will compare only the control versus sorbitol alone results.

Mayersohn et al. (1977) administered three 325 mg aspirin tablets plus 50 mL water to human subjects and, in one treatment, followed this with 100 mL of an aqueous 70% sorbitol solution. Salicylate excretion into the urine was followed for 48 hr. The control case gave 73.2% recovery of the aspirin in the urine, while the sorbitol case gave 70.2% recovery—not a large difference. The sorbitol solutions produced diarrhea in all subjects.

Van de Graaff et al. (1982) studied the effects of sorbitol and other cathartics alone or in combination with charcoal on the absorption of acetaminophen in dogs. A dose of 0.6 g/kg acetaminophen was given by orogastric tube and then followed by mannitol plus sorbitol (2 g/kg) or castor oil (3 mL/kg). Cathartics alone decreased the drug $0-\infty$ AUC by 14% (castor oil) and by 31% (mannitol plus sorbitol). Both mannitol/sorbitol and castor oil produced diarrhea.

Picchioni et al. (1982) studied the in vivo absorption of four drugs in rats and determined the effects of sorbitol. The drugs were chlorpheniramine maleate (80 mg/kg), chloroquine diphosphate (100 mg/kg), aspirin (100 mg/kg), and sodium pentobarbital (50 mg/kg). They were given by stomach tube. Immediately after, the rats were treated in one case with 20 mL/kg of 70% sorbitol. The AUC values (time spans unstated) as percentages of control values for the four drugs when sorbitol was used were 68.2, 69.0, 109.4, and 40.6%, respectively. Thus, sorbitol was effective in reducing drug absorption for all drugs except aspirin (in the same study, charcoal alone was generally much more effective and charcoal plus sorbitol was better still).

Al-Shareef et al. (1990) investigated the effects of sorbitol on the absorption of sustained-release theophylline in eight healthy female subjects. The subjects were given two 300 mg slow-release theophylline tablets, followed 2 hr later by either 400 mL water then 80 mL water every 6 hr up to 20 hr, or 50 mL 70% sorbitol followed by 80 mL water every 6 hr up to 20 hr. The 0–24 hr AUC values were 97.6 and 116.6 mg-hr/L, respectively. Thus, sorbitol increased the AUC relative to the "drug only" case whereas when charcoal or charcoal/sorbitol were used, AUC reductions of 89% or greater were found. The maximum plasma theophylline concentration with sorbitol was 30% higher than that which was seen after theophylline alone and occurred earlier, indicating a higher rate of absorption of theophylline in the presence of sorbitol. Al-Shareef and co-workers feel that this was due to changes in gastrointestinal motility.

Overall, then, the effects of sorbitol alone can be seen to be rather varied. Sometimes sorbitol has produced AUC reductions on the order of 30% (however, in such cases, the use of charcoal or charcoal plus sorbitol has invariably been much more effective) but at times it has had either no effect or a slight adverse effect. Thus, the use of sorbitol alone is inferior to charcoal or charcoal plus sorbitol (comparisons of charcoal alone and charcoal plus sorbitol are presented in Chapter 15).

The use of too much sorbitol, especially the use of multiple doses of sorbitol in multiple dose charcoal/sorbitol therapy, can be quite dangerous, as severe fluid and electrolyte imbalances can occur. The same is true with respect to the overuse of saline cathartics. These dangers are discussed more fully in Chapter 17.

Perrone et al. (1994) stated that, when the use of a cathartic is desired, they prefer sorbitol, at a dose of about 1 g/kg, because it produces rapid catharsis. They also indicate that magnesium citrate (a 10% solution, at a dose of 4 mL/kg in children and 250 mL in adults), or magnesium sulfate (250 mg/kg in children, and 15–20 g in adults) should be reserved for children or frail adults.

VIII. WHOLE-BOWEL IRRIGATION

A strategy related to the use of cathartics (the aim again being to move intestinal contents rapidly through the GI tract) is whole-bowel irrigation (WBI). Whole-bowel irrigation is the oral administration of large volumes of special lavage fluids (e.g., balanced electrolyte solutions containing polyethylene glycol) in order to promote rapid bowel emptying. Such solutions are often identified by the acronym PEG-ELS (polyethylene glycol-electrolyte lavage solution). WBI is considered to be of value for treating the ingestion of toxins not well bound to charcoal, such as iron, lead, and lithium compounds (Smith et al., 1991; Roberge and Martin, 1992), and for removing ingested drug packets or vials from "body packers" and "body stuffers." The electrolyte solution is delivered at rates up to 0.5 L/hr in children and 2 L/hr in adults, until the rectal effluent is clear (4–6 hr usually). WBI is contraindicated in the presence of ileus and any GI tract obstructions, bleeding, or perforations. One potential drawback to WBI is that, if it is combined with activated charcoal therapy, the PEG-ELS solution can compete with drugs for binding sites on the charcoal or displace previously adsorbed drugs.

Potential complications include abdominal cramping, vomiting, profuse diarrhea, and hyperchloremia. Tenenbein (1988) provided a comprehensive review of WBI as a gastro-intestinal decontamination procedure after acute poisoning, in which he states that WBI is superior to gastric lavage and induced emesis and is generally tolerated well by patients. However, he admits that it is labor intensive and time consuming. He recommends WBI particularly for ingestion of iron and overdoses with delayed-release pharmaceuticals.

Before discussing WBI further, mention will be made here of a similar technique called "drug clearance by diarrhea induction" by Porter and Baker (1985). This technique also involves passing large volumes of an electrolyte solution into the gastrointestinal tract; however, the aim is not to accelerate passage of the bowel contents per se, but rather to use the GI tract membrane as a dialysis membrane, across which an already-absorbed drug can pass. This is similar to the gastrointestinal dialysis idea discussed in Chapter 5, in which activated charcoal is the medium which induces the dialysis of an already-absorbed drug across the GI tract membrane. Porter and Baker gave dogs 30 mg/kg of phenobarbital IV and then passed a dialysate solution into their GI tracts at a rate of 40 mL/min for a total volume of 10 L (this took about 4 hr). Diarrhea continued for about 45 min after the end of the infusion period. This procedure removed 24.8% of the drug dose over 5 hr, versus elimination of only 3.1% of the drug dose in controls.

One of the earliest research articles on WBI was authored by Tenenbein (1985b), who described case reports on the effective use of WBI for treating ferrous sulfate ingestions (various other ingestions were treated also but confounding problems such as emesis made firm conclusions impossible). It was suggested that if WBI is anticipated, the prior use of syrup of ipecac should be avoided. Another report by Tenenbein (1987) dealing with WBI for treating six cases of iron poisoning suggested that the procedure was of benefit. (Iron alone does not adsorb to charcoal and so charcoal therapy can not be used in such cases; however, subsequent research (see Chapter 18) has shown that the combination of iron plus a chelating agent can cause iron adsorption to charcoal in a chelated form). However, the author admits that "because this [was] an uncontrolled descriptive study, it cannot be stated with certainty that whole blood (sic) irrigation prevented serious toxicity."

Scharman et al. (1992) described a study in which human subjects were given 10 coffee beans and then 1 hr later either metoclopramide (an antiemetic and stimulant of upper GI tract motility) or a placebo. Thirty minutes later WBI was done with 10 L of PEG-ELS administered over 5 hr, with a mean time to clear effluent of 2.8 hr. The mean number of beans recovered at the time the "clear effluent" point was reached was 3.0 with metoclopramide and 2.6 without metoclopramide. At the end of the 10 L irrigation, the mean total numbers of beans passed were 4.0 and 3.6, respectively. Thus, metoclo-pramide had no effect. But, the main conclusion reached was that, by the time the "clear effluent" point had been reached, the recovery of the beans was only 30%. Even the full 10 L irrigation gave at best 40% recovery. Whether coffee beans are representative of pharmaceutical tablets is a prime question.

Another article by Tenenbein et al. (1987) reported on the use of WBI in nine adult volunteers who had been given 5.0 g ampicillin. Comparison of blood drug level data was made to the same volunteers during a control phase. The study showed that WBI decreased ampicillin bioavailability by an average of 67%. The mean WBI duration was 234 min and the mean solution volume used was 7.7 L.

IX. ALTERING DIURETIC PROCESSES

It was stated in Chapter 5 that the elimination of drugs and their metabolites by renal excretion is the primary mechanism for their removal, in the absence of activated charcoal. Thus, methods for increasing renal elimination, such as forced diuresis, represent one way of attempting to manage a drug overdose. Acidization or alkalinization of the blood by giving the patient either ascorbic acid or sodium bicarbonate, respectively, are other methods which attempt to convert drugs to forms that are more readily excretable by the kidneys.

Simple forced diuresis is usually carried out by the IV administration of fluids sufficient to give a urine output of about 5 mL/min (300 mL/hr). However, according to Wogan (1989), its efficacy is minimal and it is considered to be of value only for a few drugs such as bromine and lithium (other studies suggest that it is not effective for lithium poisoning). Moreover, it may produce electrolyte abnormalities and precipitate congestive heart failure. The idea behind forced diuresis is that increasing the flow of fluid in the tubules of the kidneys will reduce the concentration of the drug in that fluid. In theory, this should reduce the concentration gradient for reabsorption of the drug by the kidneys and hence increase the renal elimination of the drug. However, this effect does not seem to occur as strongly as one would expect.

Alkalinizing the urine by administering a sodium bicarbonate solution IV can be of some use with drugs that are acidic (salicylates, barbiturates) because this raises the pH of the urine, under which conditions the drugs are ionized. Since only the nonionized form of a drug can undergo tubular reabsorption, this reduces the reabsorption of the drug in the kidneys and increases renal elimination. If the alkaline fluid volume is made large, the technique is often called forced alkaline diuresis; however, if urine flow is adequate, this is probably not much more effective than conventional urine alkalinization.

Similarly, ascorbic acid solutions given IV may be used to acidify the urine and increase the elimination of basic drugs (which will be ionized in the acid pH range). However, acidification is essentially no longer used clinically because basic drugs such as amphetamines and phencyclidine can produce rhabdomyolsis (breakup of muscle tissue) which under acid urine conditions can be much more toxic to the kidneys.

Kärkkäinen and Neuvonen (1984) studied the effect of urine pH on sotalol pharmacokinetics, by giving seven healthy subjects sodium bicarbonate (to alkalinize the urine) or ammonium chloride (to acidify the urine) between 1 and 64 hr after an oral dose of 160 mg sotalol (they also studied the effects of charcoal, as described in Chapter 12). The sodium bicarbonate (62 g total taken) gave urine pH values ranging from 7.2 to 8.2, whereas the ammonium chloride (35 g total taken) produced urine pH values in the range of 4.7 to 5.5. The effects are shown in Table 7.7. One can see that, while the bicarbonate and ammonium chloride had some effect in lowering C_{max} and AUC values, the amount of sotalol excreted into the urine was not increased.

Kärkkäinen and Neuvonen (1985) then did a similar study but with dextropropoxyphene (DPP) as the test drug. They monitored the excretion of DPP and its active metabolite norpropoxyphene (NP). Urinary excretion values (0–72 hr) for DPP in μmole were: for controls, 1.96; with acid urine, 11.5; and with alkaline urine, 0.096. For DPP the values were: for controls, 63.5; with acid urine, 72.0; and with alkaline urine, 35.5. Thus, the excretion of DPP was increased 6-fold by acidification and reduced 20-fold by alkalinization. The excretion of NP varied much less with pH, increasing 13% with acid urine and decreasing 44% with alkaline urine. Despite the increased excretion seen

Table 7.7 Effects of Altering Urine pH on Sotalol Elimination

	Control	Sodium bicarbonate	Ammonium chloride
C_{max} (μmol/L)	5.6	3.9	3.9
t_{max} (hr)	2.9	2.6	2.7
AUC_{0-72} (μmol-hr/L)	64.3	52.4	52.3
$t_{1/2}$ in serum (hr)	9.4	11.7	11.5
$t_{1/2}$ in urine (hr)	9.4	10.1	11.5
Excretion into urine:			
0–72 hr (μmol)	332	291	284
24–72 hr (μmol)	52	53	66

From Kärkkäinen and Neuvonen (1984). Reprinted by permission of Dustri-Verlag Dr. Karl Feistle, Deisenhofen, Germany.

with acid urine, only about 25% of the drug dose was excreted over 0–72 hr. In this same study, charcoal was also used in one phase and it reduced the drug absorption by 97–99%. Repeated doses of charcoal reduced the serum half-life of DPP from 31.1 to 21.2 hr and the serum half-life of NP from 34.4 to 19.8 hr. Thus, oral charcoal was more effective than changing the urine pH.

In a third study of the same type, Kärkkäinen and Neuvonen (1986) focused on amitriptyline (AT) as the test drug. Its active metabolite, nortriptyline (NT), was also monitored. For AT, urinary excretion values (0–72 hr) in μmoles were: for controls, 0.70; with acid urine 4.79; and with alkaline urine, 0.07. For NT, the values were: for controls, 1.16; with acid urine, 5.58; and with alkaline urine, 0.06. Thus, acidification increased AT excretion 7-fold and alkalinization decreased AT excretion 10-fold. For NT, acidification increased excretion 5-fold and alkalinization decreased excretion 19-fold. The renal clearance of AT varied with urine pH by a factor of 1000 over the pH range 4 to 8. Despite the great increase in renal excretion with acidified urine, the cumulative excretion of AT and NT over 0–72 hr only accounted for about 5% of the drug dose. Thus, altering urine pH has little effect in removing AT and its metabolite from the body. In this same study, oral charcoal was also given in one phase; it reduced AT absorption by 99% and lowered serum half-lives for AT and NT by 20 and 35%, respectively. Thus, charcoal was of much greater benefit than changing the urine pH.

Therefore, it is clear that for some drugs changing the urine pH can have a very dramatic effect on urinary excretion rates. However, the fraction of the original drug dose thus excreted over a period of, say, 72 hr is often not large. In contrast, at least for the three drugs involved in these studies, oral charcoal was far more effective—not so much in enhancing elimination but in very greatly preventing absorption in the first place.

Urinary alkalinization using $NaHCO_3$ has been compared to multiple dose charcoal therapy for enhancing phenobarbital elimination by Frenia et al. (1993). They gave 5 mg/kg phenobarbital IV to 10 volunteers, followed by either $NaHCO_3$ (enough to keep the urine pH between 7.5–8.0) or multiple dose charcoal (details in this abstract are not given). Serum drug levels were obtained over 24 hr. As compared to control tests, alkalinization increased the elimination rate constant k_e only from 0.017 to 0.021 hr^{-1}

(not statistically significant). With charcoal, k_e was raised to 0.51 hr^{-1}. The observed $t_{1/2}$ values were: 59.7 hr (control), 36.2 hr (alkalinization), and 14.4 hr (charcoal).

X. EXCHANGE TRANSFUSION

Exchange transfusion (XT) is of value only when the toxic substance is one which causes methemoglobinemia (e.g., aniline dyes or their derivatives, nitrates, nitrites, bromates, chlorates, sulfanilamide, nitrobenzene and related nitro compounds, etc.). The object is to replace the oxidized hemoglobin with hemoglobin capable of carrying oxygen. However, its technical difficulty and concerns about the transmission of blood-borne pathogens such as hepatitis B and AIDS has led to its virtual discontinuance in favor of other therapies. The administration of methylene blue or toluidine blue is easier and more commonly done (an exception is in methemoglobinemia caused by chlorate poisoning, since this often does not respond well to methylene blue or toluidine blue).

Exchange transfusion is rarely used to enhance drug elimination, since it requires repetitive exchanges to effectively remove any drug which is extensively distributed into all of the body fluids and tissues. An exception may be in serious theophylline poisoning in premature neonates, where patient size makes the risk of necrotizing enterocolitis due to oral charcoal therapy greater. Shannon et al. (1992) described the exchange transfusion of an infant iatrogenically overdosed with theophylline. A triple-volume XT reduced the serum drug level from 88.3 mg/L and calculated $t_{1/2}$ of 39.0 hr to a serum drug level of 30.8 mg/L and a calculated $t_{1/2}$ of 14 hr.

Strauss and Modanlou (1992), commenting on the study by Shannon and associates, stated that their group had successfully used repeated doses of charcoal along with magnesium citrate and IV fluids to treat a theophylline overdosed infant (see Strauss et al., 1985) and a rapid decline in theophylline levels occurred without any side effects. They also made reference to another study of their group (Ginoza et al., 1987) in which ten preterm infants with steady-state therapeutic theophylline levels were given Super-Char charcoal and showed a near doubling of theophylline clearance. Thus, they recommended charcoal as a safe alternative to exchange transfusion. In reply, Shannon and Wernovsky (1992) stated that the infant in their particular case did not have bowel sounds and thus charcoal therapy (which they had, in fact, performed on neonates in the past) was precluded.

Henry et al. (1991) also report on XT treatment of a premature infant for elevated theophylline levels caused by use of this drug to treat apnea of prematurity. A single-volume XT lowered the serum drug level from 58 to 47 mg/L, a reduction of 19%.

XI. HEMODIALYSIS AND HEMOPERFUSION

There is a very extensive literature dealing with the dialysis of poisons and drugs. For several years (1967–1973), Dr. G. E. Schreiner and coworkers at the Georgetown University Hospital (Washington, D.C.) published extensive reviews on "The Dialysis of Poisons and Drugs" in the annual volumes of the *Transactions of the American Society for Artificial Internal Organs* (TASAIO). After a gap of 4 years, a final review from Schreiner's group (see Winchester et al., 1977) was published in the TASAIO. This final review went beyond dialysis and also included a detailed survey of the removal of drugs and poisons by hemoperfusion. The 1977 review was enormous, with 953 references. Obviously, no attempt will be made here to even begin to cover the same topics, as the focus of this volume must remain with orally administered activated charcoal. Nevertheless, some comments will be made on the removal of drugs and poisons by hemodialysis and hemoperfusion, for two reasons:

1. There are situations where these techniques are logical alternatives or adjuncts to the use of oral charcoal, and it is important to delineate what these situations are. This will place the use of oral charcoal in better perspective.
2. The technique of hemoperfusion involves the adsorption of drugs and poisons to activated charcoal (although in granular rather than powdered form) and other sorbents such as resins, and thus nominally falls within the scope of the title of the present work. However, because the literature on hemoperfusion is so huge, a separate volume would be needed to adequately discuss it. Thus, we will only review the basic principles of hemoperfusion.

Before proceeding further with a discussion of these two techniques, we should recognize a basic fact concerning their applicability. Since both methods involve tapping into the patient's circulatory system (e.g., into a vein in the arm or leg) and pumping blood to the removal device (dialyzer or hemoperfusion column), then the drug must exist in a reasonable amount in the blood in order for these methods to work. Thus, if the toxic substance has only been recently ingested by the patient and is not yet in the patient's blood in any significant amount, these techniques are of no use. On the other hand, if the toxic substance had been ingested long enough prior to admission so that it is well distributed throughout the patient's body fluids and tissues—and more specifically in the blood—then these methods would be quite worthy of consideration.

Winchester et al. (1977) and Gelfand and Winchester (1980) listed several criteria which suggest the applicability of hemodialysis or hemoperfusion in treating drug overdoses. These include:

1. Progressive deterioration despite intensive supportive therapy.
2. A drug blood level which is of a potentially fatal magnitude, or ingestion and probable absorption of a potentially lethal dose.
3. Severe clinical intoxication with abnormal vital signs, especially depression of midbrain function leading to hypoventilation, hypothermia, and/or hypotension.
4. Development of complications of coma such as aspiration pneumonia, peripheral neuropathy, etc. or the presence of an underlying disease (e.g., chronic bronchitis, emphysema) which would increase the hazards of coma.
5. Impairment of normal drug excretion mechanisms (hepatic, renal, etc.) by the drug.
6. Poisoning by agents known to produce delayed toxicity (acetaminophen, paraquat, amanita phalloides, etc.).
7. Intoxication by a substance which is known to be metabolized to a more toxic form (e.g., methanol, which is converted to formaldehyde, or ethylene glycol, which is converted to oxalic acid).

Knepshield and Winchester suggested plasma drug levels above which either hemodialysis or hemoperfusion should be considered. Table 7.8 shows these.

For hemodialysis to be effective in removing a drug from a patient's blood stream, the drug must be one with: (1) a molecular weight less than about 500 daltons, (2) low protein binding, (3) low lipid solubility, and (4) low volume of distribution. Since most

Table 7.8 Plasma Levels Above Which Hemodialysis (HD) or Hemoperfusion (HP) Should be Considered

Drug	Plasma level (mg/L)	Method of choice
Phenobarbital	100	HP > HD
Other barbiturates	50	HP
Glutethimide	40	HP
Methaqualone	40	HP
Salicylates	800	HD > HP
Ethchlorvynol	150	HP
Meprobamate	100	HP
Trichloroethanol	50	HP
Paraquat	1	HP > HD
Theophylline	300	HP
Methanol	500	HD
Ethylene glycol	Unknown	HD

From Knepshield and Winchester (1982). Reprinted by permission of the J. B. Lippincott Company.
The entry "HP > HD" means both methods are viable but that hemoperfusion is the more effective of the two; "HD > HP" means that hemodialysis is the better.

drugs do not meet these criteria, hemodialysis is seldom effective. Exceptions are highly water-soluble toxins (e.g., methanol and ethylene glycol) and drugs which are not effectively treated by supportive care alone due to complications (e.g., acid-base disorders, electrolyte abnormalities, hypertension, hypotension, and thermoregulatory disorders). Examples of such drugs are salicylates, theophylline, phenobarbital, boric acid, and lithium. Other than for these specific drugs, the extra risks of hemodialysis (infection, venous thrombosis, hypotension, air embolus, and hemorrhage) are generally not acceptable.

In connection with the dialysis of drugs, both Decker et al. (1971) and Spector (1980) showed that incorporating powdered charcoal in the dialysis fluid will enhance the transfer of drugs across the dialysis membrane. The reason for this is clear: after the drug crosses the membrane and enters the dialysate, it is adsorbed by the charcoal, thus keeping its concentration in the aqueous phase very low. This maintains the fluid-phase concentration difference of the drug across the membrane at a high level and thus promotes the continued diffusion of the drug across the dialysis membrane.

Robinson et al. (1991) surveyed the use of HD and HP in an AAPCC certified poison center over a 6 year period (1/84–12/89) and found only 33 cases (19 HD, 12 HP, and 1 case where both were used). The indications for HD and HP were judged to be appropriate in only 19 (58%) of the cases. Following HD or HP, 24 patients were improved, 3 were worse, and 6 were unchanged. They noted that the average time from diagnosis to initiation of HD or HP was quite long (6.3 hr).

We turn now to a specific discussion of hemoperfusion. Hemoperfusion is a method in which activated charcoals (and other adsorbent materials) are used to treat victims of poisoning or drug overdose. Figure 7.1 shows the major features of a hemoperfusion circuit. It should be emphasized here that only *granular* charcoals can be used for hemoperfusion. This fact differentiates this topic from all of the other applications of charcoal discussed in this book, since all of the other uses involve *powdered* charcoals.

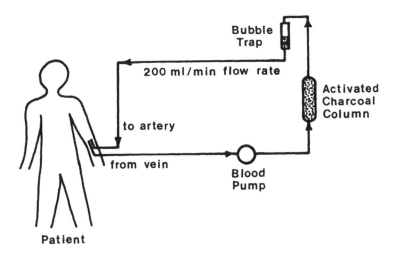

Figure 7.1 Typical hemoperfusion system.

It should also be pointed out that hemoperfusion is complementary to orally administered charcoal powder as a treatment for poisonings or drug overdoses. Powdered charcoal is mainly effective if used while the toxin is still in the stomach. Once absorption into the body as a whole occurs, hemoperfusion may then be the treatment of choice. Thus, powdered charcoal is useful at "early times" and hemoperfusion at "late times."

The published literature relating to hemoperfusion is by now quite extensive, as mentioned above. Several hundreds of papers dealing with hemoperfusion are in print (beyond the many mentioned in the 1977 review of Winchester et al., there has been a steady stream of reports over the succeeding 17 years). Readers interested in accessing these papers need only consult the annual cumulated volumes of *Index Medicus*, under the subject heading "Hemoperfusion."

The concept of flowing blood over activated charcoal granules to remove unwanted substances seems to have originated with Yatzidis (1964), who used a column of 200 g of 0.50–0.75 mm diameter granules to hemoperfuse dogs at a blood flow rate of 50–100 mL/min for up to 90 min. It was found that creatinine, uric acid, phenolic compounds, guanidine bases, and organic acids were all removed well. Yatzidis mentioned the possible use of such a system in removing endogenous toxins (as in kidney or liver failure), as well as exogenous toxins (drugs or poisons). Dunea and Kolff (1965) soon after reported on clinical tests in which uremic human subjects were treated with a system containing 200 g of 12–28 mesh size (0.59–1.65 mm diameter) activated charcoal granules at blood flow rates of 100–300 mL/min. Again, creatinine and uric acid were well removed from the blood. The only side effects noted were removal of some cells (especially platelets) from the blood and some hypotension.

Another potential problem was identified by Hagstam et al. (1966), who perfused rabbits, and upon later examination of their tissues found extensive deposits of fine charcoal particles in their lungs, spleens, livers, and kidneys (no tissue reactions were evident, however). Thus, the plugging of capillaries in various organs by fine charcoal particles, which would cause tissue necrosis if severe enough, seemed to be a definite hazard of hemoperfusion. Later work by Andrade et al. (1971, 1972) showed that, even

Table 7.9 Commercial Hemoperfusion Devices as of 1980

Name	Manufacturer	Adsorbent	Adsorbent size (mm diameter)	Adsorbent coating
Hemocol	Sandev, Ltd.	300 g charcoal	2–4	5 μm acrylic hydrogel
Adsorba 300C	Gambro, Inc.	300 g charcoal	1 × 2 mm	3–5 μm cellulose
Hemodetoxifier	Becton-Dickinson	94 g charcoal	0.3–0.84	None
XR-010	Extracorporeal Medical Specialties	312 g uncharged resin (XAD-4)	0.3–0.84	None

with very extensive preliminary washing of a charcoal bed, fine particles still continued to be generated (due to the particles rubbing together very slightly). Several investigators suggested using various polymer coatings to prevent fine particle release. Chang (1969) suggested using nylon, collodion, or heparin-complexed collodion; Denti et al. (1975) proposed cellulose acetate; and Andrade (1971, 1972) tested an acrylic hydrogel and albumin cross-linked with glutaraldehyde. All of these coatings were successful in preventing fine particle release; moreover, some of the coatings were "biocompatible" enough with blood so that blood cell losses during hemoperfusion were decreased to tolerable levels.

In the early to middle 1970s, several companies developed and began to market hemoperfusion columns commercially. These devices and their major features are listed in Table 7.9. Also included in Table 7.9 is a device which utilizes a synthetic resin rather than charcoal. The resin, Amberlite XAD-4, is a spherical polystyrene bead type of resin having no ionic groups in its structure. Lipid-soluble organic molecules (e.g., glutethimide, barbiturates) are particularly well adsorbed by this material.

These devices are now of mainly historical interest; only the Gambro Adsorba unit and a hemoperfusion column made in Japan are presently available. The reasons why hemoperfusion was never adopted to a sigificant extent in the USA (it has been, and still is, more widely used in Europe) are many and varied. The foremost reasons, perhaps, are that the technique is more expensive, more invasive, more risky, and no more effective than other therapeutic approaches, and that it can often be replaced just as effectively by hemodialysis, which is more readily available and better understood, in those cases where an extracorporeal approach is called for. Because hemoperfusion requires the use of an anticoagulant to prevent clot formation in the cartridge, and can sometimes cause substantial platelet removal from the blood, the risk of bleeding side effects is significant.

Another reason for the demise of hemoperfusion columns is essentially the same as one involved in the demise of many antidotal charcoal formulations. In both cases, the cost of research, development, manufacture, and product liability insurance has tended to exceed whatever profits could be made from the marketing of these products. Sales of hemoperfusion columns (at, say, $50 per column) were small—perhaps less than 1000 total units per year in the world. Thus, little financial incentive existed for making and selling hemoperfusion columns.

Some of the drugs which have been found to be effectively removed by hemoperfusion devices are barbiturates, glutethimide, methyprylon, meprobamate, ethchlorvynol, acetaminophen, salicylates, chloral hydrate, digitalis, imipramine, nortriptyline, meproscillaridin, and diazepam. In addition, various poisons have been effectively removed by hemoperfusion, including organophosphates, the herbicide paraquat, paralytic shellfish poisons, insecticides, carbon tetrachloride, and mushroom (*Amanita phalloidin*) toxins. Endogenous species removed well by hemoperfusion include: (1) excess thyroid hormones in cases of "thyroid storm;" (2) various uremic metabolites such as creatinine, uric acid, "middle molecular weight" molecules (identities unknown); (3) various species which accumulate during liver failure, such as bilirubin, bile acids, and certain amino acids (aromatic types); and (4) bacterial endotoxins. Excess amounts of methotrexate, an anticancer agent, have also been removed by charcoal hemoperfusion.

As a general rule, it would appear that any substance that has been found to be well adsorbed by powdered charcoal would also be adsorbed in a granular charcoal hemoperfusion column. The only exception to this would be that larger molecules (say, more than 5,000-10,000 daltons) could not adsorb in coated charcoal columns since the pores in typical polymer coatings are small enough to exclude species of that size. However, they could adsorb in an uncoated charcoal column (e.g., the Becton-Dickinson column which is no longer available).

With respect to hemoperfusion, it is common practice to give sodium heparin (an anticoagulant) IV to the patient in order to reduce the tendency for blood clots to form when the blood contacts the "foreign" surfaces in the hemoperfusion system (particularly the adsorbent column). The question thus arises as to whether heparin itself might adsorb to the charcoal in such columns, thereby reducing or nullifying the role of heparin. Thus, Cooney (1977) studied the adsorption of sodium heparin to powdered Norit A activated charcoal (although hemoperfusion columns contain granular charcoal, powdered charcoal was used because it reaches adsorption equilibrium much faster than granular charcoal and the goal of the study was to assess the potential of charcoal for adsorbing heparin). A linear adsorption isotherm fitted the data well, and had the form $Q = 0.407\,C_f$, where Q = g heparin adsorbed/g charcoal and C_f = g heparin/L of solution. Based on a level of 8 mg/L heparin in a patient's blood stream (1000 I.U./L concentration and a heparin activity of 125 I.U./mg), the Q value corresponding to this would be 0.0033 g/g. A column containing 100 g charcoal could thus potentially adsorb 0.33 g heparin or over 40,000 I.U. Since the total heparin in the hemoperfusion system at any one time is on the order of 15,000 I.U., then significant heparin removal from the blood seemed to be a real possibility.

However, because the charcoal in these systems is granular, and because heparin is a large molecule, Cooney thought that the rate of heparin adsorption in actual hemoperfusion column operation might be quite low. Thus, he carried out experiments (Cooney, 1979) in which 20 mg/L sodium heparin in a pH 7.4 phosphate buffer was pumped at 100–200 mL/min through each of four different hemoperfusion columns (without recirculation) and the concentrations of heparin in the effluent streams were monitored for 30 min. His data showed that heparin adsorption decreased sharply with time and was never higher than 7% of the inlet amount. Thus, his hypothesis was borne out.

XII.   SUMMARY

Our review of methods for treating victims of drug overdoses and poisonings can probably best be summarized by a list of conclusions:

1. Oral dilution has a role in the dilution of ingested mineral acids and bases (but, with modest amounts of water, say 200–300 mL), but otherwise should be avoided, as it increases the rate of drug absorption.

2. Gastric emptying procedures (induced emesis, gastric lavage, cathartics, whole-bowel irrigation) have generally been shown to have no significant benefit in terms of clinical outcome in poisonings and overdoses, and are currently falling into greater and greater disfavor.

3. Syrup of ipecac is the only emetic agent without significant danger. All other emetics should be avoided. Various salt "emetics" can be toxic and apomorphine often causes serious CNS depression. Moreover, although apomorphine acts more quickly than ipecac, it does not generally give greater removal of stomach contents.

4. Syrup of ipecac gives variable and generally low (often about 30%) recovery of ingested drugs and delays the use of charcoal. It has value for administration in the home but in a medical facility environment one should proceed directly to the use of charcoal. Generally, it is recommended for accidental pediatric ingestions managed immediately in the home. In a hospital environment, it is recommended for adults only when they present very early after a large and potentially severe ingestion of agents for which activated charcoal is not effective (e.g., iron, lead, and lithium compounds), and agents with delayed toxicity (e.g., sustained-release preparations).

5. Gastric lavage is uncomfortable, time-consuming, somewhat risky, and—like syrup of ipecac—gives highly variable and often low recoveries of ingested drugs (often about 30%). Lavage has generally not been found to be effective in terms of clinical outcome. Thus, the immediate use of charcoal is more effective.

6. Studies which have compared charcoal, lavage, and ipecac have either shown them to be similar in efficacy or have shown charcoal to be better, with lavage and ipecac similar and less effective. Since charcoal is easier and faster to administer, and is much more comfortable for the patient, it is to be preferred.

7. The use of sorbitol alone has given highly variable results—sometimes a drug's AUC decreases moderately (e.g., 30%) but sometimes not at all. In any event, charcoal or charcoal plus sorbitol has always been found to be more effective than sorbitol alone. The use of saline cathartics alone has not proved to be of significant value and the overuse of cathartics can lead to severe fluid and electrolyte imbalances.

8. Whole-bowel irrigation can be moderately effective for substances which do not adsorb to charcoal (e.g., ferrous sulfate, lead compounds, and lithium compounds) but it generally is less effective than charcoal. Studies with markers have given recoveries in the 30–40% range.

9. Forced diuresis is minimally effective and involves some risk (e.g., of electrolyte abnormalities). Likewise, alkalinization or acidification of the urine can greatly accelerate the urinary elimination of some drugs but usually has not significantly increased the overall rate of elimination. In all cases where changing the urine pH has been compared to the use of charcoal, charcoal has been far more efficacious.

10. Exchange transfusion must be done too many times to have any significant effect. However, it can be useful for achieving a modest but fast reduction in the level of a drug like theophylline in neonates.

11. Criteria for the use of hemodialysis and hemoperfusion have been enumerated. Hemodialysis can be useful for substances which do not adsorb to charcoal (e.g., alcohols like ethylene glycol) and both techniques can be useful when a drug is

already absorbed extensively into the patient's system and does not undergo enteroenteric/enterohepatic circulation. In such a case, its removal by oral charcoal would be impossible and, if the patient is in a serious coma, removal of the drug directly from the blood may be imperative.

One may deduce from all of the forgoing that oral dilution, induced emesis, gastric lavage, the use of cathartics alone, whole-bowel irrigation, forced diuresis, and alteration of urine pH are all inferior to the prompt use of charcoal. One exception is the use of syrup of ipecac in the home, where no other alternatives are immediately available. Another exception is the use of hemodialysis or hemoperfusion under certain specific circumstances where drug removal is essential and oral charcoal would not work.

XIII. REFERENCES

Abdallah, A. H. and Tye, A. (1967). A comparison of the efficiency of emetic drugs and stomach lavage, Am. J. Dis. Child. *113*, 571.

Adler, A. G., Walinsky, P., Krall, R. A., and Cho, S. Y. (1980). Death resulting from ipecac syrup poisoning, JAMA *243*, 1927.

Agocha, A., Reyman, L., Longmore, W., Acker, P., Sauter, D., and Goldfrank, L. (1986). Can pills really fit through the lavage tubes? (abstract), Vet. Hum. Toxicol. *28*, 494.

Albertson, T. E., Derlet, R. W., Foulke, G. E., Minguillon, M. C., and Tharratt, S. R. (1989). Superiority of activated charcoal alone compared with ipecac and activated charcoal in the treatment of acute toxic ingestions, Ann. Emerg. Med. *18*, 56.

Allport, R. B. (1959). Ipecac is not innocuous, J. Dis. Child. *98*, 786.

Al-Shareef, A. H., Buss, D. C., Allen, E. M., and Routledge, P. A. (1990). The effects of charcoal and sorbitol (alone and in combination) on plasma theophylline concentrations after a sustained-release formulation, Hum. Exp. Toxicol. *9*, 179.

Amdur, M. O., Doull, J., and Klassen, C. D., Eds. (1991). *Casarett and Doull's Toxicology: The Basic Science of Poisons*, 4th edition, Pergamon Press, New York.

Amitai, Y., Mitchell, A. A., McGuigan, M. A., and Lovejoy, F. H., Jr. (1987). Ipecac-induced emesis and reduction of plasma concentrations of drugs following accidental overdose in children, Pediatrics *80*, 364.

Andrade, J. D., Kunitomo, K., Van Wagenen, R., Kastiger, B., Gough, D., and Kolff, W. J., (1971). Coated adsorbents for direct blood perfusion: HEMA/activated carbon, Trans. Am. Soc. Artif. Int. Org. *XVII*, 222.

Andrade, J. D., et al. (1972). Coated adsorbents for direct blood perfusion II, Trans. Am. Soc. Artif. Int. Organs *XVIII*, 473.

Anonymous (1947). Value of gastric lavage in treatment of acute poisoning (editorial), JAMA *133*, 545.

Anonymous (1979). Ipecac syrup and activated charcoal for treatment of poisoning in children, The Medical Letter *21*, August 24, pp. 70-72.

Anonymous (1991). Is gastric emptying harmful?, Emerg. Med. *23*, 83.

Arena, J. M. and Drew, R. H. (1986). *Poisoning: Toxicology, Symptoms, Treatments*, 5th edition, C. C. Thomas, Springfield, Illinois.

Arnold, F. J., Jr., Hodges, J. B., Jr., Barta, R. A., Jr., Spector, S., Sunshine, I., and Wedgwood, R. J. (1959). Evaluation of the efficacy of lavage and induced emesis in treatment of salicylate poisoning, Pediatrics *23*, 286.

Askenasi, R., Abramowicz, M., Jeanmart, J., Ansay, J., and Gegaute, J. P. (1984). Esophageal perforation: An unusual complication of gastric lavage, Ann. Emerg. Med. *13*, 146.

Auerbach, P. S., Osterloh, J., Braun, O., Hu, P., Geehr, E. C., Kizer, K. W., and McKinney, H. (1986). Efficacy of gastric emptying: Gastric lavage versus emesis induced with ipecac, Ann. Emerg. Med. *15*, 692.

Bartecchi, C. E. (1974). A modification of the gastric lavage technique, JACEP *3*, 304.

Bates, T. and Grunwaldt, E. (1962). Ipecac poisoning: A report of two cases of ingestion of fluid extract of ipecac, Am. J. Dis. Child. *102*, 169.

Boeck, P., Bobbink, S., and Robertson, W. O. (1985). Dosing of ipecac (abstract), Vet. Hum. Toxicol. *27*, 317.

Bond, G. R., et al. (1989). Influence of time until emesis on the efficacy of decontamination using acetaminophen as a marker in a pediatric population (abstract), Vet. Hum. Toxicol. *31*, 336.

Borowitz, J. L., Moore, P. F., Yim, G. K. W., and Miya, T. S. (1971). Mechanism of enhanced drug effects produced by dilution of the oral dose, Toxicol. Appl. Pharmacol. *19*, 164.

Boxer, L., Anderson, F. P., and Rowe, D. S. (1969). Comparison of ipecac-induced emesis with gastric lavage in the treatment of acute salicylate ingestion, J. Pediatrics *74*, 800.

Brotman, M. C., Forbath, N., Garfinkel, P. E., and Humphrey, J. G. (1981). Myopathy due to ipecac syrup poisoning in a patient with anorexia nervosa, Can. Med. Assoc. J. *125*, 453.

Bukis, D., Kuwahara, L., and Robertson, W. O. (1978). Results of forcing fluids: Pre- versus post-ipecac, Vet. Hum. Toxicol. *20*, 90.

Burke, M. (1972). Gastric lavage and emesis in the treatment of ingested poisons: A review and a clinical study of lavage in ten adults, Resuscitation *1*, 91.

Chang, T. M. S. (1969). Removal of endogenous and exogenous toxins by a microencapsulated adsorbent, Can. J. Physiol. Pharmacol. *47*, 1043.

Charney, E. B. and Bodurtha, J. N. (1981). Intractable diarrhea associated with the use of sorbitol, J. Pediatr. *98*, 157.

Chin, L. (1971). Gastrointestinal dilution of poisons with water—An irrational and potentially harmful procedure, Am. J. Hosp. Pharm. *28*, 712.

Chin, L. (1972). Induced emesis—A questionable procedure for the treatment of acute poisoning, Am. J. Hosp. Pharm. *29*, 877.

Comstock, E. G., Faulkner, T. P., Boisaubin, E. V., Olson, D. A., and Comstock, B. S. (1981). Studies on the efficacy of gastric lavage as practiced in a large metropolitan hospital, Clin. Toxicol. *18*, 581.

Cooney, D. O. (1977). Heparin adsorption on activated charcoal, Clin. Toxicol. *11*, 569.

Cooney, D. O. (1979). Rates of heparin adsorption in hemoperfusion devices, Clin. Toxicol. *15*, 287.

Cooney, D. O. and Wijaya, J. (1986). Effect of magnesium citrate on the adsorptive capacity of activated charcoal for sodium salicylate, Vet. Hum. Toxicol. *28*, 521.

Corby, D. G., Lisciandro, R. C., Lehman, R. H., and Decker, W. J. (1967). The efficiency of methods used to evacuate the stomach after acute ingestions, Pediatrics *40*, 871.

Corby, D. G., Decker, W. J., Moran, M. J., and Payne, C. E. (1968). Clinical comparison of pharmacologic emetics in children, Pediatrics *42*, 361.

Dabbous, I. A., Bergman, A. B., and Robertson, W. O. (1965). The ineffectiveness of mechanically induced vomiting, J. Pediatr. *66*, 952.

Danel, V., Henry, J. A., and Glucksman, E. (1988). Activated charcoal, emesis, and gastric lavage in aspirin overdose, Br. Med. J. *296*, 1507.

Dean, B. S. and Krenzelok, E. P. (1985). Syrup of ipecac ... 15 mL versus 30 mL in pediatric poisonings, J. Toxicol.-Clin. Toxicol. *23*, 165.

Decker, W. J. (1971). In quest of emesis: Fact, fable, and fancy, Clin. Toxicol. *4*, 383.

Decker, W. J., Combs, H. F., Treuting, J. J., and Banez, R. J. (1971). Dialysis of drugs against activated charcoal, Toxicol. Appl. Pharmacol. *18*, 573.

Denti, E., Luboz, M. P., and Tessore, V. (1975). Adsorption characteristics of cellulose acetate coated charcoals, J. Biomed. Mater. Res. *9*, 143.

Dockstader, L. L., Lawrence, R. A., and Bresnick, H. L. (1986). Home administration of activated charcoal: Feasibility and acceptance (abstract), Vet. Hum. Toxicol. *28*, 471.

Dreisbach, R. H. and Robertson, W. O. (1987). *Handbook of Poisoning: Prevention, Diagnosis, and Treatment*, 12th edition, Appleton and Lange, Norwalk, CT.

Dunea, G. and Kolff, W. J. (1965). Clinical experience with the Yatzidis charcoal artificial kidney, Trans. Am. Soc. Artif. Int. Org. *XI*, 178.

Easom, J. M. and Lovejoy, F. H., Jr. (1979). Efficacy and safety of gastrointestinal decontamination in the treatment of oral poisoning, Pediatr. Clinics North Am. *26*, 827.

Eisenga, B. H. and Meester, W. D. (1978). Evaluation of the effect of motility on syrup of ipecac-induced emesis (abstract), Vet. Hum. Toxicol. *20*, 462.

Ellenhorn, M. J. and Barceloux, D. G., Eds. (1988). *Medical Toxicology: Diagnosis and Treatment of Human Poisoning*, Elsevier, New York.

Everson, G. W. (1986). Gastric emptying and activated charcoal (letter), Ann. Emerg. Med. *15*, 225.

Fane, L. R., Combs, H. F., and Decker, W. J. (1971). Physical parameters in gastric lavage, Clin. Toxicol. *4*, 389.

Federal Register (1982). Drug products for over-the-counter human use for the treatment of toxic ingestion: Establishment of a monograph, *47* (No. 2), 444.

Federal Register (1985). Poison treatment drug products for over-the-counter human use: Tentative final monograph, *50* (No. 10), 2244.

Ferguson, H. C. (1962). Dilution of dose and acute oral toxicity, Toxicol. Appl. Pharmacol. *4*, 759.

Fine, J. S. and Goldfrank, L. R. (1992). Update in medical toxicology, Pediatr. Clin. North Am. *39*, 1031.

Flomenbaum, N. E., Goldfrank, L. R., Weisman, R. S., Howland, M. A., Lewin, N. A., and Kulberg, A. G. (1990). General management of the poisoned or overdosed patient, in *Toxicologic Emergencies*, 4th edition, L. R. Goldfrank, Ed., Appleton & Lange, Norwalk, CT.

Frenia, M. L., Schauben, J. S., Tucker, C., Wears, R., Karlix, J., and Kunisaki, T. (1993). Multiple dose activated charcoal compared to urinary alkalinization for the enhancement of phenobarbital elimination (abstract), Vet. Hum. Toxicol. *35*, 367.

Friday, K. J., Powell, S. H., Thompson, W. L., Groden, D. L., Sunshine, I., and Neumeyer, J. L. (1980). Fluid administration in induced emesis: Efficacy independent of ingested volume (abstract), Vet. Hum. Toxicol. *22*, 365.

Gaudreault, P., Lewander, W. J., Parent, M., Chicoine, L., and Lovejoy, F. H., Jr. (1984). Ipecac syrup: Lack of dose-response effect, Vet. Hum. Toxicol. *26* (Suppl. 2), 46.

Gelfand, M. C. and Winchester, J. F. (1980). Hemoperfusion in drug overdosage: A technique when conservative management is not sufficient, Clin. Toxicol. *17*, 583.

Ginoza, G. W., Strauss, A. A., Iskra, M. K., and Modanlou, H. D. (1987). Potential treatment of theophylline toxicity by high surface area activated charcoal, J. Pediatr. *111*, 140.

Goldberg, M. J., Spector, R., Park, G. D., and Roberts, R. J. (1986). An approach to the management of the poisoned patient, Arch. Intern. Med. *146*, 1381.

Goldfrank, L. R., Ed. (1986). *Toxicologic Emergencies*, 3rd edition, Appleton-Century-Crofts, Norwalk, CT.

Grande, G. and Ling, L. (1986). Fluid volume and ipecac (abstract), Vet. Hum. Toxicol. *28*, 493.

Grande, G. A. and Ling, L. J. (1987). The effect of fluid volume on syrup of ipecac emesis time, J. Toxicol.-Clin. Toxicol. *25*, 473.

Grbcich, P. A., Lacouture, P. G., Lewander, W. J., and Lovejoy, F. H. (1986a). Does milk delay the onset of ipecac induced emesis? (abstract), Vet. Hum. Toxicol. *28*, 499.

Grbcich, P. A., Lacouture, P. G., and Lovejoy, F. H. (1986b). The effect of fluid volume on ipecac induced emesis (abstract), Vet. Hum. Toxicol. *28*, 493.

Grbcich, P. A., Lacouture, P. G., Lewander, W. J., and Lovejoy, F. H., Jr. (1987). The effect of milk on ipecac-induced emesis (abstract), Vet. Hum. Toxicol. *29* (Suppl. 2), 30.

Greensher, J., Mofenson, H. C., and Caraccio, T. R. (1988). Ascendency of the black bottle (reply to letter), Pediatrics *82*, 522.

Haddad, L. M. and Winchester, J. M., Eds. (1990). *Clinical Management of Poisoning and Drug Overdose*, 2nd edition, W. B. Saunders Co., Philadelphia.

Haddad, L. M. and Roberts, J. R. (1990). A general approach to the emergency management of poisoning, in *Clinical Management of Poisoning and Drug Overdose*, 2nd edition, L. M. Haddad and J. F Winchester, Eds., W. B. Saunders Co., Philadelphia, pp. 2-21.

Hagstam, K. E., Larsson, L. E., and Thysell, H. (1966). Charcoal deposition in internal organs after hemoperfusion with the Yatzidis technique in rabbits, Proc. Eur. Dialysis Transpl. Assoc. *3*, 352.

Hall, A. H. (1991). Gastrointestinal decontamination: Sifting through supportive therapeutic options, Emerg. Med. Reports *12*, 171.

Harris, C. R. and Kingston, R. (1992). Gastrointestinal decontamination: Which method is best?, Postgrad. Med. *92*, 116.

Harrison, R. T. (1908). Case of ipecacuanha poisoning, Lancet *2*, 536.

Harstad, E., Møller, K. O., and Simesen, M. H. (1942). Über den Wert der Magenspülung bei der Behandlung von akuten Vergiftungen, Acta Med. Scand. *112*, 478.

Hart, L. G., Guarino, A. M., and Schanker, L. S. (1969). Gastric dialysis as a possible antidotal procedure for removal of absorbed drugs, J. Lab. Clin. Med. *73*, 853.

Heath, A. (1986). Pharmacokinetic evaluation of forced diuresis, dialysis, and hemoperfusion, in *New Concepts and Developments in Toxicology*, P. L. Chambers, P. Gehring, and F. Sakai, Eds., Elsevier, Amsterdam.

Henderson, M. L., Picchioni, A. L., and Chin, L. (1966). Evaluation of oral dilution as a first aid measure in poisoning, J. Pharm. Sci. *55*, 1311.

Henry, G. C., Wax, P. M., Howland, M. A., Hoffman, R. S., and Goldfrank, L. R. (1991). Exchange transfusion for the treatment of a theophylline overdose in a premature neonate (abstract), Vet. Hum. Toxicol. *33*, 354.

Jawary, D., Cameron, P. A., Dziukas, L., and McNeil, J. J. (1992). Drug overdose: Reducing the load, Med. J. Aust. *156*, 343.

Kärkkäinen, S. and Neuvonen, P. J. (1984). Effect of oral charcoal and urine pH on sotalol pharmacokinetics, Int. J. Clin. Pharmacol. Ther. Toxicol. *22*, 441.

Kärkkäinen, S. and Neuvonen, P. J. (1985). Effect of oral charcoal and urine pH on dextropropoxyphene pharmacokinetics, Int. J. Clin. Pharmacol. Ther. Toxicol. *23*, 219.

Kärkkäinen, S. and Neuvonen, P. J. (1986). Pharmacokinetics of amitriptyline influenced by oral charcoal and urine pH, Int. J. Clin. Pharmacol. Ther. Toxicol. *24*, 326.

Kaye, S. (1988). *Handbook of Emergency Toxicology: A Guide for the Identification, Diagnosis, and Treatment of Poisoning*, 5th ed., C. C. Thomas, Springfield, Illinois.

King, W. D. (1980). Syrup of ipecac: A drug review, Clin. Toxicol. *17*, 353.

Kingston, R., Carmine, E., Skoglund, R., and Hovda, L. (1993). Neutralization as a treatment for acid ingestion (abstract), Vet. Hum. Toxicol. *35*, 332.

Klein-Schwartz, W., Gorman, R. L., Oderda, G. M., Wedin, G. P., and Saggar, D. (1984). Ipecac use in the elderly: The unanswered question, Ann. Emerg. Med. *13*, 1152.

Klein-Schwartz, W., Litovitz, T., Oderda, G., Bailey, K., and Kuba, A. (1991). Fluid administration with ipecac: Milk versus clear fluids (abstract), Vet. Hum. Toxicol. *33*, 368.

Knepshield, J. H. and Winchester, J. F. (1982). Hemodialysis and hemoperfusion for drugs and poisons, Trans. Am. Soc. Artif. Intern. Organs *28*, 666.

Knight, K. M. and Doucet, H. J. (1987). Gastric rupture and death caused by ipecac syrup, South. Med. J. *80*, 786.

Krenzelok, E. P. and Dean, B. S. (1985a). Syrup of ipecac failures: A two year review of 4,306 cases (abstract), Vet. Hum. Toxicol. *28*, 317.

Krenzelok, E. P. and Dean, B. S. (1985b). Syrup of ipecac in children less than one year of age, J. Toxicol.-Clin. Toxicol. *23*, 171.

Krenzelok, E. P. (1987). Role of sorbitol in theophylline elimination (letter), Ann. Emerg. Med. *16*, 1409.

Krenzelok, E. P. and Dunmire, S. M. (1992). Acute poisoning emergencies: Resolving the gastric decontamination controversy, Postgrad. Med. *91*, 179.

Krenzelok, E. P. (1992). The contemporary management of poisoning emergencies, J. Pract. Nurs. *42*, 24.

Kulig, K., Bar-Or, D., Cantrill, S. V., Rosen, P., and Rumack, B. H. (1985). Management of acutely poisoned patients without gastric emptying, Ann. Emerg. Med. *14*, 562.

Kulig, K. (1986). Gastric emptying and activated charcoal (reply to letter). Ann. Emerg. Med. *15*, 225.

Kulig, K. (1992). Initial management of ingestions of toxic substances, New Engl. J. Med. *326*, 1677.

Kunkel, D. B. (1985). The toxic emergency: A critical look at gut decontamination, Emerg. Med. *17*, 179.

Lamminpää, A., Vilska, J., and Hoppu, K. (1993). Medical charcoal for a child's poisoning at home: Availability and success of administration in Finland, Hum. Exp. Toxicol. *12*, 29.

LaPierre, G., Algozzine, G., and Doering, P. L. (1981). Effect of magnesium citrate on the in vitro adsorption of aspirin by activated charcoal, Clin. Toxicol. *18*, 793.

Lindberg, M. C., Cunningham, A., and Lindberg, N. H. (1992). Acute phenobarbital intoxication, South. Med. J. *85*, 803.

Litovitz, T. L., Klein-Schwartz, W., Oderda, G. M., Matyunas, N. J., Wiley, S., and Gorman, R. L. (1985). Ipecac administration in children younger than 1 year of age, Pediatrics *76*, 761.

Litovitz, T. L. (1986). Emesis versus lavage for poisoning victims (editorial), Am. J. Emerg. Med. *4*, 294.

McDougal, C. B. and Maclean, M. A. (1981). Modifications in the technique of gastric lavage, Ann. Emerg. Med. *10*, 514.

MacGregor, D. F. (1988). Ascendency of the black bottle (letter), Pediatrics *82*, 521.

MacLean, W. C., Jr. (1973). A comparison of ipecac syrup and apomorphine in the immediate treatment of ingestion of poisons, J. Pediatr. *82*, 121.

Manno, B. R. and Manno, J. E. (1977). Toxicology of ipecac: A review, Clin. Toxicol. *10*, 221.

Manoguerra, A. S. and Krenzelok, E. P. (1978). Rapid emesis from high-dose ipecac syrup in adults and children intoxicated with antiemetics or other drugs, Am. J. Hosp. Pharm. *35*, 1360.

Matthew, H., MacIntosh, T. F., Tompsett, S. L., and Cameron, J. C. (1966). Gastric aspiration and lavage in acute poisoning, Br. Med. J., May 28, p. 1333.

Matthew, H. (1970). Gastric aspiration and lavage, Clin. Toxicol. *3*, 179.

Mayersohn, M., Perrier, D., and Picchioni, A. L. (1977). Evaluation of a charcoal-sorbitol mixture as an antidote for oral aspirin overdose, Clin. Toxicol. *11*, 561.

Meester, W. D. (1980). Emesis and lavage, Vet. Hum. Toxicol. *22*, 225.

Merigian, K. S., Woodard, M., Hedges, J. R., Roberts, J. R., Steubing, R., and Rashkin, M. C. (1990). Prospective evaluation of gastric emptying in the self-poisoned patient, Am. J. Emerg. Med. *8*, 479.

Miser, J. S. and Robertson, W. O. (1978). Ipecac poisoning, West. J. Med. *128*, 440.

Noji, E. K. and Kelen, G. D., Editors (1989). *Manual of Toxicologic Emergencies*, Year Book Medical Publishers, Chicago.

Olson, K. R. (1990). Is gut emptying all washed up? (editorial), Am. J. Emerg. Med. *8*, 560.

Perrone, J., Hoffman, R. S., and Goldfrank, L. R. (1994). Special considerations in gastrointestinal decontamination, Emerg. Med. Clinics North Am. *12*, 285.

Phillips, S., Gomez, H., and Brent, J. (1993). Pediatric gastrointestinal decontamination in acute toxin ingestion, J. Clin. Pharmacol. *33*, 497.

Picchioni, A. L., Chin, L., and Gillespie, T. (1982). Evaluation of activated charcoal-sorbitol suspension as an antidote, J. Toxicol.-Clin. Toxicol. *19*, 433.

Pietrusko, R. G. (1977). Use and abuse of laxatives, Am. J. Hosp. Pharm. *34*, 291.

Pollack, M. M., Dunbar, B. S., Holbrook, P. R., and Fields, A. I. (1981). Aspiration of activated charcoal and gastric contents, Ann. Emerg. Med. *10*, 528.

Pond, S. M. (1986). A review of the pharmacokinetics and efficacy of emesis, gastric lavage and single and repeated doses of charcoal in overdose patients, in *New Concepts and Developments in Toxicology*, P. L. Chambers, P. Gehring, and F. Sakai, Eds., Elsevier, Amsterdam.

Porter, R. S. and Baker, E. B. (1985). Drug clearance by diarrhea induction, Am. J. Emerg. Med. *3*, 182.

Rauber, A. (1978). The cardiac safety of ipecac used as an emetic, Vet. Hum. Toxicol. *20*, 166.

Rauber, A. P. and Maroncelli, R. D. (1982). Two studies of the duration of emesis induced by therapeutic doses of syrup of ipecac, Vet. Hum. Toxicol. *24* (Suppl.), 60.

Reigel, J. M. and Becker, C. E. (1981). Use of cathartics in toxic ingestions, Ann. Emerg. Med. *10*, 254.

Roberge, R. J. and Martin, T. G. (1992). Whole bowel irrigation in an acute oral lead intoxication, Am. J. Emerg. Med. *10*, 577.

Robertson, W. O. (1962). Syrup of ipecac: A fast or slow emetic?, Am. J. Dis. Child *103*, 58.

Robertson, W. O. (1979). Syrup of ipecac associated fatality: A case report, Vet. Hum. Toxicol. *21*, 87.

Robinson, K. J, Martin, T. G., Wolfson, A. B., Dean, B. S., and Krenzelok, E. P. (1991). Hemodialysis and hemoperfusion in overdose victims (abstract), Vet. Hum. Toxicol. *33*, 355.

Rodgers, G. C., Jr. and Matyunas, N. J. (1986). Gastrointestinal decontamination for acute poisoning, Pediatr. Clin. North Am. *33*, 261.

Rose, N. J. (1970). Report of accidental poisoning death from fluid extract of ipecac, Ill. Med. J. *137*, 338.

Rumack, B. H. (1976). Emesis, charcoal, and cathartics, JACEP *5*, 44.

Rumack, B. H. and Rosen, P. (1981). Emesis: Safe and effective? (editorial), Ann. Emerg. Med. *10*, 551.

Rumack, B. H and Lovejoy, F. H., Jr. (1991). Clinical toxicology, in *Casarett and Doull's Toxicology: The Basic Science of Poisons*, 4th edition, M. O. Amdur, J. Doull, and C. D. Klassen, Eds., Pergamon Press, New York, pp. 924-946.

Ryan, C. F., Spigiel, R. W., and Zeldes, G. (1980). Enhanced adsorptive capacity of activated charcoal in the presence of magnesium citrate, N.F., Clin. Toxicol. *17*, 457.

Saetta, J. P. and Quinton, D. N. (1991). Residual gastric content after gastric lavage and ipecacuanha-induced emesis in self-poisoned patients: An endoscopic study, J. Royal Soc. Med. *84*, 35.

Saetta, J. P., March, S., Gaunt, M. E., and Quinton, D. N. (1991). Gastric emptying procedures in the self-poisoned patient: Are we forcing gastric content beyond the pylorus?, J. Royal Soc. Med. *84*, 274.

Saincher, A., Sitar, D., and Tenenbein, M. (1993). Efficacy of ipecac during the first hour after drug ingestion (abstract), Vet. Hum. Toxicol. *35*, 325.

Scalzo, A. J., Tominack, R. L., and Thompson, M. W. (1989). Gastric lavage in pediatric poisoning patients: Frequent tube malpositioning demonstrated radiographically (abstract), Vet. Hum. Toxicol. *31*, 332.

Scharman, E. J., Lembersky, R., and Krenzelok, E. P. (1992). Efficiency of whole bowel irrigation with and without metoclopramide pretreatment (abstract), Vet. Hum. Toxicol. *34*, 361.

Schiff, R. J., Wurzel, C. L., Brunson, S. C., Kasloff, I., Nussbaum, M. P., and Frank, S. D. (1986). Death due to chronic syrup of ipecac use in a patient with bulemia, Pediatrics *78*, 412.

Schofferman, J. A. (1976). A clinical comparison of syrup of ipecac and apomorphine use in adults, JACEP *5*, 22.

Shannon, M., Fish, S. S., and Lovejoy, F. H., Jr. (1986). Cathartics and laxatives: Do they still have a place in management of the poisoned patient?, Med. Toxicol. *1*, 247.

Shannon, M. W., Wernovsky, G., and Morris, C. (1992). Exchange transfusion in the treatment of severe theophylline poisoning, Pediatrics *89*, 145.

Shannon, M. and Wernovsky, G. (1992). Treatment of theophylline poisoning (reply to letter), Pediatrics *90*, 781.

Smith, R. P. and Smith, D. M. (1961). Acute ipecac poisoning: Report of a fatal case and review of the literature, New Engl. J. Med. *265*, 523.

Smith, S. W., Ling, L. J., and Halstenson, C. E. (1991). Whole bowel irrigation as a treatment for acute lithium overdose, Ann. Emerg. Med. *20*, 536.

Smolinske, S. C., Wruk, K. M., Knapp, G. L., Krenzelok, E. P., Dean, B. S., and Rumack, B. H. (1987). Ipecac: A comparative, collaborative study on efficacy of 15 versus 30 mL doses in children (abstract), Vet. Hum. Toxicol. *29*, 491.

Spector, T. (1980). Charcoal-facilitated dialysis, Anal. Biochem. *103*, 313.

Speer, J. D., Robertson, W. O., and Schultz, L. R. (1963). Ipecacuanha poisoning: Another fatal case, Lancet *1*, 475.

Stewart, J. J. (1983). Effects of emetic and cathartic agents on the gastrointestinal tract and the treatment of toxic ingestion, J. Toxicol.-Clin. Toxicol. *20*, 199.

Strauss, A. A., Modanlou, H. D., and Komatsu, G. (1985). Theophylline toxicity in a preterm infant: Selected clinical aspects, Pediatr. Pharmacol. *5*, 209.

Strauss, A. and Modanlou, H. D. (1992). Treatment of theophylline poisoning (letter), Pediatrics *90*, 780.

Sue, Y., Shannon, M. W., and Woolf, A. D. (1991). Efficacy of magnesium citrate in pediatric ingestions (abstract), Vet. Hum. Toxicol. *33*, 352.

Tandberg, D., Liechty, E. J., and Fishbein, D. (1981). Mallory-Weiss syndrome: An unusual complication of ipecac-induced emesis, Ann. Emerg. Med. *10*, 521.

Tandberg, D., Diven, B. G., and McLeod, B. S. (1986). Ipecac-induced emesis versus gastric lavage: A controlled study in normal adults, Am. J. Emerg. Med. *4*, 205.

Tandberg, D. and Murphy, L. C. (1989). The knee-chest position does not improve the efficacy of ipecac-induced emesis, Am. J. Emerg. Med. *7*, 267.

Tenenbein, M. (1985a). Inefficacy of gastric emptying procedures, J. Emerg. Med. *3*, 133.

Tenenbein, M. (1985b). Whole bowel irrigation for toxic ingestions, J. Toxicol.-Clin. Toxicol. *23*, 177.

Tenenbein, M. (1987). Whole bowel irrigation in iron poisoning, J. Pediatr. *111*, 142.

Tenenbein, M., Cohen, S., and Sitar, D. A. (1987). Whole bowel irrigation as a decontamination procedure after acute drug overdose (abstract), Vet. Hum. Toxicol. *29* (Suppl. 2), 81.

Tenenbein, M. (1988). Whole bowel irrigation as a gastrointestinal decontamination procedure after acute poisoning, Med. Toxicol. *3*, 77.

Thoman, M. E. and Verhulst, H. L. (1966). Ipecac syrup in antiemetic ingestion, JAMA *196*, 147.

Thompson, A. M., Robbins, J. B., and Prescott, L. F. (1987). Changes in cardiorespiratory function during gastric lavage for drug overdose, Hum. Toxicol. *6*, 215.

Timberlake, G. A. (1984). Ipecac as a cause of the Mallory-Weiss syndrome, South. Med. J. *77*, 804.

Uden, D. L., Davidson, G. J., and Kohen, D. P. (1981). The effect of carbonated beverages on ipecac-induced emesis, Ann. Emerg. Med. *10*, 79.

Underhill, T. J., Greene, M. K., and Dove, A. F. (1990). A comparison of the efficacy of gastric lavage, ipecacuanha and activated charcoal in the emergency treatment of paracetamol overdose, Arch. Emerg. Med. *7*, 148.

Vale, J. A., Meredith, T. J., and Proudfoot, A. T. (1986). Syrup of ipecacuanha: Is it really useful?, Br. Med. J. *293*, 1321.

Vale, J. A. and Meredith, T. J. (1986). Antidotal therapy: Pharmacokinetic aspects, in *New Concepts and Developments in Toxicology*, P. L. Chambers, P. Gehring, and F. Sakai, Eds., Elsevier, Amsterdam.

Vale, J. A. (1992). [Primary decontamination: Vomiting, gastric irrigation, or only medicinal charcoal?] Ther. Umsch. *49*, 102.

Van de Graaff, W. B., Thompson, W. L., Sunshine, I., Fretthold, D., Leickly, F., and Dayton, H. (1982). Adsorbent and cathartic inhibition of enteral drug absorption, J. Pharmacol. Exp. Ther. *221*, 656.

Varipapa, R. J. and Oderda, G. M. (1977). Effect of milk on ipecac-induced emesis, J. Am. Pharm. Assoc. *17*, 510.

Vasquez, T. E., Evans, D. G., and Washburn, W. L. (1988). Efficacy of syrup of ipecac-induced emesis for emptying gastric contents, Clin. Nucl. Med. *13*, 638.

Voldeng, A. N. (1976). Management of the acutely poisoned patient. Part II. Alternatives to induced emesis, J. Arkansas Med. Soc. *72*, 368.

Watson, W. A., Leighton, J., Guy, J., Bergman, R., and Garriott, J. C. (1989). Recovery of cyclic antidepressants with gastric lavage, J. Emerg. Med. *7*, 373.

Wheeler-Usher, D. H., Wanke, L. A., and Bayer, M. J. (1986). Gastric emptying: Risk versus benefit in the treatment of acute poisoning, Med. Toxicol. *1*, 142.

Winchester, J. F., Gelfand, M. C., Knepshield, J. H., and Schreiner, G. E. (1977). Dialysis and hemoperfusion of poisons and drugs—update, Trans. Am. Soc. Artif. Intern. Organs *XXIII*, 762.

Wogan, J. M. (1989). Enhancement of elimination, in *Manual of Toxicologic Emergencies*, E. K. Noji and G. D. Kelen, Eds., Year Book Medical Publishers, Chicago, pp. 48-52.

Wolowodiuk, O. J., McMicken, D. B., and O'Brien, P. (1984). Pneumomediastinum and retro-pneumoperitoneum: An unusual complication of syrup-of-ipecac-induced emesis, Ann. Emerg. Med. *13*, 1148.

Yatzidis, H. (1964). A convenient haemoperfusion micro-apparatus over charcoal for the treatment of endogenous and exogenous intoxication, Proc. Eur. Dialysis Transpl. Assoc. *1*, 83.

8

The Design of Clinical Studies and Data Treatment

The majority of the clinical studies discussed in this book involve the statistical treatment of data, to determine whether one form of treatment (e.g., nothing, supportive therapy, gastric lavage, cathartics, charcoal, charcoal plus cathartics, etc.) is more efficacious (in some sense) than another treatment. Thus, it is important to know the essential elements of what constitutes a "controlled" clinical trial and how the data obtained are analyzed statistically for significance.

I. RANDOMIZED CROSS-OVER AND OTHER TYPES OF STUDIES

In a controlled clinical trial, a number of subjects (healthy humans, sick humans, laboratory animals, etc.) are selected by some rational means to receive one form of treatment or another. In terms of determining statistically significant differences between different treatments, it is best to make the number of subjects in each treatment group as large as practical. The subjects should be chosen with respect to important variables such as age, gender, body weight, severity of toxic symptoms, and so forth, in order that the groups are as statistically similar as possible. When a rational and systematic assignment of subjects to groups so that these variables are essentially identical is difficult for practical reasons, then a totally random assignment process may be acceptable (e.g., treating patients one way if they present on odd-numbered days and another way if they present on even-numbered days).

All treatment methods are generally compared against a "control" treatment which, in the case of treating a patient for drug or poison ingestion might be simple supportive therapy. Or, in many studies involving volunteers who ingest subtoxic amounts of drugs, the "control" situation is invariably no treatment at all. The best clinical studies, especially those involving subtoxic doses, are conducted such that each subject serves as his/her own control. That is, in one trial the subject receives no treatment and the variables to be measured (blood drug concentrations, urine drug concentrations, etc.) are determined. Then, after a "washout" period during which the drug essentially totally disappears from the person's system, the drug is given again, this time with a certain treatment (e.g., oral charcoal). Presumably, if a person is somehow different from another person in the way his/her system responds to one treatment or another, a comparison of what happens to that person with one treatment as compared to what happens to the same person without any treatment provides the best comparison of the effect of that particular treatment modality. When one then averages the differences (treatment versus no treatment) effects seen for all patients in the study, a clearly significant (or insignificant) effect should be apparent.

Obviously, all conditions in each trial, such as the amount and type of food eaten prior to the test, the elapsed time between food ingestion and the drug ingestion, etc., should be identical. In practice, the best way to avoid the effects of food type, amount, and time of ingestion is to have the subjects fast for a sufficient time prior to the start of each trial phase. Additionally, the various treatment protocols, including the "control" protocol, should be done in random order. Such a study is thus called a randomized cross-over study.

The different treatments tested in an overall study are often called *phases* (e.g., one might refer to a study as a three-phase study), *limbs* (e.g., one might call the study a three-limbed study), or *arms* (e.g., one might state that "In one treatment arm, only charcoal was administered."). The term *cross-over* simply means that each subject underwent one type of treatment and then (after a sufficient washout period) "crossed over" to another type of treatment (the number of cross-overs required by any single patient in a study would be one less than the number of treatment types).

A further means, besides randomization, of achieving impartiality is to perform the study in a *double-blind* manner. In such, neither the patient nor the researcher knows which treatment protocol is being administered. In a typical study of this sort, a capsule or similar dosage form is employed which always looks, feels, and tastes the same every time it is given. In one test, the capsule might contain nothing of significance (e.g., milk sugar), while in another test it might contain some powdered charcoal alone, and in a third case it might contain a cathartic alone. The dosage form, when it contains nothing of significance, is called a *placebo*. The idea behind double-blinding a study is to prevent the subjective feelings of both the volunteers and of the researchers from having any possible influence on the results. After all, in many studies, some or all of the variables measured are subjective ones, such as the presence/absence of feelings of nausea, bloating, dizziness, thirst, etc. Objective measures of treatment effects, such as blood concentrations, blood pressures, urine volumes, and the like, are apt to be far less influenced by the subjective feelings of the study participants. Nevertheless, some slight effects are possible.

In the cases of evaluating different methods for treating drug overdoses, double-blinding is often impossible. If the treatments involve swallowing a given volume of plain water, versus the same volume of a charcoal slurry, or the same volume of a charcoal/sorbitol mixture, it is clear that preventing at least the subject, and quite possibly the researcher, from knowing which is which is essentially impossible and impractical. Thus, most of the studies described in this book are not of the double-blind type. However, many—if not most—are of the randomized cross-over variety.

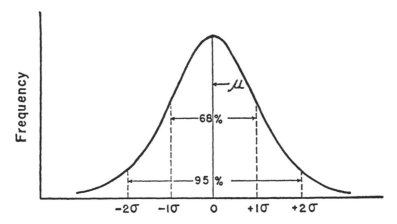

Figure 8.1 The normal probability curve, showing standard deviation units around the average. From *Remington's Pharmaceutical Sciences* (1975). Reproduced by permission of the Mack Publishing Company.

II. ASSESSMENT OF THE STATISTICAL SIGNIFICANCE OF DIFFERENCES IN TREATMENTS

When comparing objective measures of variables such as areas-under-curves (AUCs), peak blood drug levels, etc., one usually computes the mean values for all subjects in a certain treatment group (e.g., a group receiving a given amount of oral charcoal). The mean, or average, value is simply:

$$\bar{x} = \frac{x_1 + x_2 + \ldots + x_n}{n}$$

where n is the number of observations (usually the number of subjects). The standard deviation, s, of the x values is obtained by the formula

$$s = \sqrt{\frac{\sum (x_i - \bar{x})^2}{n - 1}}$$

The standard deviation is a measure of the variation of individual observed values around the average. The quantity $(n - 1)$ in the denominator is called the degrees of freedom (DF), and is one less than the number of observed values. If the sample size (i.e., number of subjects) were infinite, we would speak of the standard deviation of the infinite population as being σ. Thus, for a finite sample, s is merely an estimate of σ. The square of the standard deviation is called the variance.

The so-called normal probability curve is shown in Figure 8.1. This is a plot of the frequency of occurrence of values of a measured variable (e.g., systolic blood pressure). There are several kinds of probability distributions in nature and only one is of the kind shown (the "normal" or Gaussian probability distribution). Other kinds of distributions are the Poisson, binomial, Chi-square, and t distributions. Most of the data treated by

Table 8.1 The t Table: Distribution of t, Giving the Two-Tailed Probability According to the Degrees of Freedom (DF)

DF	p 0.60	p 0.40	p 0.20	p 0.10	p 0.05	p 0.02	p 0.01
1	0.727	1.376	3.078	6.314	12.706	31.821	63.657
5	0.559	0.920	1.476	2.015	2.571	3.365	4.032
10	0.542	0.879	1.372	1.812	2.228	2.764	3.169
15	0.536	0.866	1.341	1.753	2.131	2.602	2.947
20	0.533	0.860	1.325	1.725	2.086	2.528	2.845
30	0.530	0.854	1.310	1.697	2.042	2.457	2.750
60	0.527	0.848	1.296	1.671	2.000	2.390	2.660
120	0.526	0.845	1.289	1.658	1.980	2.358	2.617
∞	0.524	0.842	1.282	1.645	1.960	2.326	2.576

statistical means in the studies described in this book assume that the normal probability distribution applies.

Of the three types of tests of significance which are most common (the t test, the Chi-square test, and the so-called F test or analysis of variance), we will discuss here only the t test. It makes use of the t probability distribution, which depends on the sample size (number of observations) and which approaches the normal distribution as the sample size becomes large (the normal distribution is strictly correct only for a population that is infinite in size).

Table 8.1 shows values for the "two-tailed" t distribution (the table given here is incomplete and only selected entries are included to give the reader a feel for the magnitudes of t values; any standard statistics reference should be consulted for a complete table). The table shows the multiple of the standard deviation on either side of the average which would include any desired proportion of the observations (e.g., 99%, 95%, 90%, 80%, etc.). Clearly, the values of the multiples depend on the sample size n, since the values vary with DF and since DF $= n - 1$. This table says, for example, that for an infinite sample size, the probability of a value being outside of $\bar{x} - 2.326\sigma$ to $\bar{x} + 2.326\sigma$ is 2% (i.e., 1 chance out of 50), or more specifically 1 chance in 100 that it is less than $\bar{x} - \sigma$ and 1 chance in 100 that it is greater than $\bar{x} + \sigma$.

When comparing results from two different trials, we begin by computing a t value from the formula

$$t = \left(\frac{\bar{x}_1 - \bar{x}_2}{s} \right) \sqrt{ \frac{n_1 n_2}{n_1 + n_2} }$$

where $\bar{x}_1$ is the mean of the first group of n_1 observed values, and $\bar{x}_2$ is the mean of the second group of n_2 observed values. We also compute the "pooled variance" s^2 from the equation

$$s^2 = \frac{\sum x_{1i}^2 - \left(\sum x_{1i} \right)^2 \Big/ n_1 + \sum x_{2i}^2 - \left(\sum x_{2i} \right)^2 \Big/ n_2}{n_1 + n_2 - 1}$$

where

$\sum x_{1i}^2$ is the sum of the squares of the observed values in the first group

$\sum x_{1i}$ is the sum of the observed values in the first group

$\sum x_{2i}^2$ is the sum of the squares of the observed values in the second group

$\sum x_{2i}$ is the sum of the observed values in the second group

From this we take the square root of s^2 to get s. Then, based on the degrees of freedom,

$$DF = (n_1 - 1) + (n_2 - 1) = n_1 + n_2 - 2$$

and a selection of the level of probability we wish to use (e.g., $p = 0.05$), we go to Table 8.1 and find the tabulated t value. If the computed t value is greater than this, we may conclude, with the probability $p < 0.05$, that the means of the two groups are statistically different. That is, the treatment protocols which led to the two groups of data produced significantly different effects.

An example may clarify this. Let us assume that one group of seven subjects (Group 1) was given plain water after drug ingestion, and a second group of 10 subjects (Group 2) was given a certain dose of charcoal in water. Let us assume their peak blood drug concentrations were as follows:

Group 1: 39, 46, 41, 50, 35, 47, and 55 mg/L
Group 2: 34, 43, 39, 30, 36, 46, 27, 40, 32, and 41 mg/L

Figure 8.2 shows in visual form the values for the two groups and it is apparent that there is a substantial degree of "overlap" of the results. Thus, it is not clear, without calculating a t value, whether there is any significant difference between the two groups and, if so, at what level of probability significance occurs.

The means for the two groups are 44.71 and 36.70, respectively. The sum of squares of values in the first group, Σx_1^2, is 14,277 and that for the second group, Σx_2^2, is 13,872. The sum of the values for Group 1, Σx_1, is 313 and that for Group 2, Σx_2, is 368. The squares of these sums, $(\Sigma x_1)^2$ and $(\Sigma x_2)^2$, are 97,969 and 135,424, respectively). Thus, s^2, the pooled variance for the two groups, is

$$s^2 = \frac{[14,277 - (97,969/7) + 13,872 - (135,424/10)]}{(7 + 10 - 2)}$$

$$= 40.735$$

and $s = 6.382$. The value of t is thus

$$t = \frac{44.71 - 36.70}{6.382} \sqrt{\frac{(7)(10)}{7 + 10}} = 2.516$$

The degrees of freedom in the study are $7 + 10 - 2 = 15$. Looking at Table 8.1 for DF = 15 and $p = 0.05$, we see that $t = 2.131$. For $p = 0.02$ the table t value is 2.602. Since our calculated t value is greater than that corresponding to $p = 0.05$ and less than that corresponding to $p = 0.02$, this means that the observed difference between the two

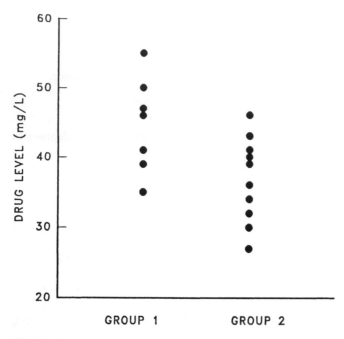

Figure 8.2 Example data on drug levels in two groups of patients.

groups would occur by chance less than one time out of 20 trials but more than one time in 50 trials. Thus, the observed difference is "significant" at the $p < 0.05$ level and "not significant" at the $p < 0.02$ level. Normally, for "significance" we desire the observed difference to be significant at $p < 0.05$ at most (i.e., not significant at a higher p value such as $p < 0.10$).

9

Some Basic Aspects of Antidotal Charcoal

Before beginning a discussion of the in vitro and in vivo effects of charcoal on specific drugs or poisons, we consider some of the more general aspects of antidotal charcoal. We compare the efficacies of charcoal given as a powder or in tablet form and consider what amount of charcoal should be given (i.e., the "optimal dose"). Then we discuss the issue of whether a drug or poison, once bound to charcoal, can desorb as the charcoal passes through the gastrointestinal tract. The next topic to be discussed is whether charcoal can lose any of its adsorption capabilities during storage in closed containers. Finally, we discuss the formulation called "universal antidote" (charcoal, tannic acid, and magnesium oxide in the ratio 2:1:1) and will show why it is inferior to plain charcoal.

I. CHARCOAL POWDER VERSUS TABLETED CHARCOAL

The great adsorption capacity of activated charcoal is primarily due to the very large internal surface area associated with the numerous tiny pores which develop in the charcoal during activation. Therefore, it is clear that for charcoal to be effective in adsorbing drugs or poisons, one should maximize the ease of access of the drug or poison molecules to the internal surface. To adsorb, a drug or poison molecule must reach the external surface of the charcoal, diffuse inside, and then travel by molecular diffusion through the pore network until a vacant adsorption "site" is found. Molecular diffusion is an extremely slow process.

By dispersing a given weight of charcoal as fine particles (as opposed to larger granules), two advantages are gained: (1) a much larger external surface area is generated, and (2) the length of the diffusion paths inside the charcoal are reduced. The net effect is that, in general, the rate of uptake of a drug or poison by charcoal will be inversely proportional to the particle size squared. Thus 325-mesh particles (0.043 mm diameter) compared to, for example, 20-mesh particles (0.833 mm diameter) will adsorb a drug

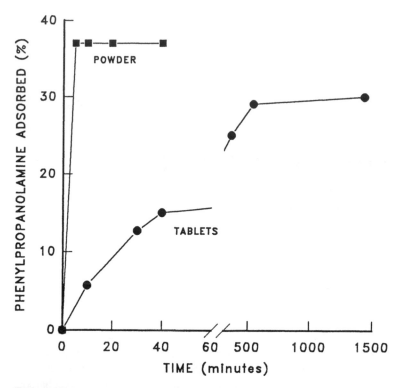

Figure 9.1 Rate of adsorption of phenylpropanolamine (125 mg in 50 mL of 0.1 N HCl) by 0.5 g powdered activated charcoal and by 1.0 g tableted activated charcoal (as three tablets) at 37°C. From Tsuchiya and Levy (1972a). Reproduced by permission of the American Pharmaceutical Association.

$(0.833/0.043)^2$ or 375 times faster. In practice, if the fluid medium is not well agitated, the difference will not be quite this large but it will still be very appreciable. Thus, one should use powdered charcoal, indeed a finely powdered grade, for best results.

Tsuchiya and Levy (1972a) have compared the in vitro and in vivo adsorption rates of a powdered charcoal and a commercially available (at that time, but perhaps no longer) tableted charcoal. The tablets, which weighed 0.44 g each, contained 0.33 g charcoal. The difference (0.11 g) represents added ingredients required to produce the tablet form. Using phenylpropanolamine as a test drug, Tsuchiya and Levy found a very great difference in adsorption rates for the powder and tablet forms, as shown in Figure 9.1. In vivo studies with human volunteers demonstrated that equal doses of charcoal powder and tablets reduced phenylpropanolamine absorption by 73 and 48%, respectively. Actually, the tablets, though definitely less effective, did surprisingly well considering their long disintegration times. A second study with aspirin showed 38 and 15% reductions in absorption with the use of powdered and tableted charcoal, respectively.

A similar study was performed by Otto and Stenberg (1973) in which the in vitro and in vivo effectiveness of powdered charcoal was compared to coated charcoal tablets which were commercially available in Sweden. The test drug was salicylamide. Their in vitro results were very similar to the ones shown in Figure 9.1. In vivo tests with

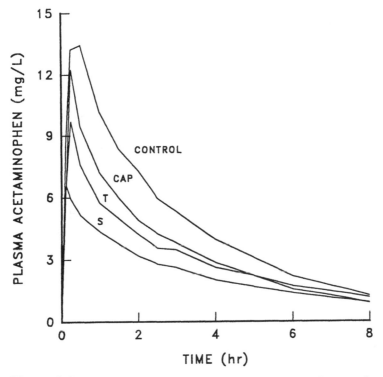

Figure 9.2 Mean acetaminophen concentrations in plasma for control tests and after treatment with activated charcoal in the form of a suspension (S), capsules (CAP), and tablets (T). From Remmert et al. (1990). Figure 1. Copyright Springer-Verlag GmbH & Co. Used with permission.

humans found that for equal doses of each charcoal form, the coated tablets and powdered charcoals reduced absorption by 3 and 31%, respectively, at 6 hr, and by 0 and 15%, respectively, over a 48 hr period.

The comparative antidotal efficacy of activated charcoal tablets, capsules, and suspension in healthy volunteers has been examined by Remmert et al. (1990). A solution of 1 g paracetamol (acetaminophen) was given orally, followed 2 min later by 5 g charcoal in 200 mL water, 5 g charcoal as tablets (forty 125-mg tablets), or 5 g charcoal in capsules (twenty-five 200-mg capsules). Neither the tablets nor capsules were chewed. Blood samples were taken over 7.5 to 480 min and plasma AUC values were determined. Their results are shown in Figure 9.2. The 0–8 hr AUC values in mg-hr/L were: control (plain water) 39.9; AC suspension, 19.5; AC tablets, 25.5; and AC capsules, 28.7. The peak plasma drug concentrations in mg/L were, in the same order, 15.3, 6.8, 10.5, and 13.5. Thus, charcoal in all forms reduced paracetamol absorption; however, the suspension was the most effective, as one would expect. The tablets were more effective than the capsules, which might not have been anticipated. However, tablet disintegration depends strongly on tableting pressure and other factors. It is clear from this study that charcoal in suspension is to be greatly preferred over charcoal in tablet or capsule form, despite some measure of convenience in handling the latter two forms.

These studies make it very clear that any tableted or granular form of charcoal would be much less efficacious than finely powdered charcoal and that coated tablets, in particular, should not be used.

II. OPTIMAL DOSE OF ACTIVATED CHARCOAL

Various bases have been recommended for determining the "optimal" dose of charcoal to administer. Two common and obvious bases are: (1) the patient's weight, and (2) the weight of drug ingested. The obvious problem in using the weight of drug as a basis (one of the more common such recommendations is to use a weight of charcoal 10 times the drug weight ingested) is that the amount of drug ingested is usually unknown. Additionally, it may not even be known what kind of drug was taken. Moreover, drugs vary enormously in their toxicities, rates of absorption, and the specific effects which overdoses produce (e.g., depression of respiration, convulsions, etc.). Some drugs adsorb very well to charcoal, whereas others do not. The idea of a fixed charcoal:drug ratio being appropriate for all drugs is clearly flawed.

Basing the charcoal dose on the patient's weight (e.g., 1 g charcoal per kg body weight, as has been suggested) may lead to doses which are too small in children. Also, measuring out a 1 g/kg dose is likely to be a messy affair.

Many authors have recommended a fixed amount of charcoal, unrelated to patient weight or the amount of drug ingested. Dordoni et al. (1973) recommend 50 g charcoal, as do Lawrence and McGrew (1975), as well as an editorial in the *Drug and Therapeutics Bulletin* (Anonymous, 1974). Levy and Gwilt (1972) suggested a figure of 30–50 g and Levy and Houston (1976) later recommended 50–100 g. Comstock (1975) suggested that no less than 100 g be administered. Hayden and Comstock (1975), noting that textbooks that physicians use for reference often recommend a standard dose of 10 g charcoal, stated that "in reality the dose of activated charcoal that should be administered to adsorb drugs from the gastrointestinal tract is in the range of 100 to 120 g." DiPalma (1979) has stated that "from a practical point of view ... the minimum dose is 30 g and the maximum dose is 120 g, which is the greatest amount that is feasible to administer."

As an example of recommendations in a standard reference, Goodman and Gilman's *The Pharmacological Basis of Therapeutics*, the 4th edition (1975) recommends a charcoal dose of 1–8 g, the 5th edition (1975) recommends 10 g, the 6th edition (1980) recommends preparing 50 g charcoal in 400 mL water and then administering this in the amount of 5 mL/kg (hence, 50 g for an 80 kg person), and the 7th edition (1985) simply recommends that the entire 50 g/400 mL slurry be given. *Remington's Pharmaceutical Sciences*, up through the 12th edition (1961) recommended 1–8 g charcoal; the 13th–15th editions (1965–1975) stated a dose range of 5–50 g, with a "usual" dose of 10 g; the 16th edition (1980) and later still has the dose range as 5–50 g but the usual dose is stated as "adults 50 g, children 25 g." Thus, both of these references show a gradual increase from 1–8 to 10 to 50 g (for adults).

Corby and Decker (1974) pointed out, however, that since charcoal is harmless, the only limiting factor is the quantity the individual is willing to accept; accordingly, the optimal dose is the maximum that can be given practically. Nevertheless, it would seem that 100 g charcoal would be sufficient for nearly all cases. For if a 10:1 ratio of charcoal to drug is really enough, then 100 g charcoal would counteract 10 g drug. This corresponds to one hundred 100 mg barbiturate capsules, for example. Palatnick and Tenen-

Table 9.1 Effect of Various Charcoal-to-Drug Ratios

| Drug | Percent reduction in tissue concentrations relative to controls (various ratios) | | | |
	1:1	2:1	4:1	8:1
Sodium pentobarbital	7.0	38.0	62.0	89.0
Chloroquine phosphate	20.0	30.0	70.0	96.0
Isoniazid	1.2	7.2	35.0	80.0

Adapted from Chin et al. (1973). Reprinted by permission of Academic Press, Inc.

bein (1992) have reviewed the optimal dose question and recommend an initial charcoal dose which is as high as the patient is likely to tolerate. Such doses, as shown by experience (they say), are roughly 25–50 g for children less than 5 years old, and 50–100 g for older children and adults.

With this preamble, we now consider a few studies which have attempted to specifically resolve the optimal dose question. Chin et al. (1973) did in vivo studies in which three drugs—sodium pentobarbital, chloroquine phosphate, and isoniazid—were given to rats, followed by charcoal in various amounts. The charcoal doses were one, two, four or eight times the weight of drug employed. Table 9.1 shows representative tissue concentrations relative to control animals. What this study suggests is a rather obvious fact: the more charcoal, the better. The data do suggest that to be highly effective, an 8:1 ratio of charcoal to drug should be used. However, many investigations which have been carried out on fasted animals using a 5:1 ratio also have shown charcoal to be quite effective.

On the other hand, other studies have shown that too little charcoal is ineffective. For example, in a study by Levy and Tsuchiya (1969) it was shown that a charcoal:drug ratio of 1.9:1.0 had little in vivo effect on aspirin absorption in fasted human subjects. A ratio of 1.7:1.0 was totally ineffective in another study (Chin et al., 1970) in reducing glutethimide absorption in fasted dogs. It would thus appear that a 5:1 ratio is roughly a minimal figure. With this in mind, many investigators have recommended that a level twice as high (a 10:1 ratio) would be a safer minimum.

Olkkola (1985) examined the effect of the charcoal:drug ratio in some in vivo studies with adult humans using p-aminosalicylic acid (PAS). The PAS was given in doses of 1, 5, 10, or 20 g. Then, 50 g activated charcoal was given in 300 mL water immediately after. Thus, the charcoal:drug ratios were 50:1, 10:1, 5:1, and 2.5:1. For the 50:1 case, less than 5% of the PAS was absorbed into the body from the GI tract, whereas at the 2.5:1 ratio, 37% of the drug was absorbed from the GI tract. In vitro studies of the adsorption capacity of charcoal for PAS correlated well to the in vivo results. Olkkola concluded that large doses (50–100 g) of activated charcoal should be used whenever possible.

Neuvonen and Olkkola (1984) brought up the important point that the direct application of in vitro data to estimate appropriate oral doses of activated charcoal has led to a wide variety of recommendations. Therefore, Neuvonen and Olkkola tested the efficacy of activated charcoal in adult human subjects using three different drugs (disopyramide, indomethacin, and trimethoprim) at several drug:charcoal ratios. The amounts of charcoal used (given 5 min after drug ingestion) were 2.5, 10, 25, or 50 g Norit A charcoal in 300 mL water, or 10 g of Amoco PX-21 "superactive" charcoal in 300 mL water. Increasing the charcoal dose from 2.5 to 50 g decreased the absorption of disopyramide and indomethacin from values of about 30–40% to values of about 3–5%. For trimethoprim, absorption fell from 10% to 1%. In vivo studies suggested that the charcoal's adsorption capacity was saturated (i.e., all used up) at charcoal:drug ratios less than about 7.5:1. Therefore, the conclusion reached from both the in vivo and in vitro studies was that the best charcoal dose is one that is "as large as possible."

To summarize the foregoing review of the optimal charcoal dose, it is important to point out that recommendations that the dose be "as large as possible" are not intended to suggest that doses of several hundred grams be given. Since the potential for vomiting and subsequent charcoal aspiration increases with larger doses, in general no more than 100 g should be given to adults, and correspondingly less to children.

Since the product information in Chapter 4 indicates that commercial pre-packaged charcoal slurries are available in bottles containing 12.5, 15, 25, 30, and 50 g charcoal without sorbitol, and 25, 30, and 50 g charcoal with sorbitol, convenience would suggest that these amounts, or multiples thereof, be used. Thus, one could easily use two 50 g bottles of charcoal, or 100 g total, for an adult. For children, either a 25-g bottle or a 50-g bottle could be used, depending on the size of the child (for very young children, the 12.5-g or 15-g sizes might be appropriate). There are no "hard and fast" rules, and the choice of how much charcoal to give should be tempered by whatever knowledge can be gained as to the type and amount of drug or poison involved. However, as we will discuss in a later chapter, the use of formulations containing sorbitol should be approached with caution. Many physicians now recommend that sorbitol, if given, be limited to the first charcoal dose. However, in evaluating this recommendation, it should be kept in mind that available charcoal/sorbitol formulations vary widely in their sorbitol contents (from 225 to 700 mg/mL, as cited in Chapter 4); thus, the use of even three bottles of a 225 mg/mL sorbitol formulation would deliver less sorbitol than one bottle of a 700 mg/mL sorbitol formulation (assuming equal bottle sizes, of course).

III. STABILITY OF THE DRUG/CHARCOAL COMPLEX

Some concern has been expressed about whether a drug or poison adsorbed to charcoal in the stomach may subsequently be desorbed from the charcoal further along the gastrointestinal tract. The possibility does indeed exist, since the equilibrium between the charcoal and the drug is intrinsically reversible, that is, it can be represented symbolically as

Free drug + charcoal $\rightleftarrows$ drug/charcoal complex

When this system is in equilibrium this simply means that the forward rate (adsorption) equals the reverse rate (desorption), and that the concentration of drug on the charcoal, Q, is in equilibrium with the concentration of free drug in solution, C_f.

The equilibrium relationship (the "isotherm") relating Q and C_f is usually determined by batch type adsorption experiments, in which different amounts of charcoal, W, are added to batches of well-stirred drug solutions of initial concentration C_o, sufficient time is allowed for equilibrium to occur, the residual free drug concentrations C_f are determined, and the Q values are calculated from the equation

$$Q = \frac{V(C_o - C_f)}{W}$$

as explained in Chapter 3. But the key question is whether the equilibrium relationship determined in such adsorption experiments is the same as one would get if one were to approach the equilibrium from the opposite direction, i.e., start with essentially unwetted charcoal to which the drug had previously been adsorbed and contact it with drug-free solution, thereby letting desorption occur until final Q and C_f values are obtained. In other words, would pairs of Q and C_f values obtained in desorption tests fall on the same isotherm line obtained by plotting Q and C_f values obtained in adsorption tests? If such pairs of Q and C_f values obtained both ways fall on the same line, then the adsorption process can be said to be "fully reversible."

To answer this question, Cooney (1992) did a study in which adsorption isotherms for sodium salicylate adsorbing from aqueous buffer solutions onto charcoal were determined at room temperature (ca. 21°C) at pH 1.2, 7.4, and 12.6. True desorption isotherms were determined by separating the residual solution from the charcoal/drug, discarding the residual solution, and adding fresh drug-free buffer of the same pH to the test container. Desorption was allowed to occur until a new equilibrium condition was attained and the concentration of desorbed salicylate in the solution was then determined. Mass balance calculations gave Q and C_f values for the desorption tests. The determination of true desorption isotherms in this way, by carrying out actual desorption tests, is fairly rare. The results (Figure 9.3) showed that the adsorption isotherms and desorption isotherms at each pH level were essentially identical, proving that, for sodium salicylate, the adsorption/desorption process is 100% reversible.

However, adsorption is not always fully reversible. Several studies done by chemical engineers and chemists have proved that substantial irreversibility can occur, if the adsorbed species has sufficient time to react (the reaction being catalyzed by the charcoal) to form higher molecular weight derivatives which adsorb more strongly than the original species. Cooney and Xi (1994) presented data which show that many aromatic chemicals undergo such reactions after adsorption. In general, such reactions (most of which are "oxidative coupling" reactions) are quite slow and would occur to an insignificant extent in the time frame involved in antidotal charcoal applications. Thus, it is the writer's belief that, for most drugs, the adsorption/desorption process is fully reversible and that there are other controlling factors which determine the actual amount of drug desorption.

Let us assume, for purposes of example, that a certain drug is not absorbed from the stomach and that an equilibrium state exists in the stomach between the free drug in the gastric fluid and the drug bound up in the drug/charcoal complex. As the stomach contents pass into the intestines, some of the free drug will be absorbed into the rest of the body. This lowers the free drug concentration in the intestinal fluid and, by the principles of chemical equilibrium, will cause a shift in the equation given above to the left until a new equilibrium situation is established. In principle, if free drug continues to be absorbed from the intestinal tract, there would be no limit to the desorption process.

The nature of the equilibrium for most systems of charcoal plus drug is as shown in Figure 3.14, which is an expression (called the adsorption "isotherm") of all possible

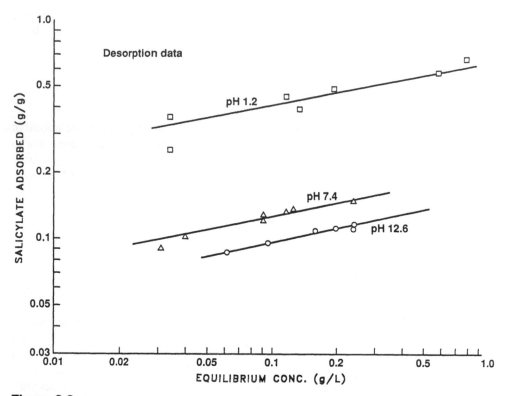

Figure 9.3 Desorption data for sodium salicylate at low, medium, and high pH levels. Data points are actual desorption data; the lines are the "best fit" lines to adsorption data at the same three pH values. From Cooney, 1992.

equilibrium states at a given temperature (other important factors, such as pH, must also be constant). Thus, for any given concentration of free drug, C_f, there is only one possible value for the amount of adsorbed drug, Q, if true equilibrium exists. Note that, as C_f falls, Q also decreases (i.e., drug absorption will cause desorption of drug from the drug/charcoal complex). However, the isotherm is typically convex in shape, as depicted in the figure. This means that, in general, the free drug concentration, C_f, would have to fall to a rather low value before much change in Q (the amount adsorbed) would occur. Another way of saying this is that the equilibrium state is strongly biased in favor of adsorption.

Another factor relates to the amount of charcoal used. If considerable charcoal is employed, then at equilibrium the residual free drug concentration, C_f, in the fluid phase will be very small. When the concentration of free drug in the GI tract fluid is very small, the rate of drug absorption from the GI tract will be very small. This, in turn, means that the rate of desorption of the drug from the drug/charcoal complex will be very low.

These two effects (the isotherm curvature in the direction favoring adsorption and low C_f values created by using sufficient charcoal) together can explain why most in vivo studies have concluded that very little desorption occurs. Such conclusions are usually drawn from curves of plasma drug concentrations or urinary drug excretion

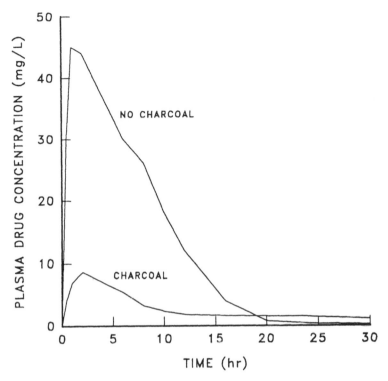

Figure 9.4 Plasma drug level curves for control and charcoal-treated subjects which show a cross-over at later times, indicating desorption of drug from the charcoal/drug complex.

amounts versus time for the cases of a drug given without charcoal and a drug given with charcoal. If the curves for the latter case always lie below the curves for the former case, then this is proof that some of the drug must remain permanently bound to the charcoal. Several figures presented in this book demonstrate this.

However, any in vivo study is always terminated at some point in time (e.g., 12, 24, 48 hr, etc.). Therefore, the possibility always exists that, had the investigators extended the study to longer time periods, ultimately the curve for "drug with charcoal" would cross over to a position above the curve for "drug without charcoal." This would signify that substantial drug desorption did occur, albeit slowly. Figure 9.4 is an example of such a cross-over. The figure is based loosely on data from an actual study but does not duplicate the data of that study.

Another factor not yet mentioned here is the effect of pH on the isotherm of a drug. As previously discussed (see Chapter 5), the increase in pH that occurs in passing down the gastrointestinal tract will tend to decrease the adsorption of an acidic drug and increase the adsorption of a basic drug. This occurs because pH affects the extent of ionization of many drugs and because nonionized species adsorb better than ionized ones. One would therefore expect the desorption of acidic drugs like barbiturates and salicylates to be enhanced as the drug/charcoal complex passes down the gastrointestinal tract.

A study which clearly shows the effect of pH on adsorption has been done by Tsuchiya and Levy (1972b). In vitro adsorption isotherms for aspirin, phenylpropanolamine, and salicylamide adsorbing to charcoal at 37°C at both pH 1 and pH 8.2 were determined. The aim was to assess how much change in the amount of drug adsorbed to the charcoal would occur for these drugs when the charcoal/drug complex passed from the stomach (pH 1) into the small intestine (pH 8.2). The results showed that one could expect significant desorption of the acidic drug (aspirin), some increase in adsorption of the basic drug (phenylpropanolamine), and a slight decrease for salicylamide, which is largely neutral over the pH 1–8.2 range (salicylamide has an acid group and a basic group, and when these two groups are both ionized, their opposite charges largely cancel each other).

Despite this potential for desorption, in practice, if sufficient charcoal is used (say, a 5:1 or greater ratio of charcoal to drug) the amount of desorption will, in any case, be small because an extremely low value of C_f in the fluid phase will exist, and this will make the rate of drug absorption very low. This is the reason why many investigators, noting that their plasma drug concentration versus time curves for "drug with charcoal" lay underneath their curves for "drug without charcoal" over the time periods of their studies, have concluded that significant desorption does not occur. Thus, Chin et al. (1970), in studying the in vivo effects of activated charcoal on the absorption of seven different drugs in dogs and rats, using charcoal:drug ratios of 5:1 and time periods up to 72 hr, noted that: "Semilogarithmic plots of the descending portions of the drug concentration curves ... indicate that the drug half-lives are nearly the same in test and control animals. Hence dissociation of the charcoal-drug complex in the digestive tract, if it occurs, appears to be inconsequential." Decker et al. (1969), after studying the effects of charcoal on aspirin absorption in human volunteers with a charcoal:drug ratio of 10:1 over a time period of 24 hr, similarly concluded that "the aspirin is not released from this complex [the charcoal/drug complex] to a clinically significant extent since serum salicylate concentrations continued to decrease in subjects receiving activated charcoal." Decker and Corby (1970), in a brief review of their various charcoal studies, reiterated the same view, stating that "it was evident in these studies that the activated charcoal-drug complexes were essentially stable throughout the gastrointestinal tract since no secondary serum peaks appeared." Picchioni (1970) stated that "experimental evidence [of his] suggests that the charcoal-poison complex remains stable throughout the gastrointestinal tract." As support for this view, he refers to two figures in his paper which show liver chloroquine levels in rats for 0–64 hr and blood pentobarbital levels in rats for 0–320 min, with charcoal-treated rat and control rat results in each case. The charcoal:drug ratio was 10:1 for both drugs.

In contrast to this, experimental evidence which suggests that substantial desorption can occur has been obtained in some studies. Levy and Tsuchiya (1972) report that aspirin adsorption to charcoal is definitely partly reversible in vivo. When aspirin was equilibrated in water with sufficient charcoal to give 50 or 99% adsorption under pH 1 and 37°C conditions (as indicated by the aspirin adsorption isotherm they determined) and these mixtures were then administered orally to nine healthy volunteers, 87 and 61%, respectively, of the total aspirin dose was recovered in the urine over 0–50 hr. They state: "... the in vivo desorption of aspirin is most likely due to the higher pH of the intestinal fluids and to competitive effects of constituents of the gastrointestinal fluids." Exactly the same types of tests were done by Tsuchiya and Levy (1972b) with phenylpropanolamine and with salicylamide. With phenylpropanolamine, the 50% preadsorbed and 99% preadsorbed charcoal/drug mixtures gave 53 and 7% recoveries of drug in the urine, respectively. With salicylamide, the recoveries in the urine were 78 and 23%, respectively. These results are as expected. For example, phenylpropanolamine is a basic drug and under intestinal pH conditions it is less ionized. Thus it adsorbs to the charcoal more strongly and in turn it is less available for absorption. Thus, the recoveries in the urine for phenylpropanolamine should be less than for aspirin and indeed they were. Salicylamide has essentially no net charge over a wide range of pH values and thus should—and did—show behavior intermediate between that of aspirin and of phenylpropanolamine.

Another in vivo study with aspirin is that of Neuvonen et al. (1978). Six adults were given 1 g aspirin followed by 50 g Norit A charcoal. When the charcoal was given immediately, the peak serum aspirin levels were reduced by 95% relative to control values. However, the total systemic absorption decreased less (by 70%). Serum concentration data taken for 0–96 hr showed higher levels during the period of 24–96 hr when charcoal was used than when it was not. This is strong evidence that some desorption occurred as the drug/charcoal complex passed down the gastrointestinal tract. Nevertheless, the serum levels were still, in an absolute sense, rather small after 24 hr in all instances.

Filippone et al. (1987) did a study with eight volunteers which was aimed specifically at determining whether aspirin pre-bound to charcoal desorbs in vivo and is then absorbed from the GI tract. Control subjects were given 1 g crushed aspirin which had been mixed with 50 mL water 15 min before. Test subjects were given a slurry of 1 g crushed aspirin plus 10 g charcoal in 50 mL water which also had been prepared 15 min earlier. This assured that the aspirin had an opportunity to become pre-bound to the charcoal. However, no assays were done on the mixture of aspirin, charcoal, and water to determine how much aspirin did indeed "pre-bind" to the charcoal. Blood samples were drawn for up to 30 hr and the 0–30 hr AUC values were found to be 476 and 89 mg-hr/L for the control and charcoal groups, respectively. The authors argued that, had aspirin not desorbed from the charcoal, none of it would have appeared in the blood and so the existence of a significant AUC for the charcoal group proved that aspirin desorption must have occurred. However, as stated earlier, it was never shown that all of the aspirin had prebound to the charcoal. Thus, desorption (which, in fact, probably did occur) is not proved by this study. Tenenbein (1989), in a letter commenting on the Filippone study, makes this same point. The other shortcoming in this work is that plasma drug concentrations were determined only up to 30 hr, at which time the plasma drug levels for the charcoal group were still measurable, and only slowly declining. To get a more accurate measure of "desorption", an AUC value over essentially 0–∞ should have been determined.

Rauws and Olling (1976) did some "desorption" experiments, in which rats were given imipramine or desipramine (50 mg/kg) orally in 0.1 N HCl with or without added charcoal (a 10% wt/vol suspension, volume unstated). Time was allowed for adsorption equilibrium to occur before giving the drug/charcoal mixture to the rats. Four hours after administration of the drug or drug/charcoal, the rats were killed and organ drug concentrations were determined. The results are shown in Table 9.2. Clearly, if the charcoal had bound most of these drugs prior to administration (however, this was not shown) there must have been significant desorption in vivo. Rauws and Olling considered these results to be very disappointing. However, it would appear that their results are not as bad as they imply (Table 9.2 shows average reductions of imipramine and desipramine in the various organs of 33 and 71%, respectively). Perhaps the use of more charcoal is all that was needed (it is impossible to determine the charcoal:drug ratio from their paper).

Table 9.2 Drug Concentration Ratios (Charcoal/No Charcoal)

Organ	Imipramine	Desipramine
Heart	0.61	0.11
Liver	0.69	0.40
Lungs	0.84	0.47
Brain	0.53	0.20

Adapted from Rauws and Olling (1976), Table 4. Copyright Springer-Verlag GmbH & Co. Used with permission.

Other evidence that substantial drug desorption can occur in vivo has been mentioned by Andersen (1948). In reviewing a study by Wiechowski in 1910, Andersen remarked that

> ... a rabbit given a lethal dose of phenol with charcoal displays no signs of poisoning, but a comparison of the amount of phenol excreted in the urine with that excreted without administration of charcoal shows the difference to be negligible. The charcoal must act by prolonging absorption, so that the threshold value for toxic concentration is not attained.

But, as he goes on to say, in any event, "it is rather unimportant whether the charcoal adsorbs the poison completely or merely prevents the attainment of a toxic concentration." Its therapeutic value is the same in either case.

Roivas and Neuvonen (1992) studied the reversibility of nicotinic acid adsorption/desorption to charcoal in vitro and showed that desorption was readily achieved. However, they did not determine desorption isotherms, so there was no proof in the study that desorption was truly 100% reversible.

In summary, it appears that desorption is possible, but if the affinity of the charcoal for the drug is reasonable and if a fairly substantial quantity of charcoal is used, then the desorption process will be very slow. Thus, drug absorption will be "stretched out" enough to prevent toxic levels in the body from occurring. Of course, if the time needed for significant desorption to occur were greater than the transit time of the charcoal in the digestive tract, then significant desorption obviously could not occur.

IV. STORAGE STABILITY OF CHARCOAL SUSPENSIONS

Although warnings have occasionally been given that an aqueous charcoal suspension must be prepared immediately before use or else its adsorption ability will be impaired, these have no known basis in fact. The chemical groups (e.g., oxides) which exist on the internal surfaces of activated charcoals have specific affinities for organic chemicals and the presence of a small polar substance like water does not destroy this. To prove this point, Picchioni et al. (1974) tested the ability of a freshly prepared charcoal suspension against that of aqueous suspensions that were 3, 6, and 12 months old for reducing strychnine toxicity in rats. The stored suspensions were, if anything, better (8%

Table 9.3 Efficacy of Stored Versus Fresh Charcoal Suspensions

Test time (months)	LD_{50} of Strychnine in Rats (mg/kg)		
	Stored AC[a] suspension	Fresh AC[a] suspension	Stored/fresh ratio
3	38.0	36.0	1.06
6	38.5	35.0	1.10
12	33.5	31.4	1.07

[a] AC, Activated charcoal
From Picchioni et al. (1974).

better on average, but this difference was not statistically significant). Their results are shown in Table 9.3.

Picchioni et al. (1982) also studied the storage stability of mixtures of charcoal in 70% w/v sorbitol. The charcoal/sorbitol mixtures were stored for 3, 6, and 12 months, and were then used, along with freshly prepared charcoal/sorbitol mixtures, to treat rats which had been given 100 mg/kg chloroquine diphosphate, 50 mg/kg sodium pentobarbital, or 100 mg/kg aspirin. Immediately after receiving the drugs orally, the rats were given 20 mL/kg of charcoal/sorbitol. Liver chloroquine levels in mg/100 g determined at 4 hr averaged 38.1 for control rats, 5.6 for rats treated with fresh charcoal/sorbitol, and 6.0, 4.3, and 3.9 for rats treated with the 3, 6, and 12-month-old suspensions, respectively. Blood pentobarbital levels in mg/100 mL determined at 30 min averaged 2.7 for control rats, 1.0 for rats treated with fresh charcoal/sorbitol, and 0.9, 1.1, and 1.2 for rats treated with the 3, 6, and 12-month-old suspensions. Mean blood aspirin values in mg/100 mL at 4 hr were 15.7 in the control tests, 8.1 when fresh charcoal/sorbitol was used, and were 7.4, 7.4, and 7.2 when 3, 6, and 12-month-old charcoal/sorbitol was used. The fresh and 3, 6, or 12-month-old suspensions were not statistically different ($p < 0.05$) from each other in any of the studies.

Therefore, it seems certain that as long as a charcoal suspension is stored in a closed container so that the adsorption of any substances from air is prevented, the potency of the suspension will be preserved indefinitely.

V. UNIVERSAL ANTIDOTE

Universal antidote, popularized many years ago by Dr. Jay Arena (Lampe, 1975), consists of 2 parts activated charcoal, 1 part magnesium oxide, and 1 part tannic acid. The principle underlying this combination is that it should be capable of very broad action, since the charcoal would adsorb many organic poisons, the magnesium oxide would neutralize ingested acids, and the tannic acid would precipitate alkaloids (and thus render them unabsorbable).

During the last 15 years or so, it has been increasingly recognized that not only is universal antidote in almost any case less effective than plain charcoal but it is downright hazardous. As Daly and Cooney (1978a) have clearly shown, the reduced effectiveness

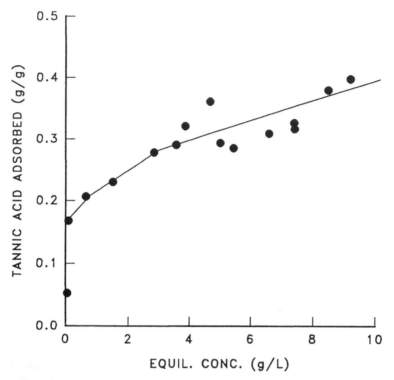

Figure 9.5 Tannic acid adsorption by activated charcoal from simulated gastric fluid. From Daly and Cooney, 1978a.

of universal antidote as compared to plain charcoal occurs because, in any aqueous medium, the tannic acid component itself adsorbs to an appreciable extent to the charcoal. Figure 9.5 shows an adsorption isotherm determined by Daly and Cooney. Since the number of adsorption sites on any given charcoal is a relatively fixed quantity, the adsorption of tannic acid uses up a significant portion of these sites and therefore leaves fewer sites available for the adsorption of whatever toxic substances are present. One example of the effect of this competition between a drug and tannic acid is shown in Figure 9.6, which shows the in vitro adsorption of sodium salicylate by charcoal and by a 2:1 mixture of charcoal and tannic acid (Daly and Cooney, 1978a).

Picchioni et al. (1966) directly compared universal antidote to charcoal alone. They showed clearly, with in vitro tests and with in vivo tests with rats, that charcoal is more effective than universal antidote in adsorbing strychnine, malathion, and pentobarbital. Their detailed results, summarized in a simpler form by Hayden and Comstock (1975), are shown in Table 9.4.

The other objection to tannic acid, beyond its competition for adsorption, is the fact that in significant amounts it can cause severe liver damage. Even in the case of alkaloid poisoning, charcoal alone is quite effective and probably is still better than universal antidote. Although tannic acid does precipitate most alkaloids, it would not prevent their gradual absorption. It is likely that charcoal would bind alkaloids tenaciously enough such that, even if some gradual desorption occurred further down the gastrointestinal

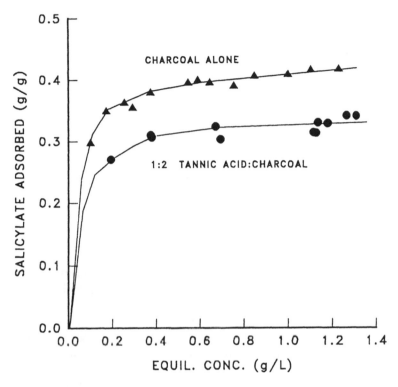

Figure 9.6 Salicylate adsorption from simulated gastric fluid by activated charcoal in the absence of tannic acid and in the presence of tannic acid. From Daly and Cooney, 1978a.

Table 9.4 Relative Efficacy of Activated Charcoal and Universal Antidote

Drug	In vitro adsorption (g/100 g)		In vivo effects, tissue concentration (% of control)	
	Activated charcoal	Universal antidote	Activated charcoal	Universal antidote
Pentobarbital	32	10	46[a]	54[a]
Malathion	31	12	34[b]	61[b]
Strychnine	22	9	28[c]	55[c]

[a]Blood level versus control.
[b]Blood cholinesterase depression versus control.
[c]Liver concentration versus control.
From Hayden and Comstock (1975).

Table 9.5 Relative Efficacy of Charcoal Alone or With Tannic Acid or Magnesium Oxide

Drug	Mean tissue concentration (mg/100 mL or mg/g)			
	Drug alone	With AC[a]	With AC[a] + TA[b]	With AC[a] + MgO
Aspirin 100 mg/kg	33.3	23.6	30.3	27.6
Chloroquine 100 mg/kg	23.5	17.0	16.3	13.6
Pentobarbital sodium 50 mg/kg	3.0	2.1	3.2	2.1

[a] AC, Activated charcoal.
[b] TA, Tannic Acid.
From Picchioni (1974).

tract, the overall rate of absorption would be slower than would occur if universal antidote were used. This idea is supported by the data of Picchioni et al. (1966) on their in vivo work with strychnine, as shown in Table 9.4.

Although magnesium oxide does not adsorb significantly to charcoal, as has been shown by Daly and Cooney (1978a), its inclusion in universal antidote does not make sense either. The majority of acid or base poisonings occur with bases (usually sodium hydroxide), which would not be neutralized by magnesium oxide. Even in cases where it is known that an acid has been ingested, it is probably better to simply dilute the acid with water or milk, since rapid neutralization can generate enough heat of reaction to severely burn gastrointestinal tissues.

Picchioni (1974) presented data on the mean tissue concentrations in rats given aspirin, or chloroquine phosphate, or pentobarbital sodium, followed by: (1) nothing, (2) charcoal alone, (3) charcoal plus tannic acid, or (4) charcoal plus magnesium oxide. For the latter two treatments, the ratio of charcoal to tannic acid or magnesium oxide was 2:1. His results are shown in Table 9.5.

The tannic acid interfered more than the MgO with the effectiveness of the charcoal, which is as one would expect. Magnesium oxide, being inorganic, probably does not adsorb to charcoal to any measurable degree, whereas tannic acid does, as shown earlier. Thus, tannic acid uses up some of the adsorption sites that would otherwise be able to bind the drug. The enhancement by MgO of the charcoal effect in the case of chloroquine is interesting and can be explained as follows. The MgO neutralizes some of the acid in the stomach, increasing the pH (when the stomach contents pass into the intestines, the pH levels there will likewise be higher than they would otherwise be in the absence of MgO). This in turn makes the chloroquine, a basic drug, less ionized, in which form it adsorbs better to charcoal.

Henschler and Kreutzer (1966) also presented much data bearing on the question of universal antidote. Their superb work (published in German) compares LD_{50} values in mice for various drugs and poisons determined using charcoal alone, tannic acid alone, MgO alone, charcoal plus tannic acid, charcoal plus MgO, and universal antidote. It is clear from their data that MgO does not interfere with charcoal, but that tannic acid does. Their paper also presents isotherms similar in nature to those published by Daly and Cooney (1978a).

Universal antidote is not the only substance which has been recommended to replace charcoal. In the late 1950s and early 1960s many "first aid" types of handbooks recommended that burnt toast (or burnt sugar) be used as a convenient household substitute for activated charcoal. Since the adsorptive capacity of activated charcoal is due to the very large internal surface area generated by careful and slow pyrolysis and activation at very high temperatures (600–900°C), and since burnt toast particles by comparison obviously have a negligible surface area for adsorption (notwithstanding the question of whether its surface has any ability to adsorb), then this notion is seen to be patently absurd.

An emphatic example of the ineffectiveness of burnt toast and burnt sugar (as if one were needed!) has been given by Lehman (1963). He conducted the USP strychnine sulfate adsorption test with the specified amount (5 g) of activated charcoal, with 5 g of burnt toast, and with 5 g of burnt sugar. The LD_{50} values of the filtrates from the three tests were determined by injecting them intraperitoneally into mice. The filtrate from the charcoal-treated strychnine sulfate solution showed no toxicity, whereas the LD_{50} values of the solutions treated with burnt toast and burnt sugar were identical to the value for untreated strychnine sulfate solution.

VI. EFFECT OF PEPSIN ON IN VITRO ADSORPTION

Having just discussed the effect of tannic acid on the adsorption of drugs and poisons to charcoal, this is an appropriate place to consider similar effects due to pepsin. Although Simulated Gastric Fluid (SGF), USP, calls for the use of 3.2 g/L pepsin, most researchers omit pepsin when performing in vitro adsorption tests with substances dissolved in SGF. The implicit assumption is that the omission of pepsin has no effect on the adsorption of the drug or poison. Recognizing that tannic acid does offer significant interference in drug/poison adsorption, Daly and Cooney (1978b) decided to determine if pepsin also offered any significant interference, and, therefore, if its omission by researchers is justified. Sodium salicylate was again used as the test drug.

The study hypothesis was that pepsin, being a large molecule (about 35,000 daltons), would not penetrate the pores of activated charcoal (its equivalent spherical diameter is about 40 Å and the pores wherein most of the surface area resides in activated charcoal are those with diameters of about 10–20 Å) and thus would not compete for adsorption with molecules the size of most drugs. Daly and Cooney, using a solution of 4 g/L pepsin in a pH 1.2 KCl-HCl buffer contacted with different amounts of Norit A charcoal for 24 hr, showed first that pepsin itself does adsorb well to charcoal. Figure 9.7 shows the adsorption isotherm for pepsin.

Next, equilibrium adsorption tests were done with a solution of 1 g/L sodium salicylate in the same KCl-HCl buffer, both without pepsin and with 4 g/L pepsin added. Figure 9.8 shows the salicylate adsorption results. The presence of pepsin decreased the fraction of salicylate adsorbed from 0.401 to 0.379 at a charcoal:salicylate ratio of 1:1 (i.e., a 5.4% reduction) and from 0.982 to 0.929 at a charcoal:salicylate ratio of 4:1 (also a 5.4% reduction). Thus, pepsin did offer some interference but not a large one. Daly and Cooney also studied whether 0.5 g/L gastric mucin, which is a component of Simulated Intestinal Fluid, USP, interfered with salicylate adsorption. In this case, no detectable interference was found. The conclusions from this study were that pepsin can

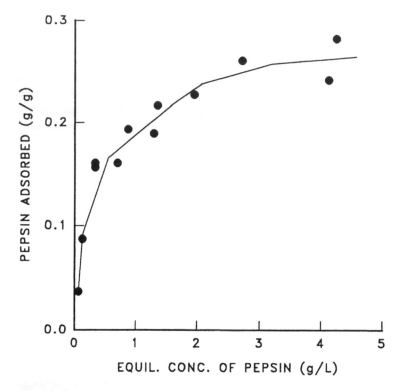

Figure 9.7 Adsorption isotherm for pepsin in a pH 1.2 KCl-HCl buffer. From Daly and Cooney (1978b). Reproduced by permission of the American Pharmaceutical Association.

produce a small (e.g., 5%) reduction in the in vitro adsorption of a drug or poison but that gastric mucin appears not to interfere.

In agreement with the Daly/Cooney pepsin study is an in vitro study of aspirin adsorption to charcoal by Neuvonen et al. (1984). In one series of tests, 0.1 M HCl was used, while in another series of tests SGF, USP (with 3.2 g/L pepsin) was employed. For a charcoal:aspirin ratio of 10:1, the percentage of unbound aspirin was 1.4% with the 0.1 M HCl and was 5.3% with the SGF. However, the difference between the percentages of aspirin adsorbed (98.6% versus 94.7%) was only about 4%. This reduction in aspirin adsorption of 4% agrees well with the 5% reduction in salicylate adsorption found by Daly and Cooney.

VII. SUMMARY

A variety of different topics related to the use of antidotal charcoal were gathered together in this chapter. We first discussed the importance of giving charcoal in powdered form, as opposed to tablet or capsule form, so that no time is lost waiting for tablets/capsules to disintegrate. In vivo studies have shown that, while tablets/capsules are reasonably effective, powdered charcoal is always significantly better.

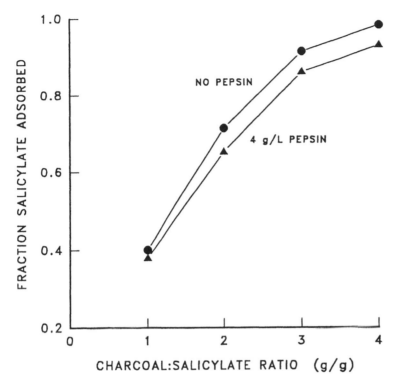

Figure 9.8 Salicylate adsorption by activated charcoal from a 1 g/L pH 1.2 solution in the absence of pepsin and in the presence of 4 g/L pepsin. From Daly and Cooney (1978b). Reproduced by permission of the American Pharmaceutical Association.

Next, the question of the optimal charcoal dose was considered. The simple answer to this is: however much the patient is willing to take. However, if one desires a specific answer, research suggests that a dose of 25–50 g for children and 50–100 g for adults would normally be quite effective. If one knows the amount of drug ingested, sufficient charcoal to give a charcoal:drug ratio of at least 5:1 should be used, although this is certainly a minimal target. A ratio of 8:1 or 10:1 is better. In using such guidelines, the surface area effect of charcoal should be kept in mind. For example, if one were to use Norit B Supra charcoal (1400 m^2/g) instead of Norit USP XXII charcoal (900 m^2/g), one could "get by with" a lesser amount, since the B Supra type should be more effective per unit weight. Many experiments with "superactive charcoal" (> 3000 m^2/g), which is no longer available, have demonstrated the effect of surface area.

Stability of the drug/charcoal complex was addressed next. Although adsorption of a drug to charcoal is usually intrinsically reversible, the use of a reasonable amount of charcoal keeps the fluid-phase drug concentration very low and thus keeps the rate of absorption of desorbed drug low. Thus, the absorption of any desorbed drug will be slowed enough to prevent toxic levels from being attained. However, it should be kept in mind that the pH gradient in the GI tract will favor the desorption of acidic drugs and increased binding of basic drugs. Thus, desorption would normally be more pronounced with acidic drugs.

The storage stability of charcoal suspensions was discussed and it was deemed quite certain that as long as the suspension is stored in a closed container (so that the adsorption of any substances from air is prevented), the potency of the suspension will be preserved indefinitely.

The concoction known as "universal antidote" was shown by many studies to be inferior to plain charcoal, as the tannic acid component adsorbs to the charcoal and interferes with its ability to adsorb drugs. Additionally, tannic acid, in sufficient amount, is quite toxic. While the magnesium oxide component of "universal antidote" does not interfere with adsorption, it could potentially neutralize acids too quickly and cause thermal injury to the stomach.

Finally, we discussed the effects of omitting pepsin, a component of SGF, USP, when doing in vitro adsorption tests in low pH media. While studies have shown that pepsin itself adsorbs reasonably well to charcoal, its omission from SGF appears to give only about 4–5% higher drug adsorption than would occur with its inclusion. It is up to each researcher to decide whether this difference is significant.

VIII. REFERENCES

Andersen, A. H. (1948). Experimental studies on the pharmacology of activated charcoal. IV. Adsorption of allylpropynal (allyl-isopropyl-barbituric acid) in vivo, Acta Pharmacol. Toxicol. *4*, 379.

Anonymous (1974), Activated charcoal in the treatment of acute poisoning, Drug Ther. Bull. *12*, March 29, pp. 27-28.

Chin, L., Picchioni, A. L., and Duplisse, B. R. (1970). The action of activated charcoal on poisons in the digestive tract, Toxicol. Appl. Pharmacol. *16*, 786.

Chin, L., Picchioni, A. L., Bourn, W. M., and Laird, H. E. (1973). Optimal antidotal dose of activated charcoal, Toxicol. Appl. Pharmacol. *26*, 103.

Comstock, E. (1975). Guide to management of drug overdose, Clin. Toxicol. *8*, 475.

Cooney, D. O. (1992). Desorption of sodium salicylate from powdered activated charcoal, unpublished paper.

Cooney, D. O. and Z. Xi (1994). Activated carbon catalyzes reactions of phenolics during liquid phase adsorption, Am. Inst. Chem. Eng. J. *40*, 361.

Corby, D. G. and Decker, W. J. (1974). Management of acute poisoning with activated charcoal, Pediatrics *54*, 324.

Daly, J. S. and Cooney, D. O. (1978a). Interference by tannic acid with the effectiveness of activated charcoal in "universal antidote," Clin. Toxicol. *12*, 515.

Daly, J. S. and Cooney, D. O. (1978b). Omission of pepsin from Simulated Gastric Fluid in evaluating activated charcoals as antidotes, J. Pharm. Sci. *67*, 1181.

Decker, W. J., Shpall, R. A., Corby, D. G., Combs, H. F., and Payne, C. E. (1969). Inhibition of aspirin absorption by activated charcoal and apomorphine, Clin. Pharmacol. Ther. *10*, 710.

Decker, W. J. and Corby, D. G. (1970). Activated charcoal as a gastrointestinal decontaminant: Experiences with experimental animals and human subjects, Clin. Toxicol. *3*, 1.

DiPalma, J. R. (1979). Activated charcoal—a neglected antidote, Am. Family Physician *20*, 155.

Dordoni, B., Willson, R. A., Thompson, R. P. H., and Williams, R. (1973). Reduction of absorption of paracetamol by activated charcoal and cholestyramine: A possible therapeutic measure, Br. Med. J. *3*, 86.

Filippone, G. A., Fish, S. S., Lacouture, P. G., Scavone, J. M., and Lovejoy, F. H., Jr. (1987). Reversible adsorption (desorption) of aspirin from activated charcoal, Arch. Intern. Med. *147*, 1390.

Goodman and Gilman's *The Pharmacological Basis of Therapeutics*, various editions, Macmillan, New York.

Hayden, J. W. and Comstock, E. G. (1975). Use of activated charcoal in acute poisoning, Clin. Toxicol. *8*, 515.

Henschler, D. and Kreutzer, P. (1966). Intoxikationsbehandling durch Binding von Giftstoffen im Magen-Darmkanal: Tierkohle oder "Universalantidot"? Deutsch Med. Wochenschr. *91*, 2241.

Lampe, K. F. (1975). Activated charcoal preparations (letter), Clin. Toxicol. *8*, 483.

Lawrence, F. H. and McGrew, W. R. (1975). Activated charcoal: A forgotten antidote, J. Maine Med. Assoc. *66*, 311.

Lehman, A. J. (1963). In universal antidote, activated charcoal should *not* be replaced by burned sugar or toast, Am. Profess. Pharm. *29*, 74.

Levy, G. and Tsuchiya, T. (1969). Effect of activated charcoal on aspirin absorption in man (abstract), Pharmacologist *11*, 292.

Levy, G. and Tsuchiya, T. (1972). Effect of activated charcoal on aspirin absorption in man, Clin. Pharmacol. Ther. *13*, 317.

Levy, G. and Gwilt, P. (1972). Activated charcoal for acute acetaminophen intoxication (letter), JAMA *219*, 621.

Levy, G. and Houston, J. B. (1976). Effect of activated charcoal on acetaminophen absorption, Pediatrics *58*, 432.

Neuvonen, P. J., Elfving, S. M., and Elonen, E. (1978). Reduction of absorption of digoxin, phenytoin, and aspirin by activated charcoal in man, Eur. J. Clin. Pharmacol. *13*, 213.

Neuvonen, P. J. and Olkkola, K. T. (1984). Effect of dose of charcoal on the absorption of disopyramide, indomethacin, and trimethoprim by man, Eur. J. Clin. Pharmacol. *26*, 761.

Neuvonen, P. J., Olkkola, K. T., and Alanen, T. (1984). Effect of ethanol and pH on the adsorption of drugs to activated charcoal: Studies in vitro and in man, Acta Pharmacol Toxicol. *54*, 1.

Olkkola, K. T. (1985). Effect of charcoal-drug ratio on antidotal efficacy of oral activated charcoal in man, Br. J. Clin. Pharmacol. *19*, 767.

Otto, U. and Stenberg, B. (1973). Drug adsorption properties of different activated charcoal dosage forms in vitro and in man, Svensk. Farm. Tids. *77*, 613.

Palatnick, W. and Tenenbein, M. (1992). Activated charcoal in the treatment of drug overdose: An update, Drug Safety *7*, 3.

Picchioni, A. L., Chin, L., Verhulst, H. L., and Dieterle, B. (1966). Activated charcoal versus "universal antidote" as an antidote for poisons, Toxicol. Appl. Pharmacol. *8*, 447.

Picchioni, A. L. (1970). Activated charcoal - a neglected antidote, Pediatr. Clin. North Am. *17*, 535.

Picchioni, A. L. (1974). Research in the treatment of poisoning, in *Toxicology Annual—1974*, pp. 27-51.

Picchioni, A. L., Chin, L., and Laird, H. E. (1974). Activated charcoal preparations—Relative antidotal efficacy, Clin. Toxicol. *7*, 97.

Picchioni, A. L., Chin, L., and Gillespie, T. (1982). Evaluation of activated charcoal-sorbitol suspension as an antidote, J. Toxicol.-Clin. Toxicol. *19*, 433.

Rauws, A. G. and Olling, M. (1976). Treatment of experimental imipramine and desipramine poisoning in the rat, Arch. Toxicol. *35*, 97.

Remington's Pharmaceutical Sciences, various editions, Mack Publishing Company, Easton, PA.

Remmert, H. P., Olling, M., Slob, W., van der Giesen, W. F., van Dijk, A., and Rauws, A. G. (1990). Comparative antidotal efficacy of activated charcoal tablets, capsules, and suspension in healthy volunteers, Eur. J. Clin. Pharmacol. *39*, 501.

Roivas, L. and Neuvonen, P. J. (1992). Reversible adsorption of nicotinic acid onto charcoal in
   vitro, J. Pharm. Sci. *81*, 917.
Tenenbein, M. (1989). Desorption of aspirin from activated charcoal (letter), Arch. Intern. Med.
   149, 717.
Tsuchiya, T. and Levy, G. (1972a). Drug adsorption efficacy of commercial activated charcoal
   tablets in vitro and in man, J. Pharm. Sci. *61*, 624.
Tsuchiya, T. and Levy, G. (1972b). Relationship between effect of activated charcoal on drug
   absorption in man and its drug adsorption characteristics in vitro, J. Pharm. Sci. *61*, 586.

10

The Classic Studies of Andersen

In the period 1944–1948, A. H. Andersen published a series of papers which not only summarized many previous studies on charcoal but, through a description of very detailed studies of his own, greatly extended previous work. Andersen's research provided a solid and extensive basis for later work by others. While much of what Andersen did was not new in concept, his research was extremely definitive in nature and explained several aspects of adsorption which were previously incompletely understood.

I. ANDERSEN'S IN VITRO ADSORPTION STUDIES

In his first study, Andersen (1944, 1946) describes the determination of equilibrium adsorption isotherms for a wide variety of drugs and poisons. Solutions of these chemicals in distilled water were shaken with various amounts of powdered Carbo medicinalis Merck at room temperature for 1 hr. After removal of the charcoal by filtration through filter paper, the residual drug or poison in the filtrate was determined by standard methods. Figure 10.1 shows a typical isotherm obtained in this way for strychnine nitrate (the initial solution concentration was 10 mmole/L). As is clear from such a figure, the isotherm tends to level off to some maximal adsorption value at higher residual concentrations in the fluid.

Table 10.1 lists a variety of such adsorption maxima determined by Andersen for various substances from his adsorption isotherm data. As these results imply, there is a great decrease in adsorption when barbital is converted from its natural acidic form to its sodium salt. The reason for this will be discussed later. Regarding some of the other substances, it is somewhat surprising that alcohol (presumably ethyl alcohol) adsorbs reasonably well. However, the amount of alcohol usually ingested before charcoal therapy would be considered is so large that charcoal as an antidote would seem to be impractical.

Andersen also demonstrated, in his first paper, that the rate of adsorption to powdered charcoal is extremely fast. His data on strychnine nitrate adsorption show an approach

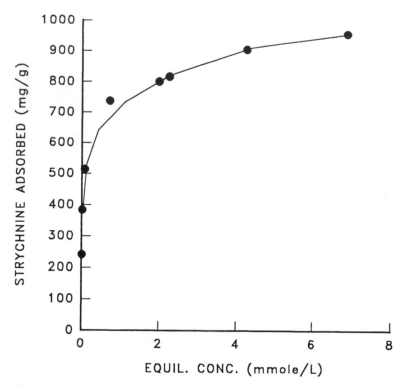

Figure 10.1 Adsorption isotherm for strychnine nitrate in distilled water. From Andersen (1946). Reprinted by permission of Munksgaard International Publishers Ltd.

Table 10.1 Maximal Amounts of Substances Adsorbed on Charcoal

Substance	Amount adsorbed (mg/g)
Mercuric chloride	1,800
Sulfanilamide	1,000
Strychnine nitrate	950
Morphine hydrochloride	800
Atropine sulfate	700
Nicotine	700
Barbital	700
Sodium barbital	150
Five other Na or Ca salts of barbiturates	300–350
Salicylic acid	550
Phenol	400
Alcohol	300
Potassium cyanide	35

From Andersen (1946). Reprinted by permission of Munksgaard International Publishers Ltd.

to essentially complete equilibrium after only 2 minutes; however, similar data of his on mercuric chloride adsorption show that small changes in the fluid-phase mercuric chloride concentration continued to occur up to a point somewhere between 30 and 60 minutes. This agrees with the observations of Cooney and Kane (1980), who found that about 30 minutes are needed for powdered charcoal to equilibrate closely with drug solutions, even if such solutions are well agitated. The reason for the exceedingly fast approach to equilibrium in Andersen's strychnine nitrate tests is unknown; however, it is obvious that the time of contact for, say, his "1 minute" sample was substantially longer in reality, for the filtration step he used to remove the charcoal from the solution would have required a reasonable time to perform.

Despite his conclusion that the rate of adsorption in his tests was quite rapid, Andersen was careful to point out that in more viscous or more complex fluids (e.g., stomach or intestinal fluids) equilibrium times could easily be very much longer. Also, he mentioned that if the charcoal is granular, is prepared in tablet form, or indeed exists in any form other than a fine powder, the attainment of equilibrium would take much longer.

One example will suffice to illustrate the charcoal particle size effect. Dedrick and Beckmann (1967) have shown that, for cylindrical charcoal pellets of 4–6 mesh size (0.43 cm diameter × 0.33 cm long) adsorbing the pesticide 2,4-dichlorophenoxyacetic acid, equilibrium was attained only after about 5 months! As a general rule-of-thumb the adsorption rate varies inversely as the square of the charcoal particle size; thus, as pointed out in Chapter 9, a 325 mesh-sized charcoal (0.043 mm) compared to a 20 mesh-sized charcoal (0.833 mm) will approach equilibrium $(0.833/0.043)^2$ or 375 times faster. The virtues of using a finely divided charcoal are therefore quite clear.

II. THE EFFECT OF pH

In his second study, Andersen (1947) considered the effect of pH on the adsorption of substances by charcoal from aqueous solutions. The first question he addressed was whether charcoal has the capacity to adsorb simple acids and bases (e.g., HCl, NaOH) and therefore directly change the pH of whatever fluid medium it finds itself in.

As Andersen indicates (and as we discussed in Chapter 3), the ability of charcoals to adsorb simple acids and bases has been clearly shown. The acid-base properties depend on the temperature used for activation. As previously discussed, those charcoals activated at low temperatures (< 400°C), called L-type charcoals, have the ability to adsorb hydroxide ions and thus give an acidic pH when immersed in pure water. High temperature (> 750°C) charcoals, called H-types, show reverse behavior (i.e., they can adsorb hydrogen ions and thus give alkaline solutions in water). Most charcoals are of the H-type and thus cause pH shifts in the alkaline direction.

Andersen goes on to explain experiments on the ability of charcoal to change the pH of buffers. He prepared the following four buffer solutions and shook 20 mL of each with 0.2 g of Merck charcoal for 1 hr.

1. Glycocoll-HCl, pH 1.64
2. Phosphate-citric acid, pH 3.50
3. Phosphate-citric acid, pH 5.46
4. Phosphate-citric acid, pH 6.95

The pHs of these were found to shift in the alkaline direction (upwards) by 0.03, 0.30, 0.12, and 0.07 pH units, respectively. Andersen suggests that the most probable explanation for these slight pH changes is the preferential adsorption of one of the buffer components, especially the citric acid component of the last three buffers. In fact, the pH changes of these buffers, made by combining different proportions of a phosphate salt solution and a citric acid solution, were in almost direct proportion to the proportions of the citric acid solution used (70, 40, and 18%, respectively). He thus suggested that, in experiments with charcoal, one should avoid using buffers containing organic compounds such as citric acid. Cooney and Wijaya (1986) have, in fact, shown conclusively that under low pH conditions where citric acid exists in undissociated form, it can adsorb well to charcoal.

The central question addressed in Andersen's second study was the effect of pH on the extent of ionization of various drugs and poisons, and the effect of the state of ionization on the degree of adsorption to charcoal. This is an important factor, since the pH in the gastrointestinal system varies from about 1–3 in the stomach to about 8 in the lower small intestine. Furthermore, many drugs involved in overdose situations are weak acids (e.g., barbituric acid derivatives), weak bases (e.g., alkaloids, amines), or the salts of each, and their adsorption has clearly been shown by many investigators to depend on pH. Hauge and Willamann (1927) demonstrated, for example, that substances capable of ionizing to give an anion adsorbed best at low pH, substances capable of ionizing to give a cation adsorbed best at high pH, electroneutral substances like glucose did not exhibit any change in adsorption with pH, and ampholytes (substances that can exist as anions, neutral species, or cations, depending on the pH) like albumin and gelatin showed maximal adsorption at their isoelectric pH values. Studies like this suggested to Andersen that the substances involved seemed to be most easily adsorbed under electroneutral conditions. He thus said:

> It may be envisaged that the effect of pH on the adsorption of a substance in the first place is the effect of the pH on the dissociation of that substance according to the following rule: changes in pH that cause a decrease in dissociation of the substance will increase the adsorption; changes in pH that cause an increase in dissociation of the substance will decrease the adsorption.

Andersen therefore conducted experiments to test this rule. He studied the adsorption of a weak base (nicotine), weak acids (phenol, diethylbarbituric acid, salicylic acid), ampholytes (sulfanilamide), neutral substances (alcohol), and weakly ionized metallic salts (mercuric chloride). He found that the bases were best adsorbed at high pHs, the acids at low pHs, and the ampholyte at its isoelectric point (point of zero net charge). Adsorption of the electroneutral substances was not affected by pH. $HgCl_2$ adsorbed least in strong acid, under which conditions the formation of complex ions is favored. Andersen's hypothesis was thus confirmed.

The experiments were carried out in the same general fashion as those reported in his first study, either at room temperature (20–21°C) or at 37°C, and pH variations were achieved by the use of HCl or NaOH solutions of different normalities. Thus all acids tested existed essentially totally undissociated at low pHs and almost totally dissociated (as anions) at high pHs. The nicotine existed predominantly as a cation at low pHs and as an undissociated base at high pHs.

Adsorption isotherms for nicotine and diethylbarbituric acid are shown in Figures 10.2 and 10.3, while variations in the adsorption of these compounds with pH are shown in Figures 10.4 and 10.5. These results are obviously in agreement with the foregoing discussion, and show the dramatic effect of pH on adsorption.

Sulfanilamide adsorption isotherms were determined at three pH levels: 1.3, 6.0, and 12.2. The adsorption isotherms were well-spaced apart, with the one for pH 6.0, which is the isoelectric point of sulfanilamide, being the highest of the three. To give an idea of the differences in the three isotherms, the equilibrium amounts adsorbed at a fluid-phase sulfanilamide concentration of 20 mmole/L were (in mg/g): 1012 at pH 6.0, 793 at pH 1.3, and 563 at pH 12.2. Thus, as expected, adsorption of this ampholytic compound was greatest at its isoelectric point.

For mercuric chloride, Andersen found that the amounts adsorbed by 1 g charcoal from four 0.05 M solutions adjusted to pH values of 0.19, 0.33, 1.33, and 3.42 were 0.40, 0.61, 1.22, and 1.42 g, respectively. The decreased adsorption with decreased pH presumably reflects the fact that the number of complex ions of the type HgCl3 and HgCl4 increases with decreasing pH. The number of electroneutral molecules in solution is reduced and, in accordance with the principles stated earlier, this tends to decrease adsorption.

Andersen further found that for the electroneutral compound alcohol, the amount adsorbed by 1 g charcoal in a neutral solution was 320 mg and in a strong acid solution was 300 mg. The effect of pH was certainly small, as one would expect, knowing that alcohol cannot ionize. The amounts of alcohol (ethanol) adsorbed are surprisingly high, considering that ethanol is extremely hydrophilic and has a vast preference for staying in aqueous solution rather than being adsorbed. Many other studies (see Chapter 11) have shown that ethanol adsorbs very poorly to charcoal.

Finally, Andersen confirmed that if an adsorption equilibrium was attained at one pH and then the pH was shifted, a new equilibrium state was soon achieved which was the same that would have been reached had the adsorption occurred entirely under conditions of the final pH. This suggests that as a drug/charcoal complex passes down the gastrointestinal tract, the degree of drug binding continually adjusts to the equilibrium state characteristic of the prevailing pH. For example, an acidic drug would be expected to desorb somewhat as its complex with charcoal passes down the gastrointestinal tract.

III. ADSORPTION FROM GASTROINTESTINAL CONTENTS

The third study by Andersen (1948a) deals with the matter of whether substances ordinarily found in the gastrointestinal tract (food, partly split derivatives of food, digestive enzymes, and various secretions) might inhibit the adsorption of drugs or poisons by charcoal in vivo. To test this, he carried out experiments on the adsorption of strychnine nitrate, mercuric chloride, and diethylbarbituric acid to Carbo medicinalis Merck charcoal in actual gastric contents. The gastric contents were obtained by withdrawing standard test meals from 15 patients, pooling them, and mixing them. The gastric contents contained 60% solids and had a pH of 1.50.

Various solutions of the three test substances were added to equal portions of the gastric fluid and mixed with 1 g charcoal (with shaking) for 1 hr at room temperature. After filtration, the fluids were assayed for the residual drug or poison. The pH values in the gastric media were about 1.95 for all three compounds. Adsorption isotherms were also determined for the three compounds in distilled water solutions, for which the

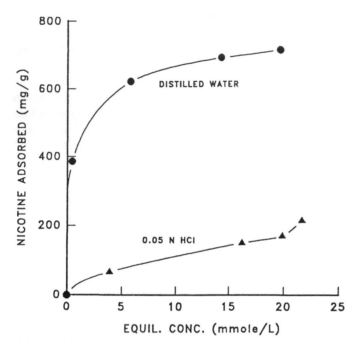

Figure 10.2 Adsorption of nicotine from distilled water solution and from 0.05 N HCl solution at 19–20°C. From Andersen (1947). Reprinted by permission of Munksgaard International Publishers Ltd.

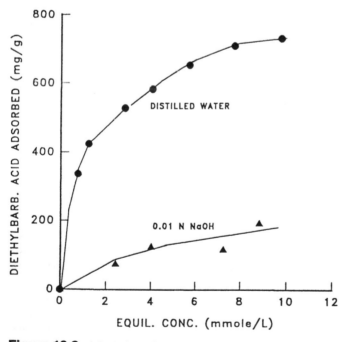

Figure 10.3 Adsorption of diethylbarbituric acid from distilled water solution and from 0.01 N NaOH solution at 20–21°C. From Andersen (1947). Reprinted by permission of Munksgaard International Publishers Ltd.

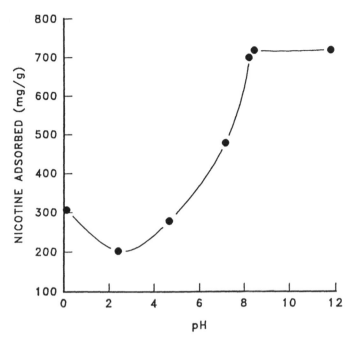

Figure 10.4 Effect of pH (adjusted using HCl) on nicotine adsorption at 19–20°C. From Andersen (1947). Reprinted by permission of Munksgaard International Publishers Ltd.

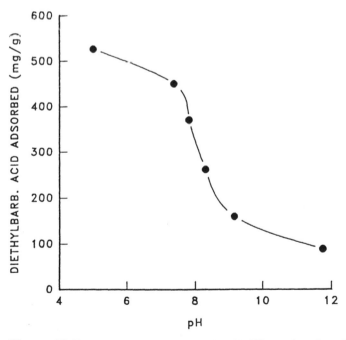

Figure 10.5 Effect of pH (adjusted using NaOH) on the adsorption of diethylbarbituric acid at 21–22°C. From Andersen (1947). Reprinted by permission of Munksgaard International Publishers Ltd.

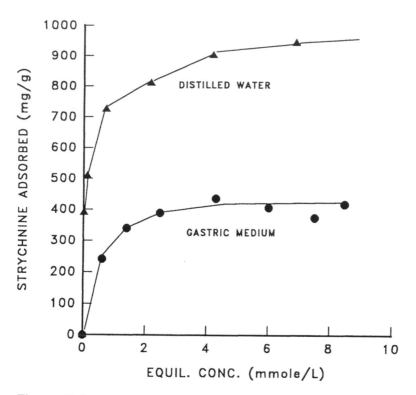

Figure 10.6 Adsorption of strychnine nitrate from distilled water and from a gastric medium. From Andersen (1948a). Reprinted by permission of Munksgaard International Publishers Ltd.

pH values were 4.91 (strychnine nitrate), 3.42 (mercuric chloride), and 5.00 (diethyl-barbituric acid). The adsorption isotherms for the three compounds in the gastric fluid environments were nearly the same in shape as those determined in distilled water solutions but were lower in magnitude by about 50%. Figure 10.6 shows the isotherms for strychnine nitrate adsorption from distilled water (pH 4.91) and from the gastric medium (pH 1.95). At a final fluid-phase concentration of 8.5 mmole/L, for example, the amount adsorbed was 960 mg/g in distilled water and 415 mg/g in the gastric medium, a decrease of 57%.

Andersen then explored whether such a decrease was actually due to the presence of the gastric material and not because the gastric test pH values were different from the pH values of the distilled water solutions. Figure 10.7 shows the adsorption of strychnine nitrate from solutions which were all prepared by combining equal volumes of a solution of strychnine nitrate in distilled water with gastric contents but with the final pH values of three of the mixtures adjusted to about 3, 4, and 5 using different amounts of NaOH. A change in pH from 1.95 to 4.91 is seen to cause the amount of strychnine nitrate adsorbed to increase from 415 mg/g to 750 mg/g. This latter value of 715 mg/g for a pH 4.91 gastric medium is the one which should be compared to the 960 mg/g value for a pH 4.91 distilled water medium shown in Figure 10.6. Thus, the decrease in adsorption due to gastric contents is really just from 960 to 750 mg/g, a

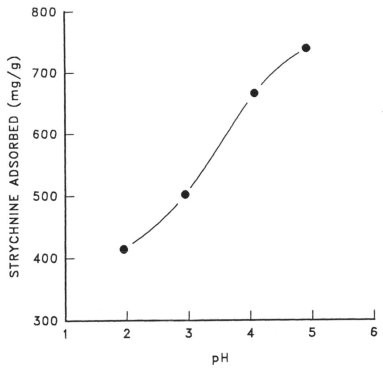

Figure 10.7 Adsorption of strychnine nitrate from gastric media with pH adjusted to different values using NaOH. From Andersen (1948a). Reprinted by permission of Munksgaard International Publishers Ltd.

22% decrease. Andersen seems not to have analyzed his strychnine nitrate results in this quantitative manner and, surprisingly, concluded that only a "moderate reduction" in adsorption occurs due to a pH change from 4.91 to 1.95, so it is still true that "adsorption in the stomach contents amounts to about half that in pure aqueous solution."

Isotherms for mercuric chloride and for diethylbarbituric acid in both distilled water solutions and in solutions with added gastric contents were also determined. The pH values in the distilled water solutions were 3.42 for mercuric chloride and 5.00 for diethylbarbituric acid, as mentioned above (these values are not given in the 1948a paper under discussion but can be found in Andersen's 1947 paper). The isotherms showed about a 50% decrease (for mercuric chloride) and about a 55% decrease (for diethylbarbituric acid) in adsorption from the media containing gastric contents, versus adsorption from distilled water.

From data given in Andersen's 1947 paper, one can easily estimate that a reduction in pH from 3.42 to 1.95 would decrease mercuric chloride adsorption by 11%. Thus, of the 50% difference in adsorption shown by the mercuric chloride isotherms, 11% can be ascribed to the pH difference and the other 39% must be due to the presence of gastric contents. Similarly, for diethylbarbituric acid, Figure 10.5 suggests that a pH change from 5.00 to 1.95 would cause virtually no change in adsorption because the

Table 10.2 Effect of Bile on Adsorption to Charcoal

Solution	pH	Bile dilution (water:bile)	Diethylbarbituric acid adsorbed (mg/g)
1	1.16	0:1	336
2		4:1	394
3		10:1	440
4		25:1	442
5	5.0	∞ (no bile)	526

From Andersen (1948a). Reprinted by permission of Munksgaard International Publishers Ltd.

diethylbarbituric acid remains almost totally undissociated over this pH range. Thus, the 55% difference in the isotherms for diethylbarbituric acid is due essentially entirely to the presence of gastric contents.

To recapitulate these gastric media results, we can state that the reduction in adsorption due to the presence of gastric contents, after accounting for effects due solely to pH changes, is fairly modest (22%) for strychnine nitrate, much stronger (39%) for mercuric chloride, and even stronger (55%) for diethylbarbituric acid. Why the effect varies with the type of compound is not clear.

Andersen also studied the adsorption of diethylbarbituric acid from duodenal juice drawn from normal individuals. The juice was clear, yellow-green and had a pH of 8.44. By comparison to adsorption from aqueous solutions of the same barbiturate concentration (0.02 M), which have a natural pH of 5.00, it was found that adsorption was 50% lower in the duodenal juice. However, as shown earlier (see Figure 10.5), a change in pH from 5.00 to 8.44 will automatically decrease adsorption of diethylbarbituric acid from about 530 to 260 mg/g, or 51%. Hence, in contrast to an implied conclusion by Andersen, the reduction in adsorption in duodenal juice can be entirely accounted for as a simple pH effect. There appears to be no "extra" effect due to other materials in the juice.

Andersen finally studied the adsorption of diethylbarbituric acid from solutions of hum an bile, collected by dissection of human subjects. The bile was from the gallbladder and was a dark green liquid of pH 5.88. Experiments were carried out with 0.02 M diethylbarbituric acid solutions to which were added equal parts of bile which had been diluted with water to various extents. The final solution pHs ranged from 1.16 to 5.0, which is a range in which pH has only a minor effect on adsorption (see Figure 10.5). Table 10.2 shows the results. In pure bile, the diethylbarbituric acid adsorption was reduced by 36.1% as compared to adsorption from distilled water. As the bile was progressively diluted with water, the interference effect diminished, as expected. Again, even though the pH varied in these tests, no correction for pH is needed in the pH range involved.

In summary, excluding effects on adsorption due to pH changes, the reductions in adsorption due to interference from (or competition with) other components in gastrointestinal fluids were found to be 22–55% for stomach contents, 36% for bile, and essentially zero for duodenal juice.

IV. FURTHER IN VIVO EXPERIMENTS

In his fourth study, Andersen (1948b) did in vivo experiments with rabbits. A 2% aqueous solution of a barbituric acid derivative, allylisopropylbarbituric acid, or allylpropynal, was introduced into the rabbits' stomachs. Because a chemical test for this substance at low concentrations was unavailable, Andersen indirectly determined the amount of drug absorbed by measuring the depth of narcosis using a corneal reflex method. It was first determined that, in the absence of charcoal, the anesthetic dose (i.e., the dose just causing loss of the corneal reflex) was 100 mg/kg. The rabbits were then given 1 g of charcoal with varying amounts of allylpropynal. By determining the smallest drug dose which just gave loss of the corneal reflex, it was estimated that the amount of drug adsorbed by 1 g of charcoal in vivo was 633 mg. This amount adsorbed was compared to an estimated amount which would be adsorbed in vitro at a pH value typical of that found in the small intestine. This value, based on Andersen's third study, was estimated as 1200 mg/g. Thus, Andersen concluded that charcoal, in vivo, adsorbs about 633/1200 or half of what it would adsorb in vitro at the same pH. This agrees with his earlier assessment of the relative degrees of adsorption in vivo versus in vitro.

In the final study of this classic series, Andersen (1948c) studied the adsorption of sulfanilamide by charcoal in vivo, using rabbits and dogs. This substance was used because: (1) its levels in blood could be determined very accurately, (2) it is not toxic enough to affect absorption, and (3) the amount in the blood was found to be proportional to the dose given (in the absence of charcoal). In both dogs and rabbits, varying amounts of charcoal were given simultaneously along with standard doses of the drug. The amounts of charcoal needed to give essentially zero appearance of the drug in the blood after a drug dose of 1 g were found to be about 4 g charcoal in dogs and about 6 g charcoal in rabbits. In vitro, the amounts of charcoal required for essentially total adsorption of sulfanilamide were estimated to be only one-third to one-half of the amounts required in vivo. Additional experiments were done with the charcoal delayed by 15, 30, and 45 minutes. These results, which will be discussed in some detail in Chapter 13, show clearly that, the greater the delay time, the less effective is the charcoal in reducing drug absorption, especially for the first ten hours.

V. REFERENCES

Andersen, A. H. (1944). Medicinal carbon, Dansk. Tids. Farm. *18*, 21.

Andersen, A. H. (1946). Experimental studies on the pharmacology of activated charcoal. I. Adsorption power of charcoal in aqueous solutions, Acta Pharmacol. Toxicol. 2, 69.

Andersen, A. H. (1947). Experimental studies on the pharmacology of activated charcoal. II. The effect of pH on the adsorption by charcoal from aqueous solutions, Acta Pharmacol. Toxicol. *3*, 199.

Andersen, A. H. (1948a). Experimental studies on the pharmacology of activated charcoal. III. Adsorption from gastro-intestinal contents, Acta Pharmacol. Toxicol. *4*, 275.

Andersen, A. H. (1948b). Experimental studies on the pharmacology of activated charcoal. IV. Adsorption of allylpropynal (allyl-isopropyl-barbituric acid) in vivo, Acta Pharmacol. Toxicol. *4*, 379.

Andersen, A. H. (1948c). Experimental studies on the pharmacology of activated charcoal. V. Adsorption of sulphanilamide in vivo, Acta Pharmacol. Toxicol. *4*, 389.

Cooney, D. O. and Kane, R. P. (1980). "Superactive" charcoal adsorbs drugs as fast as standard antidotal charcoal (letter), Clin. Toxicol. *16*, 123.

Cooney, D. O. and Wijaya, J. (1986). Effect of magnesium citrate on the adsorptive capacity of activated charcoal for sodium salicylate, Vet. Hum. Toxicol. *28*, 521.

Dedrick, R. L. and Beckmann, R. B. (1969). Kinetics of adsorption by activated carbon from dilute aqueous solution, Chem. Eng. Prog. Symp. Ser. *63* (No. 74), 68.

Hauge, S. M. and Willamann, J. J. (1927). Effect of pH on adsorption by carbons, Ind. Eng. Chem. *19*, 943.

11

Effects of Activated Charcoal on Major Classes of Drugs and Chemicals

In this chapter, we begin a detailed discussion of in vitro and in vivo studies of various specific drugs or chemicals, or classes thereof. This chapter will focus on "major" classes of drugs and poisons, and the following chapter will deal with "other" classes of drugs. Assigning certain drugs to one chapter or the other is admittedly arbitrary in many cases. The division was made to make the text more tractable (i.e., to avoid having one gigantic chapter). Included in this chapter will be common household chemicals, alkaloids, aspirin and other salicylates, hypnotics and sedatives, tricyclic antidepressants, cardiac glycosides, organic solvents, and ethanol.

This chapter will not include studies in which the charcoal was given solely according to a "multiple-dose" or "repeated-dose" regimen, as these types of studies will be considered separately in Chapter 14. Thus, the studies reported in this chapter will concern mainly single dose charcoal investigations. An exception to this will be those studies in which both single dose and multiple dose regimens were used. In this case, such studies will be considered here, with the primary focus being on the single dose aspects.

Some studies have involved resins such as cholestyramine, as well as charcoal. Such studies will be included here, since comparisons of the relative efficacies of charcoal and resins may be of interest in certain clinical situations. Studies dealing only with resins will not be considered here, as they are grouped together in Chapter 22.

Also, despite the fact that a separate part of this book will be devoted specifically to looking at the effects of delaying the administration of charcoal (Chapter 13, Section I), studies in which charcoal treatment has been initiated significantly after drug or poison ingestion (e.g., 30 min or more) will still be discussed in the present chapter, if those studies have involved major classes of drugs or poisons.

One more point needs to be made before embarking. In this chapter, in the following chapter, and indeed in many other places in this volume, many clinical studies are described in which changes in pharmacokinetic parameter values (AUCs, $t_{1/2}$ values, clearances, etc.) been determined with and without charcoal (or with and without other therapeutic interventions). What really matters, of course, are not changes in pharmacokinetic parameter values but rather the true clinical efficacies of different therapeutic approaches in terms of clinical course and improved patient outcome. However, the majority of clinically-based research articles give insufficient details regarding comparative clinical courses and patient outcomes observed with different treatment protocols, whereas all of the pharmacokinetic parameter information is normally presented. Moreover, descriptions of clinical courses and patient outcomes are, by their very nature, often subjective and qualitative, and therefore investigator-dependent. Thus, comparing several different studies in terms of clinical courses and patient outcomes is extremely difficult. Thus, most of what will be presented relative to clinical studies will, of necessity, be couched in terms of quantitative pharmacokinetic parameter information.

I. COMMON HOUSEHOLD CHEMICALS

Decker et al. (1968a) and Corby and Decker (1974) performed some basic in vitro adsorption studies on a wide range of drugs and toxic substances which are commonly found in the home. Their experiments involved adding the substances to 100 mL simulated gastric juice, then mixing in 5 g Norit A charcoal in 50 mL water, and shaking the resultant mixture at 37°C for 20 min. The charcoal was separated from the supernate, which was then analyzed for residual substance. Drugs supplied in tablet form were crushed beforehand to assure dispersal.

Table 11.1 shows the percentage of each drug adsorbed as a function of the number of tablets used. It was also determined that iodine and phenol (a constituent of calamine lotion) are well adsorbed; chlorpromazine, methyl salicylate, and 2,4 dichlorophenoxyacetic acid are moderately adsorbed; and malathion, DDT, N-methylcarbamate, and boric acid are poorly adsorbed. Mineral acids, alkalis, and compounds insoluble in aqueous acidic solution (such as tolbutamide) were not adsorbed to any measurable extent.

Gloxhuber (1968) stated that activated charcoal can be used advantageously in cases where detergents and cleaning agents have been ingested. Data on the adsorption of constituents of such materials, such as alkylbenzene sulfonate and benzalkonium chloride, onto Carbo medicinalis charcoal are presented.

II. ALKALOIDS

Kunzova (1937) reported some interesting studies with frogs. Strychnine was dissolved in a dilute salt solution, treated with various charcoals, filtered, and then injected into the lymph sacs of frogs. When the ratio of charcoal to strychnine was 65:1, convulsions appeared in 25 min; when the ratio was somewhat larger, convulsions failed to appear. Little variation between the different charcoals was observed. It appeared that the amount of charcoal needed to adsorb 1 mg strychnine ranged only between 65 and 76 mg. Some similar studies were done by Saunders et al. (1931). They shook activated charcoal with solutions of various drugs, filtered them, and injected the filtrates into dogs. Strychnine, brucine, adrenaline, histamine, and tyramine were completely inactivated. Acetylcholine and ephedrine solutions were partly inactivated.

Table 11.1 Drug Adsorption in Vitro by Activated Charcoal

Drug	Dose (mg)	Number of tablets	
		10	20
Acetylsalicylic acid	325	90	85
Amphetamine	5	94	92
Chlorpheniramine	4	96	96
Colchicine	0.5	94	92
Diphenylhydantoin	100	90	86
Ergotamine	1	92	90
Phenobarbital	32	86	45
Primaquine	25	97	94
Propoxyphene	32	100	85
Digitoxin	100	66	60
Probenecid	100	58	40
Quinacrine	325	68	26
Acetaminophen	325	23	8
Glutethimide	500	45	—
Meprobamate	400	25	—
Propylthiouracil	50	33	23
Quinidine	325	44	1
Quinine	325	32	1
Phenylbutazone	100	15	—
Ferrous sulfate	325	5	1
Chloroquine	500	6	—

[a]Values expressed as percent drug adsorbed after 20 minutes. From Corby and Decker (1974). Reproduced by permission of *Pediatrics*, American Academy of Pediatrics.

The reader may recall from Chapter 2 that various alkaloids (mostly strychnine but also morphine and nicotine) were used in the period of 1920–1940 as standard test substances for the evaluation of medicinal charcoals. As early as 1920, Joachimoglu (1920) showed that charcoal could effectively nullify the effects of strychnine given to dogs. Dingemanse and Laqueur (1926) did similar in vivo studies with pigs and Kunzova (1937) performed experiments with frogs. However, the greater portion of early work done with alkaloids was of the in vitro variety. Generally, the better charcoals were found to adsorb alkaloids quite well.

Andersen (1944, 1946) used strychnine nitrate, morphine hydrochloride, atropine sulfate, and nicotine in his in vitro studies. Maximal adsorbances for these were 950, 800, 700, and 700 mg per gram of charcoal, respectively. Andersen (1947; 1948a,b,c) also worked with nicotine and strychnine in his later in vitro studies which were described earlier in Chapter 10.

We now discuss a series of studies on strychnine compounds by a research group at the University of Arizona in the late 1960s and early 1970s. The first-named authors of the resulting papers are either Picchioni or Chin, the leading members of that group. In vitro work with strychnine sulfate was reported in a paper by Picchioni et al. (1966).

Forty milliliters of SGF, USP, containing 1.62 mg of strychnine sulfate were contacted with either 30 mg plain charcoal or 60 mg of universal antidote (which contained 30 mg charcoal). The percentages of the strychnine sulfate adsorbed were 40.6 and 35.1%, respectively. Thus, plain charcoal was superior to the same quantity of charcoal given in the form of universal antidote.

Picchioni et al. (1974) also studied the adsorption of strychnine sulfate (40 mg in 40 mL of solution) in vitro to 300 mg of each of four charcoals. The percentages adsorbed were: for Merck charcoal, 98.8; for Nuchar C charcoal, 97.5; for Norit A charcoal, 96.5; and for Darco G-60 charcoal, 66.3.

In the 1966 study by Picchioni et al., some in vivo work with strychnine-dosed rats was also done. The rats were given 20 mg/kg of strychnine sulfate and then either water, 350 mg/kg plain charcoal, or 700 mg/kg universal antidote (which contained 350 mg/kg charcoal). After 5 min, liver samples were taken and analyzed for the poison. Concentrations of strychnine sulfate in these samples in µg/g were: control (water), 28.5; plain charcoal, 8.2; and universal antidote, 15.5. Thus, both charcoal treatments were effective in reducing absorption of the strychnine, but plain charcoal was better. In other in vivo tests with rats, LD_{50} values were determined. With no charcoal treatment, the LD_{50} for strychnine sulfate was 9.9 mg/kg. With the use of 350 mg/kg charcoal, the LD_{50} was 36.5 mg/kg, and with 700 mg/kg universal antidote the LD_{50} was 28.2 mg/kg. Again, both charcoal treatments were effective, and plain charcoal was superior. Explanations for the lower efficacy of charcoal when given in the form of universal antidote were discussed in detail in Chapter 9, Section V.

Strychnine phosphate was given to rats in a study by Chin et al. (1969). The poison dose (20 mg/kg) was followed by either water, charcoal (200 mg/kg), Arizona montmorillonite (200 mg/kg), or evaporated milk (20 mL/kg). Again, liver samples obtained at 5 min were analyzed for the poison. The concentrations in µg/g were: control (water), 4.81; charcoal, 0.58; montmorillonite, 0.52; and evaporated milk, 1.06. Thus, all treatments were effective, with the charcoal and montmorillonite being the best and nearly equivalent. Again, LD_{50} values were determined; the values in mg/kg were: with water, 12.5; with 200 mg/kg charcoal, 30.9; with 200 mg/kg montmorillonite, 30.4; and with 20 mL/kg evaporated milk, 23.1. Thus, charcoal and montmorillonite were the best treatments, and again were essentially equivalent.

An extensive set of in vitro adsorption isotherms at various pH values has been presented by Henschler (1970) for nicotine, atropine, quinine, strychnine, yohimbine, aconitine, and veratrine. Adsorption values in the range of 0.5–0.8 g alkaloid per gram of charcoal were commonly found. Henschler also tested the in vivo effectiveness of charcoal in alkaloid poisoning using mice. He found that the ratio of LD_{50} values with charcoal to LD_{50} values without charcoal were 4.3, 18.2, 3.0, 2.3, 2.5, and 5.5 for six different alkaloids.

Buck et al. (1985) gave lethal doses of strychnine (65 mg/kg) to rats and then gave them single doses of three different charcoals. The 7-day survival rates were 0/10 (controls), and 3/10, 4/10, and 9/10 for the three charcoals (the amounts of charcoal used are not stated). The best charcoal was SuperChar.

Taken as a whole, the alkaloid studies described indicate that activated charcoal adsorbs this class of compounds very well, and is highly effective in preventing toxicity: if sufficient charcoal is used, and if the charcoal is given quickly (many of the alkaloids mentioned are highly toxic and can cause death within a short time if not promptly counteracted).

III. ASPIRIN AND OTHER SALICYLATES

A. In Vitro Studies

Several in vitro studies have shown that aspirin and other salicylates are well adsorbed by charcoal. Decker et al. (1968a) found that when 5 g Norit A charcoal in 50 mL water was added to simulated gastric fluid containing 10 aspirin tablets (325 mg each), 90% of the aspirin was adsorbed in 20 min, and when 20 aspirin tablets were used, 85% adsorption occurred in 20 min (these values appear in Table 11.1).

Phansalkar and Holt (1968) added 100 mg of four different charcoals to 25 mg sodium salicylate dissolved in 8 mL water plus 2 mL of an HCl-KCl buffer. After equilibration, the charcoals were found to have adsorbed the following amounts: Darco G-60, 20.0 mg; Nuchar CN, 23.0 mg; Norit A, 22.5 mg; and Norit USP, 24.5 mg.

Chin et al. (1970) studied the adsorption of aspirin by 30 mg of Merck USP charcoal from 0.1 g/L solutions of pH 1.5 and pH 8.5, and found the amounts adsorbed to be 3.78 and 2.44 mg, respectively. The lower adsorption at the higher pH, where the aspirin is ionized, is as one would expect.

Tsuchiya and Levy (1972), Sellers et al. (1977), and Boehm and Oppenheim (1977) also showed that aspirin adsorbs well to charcoal in vitro. The aspirin results of Sellers et al. are given later in Table 11.5.

B. In Vivo Studies

Decker et al. (1968b) carried out in vivo experiments with rats. When 125 mg activated charcoal was administered 30 min after a dose of 320 mg sodium salicylate, the serum salicylate concentrations at 60, 90, and 120 min were reduced by 66, 62, and 62%, respectively, compared to control animals. Additional work with dogs gave similar results. Studies by Decker et al. (1969) in humans also showed significant inhibition of aspirin absorption. Oral administration of 30 g activated charcoal to adult male volunteers 30 min after they had received 50 grains of aspirin resulted in serum salicylate concentrations roughly 50% of those in a control group.

Phansalkar and Holt (1968) found that dogs given 100 grains of aspirin plus either 60 g or 90 g activated charcoal had plasma salicylate concentrations that were very small (< 10%) compared to controls. However, it should be noted that the charcoal:drug ratios were roughly 11:1 and 14:1, respectively, so it is not surprising that the reductions in the salicylate levels were so large. Even when charcoal was delayed for 30 min, the charcoal soon halted the rise in plasma salicylate levels (the drug levels stopped rising at about 1 hr and began a steady fall).

Collombel and Perrot (1970) showed in vitro that 100 mg of various activated charcoals adsorbed 15–22 mg salicylate from a solution of 25 mg sodium salicylate in 10 mL 0.02 N HCl. In vivo experiments with rats given 100 mg/kg aspirin and 1 g/kg charcoal after different delay times (30, 60, and 120 min) showed that the charcoal was capable of lowering serum salicylate levels to varying degrees, depending on the amount of delay. Their results, shown in Table 11.2, suggest that the charcoal is quite effective at 30 min, somewhat effective at 60 min, and ineffective at 120 min.

Chin et al. (1969) reported on in vivo studies with dogs. Aspirin, 100 mg/kg in an aqueous suspension of 3 mL/kg, was administered by stomach tube to four groups of eight dogs each. One minute later three of the groups were treated with activated charcoal (500 mg/kg), Arizona montmorillonite (500 mg/kg), or evaporated milk (10 mL/kg).

Table 11.2 Salicylate Levels (mg/liter) in Rats

Hours since aspirin dose	No charcoal	Charcoal at 30 min	Charcoal at 60 min	Charcoal at 120 min
1	150	80	160	188
2	240	83	153	230
4	290	85	138	231
8	235	70	103	215
12	140	34	65	103
24	20	15	29	24

From Collombel and Perrot (1970).

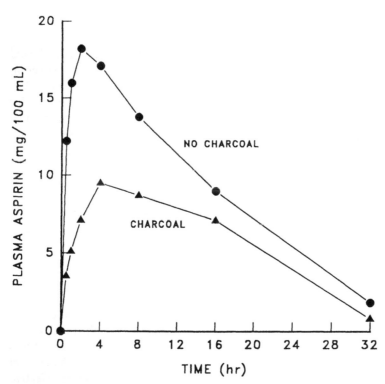

Figure 11.1 Effect of activated charcoal on plasma concentrations of aspirin in dogs. From Chin et al. (1969). Reproduced by permission of the American Pharmaceutical Association.

Blood samples taken at various times showed that only the activated charcoal was significantly effective in reducing aspirin levels. The results for the charcoal-treated dogs are shown in Figure 11.1. Chin et al. (1970) later repeated these experiments with aspirin as well as with various other drugs. The charcoal dosages were the same as in their 1969 study and the results for aspirin were very similar to those in the earlier study.

Picchioni et al. (1974) carried out in vivo aspirin studies in rats. The dosages were 100 mg/kg aspirin and 500 mg/kg of either Darco G-60 charcoal or Merck charcoal (given after a 1 min delay). Blood samples taken at 30 min showed aspirin levels of 22.3, 8.8, and 4.2 mg/100 mL for the control group, the Darco G-60 charcoal group, and the Merck charcoal group, respectively.

Levy and Tsuchiya (1969, 1972) reported on in vivo work with aspirin and human subjects. They found that charcoal, if administered promptly and in sufficient amounts, significantly inhibited the absorption of aspirin from solution, conventional tablets, enteric-coated tablets, and sustained-release tablets. The effects increased with the amount of charcoal given, decreased as the time delay between the aspirin and charcoal increased, and were decreased by the presence of food in the gastrointestinal tract. They also report that aspirin adsorption to charcoal is partly reversible in vivo. When aspirin was equilibrated in water with sufficient charcoal to give 50 or 99% adsorption, and these mixtures were then administered orally, 87 and 61%, respectively, of the total aspirin dose was recovered in the urine. The in vivo desorption of aspirin was most likely due to the higher pH of the intestinal fluid as well as to competitive effects of other constituents of the gastrointestinal fluid. Levy and Tsuchiya also showed that charcoal will reduce aspirin absorption even if administered 3 hr after aspirin ingestion, provided that the drug is still in the gastrointestinal tract at that time.

Tsuchiya and Levy (1972), in another paper, investigated the desorption question further. With in vitro measurements on aspirin adsorption to charcoal from pH 1.0 and 8.2 solutions, they showed that lower pH enhances aspirin adsorption. This suggests that aspirin will desorb from the charcoal as the charcoal/drug complex passes through the gastrointestinal system. In vivo experiments in human subjects were done which were similar to those done by Levy and Tsuchiya (1972), that is, aspirin was first adsorbed to the charcoal to varying extents before administration. Again, more aspirin was recovered in the urine than was present unbound in the mixture ingested. Thus, desorption must have occurred in vivo.

Another in vivo study with aspirin is that of Neuvonen et al. (1978). Six adults were given 1 g aspirin followed by 50 g Norit A charcoal. When the charcoal was given immediately, the peak serum aspirin levels were reduced by 95% relative to control values. However, the total systemic absorption decreased less (by 70%). Serum concentration data taken for 0–96 hr showed higher levels during the period of 24–96 hr when charcoal was used than when it was not. This is strong evidence that some desorption occurred as the drug/charcoal complex passed along the gastrointestinal tract. Nevertheless, the serum levels were still, in an absolute sense, rather small after 24 hr in all instances. The matter of desorption was discussed in considerable detail earlier in Chapter 9, Section III ("Stability of the Drug/Charcoal Complex").

The effect of a kaolin/pectin suspension as well as oral charcoal on aspirin absorption was studied by Juhl (1979). He gave 10 fasting volunteers three 325-mg aspirin tablets, followed by 240 mL water, 10 g charcoal in 240 mL water, 30 mL of a kaolin/pectin (Kaopectate) suspension with 210 mL water, 60 mL of the kaolin/pectin with 180 mL water, or 90 mL of the kaolin/pectin with 150 mL water (hence the volume given was

240 mL in all cases). The mean urine salicylate recoveries were 98.6, 69.5, 90.6, 94.6, and 95.3%, respectively, for these five treatments. Since the 5–10% reduction in aspirin absorption caused by kaolin/pectin was far less than the 30% reduction noted with charcoal, charcoal treatment was superior to kaolin/pectin.

Elonen and Neuvonen (1981) commented on an editorial in *The Lancet* which mentioned the slow absorption of enteric-coated aspirin. These authors mentioned that, after ingestion of enteric-coated aspirin, peak levels in the blood may not be reached for 60–70 hr. Therefore, to some extent, charcoal can "catch up" with overdoses due to enteric-coated aspirin and should be considered as an important treatment step.

Dillon et al. (1989) studied aspirin adsorption in vitro by two different charcoal preparations, SuperChar charcoal, and Actidose Aqua charcoal. The results showed that, at pH 1 and 8.1, the Superchar charcoal adsorbed more aspirin than did the Actidose Aqua (this is expected, as SuperChar has the higher internal surface area). They also did in vivo studies with 12 healthy volunteers who ingested 20 mg/kg aspirin, followed 1 hr later by 25 g charcoal, given as SuperChar or as Actidose Aqua. Urine was collected for up to 72 hr and analyzed for salicylates. The percentage of aspirin absorbed was 78% for the Actidose Aqua case, and 50% for the SuperChar case (the control value was 96%). Thus, in vivo the SuperChar formulation was the more effective.

McKinney et al. (1992) recently described a study in which 15 healthy volunteers were given charcoal prior to aspirin. In one phase, they were given 10 g charcoal 30 min prior to aspirin and in another phase the charcoal was given 60 min before. The aspirin dose was 975 mg in all phases. Urine was collected for 48 hr. Aspirin recovery in the urine was 88.8% for the control phase, 84.8% for the "charcoal 30 min prior" phase, and 85.8% for the "charcoal 60 min prior" phase. The volume of the charcoal suspension was 160 mL in each case and the aspirin was given as uncoated aspirin suspended in 220 mL water. These results suggest that charcoal administered 30–60 min before aspirin ingestion has little effect. The 160 mL of fluid volume of the charcoal suspension, plus the 30–60 min interval prior to drug administration, might have caused the charcoal to have emptied from the stomach before the aspirin was ingested, thus minimizing contact between the charcoal and the aspirin. A drug that is less quickly absorbed than aspirin could possibly "catch up with" the charcoal in the upper intestines, in which case an effect due to charcoal might occur.

All of the in vivo studies involving aspirin and sodium salicylate which we have just reviewed show emphatically that oral charcoal is very effective in reducing the absorption of these drugs, if the amount of charcoal employed is reasonable. Even if the charcoal is delayed for 30 min to 1 hr, it is generally still quite effective.

IV.   ACETAMINOPHEN AND INTERACTIONS WITH ACETYLCYSTEINE

Acetaminophen is an effective nonprescription, antipyretic, and mild analgesic agent which is used extensively in the United States. In European countries (where it is called paracetamol), it became popular even earlier than in the USA. Acetaminophen overdoses in both the USA and in Europe are now quite common. While acetaminophen is very safe in ordinary doses, persons who ingest large doses (15 g or more) develop liver damage; renal and myocardial necrosis have also been observed. The maximal necrotic effect appears 2–4 days after drug ingestion and death can occur between 2 and 7 days after the overdose. Other than the oral or intravenous administration of the specific antagonist N-acetylcysteine, or (less commonly) the oral administration of the antagonist D,L-methionine, only activated charcoal has been shown to be of therapeutic value in preventing hepatic necrosis. Thus, we survey here studies done to date which show that activated charcoal may be of great value in counteracting this drug.

A.   In Vitro Studies

Bainbridge et al. (1977) reported results on the adsorption of acetaminophen from a 2.5-g/L solution of this drug in simulated gastric fluid (pepsin omitted) to Norit A activated charcoal. They found that the percentages of acetaminophen adsorbed at charcoal:drug ratios of 0.8, 2.4, 4.0, 5.6, and 8.0 were 16.8, 58.2, 77.7, 94.5, and 98.3, respectively. A Langmuir isotherm of the following form was established for pH 1.2 conditions:

$$Q = \frac{C_f}{4.39C_f + 0.268}$$

where C_f is in g/L and Q is in g/g. At high drug concentrations, Q approaches 1/4.39 or 0.228 g/g. That is, the maximal adsorption capacity, Q_m, of the charcoal would be about 23% of its own weight.

Boehm and Oppenheim (1977) also reported that acetaminophen adsorbs well to activated charcoal in vitro. Three of the four charcoals they tested adsorbed more than 90% of this drug from simulated gastric fluid when a charcoal:drug ratio of 10:1 was employed. Van de Graaff et al. (1982) studied the adsorption of acetaminophen to 16 charcoals and resins in vitro at 37°C in simulated gastric fluid. Maximal binding capacities ranged from about 7.9 mole/kg (1.19 g/g) for superactive charcoal, 0.8–2.1 mole/kg (0.121–0.317 g/g) for most of the other charcoals, about 0.8 mole/kg for cholestyramine, and about 0.5 mole/kg (0.076 g/g) for two nonionic resins. When the tests were done in simulated intestinal fluid, maximal binding capacities for the charcoals fell an average of 33% compared to those for gastric fluid. Effects on adsorption caused by adding mannitol/sorbitol, N-acetylcysteine, or methionine to the test media were small.

These studies indicate that acetaminophen binding capacities for typical charcoals (excluding superactive charcoal) in simulated gastric fluids are in the range of 0.12–0.32 g/g, and such capacities are relatively good.

B.   In Vivo Studies

The first in vivo work with acetaminophen appears to be a brief study by Levy and Gwilt (1972) in which it was stated that 10 g activated charcoal administered immediately after the ingestion of 1 g acetaminophen by two human subjects resulted in 77 and 69% reductions in the amounts of drug absorbed from the gastrointestinal tract.

A year later, Dordoni et al. (1973) reported on experiments in which 2-g oral doses of acetaminophen were given to 14 human subjects. Seven of the subjects were then given 10 g activated charcoal as a suspension in methylcellulose solution (20 g/100 mL). The charcoal resulted in much lower plasma concentrations (Figure 11.2) and the bioavailability of the drug (measured as the area under the plasma concentration versus time curve over the period 0–120 min) was reduced by an average of 63% (range 32–87%). Four other subjects were given 12 g cholestyramine resin in 200 mL of water after ingesting the acetaminophen. From Figure 11.2 it is clear that this adsorbent also

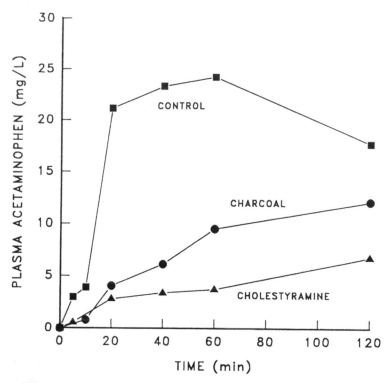

Figure 11.2 Effect of activated charcoal and cholestyramine resin on plasma concentrations of acetaminophen in human subjects. From Dordoni et al. (1973). Reproduced with permission of the BMJ Publishing Group.

was very effective; reductions in bioavailability averaged 62% (range 30–98%). In additional tests, the charcoal and cholestyramine were given 60 min after acetaminophen ingestion. In these cases, the bioavailabilities (determined for the 60–120 min period) averaged 23 and 16% less for charcoal and cholestyramine, as compared to the control groups. The adsorbents were, therefore, only about one-quarter to one-third as effective when their administration was delayed an hour.

In vivo experiments with acetaminophen have been done by Lipscomb and Widdop (1975) using pigs. The acetaminophen (10 g) was given either in the form of a suspension or tablets. One group of pigs was then given 50 g activated charcoal in 250 mL water. The results of several experiments showed that: (1) drug absorption from the suspension form was much faster than from the tablet form, (2) activated charcoal reduced blood drug concentrations considerably if given at once, and (3) even if delayed for 1/2 or 1 hr, activated charcoal still gave a very significant subsequent reduction in blood levels. One interesting fact that was noted was that, while the pigs' blood levels were often at what would be hepatotoxic levels for humans, no clinical, biochemical, or histological evidence of liver damage was noted.

Another extensive set of results on the in vivo adsorption of acetaminophen in humans is that reported by Levy and Houston (1976). One gram of acetaminophen was given to volunteers as an elixir (Tylenol), as a suspension (Liquiprin), or as tablets

Table 11.3 Effect of Charcoal Dose on Acetaminophen at Constant Drug:Charcoal Ratio of 1:10[a]

Charcoal dose (g)	Percent acetaminophen recovered in urine
5	42.5
10	34.9
20	22.6
30	14.8

[a]Drug given as elixir followed immediately by charcoal in 200 mL water (same subject in all tests). From Levy and Houston (1976). Reproduced by permission of *Pediatrics*, American Academy of Pediatrics.

(Tylenol). In the control experiments 200 mL water was then given. In the other experiments each subject was immediately given a slurry of 5 or 10 g Norit charcoal in 200 mL water. Urine samples were collected at set times up to 36 hr and were assayed for total acetaminophen (i.e., acetaminophen and its metabolites). For the experiments involving the elixir, the amounts of acetaminophen absorbed, relative to that for the elixir without charcoal, were 52.8 and 38.6% for the 5 and 10 g charcoal doses, respectively. Clearly, the charcoal was quite effective in reducing acetaminophen absorption from the elixir. When 10 g of charcoal was given after a 30 min delay, the percentages of acetaminophen absorbed from the different dosage forms were: elixir, 68.9; suspension, 49.7; and tablet, 46.0. Thus the 30 min delay of 10 g charcoal increased absorption from the elixir from 38.6 to 68.9%, relative to the case of the elixir without charcoal.

Levy and Houston (1976) also did a series of tests where the drug:charcoal ratio was held at 1:10, but different amounts of each were given. Table 11.3 shows the results. Previous studies by Levy and Tsuchiya (1972) on aspirin showed the same trend exhibited here, i.e., the effectiveness of charcoal increases with the amount of charcoal even when the ratio of drug to charcoal is held constant.

Another factor which favors the effectiveness of charcoal in acute acetaminophen overdose is that this drug inhibits gastric emptying, just as aspirin overdoses have been found to do (Levy and Tsuchiya, 1972). This in turn slows down the drug's absorption and suggests that activated charcoal administered even 2–3 hr after the acetaminophen ingestion may be significantly effective.

Levy and Houston (1976) recommended that a dose of 50–100 g activated charcoal should be administered to adults (and proportionately less to children) in acetaminophen overdose cases. Whether reducing the total amount absorbed or only slowing the rate of absorption of the drug (e.g., lowering the peak plasma concentration) would reduce liver damage is not known. It has not yet been shown whether it is the total amount of drug which is harmful or whether it is the peak concentration which is more critical. In either case, the use of activated charcoal would seem to be an effective therapeutic tool.

Van de Graaff et al. (1982) studied acetaminophen pharmacokinetics in vivo using dogs. After an oral acetaminophen dose of 0.6 mg/kg, the dogs were given: (1) water alone, (2) Norit A or Nuchar charcoal (3 g/kg) in water, (3) mannitol/sorbitol (2 g/kg), (4) castor oil (3 mL/kg), (5) charcoal plus mannitol/sorbitol, or (6) charcoal plus castor

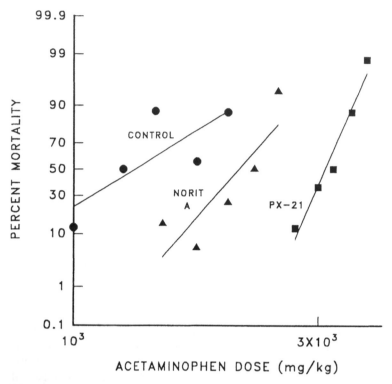

Figure 11.3 Acute lethality of acetaminophen in untreated mice and in mice given 7.4 g/kg of either Norit A activated charcoal or Amoco PX-21 activated charcoal. From Van de Graaff et al. (1982). Copyright 1982 Williams & Wilkins. Reproduced with permission of the Williams & Wilkins Company.

oil. The Norit A charcoal reduced the 0–11 hr AUC by 94.5%, and Nuchar caused a 92.7% reduction. Thus, the charcoal was highly effective in reducing acetaminophen absorption. Mannitol/sorbitol alone reduced the AUC by 31%. When combined with the charcoals, mannitol/sorbitol gave AUC values 75% greater than with the charcoals alone. Castor oil reduced the AUC by 14% compared to controls, while its combination with the charcoals caused the AUCs to more than double as compared to the AUCs determined for charcoal alone. Thus, the cathartics strongly interfered with the action of the charcoals.

Van de Graaff et al. also studied the effects of charcoal on acetaminophen lethality in mice. They found an LD_{50} of 1321 mg/kg in mice using acetaminophen alone. When 7.4 g/kg Norit A was given along with the drug, the LD_{50} increased 61% to 2128 mg/kg, and when 7.4 mg/kg of superactive charcoal (Amoco PX-21) was given, the LD_{50} was increased 136.5% to 3137 mg/kg. Figure 11.3 shows the results.

Acetaminophen pharmacokinetics were studied by Rose et al. (1991) in ten adult males who were given 5 g acetaminophen (in elixir form) and then were given 30 g charcoal administered 15 min later (in one test), 30 min later (in a second test), or 120 min later (in a third test). In a control test, the drug level peaked at 1.42 hr and the drug was 97% absorbed by a mean time of 2.05 hr. In the charcoal tests, the reductions in the recoveries of acetaminophen and its metabolites in the urine were 48, 44, and 33% for the doses given at 15, 30, and 120 min, respectively. The data taken under control conditions indicated that acetaminophen in the elixir form is fairly rapidly absorbed (absorption half-life = 0.41 hr). Hence, by the time the charcoal dose at 120 min had been given, there would have been very little drug in the GI tract; and yet there was a 33% reduction in acetaminophen recovery in the urine in this case. This suggests that there was adsorption of acetaminophen to charcoal after absorption due to diffusion of the drug through the GI tract walls or due to secretion of the drug into the bile.

Mofenson and Caraccio (1992) commented on the study by Rose et al., wondering why they did not use the solid form of acetaminophen, which is the form most commonly encountered in overdose situations. They then told about a 10-month-old boy who had ingested many 500 mg Tylenol Caplets and had a blood level of 198 μg/L 2 hr after ingestion. They gave him charcoal at 1 hr and again at 3 hr post-ingestion. The 4 hr drug level was 85 μg/L and they interpreted this as evidence that the charcoal was effective in reducing acetaminophen absorption. Based on this, Mofenson and Caraccio stated that Rose's group should have considered giving charcoal later than 2 hr after the drug. In reply, Rose et al. (1992) stated that they simply chose to study the elixir form in order to provide useful information on treating overdoses from elixirs, and that charcoal was not given beyond 2 hr because their objective was to determine the effect of charcoal given at the particular times they selected on the absorptive phase of acetaminophen.

From the studies just described, it is clear that oral activated charcoal effectively reduces acetaminophen absorption, if given in sensible amounts, and can be effective even if the charcoal is delayed for a reasonably short time (e.g., 30 min to 1 hr.)

C. Interactions with N-acetylcysteine

N-acetylcysteine (NAC), given orally or IV, is a specific antidote for acetaminophen. D,L-methionine, given orally, is also a specific antidote, but is less commonly used. The optimum required NAC dose amounts and dosage schedule are not clear. NAC is given according to several protocols: US 72-hr oral, UK 20-hr IV, and US 48-hr IV (the IV protocol is approved for use in only in few US centers since it is experimental and is presently somewhat risky, as commercial NAC preparations are not tested for sterility and nonpyrogenicity). It appears that all three protocols have about the same clinical efficacy in preventing acetaminophen hepatotoxicity (which is due to depletion of glutathione levels). It is not clear exactly how much additional NAC should be given, if any, when charcoal is also given to the patient. This question has been the subject of several studies which will soon be discussed.

Flanagan and Meredith (1991) reviewed the use of NAC in clinical toxicology. They recommend, in agreement with the US 72-hr protocol, that a loading dose of 140 mg/kg be given orally as soon as possible, followed by oral maintenance doses of 70 mg/kg every 4 hr for 17 additional doses. Oral doses are contraindicated in the presence of coma or vomiting, in which case intravenous NAC should be used. Nausea/vomiting and diarrhea are common side effects of NAC (50 and 35% rates of occurrence, respectively, have been reported). Pharmacokinetic data (Holdiness, 1991) indicate that NAC alone shows a peak serum level 1–2 hr after administration, has a volume of distribution of 0.33–0.47 L/kg, and exhibits about 50% protein-binding 4 hr after being given. Since activated charcoal is often given to counteract acetaminophen overdoses, one obvious question which arises is whether charcoal would adsorb N-acetylcysteine if both are administered to a patient.

Two brief letters (Batizy, 1980, and Greensher et al., 1980) stated without details that charcoal interferes with acetylcysteine and therefore, if charcoal has been administered, lavage must be performed before initiating treatment with acetylcysteine. Two other brief letters, one by Llera (1983) and one by Levy (1983) also mention that charcoal adsorbs acetylcysteine, and should not be given in acetaminophen overdose if acetylcysteine is to be used, although Levy points out that acetylcysteine "can be administered intravenously and may be more effective by that route." Several studies have addressed this issue in a scientific way. The studies are of two types: in vitro studies to determine how well N-acetylcysteine and methionine adsorb to charcoal, and those which have evaluated interference effects in vivo.

We shall address the in vitro studies first. Chinouth et al. (1980) studied the in vitro adsorption of N-acetylcysteine to charcoal from SIF, USP (pH 7.5) and in very dilute sodium hydroxide solutions (pH 7.5). Norit SGL charcoal was used. NAC was found to adsorb well to the charcoal, but the degree of adsorption was less in the simulated intestinal fluid environment. Since SIF contains pancreatin, which presumably can adsorb to charcoal and take up some of the adsorption sites, this result is not unexpected. They quote maximal adsorption capacities; however the value they give for the sodium hydroxide environment is 4.63 g NAC/g charcoal, which is impossibly high (adsorption capacities for charcoals which are not superactive are rarely above about 0.3–0.4 g/g).

Another in vitro study is that of Klein-Schwartz and Oderda (1981). They studied NAC and methionine (METH) adsorption to charcoal in roughly 0.1 N HCl (for METH) or in distilled water (for NAC). The ratio of NAC to charcoal was 1:3 and 1:6 and the ratio of METH to charcoal was 1:12, 1:24, and 1:40 in their samples. Adsorption in the solutions, which were agitated, was rapid. The percentage METH adsorbed was (with increasing charcoal dose) 46.9, 76.8, and 89.5%, while the percentage of NAC adsorbed was (with increasing charcoal dose) 54.6 and 96.2%. The authors concluded that NAC or METH should not be used concomitantly with charcoal in acetaminophen overdose.

Van de Graaff et al. (1982) also studied NAC and METH adsorption to several charcoals in vitro and found maximal binding capacities for NAC in gastric fluid of 0.8–1.3 mole/kg, and in gastric plus intestinal fluid of 0.2–0.5 mole/kg. For methionine, 26–36% adsorption occurred in gastric fluid and 10–24% adsorption in gastric plus intestinal fluid was found (Q_m values were not given). They also determined that NAC or METH, in the amounts they used, did not interfere significantly with acetaminophen adsorption by the charcoals. However, since NAC and METH alone adsorbed reasonably well to charcoal, some interference with acetaminophen adsorption to charcoal would be expected if the amounts of NAC or METH involved were appreciable.

Another in vitro study is that of Rybolt et al. (1986) in which the adsorption of acetaminophen and NAC was studied with Nuchar SA charcoal in SGF (pH 1.2) and in SIF (pH 7). Charcoal:drug ratios ranged from 1 to 7. When the substances were used alone, acetaminophen adsorption was greater than NAC adsorption at both pH values. When both substances were in solution together, the NAC hardly affected the acetaminophen adsorption but the acetaminophen strongly affected NAC adsorption. Maximal adsorption Q_m values for acetaminophen were decreased from 2.14 to 2.05 mmol/g at pH 7 when NAC was added and from 1.93 to 1.89 mmol/g at pH 1.2 when NAC was added. The Q_m values for NAC decreased from 3.94 to 0.49 mmol/g at pH 7 when acetaminophen was added and from 1.49 to 0.52 mmol/g at pH 1.2 when acetaminophen was

Table 11.4 Acetaminophen Lethality in Mice After 24 hr (%)

Drug dose (g/kg)	Control	Charcoal alone	METH alone	METH plus charcoal	NAC alone	NAC plus charcoal
1.5	45	17	30	0	17	12
2.0	100	50	100	75	89	71
2.5	100	85	100	100	100	100

From Van de Graaff et al. (1982). Copyright 1982 Williams & Wilkins. Reproduced with permission.

added. Thus, the authors concluded that concomitant administration of NAC and charcoal is appropriate because the charcoal would adsorb acetaminophen without significant interference and there would at least be some NAC present to provide an antidotal effect.

While in vitro studies are of some value in shedding light on the NAC/charcoal question, the major deficiencies of such studies are that: (1) the choices of how much charcoal to use, how much NAC to use, the concentrations and pH values of the solutions, and so forth, are so arbitrary that their relevance to in vivo conditions is highly uncertain, and (2) the lack of competing solutes, food, and other substances which occur in the in vivo environment makes the relevance of in vitro results even more dubious. Thus, we now turn to in vivo studies involving NAC.

North et al. (1981a) provided one of the first in vivo studies of the concomitant use of charcoal and NAC. Three subjects were each given 140 mg/kg NAC orally without charcoal and also with a 50-g oral charcoal dose 15 min prior. Charcoal lowered the 0–8 hr AUC by 8% (not statistically significant). Peak heights and peak times were also statistically the same. These data suggest that charcoal does not affect NAC absorption in vivo, a result one would not have expected based on the in vitro studies discussed above, which show that charcoal adsorbs NAC reasonably well. No explanation for this discrepancy was offered.

Van de Graaff et al. (1982) studied the concomitant use of charcoal and NAC, and of charcoal and METH, on lethality in acetaminophen-dosed mice. The results are shown in Table 11.4. Clearly, charcoal (Amoco PX-21) alone reduced lethality significantly. METH alone had a slight effect at the lowest drug dose, and NAC alone had a significant effect at the two lowest drug doses. Thus, charcoal alone > NAC alone > METH alone. When added to the charcoal, METH helped at the lowest dose and hurt at the next two higher drug dose levels. The same pattern was true for NAC/charcoal versus charcoal alone. Thus, in general, it appears that the concurrent use of METH or NAC with charcoal should be avoided, at least in mice.

Renzi et al. (1985) also carried out an in vivo study, in which ten subjects were given oral 140 mg/kg NAC (the same dose used by North's group) without charcoal and with a simultaneous 60-g charcoal dose in slurry form (Antidose, Mead-Johnson). Again it was found from data on blood NAC levels versus time over 12 hr that no significant differences existed in the AUC values, the peak heights, or the half-life values. They noted that the results were in apparent conflict with in vitro findings. In a comment on this study by Krenzelok (1986), it was pointed out that the charcoal did reduce the AUC in Renzi's study by 18%, which was not statistically significant, but that in a similar

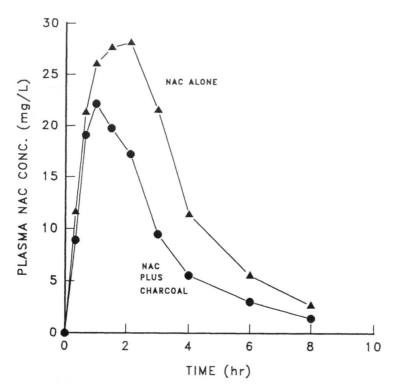

Figure 11.4 Plasma levels of N-acetylcysteine (NAC) following 140 mg/kg oral doses alone or accompanied with 100 g activated charcoal. From Ekins et al. (1987). Reproduced by permission of the W. B. Saunders Company.

study by Ekins et al., presented at a meeting in 1984, a 100-g charcoal dose (i.e., 40 g more) reduced the AUC value for NAC by 39% and the peak height by 28% (both effects were statistically significant). Figure 11.4 shows the results of Ekins et al., and it is evident that the larger charcoal dose did cause a significant reduction in NAC absorption. This study was published in 1987 (see Ekins et al. 1987). Donovan and Renzi (1986), in a reply to Krenzelok's letter, pointed out that two studies thus showed no effects with 50- and 60-g charcoal doses, and one study showed definite effects with a 100-g charcoal dose. Hence, a charcoal dose of up to 60 g can be given without apparent influence on NAC absorption. However, if one wishes to give a higher charcoal dose (e.g., 100 g), one can counteract the effect of charcoal on NAC absorption by simply giving the patient more NAC. This strategy, which is remarkably simple, was mentioned by Ekins et al. (1987), who state that with a 100-g dose of charcoal, "doses of NAC (should) be increased by 40% to compensate for the decreased oral absorption of NAC." Donovan and Renzi also mention that charcoal may have the added benefit of reducing the nausea associated with oral NAC use, thereby making NAC therapy more tolerable.

Watson and McKinney (1991) discussed the studies of North et al., Renzi et al., and Ekins et al.. They pointed out that the plasma AUC values in these three studies ranged from 0.3 to over 3000 mg-hr/L, a 10,000-fold variation. They also mention that studies have shown the measurable systemic bioavailability (plasma AUC) of NAC after oral administration to be only 6–10% (NAC has a large first-pass effect; it is metabolized to cysteine, diacetylcysteine, cystine, and taurine, and is used in the synthesis of various peptides and proteins). Thus, since the remainder of the oral dose can not be accounted for, studying the effect of charcoal on NAC AUC values is extremely uncertain. They state: "In this case, plasma AUC would be like the tip of an iceberg. Although it can be measured, it is not known whether it is an accurate reflection of what happens to the most important part of the iceberg, the large portion under the water."

Chamberlain et al. (1993) investigated the effectiveness of a "supranormal" dose of NAC in overcoming the effects of charcoal on NAC bioavailability in healthy human subjects. In Phase I of their cross-over study, subjects ingested 3 g acetaminophen and then, 1 hr later, NAC (140 mg/kg). In Phase II, the subjects again ingested 3 g of acetaminophen, followed 1 hr later by 60 g charcoal and a supranormal NAC dose of 235 mg/kg. Serum NAC levels were measured every 30 min for 6 hr, by which time the serum NAC levels had fallen to zero. For the NAC, AUC values were 4267 mg-min/L in Phase I and 5849 mg-min/L in Phase II. Peak NAC and time-to-peak were not significantly different. Serum acetaminophen was measured at 4 hr, and the levels were 19.4 mg/L in Phase I and 12.4 mg/L in Phase II. Thus, charcoal lowered the acetaminophen level in the blood by 36%. At the same time, when the larger-than-normal NAC dose was used, the bioavailability of NAC in the presence of charcoal was very good.

Spiller et al. (1994) also have determined whether charcoal interferes with NAC therapy for acetaminophen overdose. Acetaminophen overdose cases ($N = 100$) in three regional Poison Control Centers were treated with charcoal followed by NAC therapy, by standard NAC therapy alone, or by charcoal followed by increased NAC to compensate for charcoal binding of the NAC (the amounts of charcoal and NAC given in these cases are not detailed). The authors looked at maximal SGOT and acetaminophen levels, and concluded that the administration of charcoal before NAC did not alter the efficacy of NAC therapy.

Brent (1993) has questioned the need for the supranormal NAC dose recommended by Chamberlain et al.. He points out that one of the challenges of oral NAC therapy "is the frequent emesis of the Mucomyst [NAC] given its unpleasant taste and odor. To use a 40% larger dose of this agent may further promote emesis and thus ultimately compromise the treatment of the poisoning for which it is given." Brent outlines protocols for patients who present within 1–2 hr of acetaminophen overdose (he recommends charcoal first, then NAC later if the 4 hr drug level warrants it) and for patients who present more than 1–2 hr after ingestion (he recommends only NAC, if toxic levels exist). For polydrug poisoning, he states that both charcoal and NAC may be needed but that the effect of charcoal on NAC bioavailability appears to be small enough that supranormal doses of NAC are not needed.

The conclusions to be drawn from the in vivo studies just reviewed are reasonably clear: if a large charcoal dose (e.g., 100 g) is to be given with NAC, the charcoal can be expected to reduce the NAC bioavailability significantly (on the order of 40%, according to the study of Ekins' group) and a larger-than-normal NAC dose should be considered. The exact size of the supranormal NAC dose will clearly depend on the size of the charcoal dose (e.g., if 75 g charcoal is used, perhaps only a moderate increase in the NAC dose would be needed, whereas with a 100-g charcoal dose, the NAC may have to be increased substantially). Overlying all of this uncertainty is the further uncertainty that the optimal NAC dose amounts and dosage schedule, even in the absence of charcoal, have not been established. Clearly, further clinical studies are needed.

V.   HYPNOTICS AND SEDATIVES

A.   In Vitro Studies

We have already discussed the early work done by Andersen (1946, 1947, 1948a) in which it was demonstrated that barbiturates like diethylbarbituric acid adsorb well to activated charcoal, both in vitro and in vivo.

Sellers et al. (1977) determined, quantitatively, expressions for the Langmuir isotherms for glutethimide, methaqualone, secobarbital, pentobarbital, phenobarbital, amobarbital, and six other drugs, for in vitro adsorption to charcoal from pH 1.3 and 10.8 solutions. Table 11.5 shows their results. This table also includes many drugs which are not hypnotics or sedatives, but since it includes more hypnotics/sedatives than any other class of drugs we present the table here. The entries marked (—) are cases where the solubility of the drug at the given pH (10.8) was insufficient to allow determination of the adsorption isotherm accurately.

Likewise, Smith et al. (1967) also investigated the in vitro adsorption of pentobarbital and five other drugs to activated charcoal and Alaskan montmorillonite. Boehm and Oppenheim (1977) studied amobarbital adsorption by charcoal from SGF and Henschler (1970) determined the adsorption isotherm for phenobarbital from a pH 2.0 solution. Each of these studies support the general conclusion that charcoal adsorbs all of these hypnotics and sedatives to a significant extent.

Chin et al. (1970) studied the adsorption of pentobarbital sodium by 30 mg of Merck USP charcoal from solutions of 4 mg of the drug at pH 1.5 and 8.5. They found that the amounts of drug adsorbed were 3.76 and 2.92 mg, respectively. Of course, at the higher pH the pentobarbital is ionized and thus adsorbs less than at the lower pH.

Picchioni et al. (1966) described the adsorption of pentobarbital sodium from 40 mL of a 45 µg/mL solution by 30 mg of charcoal or 60 mg of universal antidote (which contains 30 mg charcoal). The plain charcoal and universal antidote adsorbed 53.3 and 35.0% of the pentobarbital, respectively. Later, Picchioni et al. (1974) did some other in vitro tests with pentobarbital sodium. Four different charcoals (300 mg) adsorbed the

Table 11.5 Maximal Binding Capacities for Various Drugs at 37°C

Drug	Q_m at pH 1.3 (mg/g)	Q_m at pH 10.8 (mg/g)
Amobarbital	51	47
Aspirin	262	141
Amitriptyline	133	—
Chlordiazepoxide	157	—
Diazepam	136	—
Methaqualone	179	—
Glutethimide	252	—
Pentobarbital	103	95
Phenobarbital	70	56
Propoxyphene HCl	127	—
Propoxyphene napsylate	137	—
Secobarbital	124	85

From Sellers et al. (1977). Reproduced by permission of the American Pharmaceutical Association.

following percentages of pentobarbital from a solution of 40 mg of drug in 40 mL: Merck, 43.8; Nuchar C, 44.5; Norit A, 43.0; and Darco G-60, 26.8.

The in vitro adsorption of phenobarbital to charcoal was studied by Javaid and El-Mabrouk (1983) using solutions of phenobarbital sodium in SGF (pH 1.2) and various buffers (pH 4, 6, and 8). Charcoal from BDH Chemicals Ltd. (Poole, England) was used. The Q_m values, based on fits to the Langmuir equation, were 292, 265, 259, and 261 mg/g at pH 1.2, 4, 6, and 8, respectively. These capacities are fairly high. Since the pK_a of phenobarbital is around 7.4, it is ionized to the extent of about 80% at pH 8, and thus the Q_m value at pH 8 would be expected to be substantially less than at the other pH values. The authors in fact state that the amount of drug adsorbed at pH 8 is "significantly lower than from solutions of pH 1.2, 4, and 6." Thus, the value of 261 mg/g at pH 8 is certainly a misprint (perhaps it should have been 161).

Two in vitro studies of pentobarbital adsorption by charcoal will be discussed here. In the first, Curd-Sneed et al. (1986) reported on sodium pentobarbital adsorption by three types of charcoals, USP (Mallinckrodt, Paris, KY), SuperChar, and Darco G-60, from aqueous solution and from buffer solutions of pH 8.1, 9, 9.7, 11, and 12. Solution pH had no effect, as one would expect since the pH values were all much higher than the drug's pK_a value, and hence the drug was in nearly the same state of ionization at all pH values employed (i.e., essentially totally ionized). SuperChar, having by far the highest surface area (2800–3500 m²/g) adsorbed the drug best, followed by USP (surface area about 1000 m²/g), and then the Darco G-60 (surface area about 650 m²/g). A second study by Curd-Sneed et al. (1987a) was much the same. Pentobarbital adsorption to the same three charcoals was examined using solutions prepared with distilled water or 70% w/v sorbitol. This time, Q_m values were determined. From water solution, the Q_m values for the SuperChar, USP, and Darco charcoals were 1141, 580, and 381 µmoles/g, respectively, and from sorbitol solution they were 716, 511, and 356 µmoles/g. The results indicate that the Q_m values are proportional to surface area and are decreased in the case of SuperChar charcoal by the presence of sorbitol. Since the contact time between the charcoal and the solution was only 10 minutes and since sorbitol makes the solution viscous, one can not tell if the sorbitol effect is due to a decreased rate of drug adsorption or due to a decreased capacity for adsorption. Had the samples been allowed to come to adsorption equilibrium, it would have been possible to distinguish between rate and equilibrium effects.

In a letter commenting on this last study, Weaver (1987), Medical Director of Gulf Bio-Systems (which marketed the SuperChar at the time) pointed out that SuperChar has about 65 wt% water, the USP charcoal has about 15 wt% water, and Darco G-60 has about 12 wt% water. Thus, unless one dries them, the amounts weighed out do not represent the true weights of the charcoal alone (he pointed out that his company adjusted for this when it bottled and labeled the SuperChar). Weaver also mentioned that his company's product (no longer available) used 26% w/v sorbitol; thus, the effects seen with a 70% w/v sorbitol solution might not carry over to the case of a 26% w/v sorbitol solution.

Curd-Sneed and Stewart (1987) replied to this letter, saying that they decided not to dry the charcoals in an oven but to use them as received, since drying is not a standard procedure in the clinical setting. They agreed that, had the SuperChar charcoal been dried beforehand, its binding capacity per unit dry weight would no doubt have been higher. The letter by Weaver and this reply make the important point that charcoals can differ greatly in water content, and so binding capacities per unit weight of "as received" charcoal, while perhaps being more realistic in a practical sense, do not always reflect the comparative binding capacities one would determine using oven-dried charcoals.

Phenobarbital adsorption from SIF and SGF by two charcoals (SuperChar and Darco KB-B) was studied by Wurster et al. (1988). In SIF, the maximal adsorption capacities were 418 mg/g for the Darco charcoal, and 1064 mg/g for the SuperChar charcoal; in SGF, these two values were 373 and 1023 mg/g, respectively. Several comments can be made. First, the SuperChar adsorbed 2.5–2.7 times as much drug as did the Darco charcoal. The ratio of surface areas is 2773/1511 or 1.8; thus, the Darco KB-B should have been more effective. A reason might be that some of the pores in the Darco charcoal were inaccessible to the drug. Secondly, at the pH of SIF (7.5) the phenobarbital would be about half dissociated and thus should have adsorbed substantially less well than at pH 1.2 (the pH of SGF). Yet, it did not. The reason for this is unclear.

A related paper by Burke et al. (1991), which is from the same group as the Wurster paper, gives further in vitro data on phenobarbital adsorption from SIF. Besides the Q_m values for SuperChar and Darco KB-B, which are now given as 383 and 980 mg/g (versus the figures of 418 and 1064 mg/g given before), Q_m values for Norit B Supra (surface area 1510 m^2/g) and Norit USP (surface area 940 m^2/g) of 483 and 184 mg/g, respectively, are quoted. The Q_m for the Norit USP charcoal seems surprisingly low, even considering that it has the lowest surface area of the four charcoals.

The large number of in vitro studies carried out with hypnotics and sedatives, especially barbiturates, attests to the significance and frequency of this class of drugs in overdose situations. Fortunately, these studies show conclusively that such drugs have high intrinsic affinities for charcoal.

B. In Vivo Studies

Except for the very early in vivo work by Andersen (1947, 1948a), the first in vivo studies with hypnotics and sedatives were performed by Picchioni's research group at the University of Arizona. Picchioni et al. (1966) did in vivo studies with rats who were given 70 mg/kg of the drug followed by 350 mg/kg of plain charcoal or 700 mg/kg of universal antidote (which contained 350 mg/kg charcoal). Ten minutes later, blood samples were drawn and analyzed. Blood drug levels in mg/100 mL were: in control tests, 5.28; with plain charcoal, 2.41; and with universal antidote, 2.87. Thus, both charcoal treatments were effective in lowering pentobarbital absorption, with the plain charcoal being slightly better. Acute toxicity tests with rats showed LD$_{50}$ values in mg/kg as follows: in control tests, 49; with 350 mg/kg plain charcoal, 123; and with 700 mg/kg universal antidote, 87. In this case, the plain charcoal was much more effective than universal antidote.

Later, Picchioni et al. (1974) did similar studies in vivo with rats using pentobarbital sodium and different brands of activated charcoals. In vivo, after a drug dose of 50 mg/kg and charcoal doses of 250 mg/kg, blood drug levels at 60 min in mg/100 mL were: in control rats, 4.1; in Merck charcoal-treated rats, 1.8; and in Darco G-60-treated rats, 2.2.

Another study from the same group (Chin et al., 1973) reported on in vivo studies in which rats were given 75 mg/kg pentobarbital sodium and then Merck charcoal in charcoal:drug ratios of 1:1, 2:1, 4:1, or 8:1. Drug blood levels at 20 min were reduced from control levels by 7, 38, 62, and 89%, respectively, for the four charcoal:drug ratios. Yet another paper (Chin et al., 1970) concerned in vivo studies using dogs and various drugs, among which were pentobarbital, barbital, and glutethimide. Figures 11.5 and 11.6

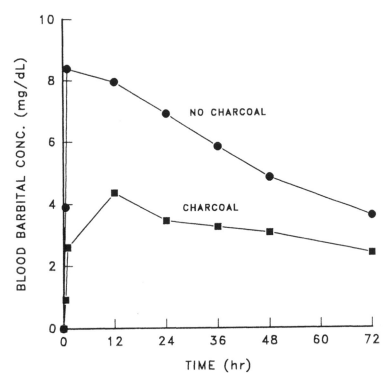

Figure 11.5 Effect of a 5:1 ratio of activated charcoal:drug on blood concentrations of barbital in dogs. The charcoal was given 1 min after the drug. From Chin et al. (1970). Reprinted with permission of Academic Press, Inc.

show blood drug concentrations versus time for the barbital and glutethimide. The drug doses were 30 and 100 mg/kg, respectively. The charcoal (given at 1 min time) amounts were five times the drug amounts. It is clear that the charcoal, at the 5:1 ratio of charcoal to drug, was effective in lowering blood levels.

Fiser et al. (1971), using dogs, found that charcoal given after 30 min was effective in reducing blood levels of secobarbital, phenobarbital, and glutethimide by an average of 53, 56, and 74%, respectively, over the period of 1–24 hr. They also investigated, for secobarbital only, the effect of waiting 1 hr before giving the charcoal. In this case, the charcoal had very little effect. The reason for this is that secobarbital is relatively quickly absorbed.

Andersen (1973) gave 40 mg/kg diethylbarbituric acid to pigs, then administered 2 g/kg activated charcoal, with delays of 0, 2, 4, 6, and 8 hr. Plasma drug levels versus time were followed for 24 hr. The smaller the delay time, the more effective was the charcoal in reducing drug absorption, as one would expect (AUC values are not given, so no quantitative indexes of the delay effect can be quoted). Even when the charcoal was delayed for 8 hr, it "considerably reduced" the absorption of the drug.

Lipscomb and Widdop (1975) carried out in vivo work in which pigs were given an amobarbital suspension, without charcoal or with charcoal 30 min later, and pigs which were given amobarbital capsules, without charcoal or with charcoal 1 hr or 4 hr

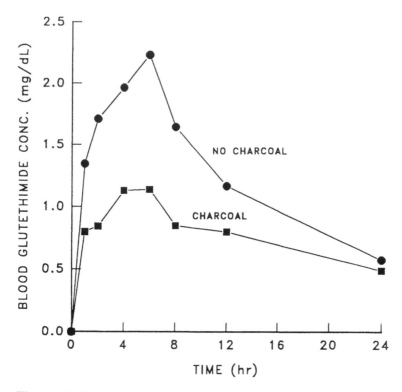

Figure 11.6 Effect of a 5:1 ratio of activated charcoal:drug on blood concentrations of gluteth-imide in dogs. The charcoal was given 1 min after the drug. From Chin et al. (1970). Reprinted with permission of Academic Press, Inc.

later. The control pigs showed the common clinical response of a lapse into coma, a gradual regaining of consciousness which increased intestinal activity (and therefore speeded up drug absorption), and subsequent relapsing into coma again. The pigs receiving charcoal, even after 4 hr, did not suffer such relapses.

Neuvonen and Elonen (1980) studied the effect of charcoal on the absorption of phenobarbital, carbamazepine, and phenylbutazone. These three drugs were given together orally (200, 400, and 200 mg, respectively) to five healthy subjects, in a randomized cross-over study. The subjects were then given 50 g charcoal after 5 min, or 50 g charcoal after 1 hr, or 50 g charcoal after 10 hr and then 17 g charcoal at 14, 24, 36, and 48 hr. The AUC values and serum half-lives of the drugs are given in Table 11.6. It can be seen that charcoal prevented the absorption of all three drugs very effectively, when the charcoal was given at 5 min. Even when started at 10 hr, the charcoal, while much less effective in terms of AUC reduction, decreased serum half lives substantially (82, 45, and 29%, respectively). The charcoal was most effective for the phenobarbital.

Gillespie et al. (1986) applied the theory of linear systems to pharmacokinetic data on phenobarbital elimination obtained with three healthy subjects. Treatments included charcoal in a sorbitol vehicle (Charcoaid) and charcoal in plain water. The drug dose was 200 mg/70 kg body weight and the charcoal doses were 30 g at 1 hr, and 15 g at 7, 13, 19, 25, and 37 hr. The analysis, which is outside of the scope of the present work,

Table 11.6 Effects of Charcoal on Phenobarbital, Carbamazepine, and Phenylbutazone Absorption

Drug	AUC (0–96 hr) % of control	Serum half-life (10–48 hr), hr
Phenobarbital		
Control	100	110
+ Charcoal at 5 min	< 3	
+ Charcoal at 1 hr	53	
+ Charcoal at 10 hr etc.	51	19.8
Carbamazepine		
Control	100	32.0
+ Charcoal at 5 min	< 5	
+ Charcoal at 1 hr	59	
+ Charcoal at 10 hr etc.	73	17.6
Phenylbutazone		
Control	100	51.5
+ Charcoal at 5 min	2	
+ Charcoal at 1 hr	70	
+ Charcoal at 10 hr etc.	95	36.7

From Neuvonen and Elonen (1980), Table 1. Copyright Springer-Verlag GmbH & Co. Used with permission.

indicated that the charcoal reduced phenobarbital absorption by 25–53% in the three subjects. Thus, even though the first charcoal dose was delayed for 1 hr, a significant effect still occurred.

Curd-Sneed et al. (1987b) carried out studies with rats given 40 mg/kg phenobarbital by gavage. The rats were then given 40 mg of either Darco G-60 (650 m^2/g), Mallinckrodt USP (1000 m^2/g), or SuperChar (3150 m^2/g) charcoal 5 min later as a slurry in 1 mL water, or as a slurry in 1 mL of 70% w/v sorbitol solution. Table 11.7 gives values for the absorption rate constants (K_a), the elimination rate constants (K_e), the peak plasma levels (C_{max}), the times to the peak plasma levels (t_{max}), and the 0–6 hr AUC values. The codes are: G-60 = Darco G-60 charcoal, USP = Mallinckrodt USP charcoal, and SC = SuperChar charcoal. Clearly, in water solution, all charcoals reduced absorption (although only SuperChar was significant at $p < 0.05$), and SuperChar was the best. In sorbitol, all charcoals were statistically effective at $p < 0.01$. An interpretation of the data suggested that the effectiveness of the charcoals in water was proportional to their surface areas and that sorbitol enhanced the effectiveness of the Darco and Mallinckrodt charcoals, but not the SuperChar charcoal.

Adler et al. (1986) investigated the effect of oral charcoal on the sleep times of mice given IV injections of phenobarbital. They found a linear relationship between the sleep times and the logarithm of the drug dose, in both control and in charcoal-treated mice. Half-lives of phenobarbital in the two groups were about 8.1 and 0.9 hr, respectively. A linear decline in sleep times with increasing charcoal dose was observed, up to a maximal charcoal dose, beyond which an increase in the charcoal caused no further reduction in sleep time. Other hypnotics were studied also. Overall, charcoal reduced

Table 11.7 Effects of Various Treatments on Pharmacokinetic Parameters in Rats Given 40 mg/kg Oral Sodium Pentobarbital

Treatment	K_a (hr^{-1})	K_e (hr^{-1})	C_{max} (mg/L)	t_{max} (hr)	0–6 hr AUC (mg-hr/L)
Control	1.9	0.48	27.6	1.00	93.4
Water/G-60	3.7	0.45	26.6	0.70	82.2
Water/USP	4.4	0.38	21.2	0.65	73.1
Water/SC	4.3	0.30	13.4	0.71	55.5
Sorbitol	3.1	0.34	21.5	0.90	89.4
Sorbitol/G-60	5.1	0.37	19.9	0.63	70.0
Sorbitol/USP	5.1	0.37	17.2	0.57	57.0
Sorbitol/SC	5.4	0.35	18.1	0.57	64.6

From Curd-Sneed et al. (1987b).

sleep times 82–88% with phenobarbital, methyprylon, glutethimide, ethchlorvynol, and methaqualone. However, charcoal had no effect with amobarbital and pentobarbital.

The in vivo studies with hypnotics and sedatives which we have reviewed in this section demonstrate clearly that oral activated charcoal is very efficacious in decreasing the absorption of these drugs, if an adequate charcoal dose is given without undue delay. In some cases, charcoal has been effective even if delayed for a considerable time (e.g., with phenobarbital), although for some hypnotics/sedatives (e.g., secobarbital) this is not the case.

VI. TRICYCLIC ANTIDEPRESSANTS

Crammer and Davies (1972) were perhaps the first investigators to seriously consider the use of adsorbents in treating overdoses of this class of drugs. They pointed out that the revival of patients poisoned by excessive doses of tricyclic antidepressants is difficult since these drugs are rapidly absorbed from the stomach and are strongly protein-bound. Thus, only negligible amounts can thereafter be removed by techniques such as forced diuresis and dialysis. Until the normal processes of metabolism and urinary excretion clear the drug (which may take several days), one is faced with having to prevent possible convulsions, hypotension, cardiac arrhythmias, and so forth.

However, within the body, these drugs undergo enterohepatic circulation. Significant amounts are secreted into the bile, carried into the duodenum, and reabsorbed from the gut into the blood. It occurred to Crammer and Davies that by use of a suitable adsorbent these drugs might be trapped in the gut and held there until excretion in the feces. Simple in vitro tests carried out by them showed that imipramine adsorbed well to activated charcoal (0.25 g/g charcoal under their conditions, which were not described). Magnesium silicate and aluminum hydroxide gel were found to be useless adsorbents but kaolin showed some adsorption, particularly when the pH was raised from 7.4 to 9.1. In response to this study report, Matthew (1972) wrote that the statement by Crammer and Davies to the effect that large amounts of tricyclic drugs are secreted into the bile needed

Table 11.8 Binding of Imipramine and Desipramine to Charcoal In Vitro

Total drug in the system; bound and free (mg/L)	Imipramine (% adsorbed)	Desipramine (% adsorbed)
10	> 99	> 99
50	92	90
100	82	62
150	62	30

From Rauws and Olling (1976). Table 1. Copyright Springer-Verlag GmbH & Co. Used with permission.

documentation, for it might encourage someone to try "biliary drainage" as an emergency measure, a technique which Matthew implies is not effective.

A. Other In Vitro Studies

Rauws and Olling (1976) dissolved various amounts of imipramine hydrochloride and desipramine hydrochloride in 20 mL of a pH 7.4 phosphate buffer and dialyzed this against 5 mL of the same buffer having 1 mg activated charcoal for 8 hr at 37°C. Table 11.8 shows their results, which indicate that both drugs are well adsorbed, with imipramine being somewhat more bindable than desipramine.

Oppenheim and Stewart (1975) did very similar in vitro studies on the adsorption of amitriptyline HCl, nortriptyline HCl, imipramine HCl, and desipramine HCl from SGF (USP, pepsin omitted) at 37°C to five different activated charcoals. The initial drug concentrations were 30 mg/L for nortriptyline HCl and 50 mg/L for the other three drugs. The percentages of the drugs adsorbed at equilibrium by the three best charcoals are shown in Table 11.9. As this table suggests, the percentage of the drug adsorbed was virtually identical for each of the four drugs. This is as one might expect, since their chemical structures are very similar. Oppenheim and Stewart also found that in their system (essentially a well-stirred batch type of system) the approach to adsorption equilibrium was very fast. Typically, 80% of the final amount adsorbed was bound in only 2 min.

Table 11.9 Percent Adsorption of Four Tricyclic Antidepressants by Three Kinds of Charcoal

Drug:charcoal ratio	Charcoal used		
	Merck	BDH	Ajax
1:10	97–98	91–92	62–68
1:5	62–70	61–66	33–36

From Oppenheim and Stewart (1975). Reproduced by permission of the Australian Pharmaceutical Publishing Company.

Further studies, by Boehm and Oppenheim (1977) on nortriptyline adsorption by charcoal from SIF have shown that, at a charcoal to drug ratio of 10:1, 97% adsorption occurred.

Crome et al. (1977) determined a Langmuir isotherm at 37°C for nortriptyline hydrochloride by combining a solution of 5 mg/mL nortriptyline hydrochloride in a pH 1.0 HCl solution with various amounts of an effervescent activated charcoal mixture (Medicoal), shaking the samples for 10 min, filtering out the charcoal, and analyzing the filtrates. They report that a maximal adsorption of 318 mg nortriptyline per gram of effervescent charcoal mixture (of which about half of this is charcoal per se) is possible. This implies a maximal adsorption of about 0.62 g/g charcoal, which is a very substantial amount.

Certainly these in vitro studies have shown that tricyclic antidepressants have high affinities for activated charcoal; thus charcoal could be expected to be an effective agent for reducing the absorption of such drugs in vivo.

B. In Vivo Studies

Rauws and van Noordwijk (1972) reported on experiments in which rats were given infusions of imipramine, either 3 mg/kg intravenously or 10 mg/kg intraperitoneally. Certain rats were pretreated with orally administered charcoal (amount unstated); other rats served as the control group. The IV and IP routes of drug administration were chosen to see if charcoal could successfully intercept enteroenteric and/or enterohepatic circulation of the drug. Mean values of the ratios of the drug concentrations found in various organs of the charcoal-treated rats compared to levels in the same organs of the control rats for the IV infusion tests were: heart, 0.71; liver, 0.80; and lungs, 0.72. For IP drug administration, the ratios were: heart, 0.80; liver, 0.99; and lungs, 0.75. Although these results show some variation, it seems clear that the charcoal was in most cases effective in moderately lowering organ imipramine levels.

In a follow-up to this study, Rauws (1974) indicated that the results of the first study led to further investigation of the kinetics and distribution of imipramine and desipramine. It was found that "the enterohepatic cycle was not as intensive as expected." However, concentrations in gastric contents of well above 200 µg/g were found and it was concluded that an intensive gastroenteral cycle was operative.

Alván (1973a) has described an in vivo study in which six healthy volunteers were given nortriptyline in doses of 0.86–1.00 mg/kg on two separate occasions. On one occasion, the drug was followed 30 min later with 5 g charcoal in a commercially available suspension (ACO, Ltd.). Blood samples were taken at 0, 2, 4, 12, 24, 32, 48, 56, and 84 hr in both tests. The charcoal reduced the 0–84 hr plasma AUC values for the six subjects by 0, 5, 19, 27, 50 and 72%, a rather large variation in effect. Peak plasma concentration reductions ranged from about 4 to 69%, but the times corresponding to the peak concentrations were unchanged in five out of the six cases. Drug half-lives were generally lengthened, by amounts ranging from 0 to about 29%. While these results are highly variable in nature, it does appear that, on average, the use of charcoal can significantly reduce nortriptyline levels, even after 30 min delay.

In a letter commenting on Alván's study, Chaput de Saintonge (1973) pointed out that the patients received nortriptyline plus charcoal in a mixture containing ethyl alcohol, sorbitol, and carboxymethylcellulose and that the controls received only nortriptyline, but clearly the controls should have received all of the constituents of the charcoal suspension without the charcoal. He states that some of these constituents are known to reduce the rate of gastric emptying. Alván (1973b) replied that his data suggest that the charcoal suspension did not delay absorption, but merely reduced it; however, he agreed that it would indeed be interesting to see if a control suspension without charcoal would affect bioavailability.

Lipscomb and Widdop (1975) and Rauws and Olling (1976) discussed additional in vivo experiments with animals. In Lipscomb and Widdop's study, 20–25 kg pigs were dosed with 2 g amitriptyline on one occasion and with 5 g amitriptyline on another occasion. Despite these massive doses, no clinical signs of toxicity were observed. The pigs metabolized the drug at such an astonishing rate that plasma levels were always very low and therefore no attempts to study the effects of modifying them with the use of charcoal were attempted (their paper did report, however, on the effects of charcoal on levels of other drugs).

Rauws and Olling (1976) extended the earlier work of Rauws and van Noordwijk (1972) on rats. By cannulating the common bile duct and then administering imipramine or desipramine to rats, Rauws and Olling showed that only negligible fractions of these drugs (less than 1%) are eliminated via the bile. Therefore, they concluded that the idea of adsorbing these drugs from the bile would not be a worthwhile strategy. They did find, however, considerable and increasing quantities (up to about 5% of the dose) of these drugs in the stomach contents, indicating that a gastroenteral cycle exists. This suggested that adsorption of these drugs by activated charcoal from the gastric contents might be a possibility. However, further experiments along the lines of those done by Rauws and van Noordwijk still failed to yield any dramatic effects due to the use of charcoal.

In some "desorption" experiments, rats were given imipramine or desipramine (50 mg/kg) orally in 0.1 N hydrochloric acid with or without added charcoal (10% w/v). Time was allowed for adsorption equilibrium to occur before giving the drug/charcoal mixture to the rats. Four hours after administration of the drug or drug/charcoal, the rats were killed and organ drug concentrations were determined. The results are shown in Table 11.10. Clearly, if the charcoal had bound most of these drugs prior to administration, there must have been significant desorption (unbinding) in vivo. Rauws and Olling considered these results to be very disappointing and suggested that the cause of this was that, in contrast to drugs like barbiturates, the antidepressants bind very strongly to the body tissues and thus are not accessible to the charcoal. However, it would appear that their results are not as poor as they imply (Table 11.9 shows average reductions of

Table 11.10 Drug Concentration Ratios (Charcoal/No Charcoal)

Organ	Imipramine	Desipramine
Heart	0.61	0.11
Liver	0.69	0.40
Lungs	0.84	0.47
Brain	0.53	0.20

From Rauws and Olling (1976). Table 4. Copyright Springer-Verlag GmbH & Co. Used with permission.

imipramine and desipramine in the various organs of 33 and 71%, respectively). Perhaps the use of more charcoal is all that is needed. It should also be recalled from the study of Lipscomb and Widdop with pigs that the extrapolation of results from animals to humans is, at least with these drugs, not apt to be valid.

Moreover, very positive results in vivo with humans were reported by Crome et al. (1977). These same results were also reported, in abbreviated form, in a paper by Braithwaite et al. (1978). They gave 75 mg nortriptyline plus a glass of water to healthy volunteers, followed 30 min later by 5 g activated charcoal contained in a 10-g packet of an effervescent charcoal preparation (Medicoal) along with 400 mL more water. Typical plasma nortriptyline concentrations versus time in control and charcoal-treated subjects showed that average reductions in the peak plasma drug concentrations were 60% (range 30–81%) and average reductions in drug bioavailability were 63% (range 32–85%). Availability was calculated as the area under the plasma concentration versus time curve for the period 0–48 hr. A report in the "Medical News" section of the *JAMA* (see Montgomery, 1978) refers to the 1977 study by Crome et al. and reviews its recommendations.

In another related study by Dawling et al. (1978), adults were given 75 mg nortriptyline then a single 13-g packet of Medicoal at 30 min, 2 hr, or 4 hr. Drug bioavailabilities were reduced by 74, 38, and 13%, while peak plasma concentrations fell by 77, 37, and 19%, respectively. Certainly these results show a substantial favorable effect of charcoal on a representative tricyclic antidepressant in humans.

The extrapolation of Crome's results to doses that are more typical of an overdose (3 g, or 120 25-mg tablets at most) represents an uncertainty. However, with an overdose, the anticholinergic effects of these drugs may significantly slow down the rate of absorption. Activated charcoal may be able to reduce absorption several hours after the overdose has occurred, as was the case with experimental barbiturate poisoning in pigs (Lipscomb and Widdop, 1975). Based on Crome's demonstration that 5 g charcoal had a sizable effect in lowering drug levels after a 75-mg dose of nortriptyline, it would appear then that for each gram of such a drug ingested one should administer 5(1/0.075) = 67 g charcoal. This is not an unduly large amount to administer, particularly down an orogastric tube to an unconscious patient. It thus appears that activated charcoal can be of significant therapeutic use in tricyclic antidepressant overdoses, if enough is employed.

Crome et al. (1983) carried out a randomized clinical trial to determine the effects of charcoal in treating tricyclic antidepressant overdose patients presenting to EDs in England. Of 48 patients enrolled in the study, 20 received supportive care plus oral charcoal (10 g) and 28 received only supportive care. The charcoal (Medicoal) was given via a lavage tube after gastric lavage, or as a drink if the patient was conscious and cooperative. Activated charcoal had no effect on either the rate of lightening of coma or on the fall in blood tricyclic levels. However, a single 10-g dose of charcoal would hardly seem to be sufficient to interrupt the enteric (mainly enterogastric) cycling of this class of drugs.

The effect of single and repeated doses of charcoal on the pharmacokinetics of doxepin in man have been studied by Scheinen et al. (1985). Doxepin is a commonly used tricyclic antidepressant, whose main metabolite is desmethyldoxepin. Eight healthy volunteers were each given two 25-mg tablets of doxepin hydrochloride, plus 15 g charcoal (Medicoal) mixed with 200 mL water at 30 min. The peak serum doxepin levels were reduced by 70% and the 0–48 hr doxepin AUC values by 49%, compared to control values. When the single charcoal dose was given 3 hr after the drug, absorption was not significantly reduced. The apparent elimination half-lives of doxepin and desmethyldox-

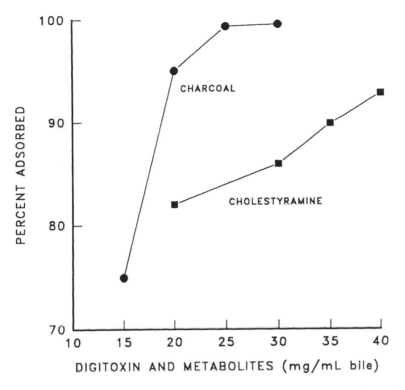

Figure 11.7 Adsorption of digitoxin and its metabolites from guinea pig bile by activated charcoal and cholestyramine resin. From Haacke et al. (1973). Reproduced by permission of F. K. Schattauer Verlagsgesellschaft mbH.

epin were prolonged by 350 and 140%, respectively, when the charcoal was given after 30 min, suggesting gradual dissociation of the drug/charcoal complex. When multiple doses of charcoal were given over 3–24 hr (15 g at 3 hr, and 10 g after 6, 9, 12, and 24 hr), the 0–48 hr AUC values for doxepin were reduced by 26.3%. Thus, both single and multiple-dose charcoal strategies were effective. A partial report on this work was given 5 years earlier in an abstract by Scheinen et al. (1980).

Shader (1986) mentioned that in one study concerning EDs, antidepressants were determined to be present in 29.5 and 21.4% of poisoning deaths in 1983 and 1984, respectively. He reviewed briefly some conflicting clinical results of tricyclic antidepressants, noting that in some cases charcoal seemed to help, while in others it did not.

Kärkkäinen and Neuvonen (1986) performed a study in which the pharmacokinetics of amitriptyline (AT) and its active metabolite nortriptyline (NT) were determined when charcoal or changing urine pH were employed. The results from changing the urine pH with sodium bicarbonate or ammonium chloride were discussed in Chapter 7, and will not be reviewed here. After a single dose of 50 g charcoal given 5 min after a 75-mg dose of AT hydrochloride, AT absorption was reduced by 99%. With repeated charcoal doses (50 g initially followed by 12.5 g every 6 hr between 6–54 hr, for a total of 150 g charcoal) the serum half-life of AT was reduced by 20% and the serum half-life of NT was reduced by 35%. Thus, charcoal was effective in substantially preventing AT absorption in the first place and in increasing the excretion rate of AT and NT to a modest extent, presumably by interrupting their enteroenteric cycles.

Hedges et al. (1987) performed a study in which nine patients who had experienced overdoses of amitriptyline were given charcoal doses ranging from 25–75 g at times ranging from 45–250 min after presentation. They correlated the drug half-life values (range of 1.6–14.3 hr) to the time between presentation and the charcoal dose (t_c) with the equation

$$t_{1/2} = 2.68 + 0.47t_c$$

Thus, a delay in giving charcoal resulted in an increase in the drug half-life. An inverse correlation between the charcoal dose and $t_{1/2}$ was also noted. Nortriptyline, the major metabolite of amitriptyline, decreased in two of three patients who received 50 g charcoal or greater within 60 min of presentation. These findings suggest that charcoal is effective for amitriptyline overdose.

A study by Hultén et al. (1988), however, suggests that charcoal is not very effective in tricyclic antidepressant overdoses. Ninety-one patients presenting to EDs in several hospitals were considered for the study. A single dose of 20 g charcoal was given to 34, while 43 served as controls, and 14 were excluded. Plasma drug concentrations were determined at 1, 2, 4, 8, and 24 hr. There was no significant difference in AUC values, peak plasma drug levels, or plasma half-lives when charcoal was used. Because most of the patients arrived at the EDs several hours after drug ingestion, the charcoal was given after drug absorption was fairly complete; thus the charcoal could act only by interrupting the enterogastric cycling of the drugs. A single dose of 20 g charcoal is most probably insufficient to produce significant interruption of the enterogastric cycle.

To summarize in vivo studies with tricyclic compounds, it appears that if a large dose of charcoal is given soon after drug ingestion, significant prevention of drug absorption can be achieved. There is evidence that many of these drugs are not significantly excreted into bile but are reasonably well excreted into the stomach (and possibly into the small intestine). Thus, if one wishes to accelerate the elimination of already-absorbed tricyclics, multiple doses of charcoal of reasonable size are required.

VII.   CARDIAC GLYCOSIDES

For the treatment of severe accidental or suicidal glycoside intoxication, adsorption to activated charcoal and other adsorbents, particularly cholestyramine resin, has been recommended. Since many cardiac glycosides undergo enterohepatic circulation, adsorbent treatment is suitable not only for oral but also for intravenous glycoside intoxication.

Haacke et al. (1973) showed that ^{3}H-labeled digitoxin and its metabolites are adsorbed well in vitro from guinea pig bile, by both activated charcoal and cholestyramine. Figure 11.7 shows the extent of such adsorption which occurred when 1 mL bile containing digitoxin and its metabolites was contacted for 1 hr with activated charcoal (15–30 mg) or cholestyramine (20–40 mg). Further data showed that the extent of adsorption of unmetabolized digitoxin versus that of a mixture of digitoxin and its metabolites was about equal (percentage-wise) with the charcoal but that the cholestyramine showed only about half as much adsorption of the digitoxin as compared to

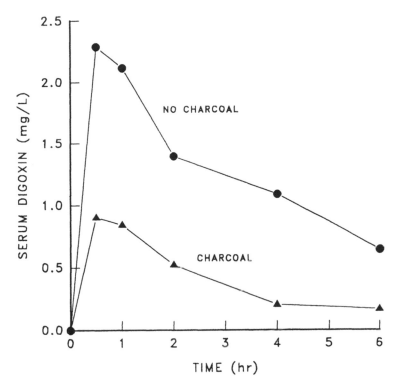

Figure 11.8 Mean serum levels of digoxin in six healthy subjects after ingestion of 0.5 mg digoxin and 0.5 mg digoxin plus 2 g activated charcoal. From Härtel et al. (1973). Copyright 1973 by The Lancet Ltd. Reproduced by permission.

the mixture of digitoxin and its metabolites. Haacke et al. estimated that a toxic dose of from 2 to 4 mg digitoxin could be satisfactorily counteracted by an oral dose of 30 g activated charcoal or 40 g cholestyramine.

In the same year, Härtel et al. (1973) reported on an in vivo study in which healthy male volunteers were given either 0.5 mg digoxin or 0.5 mg digoxin plus 2 g activated charcoal suspended in water 5 min later. Figure 11.8 shows the serum digoxin levels measured. It is apparent that the charcoal did significantly reduce the serum digoxin levels in vivo. Härtel et al. mention that charcoal may be even more effective in digitoxin overdose since this glycoside undergoes substantial enterohepatic circulation (this occurs with digoxin only to a minor degree).

In 1974, Belz et al. reported on three studies involving various glycosides. In one study (Belz and Bader, 1974) healthy volunteers were given 1 mg methyl proscillaridin IV two times (21 days apart). During one of the periods they received 2 g activated charcoal three times per day. Plasma glycoside levels over the 48 hr following the drug administration are shown in Figure 11.9. These data show that on average the drug concentrations were reduced about 40% by charcoal during the period from 10 to 48 hr. Thus, it is clear that methyl proscillaridin undergoes bile excretion and extensive enterohepatic circulation. Therefore, charcoal shows definite promise for treating overdoses of this drug.

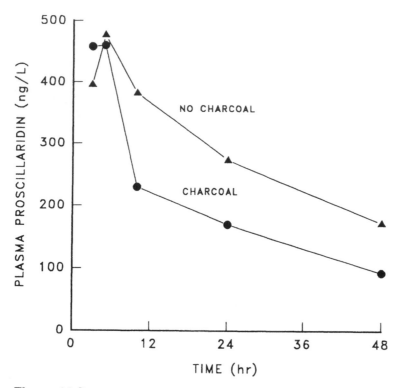

Figure 11.9 Mean plasma proscillaridin concentrations following intravenous 1-mg doses in six human subjects with and without oral activated charcoal treatment. From Belz and Bader (1974), Figure 1. Copyright Springer-Verlag GmbH & Co. Used with permission.

In an identical study involving the glycoside β-methyl digoxin, Belz (1974a) found that while slightly lower plasma levels did occur with charcoal-treated patients, the differences were not statistically significant. One explanation for the lack of effect may be that β-methyl digoxin does not undergo enterohepatic circulation to a substantial enough extent, since Belz (1974b) has proved that, in vitro, charcoal does adsorb this drug well. Belz's in vitro work concerned the adsorption of a large series of glycosides and related derivatives to activated charcoal. Figure 11.10 shows the amount of charcoal needed to adsorb 50% of the glycoside from a solution of 300 μg glycoside in 1 mL of a predominantly aqueous solution (20 vol% ethanol, 80 vol% water) versus the drug's organic/water distribution index (determined separately using carbon tetrachloride as the organic phase). One can see that glycosides do adsorb fairly well to charcoal in vitro, and that for those types having greater organic character the amount of charcoal needed is less, as one would expect.

Zajtchuk et al. (1975) reported in vivo work on the use of charcoal in digoxin-dosed dogs. Four dogs were given 0.02 mg/lb digoxin IV, four other dogs were given 0.05 mg/lb digoxin IV, and two dogs were given 0.03 mg/lb digoxin orally. Blood samples were collected for up to 96 hr. Dogs treated with charcoal (30 g Norit A charcoal 15 min before and 2 hr after drug administration) did not show toxic symptoms such as vomiting and

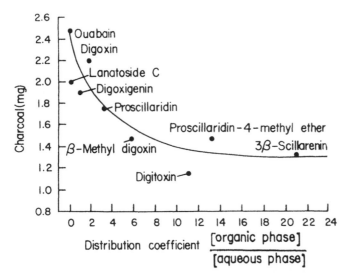

Figure 11.10 Correlation between polarity of various glycosides or glycoside derivatives and charcoal binding. From Belz (1974). Reproduced by permission of Birkhäuser Verlag AG, Basel.

arrhythmias. In dogs given the digoxin IV, charcoal lowered serum drug levels by 25–35%, while dogs given the digoxin orally showed a 75% reduction in serum drug levels.

Some additional work on a cardiac glycoside is the in vivo work of Neuvonen et al. (1978), in which six adults were given 0.5 mg digoxin, followed by 50 g Norit A charcoal. When the charcoal was given immediately, the systemic absorption of the digoxin was reduced by 98% relative to controls. Even when the charcoal was delayed one hour, a 40% reduction occurred.

Caldwell et al. (1980) studied the intestinal secretion of digoxin in the rat. Rats whose bile ducts had been ligated excreted 13.4% of parenterally-administered digoxin (determined by measuring tritium-labeled digoxin in 4 day stool collections). When charcoal was given by gastric tube, the intestinal secretion of digoxin rose to 33.4%. These results suggest that net nonbiliary intestinal secretion of digoxin can be augmented by the intraluminal binding of the drug to charcoal.

Neuvonen and Kivistö (1989) presented results of one of their studies which indicate that the absorption of a 0.25-mg dose of digoxin was 96–98% prevented by 8 g of charcoal, 30–40% reduced by 8 g cholestyramine, and not significantly reduced by 10 g colestipol. All sorbents were given simultaneously with the drug. In another part of that same study, 40 mg furosemide was used, and these same sorbent doses reduced furosemide absorption by 99.5% with charcoal, 95% with cholestyramine, and 80% with colestipol. Thus, charcoal was much more effective than the resins for digoxin, and slightly more effective for furosemide.

The in vivo studies reviewed here show that the absorption of orally-ingested digoxin and digitoxin is very effectively reduced by charcoal. With cardiac glycosides administered IV, the effects of charcoal appear to depend on whether a significant enterohepatic cycle exists for the drug. With IV methyl proscillaridin, charcoal is effective, whereas with IV β-methyl digoxin charcoal has no effect.

VIII. ORGANIC SOLVENTS

Overdoses of organic solvents such as gasoline, kerosene, lighter fluids, and cleaning fluids are of considerable concern. Especially frequent are cases of children ingesting kerosene in rural or less developed areas where kerosene is commonly used for cooking, heating, and lighting. The most common clinical manifestations of organic solvent poisoning involve the lungs and central nervous system. Although it is known that organic solvents are absorbed from the gastrointestinal tract, it is still not entirely certain whether damage to the lungs occurs primarily from the solvent absorbed into the blood or whether primary aspiration, or aspiration secondary to regurgitation or vomiting, is more important. Current opinion is that pulmonary toxicity occurs mainly by pulmonary aspiration, and not by systemic absorption from the GI tract, especially in the case of kerosene. Thus, since oral charcoal may promote vomiting, which could result in aspiration of gastric contents into the lungs, there is a strong argument for withholding charcoal altogether. Whether this should be done will require more animal studies than the few to be discussed here. The answer may depend on the substance (e.g., it may turn out that charcoal should be withheld in kerosene poisoning but should not be withheld in, say, benzene poisoning).

Fortunately, the lethal dose of organic solvents seems to be relatively large. Ashkenazi and Berman (1961) stated that kerosene poisoning in Israel, though frequent at that time, was rarely fatal. McNally (1956) also offered the view that "it is difficult to conceive of a child drinking a sufficient quantity of kerosene to produce a fatality by absorption from the gastrointestinal tract alone. For a child of 50 pounds a volume in excess of one pint would be the minimal lethal dose without aspiration." Working with rabbits, Deichman et al. (1944) found the lethal dose for this animal to be 28 mL/kg.

One of the earliest studies of organic solvent poisoning was that of Chin et al. (1969) in which 8 mL/kg (6.4 g/kg) kerosene was administered by oral intubation to groups of 50 rats each. One minute later two of the groups were treated with 3.6 g/kg activated charcoal or 3.6 g/kg Arizona montmorillonite. Figure 11.11 shows the blood concentrations when charcoal was used. It is clear that the charcoal was quite effective in reducing blood levels even though the ratio of charcoal to kerosene was only 3.6:6.4, or 0.56:1. The Arizona montmorillonite was somewhat less effective.

Another study was described by Laass (1974), who administered lethal doses of various organic solvents to rats. The doses used were 10 mL/kg of benzene, diethylaniline, tetrachloroethane, or carbon tetrachloride and 5 mL/kg diethylaniline or tetrachloroethane (only one solvent was given to each rat). Following administration of these by oral intubation, each rat was then given either 40 mL/kg physiological saline or 40 mL/kg of a suspension of 10% w/v activated charcoal in water. The lifetimes of the rats are shown in Table 11.11. Survival times were increased by an average of 166% (range 30–120%) when charcoal was used. Clearly, this is a significant effect. If more charcoal had been given it is quite likely that some of the rats would have survived.

Laass et al. (1977) later repeated this same type of study, using nitrobenzene, aniline, pyridine, n-butylamine, dichloroethane, tetrachloroethane, diethylaniline, and carbon disulfide. The results were similar except for the case where nitrobenzene was given; in this case the use of charcoal decreased the survival time slightly for some unknown reason.

Laass (1975) has also studied the effects of castor oil, heavy mineral oil, light mineral oil, physiological saline, and activated charcoal on blood levels of three solvents (benz-

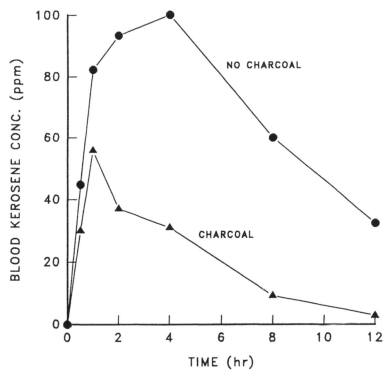

Figure 11.11 Effect of activated charcoal on blood kerosene concentration in rats. From Chin et al. (1969). Reproduced by permission of the American Pharmaceutical Association.

Table 11.11 Mean Survival Times in Hours for Rats Given Various Solvents

Solvent	Dose (mg/kg)	+ 40 mL/kg NaCl	+ 40 mL/kg 10% w/v charcoal
Benzene	10	0.98	2.63
Diethylaniline	10	5.88	10.85
Tetrachloroethane	10	0.15	0.44
Carbon tetrachloride	10	18.83	38.10
Diethylaniline	5	9.21	11.96
Tetrachloroethane	5	0.14	0.75

Adapted from Laass (1974).

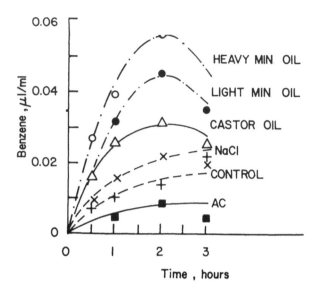

Figure 11.12 Blood levels of benzene in rats dosed with 2 mL/kg benzene and rats treated with activated charcoal, saline, or various oils. Adapted from Laass, 1975.

ene, ethylene dichloride, and diethylene dioxide) after administration of each solvent to rats. The dosages were: solvents = 2 mL/kg, castor and mineral oils and saline = 20 mL/kg, and activated charcoal = 20 mL/kg of a 10% w/v slurry in water. Figure 11.12 shows typical results, in this case for benzene. In general, the mineral and castor oils and the saline all consistently increased the blood levels of the solvents. The charcoal tests gave lower blood levels than those observed in the control tests, as expected.

Tietze and Laass (1979a, 1979b) continued studies of the effects of activated charcoal on the treatment of toxic effects due to poisoning with organic solvents. In their first paper (1979a), they presented in vitro results on the adsorption of benzene, 1,2-dichloroethane, pyridine, and 1,4-dioxane to activated charcoal from aqueous solution. Benzene adsorbed very strongly, as one would expect since it is an aromatic compound with a very low water solubility. The 1,2-dichloroethane adsorbed almost as strongly. Pyridine leveled off at a Q_m value of about 3.5 mmole/g (or about 0.277 g/g), which is quite good, and 1,2-dioxane attained a maximal adsorbance (at a solution concentration of 20 mmole/L) of about 1 mmole/g (or 0.088 g/g), which is somewhat low (a "good" Q_m value for an organic compound is on the order of 0.25–0.40 g/g). Thus, it appears that these organic compounds generally adsorb well to charcoal in vitro. As one would expect, their strengths of adsorption increase as their aqueous solubilities decrease.

Tietze and Laass (1979b) described a study in which rats were given 0.5 mL/kg benzene, carbon tetrachloride, or 1,2-dichloroethane, followed by either physiologic saline or a slurry of charcoal in physiologic saline (5 or 20 mL/kg of a 10% w/v charcoal slurry). Blood levels of each substance were determined at 30, 60, 120, and 180 min and average values were computed. The saline alone actually raised the mean blood levels somewhat. With 5 mL/kg of the charcoal slurry, the mean blood levels fell from

13.4 to 11.3 µL/L for benzene. No effect on the other two organics was noted. With 20 mL/kg of the charcoal slurry, the benzene fell from 13.4 to 8.3 µL/L, the carbon tetrachloride fell from 7.4 to 5.2 µL/L, and the 1,2-dichloroethane fell from 37.5 to 34.7 µL/L. The effects of charcoal were statistically significant ($p < 0.05$) for benzene and carbon tetrachloride at the 20 mL/kg charcoal dose. It is not surprising that there was no effect on dichloroethane, since this compound has no ring structure and would be expected to adsorb reasonably poorly to charcoal.

In a paper by Laass (1980) published a year later, Laass discussed the treatment of oral poisonings by organic solvents using combinations of charcoal and laxatives. Rats were injected with 0.2, 0.5, or 2.0 mL/kg benzene or 1,2-dichloroethane directly into the duodenum. These were followed immediately with injections of various agents (saline, charcoal slurry, two kinds of paraffins, castor oil) and combinations of these agents, and the effects on the mean levels of the organic solvents in the blood over 0–3 hr were determined. It was found that high doses of charcoal could reduce benzene levels in the blood. For 1,2-dichloroethane, charcoal produced inconsistent results. However, the combination of liquid paraffin or castor oil with charcoal gave benzene blood levels lower than those observed when charcoal alone was used. These combinations were ineffective for the 1,2-dichloroethane.

Another study on this general topic was reported by Morgan et al. (1977). They administered lindane, a chlorinated hydrocarbon pesticide (benzene hexachloride), to pigs in the form of 0.4 g dissolved in 2 mL petroleum distillates. They then administered either activated charcoal, mineral oil, or castor oil and followed the blood lindane concentrations versus time. Unfortunately, no consistent pattern of results was obtained. Note that in this case the component being monitored (lindane) was only a part of the toxic material administered (the other part being the petroleum distillates). This factor may have complicated the situation.

A study by Decker et al. (1981) concerned the in vitro adsorption of methanol, ethylene glycol, kerosene, and turpentine by charcoal as well as by a carbonized resin (a nonion exchange resin), calcium silicate, magnesium silicate, a commercial gelling agent used for laboratory spill cleanup, and hydrophobic silica. The organic solvents were mixed with SGF and the sorbent material in flasks at 37°C. The solvent doses were 1, 10, 50, and 100 mL. These were mixed with 100 mL SGF and 5 g of sorbent contained in 50 mL water. After 20 minutes shaking of each flask, the sorbent and the solution were separated by filtration and the solution analyzed for unadsorbed solvent. The results are shown in Tables 11.12 and 11.13. With the exception of the sorbents forming gels, none of the sorbents tested appeared to offer any significant advantage over activated charcoal.

Buck et al. (1985) studied the effectiveness of three charcoals in preventing toxicosis and death in rats given a supralethal dose of carbaryl (300 mg/kg). Carbaryl (1-naphthyl-N-methylcarbamate) is an insecticide (also called Sevin), which, although by itself is a crystalline solid, is marketed in a liquid vehicle. Survival over 7 days was monitored. The control rats all died. For the three charcoals, survival rates were 1/10, 10/10, and 2/10 (the high value was obtained with SuperChar charcoal).

In summary, it appears that charcoal, when given in sufficient amounts, can have a favorable effect in reducing the absorption of liquid organic substances. Other agents (e.g., liquid paraffins, mineral oil, castor oil) are ineffective alone. However, one study has shown that, in combination with charcoal, liquid paraffin and castor oil enhanced the action of charcoal. This synergistic effect deserves further scrutiny.

Table 11.12 Percent of Methanol (MeOH) and Ethylene Glycol (EG) Adsorbed by Various Adsorbents

	1 mL		10 mL		50 mL		100 mL	
Sorbent	MeOH	EG	MeOH	EG	MeOH	EG	MeOH	EG
Activated charcoal	59	68	48	40	35	20	26	2
Carbonized resin	70	68	35	48	25	17	20	2
Calcium silicate	57	68	30	28	25	18	15	2
Magnesium silicate	57	85	31	56	21	13	12	2
Commercial gel agent	64	60	44	28	27	7	8	3
Hydrophobic silica	72	40	45	19	15	8	5	4

From Decker et al. (1981). Reprinted with permission from *Veterinary and Human Toxicology*.

Table 11.13 Percent of Kerosene (Ker) and Turpentine (Tur) Adsorbed by Various Adsorbents

	1 mL		10 mL		50 mL		100 mL	
Sorbent	Ker	Tur	Ker	Tur	Ker	Tur	Ker	Tur
Activated charcoal	76	75	38	35	13	13	10	7
Carbonized resin	85	75	48	29	26	12	16	5
Calcium silicate	75	69	35	30	32	12	28	4
Magnesium silicate	80	43	18	33	28	2	21	2
Commercial gel agent	82	60	79	38	92	25	2	8
Hydrophobic silica	88	85	78	39	98	96	3	6

From Decker et al. (1981). Reprinted with permission from *Veterinary and Human Toxicology*.

IX. ETHANOL

Many in vitro studies have shown that ethanol adsorbs poorly to charcoal. This is as one would expect. Ethanol is a simple molecule (CH_3CH_2OH), which is totally water-soluble in all proportions, and which has no molecular moieties such as aromatic rings which would make it attract to carbon surfaces. Thus, there is little reason to expect that oral charcoal would be of any significant benefit in ethanol overdoses. Additionally, ethanol is absorbed from the gut rapidly, so that by the time charcoal would usually be administered in a hospital setting there would be very little ethanol left in the gut.

We first review studies in which the intent was to determine if oral charcoal can affect the absorption of ethanol; then we review some work aimed at assessing whether ethanol can affect the in vitro adsorption of other drugs, or can interfere with the ability of charcoal to reduce the absorption of other drugs.

In a letter by Jackson et al. (1980), ethanol blood level data were presented for rats which were given 800 mg/kg ethanol by orogastric tube, without charcoal and with a simultaneous dose of 500 mg/kg charcoal. These data are shown in Figure 11.13. Clearly, there is no significant effect due to the charcoal.

North et al. (1981b) gave six healthy dogs an oral dose of 2 mL/kg ethanol in a water solution. Blood ethanol concentrations were then measured at 0.5, 1, 2, 3, and 6

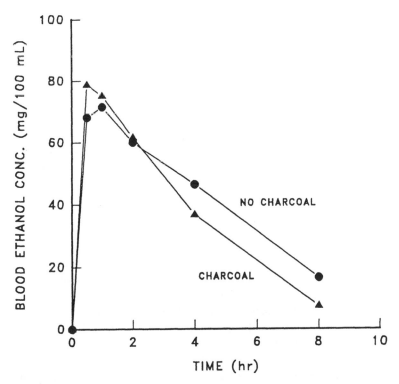

Figure 11.13 Blood ethanol concentrations in rats following 800 mg/kg ethanol by orogastric tube, given with or without activated charcoal. From Jackson et al. (1980). Reprinted with permission of the American College of Emergency Physicians.

hr. A week later the tests were repeated (same dogs), only 50 g activated charcoal in water was given just prior to the ethanol dose. The results showed that activated charcoal greatly reduced ethanol absorption, especially during the first two hours. Blood ethanol levels during 30–60 min were about 40% lower with charcoal than without charcoal, for example. These results are unexpected. It is to be noted that when the charcoal was given, it was given as a water slurry (volume unstated). Hence, the dogs which received charcoal possibly had much more water in their stomachs when they received the ethanol than those dogs which did not receive charcoal. Thus, the ethanol might have become diluted in the charcoal-treated dogs, reducing its concentration, thereby reducing its rate of absorption from the gut. In any event, some explanation other than a charcoal effect per se must exist.

Hultén et al. (1985) conducted a randomized cross-over study in which eight healthy men drank 88 g ethanol (260 mL vodka) and then, 30 min later, took either 20 g charcoal in water or just plain water (same volume). There were no significant differences in blood ethanol concentrations with or without the charcoal. The authors were trying to answer the question of whether charcoal would adsorb ethanol significantly if given after 30 min, by which time little of the ethanol was probably left in the gut. The obvious answer was confirmed.

Minocha et al. (1986) did a similar study with six healthy young adults, but in this case the charcoal (60 g superactive charcoal) was given just prior to the ethanol ingestion. No significant differences in blood alcohol versus time over 6+ hr were noted with or without charcoal.

Katona et al. (1989) studied the effect of superactive charcoal and magnesium citrate on ethanol pharmacokinetics in humans. The subjects were given 0.6 g/kg ethanol orally and then 60 g charcoal plus 300 mL of 5.8% magnesium citrate solution at 1 hr and again at 3 hr. Comparison to control tests showed no effect of the charcoal/citrate on the AUC value and the peak ethanol blood levels. These results are not surprising, since ethanol is rapidly absorbed. By the time the first charcoal dose was given at 1 hr, very little ethanol would have been left in the gastrointestinal tract. In addition, ethanol adsorbs poorly to charcoal anyway, so that even giving charcoal immediately would probably have not had a significant effect. The choice of ethanol and a charcoal/citrate delay time of 1 hr in this study made the insignificance of charcoal therapy a foregone conclusion.

Erickson (1993) studied the effects of Charcoaid (30 g charcoal in 150 mL of 70% w/v sorbitol or 20% w/v charcoal in 70% w/v sorbitol) and a type of activated charcoal called Alcosorb (Calgon Carbon Corp., Hinsdale, IL) on blood alcohol levels in rats and dogs. The rats were given either the Charcoaid or Alcosorb (prepared in a "similar concentration", presumably of both the charcoal and sorbitol) at a dose of 1.5 g charcoal/kg or a 20% w/v sorbitol solution (control) intragastrically via a rubber stomach tube and were then given doses of 2.5 g/kg ethanol intraperitoneally. Note that the control solution is incorrect, as it should have been 70% w/v sorbitol rather than 20% w/v sorbitol (it is the charcoal which is present in Charcoaid to the extent of 20% w/v). Tail-tip blood samples were drawn at 0.5, 1, 2, 4, and 6 hr and analyzed for ethanol. With IP ethanol, neither charcoal formulation produced blood level results different from control (sorbitol solution) results. Thus, charcoal given into the stomach does not affect ethanol given intraperitoneally. The experiment was then repeated with the two charcoal/sorbitol mixtures and the sorbitol alone, plus tap water as a control, and 3.5 g/kg ethanol given intragastrically. With intragastric ethanol, all three treatments (Charcoaid, Alcosorb/sorbitol, or sorbitol alone) reduced peak blood ethanol values very significantly (by

58, 74, and 75%, respectively). Why the sorbitol alone was as effective as the Alcosorb/sorbitol and better than the Charcoaid is quite unclear.

In Erickson's dog experiments, the dogs were given: (1) low doses of the two charcoals (780 mg) in capsule form, with talc capsules as the control, (2) high doses of the charcoals (20 g) given in 80 mL water, with 80 mL tap water as the control, and (3) high doses of the charcoals (20 g) given as a suspension in 20% w/v ethanol, with 80 mL of 20% w/v ethanol as the control. The first two treatments were given 5 min before an intragastric dose of 1.5 g/kg ethanol in tap water (in the third treatment, the vehicle was a 20% w/v ethanol solution, so the charcoal and ethanol were of course given simultaneously). The results on peak blood levels showed no effect of the low charcoal doses and no effect of the high dose charcoals when they were given 5 min before the ethanol. However, when the charcoals and ethanol were given simultaneously, the Alcosorb and Charcoaid charcoals (Charcoaid itself was not given; rather, the type of charcoal used to make Charcoaid was what was given) reduced peak blood levels by 28.4 and 27.0%, respectively. Erickson explained these results by hypothesizing that ethanol, mixed with charcoal prior to administration, adsorbs to the charcoal and thus is not available for absorption once the ethanol/charcoal is in the stomach. However, ethanol does not bind to charcoal to any significant degree, so the explanation must lie elsewhere.

This writer located the Calgon researcher who was involved in making and patenting the Alcosorb: Mr. Paul Tower, currently with Filtercorp, Woodinville, WA. He stated that Alcosorb was an experimental low-ash charcoal, having a surface area of 1400 m2/g, made from coconut shells obtained from the Philippines. The ash type and amount was found to make this charcoal superior to others tested for reducing the hangover symptoms associated with ethanol intoxication. The charcoal presumably worked by either adsorbing ethanol, aldehydes, higher alcohols, and other fusel-oil components in liquors prior to their absorption into the systemic circulation and/or the subsequent adsorption of these compounds and their metabolites in the GI tract by gastrointestinal dialysis.

A conversation by this writer with Dr. Erickson revealed that the tests done by Mr. Tower and collaborators involved giving human volunteers two 350-mg charcoal capsules prior to drinking, two more capsules during the drinking period, and two capsules after drinking ceased. The test liquors had 25–40 volume percent ethanol. The charcoals were remarkable in reducing the effects of alcohol, especially hangover symptoms, with the Alcosorb type of charcoal being the most effective. This research attracted the interest and financial support of several companies but this support largely vanished after MADD (Mothers Against Drunk Driving) learned of the studies and protested. To the writer's knowledge, Alcosorb has not been commercialized, probably for the same reason.

In summary, despite the results of North's group and of Erickson, there is ample evidence from in vitro tests and from other in vivo work to support the conclusion that charcoal can not significantly affect the absorption of ethanol in vivo.

We now turn to two studies which examined whether ethanol, added to aqueous test solutions, affects the in vitro adsorption of other drugs. Neuvonen et al. (1984) studied the effect of ethanol on the adsorption of aspirin, quinidine, and amitriptyline to activated charcoal in vitro at pH 1.2 and pH 7 using 0, 2.5% v/v, or 10% v/v ethanol in the drug solutions. The ethanol decreased the adsorption of all three drugs significantly. Olkkola (1984) did an in vitro study similar to the in vitro study of Neuvonen et al., using strychnine as the test drug, also at pH 1.2 and pH 7. Ten percent ethanol in solution again reduced in vitro drug adsorption significantly. However, the observed effects are not due to competitive adsorption of ethanol to the charcoal. The addition of ethanol to an aqueous solution makes the solution less polar (water is extremely polar and ethanol is polar but less so than water; hence, solutions containing some ethanol are less polar than pure water). This makes drugs having substantial organic character, such as the ones used in the studies just mentioned, prefer to adsorb less to the charcoal and remain in solution to a larger extent when ethanol is present.

Two in vivo studies have been conducted which were aimed at evaluating whether ethanol, ingested simultaneously with drugs, would affect the ability of charcoal to reduce the absorption of the drugs. Neuvonen et al. (1984) gave six human subjects 50 g charcoal 5 min after the subjects ingested 1000 mg aspirin or 200 mg quinidine sulfate. The charcoal caused a reduction in the bioavailabilities of these two drugs of 70 and 99%, respectively. The simultaneous ingestion of 50 g ethanol with these drugs only slightly affected the ability of the charcoal to reduce their absorption. Thus, even though many suicidal patients ingest ethanol along with other drugs, the ability of charcoal to help reduce drug toxicity should not be strongly affected by the ethanol (as noted above, ethanol absorbs so rapidly that by the time charcoal is given there may essentially be no significant ethanol left in the gut anyway).

In contrast to the in vivo tests just described are some in vivo tests performed with strychnine by Olkkola (1984) using mice. With charcoal (1000 mg/kg) the LD_{50} was 383 mg/kg; with added ethanol (20 mL/kg) given simultaneously, the LD_{50} was 236 mg/kg. Thus, the ethanol must have decreased the extent of binding of the strychnine to the charcoal, presumably by lowering the polarity of the GI tract fluids. However, strychnine is a relatively toxic chemical and so any change in the amount of strychnine left unadsorbed by charcoal can have a relatively large effect. Whether ethanol would affect the LD_{50} of less toxic chemicals as much as it did with strychnine is not certain.

X. SUMMARY

Summarizing the large number of studies presented in this chapter is difficult. With respect to the in vitro investigations, the choices of drug concentrations, solution volumes, pH levels, buffer types (if any), and charcoal amounts vary so widely that there is almost no basis for comparing one study to another. Indeed, in vitro results often conflict with in vivo results because the charcoal:drug ratios differed in the two types of studies and gastric/intestinal pH values were not used in the in vitro work.

In vitro studies, as a way of determining intrinsic adsorbabilities of drugs, are useful only if: (1) the solution pH values are of physiological significance (representative of gastric and/or intestinal conditions) and (2) maximal adsorption capacities (Q_m values) are determined by fitting the data to the Langmuir model. These values at least indicate the maximal possible amount of adsorption of a drug and comparing Q_m values for different drugs under the same conditions (charcoal type, solution pH, type of buffer salts used, temperature) does provide a rational basis for comparing the relative adsorbabilities of different drugs.

In many in vitro studies, an arbitrary amount of drug is dissolved in an arbitrary volume of plain distilled water, to which is added an arbitrary amount of charcoal, and then the percentage of the drug adsorbed at equilibrium is determined. Such studies are of little value, since varying any one of the arbitrarily-chosen parameters (drug amount, water volume, charcoal amount) will simply give a different result. One can obtain any degree of adsorption from 0–100% by varying the parameters widely. Thus, the in vitro results of value are the Q_m values, and the reader should focus mainly on them.

In vivo studies also suffer from arbitrary choices of the drug amount, the charcoal amount, and the delay time (if any) between drug ingestion and charcoal administration. The effects of different time delays are important, so almost any delay ranging from 0 to several hours can yield valuable information. However, comparing two studies involving different drugs and/or charcoal amounts is certainly complicated if the time delays are different. With respect to the choice of the amount of drug used, one is constrained. If too little is used, blood and urine drug levels will be difficult to determine accurately. For ethical reasons, drug amounts that cause toxic symptoms can not be used in controlled clinical studies. In this respect, actual overdose situations are more realistic. However, there are several drawbacks to actual overdose cases: (1) one often does not know how much drug was ingested, (2) food in the person's stomach can affect drug absorption and the efficacy of charcoal in nonrepeatable ways, and (3) each such case involves one person, and thus is anecdotal rather than controlled.

Thus, many if not most of the studies we have reviewed in this chapter stand alone and can not be compared to the others. With almost any substance, if enough charcoal is given without undue delay, the charcoal will be reasonably effective in reducing the absorption of that substance. Therefore, rather than to recapitulate when charcoal has been effective, we just mention the few cases when it has not been of value. These cases are few. Charcoal appears to be ineffective for treating ethanol toxicity (despite some studies which suggest that it is) because it simply does not adsorb to charcoal to any significant degree. Simple organic substances such as ethylene glycol (see Chapter 12) and dichloroethane (as used in Laass's studies with rats) also do not adsorb well to charcoal.

Before leaving this summary, we remark on one issue that was raised in the chapter: whether charcoal interferes with N-acetylcysteine (NAC) given as a chemical antidote for acetaminophen overdose. In vivo studies showed that 50, 60, and 100-g doses of charcoal lowered the AUC for NAC (a normal NAC dose was used) by 8, 18, and 39%, respectively. Thus, it is clear that charcoal can interfere with NAC absorption if the charcoal dose is high enough. However, giving a supranormal NAC dose has been shown to counteract the reduction of NAC bioavailability due to charcoal.

XI. REFERENCES

Adler, L. J., Waters, D. H., and Gwilt, P. R. (1986). The effect of activated charcoal on mouse sleep times induced by intravenously administered hypnotics, Biopharm. Drug Dispos. 7, 421.

Alván, G. (1973a). Effect of activated charcoal on plasma levels of nortriptyline after single doses in man, Eur. J. Clin. Pharmacol. 5, 236.

Alván, G. (1973b). Effect of activated charcoal on plasma levels of nortriptyline after single doses in man (reply to letter), Eur. J. Clin. Pharmacol. 6, 207.

Andersen, A. H. (1944). Medicinal carbon, Dansk. Tids. Farm. 18, 21.

Andersen, A. H. (1946). Experimental studies on the pharmacology of activated charcoal. I. Adsorption power of charcoal in aqueous solutions, Acta Pharmacol. Toxicol. 2, 69.

Andersen, A. H. (1947). Experimental studies on the pharmacology of activated charcoal. II. The effect of pH on the adsorption by charcoal from aqueous solutions, Acta Pharmacol. Toxicol. 3, 199.

Andersen, A. H. (1948a). Experimental studies on the pharmacology of activated charcoal. III. Adsorption from gastro-intestinal contents, Acta Pharmacol. Toxicol. *4*, 275.

Andersen, A. H. (1948b). Experimental studies on the pharmacology of activated charcoal. IV. Adsorption of allylpropynal (allyl-isopropyl-barbituric acid) in vivo, Acta Pharmacol. Toxicol. *4*, 379.

Andersen, A. H. (1948c). Experimental studies on the pharmacology of activated charcoal. V. Adsorption of sulphanilamide in vivo, Acta Pharmacol. Toxicol. *4*, 389.

Andersen, A. H. (1973). [Medicinal charcoal in the treatment of poisoning. Treatment of experimental barbiturate poisoning in pigs], Ugeskr. Laeger *135*, 797.

Bainbridge, C. A., Kelly, E. L., and Walking, W. D. (1977). In vitro adsorption of acetaminophen onto activated charcoal, J. Pharm. Sci. *66*, 480.

Batizy, L. (1980). Activated charcoal (letter), Ann. Emerg. Med. *9*, 111.

Belz, G. G. (1974a). Plasma concentrations of intravenous β-methyl digoxin with and without oral charcoal, Klin. Wochenschr. *52*, 749.

Belz, G. G. (1974b). Adsorption to activated charcoal and polarity of cardenolides and bufadienolides, Experientia *30*, 530.

Belz, G. G. and Bader, H. (1974). Effect of oral charcoal on plasma levels of intravenous methyl proscillaridin, Klin. Wochenschr. *52*, 1134.

Boehm, J. J. and Oppenheim, R. C. (1977). An in vitro study of the adsorption of various drugs by activated charcoal, Aust. J. Pharm. Sci. *6*, 107.

Braithwaite, R. A., Crome, P., and Dawling, S. (1978). The in vitro and in vivo evaluation of activated charcoal as an adsorbent for tricyclic antidepressants, Br. J. Clin. Pharmacol. *5*, 369.

Brent, J. (1993). Are activated charcoal-N-acetylcysteine interactions of clinical significance?, Ann. Emerg. Med. *22*, 1860.

Buck, W. B., Bratich, P. M., and Ernst, M. (1985). Experimental studies with activated charcoals and oils in preventing toxicoses (abstract), Vet. Hum. Toxicol. *27*, 311.

Burke, G. M., Wurster, D. E., Buraphacheep, V., Berg, M. J., Veng-Pedersen, P., and Schottelius, D. D. (1991). Model selection for the adsorption of phenobarbital by activated charcoal, Pharm. Res. *8*, 228.

Caldwell, J. H., Caldwell, P. B., Murphy, J. W., and Beachler, C. W. (1980). Intestinal secretion of digoxin in the rat: Augmentation by feeding activated charcoal, Naunyn-Schmiedeberg's Arch. Pharmacol. *312*, 271.

Chamberlain, J. M., Gorman, R. L., Oderda, G. M., Klein-Schwartz, W., and Klein, B. L. (1993). Use of activated charcoal in a simulated poisoning with acetaminophen: A new loading dose for N-acetylcysteine?, Ann. Emerg. Med. *22*, 1398.

Chaput de Saintonge, D. M. (1973). Effect of activated charcoal on plasma levels of nortriptyline after single doses in man (letter), Eur. J. Clin. Pharmacol. *6*, 207.

Chin, L., Picchioni, A. L., and Duplisse, B. R. (1969). Comparative antidotal effectiveness of activated charcoal, Arizona montmorillonite, and evaporated milk, J. Pharm. Sci. *58*, 1353.

Chin, L., Picchioni, A. L., and Duplisse, B. R. (1970). The action of activated charcoal on poisons in the digestive tract, Toxicol. Appl. Pharmacol. *16*, 786.

Chin, L., Picchioni, A. L., Bourn, W. M., and Laird, H. E. (1973). Optimal antidotal dose of activated charcoal, Toxicol. Appl. Pharmacol. *26*, 103.

Chinouth, R. W., Czajka, P. A., and Peterson, R. G. (1980). N-acetylcysteine adsorption by activated charcoal, Vet. Hum. Toxicol. *22*, 392.

Collombel, C. and Perrot, L. (1970). [Experimental study of the treatment of salicylate intoxication with activated carbon], Eur. J. Toxicol. *3*, 352.

Corby, D. G. and Decker, W. J. (1974). Management of acute poisoning with activated charcoal, Pediatrics *54*, 324.

Crammer, J. and Davies, B. (1972). Activated charcoal in tricyclic drug overdoses (letter), Br. Med. J., August 26, p. 527.

Crome, P., Dawling, S., Braithwaite, R. A., Masters, J., and Walkey, R. (1977). Effect of activated charcoal on absorption of nortriptyline, Lancet, December 10, p. 1203.

Crome, P., Adams, R., Ali, C., Dallos, V., and Dawling, S. (1983). Activated charcoal in tricyclic antidepressant poisoning: Pilot controlled clinical trial, Hum. Toxicol. 2, 205.

Curd-Sneed, C. D., McNatt, L. E., and Stewart, J. J. (1986). Absorption (sic) of sodium pentobarbital by three types of activated charcoal, Vet. Hum. Toxicol. 28, 524.

Curd-Sneed, C. D. and Stewart, J. J. (1987). In vitro adsorption of sodium pentobarbital by SuperChar, USP, and Darco G-60 activated charcoals (reply to letter), J. Toxicol.-Clin. Toxicol. 25, 439.

Curd-Sneed, C. D., Parks, K. S., Bordelon, J. G., and Stewart, J. J. (1987a). In vitro adsorption of sodium pentobarbital by SuperChar, USP, and Darco G-60 activated charcoals, J. Toxicol.-Clin. Toxicol. 25, 1.

Curd-Sneed, C. D., Bordelon, J. G., Parks, K. S., and Stewart, J. J. (1987b). Effects of activated charcoal and sorbitol on sodium pentobarbital absorption in the rat, J. Toxicol.-Clin. Toxicol. 25, 555.

Dawling, S., Crome, P., and Braithwaite, R. (1978). Effect of delayed administration of activated charcoal on nortriptyline absorption, Eur. J. Clin. Pharmacol. 14, 445.

Decker, W. J., Combs, H. F., and Corby, D. G. (1968a). Adsorption of drugs and poisons by activated charcoal, Toxicol. Appl. Pharmacol. 13, 454.

Decker, W. J., Corby, D. G., and Ibanez, J. D., Jr. (1968b). Aspirin adsorption with activated charcoal, Lancet, April 6, p. 754.

Decker, W. J., Shpall, R. A., Corby, D. G., Combs, H. F., and Payne, C. E. (1969). Inhibition of aspirin absorption by activated charcoal and apomorphine, Clin. Pharmacol. Ther. 10, 710.

Decker, W. J., Corby, D. G., Hilburn, R. E., and Lynch, R. E. (1981). Adsorption of solvents by activated charcoal, polymers, and mineral sorbents, Vet. Hum. Toxicol. 23 (Suppl. 1), 44.

Deichman, W. B., Kitzmuth, K. V., Witherup, S., and Joaansman, K. (1944). Kerosene intoxication, Arch. Intern. Med. 21, 803.

Dillon, E. C., Jr., Wilton, J. H., Barlow, J. C., and Watson, W. A. (1989). Large surface area activated charcoal and the inhibition of aspirin absorption, Ann. Emerg. Med. 18, 547.

Dingemanse, E. and Laqueur, E. (1926). Adsorption of poisons on charcoal. III. The distribution of poisons between stomach and intestine wall and charcoal, Biochem. Z. 169, 235.

Donovan, J. W. and Renzi, F. P. (1986). Use of activated charcoal (reply to letter), Ann. Emerg. Med. 15, 102.

Dordoni, B., Willson, R. A., Thompson, R. P. H., and Williams, R. (1973). Reduction of absorption of paracetamol by activated charcoal and cholestyramine: A possible therapeutic measure, Br. Med. J. 3, 86.

Ekins, B. R., Ford, D. C., Thompson, M. I. B., Bridges, R. R., Rollins, D. E., and Jenkins, R. D. (1987). The effect of activated charcoal on N-acetylcysteine absorption in normal subjects, Am. J. Emerg. Med. 5, 483.

Elonen, E. and Neuvonen, P. J. (1981). Charcoal in the treatment of aspirin poisoning (letter), Lancet, September 5, p. 536.

Erickson, C. K. (1993). Lowering of blood ethanol by activated carbon products in rats and dogs, Alcohol 10, 103.

Fiser, R. H., Maetz, H. M., Treuting, J. J., and Decker, W. J. (1971). Activated charcoal in barbiturate and glutethimide poisoning of the dog, J. Pediatr. 78, 1045.

Flanagan, R. J. and Meredith, T. J. (1991). Use of N-acetylcysteine in clinical toxicology, Am. J. Med. 91 (Suppl. 3C), 131S.

Gloxhuber, C. (1968). [On the treatment after intake of detergents and cleansing agents], Med. Welt. 6, 351.

Greensher, J., Mofenson, H. C., and Picchioni, A. L. (1980). Activated charcoal (reply to letter), Ann. Emerg. Med. 9, 111.

Haacke, H., Johnsen, K., and Kolenda, K. D. (1973). [On the therapy of digitalis intoxication: Another experimental indication of the efficacy of adsorbents], Med. Welt. *24*, 1374.

Härtel, G., Manninen, V., and Reissell, P. (1973). Treatment of digoxin intoxication, Lancet, July 21, p. 158.

Hedges, J. R., Otten, E. J., Schroeder, T. J., and Tasset, J. J. (1987). Correlation of initial amitriptyline concentration reduction with activated charcoal therapy in overdose patients, Am. J. Emerg. Med. *5*, 48.

Henschler, D. (1970). Antidotal properties of activated charcoal, Arh. Hig. Rada. Toksikol. *21*, 129.

Holdiness, M. R. (1991). Clinical pharmacokinetics of N-acetylcysteine, Clin. Pharmacokinet. *20*, 123.

Hultén, B. Å., Heath, A., Mellstrand, T., and Hedner, T. (1985). Does alcohol adsorb to activated charcoal?, Hum. Toxicol. *5*, 211.

Hultén, B. Å., et al. (1988). Activated charcoal in tricyclic antidepressant poisoning, Hum. Toxicol. *7*, 307.

Jackson, J. E., Picchioni, A. L., and Chin. L. (1980). Contraindications for activated charcoal use (letter), Ann. Emerg. Med. *9*, 599.

Javaid, K. A. and El-Mabrouk, B. H. (1983). In vitro adsorption of phenobarbital onto activated charcoal, J. Pharm. Sci. *72*, 82.

Joachimoglu, G. (1920). The theoretical principles of charcoal therapy, Chem.-Ztg. *44*, 780.

Juhl, R. P. (1979). Comparison of kaolin-pectin and activated charcoal for inhibition of aspirin absorption, Am. J. Hosp. Pharm. *36*, 1097.

Kärkkäinen, S. and Neuvonen, P. J. (1986). Pharmacokinetics of amitriptyline influenced by oral charcoal and urine pH, Int. J. Clin. Pharmacol. Ther. Toxicol. *24*, 326.

Katona, B. G., Siegel, E. G., Roberts, J. R., Fant, W. K., and Hassan, M. (1989). The effect of superactive charcoal and magnesium citrate solution on blood ethanol concentrations and area under the curve in humans, J. Toxicol.-Clin. Toxicol. *27*, 129.

Klein-Schwartz, W. and Oderda, G. M. (1981). Adsorption of oral antidotes for acetaminophen poisoning (methionine and N-acetylcysteine) by activated charcoal, Clin. Toxicol. *18*, 283.

Krenzelok, E. P. (1986). Use of activated charcoal (letter), Ann. Emerg. Med. *15*, 102.

Kunzova, H. (1937). The evaluation and the use of animal charcoals, Prakticky Lekar *17*, 337.

Laass, W. (1974). [Suitability of using activated charcoal for the treatment of acute oral poisoning with organic solvents], Pharmazie *29*, 728.

Laass, W. (1975). [Gas chromatographic blood level study of the effect of activated charcoal, castor oil, and heavy or light mineral oil on the intestinal absorption of toxic organic solvents], Proc. Eur. Soc. Toxicol. *16*, 297.

Laass, W., Bräter, E., and Tietze, G. (1977). [The suitability of Carbo medicinalis for the treatment of acute oral poisoning with organic solvents: Part 2. Further studies on the survival effect], Pharmazie *32*, 542.

Laass, W. (1980). Therapy of acute oral poisonings by organic solvents: Treatment by activated charcoal in combination with laxatives, Arch. Toxicol. (Suppl. 4), 406.

Levy, G. and Tsuchiya, T. (1969). Effect of activated charcoal on aspirin absorption in man (abstract), Pharmacologist *11*, 292.

Levy, G. and Tsuchiya, T. (1972). Effect of activated charcoal on aspirin absorption in man, Clin. Pharmacol. Ther. *13*, 317.

Levy, G. and Gwilt, P. (1972). Activated charcoal for acute acetaminophen intoxication (letter), JAMA *219*, 621.

Levy, G. and Houston, J. B. (1976). Effect of activated charcoal on acetaminophen absorption, Pediatrics *58*, 432.

Levy, G. (1983). Charcoal for gastrointestinal clearance of drugs (reply to letters), New Engl. J. Med. *308*, 157.

Lipscomb, D. J. and Widdop, B. (1975). Studies with activated charcoal in the treatment of drug overdosage using the pig as an animal model, Arch. Toxicol. *34*, 37.

Llera, J. L. (1983). Charcoal for gastrointestinal clearance of drugs (letter), New Engl. J. Med. *308*, 157.

McKinney, P. E., Gillilan, R., and Watson, W. A. (1992). The preadministration of activated charcoal and aspirin absorption, J. Toxicol.-Clin. Toxicol. *30*, 549.

McNally, W. D. (1956). Kerosene poisoning in children: A study of 204 cases, J. Pediatrics *48*, 296.

Matthew, H. (1972). Activated charcoal in tricyclic drug overdoses (letter), Br. Med. J., November 4, p. 298.

Minocha, A., Herold, D. A., Barth, J. T., Gideon, D. A., and Spyker, D. A. (1986). Activated charcoal in oral ethanol absorption: Lack of effect in humans, J. Toxicol.-Clin. Toxicol. *24*, 225.

Mofenson, H. C. and Caraccio, T. R. (1992). Is activated charcoal useful for acetaminophen overdose beyond two hours after ingestion? (letter), Ann. Emerg. Med. *21*, 894.

Montgomery, B. R. (1978). Tricyclic overdose? Treat with activated charcoal, JAMA *240*, 424.

Morgan, D. P., Dotson, T. B., and Lin, L. I. (1977). Effectiveness of activated charcoal, mineral oil, and castor oil in limiting gastrointestinal absorption of a chlorinated hydrocarbon pesticide, Clin. Toxicol. *11*, 61.

Neuvonen, P. J., Elfving, S. M., and Elonen, E. (1978). Reduction of absorption of digoxin, phenytoin, and aspirin by activated charcoal in man, Eur. J. Clin. Pharmacol. *13*, 213.

Neuvonen, P. J. and Elonen, E. (1980). Effect of activated charcoal on absorption and elimination of phenobarbitone, carbamazepine, and phenylbutazone in man, Eur. J. Clin. Pharmacol. *17*, 51.

Neuvonen, P. J., Olkkola, K. T., and Alanen, T. (1984). Effect of ethanol and pH on the adsorption of drugs to activated charcoal: Studies in vitro and in man, Acta Pharmacol Toxicol. *54*, 1.

Neuvonen, P. J. and Kivistö, K. T. (1989). Activated charcoal should replace the resins in the treatment of digoxin intoxication (letter), Arch. Intern. Med. *149*, 2603.

North, D. S., Peterson, R. G., and Krenzelok, E. P. (1981a). Effect of activated charcoal administration on acetylcysteine serum levels in humans, Am. J. Hosp. Pharm. *38*, 1022.

North, D. S., Thompson, J. D., and Peterson, C. D. (1981b). Effect of activated charcoal on ethanol blood levels in dogs, Am. J. Hosp. Pharm. *38*, 864.

Olkkola, K. T. (1984). Does ethanol modify antidotal efficacy of oral activated charcoal: Studies in vitro and in experimental animals, J. Toxicol.-Clin. Toxicol. *22*, 425.

Oppenheim, R. C. and Stewart, N. F. (1975). Adsorption of tricyclic antidepressants by activated charcoal. I. Adsorption in low pH conditions, Aust. J. Pharm. Sci. *NS4*, 79.

Phansalkar, S. V. and Holt, L. E., Jr. (1968). Observations on the immediate treatment of poisoning, J. Pediatr. *72*, 683.

Picchioni, A. L., Chin, L., Verhulst, H. L., and Dieterle, B. (1966). Activated charcoal versus universal antidote as an antidote for poisons, Toxicol. Appl. Pharmacol. *8*, 447.

Picchioni, A. L., Chin, L., and Laird, H. E. (1974). Activated charcoal preparations—Relative antidotal efficacy, Clin. Toxicol. *7*, 97.

Rauws, A. G. and van Noordwijk, J. (1972). Activated charcoal in tricyclic drug overdoses (letter), Br. Med. J., November 4, p. 298.

Rauws, A. G. (1974). Treatment of experimental imipramine intoxication by interrupting enteral cycles with activated charcoal, Naunyn Schmiedeberg's Arch. Pharmacol. *282* (Suppl.), p. R78.

Rauws, A. G. and Olling, M. (1976). Treatment of experimental imipramine and desipramine poisoning in the rat, Arch. Toxicol. *35*, 97.

Renzi, F. P., Donovan, J. W., Martin, T. G., Morgan, L., and Harrison, E. F. (1985). Concomitant use of activated charcoal and N-acetylcysteine, Ann. Emerg. Med. *14*, 568.

Rose, S. R., Gorman, R. L., Oderda, G. M., Klein-Schwartz, W., and Watson, W. A. (1991). Simulated acetaminophen overdose: Pharmacokinetics and effectiveness of activated charcoal, Ann. Emerg. Med. *20*, 1064.

Rose, S. R., Watson, W. A., Oderda, G. M., Gorman, R. L., and Klein-Schwartz, W. (1992). Is activated charcoal useful for acetaminophen overdose beyond two hours after ingestion? (reply to letter), Ann. Emerg. Med. *21*, 894.

Rybolt, T. R., Burrell, D. E., Shults, J. M., and Kelley, A. K. (1986). In vitro coadsorption of acetaminophen and N-acetylcysteine onto activated carbon powder, J. Pharm. Sci. *75*, 904.

Saunders, F., Lackner, J. E., and Schochet, S. S. (1931). Studies in adsorption. I. The adsorption of physiologically active substances by activated charcoal, J. Pharmacol. *42*, 169.

Scheinen, M., Virtanen, R., Iisalo, E., and Salonen, J. S. (1980). Effect of activated charcoal on the pharmacokinetics of doxepin (abstract), Naunyn-Schmiedeberg's Arch. Pharmacol. *313* (Suppl.), p. R56.

Scheinen, M., Virtanen, R., and Iisalo, E. (1985). Effect of single and repeated doses of activated charcoal on the pharmacokinetics of doxepin, Int. J. Clin. Pharmacol. Ther. Toxicol. *23*, 38.

Sellers, E. M., Khouw, V., and Dolman, L. (1977). Comparative drug adsorption by activated charcoal, J. Pharm. Sci. *66*, 1640.

Shader, R. I. (1986). Is activated charcoal helpful in the treatment of overdosages (ODs) with tricyclic antidepressants (TCAs)?, J. Clin. Psychopharmacol. *6*, 327.

Smith, R. P., Gosselin, R. E., Henderson, J. A., and Anderson, D. M. (1967). Comparison of the adsorptive properties of activated charcoal and Alaskan montmorillonite for some common poisons, Toxicol. Appl. Pharmacol. *10*, 95.

Spiller, H. A., Krenzelok, E. P., Grande, G. A., Safir, E. F., and Diamond, J. J. (1994). A prospective evaluation of the effect of activated charcoal before oral N-acetylcysteine in acetaminophen overdose, Ann. Emerg. Med. *23*, 519.

Tietze, G. and Laass, W. (1979a). [Suitability of Carbo medicinalis for the treatment of acute oral poisoning with organic solvents. Part 3. Adsorption of organic solvents by Carbo medicinalis], Pharmazie *34*, 253.

Tietze, G. and Laass, W. (1979b). [Suitability of Carbo medicinalis for the treatment of acute oral poisoning with organic solvents. Part 4. Blood level studies], Pharmazie *34*, 254.

Tsuchiya, T. and Levy, G. (1972). Relationship between effect of activated charcoal on drug absorption in man and its drug adsorption characteristics in vitro, J. Pharm. Sci. *61*, 586.

Van de Graaff, W. B., Thompson, W. L., Sunshine, I., Fretthold, D., Leickly, F., and Dayton, H. (1982). Adsorbent and cathartic inhibition of enteral drug absorption, J. Pharmacol. Exp. Ther. *221*, 656.

Watson, W. A. and McKinney, P. E. (1991). Activated charcoal and acetylcysteine absorption: Issues in interpreting pharmacokinetic data, Drug Intell. Clin. Pharm. *25*, 1081.

Weaver, W. R. (1987). In vitro adsorption of sodium pentobarbital by SuperChar, USP, and Darco G-60 activated charcoals (letter), J. Toxicol.-Clin. Toxicol. *25*, 437.

Wurster, D. E., Burke, G. M., Berg, M. J., Veng-Pedersen, P., and Schottelius, D. D. (1988). Phenobarbital adsorption from simulated intestinal fluid, U. S. P., and simulated gastric fluid, U.S.P., by two activated charcoals, Pharm. Res. *5*, 183.

Zajtchuk, R., Corby, D. G., Miller, J. G., and O'Barr, T. P. (1975). Treatment of digoxin toxicity with activated charcoal (abstract), Am. J. Cardiol. *35*, 178.

12

Effect of Charcoal on Other Classes of Drugs

In the previous chapter, we considered the effects of charcoal on major classes of drugs and chemicals, such as alkaloids, salicylates, hypnotics and sedatives, tricyclic antidepressants, cardiac glycosides, organic solvents, and ethanol. In the present chapter, we will consider a variety of other drug categories, and will group them into the following classes: antihistamines, anti-infective drugs, cardiac drugs, anticonvulsants, CNS agents, gastrointestinal drugs, respiratory relaxants, antidiabetic sulfonylureas, organic substances, and miscellaneous drugs. Some drugs fall into more than one category (e.g., phenytoin is both an anticonvulsant and an antiarrhythmic). In such cases, the drug was assigned either to the category of its most common use or was classified more or less arbitrarily.

As was the case in the previous chapter, some studies have involved the resin cholestyramine, as well as charcoal. Again, such studies will be included here, as comparisons of the relative efficacies of charcoal and resins are of value. Studies dealing only with resins are deferred to Chapter 22.

First, we review one general study which included many of the drugs to be discussed in this chapter. Sellers et al. (1977) have done a comprehensive in vitro study, using 12 drugs adsorbing at 37°C to Norit A charcoal at pH 1.3 and at pH 10.8. The data were fitted to the Langmuir equation, and the maximal binding capacities, Q_m values, in mg drug/g charcoal were as shown in Table 12.1 (these data were also given earlier in Chapter 11). The entries marked (—) are cases where the solubility of the drug at the given pH (10.8) was insufficient to allow determination of the adsorption isotherm accurately. As expected, most drugs adsorb reasonably well at pH 1.3, in the range of about 0.10–0.25 g drug/g charcoal. Only amobarbital and phenobarbital fall below this range. As expected, the values at pH 10.8—which are all for acidic drugs—are lower than the values at pH 1.3, since at higher pH values acidic drugs are ionized and thus adsorb less well than at pH 1.3, where they are essentially neutral.

Neuvonen and Olkkola (1988) presented a figure summarizing a number of studies from Neuvonen's group over the period 1978–1986 in which subtoxic doses of various drugs were given to 5–7 healthy volunteers, followed 5 min later by the administration

Table 12.1 Maximal Binding Capacities for Various Drugs at 37°C

Drug	Q_m at pH 1.3 (mg/g)	Q_m at pH 10.8 (mg/g)
Amobarbital	51	47
Aspirin	262	141
Amitriptyline	133	—
Chlordiazepoxide	157	—
Diazepam	136	—
Methaqualone	179	—
Glutethimide	252	—
Pentobarbital	103	95
Phenobarbital	70	56
Propoxyphene HCl	127	—
Propoxyphene napsylate	137	—
Secobarbital	124	85

From Sellers et al. (1977). Reproduced by permission of the American Pharmaceutical Association.

Table 12.2 Reduction of Drug Bioavailabilities by Activated Charcoal Given 5 Minutes After the Drug to Healthy Volunteers

Drug and Dose	Bioavailability (% of control)	Reference
Trimethoprim 200 mg	0.5	Neuvonen and Olkkola (1984a)
Amitriptyline 75 mg	0.5	Kärkkäinen and Neuvonen (1986)
Quinidine 200 mg	0.7	Neuvonen et al. (1984)
Phenytoin 500 mg	0.7	Neuvonen et al. (1978)
Pindolol 10 mg	0.7	Neuvonen and Olkkola (1984b)
Cimetidine 400 mg	1.1	Neuvonen and Olkkola (1984b)
Sotalol 160 mg	1.2	Kärkkäinen and Neuvonen (1984)
Digoxin 0.5 mg	1.6	Neuvonen et al. (1978)
Phenylbutazone 200 mg	1.9	Neuvonen and Elonen (1980a)
Indomethacin 50 mg	1.9	Neuvonen and Olkkola (1984a)
D-Propoxyphene 130 mg	1.9	Kärkkäinen and Neuvonen (1985)
Tetracycline 500 mg	2.4	Neuvonen et al. (1983c)
Phenobarbital 200 mg	2.6	Neuvonen and Elonen (1980a)
Disopyramide 200 mg	3.0	Neuvonen and Olkkola (1984a)
Mexiletine 200 mg (*)	3.0	Olkkola and Neuvonen (1984)
Carbamazepine 400 mg	3.0	Neuvonen and Elonen (1980a)
Atenolol 100 mg (*)	3.6	Neuvonen and Olkkola (1986)
Tolbutamide 500 mg	10.4	Neuvonen et al. (1983a)
Tolfenamic acid 400 mg (*)	12.0	Olkkola and Neuvonen (1984)
Aspirin 1000 mg	15.0	Neuvonen et al. (1978)
Acetaminophen 1000 mg	15.0	Neuvonen et al. (1983c)
Chlorpropamide 250 mg	17.4	Neuvonen and Kärkkäinen (1983)

(*) indicates only 25 g charcoal was used. Adapted from Neuvonen and Olkkola (1988). Reproduced by permission of ADIS International.

of activated charcoal. The results of these studies are presented here since only 7 of the 22 drugs involved fall into the "major classes of drugs" categories of Chapter 11.

They determined the percentage bioavailabilities with charcoal compared to control (no charcoal) cases. Their figure is in the form of a bar graph which, upon careful measurement of the lengths of the bars, gave the numerical values shown in Table 12.2 (the figure is small and so the values thus derived are not precise; however, they are probably correct to within 10% of their magnitudes). The charcoal dose (only one dose was used in these studies) was 50 g, except where noted. Clearly, the charcoal reduced the drug bioavailabilities substantially (the average of the values shown is about 4.5%).

The remainder of this chapter will be devoted to a discussion of specific drugs class-by-class. The sequence of classes will more or less follow the order of the AHFS (American Hospital Formulary Service) Pharmacologic-Therapeutic Classification. The order to be used is shown in Table 12.3.

I. ANTIHISTAMINES

A. Chlorpheniramine

Chlorpheniramine maleate is an antihistamine which has caused some overdoses in children. Picchioni et al. (1974) studied the effects of two types of charcoal on chlorpheniramine maleate absorption in vivo using rats. The rats were given 40 mg/kg of the drug, followed 1 min later by 200 mg/kg of either Merck or Darco G-60 charcoal. Liver samples were collected at 30 min and analyzed for the drug. The drug concentrations in the liver tissues in mg/100 g were: control tests, 9.1; with Darco G-60 charcoal, 3.8; and with Merck charcoal, 1.3. Thus, both charcoals were effective in reducing chlorpheniramine absorption, with the Merck charcoal being the better of the two.

Boehm et al. (1978) studied the in vitro adsorption of chlorpheniramine maleate (200 mg/L) in SGF and SIF, using Norit Medicinal charcoal at a charcoal:drug ratio of 10:1 and found 65% adsorption from SGF versus 90% adsorption from SIF (in which the drug is less ionized). In vivo experiments were done with dogs. Doses of 20 mg/kg of drug were followed 2 min later by 200 mg/kg charcoal, both given via a gastric tube. Mean chlorpheniramine blood levels are shown in Figure 12.1 for the charcoal-treated dogs and for control dogs (six animals in each group). Clearly, the charcoal had a significant effect in reducing chlorpheniramine absorption.

Chin et al. (1970) also studied chlorpheniramine maleate, both in vitro and in vivo in rats. At pH 1.5, 30 mg charcoal adsorbed 2.39 mg of this drug from a 4 mg solution. At pH 8.5, 30 mg charcoal adsorbed 3.93 mg of the drug. Since chlorpheniramine is a basic drug, it is cationic at low pHs and neutral at higher pHs. Thus, the greater amount of adsorption at the higher pH level is as expected. Rats were given 40 mg/kg chlorpheniramine maleate and then 200 mg/kg charcoal 1 min later. Figure 12.2 shows the levels of the drug in the rats' livers versus time and it is clear that the 5:1 charcoal:drug ratio had a significant effect.

These studies, although only three in number, show clearly that an adequate single dose of charcoal is effective in reducing the in vivo absorption of chlorpheniramine maleate in experimental animals (dogs, rats). Studies with human volunteers and reports of charcoal use in overdosed humans would be valuable.

Table 12.3 Outline of Substances to be Discussed

I.   Antihistamines
   A.   Chlorpheniramine
   B.   Diphenhydramine
II. Anti-infectives
   A.   Antibiotics
      1.   Streptomycin and neomycin
      2.   Gentamicin
   B.   Antituberculosis agents
      1.   Aminosalicylic acid
      2.   Isoniazid
   C.   Antimalarial agents
      1.   Chloroquine
      2.   Quinine
      3.   Sulphoxidine/pyrimethamine
   D.   Quinolones: Ciprofloxacin
   E.   Sulfones: Dapsone
   F.   Urinary anti-infectives: Trimethoprim
III. Cardiac drugs
   A.   Antiarryhthmics
      1.   Amiodarone
      2.   Disopyramide
      3.   Flecainide
      4.   Quinidine
      5.   Lidocaine, procaine, and similar compounds
      6.   Mexiletine
   B.   Beta-blockers
      1.   Nadolol
      2.   Sotalol
      3.   Propranolol
   C.   Calcium channel-blockers: Diltiazem
IV. CNS agents
   A.   Analgesics
      1.   Propoxyphene
      2.   Nefopam
      3.   Morphine
      4.   Tilidine
   B.   Anticonvulsants
      1.   Phenytoin
      2.   Carbamazepine
      3.   Valproic acid
   C.   Anti-inflammatory agents
      1.   Mefenamic acid
      2.   Piroxicam
      3.   Indomethacin
   D.   Tranquilizers
      1.   Diazepam
      2.   Lorazepam
      3.   Chlorpromazine
      4.   Meprobamate
      5.   Thioridazine
   E.   Other CNS agents: Methamphetamine

Table 12.3 (Continued)

V.		Gastrointestinal drugs
	A.	Cimetidine
	B.	Propantheline
	C.	Diphenoxylate
	D.	Nizatidine
VI.		Antidiabetic sulfonylureas
VII.		Respiratory relaxants: Theophylline
VIII.		Organic substances
	A.	Camphor
	B.	Ethylene glycol
	C.	Isopropanol and acetone
	D.	Polybrominated biphenyl
	E.	Herbicides
	F.	Insecticides
	G.	Pesticides
IX.		Other substances
	A.	Allylpropynal and sulfanilamide
	B.	Cyclosporin
	C.	Furosemide
	D.	Methotrexate
	E.	Cocaine
	F.	Phencyclidine
	G.	Miscellaneous drugs

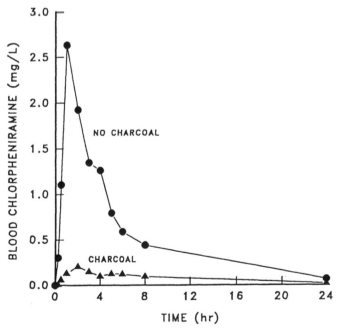

Figure 12.1 Mean levels of chlorpheniramine in whole blood of six animals after administration without activated charcoal or with activated charcoal at a charcoal:drug ratio of 10:1. From Boehm et al. (1978). Reproduced by permission of the Australian Pharmaceutical Publishing Company.

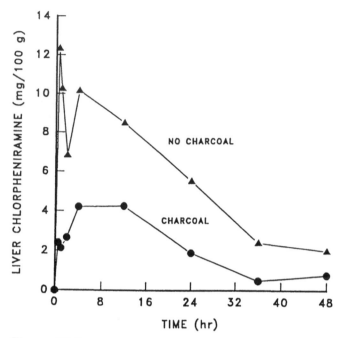

Figure 12.2 Effect of a 5:1 charcoal:drug ratio on liver chlorpheniramine levels in rats. From Chin et al. (1970). Reprinted with permission of Academic Press, Inc.

B.   Diphenhydramine

The adsorption of the antihistamine diphenhydramine (DPH) was assessed in vitro and in six healthy volunteers by Guay et al. (1984) in order to determine the effect of charcoal. When charcoal was used in vitro at a charcoal:drug ratio of 10:1 or more, greater than 85% adsorption occurred. A three-way cross-over trial in humans was done with 50 mg DPH alone, with 50 g charcoal 5 min later, or with 50 g charcoal 60 min later. These two charcoal protocols gave reductions in peak serum DPH of 94.8 and 12.3%, 0–24 hr AUC reductions of 96.9 and 20.4%, and 0–∞ AUC reductions of 90+ and 24.0%, respectively. This drug decreases GI motility due to its anticholinergic properties. Thus, charcoal, while obviously highly effective if given almost immediately, was also significantly effective if given after a 1 hr delay.

II.   ANTI-INFECTIVES

A.   Antibiotics

1.   Streptomycin and Neomycin

Ghazy et al. (1984) studied the adsorption of the antibiotics streptomycin sulfate and neomycin sulfate to activated charcoal in vitro. These drugs are aminoglycoside antibiotics which are widely used as intestinal disinfectants and in the treatment of diarrhea. Solutions were prepared in water or in buffer solutions of pH 6.5–7.0. The data were fitted to the Langmuir equation. The maximal binding capacities, Q_m, were 49 and 70 mg/g for streptomycin in water and buffer, respectively, and 53 and 78 mg/g for neomycin in water and buffer, respectively. These maximal adsorption capacities are fairly low.

2. Gentamicin

Gentamicin sulfate is an aminoglycoside antibiotic which is useful for treating a variety of bacterial infections. Because it is poorly absorbed from the GI tract, it is usually given IV or IM. Hasan et al. (1990a) gave 2 mg/kg gentamicin IV to rabbits (in an ear vein over a 1 min period) and immediately thereafter administered a single dose of 10 g charcoal in 40 mL water via gastric intubation. As compared to controls, the charcoal-treated rabbits had a drug elimination constant 1.20 times larger, a drug half-life only 0.85 times as large, and an AUC value only 0.71 times as large. Hence, the charcoal was modestly effective. This drug is known to have a small V_d, negligible binding to plasma proteins, and a low hepatic extraction ratio. Thus, one could expect its removal by gastrointestinal dialysis. Peritoneal dialysis has been shown to be effective in removing gentamicin. Only small amounts of this drug are excreted in the bile; thus, enterohepatic cycling is not important.

Another group of rabbits, which had been given uranyl nitrate to induce renal failure, was studied by Hasan and associates. The procedure was the same as before except that the charcoal was given 2.5 hr after the drug. In these rabbits, charcoal raised the elimination rate constant value and the clearance value both to 213% of the control values, and decreased $t_{1/2}$ to 53% of the control value. However, rabbits with renal failure have much lower drug elimination and clearance rates and much higher $t_{1/2}$ values. Thus, while charcoal had a more dramatic effect in the rabbits with renal failure, as compared to normal charcoal-treated rabbits, the effects were actually quantitatively smaller.

Taken together, this study and the one involving streptomycin and neomycin suggest that aminoglycoside antibiotics do not bind to charcoal in substantial amounts, and that the in vivo effects of charcoal are generally not very large, in subjects having normal renal function. Studies involving human subjects are needed to determine if these same conclusions apply to humans.

B. Antituberculosis Agents

1. Aminosalicylic Acid

Aminosalicylic acid (or, more precisely, para-aminosalicylic acid, PAS) is a synthetic antituberculosis agent. Olkkola (1985) examined the effect of the charcoal:drug ratio in some in vivo studies with adult humans using this drug. The PAS was given in doses of 1, 5, 10, or 20 g. Then, 50 g activated charcoal was given in 300 mL water immediately after. The plasma PAS values for the control and charcoal phases are shown in Figure 12.3. At a drug dose of 1 g, less than 5% of the PAS was absorbed into the body from the GI tract, whereas at a drug dose of 20 g, 37% of the drug was absorbed from the GI tract. In vitro studies of the adsorption capacity of charcoal for PAS correlated well to the in vivo results. Olkkola concluded that large doses (50–100 g) of activated charcoal should be used whenever possible, since PAS absorption is greatly reduced when the charcoal:PAS ratio is large.

2. Isoniazid

Isoniazid (isonicotinic acid hydrazide) is a synthetic antituberculosis agent which, in overdose, can cause life-threatening seizures which do not respond to standard anti-

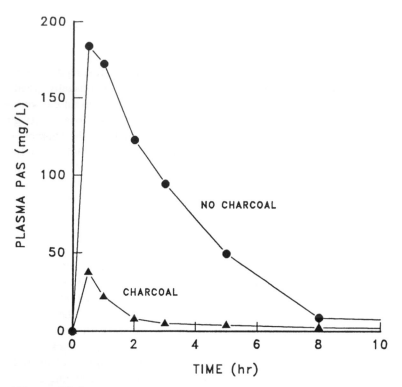

Figure 12.3 Effect of 50 g activated charcoal on the absorption of 5 g p-aminosalicylic acid (PAS). From Olkkola (1985). Reprinted by permission of Blackwell Scientific Publications Ltd.

convulsant therapy and which are accompanied by severe metabolic acidosis. Pyroxidine (vitamin B_6) is a safe, rapidly-acting systemic antidote. Alternatively, charcoal can be employed.

Orisakwe and Akintonwa (1991) studied the adsorption of isoniazid in vitro to activated charcoal and activated carbon black at pH 1.2, 7.5, and 10. For the charcoal, the Q_m values at these three pHs in mg/g were 336, 317, and 323, respectively. For the carbon black, the values were 258, 290, and 288, respectively. These results show that isoniazid adsorbs well to both adsorbents, that there is not much effect of pH (the isotherm plots show only a slightly higher adsorption at lower pH), and that the charcoal adsorbs a bit better than the carbon black.

In Chapter 9, we discussed in vivo studies of Chin et al. (1973) in which rats were given 50 mg/kg isoniazid, followed by charcoal in the amount of 1, 2, 4, or 8 times the drug dose. As was shown in Table 9.1, these charcoal doses reduced the tissue isoniazid levels at 30 min (compared to controls) by 1.2, 7.2, 35, and 80%, respectively. Thus, oral charcoal was quite effective at the higher charcoal:drug ratios.

Scolding et al. (1986) found that a 10-g dose of charcoal given 1 hr after a 600 mg oral dose of isoniazid to six healthy human subjects did not reduce the areas under plasma concentration versus time curves (evaluated over 0–8 hr) significantly. They also estimated AUC values for the zero to infinite time period. This increased the AUC values by 17.4% over those for 0–8 hr. In any event, the conclusion was that charcoal given

1 hr after an isoniazid dose does not appear to have any effect, because this drug is quite rapidly absorbed.

Siefkin et al. (1987) also studied the effect of charcoal on isoniazid pharmacokinetics. Three subjects were given 10 mg/kg isoniazid orally, followed by 60 g charcoal immediately after. The half-lives for elimination were 120, 178, and 193 min for the three subjects without charcoal. With charcoal, no measurable serum isoniazid levels were detected, so no half-lives could be determined. Thus, in contrast to the study of Scolding et al. (1986) which showed no effect of charcoal when given 1 hr after the isoniazid, this study showed that charcoal reduced isoniazid absorption to zero. This points out the need for prompt charcoal administration when a rapidly absorbed drug is involved.

Overall, these studies show that oral charcoal can be extremely effective for reducing isoniazid absorption, but only if given early, since isoniazid is rapidly absorbed.

C. Antimalarial Agents

1. Chloroquine

Chloroquine is used in the prevention and treatment of malaria, rheumatoid arthritis, and lupus erythematosus. It is quite toxic and has become a significant cause of suicide deaths. Chin et al. (1970) did both in vitro studies and in vivo studies (with rats) using chloroquine diphosphate. At pH 1.5, 30 mg charcoal adsorbed 3.68 mg of this drug from a 4 mg solution. At pH 8.5, 3.95 mg was adsorbed (thus, pH had little effect). Rats were given 100 mg/kg chloroquine diphosphate and then 500 mg/kg charcoal 1 min later. Figure 12.4 shows the drug levels in the rats' livers versus time. Clearly, the charcoal greatly reduced drug absorption.

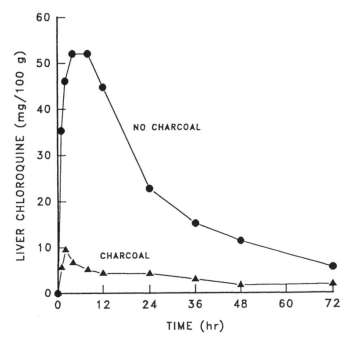

Figure 12.4 Effect of a 5:1 charcoal:drug ratio on liver chloroquine levels in rats. From Chin et al. (1970). Reprinted with permission of Academic Press, Inc.

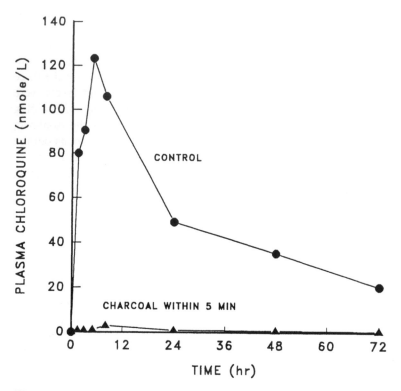

Figure 12.5 Effect of 25 g activated charcoal given within 5 min on the absorption of 500 mg chloroquine phosphate. From Neuvonen et al. (1992). Reproduced by permission of The Macmillan Press Ltd.

In Chapter 9, we discussed another in vivo study carried out by Chin et al. (1973) in which rats were given 50 mg/kg chloroquine phosphate and then were administered oral charcoal in charcoal:drug ratios of 1:1, 2:1, 4:1, or 8:1. Table 9.1 showed that the tissue levels of chloroquine at 120 min were reduced (compared to controls) by 20, 30, 70, and 96% with these charcoal doses, respectively.

Neuvonen et al. (1992) found that, in vitro at pH 1.2, 98% of chloroquine phosphate was bound by charcoal at a charcoal:drug ratio of 5:1. In vivo tests were done with human subjects, who were given 500 mg chloroquine phosphate and then 25 g charcoal 5 min later. Blood levels were followed for 192 hr. Figure 12.5 shows the data obtained. Compared to controls, the charcoal-treated patients had AUC values (0–192 hr and 0–∞) and peak drug concentrations which were all reduced by 99%. Thus, charcoal was extremely effective in reducing chloroquine absorption. These researchers summarized these striking results in a letter which commented on a study of chloroquine poisoning but failed to discuss the role of oral charcoal (see Kivistö and Neuvonen, 1993).

Laine et al. (1992) carried out experiments in rats using chloroquine diphosphate. The drug (100 mg/kg) was given subcutaneously, and 48 hr later the rats were placed on a diet containing 20 wt% charcoal. Blood samples were taken at 48, 96, 144, 216, and 264 hr. No effect of the charcoal diet on the rate of drug elimination was found, indicating that there is no significant enteroenteric or enterohepatic circulation for this drug.

These studies therefore suggest that oral charcoal is very effective in reducing chloroquine absorption if given early but may have little effect if given after the drug has become well absorbed.

2. Quinine and Quinidine

Quinine and quinidine are the levorotatory and dextrorotatory isomers of the same compound, an alkaloid extracted from the bark of the chinchona tree. As mentioned earlier, quinine is used to treat malaria, while quinidine is employed as an antiarrhythmic agent. Hayden and Comstock (1975) showed that quinine adsorbs well to charcoal in vitro. Thus, repeated-dose charcoal therapy was considered and tried by Lockey and Bateman (1989) in a study of quinine elimination using subtoxic doses in seven healthy subjects. They were each given 600 mg quinine bisulfate, then 50 g charcoal in an aqueous slurry 4 hr after the drug, and three further 50-g doses over the next 12 hr. Carbomix and Medicoal were used alternately. Blood samples were taken over 0–36 hr. Based on AUC values determined by extrapolating the blood data to infinite time, the charcoal lowered the drug absorption 31.5% and decreased the half-life from 8.23 to 4.55 hr (a 44.7% decrease). Had the charcoal not been delayed for 4 hr, even larger effects might have been noted. Since there are no experimental data to suggest that quinine undergoes enterohepatic circulation, the charcoal probably works by binding quinine which diffuses through the gut lumen.

Five patients with actual quinine overdoses were treated by Prescott et al. (1989) using repeated charcoal doses. The dosage schedule was generally 50 g charcoal (Medicoal) in 200 mL water shortly after admission, and 50 g every 4 hr thereafter up to a total of 200–400 g. Not counting one patient for which the first charcoal dose was delayed 34 hr, the quinine half-lives were 7.2, 7.6, 7.6, and 8.0 hr. In that patient for which the charcoal was delayed, the drug half-life was 33 hr prior to the charcoal and 10 hr after. A previous report from the literature suggests a half-life of about 24 hr in the absence of charcoal. Thus, repeated charcoal appears to reduce the quinine half-life from a value on the order of 24–33 hr to about 7–8 hr.

These two studies suggest that early doses of charcoal can effectively inhibit quinine absorption and that multiple doses of charcoal can accelerate quinine elimination by interruption of its enteroenteric cycle.

3. Sulphoxidine/Pyrimethamine

Akintonwa and Obodozie (1991) studied the antimalarial agent Fansidar, a combination of sulphoxidine and pyrimethamine, which in excessive amounts can cause severe cutaneous reactions and toxic epidermal necrolysis. In vitro studies by these authors showed that sulphoxidine adsorbs well to charcoal (Ultracarbon, Merck, Darmstadt). An in vivo study was performed with 20 healthy adult males who were given three tablets of Fansidar, each containing 500 mg sulphoxidine and 25 mg pyrimethamine, orally. One group then received 2 g charcoal 5 min later. Blood samples were drawn between 0 and 72 hr. The charcoal caused a sulphadoxine half-life reduction from 256 to 117 hr (a 54% decrease), and a 0–48 hr AUC reduction from 2533 to 1346 mg-hr/L (a 47% decrease). Charcoal also delayed the time to the peak sulphadoxine concentration from 5.2 hr to 6.8 hr. Clearly, even at the small charcoal dose level employed (2 g), the charcoal was effective. A larger charcoal dose, which certainly would have been easily tolerated, probably would have had even more dramatic effects.

D.   Quinolones: Ciprofloxacin

Ciprofloxacin is an antibiotic compound which has been shown to be active against the bacterium which causes Legionnaire's disease. It is generally well tolerated but in large enough doses can cause adverse GI tract or CNS effects. Torre et al. (1988) did a study in which six healthy volunteers were given 500 mg ciprofloxacin alone or together with 1 g charcoal in a cross-over study. They found no effect due to charcoal on any of the pharmacokinetic parameters they measured. However, at a charcoal:drug ratio of only 2:1, it is possible that the single oral charcoal dose was simply too small to produce an effect. Actually, the purpose of the study was not to see whether charcoal would be of use in ciprofloxacin overdose but to see whether concomitant use of a therapeutic dose of charcoal for intestinal gas would interfere with normal doses of the drug.

Sorgel et al. (1989) gave 12 healthy subjects 200 mg ciprofloxacin IV with and without concurrent oral charcoal (the amount of charcoal administered is not stated in their paper). Urine was collected over successive time intervals up to a total time of 48 hr and was assayed for unchanged ciprofloxacin. AUC values showed that renal excretion of the unchanged drug decreased by 13 mg (6.5% of the dose) when charcoal was used. Thus, it appears that the charcoal had little effect. Since the drug was given IV, this suggests that the drug does not undergo significant gastric secretion or enterohepatic circulation. However, since the amount of charcoal employed was not disclosed, it is difficult to reach firm conclusions.

Unfortunately, neither of these two studies can tell us whether oral charcoal, if given in a significant amount (e.g., at a charcoal:drug ratio of at least 5:1), would reduce oral ciprofloxacin absorption significantly. Such studies remain to be done.

E.   Sulfones: Dapsone

Dapsone is a sulfone drug used to treat leprosy and certain types of dermatitis. It is eliminated by conversion to monoacetyldapsone (MAD). Excessive levels of dapsone and MAD produce methemoglobinemia and hemolytic anemia in a dose-dependent response. Methemoglobinemia is the most obvious symptom. Dapsone has a relatively long half-life (10 to 50 hr with therapeutic doses but longer for toxic doses) and is retained in tissues, especially the liver and kidneys, for up to three weeks. It persists in the circulation for long periods, because it undergoes biliary excretion with intestinal reabsorption (i.e., it has an enterohepatic cycle). Dapsone intoxication can be effectively counteracted by the specific antagonists methylene blue and toluidine blue; however, the use of oral charcoal may accelerate the elimination of the drug and reduce the amount and duration of antagonist therapy.

Elonen et al. (1979) described the case of a 45-year-old man who had ingested about 10 g of dapsone. Because the disappearance of dapsone and MAD was observed to be slow, oral charcoal therapy (20 g, four times a day) was begun three days postingestion and was carried out for the next two days. This reduced the serum dapsone half-life from 88 to 13.5 hr and the serum half-life of MAD from 67 to 14 hr. When the charcoal was stopped, the half-lives tended to increase again. This case shows that dapsone undergoes significant enterohepatic cycling, which can be interrupted by oral charcoal.

Neuvonen et al. (1980) did an in vivo randomized cross-over study with five subjects. The subjects took a total dapsone dose of 500 mg over four days. Ten hours after the last 100-mg dose, 50 g charcoal as a slurry was given, followed by 4 consecutive doses of 17 g charcoal at 12 hr intervals. The half-life of serum dapsone was reduced by charcoal from a control period value of 20.5 hr to 10.8 hr and the half-life of serum monoacetyldapsone was decreased from 19.3 to 9.5 hr. In addition, two actual suicidal overdose patients were treated with 80 g/day charcoal for one or two days and it was found that charcoal increased the rate of elimination of dapsone by a factor of 3 to 5. Thus, the use of multiple doses of charcoal for a drug of this kind, which has a long half-life and which undergoes enterohepatic recirculation, appears to be quite effective.

Neuvonen et al. (1983b) published a later report on dapsone, in which three patients were treated after 1–10 g overdoses. Activated charcoal given orally in multiple doses (20 g, four times a day) shortened the half-life of dapsone from a value of 77 hr without charcoal to a mean of 12.7 hr with charcoal. The half-life of MAD similarly decreased, from 51 to 13.3 hr. Charcoal treatment was compared to hemodialysis, which gave dapsone and MAD half-lives of 10.4 and 10.9 hr, respectively. Thus, they concluded that oral charcoal therapy is comparable in effectiveness to hemodialysis, and is easier to perform and is much cheaper.

Reigart et al. (1982–83) reported on a case in which an 18-month-old child accidentally ingested a single 100-mg dapsone tablet. The child presented to a physician 3–4 hr later with severe cyanosis. Doses of 10 g activated charcoal were started at 20 hr postingestion and were repeated every 6 hr. The child showed rapid resolution of his symptoms and was symptom-free 64 hr postingestion. The rapid improvement was attributed to interruption of the enterohepatic circulation of the drug by the charcoal.

These reports all dealt with attempts to accelerate dapsone elimination after the drug had been well absorbed, and the reports all suggest the same conclusion: oral charcoal is quite effective in interrupting the enterohepatic circulation of dapsone and its major metabolite.

F. Urinary Anti-Infectives: Trimethoprim

Details of a study by Neuvonen and Olkkola (1984a) involving three drugs, one of which was trimethoprim (a synthetic anti-infective), are discussed immediately below in the section on antiarrhythmics, since one of the other drugs was the antiarrhythmic disopyramide.

This one study with trimethoprim indicates that oral charcoal, given 5 min after the drug, is highly effective in preventing trimethoprim absorption.

III. CARDIAC DRUGS

A. Anti-Arrhythmics

1. Amiodarone

Amiodarone hydrochloride is an antiarrhythmic agent that is slowly absorbed and has a very long half-life. It has serious effects in overdose quantities (primarily pulmonary toxicity). Kivistö and Neuvonen (1991) gave human volunteers oral doses of 400 mg amiodarone, plus 25 g charcoal immediately or 25 g charcoal at 1.5 hr. Compared to controls, immediate charcoal reduced the drug absorption by about 98% and the charcoal given at 1.5 hr reduced absorption by about 50%. Figure 12.6 shows these results. The slow absorption of this drug accounts for the significant effectiveness of charcoal.

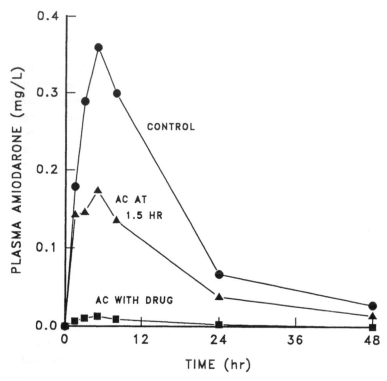

Figure 12.6 Effect of 25 g activated charcoal on the absorption of 400 mg amiodarone. The charcoal was given either with the drug or 1.5 hr later. From Kivistö and Neuvonen (1991). Reproduced by permission of The Macmillan Press Ltd.

Laine et al. (1992) gave 200 mg/kg amiodarone subcutaneously to rats and then placed the rats on a diet containing 20 wt% charcoal 48 hr later. Blood samples were taken at 48, 96, 144, 216, and 264 hr. No effect of the charcoal on the rate of amiodarone elimination was noted. This suggests that amiodarone does not undergo enteral or enterohepatic circulation.

As was the case with chloroquine (which Laine et al. also studied) these two reports suggest that oral charcoal is very effective in reducing amiodarone absorption if given early, but may have little effect if given after the drug has become well absorbed.

2. Disopyramide

Neuvonen and Olkkola (1984a) investigated the efficacy of charcoal in reducing the absorption of disopyramide, indomethacin, and trimethoprim. Disopyramide is a synthetic antiarrhythmic agent which, in toxic amounts, causes anticholinergic effects, hypotension, respiratory arrest, and cardiac problems. Indomethacin is a nonsteroidal anti-inflammatory agent which can cause GI tract and CNS disturbances in overdose amounts. Trimethoprim is a synthetic anti-infective and, in excessive doses, can cause rashes, pruritus, and GI tract problems. These drugs were given orally (separately) to six healthy subjects in doses of 200, 50, and 200 mg, respectively, and after 5 min, a slurry of 2.5, 10, 25, or 50 g of Norit A charcoal in 300 mL water or a slurry of 10 g Amoco PX-21 charcoal

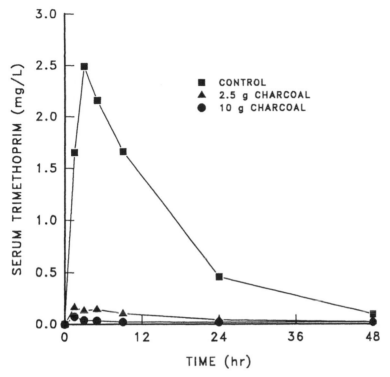

Figure 12.7 Effect of 2.5 or 10 g activated charcoal given at 5 min on the absorption of 200 mg trimethoprim in six subjects. From Neuvonen and Olkkola (1984). Figure 3a. Copyright Springer-Verlag GmbH & Co. Used with permission.

in 300 mL water was given. At a charcoal dose of 2.5 g, the absorption of these three drugs (in the same order as mentioned above) fell to 48, 53, and 8% of control values. At a dose of 10 g Norit charcoal, the absorption values were 9, 21, and 3% of the controls. For PX-21 charcoal at 10 g, the values were 4, 22, and < 1%. At a charcoal dose of 25 g, the values were 3, 11, and < 1%. Finally, at a 50 g charcoal dose, the absorption values were 3, 7, and < 1%. The results for trimethoprim are presented in Figure 12.7. Thus, charcoal was effective in reducing the absorption of these three drugs, and, as one would expect, was more effective as the charcoal amount was increased. In vitro experiments showed that adsorption of all three drugs was greater at pH 7.0 than at pH 1.2, particularly for disopyramide.

An interesting study with disopyramide was performed by Huang (1988). He separated disopyramide into two enantiomers, R- and S-disopyramide, by fractional crystallization of their bitartrate salts. The enantiomers are identical in chemical and physical properties, except in optical rotation. In vitro experiments showed that the two forms bind in identical fashion to activated charcoal. Protein-binding of the two forms is low (about 15%) and essentially equal. However, when injected separately into rabbits (into an ear vein) with or without 20 g charcoal given orally by intubation 30 min prior to the drug injection, some surprising results occurred. The 0–4 hr AUC values for the R and S forms in control rabbits (no charcoal) were 502 and 198 mg-min/L, respectively.

When charcoal was used, the 0–4 hr AUC values were 310 and 186 mg-min/L, respectively, or 38.2 and 6.1% lower than the control values. Thus, the two forms behaved differently alone and were affected by charcoal to very dissimilar extents. The differences can be explained by a difference in the hepatic extraction ratios of the two forms. The S form has a high hepatic extraction ratio, whereas that of the R form is intermediate. The higher hepatic extraction ratio for the S form causes greater removal of the S form by the liver, keeping its blood levels lower and giving it a lower AUC. The effect of charcoal was less for the S form. The effect of charcoal is masked for the S form, because any drug which escapes the GI tract goes in serial fashion to the liver where it is extracted anyway. Thus, whether charcoal is in the GI tract to bind the drug or not, the blood levels of the drug are much the same. This study shows that when considering the use of activated charcoal to enhance systemic drug elimination, the hepatic extraction ratio of the drug needs to be considered.

Another study involving disopyramide is that of Arimori and Nakano (1989). The drug was given at 20 mg/kg IV to rats, alone and with charcoal (300 mg) given as a slurry at time zero and with subsequent 150 mg charcoal doses at 1, 2, 3, and 4 hr. The half-life decreased from 1.18 to 1.05 hr, and the 0–6 hr AUC decreased from 13.8 to 11.3 mg-hr/L with the charcoal treatment. V_d was unaffected. Other tests were done in isolated perfused rat intestines. After IV disopyramide injection (10 or 30 mg/kg), bile was collected and a single-pass luminal perfusate solution (isotonic phosphate buffer) was collected at periodic intervals and both were then analyzed for the drug. The drug levels in the bile were roughly ten-fold higher than in the blood serum, showing that disopyramide is greatly excreted into the bile. Over 120 min, about 17–18% of the drug dose was exsorbed into the perfusate. These results all indicate that charcoal has the potential to accelerate the removal of disopyramide by adsorbing the enterally and enterohepatically cycled drug.

Arimori et al. (1989) then turned to using human subjects in a following study. They gave disopyramide (200 mg) orally to six healthy subjects followed by 40 g of charcoal in a suspension containing 250 mL of a magnesium citrate solution (34 g magnesium citrate content) at 4 hr, and charcoal doses of 20 g with 150–200 mL water at 6, 8, and 12 hr. The charcoal treatment decreased serum half life and 0–∞ AUC values by 33 and 19%, respectively, as compared to control values. Total body clearance increased to 122% of the control value. Serum drug level data showed that the peak drug levels occurred between 1–4 hr in the subjects; hence, even though the first charcoal dose was not given until 4 hr, there was still a significant charcoal effect.

These studies make it clear that oral charcoal is very effective in reducing disopyramide absorption when given early, and is also effective when given after the drug has been absorbed, due to interruption of the enteroenteric and enterohepatic cycles of the drug.

3.   Flecainide

Flecainide is an antiarrhythmic drug which, in toxic amounts, can cause severe cardiac and other organ manifestations. Nitsch et al. (1987) studied the effect of oral charcoal on the absorption and elimination of flecainide in eight male subjects. Each was given a single oral dose (200 mg) of flecainide, with or without a concurrent dose of 30 g charcoal. Plasma flecainide concentrations under "no charcoal" conditions were 266, 232, and 231 µg/L at 2, 4, and 6 hr after drug ingestion. When charcoal was used, plasma flecainide levels were unmeasurably low at the same times. In other tests, the charcoal was given 90 min after the drug. In this case, plasma flecainide levels were 231, 199, and 155 µg/L at 2, 4, and 6 hr. Thus, there was some effect of charcoal even when it was given after 90 min delay. Since flecainide apparently does not undergo enterohepatic circulation, charcoal should be given as early as possible.

4. Quinidine

Hasan et al. (1990b) studied the pharmacokinetics of quinidine and quinine administered intravenously to rabbits at a dose of 10 mg/kg. Quinidine is an antiarrhythmic drug, whereas quinine is an anti-malarial agent. Control experiments (no charcoal) showed that quinine had a larger volume of distribution, V_d, a larger elimination rate constant, k_e, a smaller half-life, $t_{1/2}$, and a smaller AUC than quinidine. When 15 g oral charcoal was given, no effect on the pharmacokinetic parameters for quinine occurred. However, charcoal significantly increased the rate of elimination of quinidine (k_e rose from 0.33 to 0.46 hr^{-1}), decreased its $t_{1/2}$ from 1.22 to 0.86 hr, and decreased the 0–∞ AUC from 1.53 to 1.03 mg-hr/L. Because quinine is eliminated at about three times the rate of quinidine, and because the V_d of quinine is about twice that of quinidine, these factors probably masked the effect of charcoal on quinine. Thus, it appears that oral charcoal can significantly enhance the elimination of quinidine, presumably by interruption of enteroenteric or enterohepatic circulation of the drug.

5. Lidocaine, Procaine, and Similar Compounds

Judis (1985) studied the adsorption of lidocaine, methadone, pilocarpine, and procaine to charcoal in vitro, at various pH values. We include this study in this section (arbitrarily) since lidocaine is an antiarrhythmic agent. Q_m values were not determined, so it is difficult to assess the strength of adsorption. However, for the charcoal:drug ratios used, the percent adsorption values were about 93–98% for lidocaine and procaine, were about 78–79% for methadone, and were about 45–48% for pilocarpine. There was almost no effect of pH over the range of pH 6.0–10.2. This is surprising, as the drug pK_a values ranged from 7.05 to 8.25, and these drugs (all bases) would be expected to adsorb more strongly for pH > pK_a and adsorb less strongly for pH < pK_a. The reason for the lack of a pH effect in this study is unknown. In general, pH in other studies has had the effects expected. Unfortunately, in vitro studies alone provide little information about the effects of charcoal in vivo.

6. Mexiletine

The adsorption of the antiarrhythmic drug mexiletine to charcoal from a macrogol (polyethylene glycol)-electrolyte solution (PEG-ELS) and a solution called JP XII (a disintegration medium in the Japanese Pharmacopeia) was studied by Arimori et al. (1993). Q_m values for mexiletine adsorbing from the PEG-ELS (pH 8.5) and JP XII (pH 6.8) media were 328 and 284 mg/g, respectively. The higher value at pH 8.5 is as expected since mexiletine is a basic compound (pK_a 9.1) and is less ionized at pH 8.5 than at pH 6.8. Omitting polyethylene glycol, sodium sulfate, and sodium bicarbonate (each one separately) from the PEG-ELS medium decreased the adsorption of the drug. The decrease due to sodium bicarbonate omission is easily explained since its absence lowered the PEG-ELS pH to 5.9, at which value the mexiletine is more ionized and therefore less adsorbable. The effect of removing a simple salt like sodium sulfate has been seen by others and is explained in Chapter 15. On the other hand, the decrease in adsorption in the absence of polyethylene glycol is more difficult to explain. As Arimori et al. hypothesize, the reason is probably that PEG reduces the solubility of the drug and makes it adsorb better when PEG is present.

The Q_m values of 284 and 328 mg/g indicate reasonably strong adsorption of mexiletine to charcoal but obviously in vivo studies are needed to determine if oral charcoal would be effective in mexiletine overdoses.

B.   Beta-Blockers

1.   Nadolol

Nadolol is a beta-adrenergic blocker used in angina, hypertension, and cardiac arrhythmias. About 20% binds to plasma proteins. It has a half-life roughly five times longer than that of propranolol, another beta-adrenergic blocker. Du Souich et al. (1983) performed studies aimed at determining whether nadolol undergoes enterohepatic circulation. They gave 80 mg nadolol to eight healthy subjects orally and (in one phase) followed this by giving charcoal tablets orally according to the following schedule: 500 mg at 3 hr, 500 mg at 4 hr, and 250 mg each hour for the next 8 hr (total charcoal 3000 mg). The charcoal reduced the 0–∞ AUC for nadolol from 2455 to 1355 µg-hr/L (a 45% decrease) and the half-life from 17.3 to 11.8 hr (a 32% decrease). These results suggest that enterohepatic circulation occurs with this drug and that charcoal can effectively interrupt such circulation.

2.   Sotalol

Kärkkäinen and Neuvonen (1984) examined the effects of oral charcoal on sotalol pharmacokinetics in seven subjects, in a randomized cross-over study. Sotalol hydrochloride (a beta-adrenergic blocker) was given (160 mg), followed 5 min later by 50 g oral charcoal. The charcoal reduced sotalol absorption by 99% compared to controls. When additional doses were given at 6 hr (50 g) and each 6 hr thereafter (12.5 g) up to a cumulative charcoal dose of 150 g at 54 hr, the serum sotalol half-life was shortened from 9.4 to 7.6 hr. It was concluded that early charcoal can dramatically inhibit sotalol absorption and that charcoal given later significantly interrupts the enterohepatic or enteroenteric circulation of the drug.

3.   Propranolol

Al-Meshal et al. (1993) studied the pharmacokinetics of propranolol, a nonselective beta-adrenergic blocking agent, after IV administration to rabbits. They gave rabbits 1 mg/kg of the drug IV over 1 min in an ear and then either 40 mL water or 10 g charcoal in 40 mL water (by gastric tube). Blood samples were taken over the next 360 min. Charcoal treatment decreased the drug half-life by 16.6%, increased the systemic clearance by 17%, and decreased the 0–360 min AUC by 14%. A two-compartment model adequately described the pharmacokinetics. This study demonstrates that propranolol undergoes enteroenteric or enterohepatic circulation, and that charcoal has a modest effect in interrupting such circulation.

C.   Calcium Channel-Blockers: Diltiazem

Roberts et al. (1991) described the case of one 38-year-old female who had ingested 900 mg diltiazem, a calcium channel-blocking agent. Multiple-dose charcoal therapy was started at 7 hr postadmission and consisted of 50 g charcoal in water (Charcodote Aqua, Pharmascience, Montreal) every 6 hr from 7–31 hr via nasogastric tube. Plasma concentrations versus time were determined over the 7–71 hr period, and the half-life for elimination was determined and its value compared to historical controls. This patient's drug half-life (10.2 hr) was longer than half-lives obtained for several previous patients who had received supportive therapy and was higher than most published $t_{1/2}$ values for this drug. Thus, it was concluded that the charcoal was not effective in this case. This is not surprising, because of the long delay before charcoal was given and because this drug is strongly protein-bound and has a large volume of distribution.

IV. CNS AGENTS

A. Analgesics

1. Propoxyphene

Propoxyphene HCl (the dextrorotatory form, dextropropoxyphene, is used), a frequently prescribed pain killer, has been involved in significant numbers of overdoses. Dextropropoxyphene (DPP) overdose can cause rapid development of coma, respiratory depression, and cardiovascular collapse. Deaths have occurred within 1 hr of ingestion.

Corby and Decker (1968a, 1974) reported that propoxyphene is very well adsorbed from simulated gastric juice by Norit A activated charcoal. They stated that they found 99% adsorption of 10 capsules and 70% adsorption of 30 capsules (each capsule = 32 mg) after 20 min in a 150 mL solution to which 5 g activated charcoal had been added.

Corby and Decker (1968b) demonstrated that the timely administration of activated charcoal prevented clinical symptoms of propoxyphene poisoning in dogs. The dogs did not have convulsions, nor did they appear to have muscle fasciculations or respiratory depression. Correlation of serum drug levels with the clinical state was, however, not possible. This was ascribed to the known fact that blood levels of this drug always tend to be very low and do not reflect tissue concentrations well.

Chernish et al. (1972) did studies with propoxyphene. In vitro, they found that 500 mg activated charcoal adsorbed 95% of 100 mg propoxyphene from 31 mL of a 0.1 N HCl solution, and 700 mg charcoal adsorbed 99%. In vivo studies were done with six adult men. They were given 130 mg propoxyphene hydrochloride orally with or without 4 g activated charcoal in a cross-over study. Plasma propoxyphene levels are shown in Figure 12.8. It can be seen that more than half of the drug was prevented from absorbing (Chernish estimates that 80 mg of the 130-mg dose was effectively bound). It appears, then, that charcoal is effective with this drug if enough is used (note that the charcoal:drug ratio was 4:0.13, or 30.7:1 in this study).

Glab et al. (1982) showed that the mortality of rats given DPP followed by charcoal 30 min later was significantly reduced. They gave propoxyphene hydrochloride (350 mg/kg) or propoxyphene napsylate (825 mg/kg) to rats orally and then administered charcoal equal to 10 times the drug dose 30 min later. The charcoal reduced rat mortality by approximately 50%.

Kärkkäinen and Neuvonen (1985) studied the effects of oral charcoal on the pharmacokinetics of DPP and its active metabolite norpropoxyphene (NP). They also studied the effect of changing the urine pH using sodium bicarbonate or ammonium chloride on the renal excretion of DPP and NP. These latter results were summarized in Chapter 7 and will not be repeated here. The study involved giving 130 mg DPP hydrochloride to six healthy subjects, followed by 50 g charcoal at 5 min. DPP absorption was reduced

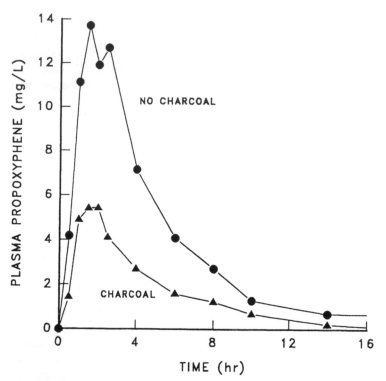

Figure 12.8 Average plasma propoxyphene levels for six human subjects, with and without activated charcoal. From Chernish, 1972.

by 97–99% as a result. When charcoal was given in repeated doses (50 g initially and 12.5 each 6 hr later between 6–54 hr, for a total of 150 g charcoal), the serum half-life of DPP was reduced from 31.1 to 21.2 hr and the serum half-life of NP was reduced from 34.4 to 19.8 hr.

These studies show that charcoal is effective in preventing DPP absorption and in interrupting the enterohepatic or enteroenteric circulation of DPP and NP.

2. Nefopam

Nefopam (fenazoxine) is a nonnarcotic analgesic. It is given in the hydrochloride form at an oral dose of about 60–180 mg/day. When amounts in the range of 1.5–2 g are ingested, grand mal convulsions can occur. Neuvonen et al. (1983–84) studied nefopam adsorption by charcoal in vitro and in vivo. It was found that, in vitro, adsorption was significant. Q_m values for Norit A charcoal were 0.55 g/g at pH 1.2 and 0.83 g/g at pH 7.4. For Amoco PX-21 charcoal (superactive charcoal) the Q_m values were 0.87 and 1.00 g/g, respectively. These are all excellent values. Nefopam is a weak base and therefore is more nonionized at higher pH values. This explains the higher Q_m values at pH 7.4 as compared to pH 1.2. As expected, the Q_m values for the very high surface area PX-21 charcoal were greater than those for the Norit A charcoal. In mice, the administration of Norit A and PX-21 charcoal (1700 mg/kg) increased the LD_{50} of nefopam by factors of 4.3 and 5.8, respectively. Figure 12.9 shows the relationships between the percentages of deaths

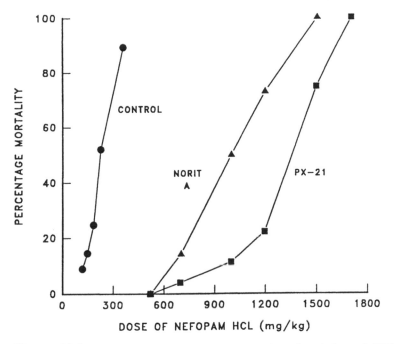

Figure 12.9 The effect of Norit A and Amoco PX-21 activated charcoals (1700 mg/kg) on the acute oral toxicity of nefopam hydrochloride in mice. The points indicate the percentage of deaths at each dose when nefopam was administered to 10–40 mice. From Neuvonen et al., 1983–84.

and the nefopam doses for the control, Norit-treated, and PX-21-treated mice. In other in vivo studies, in rats, higher charcoal:drug ratios were used, and the charcoal was even more effective. Thus, immediate use of oral charcoal is recommended for nefopam overdoses.

3. Morphine

El-Sayed and Hasan (1990) showed that a single oral dose of charcoal (10 g) given immediately reduced the half-life of morphine (1 mg/kg) given intravenously to rabbits from 1.02 to 0.70 hr. Serum morphine data are shown in Figure 12.10. A 30% decrease in the 0–∞ AUC value was noted. Systemic clearance increased 40% in the charcoal-treated group. However, the volume of distribution was unaffected by charcoal (this indicates that morphine adsorption to charcoal is essentially irreversible). A two-compartment pharmacokinetic model was used to interpret the results. The model showed that the rate of transfer of morphine from the tissue compartment to the central compartment was accelerated by the charcoal. The effectiveness of charcoal for countering IV morphine is due to interruption of the enterohepatic cycle of morphine.

4. Tilidine

Tilidine HCl, a narcotic analgesic, was studied in vitro by Cordonnier et al. (1986–87). Adsorption isotherms were determined using tilidine solutions at pH 1.2 and 7.5 (SGF without pepsin, and SIF without pancreatin, respectively). Norit A charcoal was the adsorbent. The maximal adsorption capacities (Q_m values) were 161.3 mg/g at pH 1.2 and 176.8 mg/g at pH 7.5. These are good, but not high, capacities. Thus, when tilidine

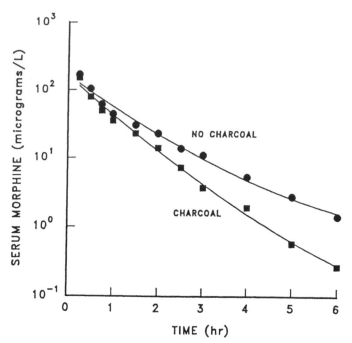

Figure 12.10 Serum morphine levels in rabbits after IV administration of 1 mg/kg morphine, with and without pretreatment with oral activated charcoal. From El-Sayed and Hasan (1990). Reprinted with permission of the Royal Pharmaceutical Society of Great Britain.

passes from the stomach to the intestines, one would expect a possible slight increase in tilidine binding to occur (other factors being neglected). The effects of added sorbitol, sucrose, cacao powder, milk, and starch on tilidine adsorption were evaluated. At a 10:1 charcoal:drug ratio, tilidine adsorption was 72.9% and decreased to the following values with addition of these materials to 10 mL of solution: 34.0% with 300 mg cacao powder, 49.0% with 300 mg sucrose, 60.0% with 300 mg potato starch, 69.1% with 300 mg corn starch, 84.2% with 10 mL milk, and 59.8, 56.5, 58.0, and 58.1% with 300, 600, 900, and 1200 mg sorbitol, respectively. Ethanol at 10% v/v was also added in other tests and was found to decrease tilidine adsorption significantly. Finally, two other charcoals (AC-Merck and AC-Federa) were tested without additives and compared to the results for the Norit A charcoal. At pH 7.5, all three brands adsorbed 100% of the tilidine in solution, but at pH 1.2 (at which pH the drug is more ionized and thus less able to adsorb) the percent tilidine adsorbed was 49, 100, and 73% for the Merck, Federa, and Norit charcoals, respectively. Thus, the Federa brand was the best.

In vivo tests with tilidine HCl were done by Cordonnier et al. (1987) and comparisons to the use of ipecac were made. The drug (50 mg) was given orally to healthy volunteers and then after 3 or 25 min either 20 g charcoal or 20 mL syrup of ipecac was given. With ipecac, emesis occurred in every subject with first emesis at a mean time of 14.9 min (range 13.6–15.4 min). Urine was collected for 48 hr and analyzed for the drug. Charcoal at 3 min reduced tilidine absorption by 89%, while charcoal at 25 min reduced the absorption by 66%. The syrup of ipecac at 3 min reduced tilidine absorption by 56% but when given at 25 min the ipecac had no significant effect

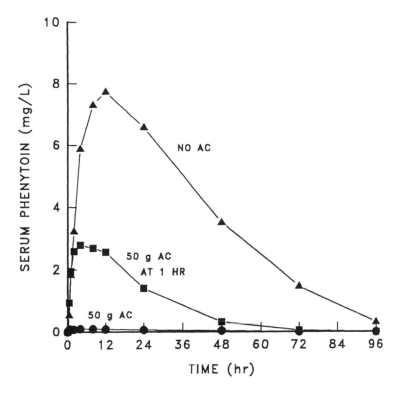

Figure 12.11 Phenytoin levels versus time after a 0.5-g dose of the drug alone, with 50 g activated charcoal at the same time, or with 50 g activated charcoal 1 hr later. From Neuvonen et al. (1978). Figure 3. Copyright Springer-Verlag GmbH & Co. Used with permission.

(absorption was reduced by 1.1%). Thus, when given at the same time (3 min), charcoal was much more effective than ipecac.

Thus, charcoal is very effective in reducing tilidine absorption, especially if given early. However, even if delayed for roughly a half hour, charcoal is still quite useful.

B. Anticonvulsants

1. Phenytoin

Phenytoin (diphenylhydantoin) is one of the most widely prescribed anticonvulsants and is occasionally used as a myocardial antiarrhythmic. Phenytoin poisoning is relatively common and presents a problem because this drug is slowly absorbed, metabolized, and excreted. It has a narrow therapeutic range and therefore can easily cause toxicity. Following absorption, the drug is 90% protein-bound and has a volume of distribution of 0.5–0.8 L/kg. It is metabolized in the liver by a saturable system according to Michaelis-Menten kinetics.

Neuvonen et al. (1978) showed that when adults were given 0.5 g phenytoin followed by 50 g Norit A charcoal, the systemic absorption of the drug was reduced (relative to control tests) by 98–99% if the charcoal was given immediately and by 80% even if the charcoal was delayed for 1 hr, as shown in Figure 12.11. The great effectiveness of

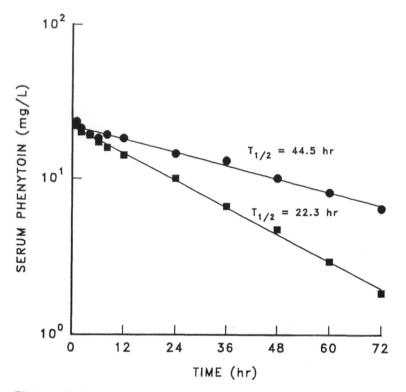

Figure 12.12 Average serum phenytoin concentrations versus time, plotted to show half-life determinations, with (squares) and without (circles) activated charcoal. From Mauro et al. (1987). Reprinted with permission of the American College of Emergency Physicians.

charcoal even after an hour's delay undoubtedly stems from the reduction in gastric motility that occurs, plus the fact that phenytoin is only sparingly soluble and thus dissolves slowly.

A cross-over study with seven healthy subjects and subtoxic doses of phenytoin was performed by Mauro et al. (1987). Each subject was given 15 mg/kg sodium phenytoin IV and then SuperChar charcoal (60 g in 470 mL 26% w/v sorbitol at the end of the infusion, and eight additional 30-g doses of charcoal at 2, 4, 8, 12, 24, 30, 36, and 48 hr either in 235 mL water or in 235 mL 26% w/v sorbitol, in such a manner "to produce one to two bowel movements per day"). Serum drug levels were measured over 0–72 hr. The charcoal reduced the 0–∞ AUC (obtained by extrapolation) by 50.2% and the drug half-life from 44.5 hr to 22.3 hr. Figure 12.12 shows the data plotted on semilogarithmic coordinates, with the respective half-lives indicated. The elimination rate constant was increased from 0.0167 to 0.0332 hr^{-1}, a 98.8% increase. Thus, multiple-dose charcoal appears to greatly enhance phenytoin elimination. The small volume of distribution of phenytoin and the fact that phenytoin is known (from studies with rats) to be secreted into the GI tract by means other than biliary excretion (which is small) would explain the effectiveness of charcoal.

Krenzelok and Lopez (1988) wrote a letter about the Mauro study, cautioning that the use of too much sorbitol can have adverse effects. They stated that sorbitol in purgative doses commonly produces massive catharsis, with up to 15 stools from a single dose, and persistence as long as 33 hours. They doubted that stooling could be titrated to one to two episodes per day, as Mauro's group indicated they had done. In reply, Mauro (1988) stated that the primary case quoted by Krenzelok and Lopez involved an 8 kg infant who received 110 g sorbitol. She pointed out that sorbitol in adults is less of a problem but agreed that close monitoring of electrolyte and fluid balances should be done. She also stated that some sorbitol can be of benefit in preventing constipation.

In summary, both early single doses of charcoal and multiple doses of charcoal are very efficacious in phenytoin-dosed subjects, due to the slow absorption of this drug.

2. Carbamazepine

Carbamazepine is commonly used to treat childhood seizure disorders and is used also for the relief of pain associated with trigeminal neuralgia (tic douloureux). It is relatively free of side effects but due to its increased use has become more frequently involved in acute overdoses or acute-on-chronic overdoses (i.e., acute effects subsequent to chronic therapy). Carbamazepine overdose can cause dangerous respiratory depression, prolonged coma, hypotension, cardiac arrhythmias, and a variety of neurological disturbances. The drug is highly protein-bound and is thus poorly removed by hemodialysis or peritoneal dialysis. Thus, oral charcoal therapy is of considerable value.

Neuvonen and Elonen (1980a) studied the effect of charcoal on carbamazepine, phenobarbital, and phenylbutazone absorption in man (this same study was published in shorter form in a different journal by Neuvonen and Elonen, 1980b). These three drugs were given separately in oral doses of 400, 200, and 200 mg, respectively, to five healthy subjects using a randomized cross-over design. When 50 g charcoal was given within 5 min, absorption of all three drugs was more than 95% prevented. When the charcoal was given after 1 hr, the absorption of the three drugs was decreased 41, 47, and 30% (in the same order as mentioned above), based on 0–96 hr AUC values. In other tests, multiple dose regimens were done with 50 g charcoal at 10 hr, and 17 g at 14, 24, 36, and 48 hr. The reductions in absorption of the three drugs were then 27, 49, and 5% (same order as before), again based on 0-96 hr AUC values. In these multiple dose tests, the half-life values, determined over 10–48 hr, decreased from 32.0 to 17.6 hr, from 110.0 to 19.8 hr, and from 51.5 to 36.7 hr for the three drugs (reductions of 45, 82, and 29%, respectively). Thus, the charcoal was effective, particularly in the case of phenobarbital.

Additional results on carbamazepine have been given in a paper by Neuvonen et al. (1988) in which results on digoxin and furosemide were also presented. In a four-phase cross-over study with six healthy volunteers, single doses of charcoal (8 g), cholestyramine (8 g), colestipol hydrochloride (10 g), or water only were given immediately after the simultaneous ingestion of 0.25 mg digoxin, 400 mg carbamazepine, and 40 mg furosemide. Plasma and urine drug concentrations were determined up to 72 hr. Reductions in absorption of the three drugs by the charcoal, cholestyramine, and colestipol, respectively were: for digoxin, 98, 40, and 3%; for carbamazepine, 92, -8, and 10%; and for furosemide, 99, 94, and 79%. Thus, charcoal was highly effective for all three drugs. Cholestyramine and colestipol were highly effective only for furosemide.

Boldy et al. (1987), in a letter, summarized some of their data on carbamazepine overdose cases. Fifteen patients received supportive therapy plus 12.5–100 g charcoal in slurry form every 1–6 hr (these were actual overdose cases and the treatment protocols varied widely). They report a mean drug half-life value, after charcoal administration, of 8.6 hr. This can be compared to a reported average value of about 19.0 hr in 12 overdose cases in which only supportive therapy was used.

Edge and Edmonds (1992) reported a case of a significant drop in serum sodium following carbamazepine overdose in a 2.5-year-old child. They recommend careful monitoring of electrolyte levels and judicious administration of fluids in carbamazepine overdose.

The studies just discussed show that charcoal given early can greatly reduce carbamazepine absorption, and multiple-dose charcoal therapy initiated subsequent to carbamazepine absorption can reduce the half-life of the drug by more than 50%, presumably by interruption of its enteroenteric and/or enterohepatic circulation.

3.   Valproic Acid

Valproic acid intoxication of a 26-month-old boy has been described by Farrar et al. (1993). Soon after admission, continuous nasogastric infusion of a 3 g/50 mL slurry of charcoal in normal saline was initiated at a rate of 50 mL/hr (giving 0.25 g/kg-hr). Sorbitol was included until a charcoal-containing stool was passed. At 6 hr postingestion the total serum drug level (free plus protein-bound) was 815 mg/L and at about 25 hr postingestion it was 56 mg/L (charcoal therapy was started at about 1.5 hr postingestion). The patient was essentially back to normal at the 25 hr point. Although valproic acid is about 90% protein-bound, charcoal was effective in enhancing elimination. The half-life during charcoal therapy was 4.8 hr versus expected values in children of 10–16 hr.

C.   Anti-Inflammatory Agents

1.   Mefenamic Acid

Mefenamic acid is a widely used nonsteroidal anti-inflammatory agent which has become increasingly involved in overdose instances. Convulsions often occur as a result of excessive ingestion. El-Bahie et al. (1985) gave 500 mg mefenamic acid orally to nine healthy subjects. On one occasion, 2.5 g charcoal was administered 1 hr after the drug. Plasma drug levels were followed for up to six hr after drug ingestion. Charcoal reduced the 0–∞ AUC values, compared to control-phase values, by 36%. To simulate the delayed gastric emptying that is reported to occur in mefenamic acid overdose, the anticholinergic agent hyoscine butylbromide was given intramuscularly (20 mg) immediately after the drug in some repeat tests. Hyocine alone delayed the time to peak mefenamic acid concentration but had no effect on the AUC. However, the combination of hyoscine and charcoal reduced the AUC (versus the control value) by 42%, that is, somewhat more than with charcoal alone. Because mefenamic acid causes delayed gastric emptying, oral charcoal was effective even after a 1 hr delay.

Allen et al. (1987) administered mefenamic acid suppositories (500 mg) rectally to eight healthy adults, followed by a charcoal (5 g) suspension in water orally each hour for 7 hr. Controls received only the water. There was no significant difference between the two groups and it was concluded that the benefits of charcoal seen in other studies were due to interference with absorption rather than due to enhancement of elimination. Thus, this drug appears not to undergo significant enteroenteric or enterohepatic circulation.

Thus, while charcoal given as late as 1 hr after mefenamic acid can be effective, because of the slow absorption of this drug, once the drug is well absorbed charcoal would be of little use.

2. Piroxicam

Piroxicam is a nonsteroidal anti-inflammatory drug which is moderately well tolerated in excess but worth studying nevertheless. This drug is known to have a low clearance with a mean elimination half-life of at least 45 hr. Laufen and Leitold (1986) did a cross- over study with six healthy volunteers. The subjects were given single oral doses (20 mg) of piroxicam. Charcoal (50 g) was given orally 5 min later. The charcoal reduced piroxicam bioavailability by 98%. In other tests, charcoal was given in a multiple dose format (50 g at 10 hr, and seven more doses of 20 or 30 g given over the next 48 hr). The drug bioavailability decreased 41.7% and the half-life decreased from 40.2 to 19.6 hr. In a repeat test, 20 mg piroxicam was given rectally, followed by multiple-dose charcoal orally (30 g at 2 hr, 20 g at 8 hr, and seven more 20 or 30-g doses over the next 50 hr). Bioavailability decreased 48.8% with charcoal and the drug half-life went from a control value of 40.7 hr to 21.6 hr. It was inferred from the effectiveness of delayed multiple-dose charcoal therapy that piroxicam undergoes enteral circulation of some type.

Further studies on piroxicam were done by Ferry et al. (1990). They gave 20 mg oral doses of the drug to eight healthy young adults. Then 5 g charcoal (Medicoal) was given starting 24 hr after the drug dose and further 5-g doses were given 3–4 times/day thereafter for several days. The charcoal therapy reduced the piroxicam half-life from 53.1 to 40.0 hr. In other tests, cholestyramine was given in 4 g amounts according to the same schedule as for the charcoal and the drug half-life was decreased even more, to 29.6 hr. Again, even though sorbent therapy was delayed for 24 hr, it was effective. This proves that enteral circulation occurs for this drug.

Charcoal therapy is thus very effective for piroxicam, whether initiated early or well after the drug has been absorbed.

3. Indomethacin

A study by Neuvonen and Olkkola (1984a) involving three drugs, one of which was the anti-inflammatory agent indomethacin, was discussed in the earlier section on the anti-arrhythmic drug disopyramide (which was one of the other drugs).

El-Sayed et al. (1990) studied the effect of oral charcoal on the systemic clearance and other pharmacokinetic parameters of IV indomethacin (2 mg/kg) given to rabbits. A single 10-g dose of charcoal in 40 mL water was given by gastric tube just prior to the drug infusion. Blood samples were taken at several times up to 6 hr. Control rabbits received only the 40 mL of water. Serum indomethacin values are shown in Figure 12.13. The charcoal treatment reduced the serum half-life from 1.26 to 0.82 hr, decreased the 0–∞ AUC from 17.56 to 10.34 mg-hr/L (a 41% decrease), and increased the systemic clearance from 1.92 to 3.23 mL/min-kg (a 68% increase). Interpretation of the data using a two-compartment model showed that the charcoal significantly increased the rate of drug transfer from the tissue compartment into and out of the central compartment. The results suggest that charcoal interrupts the enterohepatic circulation of indomethacin.

D. Tranquilizers

This section deals with drugs used in various ways as tranquilizers, that is, drugs with antianxiety effects, antipsychotic effects, and/or general sedative effects. We begin by discussing drugs of the benzodiazepine class (these drugs are widely used to relieve anxiety). We first review one study involving a variety of benzodiazepines.

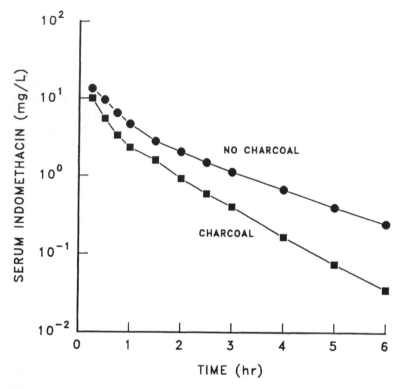

Figure 12.13 Serum indomethacin levels in rabbits after IV administration of 2 mg/kg indo-methacin, with and without pretreatment with oral activated charcoal. From El-Sayed et al. (1990). Reproduced by permission of Elsevier Science Publishers BV.

El-Khordagui et al. (1987) studied the adsorption of seven benzodiazepines to Merck charcoal in vitro at pH 1.3 (HCl solution) and 37°C. They tried to correlate the Langmuir constant product KQ_m (the use of Q_m alone would have been better) for the seven compounds versus: log octanol/water partition coefficient, log chloroform/water partition coefficient, percent protein (BSA) binding, a ranking of potency in humans, and rankings of potency in mice, rats, and cats. KQ_m correlated well with the octanol/water partition coefficients, with percent protein binding, and all of the in vivo potency indices. All correlations were positive in the sense that KQ_m increased as each of the correlating parameters increased. These results suggest that molecules of this class of drugs which are less water-soluble adsorb better to charcoal (as one would expect) and are more potent in vivo.

1.   Diazepam

Kortilla et al. (1976) studied the adsorption of diazepam, a benzodiazepine, in vitro by charcoal at both pH 1.2 and pH 6.5, at a temperature of 37°C. They found that for the amounts of drug and charcoal used, the extent of diazepam adsorption ranged from about 22 to 72%. They then did in vivo studies in which six subjects were given 0.3 mg/kg diazepam intravenously along with oral charcoal doses of "5 mg" at 60 min intervals, starting 30 min after the drug injection. It must be assumed that the quoted charcoal dose of "5 mg" is a typographical error, as such a tiny dose is absurd; it perhaps should have read "5 g." Even a 5-g dose every 60 min is quite a small dose. In any event, the charcoal had no statistically significant effect on diazepam disappearance from the blood; however, the serum diazepam curves with charcoal and without charcoal do show the curve for charcoal to be definitely lower over about the 2–6 hr time period, by as much as about 15% near the 2 hr mark.

Traeger and Haug (1986) reported on a case of prolonged coma which occurred after treatment of alcohol withdrawal seizures and delirium with IV phenobarbital and diazepam. Forty grams charcoal (Liquid Antidose, Bowman Pharmaceutical, Canton, Ohio) followed by 60 mL 6% magnesium citrate was given every 4 hr nasogastrically for a total of 6 doses on days 13 and 14. Coma was completely reversed within 12 hr and the serum half-life of diazepam was reduced from 195 to 18 hr. Because of severe liver disease, the patient was hypoalbuminemic. This caused more of the diazepam to be unbound and able to cross the gastrointestinal membrane barrier, thus making the charcoal more effective in binding the diazepam.

Controlled clinical studies with animals or humans are needed to better define the role of charcoal with respect to diazepam.

2. Lorazepam

Lorazepam is a benzodiazepine which has the characteristic antianxiety properties of that class of compounds. Excessive CNS depression occurs with overdose. The in vitro binding of lorazepam by charcoal, and by the resins cholestyramine and colestipol, was studied by Herman and Chaudhary (1991). The binding of its major metabolite, lorazepam glucuronide, was also examined. The solution pH was 7.4 in every case. For the amounts of adsorbents and the volumes and concentrations of solutions employed, the percent lorazepam bound was: 100% for charcoal, 23.7% for cholestyramine, and 11.3% for colestipol. The glucuronide binding values were 100, 74.3, and 20.8%, respectively. Since it is unclear that the three adsorbents were evaluated on the same basis, these results are only of qualitative value. Clearly, the charcoal adsorbed best. In vivo studies are required before it can be stated whether charcoal would be useful for lorazepam overdose. However, the in vitro results are encouraging.

3. Chlorpromazine

Chlorpromazine is a phenothiazine antipsychotic agent. Chin et al. (1970) did in vivo studies on rats with this drug. The rats were given 150 mg/kg chlorpromazine HCl and then, 1 min later, 750 mg/kg charcoal. Figure 12.14 shows the levels of the drug in the livers of the rats as a function of time and it is clear that the charcoal very significantly reduced the absorption of the drug. Studies with human subjects are needed.

4. Meprobamate

The use of multiple-dose charcoal therapy in two patients with meprobamate overdoses has been discussed by Hassan (1986). Meprobamate is a nonbarbiturate sedative-hypnotic, available in the U.S.A. since the mid 1950s. The patients were a 32-year-old woman and a 28-year-old woman. The first patient was given 50 g charcoal every 4 hr for five doses and the second patient was given 50 g charcoal every 6 hr for five doses. Both patients had a drug elimination half-life of about 4.5 hr with the charcoal therapy, which can be compared to an average half-life of 11.3 hr from one study in the literature

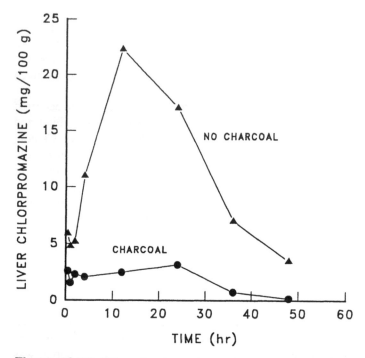

Figure 12.14 Effect of a 5:1 charcoal:drug ratio on liver chlorpromazine levels in rats. From Chin et al. (1970). Reprinted with permission of Academic Press, Inc.

involving 12 healthy subjects who received no charcoal. Thus, it appears that multiple doses of charcoal can effectively interrupt enteral cycles of meprobamate.

5. Thioridazine

Boehm and Oppenheim (1977), in addition to studying the in vitro adsorption of chlorpheniramine maleate (discussed earlier in this chapter) and some other common drugs (aspirin, amylobarbital, acetaminophen, nortriptyline HCl), also studied the tranquilizer thioridazine HCl. This weakly basic drug (pK_a = 9.5), in SGF, adsorbed to the extent of 90, 95, and 100% to three different charcoals, at charcoal:drug ratios of 10:1, 5:1, and 10:1, respectively. In SIF and with a charcoal:drug ratio of 10:1, it adsorbed 100% to the one charcoal tested. These positive results suggest that oral charcoal may have a role in thioridazine overdoses.

E.   Other CNS Agents: Methamphetamine

McKinney et al. (1991) studied the effectiveness of oral charcoal in reducing methamphetamine toxicity in mice. The mice were given 100 mg/kg of the drug orally in 0.5 mL water. Then some mice were given 1 g/kg charcoal in slurry form (Actidose Aqua) 1 min later. Mortality at 1 hr and at 24 hr was determined. With control mice, mortality at 1 hr was 10/20 and at 24 hr was 12/20. With the mice which had received charcoal, mortality at 1 hr was 1/20 and at 24 hr was 5/20. Mean times to piloerection, agitation, and tremor were 256 sec in the controls and 332 sec in the charcoal-treated group (this difference is significant at $p < 0.002$). Thus, charcoal significantly reduced the onset of symptoms and reduced mortality. Clearly, experience with charcoal in human subjects would be both valuable and interesting for this drug.

V. GASTROINTESTINAL DRUGS

A. Cimetidine

The adsorption of cimetidine, a drug used to treat duodenal ulcers and gastric hypersecretion, to charcoal and other adsorbents (kaolin, talc, magnesium trisilicate) was studied in vitro by Ganjian et al. (1980) at pH 5.0. Their Q_m value for charcoal appears to be around 0.041 g drug/g charcoal, which is rather low. For the other adsorbents, the Q_m values were much less. Thus, of the adsorbents tested, only charcoal had potential to adsorb cimetidine significantly.

No in vivo studies with cimetidine have appeared in the literature but the in vitro results of Ganjian's group suggest that charcoal would not be very effective.

B. Propantheline

Propantheline, a quaternary ammonium antimuscarinic (anticholinergic) agent is used for treating peptic ulcers. Its effect derives mainly from its ability to decrease gastric motility. Chaput de Saintonge and Herxheimer (1971) described studies in which adult volunteers were given 45 mg propantheline plus 200 mL water. In one test, nothing else was given, but in a second test 5 g activated charcoal was given at the same time as the drug. Pulse rates and saliva secretion rates were followed for several hours. After about 50 min, tachycardia appeared in all cases. Without charcoal this persisted throughout the experiment, whereas with charcoal the pulse rate fell steadily. Saliva flow began to fall after about 30 min in all tests but the fall was greater when charcoal was not used. These results suggest that the charcoal significantly decreased the anticholinergic effects of the drug, presumably by reducing its absorption. Further tests, involving the determination of drug blood levels versus time, are needed.

C. Diphenoxylate

Sanvordeker and Dajani (1975) studied the adsorption of diphenoxylate hydrochloride, a potent antidiarrheal agent, to activated charcoal in vitro and found that substantial adsorption occurred. In vivo tests were then done with mice and these indicated that the antipropulsive activity of the drug was strongly inhibited by activated charcoal given 30 min after the drug. In fact, six times as much drug was needed to yield the same gastrointestinal transit time when charcoal was used as was obtained without charcoal.

As in the case of propantheline, studies in which drug blood levels are measured as a function of time should be carried out, since indices such as saliva secretion rates (propantheline) and GI transit times (diphenoxylate) are less definitive measures of drug absorption.

D. Nizatidine

Nizatidine, a potent H_2-receptor antagonist that decreases gastric acid production, was studied by Knadler et al. (1987) using healthy male subjects. The drug (150 mg) was given orally, followed 60 min later by a slurry of 2 g charcoal. The 0–12 hr AUC value based on serum nizatidine concentrations was 22.8% lower with charcoal than without charcoal. This is actually a surprisingly large effect, considering the small size of the charcoal dose (2 g) and the considerable delay (60 min) in administering it.

VI. ANTIDIABETIC SULFONYLUREAS

Sulfonylureas comprise one class of antidiabetic agents. Although poisoning due to oral hypoglycemic agents like these is not common, some deaths and cerebral damage have been reported. The adsorption of sulfonylureas has been difficult to show in vitro at low pH conditions, because of the very low solubilities of these compounds at low pH. In vivo, charcoal can be very effective in reducing the absorption of such compounds, for their low solubilities at gastric pH cause their concentrations in the gastric fluid to be low. This, in turn, means that their rates of absorption from the gastric region will be quite low. This gives charcoal more time in which to bind such compounds. Of course, once such drugs pass out of the stomach and into the higher pH milieu of the intestines, their solubilities will be much higher and their rates of absorption will be potentially much higher. However, if charcoal is given early enough, then by the time such drugs pass into the intestines they would very likely be well bound to the charcoal.

Kannisto and Neuvonen (1984) studied the in vitro binding of several sulfonylureas (carbutamide, chlorpropamide, tolazamide, tolbutamide, glibenclamide [glyburide], and glipizide) to activated charcoal from buffer solutions at pH 4.9 and 7.5. Q_m values at pH 7.5, determined by fitting the data to the Langmuir isotherm equation, were in the range of 0.42–0.62 g/g. These are excellent maximal binding capacities. Q_m values for two compounds tested at pH 4.9 were even higher (0.78 and 0.91 g/g). These two compounds are weak acids, are therefore less ionized at lower pH values, and thus adsorb more strongly at lower pH. Tests were tried at even lower pH values (around 1) but the compounds became so nonionized that they were very poorly soluble under these conditions. These in vitro results suggest that charcoal would be very effective in sulfonylurea overdoses.

Chlorpropamide is a long-used sulfonylurea antidiabetic agent. Neuvonen and Kärk-käinen (1983) studied the effect of charcoal on chlorpropamide kinetics in man. Six healthy subjects were given 250 mg chlorpropamide orally and then a single 50-g dose of charcoal 5 min later, or 50 g charcoal 6 hr after the drug followed by 12.5 g at six hour intervals until a total of 150 g charcoal had been taken. On one occasion the urine was acidified with ammonium chloride and on another it was made alkaline with sodium bicarbonate. The single dose of charcoal reduced the 0–72 hr AUC for chlorpropamide by 90% but for the case of repeated charcoal doses begun at 6 hr the AUC was not significantly different from the control value. The 0–72 hr urinary excretion of the drug was increased four-fold by alkalinization (accounting for 85% of the drug dose) and decreased to about 5% of the control value by acidification. Thus, charcoal is recommended for chlorpropamide overdose if given early and alkalinization of the urine accelerates its renal elimination.

Kivistö and Neuvonen (1990a) studied the effect of activated charcoal and cholestyramine on the absorption of the sulfonylurea antidiabetic drug glipizide. This was a three-phase cross-over study in which single oral 8-g doses of charcoal (Carbomix, Medica, Finland), single oral 8-g doses of cholestyramine, or water only, were given to six healthy subjects. Figure 12.15 shows the plasma glipizide data for the three phases of the study. The absorption of glipizide, as determined by 0–10 hr AUC values, was

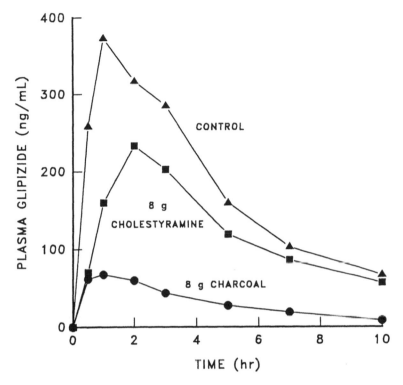

Figure 12.15 Effect of activated charcoal (8 g) and cholestyramine resin (8 g) on plasma glipizide concentrations. From Kivistö and Neuvonen (1990a). Reprinted by permission of Blackwell Scientific Publications Ltd.

reduced 81% by charcoal and 29% by cholestyramine. Charcoal reduced the peak plasma glipizide levels by 79%, compared to 33% for cholestyramine.

Neuvonen et al. (1983a) studied the effects of charcoal on tolbutamide absorption in man. Sodium valproate—a fatty acid antiepileptic—was also examined in this study. They were given to six healthy volunteers in simultaneous oral doses of 500 mg and 300 mg, respectively. In one phase of this randomized cross-over study, 50 g activated charcoal was given orally within 5 min. The tolbutamide 0–48 hr AUC was reduced 90% by the charcoal, while the 0–48 hr AUC for valproate was reduced by 65%. Thus, it is clear from the in vivo results of this study that charcoal is very effective in binding tolbutamide in vivo.

The sulfonylurea studies just reviewed, all of which were done by Neuvonen's group in Finland, consistently demonstrated that oral charcoal, given early, greatly reduces the absorption of this class of drugs.

VII. RESPIRATORY RELAXANTS: THEOPHYLLINE

Theophylline produces bronchodilation by the inhibition of phosphodiesterase and is widely used to treat diseases of reversible airway obstruction (e.g., asthma). It is frequently used in the form of aminophylline, a compound of theophylline and 1,2-ethanediamine in a molecular ratio of 1:2. Aminophylline dissociates immediately upon dissolution to produce free theophylline. Because of the ready availability and increased usage of theophylline, plus the fact that it has a narrow therapeutic index, frequent acute overdoses occur, especially among children.

A.   In Vitro Studies

Sintek et al. (1978) studied theophylline adsorption by activated charcoal in vitro. In their tests, 10 g activated charcoal was shaken for 4 min with 200 mL of solutions of 2, 10, and 20 g/L theophylline concentration. The mean percentages of drug adsorption from these three solutions were 94, 96, and 70%, respectively. An upper limit for adsorption of about 0.3 g drug per g charcoal was suggested by the data.

B.   Animal Studies

The pharmacokinetics of theophylline following oral charcoal administration in rabbits have been studied by Huang (1987). Rabbits were given 2.12 mg/hr theophylline by continuous infusion and at 4 hr charcoal (20 g) in a slurry was given by intubation. Serum drug levels thereafter gradually decreased. By comparing the steady-state drug levels in control and treated animals, it was found that the total body clearance was increased from 94.4 to 210 mL/hr-kg by the charcoal.

De Vries et al. (1989) confirmed the results of McKinnon et al. (1987) with respect to the ability of theophylline to diffuse across the intestinal wall. De Vries's group isolated the small intestines of rats and perfused the vascular system with a recirculating theophylline solution. The lumen side was perfused with a recirculating solution that was originally theophylline-free. Over the course of about 200 min, the theophylline concentration in the vascular fluid fell and that in the luminal perfusate rose, until they reached the same value. This proved that theophylline can cross the intestinal wall. When the study was repeated with a charcoal slurry as the luminal perfusate, the vascular solution concentrations continued to fall with time, since the theophylline that diffused into the lumen was continually adsorbed by the charcoal, keeping the luminal fluid concentration of free theophylline low.

C.   Human Volunteer Studies

Sintek et al. (1979) did an in vivo study in which they gave adult men and women 500–600 mg of theophylline plus 30 g activated charcoal in slurry form 30 min later. Absorption "then appeared to stop abruptly." By comparison to control tests (no charcoal given) it was found that the charcoal decreased theophylline absorption by an average of 59%. However, the decreases ranged from 26 to 96% for the five subjects, probably because of a great variability in the amount of drug absorbed in the 30 min prior to the charcoal administration. On balance it is clear that the dose of charcoal delivered (30 g) was quite effective, even after a delay of 30 min.

Helliwell and Berry (1981) gave 675 mg of a sustained-release formulation of theophylline to six volunteers and then single and multiple doses of the effervescent charcoal preparation Medicoal. First, however, they did an in vitro study which showed that, at pH 1, one gram of Medicoal (which contains 500 mg charcoal) adsorbed 210 mg theophylline from 50 mL of solutions containing 100–140 mg theophylline initially.

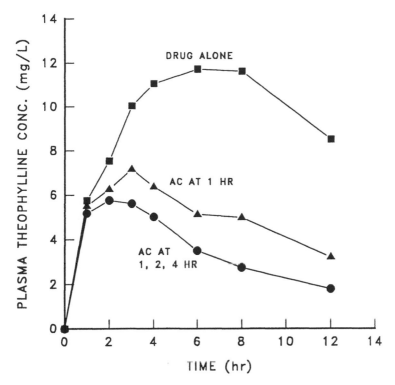

Figure 12.16 Mean plasma concentrations with theophylline alone, with activated charcoal at 1 hr, or with activated charcoal at 1, 2, and 4 hr. From Helliwell and Berry (1981). Reproduced by kind permission of Cambridge Medical Publications.

This report obviously contains an error as it states that more theophylline was adsorbed than was in the solution to begin with! The in vivo study involved giving a 10 g sachet of Medicoal (5 g charcoal) at 1 hr, or 10 g sachets at 1, 2, and 4 hr. In the single charcoal dose study, theophylline absorption over 0–12 hr was reduced by 51%. Figure 12.16 shows the mean plasma theophylline data. With the three sachet regimen, the reduction was 66%. Thus, the charcoal was quite effective.

Lim et al. (1986) studied the effect of oral charcoal on the first 12 hr of absorption and elimination of sustained-release theophylline in 20 normal children, ages 8 to 18 years. The drug dose was 10 mg/kg. After receiving the drug, the children were given from one to four doses of 1 g/kg charcoal (InstaChar) with 30 mL 70% w/v sorbitol. Group 1 was given a single dose at 1 hr; Group 2 received four doses at 1, 4, 7, and 10 hr; Group 3 was administered four doses at 3, 6, 9, and 12 hr; and Group 4 received three doses at 6, 9, and 12 hr. Blood samples were drawn at 0, 2, 4, 6, 8, 10, 12, and 24 hr. Reductions of the 0–12 hr AUC values, compared to control values, were: Group 1, 61%; Group 2, 68%; Group 3, 37%; and Group 4, 18%. The results indicate that early charcoal administration can inhibit theophylline absorption significantly and enhance elimination of previously absorbed theophylline.

Ginoza et al. (1987), working with ten low-birth-weight infants who had been given theophylline daily for treatment of apnea, gave them each a series of three IV infusions of theophylline as aminophylline dihydrate. Then, a slurry of SuperChar charcoal (1 g/kg) was given via orogastric tube, 1 hr later. A second charcoal dose was given at 4 hr. The charcoal treatment increased the drug clearance from a mean of 22.7 to a mean of 44.6 mL/kg-hr, an increase of 96.4%. Thus, high surface area charcoal given orally appears to be effective for enhancing theophylline elimination in neonates.

D. Human Overdose Studies

In a letter, Weinberger (1983) recommended syrup of ipecac followed by charcoal and a saline cathartic for the treatment of theophylline overdose in children. More specifically, he recommended repeated doses of charcoal whenever serum theophylline concentrations exceed 40 μg/mL, continuing until drug levels fall below this level and all symptoms and signs of toxicity are gone. In a brief reply, Gaudreault and Lovejoy (1983) basically agreed, stating that charcoal "be repeated probably every 4 to 6 hours until the theophylline serum concentrations are below 40 to 60 μg/mL."

Strauss et al. (1985) described the administration of a charcoal slurry to a preterm infant with parent-induced theophylline toxicity. The slurry was 2 g/kg (12 mL volume) given by gavage tube every 4 hr for three doses, starting at 12 hr. Magnesium citrate (7.5 mL) was given with the first and third charcoal doses. During the charcoal therapy, the theophylline half-life was about 10.3 hr, as compared to 23.0 hr prior to the charcoal. Thus, the charcoal considerably enhanced theophylline elimination.

A report involving charcoal given to a premature infant on the second day of life was presented by Jain and Tholl (1992). A 1,700-g premature female (33 wk gestation) was given aminophylline for apnea and became extremely jittery and hyperactive. A single small dose (1 g) of charcoal was administered to lower the infant's serum theophylline level (72 μg/mL). This single dose lowered the drug half-life from 62.5 hr to 17.8 hr. The serum drug concentration showed a plateau between 8 and 16 hr after charcoal administration but thereafter fell again with a half-life of 36 hr.

D'Angio and Sabatelli (1987) described the case of a 32-year-old man, intentionally overdosed with Theo-Dur, whom they treated with oral charcoal. Three doses of 50 g charcoal with sorbitol were administered by nasogastric tube (the times of administration are not stated). The initial theophylline level of 250 μmol/L (45 mg/L) was reduced to 31 μmole/L (5.6 mg/L) in only 12 hr. The patient was stable one day after ingesting the drug.

Amitai and Lovejoy (1987) wrote about the problem of vomiting in theophylline overdose and its interference with multiple-dose charcoal therapy. They studied 26 patients with theophylline overdose. Twenty-five of these vomited, with the vomiting extending over 63% of the drug's absorptive phase (the period between ingestion and the peak drug level in the blood). The duration of vomiting correlated with both the peak drug levels and with the duration of the drug's toxicity. Patients with peak drug levels less than 70 μg/mL were able to accept larger amounts of charcoal (113 g average) than patients with peak drug levels greater than 70 μg/mL (57 g average).

Sessler (1987) has also written about the vomiting problem with theophylline overdose. He reviewed records of 33 patients who presented to an ED with theophylline overdose and who were given oral charcoal. Seventeen (22%) of 76 oral charcoal doses were vomited. Six patients with acute toxicity vomited all 11 doses given, whereas 27 patients with chronic toxicity vomited only six of 65 doses given. Vomited doses were associated with higher serum drug levels.

E.   Conclusions About Theophylline

The theophylline studies we have discussed, though varying widely in nature, are quite consistent in showing that oral charcoal can strongly reduce theophylline absorption (typical reductions are on the order of 50%) if given early and that multiple-dose charcoal can greatly enhance the elimination of already-absorbed theophylline (typical clearance increases are on the order of 100%).

VIII.   ORGANIC SUBSTANCES

This section will deal with a variety of organic substances having widely different uses. We discuss camphor, alcohols (ethylene glycol, isopropanol), acetone, polybrominated biphenyl, three herbicides, two insecticides, and one pesticide in arbitrary order.

A.   Camphor

Dean et al. (1992) studied the effects of oral charcoal on camphor pharmacokinetics. Camphor is a cyclic ketone of the aromatic terpene group and is an ingredient in many over- the-counter products. It produces gastrointestinal and CNS symptoms following toxic ingestion. In this study, 100 rats were given camphor (40% w/w in cottonseed oil, 1 g camphor/kg dose) by oral gavage. Then, an aqueous charcoal suspension (2 g/kg) was given. Serum camphor determinations over 10 hr showed no differences between controls and charcoal-treated rats with respect to AUC, k_a, k_e, and $t_{1/2}$. The lack of effect of charcoal might have been because only a 2:1 ratio of charcoal:drug was used, which is not very large.

B.   Ethylene Glycol

Permanent antifreeze is roughly 95% ethylene glycol and a significant number of deaths each year result from its accidental or suicidal ingestion (it is also commonly ingested in overdose amounts by household pets because it is often stored in accessible places and because its sweet taste is attractive to pets). Szabuniewicz et al. (1975a,b) reported on studies in which 35 dogs were given 10 mL/kg antifreeze. Fifteen were then given 5 g/kg activated charcoal, 15 were treated with ethanol or sodium bicarbonate, and five were left untreated. In these three groups the survivors were 15/15, 3/15, and 0/5, respectively. Clearly, the charcoal was effective. This is somewhat surprising since, as Cooney (1977) has pointed out, ethylene glycol adsorbs poorly to charcoal. Based on in vitro adsorption tests, he estimated that only 3.6% of the ethylene glycol would be bound by the charcoal. The explanation of charcoal's great effect may be related more to ethylene glycol's metabolites, particularly oxalic acid, which may adsorb much better to charcoal. Some type of circulation of the metabolites back into the digestive tract would be required, of course, for contact with the charcoal to occur. Further studies are obviously needed.

C.   Isopropanol and Acetone

Burkhart and Martinez (1992) studied the adsorption of isopropanol to charcoal in HCl solutions and in plain water, and the adsorption of acetone to charcoal in plain water alone. Charcoal:solvent ratios used were 1, 2.5, 5, 10, and 20 to 1. Table 12.4 shows their results. It is clear that the adsorption patterns were similar in all cases. This study shows that charcoal, in high enough doses, can adsorb these two solvents in vitro.

Table 12.4 Percent Adsorption of Isopropanol and Acetone by Activated Charcoal

Charcoal:solvent ratio (g/g)	Isopropanol in HCl	Isopropanol in water	Acetone in water
1.0	11.4	7.4	14.8
2.5	24.3	22.1	28.8
5.0	42.0	42.3	43.8
10.0	68.1	63.3	75.6
20.0	89.1	86.7	91.5

From Burkhart and Martinez (1992).

However, there is reason to doubt that sufficient oral charcoal could be delivered in actual overdose cases to be of significant value.

D.   Polybrominated Biphenyl

McConnell et al. (1980) studied the effect of activated charcoal and cholestyramine in reducing polybrominated biphenyl (PBB) in weanling male rats. PBBs are a class of compounds which have heat-resistant and flame-retardant properties and were used to treat synthetic polymer garments until concern was raised about their toxicity. A particular PBB was added to the diets of the rats at 1 mg/kg per day for six months. This resulted in the deposition of PBB in their body tissues (primarily the fatty tissues, since PBB is highly lipid-soluble). They were then placed on a normal diet for four months. Following this, they were placed on diets containing either activated charcoal or cholestyramine for six months. Periods of restricted caloric intake were tried in order to try to induce PBB desorption from the fatty tissues. Neither adsorbent nor the restricted caloric intake were found to be effective in reducing tissue PBB levels, but the cholestyramine was useful in preventing the course of chronic progressive nephropathy, a spontaneous lesion of aging rats. This study confirms a general principle concerning highly organic substances, namely, that once they have been deposited into fatty tissues, oral sorbent therapy is largely ineffective in removing them.

E.   Herbicides

Paraquat is a bipyridyl herbicide which is quite toxic (as little as 10 mL can be fatal). Death results in hours or days from pulmonary edema and hemorrhage.

The in vitro adsorption of paraquat to activated charcoal, with and without added magnesium citrate, was studied by Gaudreault et al. (1985). Three milligrams of paraquat mixed with 30 or 60 mg activated charcoal in water gave, at equilibrium, 37 and 56% adsorption, respectively. Fuller's earth in the same amounts gave only 24 and 45% adsorption, respectively. Clearly, the charcoal was better. Using a magnesium citrate solution instead of water in repeat tests, the adsorption to charcoal was significantly increased—to 53 and 86%—and the adsorption to Fuller's earth was slightly decreased—to 20 and 41%.

In vivo work on paraquat was done by these same authors, using mice and a dose of 200 mg/kg paraquat followed 30 minutes later by: (1) water, (2) magnesium citrate solution, (3) charcoal in water, (4) Fuller's earth in water, or (5) charcoal in magnesium citrate solution. The numbers of mice surviving out of 16 in each group were 5, 11, 10,

10, and 15, respectively. Surprisingly, magnesium citrate solution was as effective as charcoal or Fuller's earth (these latter two were identical). The best treatment was activated charcoal in magnesium citrate solution. The enhancement of the in vivo action of charcoal by magnesium citrate could be due to at least two effects: (1) enhanced adsorption of paraquat to charcoal, as shown by the in vitro tests discussed in the previous paragraph, and (2) the cathartic action of magnesium citrate (note that the in vivo tests showed that magnesium citrate alone was better than water alone).

Clark (1971) showed in animals that Fuller's earth and bentonite can bind paraquat effectively in the gut if given soon enough and Smith et al. (1974) showed in rats that gastric lavage plus bentonite/purgatives given every 2–3 hr can prevent paraquat absorption. Vale et al. (1977) treated ten human patients intoxicated with large overdoses of paraquat using gastric lavage, Fuller's earth via nasogastric tube, and charcoal hemoperfusion. However, nine of the ten patients died.

The apparent conclusion to be reached regarding the paraquat studies discussed above is that, if charcoal or a suitable clay (Fuller's earth) is administered soon enough, paraquat absorption can be significantly reduced; however, if the sorbent is delayed, the likelihood of a positive outcome is low.

The herbicide linuron was studied in vitro by Bosetto et al. (1991). Linuron in aqueous solution was contacted with two different Norit charcoals and the equilibrium data were fitted to the Langmuir equation. The Q_m values for the two charcoals were 518 and 581 µmole/g at 5°C, 567 and 683 µmole/g at 25°C, and 621 and 779 µmole/g at 40°C. The first of the two values was always smaller because the charcoal surface areas were 560 and 610 m^2/g, respectively. Since linuron has a molecular weight of 249.1, a Q_m value of 600 µmole/g, for example, corresponds to a Q_m of 0.15 g linuron/g charcoal—a respectable binding capacity.

Guven et al. (1993) studied the in vitro adsorption of parathion (an organophosphate herbicide) to charcoal, at pH 1.2 (in HCl solution) and pH 7 (plain water). Charcoal:parathion ratios ranged from 1:1 to 20:1. Over this range, the percent of the parathion adsorbed at pH 1.2 ranged from 26.3 to 59.3%, while at pH 7 the percent adsorbed ranged from 25.3 to 64.5%. Thus, pH had no significant effect. However, this report does not state whether the variation of the charcoal:parathion ratio was achieved by varying the amount of charcoal or by varying the amount of parathion. The results one gets for percent adsorbed depend significantly on which approach is taken. As discussed in Chapter 3, the percent adsorbed is not a unique function of the charcoal:toxin ratio. In this study, it appears that what was varied was the amount of parathion, because the Q values range from fairly high (0.25–0.26 g/g) at a charcoal:parathion ratio of 1:1 to quite low (0.030–0.032 g/g) at a charcoal:parathion ratio of 20:1. The only way that the Q value could be so low at a charcoal:parathion ratio of 20:1 is if C_f were very low; this would mean that the amount of parathion must have been decreased by a factor of 20 (compared to the 1:1 ratio case) and not that the amount of charcoal was increased by a factor of 20.

Obviously, animals studies need to be done with linuron and parathion. The in vitro results discussed, while encouraging, are insufficient to predict what might occur in vivo.

F. Insecticides

Picchioni et al. (1966) studied the adsorption of malathion (a very toxic organic phosphate insecticide) to charcoal in vitro from SGF, USP. Forty mL of solution, containing 1.0 mg of malathion, was contacted either with 30 mg of charcoal or with 30 mg of charcoal in the form of universal antidote. The plain charcoal adsorbed 0.93 mg (i.e., 93%) of the malathion, whereas the universal antidote adsorbed 0.60 mg (i.e., 60%) of the malathion. In vivo tests were done with rats who were given 250 mg/kg malathion followed 1 min later by 500 mg/kg charcoal, 1000 mg of universal antidote, or plain water. One hour later, blood samples were obtained and analyzed for cholinesterase activity. Cholinesterase activity depressions were 15.0, 27.3, and 44.0% for the three treatments, respectively. In acute toxicity tests with other rats, different doses of malathion covering a mortality range of 0–100% in control rats were used. The control LD_{50} was found to be 49 mg/kg. When rats were treated with 500 mg/kg charcoal the LD_{50} rose to 123 mg/kg and when 1000 mg/kg universal antidote was given the LD_{50} was 87 mg/kg. Thus, it was shown that charcoal was effective in malathion toxicity and better than universal antidote (a detailed discussion of universal antidote was presented in Chapter 9).

Orisakawe and Obi (1993) studied the in vitro adsorption of diazinon, an organophosphate insecticide, from solutions in distilled water. A carbon black locally produced in Nigeria and an activated charcoal were used. The charcoal was only slightly better than the carbon black, which is surprising because their surfaces areas are totally different. The amounts needed to adsorb 50% of the diazinon from 5 mL of a 10 µg/mL diazinon solution were 294 mg for the charcoal and 315 mg for the carbon black. In vivo tests were done with diazinon-dosed mice. Two types of charcoals and one type of carbon black were evaluated. The percentage mortalities at adsorbent:diazinon ratios of 1:1, 2:1, 4:1, and 8:1 were: for BDH charcoal, 57, 43, 28, and 14; for the carbon black, 57, 43, 29, and 0; and for the Merck charcoal, 29, 29, 0, and 0. Thus, the carbon black and the BDH charcoal were essentially equivalent and the Merck charcoal superior. Again, it is surprising that the carbon black was as good as one of the charcoals.

G. Pesticides

Atkinson and Azarnoff (1971) gave 300 mg/kg of the pesticide 2,2,2-trichlor-1-hydroxyethyl phosphonate (Neguvon) to rats followed by 250 mg of charcoal at 1 min, 250 mg of attapulgite clay at 1 min, 500 mg of charcoal at 15 min, or 500 mg of attapulgite clay at 15 min. At 60 min, blood, brain, and liver were removed and assayed for cholinesterase activity (inhibition of cholinesterase activity was used as an indirect index of Neguvon absorption). Table 12.5 shows the cholinesterase activities in liters CO_2 per

Table 12.5 Effect of Adsorbents on Cholinesterase Activity

Treatment	Brain	Liver	RBC	Serum
None	80.3	21.8	8.9	3.8
Neguvon only	11.9	3.5	1.3	1.1
Neg + charcoal				
250 mg, 1 min	27.6	7.9	2.6	1.4
500 mg, 15 min	12.2	4.3	1.0	1.3
Neg + attapulgite				
250 mg, 1 min	23.7	6.3	2.1	1.8
500 mg, 15 min	17.1	6.8	1.9	1.9

From Atkinson and Azarnoff (1971).

10 min per 5 mg of tissue. One can see that both adsorbents were mildly effective in reducing the inhibition of cholinesterase activity by Neguvon. They were essentially equivalent in efficacy. The smaller doses (250 mg), when given early, were much more effective than the larger doses (500 mg) given after 15 min.

IX. OTHER SUBSTANCES

There have been a number of studies dealing with substances of various types which are essentially single representatives of miscellaneous drug or poison classes. We thus bring them all together in this general section on other substances.

A. Allylpropynal and Sulfanilamide

We have already discussed the in vivo studies of Andersen (1948a) with allylpropynal and the in vivo studies of Andersen (1948b) with sulfanilamide (see Chapter 10) and merely mention the major results here. With respect to allylpropynal, Andersen found that charcoal adsorbed 1200 mg/g (a large amount) in vitro and 633 mg/g in vivo (still a large amount), in rabbits. With respect to sulfanilamide, it was found that the amounts of charcoal required for essentially total adsorption of sulfanilamide in vitro were about one-third to one-half of the amounts required in vivo, in rabbits and dogs. Experiments were done with sulfanilamide-dosed animals, with the charcoal delayed by 15, 30, and 45 minutes. The greater the delay time, the less effective was the charcoal in reducing drug absorption, especially for the first ten hours.

B. Cyclosporin

Honcharik and Anthone (1985) reported one case involving an adult male who had received a renal transplant and who was given cyclosporin as an immunosuppressant. Due to a mistake, the first daily dose of 750 mg cyclosporin was actually 5000 mg; the mistake was discovered 3 hr later. To counteract this overdose, oral activated charcoal was given as follows: 60 g at the time the mistake was discovered, 30 g 4 hr later, and 30 g an additional 4 hr later. Plasma cyclosporin concentrations were 6700 µg/L 4 hr after drug ingestion, 675 µg/L at 13 hr, and 50 µg/L at 47 hr. The calculated cyclosporin half-life was 2.7 hr during charcoal administration and 9.0 hr after the charcoal was stopped. It appears, therefore, that oral charcoal is effective in cases of cyclosporin overdose.

C. Furosemide

Kivistö and Neuvonen (1990b) showed that the diuretic effect of furosemide can be counteracted by orally administered charcoal. Healthy subjects (61 in number) were given 40 mg furosemide with water, water only, or 40 mg furosemide and 8 g activated charcoal immediately or at different time intervals. The diuretic effect of the drug was totally prevented when the charcoal was given immediately but became stronger as the charcoal delay time increased. The average urine volumes over 0–3 hr were: 390 mL (no drug or charcoal), 340 mL (drug and immediate charcoal), and 1030 mL (drug only). When charcoal was delayed the volumes increased until, at a delay of 1 hr, no effect of charcoal occurred.

D. Methotrexate

The effects of activated charcoal on the pharmacokinetics of high-dose methotrexate (a chemotherapeutic agent) have been studied by Gadgil et al. (1982). Serum methotrexate (MTX) concentrations versus time were followed in two groups of patients who had received 6 hr infusions of MTX (1 g/m^2). One group was given 25 g charcoal orally at 12, 18, 24, 36, and 48 hr. There was a significant reduction (36.7%) in the MTX AUC from 18 hr onward. The data suggest that charcoal aids in the elimination of MTX from the body by interruption of enterohepatic circulation of the drug.

Scheufler and Bos (1983) also studied the effects of oral charcoal on intravenously administered methotrexate in rats. Although the MTX was given intravenously, 31.6% of the dose was subsequently found in the intestinal lumen in the control group (this confirms that enterohepatic circulation of MTX occurs). When charcoal was given, the free MTX that subsequently appeared in the intestinal lumen was only 1% as large (i.e., most of the MTX was bound in the lumen by the charcoal and hence was not free). However, charcoal did not decrease toxic MTX in the intestinal lumen, as determined by villus height decreases and counts of mitotic figures in crypt cells. This suggests that the MTX exerted its toxic effects prior to being adsorbed by the charcoal. Nor did charcoal affect the plasma MTX concentration versus time curve. The only explanation for this surprising result is that MTX is not well absorbed from the intestines; thus, whether the MTX which is carried into the gut by bile is adsorbed by charcoal or remains free is not of consequence to what occurs in the blood plasma, since the MTX in free form is not absorbed from the gut at any significant rate.

E. Cocaine

Makosiej et al. (1990) studied the adsorption of cocaine to charcoal in vitro at pH 1.2 and at pH 7.0, with charcoal:drug ratios ranging from 1:1 to 10:1. The data were fitted to the Langmuir equation. Adsorption was reasonably strong, with the maximal adsorption values, Q_m, being 132 mg/g at pH 1.2 and 204 mg/g at pH 7.0. Because cocaine is basic and is less ionized as pH rises, the Q_m value at pH 7.0 is greater than that at pH 1.2. The Q_m value of 204 mg/g at pH 7.0 is quite good since it means that the charcoal adsorbs cocaine to the extent of 20% of its own weight.

In a longer report on this subject, Makosiej et al. (1993) gave Q_m values of 212 mg/g at pH 1.2 and 273 mg/g at pH 7.0. Since their 1990 report was just an abstract, it does not contain enough detail to resolve the discrepancy in the two sets of Q_m values. The 1993 report included results of additional work, in which desorption of cocaine by adding polyethylene glycol electrolyte lavage solution (PEG-ELS) was studied. The addition of PEG-ELS to charcoal/cocaine slurries at pH 1.2 resulted in significantly decreased adsorption of cocaine. The reductions at charcoal:drug ratios of 1:1, 3:1, 5:1, 7:1, and 10:1 were 76.4, 22.7, 64.8, 62.4, and 56.1%, respectively. At pH 7.0, the reductions were 14.9, 29.1, 32.0, 24.1, and 17.1%, respectively. The lack of consistency, or of a consistent trend, in the PEG-ELS effect seems strange. When PEG-ELS was added to the charcoal first and then the cocaine was added, the percent reductions at pH 1.2 at the same charcoal:drug ratios were 82.4, 86.6, 90.9, 88.0, and 84.9%, respectively. At pH 7.0, the reductions were 44.0, 45.4, 48.6, 35.4, and 25.4%, respectively. In the "PEG-ELS first" cases, the results are less variable. Thus, adding PEG-ELS first interfered with cocaine adsorption much more than adding it after the cocaine had adsorbed to the charcoal; the effect was particularly severe at pH 1.2. It appears that the high molecular weight polyethylene glycol, if added first, adsorbs and blocks pores, making them inaccessible to cocaine. However, when the PEG-ELS is added second, the cocaine already adsorbed in the very smallest pores (which the polyethylene glycol can not penetrate) is not desorbed; thus the PEG-ELS effect is much smaller. Thus, if charcoal and PEG-ELS therapies are both to be used, the charcoal should be used first.

Tomaszewski et al. (1992) also have studied the adsorption of cocaine to charcoal in vitro. At pH 1.2, the percentages of cocaine adsorbed at charcoal:drug ratios of 1:1, 2.5:1, and 5:1 were 40, 92, and 99%, respectively. At pH 8.0, the values were 78, 98, and 99%, respectively. Thus, cocaine adsorbed well at both low and high pH but better at the higher pH, as was the case in the study by Makosiej et al.

Tomaszewski et al. (1993) also studied the effects of charcoal on cocaine in vivo using mice. Fasted mice were given 100 mg/kg cocaine HCl by gavage tube and then either 1 or 2 g/kg charcoal in slurry form (InstaChar) at 1 min. Control animal seizures were 20/20 and deaths were 16/20. With 1 g/kg charcoal, seizures were 4/10 and deaths were 1/10. With 2 g/kg charcoal, seizures were 5/10 and deaths were 3/10. Clearly, charcoal significantly reduced seizures and mortality but strangely the 1 g/kg dose was as good (or better) than the 2 g/kg dose. The reason for this latter result is unknown. Repeated or further studies should be done to resolve this matter. The mean times to seizure (4.4, 4.2, and 5.0 min, respectively) were not significantly different.

Overall, the in vitro and in vivo results for cocaine suggest that charcoal, in a sufficient amount, would be effective in human overdoses, if given early enough. Studies are needed to determine if charcoal given after a significant delay might also be effective.

F.   Phencyclidine

Phencyclidine (PCP) is a drug of abuse which is taken by several routes: orally, by IV injection, or by smoking. It became popular starting in the early 1970s and is now a common cause of severe morbidity and mortality in adults and children. PCP was studied by Picchioni and Consroe (1979) in vivo with dogs and rats. The dogs were given 25 mg/kg phencyclidine by mouth and then were given either an activated charcoal slurry (500 mg charcoal/kg) or plain water. Of six charcoal-treated dogs, only two had convulsions and none died. Of ten dogs receiving no charcoal, nine experienced convulsions and six died. The rat tests involved determining LD_{50} values for pure PCP and for a mixture of 10 parts charcoal per part of PCP. With charcoal added to PCP in this proportion the LD_{50} rose from 135 to 225 mg/kg. Clearly, with both dogs and rats, charcoal was quite effective in reducing the gastrointestinal absorption of PCP.

Lyddane et al. (1988) studied the effect of charcoal in treating PCP overdose. Rats were gavaged once with PCP (10, 25, or 50 mg/kg in 2 mL/kg water) and then given charcoal (1 g/kg) orally once per hour, starting 45 min after drug administration. The AUC value (determined from 45 min until the "time of recovery") was reduced by charcoal in the 10 mg/kg PCP group by 30%, in the 25 mg/kg PCP group by 4% and not at all in the 50 mg/kg PCP group. Thus, charcoal is moderately effective if given in a large enough charcoal:drug ratio. Obviously, had the charcoal been given immediately, it would have had a greater effect (but then this would not mimic the delay which is typical in treating overdose victims). Other agents employed in this study were ammonium chloride, liquid paraffin, and cholestyramine. The combination of NH_4Cl and charcoal was more effective than either alone, giving AUC reductions of 40% in the 10

Table 12.6 Other Drugs or Poisons Studied

Investigators	Drug or poison	In vitro/in vivo
De Souza et al. (1973)	Chlorpheniramine maleate	In vitro
Tsuchiya and Levy (1972a,b)	Phenylpropanolamine salicylamide	In vitro/in vivo
Otto and Stenberg (1973)	Salicylamide	In vitro/in vivo
Ivan (1972a)	Sulfanilamides	In vitro
Ivan (1972b)	Sulfonamides	In vivo
Smith et al. (1967)	D-Amphetamine, tripelennamine, ferrous sulfate, ethanol	In vitro
Cheldelin and Williams (1942)	Amino acids, vitamins	In vitro
Sorby (1961, 1965, 1966)	Phenothiazine derivatives	In vitro
Glazko (1967)	Mefenamic acid	In vivo

mg/kg PCP group, 16% in the 25 mg/kg PCP group, and 21% in the 50 mg/kg PCP group. Cholestyramine and liquid paraffin had no significant effect.

G. Miscellaneous Drugs

Other in vitro and/or in vivo studies of drugs or poisons, mostly less common substances, which have been published but which will not be discussed here are listed in Table 12.6. We merely cite them for the sake of completeness.

X. SUMMARY

We have considered a wide variety of drugs in this chapter: antidiabetics, antihistamines, anti-infectives, antiarrhythmics, beta-blockers, calcium channel-blockers, analgesics, anticonvulsants, anti-inflammatory agents, tranquilizers, gastrointestinal drugs, respiratory relaxants, and organic substances. In vivo studies in which such drugs were given orally, followed soon after by a reasonable dose of charcoal, have invariably shown charcoal to be very effective in reducing absorption. The results shown in Table 12.2, which show reductions in bioavailability of 82.6–99.5%, are typical.

Even studies in which such drugs were given intravenously have shown that oral charcoal can reduce AUC values significantly (30% reductions are typical) if the drugs undergo enteroenteric or enterohepatic cycling.

A few substances, such as ethylene glycol, which are simple and very hydrophilic, and lack moieties which attract to the surface of charcoal (like ring structures) do not adsorb significantly to charcoal. Hence, charcoal is not effective in vivo in reducing the absorption of such substances.

XI. REFERENCES

Akintonwa, A. and Obodozie, O. (1991). Effect of activated charcoal on the disposition of sulphadoxine, Arch. Int. Pharmacodyn. *309*, 185.

Al-Meshal, M. A., El-Sayed, Y. M., Al-Angary, A. A., and Al-Dardiri, M. M. (1993). Effect of oral activated charcoal on propranolol pharmacokinetics following intravenous administration to rabbits, J. Clin. Pharm. Ther. *18*, 39.

Allen, E. M., Buss, D. C., Williams, J., and Routledge, P. A. (1987). The effect of charcoal on mefenamic acid elimination, Br. J. Clin. Pharmacol. *24*, 830.

Amitai, Y. and Lovejoy, F. H., Jr. (1987). Characteristics of vomiting associated with acute sustained release theophylline poisoning: Implications for management with oral activated charcoal, J. Toxicol.-Clin. Toxicol. *25*, 539.

Andersen, A. H. (1948a). Experimental studies on the pharmacology of activated charcoal. IV. Adsorption of allylpropynal (allyl-isopropyl-barbituric acid) in vivo, Acta Pharmacol. Toxicol. *4*, 379.

Andersen, A. H. (1948b). Experimental studies on the pharmacology of activated charcoal. V. Adsorption of sulphanilamide in vivo, Acta Pharmacol. Toxicol. *4*, 389.

Arimori, K. and Nakano, M. (1989). Study on the transport of disopyramide into the intestinal lumen aimed at gastrointestinal dialysis by activated charcoal in rats, J. Pharm. Pharmacol. *41*, 445.

Arimori, K., Kawano, H., and Nakano, M. (1989). Gastrointestinal dialysis of disopyramide in healthy subjects, Int. J. Clin. Pharmacol. Ther. Toxicol. *27*, 280.

Arimori, K., Deshimaru, M., Furukawa, E., and Nakano, M. (1993). Adsorption of mexiletine onto activated charcoal in macrogol-electrolyte solution, Chem. Pharm. Bull. *41*, 766.

Atkinson, J. P. and Azarnoff, D. L. (1971). Comparison of charcoal and attapulgite as gastrointestinal sequestrants in acute drug ingestions, Clin. Toxicol. *4*, 31.

Boehm, J. J. and Oppenheim, R. C. (1977). An in vitro study of the adsorption of various drugs by activated charcoal, Aust. J. Pharm. Sci. *6*, 107.

Boehm, J. J., Brown, T. C. K., and Oppenheim, R. C. (1978). Reduction of pheniramine toxicity using activated charcoal, Clin. Toxicol. *12*, 523.

Boldy, D. A. R., Heath, A., Ruddock, S., Vale, J. A., and Prescott, L. F. (1987). Activated charcoal for carbamazepine poisoning, Lancet, May 2, p. 1027.

Bosetto, M., Arfaioli, P., and Fusi, P. (1991). Adsorption and desorption of linuron by activated charcoals, Bull. Environ. Contam. Toxicol. *46*, 37.

Burkhart, K. K. and Martinez, M. A. (1992). The adsorption of isopropanol and acetone by activated charcoal, J. Toxicol.-Clin. Toxicol. *30*, 371.

Chaput de Saintonge, D. M., and Herxheimer, A. (1971). Activated charcoal impairs propantheline absorption, Eur. J. Clin. Pharmacol. *4*, 52.

Cheldelin, V. H. and Williams, R. J. (1942). Adsorption of organic compounds. I. Adsorption of ampholytes on an activated charcoal, J. Am. Chem. Soc. *64*, 1513.

Chernish, S. M., Wolen, R. L., and Rodda, B. E. (1972). Adsorption of propoxyphene hydrochloride by activated charcoal, Clin. Toxicol. *5*, 317.

Chin, L., Picchioni, A. L., and Duplisse, B. R. (1970). The action of activated charcoal on poisons in the digestive tract, Toxicol. Appl. Pharmacol. *16*, 786.

Chin, L., Picchioni, A. L., Bourn, W. M., and Laird, H. E. (1973). Optimal antidotal dose of activated charcoal, Toxicol. Appl. Pharmacol. *26*, 103.

Clark, D. G. (1971). Inhibition of the absorption of paraquat from the gastrointestinal tract by adsorbents, Br. J. Ind. Med. *28*, 186.

Cooney, D. O. (1977). The treatment of ethylene glycol poisoning with activated charcoal, IRCS Med. Sci. *5*, 265.

Corby, D. G. and Decker, W. J. (1968a). An antidote for propoxyphene HCl (letter), JAMA *203*, 1074.

Corby, D. G. and Decker, W. J. (1968b). Treatment of propoxyphene poisoning (letter), JAMA *205*, 250.

Corby, D. G. and Decker, W. J. (1974). Management of acute poisoning with activated charcoal, Pediatrics *54*, 324.

Cordonnier, J. A., Van den Heede, M. A., and Heyndrickx, A. M. (1986-87). In vitro adsorption of tilidine HCl by activated charcoal, J. Toxicol.-Clin. Toxicol. *24*, 503.

Cordonnier, J., Van den Heede, M., and Heyndrickx, A. (1987). Activated charcoal and ipecac syrup in prevention of tilidine absorption in man, Vet. Hum. Toxicol. *29* (Suppl. 2), 105.

D'Angio, R. and Sabatelli, F. (1987). Management considerations in treating metabolic abnormalities associated with theophylline overdose, Arch. Intern. Med. *147*, 1837.

De Vries, M. H., Rademaker, C. M. A., Geerlings, C., van Dijk, A., and Noordhoek, J. (1989). Pharmacokinetic modelling of the effect of activated charcoal on the intestinal secretion of theophylline, using the isolated vascularly perfused rat small intestine, J. Pharm. Pharmacol. *41*, 528.

De Souza, J. J. V., Mitra, A. K., Gupta, S., and Gupta, B. K. (1973). Adsorption of chlorpheniramine maleate from aqueous solution by activated charcoal, Indian J. Pharm. *35*, 167.

Dean, B. S., Burdick, J. D., Geotz, C. M., Bricker, J. D., and Krenzelok, E. P. (1992). In vivo evaluation of the adsorptive capacity of activated charcoal for camphor, Vet. Hum. Toxicol. *34*, 297.

Du Souich, P., Caillé, G., and Larochelle, P. (1983). Enhancement of nadolol elimination by activated charcoal and antibiotics, Clin. Pharmacol. Ther. *33*, 585.

Edge, W. and Edmonds, J. (1992). Serum sodium and carbamazepine overdose (letter), J. Toxicol.-Clin. Toxicol. *30*, 479.

El-Bahie, N., Allen, E. M., Williams, J., and Routledge, P. A. (1985). The effect of activated charcoal and hyoscine butylbromide alone and in combination on the absorption of mefenamic acid, Br. J. Clin. Pharmacol. *19*, 836.

El-Khordagui, L. K., Saleh, A. M., and Khalil, S. A. (1987). Adsorption of benzodiazepines on charcoal and its correlation with in vitro and in vivo data, Pharm. Acta Helv. *62*, 28.

El-Sayed, Y. M. and Hasan, M. M. (1990). Enhancement of morphine clearance following intravenous administration by oral activated charcoal in rabbits, J. Pharm. Pharmacol. *42*, 538.

El-Sayed, Y. M., Al-Meshal, M. A., Al-Angary, A. A., Lutfi, K. M., and Gouda, M. W. (1990). Accelerated clearance of intravenous indomethacin by oral activated charcoal in rabbits, Int. J. Pharmaceut. *64*, 109.

Elonen, E., Neuvonen, P. J., Halmekoski, J., and Mattila, M. J. (1979). Acute dapsone intoxication: A case with prolonged symptoms, Clin. Toxicol. *14*, 79.

Farrar, H. C., Herold, D. A., and Reed, M. D. (1993). Acute valproic acid intoxication: Enhanced drug clearance with oral-activated charcoal, Crit. Care Med. *21*, 299.

Ferry, D. G., Gazeley, L. R., Busby, W. J., Beasley, D. M. G., Edwards, I. R., and Campbell, A. J. (1990). Enhanced elimination of piroxicam by administration of activated charcoal or cholestyramine, Eur. J. Clin. Pharmacol. *39*, 599.

Gadgil, S. D., Damle, S. R., Advani, S. H., and Vaidya, A. B. (1982). Effect of activated charcoal on the pharmacokinetics of high-dose methotrexate, Cancer Treat. Rep. *66*, 1169.

Ganjian, F., Cutie, A. J., and Jochsberger, T. (1980). In vitro adsorption studies of cimetidine, J. Pharm. Sci. *69*, 352.

Gaudreault, P. and Lovejoy, F. H., Jr. (1983). Treatment of theophylline overdose (reply to letter), J. Pediatr. *103*, 1004.

Gaudreault, P., Friedman, P. A., and Lovejoy, F. H., Jr. (1985). Efficacy of activated charcoal and magnesium citrate in the treatment of oral paraquat intoxication, Ann. Emerg. Med. *14*, 123.

Ghazy, F. S., Kassem, A. A., and Shalaby, S. H. (1984). Adsorption characteristics of certain antibiotics to veegum and activated charcoal, Pharmazie *39*, 821.

Ginoza, G. W., Strauss, A. A., Iskra, M. K., and Modanlou, H. D. (1987). Potential treatment of theophylline toxicity by high surface area activated charcoal, J. Pediatr. *111*, 140.

Glab, W. N., Corby, W. G., Decker, W. J., and Coldiron, V. R. (1982). Decreased absorption of propoxyphene by activated charcoal, J. Toxicol.-Clin. Toxicol. *19*, 129.

Glazko, A. J. (1967). In Pharmacology of the fenamates: III. Metabolic disposition, Ann. Phys. Med. *9* (Suppl.), 24.

Guay, D. R. P., Meatherall, R. C., Macauley, P. A., and Yeung, C. (1984). Activated charcoal adsorption of diphenhydramine, Int. J. Clin. Pharmacol. Ther. Toxicol. *22*, 395.

Guven, H., et al. (1993). The adsorption of parathion by activated charcoal in vitro (abstract), Vet. Hum. Toxicol. *35*, 359.

Hasan, M. M., El-Sayed, Y. M., and Abdelaziz, A. A. (1990a). The effect of oral activated charcoal on the systemic clearance of gentamicin in rabbits with acute renal failure, J. Pharm. Pharmacol. *42*, 85.

Hasan, M. M., Hassan, M. A., and Rawashdeh, N. M. (1990b). Effect of oral activated charcoal on the pharmacokinetics of quinidine and quinine administered intravenously to rabbits, Pharmacol. Toxicol. *67*, 73.

Hassan, E. (1986). Treatment of meprobamate overdose with repeated oral doses of activated charcoal, Ann. Emerg. Med. *15*, 73.

Hayden, J. W. and Comstock, E. G. (1975). Use of activated charcoal in acute poisoning, Clin. Toxicol. *8*, 515.

Helliwell, M. and Berry, D. (1981). Theophylline absorption [sic] by effervescent activated charcoal (Medicoal), J. Int. Med. Res. *9*, 222.

Herman, R. J. and Chaudhary, A. (1991). In vitro binding of lorazepam and lorazepam glucuronide to cholestyramine, colestipol, and activated charcoal, Pharm. Res. *8*, 538.

Honcharik, N. and Anthone, S. (1985). Activated charcoal in acute cyclosporin overdose (letter), Lancet, May 4, p. 1051.

Huang, J. D. (1987). Kinetics of theophylline clearance in gastrointestinal dialysis with charcoal, J. Pharm. Sci. *76*, 525.

Huang, J. D. (1988). Stereoselective gastrointestinal clearance of disopyramide in rabbits treated with activated charcoal, J. Pharm. Sci. *77*, 959.

Ivan, J. (1972a). Adsorption of sulphanilamides on activated carbon, Acta Pharm. Hung. *42*, 97.

Ivan, J. (1972b). Influence of activated carbon on the absorption of sulfonamides, Acta Pharm. Hung. *42*, 103.

Jain, R. and Tholl, D. A. (1992). Activated charcoal for theophylline toxicity in a premature infant on the second day of life, Dev. Pharmacol. Ther. *19*, 106.

Judis, J. (1985). Effect of pH on charcoal adsorption of lidocaine, methadone, pilocarpine, and procaine, J. Pharm. Sci. *74*, 476.

Kannisto, H. and Neuvonen. P. J. (1984). Adsorption of sulfonylureas onto activated charcoal in vitro, J. Pharm. Sci. *73*, 253.

Kärkkäinen, S. and Neuvonen, P. J. (1984). Effect of oral charcoal and urine pH on sotalol pharmacokinetics, Int. J. Clin. Pharmacol. Ther. Toxicol. *22*, 441.

Kärkkäinen, S. and Neuvonen, P. J. (1985). Effect of oral charcoal and urine pH on dextropropoxyphene pharmacokinetics. Int. J. Clin. Pharmacol. Ther. Toxicol. *23*, 219.

Kärkkäinen, S. and Neuvonen, P. J. (1986). Pharmacokinetics of amitriptyline influenced by oral charcoal and urine pH, Int. J. Clin. Pharmacol. Ther. Toxicol. *24*, 326.

Kivistö, K. T. and Neuvonen, P. J. (1990a). The effect of cholestyramine and activated charcoal on glipizide absorption, Br. J. Clin. Pharmacol. *30*, 733.

Kivistö, K. T. and Neuvonen, P. J. (1990b). Effect of activated charcoal on frusemide induced diuresis: A human class experiment for medical students, Br. J. Clin. Pharmacol. *30*, 496.

Kivistö, K. T. and Neuvonen, P. J. (1991). Effect of activated charcoal on the absorption of amiodarone, Hum. Exp. Toxicol. *10*, 327.

Kivistö, K. T. and Neuvonen, P. J. (1993). Activated charcoal for chloroquine poisoning (letter), Br. Med. J. *307*, 1068.

Knadler, M. P., Bergstrom, R. F., Callaghan, J. T., Obermeyer, B. D., and Rubin, A. (1987). Absorption studies of the H_2-blocker nizatidine, Clin. Pharmacol. Ther. *42*, 514.

Kortilla, K., Mattila, M. J., and Linnoila, M. (1976). Prolonged recovery after diazepam sedation: The influence of food, charcoal ingestion and injection rate on the effects of intravenous diazepam, Br. J. Anaesth. *48*, 333.

Krenzelok, E. P. and Lopez, G. P. (1988). Multiple-dose activated charcoal: An ever-expanding role (letter), Ann. Emerg. Med. *17*, 1134.

Laine, K., Kivistö, K. T., and Neuvonen, P. J. (1992). Failure of activated charcoal to accelerate the elimination of amiodarone and chloroquine, Hum. Exp. Toxicol. *11*, 491.

Laufen, H. and Leitold, M. (1986). The effect of activated charcoal on the bioavailability of piroxicam in man, Int. J. Clin. Pharmacol. Ther. Toxicol. *24*, 48.

Lim, D. T., Singh, P., Nourtsis, S., and Dela Cruz, R. (1986). Absorption inhibition and enhancement of elimination of sustained-release theophylline tablets by oral activated charcoal, Ann. Emerg. Med. *15*, 1303.

Lockey, D. and Bateman, D. N. (1989). Effect of oral activated charcoal on quinine elimination, Br. J. Clin. Pharmacol. *27*, 92.

Lyddane, J. E., Thomas, B. F., Compton, D. R., and Martin, B. R. (1988). Modification of phencyclidine intoxication and biodisposition by charcoal and other treatments, Pharmacol. Biochem. Behav. *30*, 371.

Makosiej, F., Hoffman, R. S., Howland, M. A., Verebey, K., Weisman, R. S., and Goldfrank, L. R. (1990). Cocaine adsorption to activated charcoal: The effects of pH (abstract), Vet. Hum. Toxicol. *32*, 350.

Makosiej, F., Hoffman, R. S., Howland, M. A., and Goldfrank, L. R. (1993). An in vitro evaluation of cocaine hydrochloride adsorption by activated charcoal and desorption upon addition of polyethylene glycol electrolyte lavage solution, J. Toxicol.-Clin. Toxicol. *31*, 381.

Mauro, L. S., Mauro, V. F., Brown, D. L., and Somani, P. (1987). Enhancement of phenytoin elimination by multiple-dose activated charcoal, Ann. Emerg. Med. *16*, 1132.

Mauro, L. S. (1988). Multiple-dose activated charcoal: An ever-expanding role (reply to letter), Ann. Emerg. Med. *17*, 1134.

McConnell, E. E., Harris, M. W., and Moore, J. A. (1980). Studies on the use of activated charcoal and cholestyramine for reducing the body burden of polybrominated biphenyls, Drug. Chem. Toxicol. *3*, 277.

McKinney, P., Tomaszewski, C., Phillips, S., Brent, J., Kulig, K., and Rumack, B. (1991). Prevention of methamphetamine toxicity by activated charcoal in mice (abstract), Vet. Hum. Toxicol. *33*, 386.

McKinnon, R. S., et al. (1987). Studies on the mechanisms of action of activated charcoal on theophylline pharmacokinetics, J. Pharm. Pharmacol. *39*, 522.

Neuvonen, P. J., Elfving, S. M., and Elonen, E. (1978). Reduction of absorption of digoxin, phenytoin, and aspirin by activated charcoal in man, Eur. J. Clin. Pharmacol. *13*, 213.

Neuvonen, P. J. and Elonen, E. (1980a). Effect of activated charcoal on absorption and elimination of phenobarbitone, carbamazepine, and phenylbutazone in man, Eur. J. Clin. Pharmacol. *17*, 51.

Neuvonen, P. J. and Elonen, E. (1980b). Phenobarbitone elimination rate after oral charcoal, Br. Med. J., March 15, p. 762.

Neuvonen, P. J., Elonen, E., and Mattila, M. J. (1980). Oral activated charcoal and dapsone elimination, Clin. Pharmacol. Ther. *27*, 823.

Neuvonen, P. J. and Kärkkäinen, S. (1983). Effects of charcoal, sodium bicarbonate, and ammonium chloride on chlorpropamide kinetics, Clin. Pharmacol. Ther. *33*, 386.

Neuvonen, P. J., Kannisto, H., and Hirvisalo, E. L. (1983a). Effect of activated charcoal on absorption of tolbutamide and valproate in man, Eur. J. Clin. Pharmacol. *24*, 243.

Neuvonen, P. J., Elonen, E., and Haapanen, E. J. (1983b). Acute dapsone intoxication: Clinical findings and effect of oral charcoal and haemodialysis on dapsone elimination, Acta. Med. Scand. *214*, 215.

Neuvonen, P. J., Vartiainen, M., and Tokola, O. (1983c). Comparison of activated charcoal and ipecac syrup in prevention of drug absorption. Eur. J. Clin. Pharmacol. *24*, 557.

Neuvonen, P. J., Kannisto, H., and Lankinen, S. (1983-84). Capacity of two forms of activated charcoal to adsorb nefopam in vitro and to reduce its toxicity in vivo, J. Toxicol.-Clin. Toxicol. *21*, 333.

Neuvonen, P. J. and Olkkola, K. T. (1984a). Effect of dose of charcoal on the absorption of disopyramide, indomethacin, and trimethoprim by man, Eur. J. Clin. Pharmacol. *26*, 761.

Neuvonen, P. J. and Olkkola, K. T. (1984b). Activated charcoal and syrup of ipecac in prevention of cimetidine and pindolol absorption in man after administration of metoclopramide as an antiemetic agent, J. Toxicol.-Clin. Toxicol. *22*, 103.

Neuvonen, P. J., Olkkola, K. T., and Alanen, T. (1984). Effect of ethanol and pH on the adsorption of drugs to activated charcoal: Studies in vitro and in man, Acta Pharmacol. Toxicol. *54*, 1.

Neuvonen, P. J. and Olkkola, K. T. (1986). Effect of purgatives on antidotal efficacy of oral activated charcoal, Hum. Toxicol. *5*, 255.

Neuvonen, P. J. and Olkkola, K. T. (1988). Oral activated charcoal in the treatment of intoxications: Role of single and repeated doses, Med. Toxicol. *3*, 33.

Neuvonen, P. J., Kivistö, K., and Hirvisalo, E. L. (1988). Effects of resins and activated charcoal on the absorption of digoxin, carbamazepine, and frusemide, Br. J. Clin. Pharmacol. *25*, 229.

Neuvonen, P. J., Kivistö, K. T., Laine, K., and Pyykkö, K. (1992). Prevention of chloroquine absorption by activated charcoal, Hum. Exp. Toxicol. *11*, 117.

Nitsch, J., Köhler, U., Neyses, L., and Lüderitz, B. (1987). Inhibition of flecainide absorption by activated charcoal, Am. J. Cardiol. *60*, 753.

Olkkola, K. T. and Neuvonen, P. J. (1984). Do gastric contents modify antidotal efficacy of oral activated charcoal?, Br. J. Clin. Pharmacol. *18*, 663.

Olkkola, K. T. (1985). Effect of charcoal-drug ratio on antidotal efficacy of oral activated charcoal in man, Br. J. Clin. Pharmacol. *19*, 767.

Orisakwe, O. E. and Akintonwa, A. (1991). In-vitro adsorption studies of isoniazid, Hum. Exp. Toxicol. *10*, 133.

Orisakwe, O. E. and Obi, N. (1993). In vitro and in vivo adsorption studies of diazinon, Hum. Exp. Toxicol. *12*, 301.

Otto, U. and Stenberg, B. (1973). Drug adsorption properties of different activated charcoal dosage forms in vitro and in man, Svensk. Farm. Tids. *77*, 613.

Picchioni, A. L., Chin, L., Verhulst, H. L., and Dieterle, B. (1966). Activated charcoal versus universal antidote as an antidote for poisons, Toxicol. Appl. Pharmacol. *8*, 447.

Picchioni, A. L., Chin, L., and Laird, H. E. (1974). Activated charcoal preparations—Relative antidotal efficacy, Clin. Toxicol. *7*, 97.

Picchioni, A. L., and Consroe, P. F. (1979). Activated charcoal: A phencyclidine antidote, or hog, in dogs, New Engl. J. Med. *300*, 202.

Prescott, L. F., Hamilton, A. R., and Heyworth, R. (1989). Treatment of quinine overdosage with repeated oral charcoal, Br. J. Clin. Pharmacol. *27*, 95.

Reigart, J. R., Trammel, H. L., Jr., and Lindsey, J. M. (1982-83). Repetitive doses of activated charcoal in dapsone poisoning in a child, J. Toxicol.-Clin. Toxicol. *19*, 1061.

Roberts, D., Honcharik, N., Sitar, D. S., and Tenenbein, M. (1991). Diltiazem overdose: Pharmacokinetics of diltiazem and its metabolites and effect of multiple-dose charcoal therapy, J. Toxicol.-Clin. Toxicol. *29*, 45.

Sanvordeker, D. R., and Dajani, E. Z. (1975). In vitro adsorption of diphenoxylate hydrochloride on activated charcoal and its relation to pharmacological effects of drug in vivo, J. Pharm. Sci. *64*, 1877.

Scheufler, E. and Bos, I. (1983). Influence of peroral charcoal on pharmacokinetics and intestinal toxicity of intravenously given methotrexate, Arch. Int. Pharmacodyn. *261*, 180.

Scolding, N., Ward, M. J., Hutchings, A., and Routledge, P. A. (1986). Charcoal and isoniazid pharmacokinetics, Hum. Toxicol. *5*, 285.

Sellers, E. M., Khouw, V., and Dolman, L. (1977). Comparative drug adsorption by activated charcoal, J. Pharm. Sci. *66*, 1640.

Sessler, C. N. (1987). Poor tolerance of oral activated charcoal with theophylline overdose, Am. J. Emerg. Med. *5*, 492.

Siefkin, A. D., Albertson, T. E., and Corbett, M. G. (1987). Isoniazid overdose: Pharmacokinetics and effects of oral charcoal in treatment, Hum. Toxicol. *6*, 497.

Sintek, C., Hendeles, L., and Weinberger, M., (1978). Activated charcoal adsorption of theophylline in vitro, Drug Intell. Clin. Pharm. *12*, 158.

Sintek, C., Hendeles, L., and Weinberger, M. (1979). Inhibition of theophylline absorption by activated charcoal, J. Pediatrics *94*, 314.

Smith, L. L., Wright, A., Wyatt, I., and Rose, M. S. (1974). Effective treatment for paraquat poisoning in rats and its relevance to the treatment of paraquat poisoning in man, Br. Med. J. *4*, 569.

Smith, R. P., Gosselin, R. E., Henderson, J. A., and Anderson, D. M. (1967). Comparison of the adsorptive properties of activated charcoal and Alaskan montmorillonite for some common poisons, Toxicol. Appl. Pharmacol. *10*, 95.

Sorby, D. L. and Plein, E. M. (1961). Adsorption of phenothiazine derivatives by kaolin, talc, and Norit, J. Pharm. Sci. *50*, 355.

Sorby, D. L. (1965). Effect of adsorbents on drug absorption: I. Modification of promazine absorption by activated attapulgite and activated charcoal, J. Pharm. Sci. *54*, 677.

Sorby, D. L., Plein, E. M., and Benmaman, J. D. (1966). Adsorption of phenothiazine derivatives by solid adsorbents, J. Pharm. Sci. *55*, 785.

Sorgel, F., Naber, K. G., Jaehde, U., Reiter, A., Seelmann, R., and Sigl, G. (1989). Gastrointestinal secretion of ciprofloxacin: Evaluation of the charcoal model for investigations in healthy volunteers, Am. J. Med. *87* (Suppl. 5A), 62S.

Strauss, A. A., Modanlou, H. D., and Komatsu, G. (1985). Theophylline toxicity in a preterm infant: Selected clinical aspects, Pediatr. Pharmacol. *5*, 209.

Szabuniewicz, M., Bailey, E. M., and Wiersig, D. O. (1975a). A new regimen for the treatment of ethylene glycol poisoning, IRCS Med. Sci. (Pharmacol.; Vet. Sci.) *3*, 102.

Szabuniewicz, M., Bailey, E. M., and Wiersig, D. O. (1975b). A new approach to the treatment of ethylene glycol poisoning in dogs, Southwest. Vet. *28*, 1.

Tomaszewski, C., Voorhees, S., Wathen, J., Brent, J., and Kulig, K. (1992). Cocaine adsorption to activated charcoal in vitro, J. Emerg. Med. *10*, 59.

Tomaszewski, C., McKinney, P., Phillips, S., Brent, J., and Kulig, K. (1993). Prevention of toxicity from oral cocaine by activated charcoal in mice, Ann. Emerg. Med. *22*, 1804.

Torre, D., Sampietro, C., Quadrelli, C., Bianchi, W., and Maggiolo, F. (1988). Effects of orally administered activated charcoal on ciprofloxacin pharmacokinetics in healthy volunteers, Chemioterapia *7*, 382.

Traeger, S. M. and Haug, M. T. (1986). Reduction of diazepam serum half life and reversal of coma by activated charcoal in a patient with severe liver disease, J. Toxicol.-Clin. Toxicol. *24*, 329.

Tsuchiya, T. and Levy, G. (1972a). Drug adsorption efficacy of commercial activated charcoal tablets in vitro and in man, J. Pharm. Sci. *61*, 624.

Tsuchiya, T. and Levy, G. (1972b). Relationship between effect of activated charcoal on drug absorption in man and its drug adsorption characteristics in vitro, J. Pharm. Sci. *61*, 586.

Vale, J. A., Crome, P., Volans, G. N., Widdop, B., and Goulding, R. (1977). The treatment of paraquat poisoning using oral sorbents and charcoal haemoperfusion, Acta Pharmacol. Toxicol. *41* (Suppl. 2), 109.

Weinberger, M. (1983). Treatment of theophylline overdose (letter), J. Pediatr. *103*, 1004.

13

Effect of Administration Time, Food, and Gastric pH

The efficacy of activated charcoal in reducing the absorption of a drug or poison is obviously affected by how soon after the drug/poison ingestion the charcoal is administered. Also, it seems likely that whether food is present in a person's stomach and, if so, the type and amount of food should affect the action of charcoal. However, the effect of food has been studied only rarely. Another factor which would seem to be important is the pH profile in the gastrointestinal tract, particularly the pH in the stomach (since that is where the drug and charcoal first interact). Since the pH of stomach contents can be altered (e.g., by antacids), the possibility exists that the simultaneous use of charcoal and agents which can change the stomach pH might be of some benefit. Here again we find that this strategy has rarely been considered.

I. EFFECT OF A DELAY IN ADMINISTRATION

The effect of a delay in time between ingestion of a drug or poison and the administration of activated charcoal depends on the balance between two factors: (1) the duration of the time delay, and (2) the rate of absorption of the drug. The latter process, in turn, will depend on how much food is in the digestive tract, the solubility of the drug or poison in the gastrointestinal fluids, the effect of the drug or poison on the gastric emptying rate and on gut motility, and the dosage form of the drug or poison (tablets, liquid, suspension, etc.).

In general, charcoal should be administered as soon as possible. Several investigators have suggested that a time of 30 min is a rough limit beyond which charcoal will be decidedly less effective. It should be mentioned that many drugs, such as sedatives, hypnotics, and tricyclic antidepressants, tend to reduce gastric motility and, therefore, the rate of drug absorption. In such cases, charcoal given after 30 min can still be of great benefit.

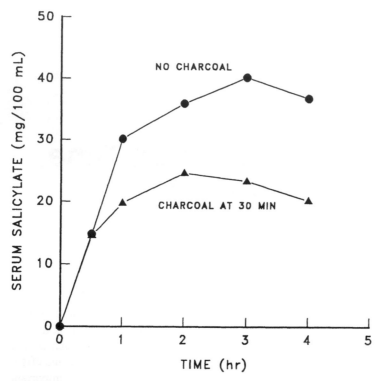

Figure 13.1 Effect of activated charcoal given at 30 min on serum salicylate concentrations in dogs. From Corby et al. (1970). Reproduced by permission of the W. B Saunders Company.

Drugs which undergo enterohepatic recycling, such as glutethimide, some cardiac glycosides, and tricyclic antidepressants can be effectively treated by delayed or additional repeated doses of activated charcoal.

A. Aspirin and Other Salicylates

Decker et al. (1968) carried out in vivo experiments with rats. When 125 mg activated charcoal was administered 30 min after a dose of 320 mg sodium salicylate, the serum salicylate concentrations at 60, 90, and 120 min were reduced by 66, 62, and 62%, respectively, compared to control animals. Additional work with dogs (Corby et al., 1970) gave similar results. Figure 13.1 presents the results of the studies with dogs, which show that, even after a delay of 30 min, the charcoal had a significant effect.

Phansalkar and Holt (1968) found that dogs given 100 grains of aspirin plus either 60 g or 90 g activated charcoal had plasma salicylate concentrations that were less than 10% of control test levels. However, it should be mentioned that the charcoal:drug ratios were roughly 11:1 and 14:1, respectively, so it is not surprising that the reductions in the salicylate levels were so large. Even when charcoal was delayed for 30 min, the charcoal halted the rise in plasma salicylate (at about the 1 hr point) and caused it to start falling.

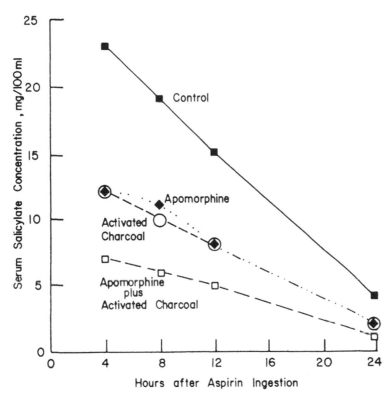

Figure 13.2 Effect of activated charcoal, apomorphine, and activated charcoal/apomorphine given at 30 min on serum salicylate in volunteer human subjects receiving aspirin. From Decker et al. (1969). Reprinted by permission of Mosby-Year Book Inc.

Studies by Decker et al. (1969) in humans also showed a significant inhibition of aspirin absorption. Oral administration of 30 g activated charcoal to adult male volunteers 30 min after they had received 50 grains of aspirin resulted in serum salicylate concentrations roughly 50% of those in a control group. They also showed that the simultaneous intramuscular administration of apomorphine (0.03 mg per pound of body weight) enhanced this effect, as shown in Figure 13.2. Thus, the combined regimen of charcoal adsorption and emesis was more efficient than either method alone (it should be mentioned that, because apomorphine can cause CNS and respiratory depression, it is essentially no longer used in human medicine). Despite the better results with the charcoal/emetic combination, most physicians today would probably choose to forego the emetic and simply use a larger dose of charcoal.

Levy and Tsuchiya (1969, 1972) reported on in vivo work with aspirin and human subjects. They found that charcoal, if administered promptly and in sufficient amounts, significantly inhibited the absorption of aspirin from solution, conventional tablets, enteric-coated tablets, and sustained-release tablets. The effects increased with the amount of charcoal given, decreased as the time delay between the aspirin and charcoal increased, and were decreased by the presence of food in the gastrointestinal tract (the effect of food is discussed in detail later in this chapter).

Table 13.1   Effect of Delayed Administration of Activated Charcoal on the Absorption of Aspirin Given as Enteric-Coated Tablets

Subject	Time delay (hr)	Dose recovered in the urine (%)	
		Tablets alone	With charcoal
A	3	90.5	38.0
B	3	85.5	68.8
C	3	97.7	68.5
D	2	101.6	60.8
Mean		93.8	59.0

From Levy and Tsuchiya (1972). Reprinted by permission of Mosby-Year Book Inc.

Levy and Tsuchiya found that aspirin given in solution form was very rapidly absorbed in vivo. In such a case, charcoal administration after even 1 or 2 hr would probably have little effect. However, they also observed that when aspirin is administered in the form of enteric-coated tablets, charcoal is effective even if given several hours after the aspirin, as shown in Table 13.1.

Collombel and Perrot (1970) carried out studies on sodium salicylate absorption versus charcoal delay time in rats. Their results (Table 13.2) suggest that the charcoal is quite effective at 30 min, somewhat effective at 60 min, and ineffective at 120 min.

Atkinson and Azarnoff (1971) gave 4.0 g sodium salicylate to fasted dogs, followed in one case by 90 g charcoal at 60 min and in another case by 60 g charcoal at 90 min. Figure 13.3 shows that the 90-g dose at 60 min was quite effective in reducing the drug absorption, while the 60-g dose at 90 min was somewhat effective.

The fact that various drugs are affected differently by delays in charcoal administration is shown clearly by the results of Neuvonen et al. (1978). When charcoal (50 g) was given immediately after three drugs—digoxin (0.5 mg), phenytoin (0.5 g), and aspirin (1 g)—systemic absorption decreased by 98, 98, and 70%, respectively, relative to control amounts. When the charcoal was delayed for 1 hr, the reductions fell to 40, 80, and 10% respectively. Thus, the delay had relatively little effect on phenytoin, a drug which

Table 13.2   Salicylate Levels (mg/liter) in Rats

Hours since aspirin dose	No Charcoal	Charcoal at 30 min	Charcoal at 60 min	Charcoal at 120 min
1	150	80	160	188
2	240	83	153	230
4	290	85	138	231
8	235	70	103	215
12	140	34	65	103
24	20	15	29	24

From Collombel and Perrot (1970).

is absorbed slowly by the body. However, for aspirin, which is absorbed fairly rapidly, the delay had a great effect on the efficacy of the charcoal.

Dillon et al. (1989) gave 12 healthy volunteers 20 mg/kg aspirin orally and, 1 hr later, gave them either 25 g SuperChar charcoal or 25 g Actidose Aqua. Urine was collected up to 72 hr and analyzed for salicylates. The percentage of the aspirin absorbed was 78% for the Actidose Aqua case and 50% for the SuperChar case. The control value was 96%. Thus, the Actidose Aqua charcoal was moderately effective (18% reduction in aspirin absorption) after the 1 hr delay, and the SuperChar charcoal (which has a much higher internal surface area) was quite effective (46% reduction in aspirin absorption) after the 1 hr delay.

The studies with aspirin and sodium salicylate just discussed show that, in general, charcoal can still be reasonably effective when delayed for 30 min, even though these drugs are fairly rapidly absorbed. However, with a delay of 1 hr or more, the efficacy of charcoal usually is greatly reduced. The exceptions to this have been in cases where superactive charcoal has been employed, or where enteric-coated aspirin tablets were involved.

B. Acetaminophen

Dordoni et al. (1973) described experiments in which 2-g oral doses of acetaminophen were given to 14 human subjects. Seven of the subjects were then given 10 g activated charcoal as a suspension in methylcellulose solution (20 g/100 mL). The charcoal resulted in much lower plasma concentrations, as shown in Figure 11.2, and the 0–2 hr AUC values were reduced by an average of 63% (range 32–87%). Four other subjects were given 12 g cholestyramine resin in 200 mL water after ingesting the acetaminophen. As shown in Figure 11.2, it is clear that this adsorbent also was very effective; reductions in bioavailability averaged 62% (range 30–98%). In addition, the charcoal and cholestyramine were given 60 min after acetaminophen ingestion. In these cases, the bioavailabilities (determined for the 60–120 min period) averaged 23 and 16% less for charcoal and cholestyramine, respectively, as compared to the control values. The adsorbents were, therefore, only about one-quarter to one-third as effective when their administration was delayed an hour.

Similarly, studies by Levy and Houston (1976) showed that when acetaminophen was given in elixir form, with 10 g charcoal being administered immediately after, the bioavailability of this drug was reduced by 61.4%. However, when the charcoal was delayed for 30 min, the bioavailability was reduced by only 31.1%. Therefore, with acetaminophen, a delay of even 30 min results in much reduced charcoal efficacy. For a less rapidly absorbed dosage form (e.g., tablets) the loss of efficacy would probably not be as large as in the case of the elixir.

Additional in vivo experiments with acetaminophen were done using pigs by Lipscomb and Widdop (1975). The acetaminophen (10 g) was given either in the form of a suspension or tablets. One group of pigs was then given suspensions of 50 g activated charcoal in 250 mL water, while a second group served as controls. The results of several experiments showed that: (1) drug absorption from the suspension form was much faster than from the tablet form, (2) activated charcoal reduced blood drug concentrations considerably if given at once, and (3) even if delayed for one half to one 1 hour, activated charcoal still gave a very significant subsequent reduction in blood levels.

Acetaminophen pharmacokinetics were studied by Rose et al. (1991) in ten adult males who were given 5 g acetaminophen (in elixir form) and then were given 30 g

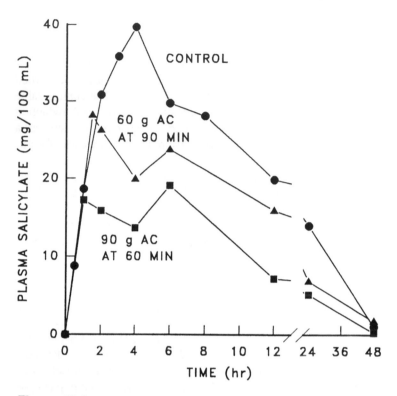

Figure 13.3 Effect of activated charcoal after administration of 4.0 g sodium salicylate to dogs. From Atkinson and Azarnoff, 1971.

charcoal administered 15 min later (in one test), 30 min later (in a second test), or 120 min later (in a third test). In a control test, the drug level peaked at 1.42 hr and the drug was 97% absorbed by a mean time of 2.05 hr. In the charcoal tests, the reductions in the recoveries of acetaminophen and its metabolites in the urine were 48, 44, and 33% for the doses given at 15, 30, and 120 min, respectively. The data taken under control conditions indicated that acetaminophen in the elixir form is fairly rapidly absorbed (absorption half-life = 0.41 hr). Hence, by the time the charcoal dose at 120 min had been given, there would have been very little drug in the GI tract and yet there was a 33% reduction in acetaminophen recovery in the urine in this case. This suggests that there was adsorption of acetaminophen to charcoal after absorption due to diffusion of the drug through the GI tract walls or due to secretion of the drug into the bile.

Mofenson and Caraccio (1992) commented on the study by Rose et al., wondering why they did not use the solid form of acetaminophen, which is the form most commonly encountered in overdose situations. They then told about a 10-month-old boy who had ingested many 500 mg Tylenol Caplets and had a blood level of 198 µg/L 2 hr after ingestion. They gave him charcoal at 1 hr and again at 3 hr post-ingestion. The 4 hr drug level was 85 µg/L, and they interpreted this as evidence that the charcoal was effective in reducing acetaminophen absorption. Based on this, Mofenson and Caraccio stated that Rose's group should have considered giving charcoal later than 2 hr after the drug. In reply, Rose et al. (1992) stated that they simply chose to study the elixir form in order to provide useful information on treating overdoses from elixirs and that charcoal was not given beyond 2 hr because their objective was to determine the effect of charcoal given at the particular times they selected on the absorptive phase of acetaminophen.

The studies by Dordoni's group and by Levy and Houston are the most definitive in terms of human subjects, and indicate that delays of 30 and 60 min in administering charcoal cause the charcoal's effect, in reducing acetaminophen absorption, to decrease by about one-half and three-quarters, respectively.

C. Barbiturates

Chin et al. (1970) studied the effect of delayed and repeated doses of charcoal on blood barbital concentrations in dogs. The drug dose was 80 mg/kg and each charcoal dose was 400 mg/kg in suspension. Their results, shown in Table 13.3, clearly show that, since barbital is a rapidly absorbed drug, a time delay of 30 min seriously reduced the efficacy of the charcoal. The data also show that the additional doses at 12 hr and 24 hr really did not have much, if any, added effect. Perhaps additional doses given earlier (say, at 1 hr, 2 hr, etc.) would have made more of a difference.

Fiser et al. (1971), using dogs, found that charcoal given after 30 min was effective in reducing blood levels of secobarbital, phenobarbital, and glutethimide by an average of 53, 56, and 74%, respectively, over the period of 1–24 hr. They also investigated, for secobarbital only, the effect of waiting 1 hr before giving the charcoal. In this case, the charcoal had very little effect. The reason for this is that secobarbital is relatively quickly absorbed.

Andersen (1973) gave 40 mg/kg diethylbarbituric acid to pigs and then administered 2 g/kg activated charcoal, with delays of 0, 2, 4, 6, and 8 hr. Plasma drug levels versus time were followed for 24 hr. The smaller the delay time, the more effective was the charcoal in reducing drug absorption, as one would expect. However, AUC values are not given by Andersen, so no quantitative measures of the delay effects can be stated. The data suggest that the drug was fairly rapidly absorbed, and that reductions in serum drug concentrations due to charcoal were the result of interruption of enteral cycling of the drug. Smaller delays in administering the charcoal caused the charcoal to become available for this process sooner, and thus led to larger decreases in the serum drug levels.

Table 13.3 Effect of Single and Repeated Charcoal Doses on Blood Barbital Concentrations (mg/100 mL) in Dogs

Charcoal treatment	Time after drug was given (hr)				
	1	12	24	48	72
None	9.5	8.5	8.0	5.5	4.4
30 min	9.1	7.7	7.1	5.5	4.2
1 min	2.7	4.5	3.5	3.0	2.2
1 min + 12 hr	3.8	4.9	3.7	2.6	1.6
1 min, 12 hr, 24 hr	3.4	5.0	3.9	2.2	2.3

Adapted from Chin et al. (1970). Reprinted by permission of Academic Press, Inc.

Table 13.4 Effect of Charcoal Delay for Three Drugs

Drug	AUC (% of control)	Serum half-life (hr)
Phenobarbital		
Control	100	110
+ Charcoal at 5 min	< 3	
+ Charcoal at 1 hr	53	
+ Charcoal at 10 hr etc.	51	19.8
Carbamazepine		
Control	100	32.0
+ Charcoal at 5 min	< 5	
+ Charcoal at 1 hr	59	
+ Charcoal at 10 hr etc.	73	17.6
Phenylbutazone		
Control	100	51.5
+ Charcoal at 5 min	2	
+ Charcoal at 1 hr	70	
+ Charcoal at 10 hr etc.	95	36.7

From Neuvonen and Elonen (1980). Table 1. Copyright Springer-Verlag GmbH & Co. Used with permission.

Lipscomb and Widdop (1975) carried out in vivo work in which pigs were given an amobarbital suspension, without charcoal or with charcoal 30 min later, and pigs which were given amobarbital capsules, without charcoal or with charcoal 1 hr or 4 hr later. The control pigs showed the common clinical response of a lapse into coma, a gradual regaining of consciousness which increased intestinal activity (and, therefore, speeded up drug absorption), and subsequent relapsing into coma again. The pigs receiving charcoal, even after 4 hr, did not suffer such relapses.

Neuvonen and Elonen (1980) studied the effects of charcoal on the absorption of phenobarbital, carbamazepine, and phenylbutazone. These three drugs were given together orally (200, 400, and 200 mg, respectively) to five healthy subjects, in a randomized cross-over study. The subjects were then given 50 g charcoal after 5 min, or 50 g charcoal after 1 hr, or 50 g charcoal after 10 hr and then 17 g charcoal at 14, 24, 36, and 48 hr. The AUC values and serum half-lives of the drugs are given in Table 13.4.

It can be seen that charcoal prevented the absorption of all three drugs very effectively, when the charcoal was given at 5 min. Even when started at 10 hr, the charcoal, while much less effective in terms of AUC reduction, decreased serum half-lives substantially (82, 45, and 29%, respectively). The charcoal was most effective for the phenobarbital (even with a 10 hr delay, charcoal reduced phenobarbital absorption by nearly half of the control amount).

Gillespie et al. (1986) applied the theory of linear systems to pharmacokinetic data on phenobarbital elimination obtained with three healthy subjects. Treatments included charcoal in a sorbitol vehicle (Charcoaid) and charcoal in plain water. The drug dose was 200 mg/70 kg body weight and the charcoal doses were 30 g at 1 hr, and 15 g at 7, 13, 19, 25, and 37 hr. The analysis, which is outside of the scope of the present work, indicated that the charcoal reduced phenobarbital absorption by 25–53% in the three subjects. Thus, even though the first charcoal dose was delayed for 1 hr, a significant effect still occurred.

Summarizing charcoal delay studies with barbiturates is difficult, as several types of barbiturates were employed and several kinds of subjects (dogs, pigs, humans) were used. The results obtained seem to be quite diverse. For example, with secobarbital, charcoal delayed for 1 hr was ineffective in dogs, whereas for phenobarbital a 10 hr delay reduced the efficacy of charcoal only by half in human subjects. The key factor appears to be whether the barbiturate involved was rapidly or slowly absorbed.

D. Tricyclic Antidepressants

Alván (1973) has described an in vivo study in which six healthy volunteers were given nortriptyline in doses of 0.86–1.00 mg/kg on two separate occasions. On one occasion, the drug was followed 30 min later with 5 g charcoal in a commercially available suspension (ACO, Ltd.). Blood samples were taken at 0, 2, 4, 12, 24, 32, 48, 56, and 84 hr in both tests. The charcoal reduced the 0–84 hr plasma AUC values for the six subjects by 0, 5, 19, 27, 50 and 72%, a rather large variation in effect. Peak plasma concentration reductions ranged from about 4 to 69% but the times corresponding to the peak concentrations were unchanged in five out of the six cases. Drug half-lives were generally lengthened by amounts ranging from 0 to about 29%. While these results are highly variable in nature, it does appear that, on average, the use of charcoal can significantly reduce nortriptyline levels, even after 30 min delay.

Delayed charcoal studies with a tricyclic antidepressant (nortriptyline) in vivo with humans have also been reported by Crome et al. (1977). They gave 75 mg nortriptyline plus a glass of water to healthy volunteers, followed 30 min later by 5 g activated charcoal contained in a 10-g packet of an effervescent charcoal preparation (Medicoal) along with 400 mL more water. Plasma nortriptyline concentrations versus time for control and charcoal-treated subjects showed that the charcoal caused average reductions in the peak plasma drug concentrations of 59% (range 30–81%) and average reductions in the plasma 0–48 hr AUC values of 59% (range 32–85%).

In another related study, by Dawling et al. (1978), adults were given 75 mg nortriptyline, followed by a single 15-g packet of Medicoal at 30 min, 2 hr, or 4 hr. Figure 13.4 shows the plasma drug levels versus time for these protocols. Drug bioavailabilities were reduced by 74, 38, and 13%, while peak plasma concentrations fell by 77, 37, and 19%, respectively. Certainly these results show a substantial favorable effect of charcoal on a representative tricyclic antidepressant in humans, even with delays of 0.5–4 hr.

Hedges et al. (1987) performed a study in which nine patients who had experienced an overdose of amitriptyline were given charcoal doses ranging from 25–75 g at times ranging from 45–250 min after presentation. They correlated the drug half-life values (range of 1.6–14.3 hr) to the time between presentation and the charcoal dose, t_c, with the equation

$$t_{1/2} = 2.68 + 0.47t_c$$

Thus, a delay in giving charcoal resulted in an increase in the drug half-life. An inverse correlation between the charcoal dose and $t_{1/2}$ was also noted. Nortriptyline, the major metabolite of amitriptyline, decreased in two of three patients who received 50 g charcoal or greater within 60 min of presentation. These findings suggest that charcoal is effective for amitriptyline overdose.

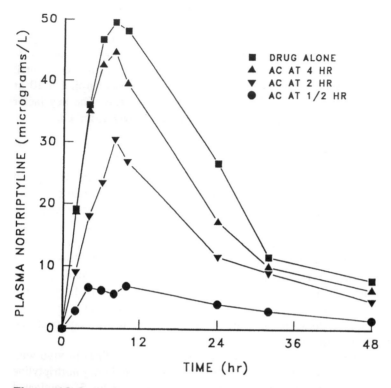

Figure 13.4 Plasma nortriptyline levels in one subject following the drug alone and the drug plus charcoal at 30 min, charcoal at 2 hr, or charcoal at 4 hr. From Dawling et al. (1978). Figure 1. Copyright Springer-Verlag GmbH & Co. Used with permission.

The effects of single and repeated doses of charcoal on the pharmacokinetics of doxepin in man have been studied by Scheinen et al. (1985). Doxepin is a commonly used tricyclic antidepressant, whose main metabolite is desmethyldoxepin. Eight healthy volunteers were each given two 25-mg tablets of doxepin hydrochloride, plus 15 g charcoal (Medicoal) mixed with 200 mL water at 30 min. The peak serum doxepin levels were reduced by 70% and the 0–48 hr doxepin AUC values by 49%, compared to control values. When the single charcoal dose was given 3 hr after the drug, absorption was not significantly reduced.

Since tricyclic antidepressants reduce gastric motility, and hence the rate of drug absorption, it is no surprise that the studies just discussed indicate that delaying charcoal by 30 min (or, in some cases, somewhat longer) does not prevent it from being significantly effective.

E. Cardiac Glycosides

We have already mentioned a study involving digoxin, as one of three drugs investigated by Neuvonen et al. (1978). This study was discussed in Section I.A. The mean absorption of digoxin was only 2% of the control value when charcoal was given immediately but was 60% of the control value when charcoal was delayed for 1 hr.

Thus, a delay of 1 hr had a significant negative effect, indicating that digoxin is reasonably rapidly absorbed.

F. Theophylline

Sintek et al. (1979) studied theophylline adsorption by activated charcoal in vivo. These investigators gave adult men and women 500–600 mg of theophylline plus 30 g activated charcoal in slurry form 30 min later. Absorption "then appeared to stop abruptly." By comparison to control tests (no charcoal given) it was found that the charcoal decreased theophylline absorption by an average of 59%. However, the decreases ranged from 26 to 96% for the five subjects, probably because of a great variability in the amount of drug absorbed in the 30 min prior to the charcoal administration. On balance, it is clear that the dose of charcoal delivered (30 g) was quite effective, even after a delay of 30 min.

Lim et al. (1986) studied the effect of oral charcoal on the first 12 hr of absorption and elimination of sustained-release theophylline in 20 normal children, ages 8 to 18 years. The drug dose was 10 mg/kg. After receiving the drug, the children were given from one to four doses of 1 g/kg charcoal (InstaChar, a charcoal suspension in water) with 30 mL 70% w/v sorbitol. Group 1 was given a single dose at 1 hr; group 2 received four doses at 1, 4, 7, and 10 hr; group 3 was administered four doses at 3, 6, 9, and 12 hr; and group 4 received three doses at 6, 9, and 12 hr. Blood samples were drawn at 0, 2, 4, 6, 8, 10, 12, and 24 hr. Reductions of the 0–12 hr AUC values, compared to control values, for the four groups were: 61, 68, 37, and 18%, respectively. The results indicate that charcoal can inhibit theophylline absorption significantly even if it is delayed for 1 hr, and can enhance the elimination of previously absorbed theophylline by interruption of its enteral cycle.

These two studies suggest that charcoal delayed for 30 min to 1 hr can significantly reduce the initial absorption of theophylline.

G. Propoxyphene

Glab et al. (1982) showed that the mortality of rats given propoxyphene (i.e., dextropropoxyphene) followed by charcoal 30 min later was significantly reduced. They gave propoxyphene hydrochloride (350 mg/kg) or propoxyphene napsylate (825 mg/kg) to rats orally and then administered charcoal equal to 10 times the drug dose 30 min later. As shown in Table 13.5, the charcoal reduced rat mortality by approximately 50%.

Kärkkäinen and Neuvonen (1985) performed a study in which six human subjects were given 130 mg dextropropoxyphene (DPP) followed 5 min later by 50 g charcoal. DPP absorption was reduced by 93–94% (based on 0–72 hr AUC and 0–∞ AUC values, respectively). In other tests, the charcoal was given as one 50-g dose at 6 hr, followed by 12.5 g at six hour intervals until a total amount of 150 g had been attained at 54 hr. In this case, DPP absorption was not significantly reduced (based on the AUC results) but the half-life of DPP did decrease from 31.1 to 21.2 hr. This suggests that DPP undergoes some enterohepatic or enteroenteric circulation. Overall, these studies suggest that charcoal given after 30 min still has a significant effect in reducing DPP absorption but if the charcoal is delayed for a long time (6 hr) it has no effect. What might occur for intermediate delay times is unknown.

Table 13.5 Times of Death and Mortality for Rats Given Propoxy-phene Hydrochloride (P-HCl) or Propoxyphene Napsylate (P-N) With or Without Activated Charcoal (AC)

	Time of death (hr)				
	0–1	1–2	2–4	4–8	Mortality
P-HCl	15	0	2	2	19/30 (63%)
P-HCl + AC	7	0	2	0	9/30 (30%)
P-N	0	1	3	6	10/30 (33%)
P-N + AC	1	0	1	3	5/30 (17%)

Numbers under "Time of death" are the numbers of rats which died in each time interval.
From Glab et al. (1982).

H.   Other Drugs

1.   Amiodarone

Kivistö and Neuvonen (1991) gave human volunteers oral doses of 400 mg amiodarone plus 25 g charcoal immediately or 25 g charcoal at 1.5 hr. Compared to controls, immediate charcoal reduced the drug absorption by about 98% and charcoal given at 1.5 hr reduced absorption by about 50%.

2.   Chlorpropamide

Neuvonen and Kärkkäinen (1983) studied the effects of charcoal on chlorpropamide kinetics in man. Six healthy subjects were given 250 mg chlorpropamide orally and then a single 50-g dose of charcoal 5 min later, or 50 g charcoal 6 hr after the drug followed by 12.5 g at six hour intervals until a total of 150 g of charcoal had been given. The single dose of charcoal reduced the 0–72 hr AUC value for chlorpropamide by 90% but for the case of repeated charcoal doses begun at 6 hr the AUC was not significantly different from the control value. Thus, a delay of 6 hr was obviously too long to permit the charcoal to be effective for this drug, and also indicates a lack of enteric cycling of the drug.

3.   Diphenhydramine

The adsorption of the antihistamine diphenhydramine (DPH) was assessed in six healthy volunteers by Guay et al. (1984). A three-way cross-over trial in humans was done with 50 mg DPH alone, with 50 g charcoal 5 min later, or with 50 g charcoal 60 min later. These two charcoal protocols gave reductions in peak serum DPH of 94.8 and 12.3%, AUC (0–24 hr) reductions of 96.9 and 20.4%, and AUC (0–∞) reductions of 90+ and 24.0%, respectively. This drug decreases GI motility due to its anticholinergic properties. Thus, charcoal, while obviously highly effective if given almost immediately, was also significantly effective if given after a 1 hr delay.

4.   Diphenoxylate

Sanvordeker and Dajani (1975) studied the in vivo absorption of diphenoxylate hydrochloride in mice. They found that the antipropulsive activity of the drug was strongly inhibited by activated charcoal given 30 min after the drug. In fact, six times as much drug was needed to yield the same gastrointestinal transit time when charcoal was used, as was obtained without charcoal.

5. Ethanol

Hultén et al. (1985) conducted a randomized cross-over study in which eight healthy men drank 88 g ethanol (260 mL vodka) and then, 30 min later, took either 20 g charcoal in water or just plain water (same volume). There were no significant differences in blood ethanol concentrations with or without the charcoal. The authors were trying to answer the question of whether charcoal would adsorb ethanol significantly if given after 30 min, by which time little of the ethanol was probably left in the stomach (additionally, ethanol adsorbs insignificantly to charcoal anyway). The obvious answer was confirmed.

6. Flecainide

Nitsch et al. (1987) studied the effects of oral charcoal on the absorption and elimination of flecainide in eight male subjects. Each was given a single oral dose (200 mg) of flecainide without charcoal. Plasma flecainide concentrations were 266, 232, and 231 µg/L at 2, 4, and 6 hr after drug ingestion. When 30 g charcoal was given concurrently with the drug, plasma flecainide levels were unmeasurably low at the same times. In other tests, the charcoal was given 90 min after the drug. In this case, plasma flecainide levels were 231, 199, and 155 µg/L at 2, 4, and 6 hr (86.8, 85.8, and 67.1% of the values observed without charcoal). Thus, there was some effect of charcoal even when it was given after 90 min delay. Since flecainide apparently does not undergo enteroenteric or enterohepatic circulation, charcoal should be given as early as possible.

7. Furosemide

Kivistö and Neuvonen (1990) showed that the diuretic effect of furosemide can be counteracted by orally administered charcoal. Healthy subjects (61) were given 40 mg furosemide with water, water only, or 40 mg furosemide and 8 g activated charcoal immediately or at different time intervals. The diuretic effect of the drug was totally prevented when the charcoal was given immediately but became stronger as the charcoal delay time increased. The average urine volumes over 0–3 hr were: 390 mL (no drug or charcoal), 340 mL (drug and immediate charcoal), and 1030 mL (drug only). When charcoal was delayed, the volumes increased until, at a delay of 1 hr, no effect of charcoal occurred.

8. Isoniazid

Scolding et al. (1986) found that a 10-g dose of charcoal given 1 hr after a 600 mg oral dose of isoniazid to six healthy human subjects did not reduce the areas under plasma concentration versus time curves (evaluated over 0–8 hr) significantly. Siefkin et al. (1987) also studied the effects of charcoal on isoniazid pharmacokinetics. Three subjects were given 10 mg/kg isoniazid orally, followed by 60 g charcoal immediately after. The half-lives for elimination were 120, 178, and 193 min for the three subjects without charcoal. With charcoal, no measurable serum isoniazid levels were detected, so no half-lives could be determined. Thus, in contrast to the study of Scolding et al. (1986), which showed no effect of charcoal given 1 hr after isoniazid, this study showed that if the charcoal is given immediately, isoniazid absorption is reduced to zero. This points out the need for prompt charcoal administration when a rapidly absorbed drug is involved.

9.   Mefenamic Acid

El-Bahie et al. (1985) gave 500 mg mefenamic acid orally to nine healthy subjects. On one occasion, 2.5 g charcoal was administered 1 hr after the drug. Plasma drug levels were followed for up to 8 hr after drug ingestion. The charcoal reduced the 0–8 hr AUC value, compared to control trials, by 36%. Thus, charcoal was effective even after a 1 hr delay.

10.   Nadolol

Du Souich et al. (1983) performed studies aimed at determining whether nadolol undergoes enteroenteric or enterohepatic circulation. They gave 80 mg nadolol to eight healthy subjects orally, and in one phase followed this by giving charcoal tablets orally according to the schedule: 500 mg at 3 hr, 500 mg at 4 hr, and 250 mg each hour for the next 8 hr (total charcoal 3000 mg). Plasma drug levels were determined up to 48 hr, and 0–∞ AUC values were estimated by an extrapolation procedure. The charcoal reduced the AUC for nadolol from 2455 to 1355 μg-hr/L (a 45% decrease) and the half-life from 17.3 to 11.8 hr (a 32% decrease). These results suggest that enteroenteric or enterohepatic circulation occurs with this drug.

11.   Phencyclidine

Lyddane et al. (1988) studied the effect of charcoal in treating phencyclidine (PCP) overdose. Rats were gavaged once with PCP (10, 25, or 50 mg/kg in 2 mL/kg water) and then were given charcoal (1 g/kg) orally once per hour, starting 45 min after drug administration. AUC values were computed from 45 min "until the time of recovery" from the PCP effects. Charcoal treatment reduced the AUC values, compared to the average control AUC value, by 30% in the 10 mg/kg PCP group, by 4% in the 25 mg/kg PCP group, and not at all in the 50 mg/kg PCP group. Thus, charcoal is effective for PCP intoxication if given in a large enough amount. Obviously, had the first dose of charcoal been given immediately, it probably would have had a greater effect.

12.   Phenytoin

Neuvonen et al. (1978) showed that when adults were given 0.5 g phenytoin followed by 50 g Norit A charcoal, the systemic absorption of the drug was reduced (relative to control tests) by 98–99% if the charcoal was given immediately and by 80% even if the charcoal was delayed for 1 hr. The great effectiveness of charcoal even after an hour's delay undoubtedly stems from the reduction in gastric motility that phenytoin causes, plus the fact that phenytoin is only sparingly soluble and thus dissolves slowly.

13.   Piroxicam

Studies on piroxicam were done by Ferry et al. (1990). They gave 20-mg oral doses of the drug to eight healthy young adults. Then 5 g charcoal (Medicoal) was given starting 24 hr after the drug dose and further 5-g doses were given 3–4 times/day thereafter for several days. The charcoal therapy reduced the piroxicam half-life from 53.1 to 40.0 hr. In other tests, cholestyramine was given in 4-g amounts according to the same schedule as for the charcoal and the drug half-life was decreased even more to 29.6 hr. Again, even though sorbent therapy was delayed for 24 hr, it was effective. This proves that enteroenteric or enterohepatic circulation occurs for this drug.

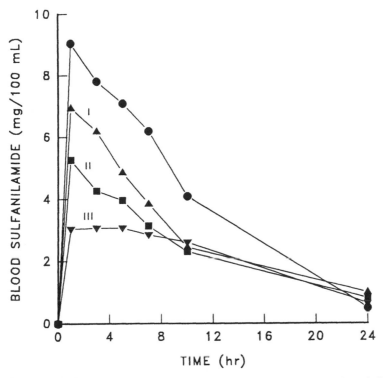

Figure 13.5 Sulfanilamide concentrations in the blood of a dog after administration of 2 g of the drug (upper curve) and 2 g of the drug with 8 g activated charcoal at 60 min (I), 45 min (II), or at 15 min (III). From Andersen (1948a). Reprinted by permission of Munksgaard International Publ. Ltd.

14. Quinine

Repeated-dose charcoal therapy was considered and tried by Lockey and Bateman (1989) in a study of quinine elimination using subtoxic doses in seven healthy subjects. They were each given 600 mg quinine bisulfate, then 50 g charcoal in an aqueous slurry 4 hr after the drug, and three further 50-g doses over the next 12 hr. Carbomix and Medicoal were used alternately. Blood samples were taken over 0–36 hr. Based on AUC values determined by extrapolating the blood data to infinite time, the charcoal lowered the drug absorption 31.5% and shortened the half-life from 8.23 to 4.55 hr (a 44.7% decrease). Had the charcoal not been delayed for 4 hr, even larger effects might have been noted. Since there are no experimental data to suggest that quinine undergoes enterohepatic circulation, the charcoal probably works by binding quinine which diffuses through the gut lumen.

15. Sulfanilamide

One of the first studies in which the effect of delaying charcoal was investigated is that of Andersen (1948a). He gave 2 g sulfanilamide to a dog, followed by 8 g activated charcoal with a delay time of 15, 45, or 60 min. Figure 13.5 shows blood sulfanilamide concentrations versus time. These results show clearly that charcoal given sooner has positive benefits, particularly over the first 10 hr.

II. EFFECT OF FOOD

Considering all of the in vivo studies (several hundred) which researchers have conducted over the years dealing with the effects of charcoal and other adsorbents on drug/toxin absorption, it is astonishing that all but about three such studies have been done without food as a complicating factor (i.e., done under fasting conditions). This is all the more surprising because medical personnel are fully aware that a large fraction of patients presenting to EDs possess significant food in their stomachs (a typical patient does not usually fast for a significant time period prior to ingesting an overdose of drugs or accidentally ingesting a poison/toxin). In particular, patients suffering from ethanol toxicity and, very commonly, toxicity due to combinations of ethanol and various other drugs, are especially likely to have significant stomach contents.

But obviously the amount and type of food in the stomach (and upper small intestine as well), and how long it has been there, are factors that must be taken into account. It is well known that the type of food eaten has a large effect on gastric emptying. Fats can delay the emptying of a fatty meal for as long as three to six hours; proteins have an intermediate effect, and carbohydrates have a lesser effect.

On one hand, it can be expected that food would reduce the effectiveness of charcoal because (1) it might prevent good contact between the drug and the charcoal (i.e., the food might constitute a physical barrier separating the two), and (2) food decomposition products might compete with the drug for adsorption sites on the charcoal. On the other hand, the presence of food would undoubtedly slow down the rate of drug absorption into a person's system. One way it would do this is by providing competition for absorption through the GI tract walls (the food components and the drug would be absorbing simultaneously, and therefore competing for such absorption). Another way in which food would slow drug absorption results from the above-mentioned effect of food on the rate of gastric emptying (this would give the charcoal more time in which to contact and adsorb the drug in the stomach). How these competing effects balance will depend on the individual, the type and amount of food eaten, the amounts of drug and charcoal ingested, the amount of time between food and drug ingestion, the amount of time between drug ingestion and charcoal administration, the chemical characteristics of the drug (whether it is a quickly or slowly absorbed drug), and, most importantly, the effect of the drug on gastric motility and emptying. The net result could easily be either an increased or decreased drug bioavailability. Many more studies than the few that have been done so far will be needed to define the role of food/drug/charcoal interactions.

Levy and Tsuchiya (1972) showed that the percentage of a 1 g aspirin dose (given dissolved in 200 mL water) recovered in the urine of human subjects, when 10 g activated charcoal was given immediately after the aspirin, was 62.6% for fasted subjects and 75.9% for subjects who had eaten a standard breakfast (2 ounces of corn flakes with sugar and 500 mL milk) 15 min before the test. Control subjects gave 99.7% aspirin recovery in the urine (in all cases, urine samples were taken until no salicylate could be detected—usually 36–48 hr). Therefore, the action of the charcoal was inhibited by the presence of food in this study.

Atkinson and Azarnoff (1971), however, claim to have observed the opposite effect of food in salicylate-dosed dogs. The dogs were given 4 g sodium salicylate, followed 45 min later by 30 g attapulgite (clay), or by 30, 60, or 90 g charcoal. In one case, one dog was allowed to eat 16 oz of canned food 60 min prior to receiving the test drug. The grams of salicylate excreted in the urine over 48 hr are shown in Table 13.6.

Table 13.6 Effect of Food on Salicylate Absorption in Dogs

Treatment	Urinary salicylate (0–48 hr)
None	2.45
Attapulgite, 30 g	2.5
Charcoal, 30 g	1.5
Charcoal, 60 g	1.2
Charcoal, 90 g	1.1
Charcoal, 90 g, plus food	1.0

From Atkinson and Azarnoff (1971).

Apart from the obvious ineffectiveness of the clay and the significant effectiveness of charcoal (which increases somewhat with the amount given), these results show that food 60 min prior to the drug had no significant effect on the action of the charcoal (the claim by the authors that food had an effect must have been based on the difference between the values 1.1 and 1.0, which is clearly not significant). However, the dogs involved were all fasted overnight, so it is quite possible that a dog ingesting food 60 min prior to the drug would have had virtually all of the food passed out of the stomach and into the small intestine by the time the drug was given, thus leaving the stomach as empty as in the "no food" case. Thus, this study is not a strong test of the effect of food.

Perhaps the finest study of the effect of gastrointestinal contents on the action of charcoal in reducing drug absorption was that described in the third paper in Andersen's classic series (Andersen, 1948b). This study was described in considerable detail in Chapter 10 and we refer the reader back to that chapter for details; we only summarize the main features of the study here. The study dealt with the matter of whether substances ordinarily found in the gastrointestinal tract (food, partly split derivatives of food, digestive enzymes, and various secretions) might inhibit the adsorption of drugs or poisons by charcoal in vivo. To test this, Andersen carried out experiments on the adsorption of strychnine nitrate, mercuric chloride, and diethylbarbituric acid to Carbo medicinalis Merck charcoal in actual gastric contents. The gastric contents were obtained by withdrawing standard test meals from 15 patients, pooling them, and mixing them. The final mixture contained 60% solids and had a pH of 1.50.

Various solutions of the three test substances were added to equal portions of the gastric fluid and contacted with 1 g charcoal (with shaking) for 1 hr at room temperature. After filtration, the fluids were assayed for the residual drug or poison. The adsorption isotherms for the three compounds in the gastric fluid environments were nearly the same in shape as those determined in distilled water solutions but were lower in magnitude by about 50%.

Andersen then explored whether such decreases were actually due to the presence of the gastric material and not because the gastric test pH values (which were all about 1.95, after the gastric contents were combined with the drug solutions) were different from the pH values of the distilled water solutions (which were 4.91 for strychnine nitrate, 3.42 for mercuric chloride, and 5.00 for diethylbarbituric acid). He therefore studied the adsorption of strychnine nitrate from three test media which were prepared by combining equal volumes of a solution of strychnine nitrate in distilled water with gastric contents but with the final pH values of the three mixtures adjusted to about 3, 4, and 5 using different amounts of NaOH. By plotting the gastric media test results for pH 1.95, 3, 4, and 5, it could be seen that a change in pH from 4.91 to 1.95 alone caused the amount of strychnine nitrate adsorbed to decrease significantly. When one subtracts the decrease caused solely by a pH change from 4.91 to 1.95 from the difference between the strychnine nitrate isotherm for the pH 4.91 distilled water medium and the strychnine nitrate isotherm for the pH 1.95 gastric medium (as shown in Chapter 10), it turns out that the decrease in adsorption due to gastric contents alone was actually 22%.

As mentioned previously, the isotherms for mercuric chloride and diethylbarbituric acid determined in both distilled water and in gastric media also showed about a 50% spread (the distilled water isotherms lay above those obtained in the gastric media by about a factor of 2). From data given in an earlier paper by Andersen (1947), one can easily estimate that a reduction in pH from 3.42 to 1.95 would decrease mercuric chloride adsorption by 11%. Thus, of the 50% difference in adsorption shown by the mercuric chloride isotherms, 11% can be ascribed to the pH difference and the other 39% must be due to the presence of gastric contents. Similarly, for diethylbarbituric acid, other data in Andersen's earlier paper (1947) suggest that a pH change from 5.00 to 1.95 would cause virtually no change in adsorption, because the diethylbarbituric acid remains almost totally undissociated over this pH range. Thus, the difference in the isotherms for diethylbarbituric acid (55%) is due essentially entirely to the presence of gastric contents.

To recapitulate Andersen's gastric media results, we can state that the reduction in adsorption due to the presence of gastric contents, after accounting for effects due solely to pH changes, was fairly modest (22%) for strychnine nitrate, much stronger (39%) for mercuric chloride, and even stronger (55%) for diethylbarbituric acid. Why the effect varies with the type of compound is not clear.

Other experiments done by Andersen and reported in the same paper (Andersen, 1948b) involved in vitro drug adsorption from actual bile; these studies showed that bile reduced diethylbarbituric acid adsorption by 36%. Additional experiments done with duodenal juice showed very little interference with adsorption.

The effect of food on the antidotal efficacy of charcoal was studied by Olkkola and Neuvonen (1984a). Six healthy subjects were given 100 mg aspirin, 200 mg mexiletine, and 400 mg tolfenamic acid either on an empty stomach or after a standard meal (meat balls weighing about 150 g and one roll with cheese). Then, 25 g charcoal in water was given either 5 or 60 min later. The 0–48 hr AUC values determined in this study are presented in Table 13.7.

The first thing to notice about these data is that the degrees of drug absorption without charcoal were 6, 38, and 60% greater for the three drugs (aspirin, mexiletine, and tolfenamic acid, respectively) when food was given than when it was not. Secondly, comparing just the "percent of control" values for charcoal given at 5 min, it is seen that food caused the drug absorption values to be higher (44 vs 23%, 5 vs 4%, and 54 vs. 12%). Thus, the food interfered with the charcoal. Thirdly, comparing the "percent of control" values for charcoal given at 60 min, it is seen that food caused greater drug absorption for the tolfenamic acid (52 vs 38%) but less drug absorption for aspirin (51 vs 75%) and for mexiletine (17 vs 94%). Thus, food had no consistent effect on charcoal given at 60 min.

Table 13.7 AUC Values (mg-hr/L) With and Without Food

Aspirin/no food (control)	940
Aspirin/food (control)	995
Aspirin/no food/AC at 5 min	218 (23.2% of control)
Aspirin/food/AC at 5 min	438 (44.0% of control)
Aspirin/no food/AC at 60 min	701 (74.6% of control)
Aspirin/food/AC at 60 min	503 (50.6% of control)
Mexiletine/no food (control)	2.82
Mexiletine/food (control)	3.88
Mexiletine/no food/AC at 5 min	0.10 (3.5% of control)
Mexiletine/food/AC at 5 min	0.20 (5.2% of control)
Mexiletine/no food/AC at 60 min	2.66 (94.3% of control)
Mexiletine/food/AC at 60 min	0.64 (16.5% of control)
Tolfenamic/no food (control)	18.6
Tolfenamic/food (control)	29.8
Tolfenamic/no food/AC at 5 min	2.29 (12.3% of control)
Tolfenamic/food/AC at 5 min	16.1 (54.0% of control)
Tolfenamic/no food/AC at 60 min	7.02 (37.7% of control)
Tolfenamic/food/AC at 60 min	15.4 (51.7% of control)

From Olkkola and Neuvonen (1984a). Reprinted by permission of Blackwell Scientific Publications Ltd.

Overall, the data suggest that food itself causes greater drug absorption in the absence of charcoal, interferes with the effect of charcoal when the charcoal is given soon after the drug, and has significant but inconsistent effects for charcoal given 1 hr later. It appears that the type and amount of drug involved may be very important in determining the effect of food (as the inconsistent nature of the "60 min AC" results show). This underscores the need for many more studies of food/charcoal interactions with many more test drugs.

III. EFFECT OF GASTRIC pH

Olkkola and Neuvonen (1984b) noted from in vitro experiments with aspirin and disopyramide that, for charcoal:drug ratios in the range of 5:1 to 12.5:1, the fraction of aspirin absorbed changed from 98.1–99.6% at pH 1.2 to 57–83% at pH 7. For disopyramide, the fraction absorbed varied between 62–83% at pH 1.2 and between 88–98.7% at pH 7. Thus, the acidic drug (aspirin) adsorbed much more strongly at low pH, where it is nonionized, and the basic drug (disopyramide) adsorbed much more strongly at the higher of the two pH's, at which value it also is largely nonionized (this pH effect has been mentioned often in this book, especially in Chapter 10, where the results of Andersen's classic 1947 study on the pH effect were described in great detail).

Based on this, Olkkola and Neuvonen wondered if such a pH effect would occur in vivo. Thus, six subjects were given 500 mg aspirin and 200 mg disopyramide, plus 200 mg tolfenamic acid (this drug could not be studied in vitro at pH 1.2 since it was

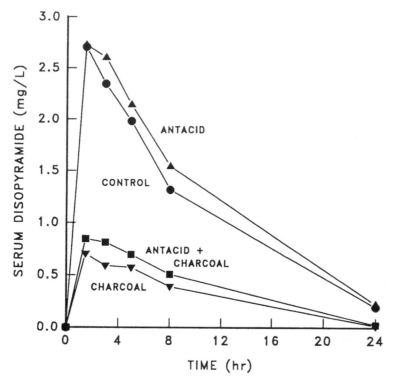

Figure 13.6 Effect of 2.5 g activated charcoal and 20 mL of 8.5% magnesium hydroxide, given alone or in combination, on the absorption of disopyramide. From Olkkola and Neuvonen (1984). Reprinted by permission of Dustri-Verlag Dr. Karl Feistle, Deisenhofen, Germany.

essentially insoluble at this pH). A small dose of 2.5 g charcoal given immediately after the drugs reduced the absorption of aspirin by 30–40% and the absorption of the other two drugs by 70–80%. When an antacid (20 mL of 8.5% magnesium hydroxide) was given with the charcoal, no significant added effects were noted. Figure 13.6 gives results for one of the drugs (disopyramide), and it is clear that the effects of magnesium hydroxide are not large. It was concluded that changing the gastric pH did not change the effectiveness of charcoal in vivo. However, gastric pH was never determined in the study; thus, it is unknown whether one 20 mL dose of antacid changed the gastric pH and, if so, for how long. Drug absorption was evaluated based on 0–48 AUC data; hence if the antacid had only a short-term effect on gastric pH, it would not be expected to affect data taken over a period as long as 48 hr.

IV. SUMMARY

The effect of a delay in giving charcoal on the absorption of a drug depends on many factors. For the moment, we will omit the effect of food, since most studies in which the effects of delayed charcoal were assessed involved fasted subjects. The dosage form (suspensions, capsules, tablets) of a drug can be very important, as it can determine how fast the drug dissolves. In particular, prolonged-release dosage forms obviously slow down the rate of drug absorption, for that is their purpose. The physicochemical properties of a drug are important, especially its solubility in gastric and intestinal fluids and its pK_a value (which determines if it is ionized at the prevailing pH values in the stomach and intestines). These factors were mentioned before in Chapter 5. Another drug property which is crucial is whether the drug has anticholinergic properties and thus suppresses movement of material down the GI tract. Drugs with such action tend to be absorbed slowly. Charcoal can often "catch up with" such drugs.

When one is using multiple-dose charcoal therapy (see Chapter 14), the important consideration is whether the drug undergoes enteroenteric and/or enterohepatic circulation. If it does, then charcoal given in regular doses, even if started well after the drug ingestion, can be very effective. However, the issue we wish to focus on here is whether a single charcoal dose can be effective if given well after a drug has been ingested.

The in vivo studies reviewed in this chapter show, in general, that with most drugs a delay of 30 min to 1 hr in giving a single dose of charcoal does decrease the effectiveness of the charcoal somewhat but that the effectiveness is still substantial (often 1/4, 1/3, or 1/2 as effective as charcoal given immediately). When the delay time is on the order of 2–3 hr or longer, the charcoal is often ineffective. However, with aspirin, certain barbiturates, tricyclic antidepressants, and quinine, for example, delays of even 4 hr still allow the charcoal to be significantly effective. For rapidly absorbed drugs, like isoniazid and furosemide, a 1 hr delay normally makes the charcoal useless.

Regarding the effect of food, one study (Levy and Tsuchiya, 1972) found that food definitely interfered with the action of charcoal in suppressing aspirin absorption. Anderson found in a series of experiments that gastric contents reduced the in vitro adsorption of three different drugs by 22, 39, and 55%. However, the study of Olkkola and Neuvonen (1984a) suggested that food itself causes greater drug absorption in the absence of charcoal, interferes with the effect of charcoal when the charcoal is given soon after the drug, and has significant but inconsistent effects for charcoal given 1 hr later. Many more studies of food/charcoal interactions are needed.

V. REFERENCES

Alván, G. (1973). Effect of activated charcoal on plasma levels of nortriptyline after single doses in man, Eur. J. Clin. Pharmacol. *5*, 236.

Andersen, A. H. (1947). Experimental studies on the pharmacology of activated charcoal. II. The effect of pH on the adsorption by charcoal from aqueous solutions, Acta Pharmacol. Toxicol. *3*, 199.

Andersen, A. H. (1948a). Experimental studies on the pharmacology of activated charcoal. V. Adsorption of sulphanilamide in vivo, Acta Pharmacol. Toxicol. *4*, 389.

Andersen, A. H. (1948b). Experimental studies on the pharmacology of activated charcoal. III. Adsorption from gastro-intestinal contents, Acta Pharmacol. Toxicol. *4*, 275.

Andersen, A. H. (1973). [Medicinal charcoal in the treatment of poisoning. Treatment of experimental barbiturate poisoning in pigs], Ugeskr. Laeger *135*, 797.

Atkinson, J. P. and Azarnoff, D. L. (1971). Comparison of charcoal and attapulgite as gastrointestinal sequestrants in acute drug ingestions, Clin. Toxicol. *4*, 31.

Chin, L., Picchioni, A. L., and Duplisse, B. R. (1970). The action of activated charcoal on poisons in the digestive tract, Toxicol. Appl. Pharmacol. *16*, 786.

Collombel, C., and Perrot, L. (1970). [Experimental study of the treatment of salicylate intoxication with activated carbon], Eur. J. Toxicol. *3*, 352.

Corby, D. G., Fiser, R. H., and Decker, W. J. (1970). Re-evaluation of the use of activated charcoal in the treatment of acute poisoning, Pediatr. Clin. N. Am. *17*, 545.

Crome, P., Dawling, S., Braithwaite, R. A., Masters, J., and Walkey, R. (1977). Effect of activated charcoal on absorption of nortriptyline, Lancet, December 10, p. 1203.

Dawling, S., Crome, P., and Braithwaite, R. (1978). Effect of delayed administration of activated charcoal on nortriptyline absorption, Eur. J. Clin. Pharmacol. *14*, 445.

Decker, W. J., Combs, H. F., and Corby, D. G. (1968). Adsorption of drugs and poisons by activated charcoal, Toxicol. Appl. Pharmacol. *13*, 454.

Decker, W. J., Shpall, R. A., Corby, D. G., Combs, H. F., and Payne, C. E. (1969). Inhibition of aspirin absorption by activated charcoal and apomorphine, Clin. Pharmacol. Ther. *10*, 710.

Dillon, E. C., Jr., Wilton, J. H., Barlow, J. C., and Watson, W. A. (1989). Large surface area activated charcoal and the inhibition of aspirin absorption, Ann. Emerg. Med. *18*, 547.

Dordoni, B., Willson, R. A., Thompson, R. P. H., and Williams, R. (1973). Reduction of absorption of paracetamol by activated charcoal and cholestyramine: A possible therapeutic measure, Br. Med. J. *3*, 86.

Du Souich, P., Caillé, G., and Larochelle, P. (1983). Enhancement of nadolol elimination by activated charcoal and antibiotics, Clin. Pharmacol. Ther. *33*, 585.

El-Bahie, N., Allen, E. M., Williams, J., and Routledge, P. A. (1985). The effect of activated charcoal and hyoscine butylbromide alone and in combination on the absorption of mefenamic acid, Br. J. Clin. Pharmacol. *19*, 836.

Ferry, D. G., Gazeley, L. R., Busby, W. J., Beasley, D. M. G., Edwards, I. R., and Campbell, A. J. (1990). Enhanced elimination of piroxicam by administration of activated charcoal or cholestyramine, Eur. J. Clin. Pharmacol. *39*, 599.

Fiser, R. H., Maetz, H. M., Treuting, J. J., and Decker, W. J. (1971). Activated charcoal in barbiturate and glutethimide poisoning of the dog, J. Pediatr. *78*, 1045.

Gillespie, W. R., Veng-Pedersen, P., Berg, M. J., and Schottelius, D. D. (1986). Linear systems approach to the analysis of an induced drug removal process. Phenobarbital removal by oral activated charcoal, J. Pharmacokin. Biopharm. *14*, 19.

Glab, W. N., Corby, W. G., Decker, W. J., and Coldiron, V. R. (1982). Decreased absorption of propoxyphene by activated charcoal, J. Toxicol.-Clin. Toxicol. *19*, 129.

Guay, D. R. P., Meatherall, R. C., Macauley, P. A., and Yeung, C. (1984). Activated charcoal adsorption of diphenhydramine, Int. J. Clin. Pharmacol. Ther. Toxicol. *22*, 395.

Hedges, J. R., Otten, E. J., Schroeder, T. J., and Tasset, J. J. (1987). Correlation of initial amitriptyline concentration reduction with activated charcoal therapy in overdose patients, Am. J. Emerg. Med. *5*, 48.

Hultén, B. Å., Heath, A., Mellstrand, T., and Hedner, T. (1985). Does alcohol adsorb to activated charcoal?, Hum. Toxicol. *5*, 211.

Kärkkäinen, S. and Neuvonen, P. J. (1985). Effect of oral charcoal and urine pH on dextropropoxyphene pharmacokinetics, Int. J. Clin. Pharmacol. Ther. Toxicol. *23*, 219.

Kivistö, K. T. and Neuvonen, P. J. (1990). Effect of activated charcoal on frusemide induced diuresis: A human class experiment for medical students, Br. J. Clin. Pharmacol. *30*, 496.

Kivistö, K. T. and Neuvonen, P. J. (1991). Effect of activated charcoal on the absorption of amiodarone, Hum. Exp. Toxicol. *10*, 327.

Levy, G. and Tsuchiya, T. (1969). Effect of activated charcoal on aspirin absorption in man (abstract), Pharmacologist *11*, 292.

Levy, G. and Tsuchiya, T. (1972). Effect of activated charcoal on aspirin absorption in man, Clin. Pharmacol. Ther. *13*, 317.

Levy, G. and Houston, J. B. (1976). Effect of activated charcoal on acetaminophen absorption, Pediatrics *58*, 432.

Lim, D. T., Singh, P., Nourtsis, S., and Dela Cruz, R. (1986). Absorption inhibition and enhancement of elimination of sustained-release theophylline tablets by oral activated charcoal, Ann. Emerg. Med. *15*, 1303.

Lipscomb, D. J., and Widdop, B. (1975). Studies with activated charcoal in the treatment of drug overdosage using the pig as an animal model, Arch. Toxicol. *34*, 37.

Lockey, D. and Bateman, D. N. (1989). Effect of oral activated charcoal on quinine elimination, Br. J. Clin. Pharmacol. *27*, 92.

Lyddane, J. E., Thomas, B. F., Compton, D. R., and Martin, B. R. (1988). Modification of phencyclidine intoxication and biodisposition by charcoal and other treatments, Pharmacol. Biochem. Behav. *30*, 371.

Mofenson, H. C. and Caraccio, T. R. (1992). Is activated charcoal useful for acetaminophen overdose beyond two hours after ingestion? (letter), Ann. Emerg. Med. *21*, 894.

Neuvonen, P. J., Elfving, S. M., and Elonen, E. (1978). Reduction of absorption of digoxin, phenytoin, and aspirin by activated charcoal in man, Eur. J. Clin. Pharmacol. *13*, 213.

Neuvonen, P. J. and Elonen, E. (1980). Effect of activated charcoal on absorption and elimination of phenobarbitone, carbamazepine, and phenylbutazone in man, Eur. J. Clin. Pharmacol. *17*, 51.

Neuvonen, P. J. and Kärkkäinen, S. (1983). Effects of charcoal, sodium bicarbonate, and ammonium chloride on chlorpropamide kinetics, Clin. Pharmacol. Ther. *33*, 386.

Nitsch, J., Köhler, U., Neyses, L., and Lüderitz, B. (1987). Inhibition of flecainide absorption by activated charcoal, Am. J. Cardiol. *60*, 753.

Olkkola, K. T. and Neuvonen, P. J. (1984a). Do gastric contents modify antidotal efficacy of oral activated charcoal? Br. J. Clin. Pharmacol. *18*, 663.

Olkkola, K. T. and Neuvonen, P. J. (1984b). Effect of gastric pH on antidotal efficacy of activated charcoal in man, Int. J. Clin. Pharmacol. Ther. Toxicol. *22*, 565.

Phansalkar, S. V. and Holt, L. E., Jr. (1968). Observations on the immediate treatment of poisoning, J. Pediatr. *72*, 683.

Rose, S. R., Gorman, R. L., Oderda, G. M., Klein-Schwartz, W., and Watson, W. A. (1991). Simulated acetaminophen overdose: Pharmacokinetics and effectiveness of activated charcoal, Ann. Emerg. Med. *20*, 1064.

Rose, S. R., Watson, W. A., Oderda, G. M., Gorman, R. L., and Klein-Schwartz, W. (1992). Is activated charcoal useful for acetaminophen overdose beyond two hours after ingestion? (reply to letter), Ann. Emerg. Med. *21*, 894.

Sanvordeker, D. R. and Dajani, E. Z. (1975). In vitro adsorption of diphenoxylate hydrochloride on activated charcoal and its relation to pharmacological effects of drug in vivo, J. Pharm. Sci. *64*, 1877.

Scheinen, M., Virtanen, R., and Iisalo, E. (1985). Effect of single and repeated doses of activated charcoal on the pharmacokinetics of doxepin, Int. J. Clin. Pharmacol. Ther. Toxicol. *23*, 38.

Scolding, N., Ward, M. J., Hutchings, A., and Routledge, P. A. (1986). Charcoal and isoniazid pharmacokinetics, Hum. Toxicol. *5*, 285.

Siefkin, A. D., Albertson, T. E., and Corbett, M. G. (1987). Isoniazid overdose: Pharmacokinetics and effects of oral charcoal in treatment, Hum. Toxicol. *6*, 497.

Sintek, C., Hendeles, L., and Weinberger, M. (1979). Inhibition of theophylline absorption by activated charcoal, J. Pediatr. *94*, 314.

14

Effect of Multiple Doses of Charcoal

In this chapter, we review those studies in which "repeated-dose" or "multiple-dose" charcoal therapies were used. Most of the investigations involved are of the controlled clinical study type, using human volunteers and subtoxic doses of the drug. However, we also include here many reports of the use of multiple-dose therapy in actual overdose situations. We have excluded studies involving animals; some repeated-dose examples involving animals may be found in Chapter 20.

Many of the studies in this chapter are the same as those mentioned in Chapters 11 and 12, but with a focus on the multiple-dose aspects. Thus, in some studies where both single and multiple-dose therapies were evaluated, we will generally only concern ourselves here with the multiple-dose parts.

The structure of this chapter will be very similar to that of Chapter 11. That is, the multiple-dose studies to be discussed are grouped according to major drugs or drug classes, followed in the end by a section which mostly includes single reports on other drugs.

I. ASPIRIN

The treatment of aspirin overdose with repeated oral charcoal was examined by Hillman and Prescott (1985). They treated four patients, first with gastric lavage, then with 75 g charcoal (Medicoal) suspended in 200 mL water. Additional doses of 50 g charcoal were given every 4 hr until toxic symptoms had been relieved. Plasma salicylate concentrations were compared against those of six control patients with mild aspirin poisoning who were treated with oral fluids alone. The average aspirin half-life in the control patients was 27 hr, whereas that in the charcoal-treated patients was 3.2 hr. The use of repeated charcoal was clearly effective, due to binding of the aspirin which diffuses from the circulation into the gut lumen ("gastrointestinal dialysis"). Acidic drugs like aspirin have small volumes of distribution and tend to diffuse well into the gut lumen.

In response to this report, Boldy and Vale (1986) gave an account of treating two aspirin-overdosed patients with repeated oral charcoal. These patients were given charcoal (Carbomix) in 50-g doses. One patient received charcoal every 4 hr (total 150 g) and the other patient received charcoal every 2.5 hr (total 250 g). The elimination half-lives in these cases were 8 hr and 15.9 hr, respectively. The reason why these half-lives were larger than in the Hillman and Prescott study was probably due to higher initial aspirin blood levels and/or differences in the two charcoal formulations, according to Boldy and Vale. They also stated that, in their experience, Carbomix tends to cause constipation, whereas Medicoal tends to cause diarrhea. Thus, they subsequently alternated 50-g doses of each in later treatment protocols.

Mofenson et al. (1985) described their use of multiple-dose charcoal for treating two aspirin and three phenobarbital overdoses in adolescents (13–19 years old). Their protocol was to give either 60 or 90-g doses of charcoal along with 30 g magnesium sulfate every four hours (the total numbers of doses in the five cases were 3, 5, 5, 6, and 12). Gastric lavage was performed prior to charcoal administration and alkaline diuresis was also employed. All patients did well and were alert and oriented within 16–24 hr postingestion. Limited drug blood level data suggested considerable shortening of the drug half-lives.

Wogan et al. (1986, 1987) used multiple-dose charcoal therapy to enhance the elimination of sodium salicylate (NaSal) which had been given IV (175 mg/kg) to dogs. After drug infusion, 60 g charcoal was given orally and repeated at 1 hr. At 2, 3, 4, and 5 hr 30-g doses were given. The 0–24 serum salicylate mean AUC value was 59.4 (units unstated) versus 97.8 for the control tests (a 39.3% difference). The elimination $t_{1/2}$ fell from 10.3 hr (control) to 6.3 hr and the mean clearance rose from 28.4 to 46.0 mL/min. Thus, it appears that multiple-dose charcoal significantly enhances the elimination of pre-absorbed salicylate in dogs.

A study which was very similar to the Wogan study, except involving rabbits, is one by Douidar et al. (1992). Male rabbits were given 250 mg/kg NaSal IV and then were given 3 doses of charcoal (0.5–1.0 g/kg) orally at 0, 4, and 8 hr. Blood samples were taken over 24 hr and urine samples were collected over 12 hr. There was no difference in urinary excretion as compared to controls but the serum AUC value was reduced 26.5% by the charcoal. This result was said to be "not significant", but the p level used to test for significance was not stated. The data do suggest a moderate effect of charcoal in enhancing the excretion of already absorbed salicylate in rabbits.

An interesting report in abstract form on multiple-dose charcoal therapy for two aspirin overdose cases and one acetaminophen case was given by Augenstein et al. (1987). All three patients showed a delayed rise in serum drug levels despite multiple-dose charcoal, and after charcoal stools had been passed. All three patients were acutely overdosed, were treated with gastric emptying procedures, were given an initial charcoal dose within 6 hr of ingestion, were given further doses of charcoal, and passed multiple charcoal stools between 9–20 hr after drug ingestion. Two of the patients died and it is believed that the late serum drug level rises contributed to the fatalities. This report does not state when charcoal therapy was terminated and it must be assumed that the late rises in serum drug levels were due to further drug absorption from the GI tract after charcoal was discontinued. Thus, the point to be made is that charcoal therapy should not be discontinued too early.

A randomized cross-over study of multiple-dose charcoal for subjects who had taken aspirin was reported by Barone et al. (1988). They gave twenty-four 81-mg aspirin tablets to 13 healthy subjects, followed by 50 g charcoal for one, two, or three doses (separated by 4 hr). Urine was collected for 48 hr and analyzed for salicylates. The recoveries of salicylates in the urine were: control, 91.0%; one-dose charcoal, 68.3%; two-dose charcoal, 65.9%; and three-dose charcoal, 49.2%. They concluded that multiple-dose charcoal was effective in reducing aspirin absorption, especially if given three times as 50-g doses 4 hr apart.

Still another in vivo study of aspirin elimination with multiple-dose charcoal (as well as single-dose charcoal) was described by Yeakel et al. (1988) in an abstract of a meeting paper. They gave six healthy adults 650 mg crushed aspirin with water every 4 hr for three days in order to establish a steady state blood aspirin concentration. Then, in one phase, they gave a single dose of 80 g charcoal on day four. In another phase, they gave additional 40 g charcoal doses 2, 4, 6, and 10 hr later. Blood samples were drawn over 0–24 hr on the fourth day. The authors state that the AUC values thus obtained showed that the multiple-dose therapy reduced aspirin absorption significantly as compared to the single-dose regimen; however, no AUC values were given in this brief abstract.

Another study involving aspirin was done by Ho et al. (1989). They gave six healthy volunteers 1300 mg aspirin (four 325-mg tablets crushed and mixed with 250 mL water). Charcoal was then given as 25 g immediately after, then as three 10-g doses at 2 hr intervals. The blood salicylate 4 hr–∞ AUC value was reduced only 6.2% by the charcoal, as compared to the control value. The aspirin elimination half-life fell by 10.8%. Neither change was significant at $p > 0.05$. Note, however, that the last three of the four charcoal doses were rather small (10 g). These are much lower than the 40 g or larger doses used in studies where definite charcoal effects have been observed.

Multiple-dose therapy for aspirin overdose has also been studied by Kirschenbaum et al. (1990). Ten human volunteers participated in a randomized cross-over study with two phases. The subjects were given 2880 mg (36 × 80 mg) aspirin and then either nothing or 25 g charcoal at 4, 6, 8, and 10 hr. Serum salicylate concentrations were determined over 0–48 hr. Urinary salicylate excretion was also monitored. When the AUCs were calculated for the entire 0–48 hr period, they were not statistically different. However, when the AUC values were computed for the 4–48 hr period (the charcoal was started at 4 hr), the values were 2139 and 1950 mg-hr/L-kg for the control and charcoal-treatment phases. The difference (8.5%) was statistically significant ($p < 0.05$). The total urinary excretion of salicylate was reduced from 9.1 to 7.5 mmole, a decline of 17.6%. Thus, the multiple-dose charcoal treatment was only modestly effective. This study appears to be the same as that reported in abstract form by Tenenbein et al. (1989), although the 1989 report refers to 20 volunteers rather than ten.

Repeated charcoal for aspirin overdose in young children was described by Vertrees et al. (1990). Two case reports were given for 16 and 23-month-old male infants who had ingested a large amount of aspirin and 56 g sodium salicylate, respectively. The first infant was given 10 g charcoal and then 1 g/kg charcoal and 4 mL/kg magnesium citrate every 4 hr by nasogastric tube. The other child was given 30 g charcoal plus 15 mL magnesium citrate, then 30 g charcoal every 4 hr for the next 36 hr, by nasogastric tube. The elimination half-lives fell from an initial value of 37 hr to 5.9 hr, and from an unspecified value to 7.0 hr, respectively, for these two patients. These values are considerably less than half-life values reported for other patients who were not treated with charcoal. The patients in this study were treated with concurrent urinary alkalinization and this should be borne in mind in interpreting the charcoal effects.

The multiple-dose studies on aspirin and sodium salicylate just reviewed have produced rather varied results but this appears to be due to the wide diversity of the charcoal doses employed. In those studies where several charcoal doses of 40 g or more were used, significant AUC and elimination half-life reductions have been observed. Even with salicylates administered IV, charcoal has been effective in increasing drug elimination, if the charcoal doses were of reasonable size. This provides evidence that charcoal can interrupt the enteroenteric cycling of salicylates.

II. ACETAMINOPHEN

Mofenson and Caraccio (1992) described the case of a 10-month-old boy who had ingested many 500-mg Tylenol caplets and had an acetaminophen blood level of 198 µg/L 2 hr after ingestion. They gave him charcoal at 1 hr and again at 3 hr postingestion. The 4 hr drug level was 85 µg/L and they interpreted this as evidence that the charcoal was effective in reducing acetaminophen absorption.

III. PHENOBARBITAL

Neuvonen and Elonen (1980) studied the effects of charcoal on phenobarbital, carbamazepine, and phenylbutazone absorption in man. These three drugs were given simultaneously in oral doses of 400, 200, and 200 mg, respectively, to five healthy subjects using a randomized cross-over design. When 50 g charcoal was given within 5 min, absorption of all three drugs was more than 95% prevented. When the charcoal was given at 1 hr, the absorption of the three drugs was decreased 47, 41, and 30% (in the same order as mentioned above). These reductions are based on 0–96 hr AUC values.

Multiple-dose charcoal regimens also were employed, with 50 g charcoal at 10 hr, and 17 g at 14, 24, 36, and 48 hr. The reductions in absorption of the three drugs were then 49, 27, and 5% (same order as before), again based on 0-96 hr AUC values. In these multiple-dose tests, the half-life values, determined over 10–48 hr, decreased from 110.0 to 19.8 hr, from 32.0 to 17.6 hr, and from 51.5 to 36.7 hr for the three drugs (reductions of 82, 45, and 29%, respectively). Thus, the charcoal was effective, particularly in the case of phenobarbital. Charcoal prevented the absorption of all three drugs very effectively when it was given at 5 min, and moderately well when it was given at 1 hr. Even when started at 10 hr, the charcoal, while much less effective in terms of AUC reduction, decreased serum half-lives substantially.

Berg et al. (1982) studied the effect of multiple oral doses of charcoal on the body clearance of phenobarbital sodium given to six healthy men. The drug was administered IV at 200 mg per 70 kg of body weight. Charcoal was given as 40 g at time zero, and 20 g at 6, 12, 18, 24, 30, 42, and 66 hr. A pharmacokinetic analysis showed that the charcoal decreased the serum half-life of the phenobarbital from 110 to 45 hr, increased the total body clearance from 4.4 to 12.0 mL/kg-hr, and increased the nonrenal clearance from 52 to 80% of the total body clearance. Thus, charcoal enhanced the nonrenal clearance of phenobarbital.

Patients with phenobarbital overdose may remain comatose for several days because of the long elimination half-life of this drug. Thus, Goldberg and Berlinger (1982) tried multiple doses of charcoal given by nasogastric tube to treat two overdose patients. After endotracheal intubation and gastric lavage, the two patients were given charcoal and sodium sulfate or magnesium citrate via a nasogastric tube. The dosage schedules were a bit different for the two patients but consisted of 30 or 40 g charcoal doses with 20 or 30 g cathartic doses about every 4–6 hr. The serum half-lives of the drug were estimated to be less than 24 hr. By comparison to literature values of about 110 hr, this represents greater than a four-fold reduction in the drug half-life.

Linden et al. (1983) stated (in a brief abstract without details) that they treated a group of patients with grade II coma or deeper (Reed scale) due to intentional phenobarbital overdose using multiple doses of charcoal (at 3–6 hr intervals). A second group of patients received only a single dose of charcoal. They found that the multiple-dose therapy "hastens the elimination of toxic amounts of phenobarbital and shortens the duration of phenobarbital-induced coma."

In contrast to reports that multiple-dose charcoal has significant benefits for patients with phenobarbital overdose is a paper by Pond et al. (1984). In their study, ten such overdosed patients were treated, five with repeated doses of charcoal plus sorbitol and five who received only one dose of charcoal plus sorbitol (they served as controls). The drug serum half-life during repeated charcoal therapy was 36 hr versus 93 hr after charcoal was discontinued and 93 hr for the single charcoal dose group. The length of time that the patients required mechanical ventilation was not significantly different (39 hr for the single-dose group, and 48 hr for the repeated-dose group), nor was the time spent in the hospital different. Thus, although charcoal significantly increased the rate of phenobarbital elimination, it did not seem to affect the patients' clinical course. In response to this paper, Goldberg et al. (1985a) stated that it appeared that the patients receiving the multiple doses of charcoal were simply more ill than those in the single-dose group. However, Pond et al. (1985) defended their original conclusions as being valid and remarked that, based on other observations, "tolerance to the drug is probably more important in determining coma time than absolute concentration or rate of phenobarbital elimination."

Boldy et al. (1986) also have used repeated-dose charcoal therapy for phenobarbital overdoses. Six patients with moderate to severe phenobarbital intoxication were given initial charcoal doses of 50 or 100 g, followed by 50-g doses every 4 hr. The total time of charcoal treatment was 16–29 hr and the total charcoal given was 225–375 g, in the six cases. The mean drug half-life was 6.2 hr (versus a normal half-life of 3–5 days) and 62–93% of the absorbed drug was eliminated within 24 hr. The mean total body clearance was 84 mL/min with charcoal treatment. Boldy et al. concluded that the use of repeated oral charcoal doses was simple, safe, and effective.

The treatment of phenobarbital poisoning with multiple-dose charcoal in an infant was described by Amitai and Degani (1990). A 28-day-old infant was treated for epilepsy with phenobarbital but, through a pharmacist's error, was given too much drug. Upon admission his serum phenobarbital level was 103 mg/L. The infant was treated with 3 g (1 g/kg) SuperChar charcoal followed by three more doses of 1.5 g each, four hours apart, and one dose of 3 mL 20% sorbitol, all given by nasogastric tube. The serum drug level after 21 hr was 28 mg/L, upon which a calculated elimination half-life value of 11.2 hr was estimated. The patient recovered uneventfully.

The use of charcoal to treat elevated serum phenobarbital in a neonate also has been described by Veerman et al. (1991). The neonate was given phenobarbital to control seizures but the serum drug concentrations became too high to permit EEG and apnea tests to be performed. Thus, six 2-g doses of charcoal (0.7 g/kg) were given by orogastric tube at six hour intervals in an effort to reduce the serum drug level. The serum drug level fell from 79 to 53 mg/L after two doses and to 22 mg/L after six doses. The drug half-life was reduced from about 250 to 22 hr. Subsequent to this case, two neonates were treated for elevated drug levels (charcoal dose details unstated). In one case, charcoal reduced theophylline half-life from 15 to 8 hr and in the second case charcoal again reduced phenobarbital half-life, this time from 250 to 30 hr.

While there is little doubt that multiple-dose charcoal therapy can significantly reduce phenobarbital AUC and $t_{1/2}$ values, the report by Pond's group which indicates that patients' clinical courses are not affected is unsettling. They state that patient "tolerance" of the drug is the determining factor in the clinical course. But, does this mean that multiple-dose charcoal therapy is of no benefit for any given patient? Further clinical studies need to be done with patients who are closely-matched with respect to phenobarbital tolerance, some being treated with multiple-dose charcoal and others not, to reassess this matter.

IV. DIGOXIN AND DIGITOXIN

Multiple doses of activated charcoal have often been employed to accelerate the elimination of cardiac glycosides. We first discuss controlled clinical studies in which subtoxic doses of glycosides were given to normal human subjects, and then consider actual overdose case reports.

A. Human Volunteer Studies with Subtoxic Doses

In an investigation reported by Reissell and Manninen (1982), patients were given digitoxin and digoxin at constant daily doses until the plasma levels of both drugs reached steady therapeutic values. Then the patients were given either charcoal (2 g, three times per day) or a fiber product (10 g, three times per day). Blood glycoside levels were measured on days 10, 11, and 12 after the start of such treatment. The digitoxin and digoxin blood levels fell by 18.3 and 31.2%, respectively, with the charcoal therapy, as compared to the pre-charcoal values. The fiber had no significant effect on glycoside blood levels. This study suggests that digitoxin and digoxin both undergo enteral recirculation of some sort and that charcoal is able to significantly interrupt such cycles.

Digoxin pharmacokinetics with and without oral charcoal have been studied by Lalonde et al. (1985). Digoxin was given IV to ten healthy subjects (10 μg/kg) over 5 min. In one phase of this randomized cross-over study, the subjects were then given 25 g charcoal orally at 0, 4, 8, 12, 16, 22, 28, 34, and 40 hr. A noncompartmental kinetic analysis was used to interpret data on serum drug levels versus time. Digoxin clearance increased by an average of 47% with charcoal treatment and the drug half-life decreased from 36.5 to 21.5 hr. The mean residence time fell from 41.1 to 19.9 hr. Since a digoxin study by another group suggests that up to 30% of a digoxin dose is excreted in the bile and potentially resorbed each day, the charcoal presumably was effective by interrupting the enterohepatic cycle.

Another pharmacokinetic study involving multiple doses of charcoal and cardiac glycosides is that of Park et al. (1985). Normal subjects were given IV infusions of digoxin (0.75 mg/kg) or digitoxin (1 mg/kg) followed by either water alone or by water

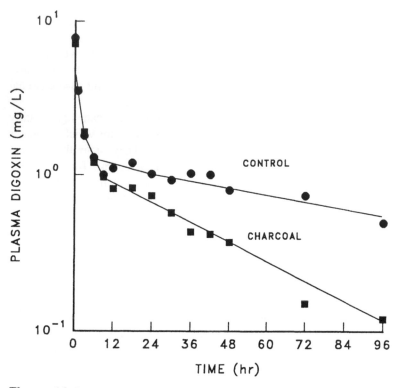

Figure 14.1 Plasma digoxin concentrations in a renal failure patient, with and without multiple oral doses of activated charcoal. From Park et al. (1985). Reprinted by permission of the Harvey Whitney Books Company, Cincinnati, OH.

with SuperChar charcoal (20 g in 300 mL water, at the end of the IV infusion, every 4 hr for 36 hr, and a final dose at 48 hr.) Blood samples were collected for 168 hr. The charcoal treatment increased digoxin clearance from 16.8 to 22.7 L/hr and increased digitoxin clearance from 0.24 to 0.47 L/hr. In one subject with chronic renal failure, the charcoal increased digoxin clearance from 3.6 to 10.1 L/hr. The data for this patient are shown in Figure 14.1. Thus, charcoal appears to be useful for digitoxin toxicity, and for digoxin toxicity in patients with prolonged elimination caused by renal dysfunction.

The studies on digitoxin and digoxin just described indicate that multiple-dose charcoal can accelerate the elimination of these drugs after they have already been absorbed (the single charcoal dose studies reviewed in Chapter 11 indicate that charcoal is also effective in reducing digoxin and digitoxin absorption in the first place, that is, when given with or soon after the drug). Thus, these drugs must undergo enteral cycling which can be interrupted by charcoal.

B. Overdose Case Reports

One case of a patient who was treated for a digitoxin overdose with oral charcoal was described by Pond et al. (1981). A 66-year-old woman was given oral doses of charcoal (60 g) and magnesium citrate (250 mL) every 8 hr for 72 hr. The digitoxin half-life was lowered from 162 hr (after 72 hr, when the charcoal was not given) to 18 hr during the charcoal treatment phase. The serum digitoxin level in this patient upon admission (264 μg/L) is the highest reported for any survivor of digitoxin overdose (she had taken about one hundred 0.1-mg digitoxin tablets). It was concluded that multiple-dose charcoal is effective for digitoxin overdose.

Moulin et al. (1982) described the case of a 17-year-old person who had ingested 10 mg digitoxin. The patient was given charcoal (5 g every 4 hr) by gastric intubation, starting 2 hr after digitoxin ingestion. The serum half-life of the drug was reduced from 138.6 hr to 53.3 hr by the charcoal.

A case report involving digoxin was described by Lake et al. (1984). A 71-year-old woman was treated in the hospital for severe congestive heart failure with diuretics and digoxin. The patient's renal function deteriorated and her serum digoxin levels became elevated. Digoxin was withdrawn and charcoal (50 g initially plus 25 g at 6 hr intervals thereafter for 8 doses total) was given by nasogastric tube in an effort to reduce her plasma digoxin. The digoxin half-life was 7.3 days prior to charcoal therapy, 6.3 days after charcoal was stopped, and 1.4 days during the charcoal period. Thus, charcoal was effective in accelerating the elimination of digoxin in this patient, who had seriously impaired renal function. In normal patients, the effect of charcoal would probably be less, since renal clearance accounts for 75% of total body clearance in normal patients.

The use of multiple doses of oral charcoal in digoxin poisoning has also been discussed by Boldy et al. (1985). A 69-year-old man who had ingested 6.5 mg digoxin presented 13.5 hr later with nausea, vomiting, and blurred vision. Because the patient did not vomit during the first 2 hr after admission, he was given 100 g Carbomix charcoal orally over the next hour and 50 g each 4 hr later for a further seven doses (total 450 g). His recovery was uneventful. The terminal digoxin half-life was 14 hr.

Vicas (1987) wrote an abstract describing the case of a 72-year-old woman who presented with intoxication due to 3.75 mg digoxin, 500 mg warfarin, 12.5 mg halcion, and 50 mg diazepam. Multiple-dose charcoal therapy was carried out over 12 hr (the exact details are not given) and the digoxin half-life was found to be 8 hr during the charcoal therapy versus 40 hr after the charcoal was discontinued.

Hantson et al. (1991) described the case of a 48-year-old man, chronically treated with digitoxin and verapamil for prevention of atrial fibrillation, who overdosed on digitoxin (2.2 mg). Following gastric aspiration and lavage, he was given 30 g charcoal every 6 hr (combined every other dose with 20% mannitol) plus 4 g cholestyramine every 6 hr over a 175 hr period. The patient's digitoxin level fell steadily and the terminal half-life was 109 hr. By comparing this to reported terminal half-life values of 2.9–6.2 days (i.e., 70–149 hr) in other overdosed patients, it was concluded that the multiple-dose charcoal and cholestyramine therapy did not accelerate digitoxin elimination in this patient. However, comparing one overdosed patient to other overdosed patients is difficult, as individual responses are so variable. Thus, their conclusion is rather fragile.

V. TRICYCLIC ANTIDEPRESSANTS

Again, to provide structure to our discussion, we first consider controlled clinical trials involving volunteers who were given subtoxic drug doses and then we review two actual overdose cases.

A.   Human Volunteer Studies with Subtoxic Doses

Very positive results with multiple-dose charcoal therapy for reducing tricyclic anti-depressant (TCA) absorption in humans have been reported by Crome et al. (1977). These same results were also reported, in abbreviated form, in a paper by Braithwaite et al. (1978). They gave 75 mg nortriptyline (NT) plus a glass of water to healthy volunteers, followed 30 min later by 5 g activated charcoal contained in a 10 g packet of an effervescent charcoal preparation (Medicoal) along with 400 mL more water. In a similar study, multiple charcoal doses of this same size were given at 30, 120, 240, and 360 min after drug administration. Typical plasma nortriptyline concentrations versus time in control and charcoal-treated subjects showed that, for the single charcoal dose study, average reductions in the peak plasma drug concentrations were 59% (range 30–81%) and average reductions in drug bioavailability were 59% (range 32–85%). Bioavailability was calculated as the area under the plasma concentration versus time curve for the period 0–48 hr. For the multiple charcoal dose studies, the average peak plasma concentration reduction was 72%, (range 62–78%) and the average bioavailability reduction was 70%, (range 58–76%), as compared to controls. A report in the "Medical News" section of the *JAMA* (see Montgomery, 1978) refers to the 1977 study by Crome et al. and reviews its recommendations.

Kärkkäinen and Neuvonen (1986) performed a study in which the pharmacokinetics of amitriptyline (AT) and its active metabolite nortriptyline were determined when charcoal or changing the urine pH were employed. The results obtained by changing the urine pH with sodium bicarbonate or ammonium chloride were discussed in Chapter 7 and will not be reviewed here. After a single dose of 50 g charcoal given 5 min after a 75-mg dose of AT hydrochloride, AT absorption was reduced by 99%. With repeated charcoal doses (50 g initially followed by 12.5 g every 6 hr between 6–54 hr, for a total of 150 g charcoal) the serum half-life of AT was reduced by 20% and the serum half-life of NT was reduced by 35%. Thus, charcoal was effective in substantially preventing AT absorption and then in increasing the elimination rates of AT and NT to a modest extent, presumably by interrupting their enterohepatic and/or enteroenteric cycles.

The effect of single and repeated doses of charcoal on the pharmacokinetics of doxepin in man have been studied by Scheinen et al. (1985). Doxepin is a commonly used tricyclic antidepressant, whose main metabolite is desmethyldoxepin. Eight healthy volunteers were each given two 25-mg tablets of doxepin hydrochloride, plus 15 charcoal (Medicoal) mixed with 200 mL water at 30 min. Peak serum doxepin levels were reduced by 70% and the 0–8 hr doxepin AUC values were reduced by 49%, compared to control values. When multiple doses of charcoal were given over 3–4 hr (15 g at 3 hr, and 10 g after 6, 9, 12, and 24 hr), the 0–48 hr AUC values for doxepin were reduced by 26.3%. Thus, both single and multiple-dose charcoal strategies were effective.

One study in which no effect of charcoal on tricyclics was noted is that of Goldberg et al. (1985b). They carried out a randomized cross-over trial with four normal men. Imipramine was given IV at a dose of 12.5 mg/70 kg body weight. In one phase, this was followed by SuperChar charcoal given as a slurry at 0, 2, 4, 6, 9, 12, 16, 20, and 24 hr (20 g charcoal each time). Serum imipramine data were collected over 0–24 hr. Values of pharmacokinetic parameters with and without charcoal were, respectively: half-life, 10.9 and 9.0 hr; clearance, 930 and 990 mL/min/70 kg; and volume of distribution, 12.4 and 11.2 L/kg. Thus, charcoal had no effect. These results suggest that multiple-dose charcoal does not increase imipramine elimination. A possible reason for this is that imipramine has a very large volume of distribution and it has an extensive tissue distribution; thus, there is little left in the blood to undergo gastrointestinal dialysis across the gut wall or excretion into the bile. Hence, enterogastric and enterohepatic cycling of imipramine is insignificant.

In summary, it appears that for TCAs such as amitriptyline, nortriptyline, doxepin, and dothiepin, multiple-dose charcoal therapy can significantly increase elimination rates, by interruption of enteroenteric and/or enterohepatic cycles. However, for imipramine, such cycles are not significant and thus charcoal is not of value.

B. Overdose Case Reports

Swartz and Sherman (1984) used repeated doses of charcoal to treat three patients who were in comas as a result of TCA overdoses. After gastric lavage, an initial charcoal dose of 40–50 g was given via nasogastric tube and later doses of 20–25 g charcoal were given as frequently as possible without regurgitation through the tube. The TCA (amitriptyline in these cases) plasma half-lives were reduced below 10 hr for each patient to as low as 4 hr. This is to be contrasted with reported amitriptyline half-lives averaging 36.8 hr and often exceeding 60 hr for patients not treated with charcoal. Repeated doses of charcoal appear to have effectively interrupted the enterohepatic circulation of this drug.

Ilett et al. (1991) studied the effects of repeated-dose charcoal (Carbosorb, Delta West, Ltd., Perth, Western Australia) on the elimination of the tricyclic antidepressant dothiepin in three overdose patients. In three patients who received multiple doses of charcoal (50 g every 12 hr for 2 doses for the first patient, and 50 g every 4 hr for 4 doses for the other two patients), dothiepin half-lives were reduced to 10.6, 12.5, and 13.1 hr, compared to literature values for non-charcoal-treated patients of 18.5–24 hr.

VI. THEOPHYLLINE

Since a very large number of studies have been conducted with theophylline, they will be untractable unless they are divided into several categories. Thus, they are organized into the following subsections: (1) animal studies, (2) human volunteer studies with subtoxic oral doses, (3) human volunteer studies with subtoxic IV doses, and (4) human overdose case reports. In many of the studies we discuss, the theophylline was used in the form of aminophylline, a compound of theophylline and 1,2-ethanediamine which dissociates upon dissolution to give free theophylline.

A. Animal Studies

A study by Chyka et al. (1993) evaluated female pigs as an animal model for multiple-dose charcoal therapy in theophylline overdose. The pigs were given 8.9 mg/kg aminophylline IV in an arm over 12 min, followed by 25 g charcoal at 0, 2, 4, 6, 12, 18, 24, and 30 hr (by orogastric tube). Blood samples were taken over 36 hr. Pharmacokinetic results for the pigs were compared to those for humans (Campbell and Chyka, 1992) and showed the following differences (pigs vs. humans): reduction in AUC compared to controls = 67% vs. 55%, reduction in $t_{1/2}$ compared to controls = 68% vs. 47%, $t_{1/2}$ = 10 hr vs. 9 hr, clearance = 1.0 mL/kg-min vs. 0.7 mL/kg-min, volume of distribution = 0.8 L/kg vs. 0.5 L/kg, and protein binding = 16% vs. 56%. Thus, for many parameters,

328   **Chapter 14**

the differences between pigs and humans were not unduly large and therefore, pigs appear to be a suitable substitute for humans in the case of theophylline.

Brashear et al. (1985) gave theophylline IV (28 mg/kg) to each of six dogs, followed by 30 g charcoal in water by nasogastric tube every 2 hr for a total of four doses. The charcoal reduced the $t_{1/2}$ from 4.0 to 2.8 hr and the clearance increased from 0.123 to 0.150 L/hr-kg but many of the other pharmacokinetic parameter values (t_{max}, C_{max}, AUC) did not change significantly. The time required for the serum drug level to fall below 20 mg/L (i.e., below the toxic level) was unchanged (the time required was about 5 hr). Since studies with intravenously-administered theophylline in human subjects have shown charcoal to be very effective (e.g., Park et al., Berlinger et al., Mahutte et al., and Radomski et al., discussed below), one must conclude that dogs respond somewhat less well than humans to charcoal therapy after IV theophylline.

Multiple-dose charcoal for theophylline poisoning was studied in an animal model by Kulig et al. (1987). They gave large IV aminophylline doses to five dogs, with and without charcoal. When charcoal was used, they gave 50-g doses of charcoal via a nasogastric tube every hour for a total of eight doses. The charcoal reduced the serum AUC values by an average of 34.4% and half-life values by an average of 39.4%. The effects were not seen when the nasogastric tube was put into the stomach instead of the small bowel because the charcoal administered did not pass beyond the pylorus. In a separate experiment in which bile theophylline concentrations were measured, it was shown that enhanced elimination was not from interruption of enterohepatic circulation of theophylline. This suggests that the effect of charcoal is due to the interruption of enteroenteric cycling of the drug.

These three animal studies show that multiple-dose charcoal therapy is effective in enhancing the elimination of IV theophylline by interrupting its enteroenteric circulation.

B. Human Volunteer Studies with Subtoxic Oral Doses

Lim et al. (1986) studied the effect of oral charcoal on the first 12 hr of absorption and elimination of sustained-release theophylline in 20 normal children, ages 8 to 18 years. The drug dose was 10 mg/kg. After receiving the drug, the children were given from one to four doses of 1 g/kg charcoal (InstaChar) with 30 mL 70% sorbitol. Group 1 was given a single dose at 1 hr; Group 2 received four doses at 1, 4, 7, and 10 hr; Group 3 was administered four doses at 3, 6, 9, and 12 hr; and Group 4 received three doses at 6, 9, and 12 hr. Blood samples were drawn at 0, 2, 4, 6, 8, 10, 12, and 24 hr. Reductions of the 0–12 hr AUC values, compared to control values, were: Group 1, 61%; Group 2, 68%; Group 3, 37%; and Group 4, 18%. The results indicate that early charcoal administration can inhibit theophylline absorption significantly and enhance elimination of previously absorbed theophylline.

C. Human Volunteer Studies with Subtoxic IV Doses

The effect of size and frequency of oral doses of charcoal on theophylline pharmacokinetics was examined by Park et al. (1983). Six healthy volunteers were given 6 mg/kg infusions of aminophylline over 1 hr, followed by the following charcoal doses: no charcoal (control), 5 g every 2 hr for 6 doses, 10 g every 2 hr for 6 doses, 10 g every hr for 12 doses, 20 g every 2 hr for 6 doses, or 40 g every 4 hr for 3 doses. The serum $t_{1/2}$ values that were determined were: 9.1, 5.6, 5.5, 4.3, 4.3, and 5.4 hr, respectively. The 0–24 hr serum AUC values were 123, 79, 72, 60, 62, and 73 mg-hr/L, respectively. These

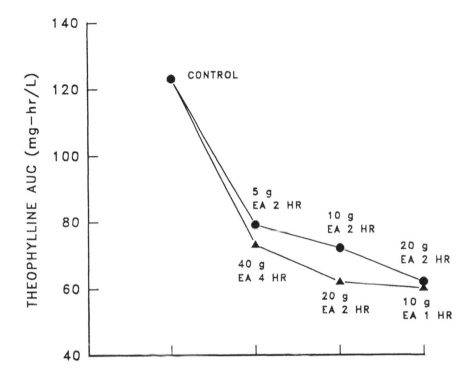

Figure 14.2 Mean theophylline AUC values for different activated charcoal doses. Upper line: different charcoal doses given every 2 hr. Lower line: increasing frequency of charcoal dose with the same total amount of charcoal (120 g). From Park et al. (1983). Reprinted by permission of Mosby-Year Book Inc.

values are shown in Figure 14.2. The effect of larger dose sizes, given every 2 hr, is small and the effect of increasing dosage frequency at a constant total dose size is also small. However, all charcoal regimens were effective as compared to the control situation.

Park et al. (1984) evaluated the effects of using standard charcoal or superactive charcoal on theophylline pharmacokinetics. Eight healthy men were given either amino-phylline (6 mg/kg) IV or theophylline (5 mg/kg) IV, over 1 hr, followed by 5 g standard (Norit USP XX) charcoal (surface area 950 m^2/g) every two hours, 20 g Norit USP XX every 2 hr, or 5 g superactive (Amoco PX-21) charcoal (surface area 3,376 m^2/g) every 2 hr. The $t_{1/2}$ and AUC values with these treatments were as shown in Table 14.1.

For equal weights (5 g), the superactive charcoal was somewhat more effective. Since the ratio of surface areas of the charcoal was 3.55, then—if surface area is the controlling factor—5 g superactive charcoal and 17.75 g (3.55 × 5) Norit charcoal should be equivalent. When 20 g Norit charcoal was used, the effect was somewhat larger than with 5 g PX-21 charcoal, as one would expect.

Hence, the results are consistent with the notion that the effect of a charcoal is proportional to its surface area. It is unfortunate that a "no charcoal" control was not included in this study; it would have been easy to do and would have shown the relative effects of each charcoal protocol versus no treatment.

Table 14.1 Effect of Charcoal on Theophylline Pharmacokinetics

Charcoal and dose	$t_{1/2}$ (hr)			AUC (mg-hr/L)		
	Norit 5 g	PX-21 5 g	Norit 20 g	Norit 5 g	PX-21 5 g	Norit 20 g
Value	6.3	5.3	4.9	88.9	74.4	67.7

From Park et al. (1984). Reprinted by permission of the J. B. Lippincott Company.

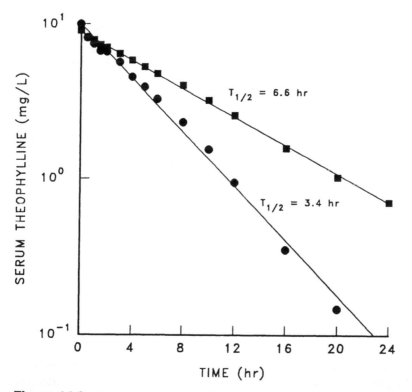

Figure 14.3   Mean serum theophylline levels after intravenous aminophylline, with (circles) and without (squares) activated charcoal. From Berlinger et al. (1983). Reprinted by permission of Mosby-Year Book Inc.

Berlinger et al. (1983) performed a randomized cross-over test in which six normal males were given aminophylline (6 mg/kg) IV and then charcoal (40 g at 0 hr, and 20 g at 2, 4, 6, 9, and 12 hr). Serum drug levels were measured over 0–24 hr. Treatment with the charcoal reduced the serum AUC from 78 to 42 mg-hr/L (a decrease of 46%) and lowered the serum $t_{1/2}$ from 6.6 to 3.4 hr (a decrease of 48%). Figure 14.3 shows the determination of half-life values, with and without charcoal. Thus, it is clear that charcoal can bind theophylline secreted into the GI tract.

Mahutte et al. (1983) did a similar study in which seven normal subjects were given aminophylline (8 mg/kg) IV and then 30 g charcoal doses at 0, 2, 4, and 6 hr. With charcoal, as compared to control results, the drug half-life fell from 10.2 to 4.6 hr, and the clearance increased from 35.6 to 72.6 mL/kg-hr. In one additional test with an actual overdose victim, this same charcoal regimen reduced the patient's theophylline half-life from 34.4 to 5.7 hr.

Byatt and Henry (1984) commented on the Mahutte study, stating that the half-life value of 34.4 hr prior to charcoal therapy was based on just two data points (one at the time of admission and a second 4 hr later, at which time the first charcoal dose was given) and because the theophylline preparation was of the slow-release type, that the change in the slope of serum theophylline versus time which occurred after 4 hr could have resulted simply from cessation of the absorptive phase of the drug and not because of the charcoal therapy. Mahutte et al. (1984) replied, stating that the 34.4 hr half-life was only meant to be approximate and that, in their view, absorption was complete at the time the patient was admitted.

Radomski et al. (1984) gave 6 mg/kg aminophylline IV to six subjects having cirrhosis and then gave them 40 g charcoal (ActaChar) at 0 hr, and 20 g at 2, 4, 6, 9, and 12 hr. The $t_{1/2}$ values fell from an average of 12.7 hr to an average of 4.0 hr. In a separate study with five patients having moderate theophylline poisoning, a charcoal regimen of 20 or 40 g initially plus 20 g each 2 hr until serum drug levels were under 25 mg/L resulted in a mean $t_{1/2}$ of 4.9 hr. This can be compared to still other control studies which gave a mean $t_{1/2}$ of 7.9 hr in normal subjects not treated with charcoal.

Ginoza et al. (1987), working with ten low-birth-weight infants who had been given aminophylline daily for treatment of apnea, gave them each a series of three IV infusions of theophylline as aminophylline dihydrate. Then, a slurry of SuperChar charcoal (1 g/kg) was given via orogastric tube, 1 hr later. A second charcoal dose was given at 4 hr. The charcoal treatment increased the drug clearance from a mean of 22.7 to a mean of 44.6 mL/kg-hr, an increase of 96.4%. Thus, high surface area charcoal given orally appears to be effective for enhancing theophylline elimination in neonates.

Theophylline pharmacokinetics have also been examined by McKinnon et al. (1987). Aminophylline was intravenously administered to humans (3 mg/kg, containing 2.37 mg/kg theophylline, given over 30 min) and charcoal was given in a 10% water slurry at the same time (400 mL) and at 2, 4, and 6 hr (200 mL). Studies were also done with dogs with an intravenous infusion of 10 mg/kg of drug plus a 10% charcoal slurry delivered into the duodenum at 0 hr (200 mL) and at 2, 4, and 6 hr (50 mL). In the human subjects, the charcoal therapy increased the rate of theophylline disappearance from the blood by 73% and reduced its half-life by 28%. The volume of distribution of the drug was unchanged by charcoal. In the dogs, the rate of theophylline clearance was increased by 31% and the half-life for elimination decreased by 31%. Thus, it appears that theophylline entering the gut was significantly bound by charcoal. However, collection of bile after theophylline infusion showed that only 0.28% of the dose entered the gut via bile but samples of jejunal aspirate showed theophylline concentrations very similar to those in the blood. Thus, it was shown that theophylline enters the gut by diffusion across the intestinal wall and not to any significant extent via the bile.

Ilkhanipour et al. (1990, 1992) carried out a randomized cross-over study with five healthy subjects who were given aminophylline IV (8 mg/kg). They were immediately given 50 g charcoal and then either: (1) 12.5 g every hr, (2) 25 g every 2 hr, or (3) 50 g every 4 hr, over an 8 hr period (total charcoal 150 g in all cases). Ten blood samples were taken over 12 hr and analyzed for the drug. AUC values for $0-\infty$ (control = 223 mg-hr/L) were 104, 121, and 122 mg-hr/L for the three regimens, respectively, and plasma $t_{1/2}$ values (control = 9.89 hr) were 4.87, 5.39, and 5.41 hr for the three regimens, respectively. Thus, all charcoal treatments were effective in reducing absorption and in increasing the drug elimination rate (since the drug was given IV, the charcoal must have interrupted the enteral cycle of the drug). The three charcoal treatment values were not statistically different ($p < 0.4$), but there is at least a suggestion in the results that the "12.5 g every hour" regimen was somewhat better. It was concluded that the "traditional every 4 hour therapy" is as effective as the other therapies.

In summary, there is strong evidence from animal studies that theophylline is secreted into the gut across the lumen wall and is not excreted to any significant extent in the bile. Moreover, multiple doses of charcoal are quite effective in binding the theophylline which is secreted into the gut, thereby significantly increasing its rate of elimination.

D. Human Overdose Case Reports

There are a large number of overdose case reports in the literature and, like most case reports, they vary widely in terms of patient characteristics, overdose quantities, and the charcoal dose schedules employed to treat the patients. Thus, it is difficult to construct some sort of logical pattern to a discussion of such reports; however, they do possess one thing in common: they all attest to the effectiveness of multiple-dose charcoal therapy in accelerating theophylline elimination.

A case report on theophylline overdose was given by Gal et al. (1984). A 23-year-old woman with an overdose of theophylline (Theo-Dur) was given 50 g charcoal every 6 hr between 18–48 hr postadmission. Her drug half-life was 17.2 hr before the charcoal and 5.9 hr during the charcoal treatment.

Another case report, this time involving a 53-day-old premature infant who inadvertently received an oral dose of 107 mg theophylline, was described briefly by Bronstein et al. (1984). The infant was given 1 g charcoal plus 250 mg magnesium sulfate every 2 hr, beginning 5 hr after drug ingestion, for five doses. Charcoal therapy reduced the elimination half-life from 30.4 to 5.5 hr and increased the clearance from 0.38 to 1.89 mL/kg-min. Serial drug levels at 5, 8, 12, and 17 hr postingestion were 42, 33, 17, and 7.9 mg/L, respectively. Thus, the multiple-dose regimen was quite effective.

Slaughter and DuPuis (1985), in a letter commenting on a study by Greenberg et al. (1984) involving the treatment of theophylline toxicity with charcoal hemoperfusion, described the treatment of two patients for theophylline toxicity with oral charcoal given as 30 g every 2–4 hr for four doses. The drug half-lives after charcoal treatment were 6 and 12 hr in the two patients. The half-lives after charcoal was discontinued were 12 and 56 hr, respectively. Clearance values indicated that the use of charcoal increased clearances two- to three-fold. Greenberg (1985), in a brief reply, mentioned that the theophylline half-life during their hemoperfusion study was 4.7 hr, comparable to the 4.3 hr half-life observed by Park et al. (1983) with oral charcoal.

Four patients with theophylline toxicity were treated by True et al. (1984) using a regimen which, in general, consisted of an initial charcoal dose of 30 g (in water) and three more identical doses spaced 2 hr apart. The half-life reductions in these four patients were: from 16.2 to 5.9 hr, from 34.4 to 5.7 hr, from 23.3 to 13.9 hr, and from 19.3 to 6.5 hr.

Fourteen patients with toxic levels of theophylline were treated by Sessler et al. (1985) using 30-g doses of oral charcoal administered about every 2 hr for a total of 2–4 doses. Ten of the patients tolerated the charcoal well and had their serum drug half-lives reduced to an average of 5.6 hr. The four patients with the highest initial theophylline levels vomited all doses of the charcoal and were treated by charcoal hemoperfusion. This reduced their drug half-lives to an average of 5.2 hr.

A case report involving a 27-year-old woman who presented with a toxic level of theophylline was described by Davis et al. (1985). They gave her 15 g charcoal orally, roughly every 2 hr, for a total of 4 doses. Magnesium citrate (300 mL) was also given. The half-life of the drug was reduced from 7.5 to 2.2 hr and the drug clearance was increased from 1.2 to 4.7 L/hr with the charcoal therapy.

Ohning et al. (1986) described the cases of two adolescents who presented to an ED with serum theophylline levels in excess of 100 mg/L. In one patient, several attempts to instill 30 g charcoal or magnesium citrate via nasogastric tube were met with immediate and repeated emesis. The other patient was given 25 g charcoal followed by 250 mL magnesium citrate by nasogastric tube and she vomited once soon after. In both cases, continuous nasogastric infusion of a charcoal slurry (10 and 50 g charcoal/hr in the two cases) was initiated and no further emesis occurred. The infusions were performed for roughly 20 hr. The theophylline elimination half-lives for the two patients were 7.7 and 13.5 hr, respectively, during the first several hours of charcoal infusion and fell to 2.6 and 3.3 hr, respectively, during the latter part of the charcoal infusion period. Thus, it appears that continuous nasogastric charcoal administration is both safe and effective in theophylline overdose.

Rygnestad et al. (1986) wrote a case report about a male patient with moderate to severe theophylline poisoning who was given gastric lavage and 45 g oral charcoal initially. Then, about 2 hr later and after his condition appeared to worsen, he was given 30 g more charcoal and three subsequent 30-g doses spaced 2 hr apart (total of 165 g charcoal). The theophylline elimination half-life was reduced to 2 hr. Tests done several months later using a subtoxic drug dose in the same patient without any charcoal gave a half-life value of 24 hr. Thus, in this patient, repeated doses of oral charcoal had a large effect.

Two additional case reports have been presented by Amitai et al. (1986). They treated a 34-year-old woman and a 5-month-old male infant for theophylline overdose. The woman was given 15 g charcoal and 60 g magnesium citrate hourly from about 14–22 hr postadmission and her drug elimination half-life fell to 3.7 hr with this treatment. The infant was given 10 g charcoal and 30 mL magnesium citrate nasogastrically at 4.5 hr, 5 g charcoal at about 8 hr, and 2.5 g charcoal at about 11 hr. His drug elimination half-life fell from 19 hr prior to the charcoal to 2.4 hr after the charcoal took effect.

Multiple-dose charcoal therapy for theophylline overdose in young infants has been discussed by Shannon et al. (1987). They reviewed five cases of theirs in which infants

Table 14.2 Half-lives for Different Theophylline Overdose Therapies

	Untreated	MDAC	HD	HP
First $t_{1/2}$, hr	22.8	8.9	3.7	3.2
Terminal $t_{1/2}$, hr	8.4	5.1	6.4	6.0

Adapted from Shannon and Woolf (1992). Reprinted with permission from *Veterinary and Human Toxicology*.

of 6-months age or less were treated with repeated charcoal doses. The regimens varied and can not be summarized in any brief way. However, suffice it to say that the theophylline half-lives observed were 11.6, 2.4, 6.5, and 12.6 hr in the four cases for which they were evaluated (usually based on only two serum drug level points). It is known that in infants theophylline half-lives are in the range of 14–30 hr and thus the repeated charcoal therapy seems to have accelerated theophylline elimination in these cases. The treatment was well tolerated, without any complications, despite the young age of the patients (one was only 2 wk old).

Lopez-Herce et al. (1991) treated a 1-month-old girl for theophylline toxicity with repeated oral doses of charcoal. At 6 hr after admission to their pediatric ICU, charcoal was administered via nasogastric tube at a dosage of 1 g/kg every 2 hr for 20 hr. A rapid fall in serum theophylline levels was noted, along with disappearance of clinical signs of toxicity. The drug half-life fell from 20 hr (before charcoal) to 3.8 hr during charcoal treatment. Drug clearance rose from 15.3 to 81.6 mL/kg-hr. This case confirms that repeated charcoal can be effective for theophylline toxicity in infants.

A report by Shannon and Woolf (1992) has compared multiple-dose activated charcoal (MDAC) with hemodialysis (HD) and hemoperfusion (HP) for treating theophylline overdoses. They determined multiphasic $t_{1/2}$ values for untreated, MDAC, HD, and HP groups. The values are shown in Table 14.2. These results show that MDAC, HD, and HP all significantly enhance theophylline elimination but that, in terms of the first $t_{1/2}$, hemodialysis and hemoperfusion are superior to MDAC.

VII. QUININE

Quinine poisoning is relatively uncommon but potentially serious. Hayden and Comstock (1975) showed that quinine adsorbs well to charcoal in vitro. Thus, repeated-dose charcoal therapy was considered and tried by Lockey and Bateman (1989) in a study of quinine elimination using subtoxic doses in seven healthy subjects. They were each given 600 mg quinine bisulfate, then 50 g charcoal in an aqueous slurry 4 hr after the drug, and three further 50-g doses over the next 12 hr. Carbomix and Medicoal were used alternately. Blood samples were taken over 0–36 hr. Based on AUC values determined by extrapolating the blood data to infinite time, the charcoal lowered drug absorption 31.5% and decreased the half-life from 8.23 to 4.55 hr (a 44.7% decrease). The data plotted to allow half-life determinations are shown in Figure 14.4. Had the charcoal not been delayed for 4 hr, even larger effects might have been noted. Since there are no

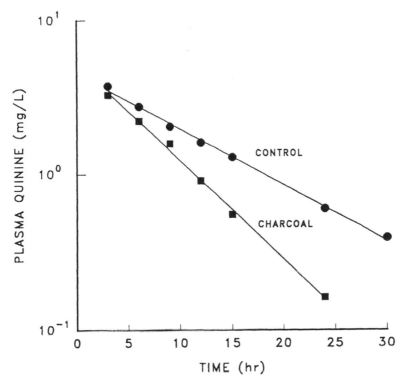

Figure 14.4   Plasma quinine concentrations following 600 mg oral quinine bisulfate in seven volunteers, with and without activated charcoal. From Lockey and Bateman (1989). Reprinted by permission of Blackwell Scientific Publications Ltd.

experimental data to suggest that quinine undergoes enterohepatic circulation, the charcoal probably works by binding quinine which diffuses through the gut lumen.

Five patients with actual quinine overdoses were treated by Prescott et al. (1989) using repeated charcoal doses. The dosage schedule was generally 50 g charcoal (Medicoal) in 200 mL water shortly after admission, and 50 g every 4 hr thereafter up to a total of 200–400 g. Not counting one patient for which the first charcoal dose was delayed 34 hr, the quinine half-lives were 7.2, 7.6, 7.6, and 8.0 hr. In the patient for which the charcoal was delayed, the drug half-life was 33 hr prior to the charcoal and 10 hr thereafter. A previous report from the literature suggests a half-life of about 24 hr in the absence of charcoal. Thus, repeated charcoal appears to have reduced the quinine half-life from a value on the order of 24–33 hr to about 7–8 hr.

VIII.   ANTIBIOTICS

Doxycycline, a broad-spectrum antibiotic of the tetracycline type, was studied by Venho et al. (1978). They first found that activated charcoal completely adsorbed doxycycline in vitro from artificial small intestinal fluid, pH 6.5. However, when 4-g charcoal doses were given at 0, 3, 8, and 12 hr to adults that had ingested 100 or 200-mg doses of doxycycline at 0 and 7 hr, there was no impact on either serum blood levels or the urinary excretion of the drug during the 24 hr period following the start of the tests. However, analysis of their protocols shows that the ratio of charcoal to drug was 400 or 200 to one in their in vitro studies and 40 or 20 to one in their in vivo studies (based on the initial doses). This may explain part of the differences observed. In any event, this study is useful in demonstrating that multiple-dose charcoal may not have a significant effect on systemic absorption of certain drugs.

Vancomycin overdose can cause nephrotoxicity and ototoxicity. Vancomycin adsorption was studied by Davis et al. (1987). Six healthy subjects were used in a randomized cross-over study. They were given 1 g vancomycin IV over a 1 hr period, with and without charcoal. The charcoal therapy involved 50 g oral charcoal immediately before the vancomycin infusion and 15-g doses at 2, 4, 6, and 8 hr. There was no statistical difference in any pharmacokinetic parameter values (clearance, half-life, urinary recovery) between the tests with charcoal and those without charcoal. It appears, therefore, that vancomycin does not undergo an enteroenteric or an enterohepatic cycle.

The lack of enteroenteric/enterohepatic cycling of vancomycin is further supported by the results of El-Sayed et al. (1993), who studied the effect of charcoal on vancomycin clearance in normal rabbits and in rabbits having induced acute renal failure (ARF). Vancomycin (7.5 mg/kg) was given IV, followed by a single oral dose of 10 g charcoal (although this study did not involve multiple-dose charcoal, we discuss it here, as it relates closely to the just-presented vancomycin findings of Davis's group). While the pharmacokinetic parameter values for the normal and ARF rabbits were different, as one would expect, charcoal did not significantly affect their values in either case.

Multiple-dose charcoal for vancomycin overdose was described by Burkhart et al. (1992). The inadvertent administration of an incorrect dose of vancomycin solution to a 47-day-old premature infant caused very high drug levels. A 1.5 volume exchange transfusion did not have any effect. Multiple doses of charcoal (1 g/kg) were given as follows: one before the transfusion and nine after the transfusion at 4 hr intervals (via nasogastric tube). The calculated drug half-life before the transfusion was 35 hr. After the exchange transfusion and during the multidose charcoal therapy the half-life was 12 hr. The results of this study are in direct contrast to those of Davis et al., discussed in the previous paragraph, which showed a complete lack of any multiple-dose charcoal effect. The reason for the difference is unknown but may be due to a possible error in the calculated 35-hr drug half-life value.

Tobramycin is an aminoglycoside and, as such, is a polar, cationic, water-soluble compound that is not absorbed significantly from the GI tract. It does not undergo significant enterohepatic circulation, as less than 1% of a dose is found in the bile. Watson et al. (1987) studied tobramycin adsorption to charcoal in vitro and also studied whether repeated oral doses of charcoal would increase the clearance of tobramycin in vivo. They found that tobramycin adsorbed to charcoal at pH 5.6 (13.6, 11.7, and 34.9% bound at drug:charcoal ratios of 1:5, 1:10, and 1:20, respectively) but not at all at pH 2.6 (for these same drug:charcoal ratios). Tobramycin is ionized at acidic pH and this accounts for its zero binding to charcoal in vitro at pH 2.6. At pH 5.6, some of the drug is not ionized and the nonionized part can bind somewhat well to charcoal.

In the in vivo tests, five healthy subjects were given 1.5 mg/kg tobramycin intravenously over 5 min. SuperChar charcoal was given in 10-g oral doses 2 hr before the drug, at the time of the drug injection, and at 2, 6, and 8 hr after the drug injection. The 0—∞ AUC mean values were 14.8 mg-hr/L with charcoal and 15.3 mg-hr/L without charcoal. The total body clearance was 7.1 L/hr with charcoal and 6.3 L/hr without charcoal (the renal clearances were 5.1 and 4.4 L/hr, respectively). However, the elimination half-life values were 2.3 hr with charcoal and 1.9 hr without charcoal. Thus, charcoal had no significant effect and it was concluded that tobramycin does not diffuse back into the gut to any reasonable extent.

Davis et al. (1988) also did an in vivo study with tobramycin, in which six volunteers were given 2.5 mg/kg tobramycin IV over 30 min, with 50 g oral charcoal given just before the drug, and 15 g charcoal doses given at 2, 4, and 6 hr. Blood and urine samples were collected over 8 and 24 hr, respectively. The mean total body clearances were 7.9 L/hr with charcoal and 8.5 L/hr without charcoal (the renal clearances were 6.9 and 8.2 L/hr, respectively). The $t_{1/2}$ values for the elimination phase were 2.1 hr both with and without charcoal. Thus, just as in the very similar study by Watson and colleagues, it was concluded that charcoal had no effect on tobramycin clearance.

Taken as a whole, these studies on various antibiotics show that, with the exception of the study by Burkhart et al., such compounds are not affected by multiple-dose charcoal therapy, presumably because they are not excreted/secreted either into the bile or into the intestinal lumen to a significant extent. Thus, charcoal is normally ineffective if given after IV antibiotic infusion or well after an oral dose has been absorbed. However, if sufficient charcoal is given soon after an oral antibiotic dose, it can be of some value.

IX. PROPOXYPHENE

Kärkkäinen and Neuvonen (1985) studied the effects of oral charcoal on the pharmacokinetics of dextropropoxyphene (DPP) and its active metabolite norpropoxyphene (NP). They also studied the effect of changing the urine pH using sodium bicarbonate or ammonium chloride on the renal excretion of DPP and NP. These latter results were summarized in Chapter 7 and will not be repeated here. The study involved giving 130 mg DPP hydrochloride to six healthy subjects, followed by 50 g charcoal at 5 min. DPP absorption was reduced by 92–96% as a result. When charcoal was delayed and then given in repeated doses (50 g at 6 hr, followed by eight doses of 12.5 g each given every 6 hr thereafter, for a total of 150 g charcoal), DPP absorption was not significantly affected, compared to controls, but the DPP terminal serum half-life was reduced from 31.1 to 21.2 hr and the terminal serum half-life of NP was reduced from 34.4 to 19.8 hr. Thus, charcoal was effective in preventing absorption and in interrupting the enterohepatic or enteroenteric circulation of DPP and NP. The delayed charcoal results suggest that, with this rapidly absorbed drug, it is important to administer charcoal as quickly as possible.

X. DAPSONE

Neuvonen et al. (1980) did an in vivo randomized cross-over study with five subjects. The subjects took a total dapsone dose of 500 mg over four days. Ten hr after the last 100-mg dose, 50 g charcoal as a slurry was given, followed by 4 consecutive doses of 17 g charcoal at 12 hr intervals. The half-life of serum dapsone was reduced by charcoal from a control value of 20.5 hr to 10.8 hr and the half-life of serum monoacetyldapsone (the main metabolite of dapsone) was decreased from 19.3 to 9.5 hr. In addition, two actual suicidal overdose patients were treated with 80 g/day charcoal for 1 or 2 days and it was found that the charcoal increased the rate of elimination of dapsone by a factor of 3 to 5. Thus, the use of multiple doses of charcoal for a drug of this kind, which has a long half-life and which undergoes secretion into the gut, appears to be quite effective.

Neuvonen et al. (1983) also presented in vivo results for three patients who were treated for 1–10 g dapsone overdoses. Activated charcoal given orally in multiple doses (20 g, four times/day) shortened the half-life of dapsone from a value of 77 hr without charcoal to a mean of 12.7 hr with the charcoal. The half-life of monoacetyldapsone similarly decreased from 51 to 13.3 hr. Charcoal treatment was compared to hemodialysis, which gave dapsone and monoacetyldapsone half-lives of 10.4 and 10.9 hr, respectively. Thus, charcoal therapy is comparable in effectiveness to hemodialysis and is easier to perform and is much cheaper.

Reigart et al. (1982–83) reported on a case in which an 18-month-old child accidentally ingested a single 100-mg dapsone tablet. The child presented to a physician 3–4 hr later with severe methemoglobinemia. Doses of 10 g activated charcoal were started at 20 hr postingestion and were repeated every 6 hr. The child showed rapid resolution of his symptoms and was symptom-free 64 hr postingestion. The rapid improvement was attributed to interruption of the enterohepatic circulation of the drug by the charcoal.

XI. CARBAMAZEPINE

All reports in the literature concerning this drug have involved human overdoses. Boldy et al. (1987), in the form of a letter to the editor, summarized some of their data on carbamazepine overdose cases. Fifteen patients received supportive therapy plus 12.5–100 g charcoal in slurry form every 1–6 hr (these were actual overdose cases and the treatment protocols varied widely). They report a mean drug half-life value, after charcoal administration, of 8.6 hr. This can be compared to a reported average value of about 19.0 hr in 12 overdose cases in which only supportive therapy was used.

Two other studies were reported in 1987 dealing with multiple-dose charcoal for carbamazepine overdose. One (Heath and Van Loo, 1987) described four patients with carbamazepine overdoses who were given 50 g charcoal orally on admission and further 50-g doses each 6 hr until neurological symptoms had subsided. The mean elimination half-life in this group was 10.5 hr. The authors state that the charcoal increased carbamazepine elimination "by a factor of at least one-third." The other study (Vale et al., 1987) described carbamazepine-overdosed patients who were given 50–100 g charcoal initially, then 50 g every 4 hr. This reduced the carbamazepine half-life from 17 to 7 hr and gave a mean total body clearance of 123 mL/min compared to a value of 100 mL/min obtained with hemodialysis. This same report gives results for a phenobarbital overdose in which the same multiple-dose therapy lowered the drug's half-life from 110 to 12 hr (the total body clearance was 84 mL/min versus a value of 74 mL/min obtained with hemodialysis and a value of 79 mL/min obtained with hemoperfusion).

The successful treatment of a massive carbamazepine overdose was reported by Sethna et al. (1989). A 36-year-old comatose woman was treated with gastric lavage and then with charcoal. The blood drug level fell and then began to rise again. More charcoal was administered, as were cathartics; these produced a rapid and complete recovery of the patient.

A study involving the use of multiple doses of charcoal for carbamazepine overdose was described by Wason et al. (1992). Five children were involved, two with acute overdose and three with acute-on-chronic overdose. In the acute cases the drug half-lives averaged 8.6 hr and in the acute-on-chronic cases they were 12.0 hr, without charcoal. Each patient was treated with different numbers and sizes of charcoal doses. The half-life values were: no charcoal, 23.3 hr; 30–50 g total charcoal, 10.2 hr; and 60–90 g total charcoal, 7.2 hr. Thus, the half-life of the drug decreased with the total amount of charcoal given. Yet, there was no relationship between the time to complete recovery and the use of multiple doses of charcoal. A similar finding was reported by Pond et al. (1984) for the use of multiple-dose charcoal in phenobarbital overdose. Vale and Heath (1992), commenting on a statement in the Wason paper to the effect that no prior work on charcoal in carbamazepine overdose had been reported, referred to their groups' 1987 letter (Boldy et al., 1987, mentioned above), and stated that the two groups' findings were in accord.

These overdose reports involving carbamazepine have shown that multiple-dose charcoal therapy reduced elimination half-life values, in general, by somewhat more than a factor of 2, depending on the charcoal dose size. It would be interesting for someone to conduct controlled clinical trials, either with human volunteers given subtoxic carbamazepine doses or with animals given toxic carbamazepine doses.

XII. PHENYTOIN

Phenytoin poisoning is relatively common and presents a problem because this drug is slowly absorbed, metabolized, and excreted. We discuss human trials with subtoxic IV drug doses first and then human overdose cases.

A. Human Studies with Subtoxic IV Doses

A cross-over study with seven healthy subjects and subtoxic doses of phenytoin was performed by Mauro et al. (1987). Each subject was given 15 mg/kg sodium phenytoin IV and then SuperChar charcoal (60 g in 470 mL 26% sorbitol at the end of the infusion and eight additional 30-g doses of charcoal at 2, 4, 8, 12, 24, 30, 36, and 48 hr either in 235 mL water or in 235 mL 26% sorbitol, in such a manner "to produce one to two bowel movements per day"). Serum drug levels were measured over 0–72 hr. The charcoal reduced the $0-\infty$ AUC (obtained by extrapolation) by 50.2% and the drug half-life from 44.5 hr to 22.3 hr. The elimination rate constant was increased from 0.0167 to 0.0332 hr^{-1}, a 98.8% increase. Thus, multiple-dose charcoal appears to greatly enhance phenytoin elimination. This drug is strongly protein-bound (93%) but has a small volume of distribution (0.6–0.7 L/kg). The small V_d and the fact that phenytoin is known (from studies with rats) to be secreted into the GI tract by means other than biliary excretion (which is small) would explain the effectiveness of charcoal.

Rowden et al. (1990) gave human subjects phenytoin IV (15 mg/kg given in 250 mL of solution over a 1 hr period) and then administered oral charcoal at the end of infusion (40 g) and at 2, 4, 6, 8, and 10 hr (20 g each time). The charcoal gave AUC values (0–72 hr) that were about 25% lower than those of the same subjects under control conditions. Clearance values rose 38.3% and half-lives decreased from 25.5 to 23.6 hr with charcoal. When sorbitol was added to the charcoal, no additional change in any parameter values occurred but the sorbitol did cause abdominal cramps in many of the subjects.

These two reports show that multiple doses of charcoal can effectively interrupt the enteroenteric circulation phenytoin that is already systemically distributed.

B. Human Overdose Reports

A 38-year-old woman who had ingested at least 10 g phenytoin was treated by Weichbrodt et al. (1987) with multiple doses of charcoal. A single charcoal dose of 30 g with 180 mL magnesium citrate was given initially and then 24 hr later serial dosing was begun with 30 g charcoal plus 180 mL magnesium citrate at 6 hr intervals for several days. The phenytoin blood level fell from a peak of 52 mg/L 42.5 hr after drug ingestion to 7 mg/L eight days after drug ingestion. The authors concluded that the charcoal therapy was effective; however, no pharmacokinetic parameters (e.g., half-life values) were deter- mined and no control values were cited either. Thus, the effect of the charcoal was not quantified.

Ros et al. (1989) reported the case of a 17-year-old male with a chronic seizure disorder who was given 300 mg phenytoin daily. He gradually developed headache, dizziness, and an unsteady gait; then vomiting and a shaking episode occurred. He was admitted and had a serum phenytoin level of 69 mg/L 24 hr later. Charcoal (30 g every 4 hr) and magnesium citrate (200 mL with each charcoal dose) were given orally. Nine doses in all were given. The drug level fell 25% in the first 2 hr following the initial charcoal dose and fell to 22 mg/L over the next 36 hr. A rebound to 33 mg/L occurred after charcoal was stopped but then the drug level began to decline again. The patient fully recovered.

Weidle et al. (1991) reported on the case of a 67-year-old woman with severe liver disease who was given 300 mg phenytoin sodium orally twice daily and who developed signs of phenytoin toxicity. Phenytoin was discontinued and she was given 30 g charcoal in 70% sorbitol by nasogastric tube every 4 hr for a total of 10 doses. Her serum phenytoin level fell from 45.2 mg/L before charcoal therapy to 16.6 mg/L on the second day after charcoal was started (the time interval between these two values was about 32 hr). The patient made a full recovery. It was concluded that charcoal was highly effective in increasing the clearance of phenytoin in this patient.

Dolgin et al. (1991) reported on the effects of multiple-dose charcoal on phenytoin pharmacokinetics in two pediatric cases. In one case, a 42-month-old boy was given 25 g charcoal in sorbitol but he immediately vomited. Then, charcoal was infused continuously as a slurry via a nasogastric tube at a rate of about 6 g/hr until 25 g had been given. No emesis occurred during this. Then 15 g charcoal in sorbitol was given in four oral doses, every 4 hr, followed by 15 g charcoal without sorbitol every 4 hr for four doses. In the second case, a 36-month-old boy was given 15 g charcoal with sorbitol orally every 4 hr for four doses, then 15 g charcoal without sorbitol every 4 hr for 12 hr, then 7.5 g charcoal without sorbitol every 4 hr for 16 more hr. Pharmacokinetic modeling of serum phenytoin values versus time was difficult but suggested that the drug could be characterized by first-order elimination with a rate constant of 0.02–0.04 hr^{-1}. The elimination of the drug appeared to be more rapid than reports in the literature would have suggested. However, in this study, involving two actual cases without any controls, such a conclusion can only be qualitative in nature.

Howard et al. (1994) described the case of a 36-year-old man who was treated twice for chronic phenytoin toxicity. In the first episode, only supportive measures were used and the patient's serum phenytoin fell slowly over the next six days from 150 to 135 μmol/L. In the second episode, 50-g doses of charcoal were given (Actidose with Sorbitol) upon admission and at 6 hr intervals for a total of four doses. With this treatment, the patient's serum phenytoin decreased from 187 µmol/L to less than 80 µmol/L within 36 hr. Although no $t_{1/2}$ values were determined, it is clear that the multiple-dose charcoal therapy considerably increased the rate of elimination of phenytoin.

Although these reports describe six patients, in no case were drug half-life values determined before, after, or during multiple-dose charcoal therapy. Instead, the reports cite primarily blood drug levels at different times. Thus, this set of overdose reports lacks pharmacokinetic quantification. Nevertheless, it does appear that charcoal did decrease drug blood levels much more than would have occurred without the use of charcoal.

XIII. PIROXICAM

Laufen and Leitold (1986) did a cross-over study with six healthy volunteers. The subjects were given single oral doses (20 mg) of piroxicam. Charcoal (50 g) was given orally 5 min later. The charcoal reduced piroxicam bioavailability by 98%. In other tests, charcoal was given in a multiple-dose format (50 g at 10 hr and seven more doses of 20 or 30 g given over the next 48 hr). The drug bioavailability decreased 41.7% and the half-life decreased from 40.2 to 19.6 hr. In a repeat test, 20 mg piroxicam was given rectally, followed by multiple-dose charcoal orally (30 g at 2 hr, 20 g at 8 hr and seven more 20 or 30-g doses over the next 50 hr). Bioavailability decreased 48.8% with charcoal and the drug half-life went from a control value of 40.7 hr to 21.6 hr. It was inferred from the effectiveness of delayed multiple-dose charcoal therapy that piroxicam undergoes enteral circulation of some type.

Further studies on piroxicam were done by Ferry et al. (1990). They gave 20 mg oral doses of the drug to eight healthy young adults. Then 5 g charcoal (Medicoal) was given starting 24 hr after the drug dose and further 5 g Medicoal doses were given 3–4 times/day thereafter for several days. The charcoal therapy reduced the piroxicam half-life from 53.1 to 40.0 hr. In other tests, cholestyramine was given in 4 g amounts according to the same schedule as for the charcoal and the drug half-life was decreased even more, to 29.6 hr. Again, even though sorbent therapy was delayed for 24 hr, it was effective. This second report also proves that enteral circulation occurs for this drug and that it can be interrupted by multiple doses of charcoal.

XIV. OTHER DRUGS

A. Glutethimide and Barbital

Repeated administration of activated charcoal after an adequate initial dose sometimes does not cause significant additional inhibition of absorption. Chin et al. (1969) showed, for example, that repeated doses of activated charcoal in dogs at two 1 hr intervals after the administration of glutethimide or at two 12 hr intervals after administration of barbital were no more effective than a single dose of activated charcoal. These drugs apparently do not undergo sufficient enterohepatic circulation for multiple-dose therapy to have a significant effect.

B. Phenobarbital and Diazepam

Traeger and Haug (1986) reported on a case of prolonged coma which occurred after treatment of alcohol withdrawal seizures with IV phenobarbital and diazepam. Charcoal

(40 g) was given every 4 hr nasogastrically for a total of six times. Coma was completely reversed within 12 hr and the serum half-life of diazepam was reduced from 195 to 18 hr. Because of severe liver disease, the patient was hypoalbuminemic. This caused more of the diazepam to be unbound and able to cross the gastrointestinal membrane barrier, thus making the charcoal more effective in binding the diazepam.

C.   Meprobamate

Linden and Rumack (1984) used multiple-dose charcoal to treat three females for mepro-bamate overdose. Meprobamate is a nonbarbiturate sedative-hypnotic, available in the USA since the mid-1950s. Peak drug levels were 221, 91, and 80.5 mg/L. Multiple-dose charcoal lowered the drug half-lives in the three cases to 4, 4.5, and 5 hr, respectively, which the authors state are significantly less than half-lives observed in the absence of charcoal (half-lives of therapeutic doses of meprobamate are on the order of 10–11 hr, according to the *AHFS Drug Information '92* reference book). This report (a brief abstract) does not give details of the charcoal therapy (e.g., amounts, times of adminis-tration).

The use of multiple-dose charcoal therapy in two patients with meprobamate over-doses was discussed by Hassan (1986). The patients were a 32-year-old woman and a 28-year-old woman. These patients were given 50-g doses of charcoal every 4 hr for five doses and 50 g charcoal doses every 6 hr for five doses, respectively. Both patients had a drug elimination half-life of about 4.5 hr with the charcoal therapy, which can be compared to an average half-life of 11.3 hr without charcoal from one study in the literature involving 12 healthy subjects.

Thus, these two reports both suggest that multiple-dose charcoal therapy reduces the half-life of meprobamate from around 10–11 hr to 4–5 hr.

D.   Diltiazem

Roberts et al. (1991) described the case of one 38-year-old female who had ingested 900 g diltiazem. Multiple-dose charcoal therapy was started at 7 hr postadmission and consisted of 50 g charcoal in water (Charcodote Aqua, Pharmascience, Montreal) every six hours from 7–31 hr via nasogastric tube. Plasma concentrations versus time were determined over 7–71 hr and the half-life for elimination was determined and its value compared to historical controls. This patient's drug half-life (10.2 hr) was longer than half-lives obtained for several previous patients who had received supportive therapy and was higher than most published $t_{1/2}$ values for this drug. Thus, it was concluded that the charcoal was not effective in this case. This is not surprising because of (1) the long delay before charcoal was given, and (2) this drug is strongly protein-bound and has a large volume of distribution.

E.   Nadolol and Sotalol

Du Souich et al. (1983) performed studies aimed at determining whether nadolol un-dergoes enterohepatic circulation. They gave 80 mg nadolol to eight healthy subjects orally and, in one phase, followed this with charcoal tablets given orally according to the following schedule: 500 mg at 3 hr, 500 mg at 4 hr, and 250 mg each hour for the next 8 hr (total charcoal = 3000 mg). The charcoal reduced the AUC for nadolol from

2455 to 1355 µg-hr/L (a 45% decrease) and the half-life from 17.3 to 11.8 hr (a 32% decrease). These results suggest that enterohepatic circulation occurs with this drug.

Kärkkäinen and Neuvonen (1984) examined the effects of oral charcoal on sotalol pharmacokinetics in seven subjects, in a randomized cross-over study. Sotalol hydrochloride (160 mg) was given, followed 5 min later by 50 g oral charcoal. The charcoal reduced sotalol absorption by 99% compared to controls. When additional charcoal doses were given at 6 hr (50 g) and each 6 hr thereafter (12.5 g) up to a cumulative charcoal dose of 150 g at 54 hr, the serum sotalol half-life was shortened from 9.4 to 7.6 hr. It was concluded that charcoal interrupts the enterohepatic or enteroenteric circulation of sotalol.

F. Chlorpropamide

Chlorpropamide is a long-used antidiabetic agent. Neuvonen and Kärkkäinen (1983) studied the effects of charcoal on chlorpropamide kinetics in man. Six healthy subjects were given 250 mg chlorpropamide orally and then a single 50-g dose of charcoal 5 min later, or 50 g charcoal 6 hr after the drug followed by 12.5 g at six hour intervals until a total of 150 g of charcoal had been taken. The single dose of charcoal reduced the 0–72 hr AUC for chlorpropamide by 90%; however, in the case of repeated charcoal doses begun at 6 hr, the AUC was not significantly different from the control value. Thus, early charcoal administration is recommended for chlorpropamide overdose.

G. Cyclosporin

Honcharik and Anthone (1985) report on one case involving an adult male who had received a renal transplant and who was given cyclosporin as an immunosuppressant. Due to a mistake, the first daily dose of 750 mg cyclosporin was actually 5000 mg; the mistake was discovered 3 hr later. To counteract this overdose, oral activated charcoal was given as follows: 60 g at the time the mistake was discovered, 30 g 4 hr later, and 30 g an additional 4 hr later. Plasma cyclosporin concentrations were 6700 µg/L 4 hr after drug ingestion, 675 µg/L at 13 hr, and 50 µg/L at 47 hr. Calculated cyclosporin half-lives were 2.7 hr during charcoal administration and 9.0 hr after charcoal was stopped. It appears, therefore, that oral charcoal is effective in cases of cyclosporin overdose.

H. Phencyclidine

Lyddane et al. (1988) studied the effect of charcoal in treating phencyclidine (PCP) overdose. Rats were gavaged once with PCP (10, 25, or 50 mg/kg in 2 mL/kg water) and then were given charcoal (1 g/kg) orally once per hour, starting 45 min after drug administration. The AUC value was reduced by charcoal in the 10 mg/kg PCP group by 30%, in the 25 mg/kg PCP group by 4%, and not at all in the 50 mg/kg PCP group. Thus, charcoal is effective if given in a large enough amount. Obviously, had the charcoal initially been given immediately, it would have had a greater effect (but then this would not mimic the delay which is typical in treating overdose victims).

I. Paroxetine

Greb et al. (1989) showed that charcoal (20 g at 20 and 40 min) can reduce the absorption of 60 mg paroxetine (an antidepressant) given to human volunteers. In 13 male adults, the average 0–∞ AUC value for the drug alone was 1005 µg-hr/L. When charcoal was given, paroxetine could not be detected in the patients' plasma over the 56 hr that blood sampling was carried out. When paroxetine was given alone, many symptoms of toxicity (nausea, tiredness, flatulence, headache, vertigo) occurred, but when charcoal was used symptoms were far fewer (mainly tiredness and headache) and much less severe. However, since symptoms were still felt to some degree, the lack of paroxetine in the plasma seems strange. Perhaps the sensitivity of the drug analysis (extraction with toluene followed by gas chromatography) was poor.

J. Disopyramide

Arimori et al. (1989) gave disopyramide (200 mg) orally to six healthy subjects followed by 40 g of charcoal in a suspension containing 250 mL of a magnesium citrate solution (34 g magnesium citrate content) at 4 hr and charcoal doses of 20 g with 150–200 mL water at 6, 8, and 12 hr. The charcoal treatment decreased serum half life and 0–∞ AUC values by 33 and 19%, respectively, as compared to control values. Total body clearance increased to 122% of the control value.

A second study involving disopyramide from Arimori's research group is that of Arimori and Nakano (1989). The drug was given at 20 mg/kg IV to rats, alone and with charcoal (300 mg) given as a slurry at time zero and with subsequent 150 mg charcoal doses at 1, 2, 3, and 4 hr. The half-life decreased from 1.18 to 1.05 hr and the 0–6 hr AUC decreased from 13.8 to 11.3 mg-hr/L with the charcoal treatment. V_d was unaffected.

These studies indicate that charcoal has the potential to modestly increase the elimination of disopyramide by interrupting its enteroenteric and/or enterohepatic cycles.

K. Methotrexate

The effect of activated charcoal on the pharmacokinetics of high-dose methotrexate (a chemotherapeutic agent) was studied by Gadgil et al. (1982). Serum methotrexate (MTX) concentrations versus time were followed in two groups of patients who had received 6 hr infusions of MTX (1 g/m^2). One group was given 25 g charcoal orally at 12, 18, 24, 36, and 48 hr. There was a significant reduction (36.7%) in the MTX AUC from 18 hr onward. The data suggest that charcoal aids in elimination of MTX from the body by interruption of enterohepatic circulation of the drug.

XV. LIMITATIONS ON MULTIPLE-DOSE CHARCOAL THERAPY

Turk et al. (1993) presented an analysis which determined the maximal fraction of an absorbed dose of a drug or toxin (F_R) which can be removed from a patient's systemic circulation by either 4 hr of hemoperfusion or 24 hr of repeated oral administration of activated charcoal. For the case of hemoperfusion, they showed that F_R can never exceed $1/V_d$, and in actual practice will be no greater than about $0.5/V_d$, where V_d is the volume of distribution of the drug or toxin in liters per kg of body weight.

For the case of 24 hr of repeated oral charcoal, the maximal value of F_R was shown to be approximately $1.17/V_d$. Thus, drugs like theophylline (V_d = 0.5 L/kg), phenobarbital (V_d = 0.8 L/kg), and phenytoin (V_d = 0.6 L/kg) are relatively well removed by repeated oral charcoal (the fractions removed in 24 hr were computed to be 0.47, 0.21, and 0.26, respectively, by Turk et al.). For tricyclic antidepressants (V_d = 20–50 L/kg), removal would be very small (less than 0.02).

They then mentioned that Radomski et al. (1984) developed a model for theophylline removal by repeated oral charcoal which assumes that the total drug clearance, CL_T, is equal to the sum of charcoal-induced GI tract clearance, CL_C, and endogenous clearances, CL_E, such as renal excretion, metabolization, etc. This is equivalent to saying that the total half-life of drug elimination, T_T, and the half-lives of drug elimination by charcoal, T_C, and by all endogenous processes, T_E, are related by the equation

$$\frac{1}{T_T} = \frac{1}{T_C} + \frac{1}{T_E}$$

By determining the half-life of IV theophylline without charcoal (for which case $T_T = T_E$), and for IV theophylline with repeated oral charcoal, Radomski et al. found that for their group of subjects T_E averaged 9.8 hr and T_C averaged 7.1 hr. Thus, oral repeated charcoal would be predicted by the T_T equation just cited to fall from 9.8 hr to

$$\frac{1}{(1/9.8) + (1/7.1)} = 4.12 \text{ hr}$$

Actual data obtained in the Radomski study gave a value of 4.0 hr, which is in excellent agreement.

Turk et al. used the T_C data of Radomski's group to estimate the theophylline clearance due to charcoal, CL_C, to be 56 mL/min for a 70 kg adult. Then, using a literature value of 50 mL/min for theophylline CL_E, they computed a total clearance with charcoal of 106 mL/min, an increase over the endogenous clearance alone (50 mL/min) by a factor of 2.12. They compared this to the results of Berlinger et al. (1983) which showed a theophylline half-life reduction from 6.4 hr to 3.3 hr (a factor of 1.94) when repeated oral charcoal was employed.

An estimate of CL_C (11 mL/min) was also made by Turk et al. for phenobarbital. Since CL_E was noted (from studies in the literature) to be about 8.5 mL/min, the use of charcoal should reduce the half-life of phenobarbital by a factor of $(11 + 8.5)/8.5 = 2.29$. This compares well with the results of Pond et al. (1984) which showed a phenobarbital half-life decrease from 93 to 36 hr when repeated oral charcoal was used, a fall by a factor of 2.58.

Turk et al. go on to state that some reports suggest that repeated oral charcoal accelerates the removal of tricyclic antidepressants (TCA's). However, they noted that the observed effects of charcoal in these studies appear to have been due to the charcoal reducing TCA absorption in the first place and not due to an enhancement of TCA removal from the systemic circulation. This points out the importance of using oral charcoal as early as possible for drugs like TCA's, whose removal by charcoal after absorption is very small.

XVI. SUMMARY

Several review articles on multiple-dose charcoal therapy have appeared, including articles by Pond (1986a, 1986b), Jones et al. (1987), Neuvonen and Olkkola (1988), an editorial (Anonymous, 1987), and McLuckie et al. (1990). In the Pond (1986a), Jones, and McLuckie articles, there are excellent summaries of studies which, up to the time of the articles, involved multiple-dose therapy. Jones' summary, which gives values for

Table 14.3 Studies Involving Multiple-Dose Charcoal Therapy

Drug	Estimated increase in total body clearance, percent	Reference
Amitriptyline	496	Swartz and Sherman (1984)
Carbamazepine	82	Neuvonen and Elonen (1980)
Chlordecone	106	Guzelian (1981)
Cyclosporin	233	Honcharik and Anthone (1985)
Dapsone	90	Neuvonen et al. (1980)
Digitoxin	800	Pond et al. (1981)
Digoxin	421	Lake et al. (1984)
Digoxin (IV)	47	Lalonde et al. (1985)
Meprobamate	151	Hassan (1986)
Methotrexate (IV)	58	Gadgil et al. (1982)
Nadolol	34	Du Souich et al. (1983)
Phenobarbital	158	Goldberg and Berlinger(1982)
Phenobarbital (IV)	173	Berg et al. (1982)
Phenylbutazone	40	Neuvonen and Elonen (1980)
Salicylate (IV)	62	Wogan et al. (1986
Theophylline	191	Gal et al. (1984)
Theophylline (IV)	104	Mahutte et al. (1983)

Adapted from Jones et al. (1987). Reproduced by permission of the W. B. Saunders Company.

Table 14.4 Additional Studies Involving Multiple-Dose Charcoal Therapy

Drug	Estimated increase in total body clearance, percent	Reference
Digitoxin	96	Park et al. (1985)
Digoxin	35	Park et al. (1985)
Disopyramide	22	Arimori et al. (1989)
Doxepin	32	Scheinen et al. (1985)
Imipramine	−6	Goldberg et al. (1985b)
Quinine	56	Lockey and Bateman (1989)
Phenytoin (IV)	38	Rowden et al. (1990)
Tobramycin (IV)	13	Watson et al. (1987)
Tobramycin (IV)	−7	Davis et al. (1988)
Theophylline	397	Bronstein et al. (1984)
Theophylline	292	Davis et al. (1986)
Theophylline	96	Ginoza et al. (1987)
Theophylline	73	McKinnon et al. (1987)
Theophylline	433	Lopez-Herce et al. (1991)

Adapted and expanded from McLuckie et al. (1990). Reprinted by permission of the Australian Society of Anaesthetists.

Table 14.5 Effects of Multiple-Dose Charcoal on Drug Half-Lives

| | Half-life (hr) | | |
Drug	Without charcoal	With charcoal	Reference
Normal volunteers			
Carbamazepine	32.0	17.6	Neuvonen and Elonen (1980)
Chlorpropamide	49.7	46.9	Neuvonen and Kärkkäinen (1983)
Dapsone	20.5	10.8	Neuvonen et al. (1980)
Digoxin	23.1	17.0	Park et al. (1985)
Digoxin	36.5	21.5	Lalonde et al. (1985)
Digitoxin	110.6	51.1	Park et al. (1985)
Doxepin	17.9	16.2	Scheinen et al. (1985)
Imipramine	9.0	10.9	Goldberg et al. (1985b)
Nadolol	17.3	11.8	Du Souich et al. (1983)
Phenobarbital	110	19.8	Neuvonen and Elonen (1980)
Phenobarbital	110	45	Berg et al. (1982)
Phenylbutazone	51.5	36.7	Neuvonen and Elonen (1980)
Propoxyphene	31.1	21.2	Kärkkäinen and Neuvonen (1985)
Sotalol	9.4	7.6	Kärkkäinen and Neuvonen (1984)
Overdose cases			
Aspirin	27	3.2	Hillman and Prescott (1985)
Cyclosporin	9.0	2.7	Honcharik and Anthone (1985)
Dapsone	52.7	11.2	Neuvonen et al. (1980)
Digoxin	7.3	1.4	Lake et al. (1984)
Digitoxin	162	18	Pond et al. (1981)
Methotrexate	8.4	7.6	Gadgil et al. (1982)
Phenobarbital	93	36	Pond et al. (1984)

Adapted from Pond (1986a). Reproduced by permission of Elsevier Science Publishers BV.

the percentage increases in total body clearance achieved with multiple-dose activated charcoal (compared to controls) is presented in Table 14.3. Quite clearly, multiple-dose therapy significantly (and sometimes tremendously) increases drug clearances. Some additional results on clearance increases besides those given by Jones have been summarized by McLuckie et al. (1990). These are shown in Table 14.4.

The summary by Pond (1986a) on the effects of multiple-dose oral charcoal on the elimination half-lives of various drugs in humans is given in Table 14.5 (similar data were presented by McLuckie et al. (1990); their summary appears to be that of Pond with additions). The additions made by McLuckie et al. are presented in Table 14.6. Again, it is clear that in most cases the use of multiple-dose charcoal significantly decreased drug elimination half-lives.

In the cases of Tables 14.4 and 14.6 (from McLuckie et al., 1990) some additional listings have been added by the writer, based on literature reports which either appeared after the tables were prepared or were not included by McLuckie et al. (these added studies have all been discussed earlier in this chapter).

Table 14.6 Effects of Multiple-Dose Charcoal on Drug Half-Lives

| | Half-life (hr) | | |
Drug	Without charcoal	With charcoal	Reference
Amitriptyline	36.8	4	Swartz and Sherman (1984)
Amitriptyline	27.4	21.1	Kärkkäinen and Neuvonen (1986)
Aspirin	5.4	4.8	Ho et al. (1989)
Aspirin	37	5.9	Vertrees et al. (1990)
Carbamazepine	19.0	8.6	Boldy et al. (1987)
Carbamazepine	17	7	Vale et al. (1987)
Carbamazepine	23.3	10.2	Wason et al. (1992)
Carbamazepine	23.3	7.2	Wason et al. (1992)
Dapsone	88	13.5	Neuvonen et al. (1980)
Dapsone	33	11.4	Neuvonen et al. (1980)
Dapsone	77	12.7	Neuvonen et al. (1983)
Diazepam	195	18	Traeger and Haug (1986)
Digitoxin	139	53	Moulin et al. (1982)
Digoxin	175	33.6	Lake et al. (1984)
Digoxin	40	8	Vicas (1987)
Disopyramide	6.1	4.1	Arimori et al. (1989)
Dothiepin	(18–24)	12.1	Ilett et al. (1991)
Meprobamate	(10–11)	4.5	Linden and Rumack (1984)
Meprobamate	(11.3)	4.5	Hassan (1986)
Nortriptyline	37.5	24.5	Kärkkäinen and Neuvonen (1986)
Phenobarbital	(110)	< 24	Goldberg and Berlinger (1982)
Phenobarbital	69.3	11.8	Mofenson et al. (1985)
Phenobarbital	(110)	6.2	Boldy et al. (1986)
Phenobarbital	110	12	Vale et al. (1987)
Phenobarbital	(110)	11.2	Amitai and Degani (1990)
Phenobarbital	250	22, 30	Veerman et al. (1991)
Phenytoin	44.5	22.3	Mauro et al. (1987)
Phenytoin (IV)	25.5	23.6	Rowden et al. (1990)
Piroxicam	40.2	19.6	Laufen and Leitold (1987)
Piroxicam (rectal)	40.7	21.6	Laufen and Leitold (1987)
Piroxicam	53.1	40.0	Ferry et al. (1990)
Quinine	33	10	Prescott et al. (1989)
Quinine	8.2	4.6	Lockey and Bateman (1989)
Salicylate (IV)	10.3	6.3	Wogan et al. (1986,87)
Tobramycin (IV)	1.9	2.3	Watson et al. (1987)
Tobramycin	2.1	2.1	Davis et al. (1988)
Vancomycin	35	12	Burkhart et al. (1992)

Adapted and expanded from McLuckie et al. (1990). Reproduced by permission of the Australian Society of Anaesthetists. Figures in parentheses are estimates taken from other literature sources.

Table 14.7 Multiple-Dose Charcoal Effect on
Theophylline Half-Life

Half-life (hr)		
Without charcoal	With charcoal	Reference
6.4	3.3	Berlinger et al. (1983)
10.2	4.6	Mahutte et al. (1983)
34.4	5.7	Mahutte et al. (1983)
16.2	5.9	True et al. (1984)
34.4	5.7	True et al. (1984)
23.3	13.9	True et al. (1984)
19.3	6.5	True et al. (1984)
12.7	4.0	Radomski et al. (1984)
17.2	5.9	Gal et al. (1984)
30.4	5.5	Bronstein et al. (1984)
—	5.6	Sessler et al. (1985)
7.5	2.2	Davis et al. (1986)
7.7	2.6	Ohning et al. (1986)
13.5	3.3	Ohning et al. (1986)
24	2	Rygnestad et al. (1986)
19	2.4	Amitai et al. (1986)
5.5	4.0	McKinnon et al. (1987)
20	3.8	Lopez-Herce et al. (1991)
15	8	Veerman et al. (1991)
9.9	5.2	Ilkhanipour et al. (1990,92)

There have been a particularly large number of multiple-dose charcoal studies performed with theophylline which have focused on determining the effect on the half-life of the drug. These theophylline half-life investigations are presented separately in Table 14.7.

Campbell and Chyka (1992) wondered whether a drug's physicochemical properties are associated with enhanced removal by multiple-dose charcoal therapy. Thus, they searched the literature dealing with multiple-dose studies. Percent reductions in plasma half-life values were calculated and correlated with variables such as the drug molecular weight, the drug pK_a, plasma protein-binding by the drug, intrinsic plasma half-life (i.e., in the absence of charcoal), and volume of distribution. Step-wise multiple regression analysis was employed. No correlations were identified ($p < 0.1$). The data base involved 34 studies, of which 28 met the criteria which they set for inclusion in their analysis. The analysis did suggest that charcoal therapy was more effective for drugs having a longer intrinsic half-life ($p < 0.04$), as shown in Figure 14.5. It does seem logical that repeated-dose charcoal therapy would be more effective for drugs which tend to be eliminated from the plasma more slowly, since this would give the charcoal "a chance to work."

Tenenbein (1991) wrote a short article calling for the reappraisal of multiple-dose charcoal therapy. He states that multiple-dose charcoal is regarded as simple, inexpensive,

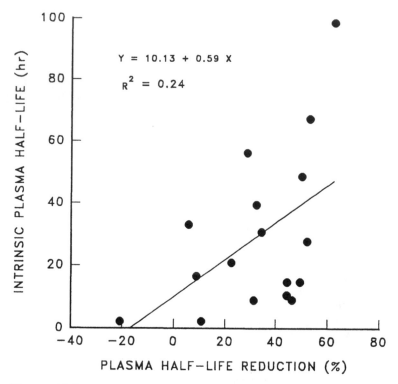

Figure 14.5   Linear regression analysis of plasma half-life reduction after repeat-dose activated charcoal and intrinsic plasma half-life of 17 drugs. From Campbell and Chyka (1992). Reproduced by permission of the W. B. Saunders Company.

effective, and safe, but that reports of its efficacy have been mostly anecdotal case reports of acutely poisoned patients or have involved controlled studies with subtoxic drug levels and that several controlled studies have shown no benefit. He points out the growing incidence of complications (aspiration, bowel obstruction, etc.).

XVII. REFERENCES

AHFS Drug Information '92, American Hospital Formulary Service, American Society of Hospital Pharmacists, Bethesda, Maryland, 1992.

Amitai, Y., Yeung, A. C., Moye, J., and Lovejoy, F. H., Jr. (1986). Repetitive oral activated charcoal and control of emesis in severe theophylline toxicity, Annals Intern. Med. *105*, 386.

Amitai, Y. and Degani, Y. (1990). Treatment of phenobarbital poisoning with multiple dose activated charcoal in an infant, J. Emerg. Med. *8*, 449.

Anonymous (1987). Repeated oral activated charcoal in acute poisoning (editorial), Lancet, May 2, p. 1013. Erratum, Lancet, May 16, p. 1160.

Arimori, K. and Nakano, M. (1989). Study on transport of disopyramide into the intestinal lumen aimed at gastrointestinal dialysis by activated charcoal in rats, J. Pharm. Pharmacol. *41*, 445.

Arimori, K., Kawano, H., and Nakano, M. (1989). Gastrointestinal dialysis of disopyramide in healthy subjects, Int. J. Clin. Pharmacol. Ther. Toxicol. *27*, 280.

Augenstein, W. L., Kulig, K. W., and Rumack, B. H. (1987). Delayed rise in serum drug levels in overdose patients despite multiple-dose charcoal and after charcoal stools (abstract), Vet. Hum. Toxicol. *29*, 491.

Barone, J. A., Raia, J.J., and Huang, Y. C. (1988). Evaluation of the effects of multiple-dose activated charcoal on the absorption of orally administered salicylate in a simulated toxic ingestion model, Ann. Emerg. Med. *17*, 34.

Berg, M. J., Berlinger, W. G., Goldberg, M. J., Spector, R., and Johnson, G. F. (1982). Acceleration of the body clearance of phenobarbital by oral activated charcoal, New Engl. J. Med. *307*, 642.

Berlinger, W. G., Spector, R., Goldberg, M. J., Johnson, G. F., Quee, C. K., and Berg, M. J. (1983). Enhancement of theophylline clearance by oral activated charcoal, Clin. Pharmacol. Ther. *33*, 351.

Boldy, D. A. R., Smart, V., and Vale, J. A. (1985). Multiple doses of charcoal in digoxin poisoning (letter), Lancet, November 9, p. 1076.

Boldy, D. and Vale, J. A. (1986). Treatment of salicylate poisoning with repeated oral charcoal (letter), Br. Med. J. *292*, 136.

Boldy, D. A. R., Vale, J. A., and Prescott, L. F. (1986). Treatment of phenobarbitone poisoning with repeated oral administration of activated charcoal, Quart. J. Med. *61* (new series), 997.

Boldy, D. A. R., Heath, A., Ruddock, S., Vale, J. A., and Prescott, L. F. (1987). Activated charcoal for carbamazepine poisoning, Lancet, May 2, p. 1027.

Braithwaite, R. A., Crome, P., and Dawling, S. (1978). The in vitro and in vivo evaluation of activated charcoal as an adsorbent for tricyclic antidepressants, Br. J. Clin. Pharmacol. *5*, 369.

Brashear, R. E., Aronoff, G. R., and Brier, R. A. (1985). Activated charcoal in theophylline intoxication, J. Lab. Clin. Med. *106*, 242.

Bronstein, A. C., Sawyer, D. R., Rumack, B. H., and Black, E. M. (1984). Theophylline intoxication in a premature infant: Multiple dose activated charcoal therapy (abstract), Vet. Hum. Toxicol. *26*, 404.

Burkhart, K. K., et al. (1992). Exchange transfusion and multidose activated charcoal following vancomycin overdose, J. Toxicol.-Clin. Toxicol. *30*, 285.

Byatt, C. M. and Henry, J. A. (1984). Increased serum theophylline clearance with orally administered activated charcoal (letter), Am. Rev. Respir. Dis. *130*, 147.

Campbell, J. W. and Chyka, P. A. (1992). Physicochemical characteristics of drugs and response to repeat-dose activated charcoal, Am. J. Emerg. Med. *10*, 208.

Chin, L., Picchioni, A. L., and Duplisse, B. R. (1969). Comparative antidotal effectiveness of activated charcoal, Arizona montmorillonite, and evaporated milk, J. Pharm. Sci. *58*, 1353.

Chyka, P. A., Mandrell, T. D., Holley, J. E., and Sugathan, P. (1993). Evaluation of a porcine model to study repeat-dose activated charcoal therapy (abstract), Vet. Hum. Toxicol. *35*, 367.

Crome, P., Dawling, S., Braithwaite, R. A., Masters, J., and Walkey, R. (1977). Effect of activated charcoal on absorption of nortriptyline, Lancet, December 10, p. 1203.

Davis, R., Ellsworth, A., Justus, R. E., and Bauer, L. A. (1985). Reversal of theophylline toxicity using oral activated charcoal, J. Fam. Prac. *20*, 73.

Davis, R. L., Roon, R. A., Koup, J. R., and Smith, A. L. (1987). Effect of orally administered activated charcoal on vancomycin clearance, Antimicrob. Agents Chemother. *31*, 720.

Davis, R. L., Koup, J. R., Roon, R. A., Opheim, K. E., and Smith, A. N. (1988). Effect of oral activated charcoal on tobramycin clearance, Antimicrob. Agents Chemother. *32*, 274.

Dolgin, J. G., Nix, D. E., Sanchez, J., and Watson, W. A. (1991). Pharmacokinetic simulation of the effect of multiple-dose activated charcoal in phenytoin poisoning: Report of two pediatric cases, Drug Intell. Clin. Pharm.-Ann. Pharmacother. *25*, 646.

Douidar, S. M., Hale, T. W., Trevino, D., and Habersang, R. (1992). The effect of multiple dose activated charcoal (MDAC) on the elimination of intravenous (IV) sodium salicylate (NaSA) in rabbits (abstract), Vet. Hum. Toxicol. *34*, 362.

Du Souich, P., Caillé, G., and Larochelle, P. (1983). Enhancement of nadolol elimination by activated charcoal and antibiotics, Clin. Pharmacol. Ther. *33*, 585.

El-Sayed, Y. M., Niazy, E. M., Tawfik, A. F., and Al-Dardiri, M. M. (1993). Effect of oral activated charcoal on vancomycin clearance in rabbits with acute renal failure, Drugs Exptl. Clin. Res. *19*, 19.

Ferry, D. G., Gazeley, L. R., Busby, W. J., Beasley, D. M. G., Edwards, I. R., and Campbell, A. J. (1990). Enhanced elimination of piroxicam by administration of activated charcoal or cholestyramine, Eur. J. Clin. Pharmacol. *39*, 599.

Gadgil, S. D., Damle, S. R., Advani, S. H., and Vaidya, A. B. (1982). Effect of activated charcoal on the pharmacokinetics of high-dose methotrexate, Cancer Treat. Rep. *66*, 1169.

Gal, P., Miller, A., and McCue, J. D. (1984). Oral activated charcoal to enhance theophylline elimination in an acute overdose, JAMA *251*, 3130.

Ginoza, G. W., Strauss, A. A., Iskra, M. K., and Modanlou, H. D. (1987). Potential treatment of theophylline toxicity by high surface area activated charcoal, J. Pediatr. *111*, 140.

Goldberg, M. J. and Berlinger, W. G. (1982). Treatment of phenobarbital overdose with activated charcoal, JAMA *247*, 2400.

Goldberg, M. J., Berlinger, W. G., and Park, G. D. (1985a). Activated charcoal in phenobarbital overdose (letter), JAMA *253*, 1120.

Goldberg, M. J., Park, G. D., Spector, R., Fischer, L. J., and Feldman, R. D. (1985b). Lack of effect of oral activated charcoal on imipramine clearance, Clin. Pharmacol. Ther. *38*, 350.

Greb, W. H., Buscher, G., Dierdorf, H. D., von Schrader, H. W., and Wolf, D. (1989). Ability of charcoal to prevent absorption of paroxetine, Acta Psychiatr. Scand. *80* (Suppl. 350) 156.

Greenberg, A., Piraino, B. H., Kroboth, P. D., and Weiss, J. (1984). Severe theophylline toxicity: Role of conservative measures, antiarrhythmic agents, and charcoal hemoperfusion, Am. J. Med. *76*, 854.

Greenberg, A. (1985). Severe theophylline toxicity (reply to letter), Am. J. Med. *78*, A82.

Guzelian, P. S. (1981). Therapeutic approaches for chlordecone poisoning in humans, J. Toxicol. Environ. Health *8*, 757.

Hantson, P., Vandenplas, O., Mahieu, P., Wallemacq, P., and Hassoun, A. (1991). Repeated doses of activated charcoal and cholestyramine for digitoxin overdose: Pharmacokinetic data and urinary elimination, J. Toxicol. Clin. Exp. *11*, 401.

Hassan, E. (1986). Treatment of meprobamate overdose with repeated oral doses of activated charcoal, Ann. Emerg. Med. *15*, 73.

Hayden, J. W. and Comstock, E. G. (1975). Use of activated charcoal in acute poisoning, Clin. Toxicol. *8*, 515.

Heath, A. and Van Loo, T. (1987). Multiple dose activated charcoal therapy in carbamazepine overdose (abstract), Vet. Hum. Toxicol. *29* (Suppl. 2), 44.

Hillman, R. J. and Prescott, L. F. (1985). Treatment of salicylate poisoning with repeated oral charcoal, Br. Med. J. *291*, 1472.

Ho, J. L., Tierney, M. G., and Dickinson, G. E. (1989). An evaluation of the effect of repeated doses of oral activated charcoal on salicylate elimination, J. Clin. Pharmacol. *29*, 366.

Honcharik, N. and Anthone, S. (1985). Activated charcoal in acute cyclosporin overdose (letter), Lancet, May 4, p. 1051.

Howard, C. E., Roberts, R. S., Ely, D. S., and Moye, R. A. (1994). Use of multiple-dose activated charcoal in phenytoin toxicity, Ann. Pharmacother. *28*, 201.

Ilett, K. F., Hackett, L. P., Dusci, L. J., and Paterson, J. W. (1991). Disposition of dothiepin after overdose: Effects of repeated-dose activated charcoal, Ther. Drug Monit. *13*, 485.

Ilkhanipour, K., Yealy, D. M., and Krenzelok, E. P. (1990). The comparative efficacy of various multiple dose activated charcoal regimens (abstract), Ann. Emerg. Med. *19*, 453.

Ilkhanipour, K., Yealy, D. M., and Krenzelok, E. P. (1992). The comparative efficacy of various multiple-dose activated charcoal regimens, Am. J. Emerg. Med. *10*, 298.

Jones, J., McMullen, M. J., Dougherty, J., and Cannon, L. (1987). Repetitive doses of activated charcoal in the treatment of poisoning, Am. J. Emerg. Med. *5*, 305.

Kärkkäinen, S. and Neuvonen, P. J. (1984). Effect of oral charcoal and urine pH on sotalol pharmacokinetics, Int. J. Clin. Pharmacol. Ther. Toxicol. *22*, 441.

Kärkkäinen, S. and Neuvonen, P. J. (1985). Effect of oral charcoal and urine pH on dextro-propoxyphene pharmacokinetics, Int. J. Clin. Pharmacol. Ther. Toxicol. *23*, 219.

Kärkkäinen, S. and Neuvonen, P. J. (1986). Pharmacokinetics of amitriptyline influenced by oral charcoal and urine pH, Int. J. Clin. Pharmacol. Ther. Toxicol. *24*, 326.

Kilgore, T. L. and Lehmann, C. R. (1982). Treatment of digoxin intoxication with colestipol, South. Med. J. *75*, 1259.

Kirschenbaum, L. A., Mathews, S. C., Sitar, D. S., and Tenenbein, M. (1990). Does multiple-dose charcoal therapy enhance salicylate excretion?, Arch. Intern. Med. *150*, 1281.

Kulig, K. W., Bar-Or, D., and Rumack, B. H. (1987). Intravenous theophylline poisoning and multiple-dose charcoal in an animal model, Ann. Emerg. Med. *16*, 842.

Lake, K. D., Brown, D. C., and Peterson, C. D. (1984). Digoxin toxicity: Enhanced systemic elimination during oral activated charcoal therapy, Pharmacotherapy *4*, 161.

Lalonde, R. L., Deshpande, R., Hamilton, P. P., McLean, W. M., and Greenway, D. C. (1985). Acceleration of digoxin clearance by activated charcoal, Clin. Pharmacol. Ther. *37*, 367.

Laufen, H. and Leitold, M. (1986). The effect of activated charcoal on the bioavailability of piroxicam in man, Int. J. Clin. Pharmacol. Ther. Toxicol. *24*, 48.

Lim, D. T., Singh, P., Nourtsis, S., and Dela Cruz, R. (1986). Absorption inhibition and enhancement of elimination of sustained-release theophylline tablets by oral activated charcoal, Ann. Emerg. Med. *15*, 1303.

Linden, C. H., Lewis, P. K., and Rumack, B. H. (1983). Phenobarbital overdosage: Treatment with multiple dose activated charcoal (abstract), Vet. Hum. Toxicol. *25*, 270.

Linden, C. H. and Rumack, B. H. (1984). Enhanced elimination of meprobamate by multiple doses of activated charcoal (abstract), Vet. Hum. Toxicol. *26*, 404.

Lockey, D. and Bateman, D. N. (1989). Effect of oral activated charcoal on quinine elimination, Br. J. Clin. Pharmacol. *27*, 92.

Lopez-Herce, J., Garcia Teresa, M. A., Ruiz Beltran, A., Valdivielso, A., and Casado, J. (1991). Severe theophylline toxicity treated with oral activated charcoal, Intensive Care Med. *17*, 244.

Lyddane, J. E., Thomas, B. F., Compton, D. R., and Martin, B. R. (1988). Modification of phencyclidine intoxication and biodisposition by charcoal and other treatments, Pharmacol. Biochem. Behav. *30*, 371.

McKinnon, R. S., et al. (1987). Studies on the mechanisms of action of activated charcoal on theophylline pharmacokinetics, J. Pharm. Pharmacol. *39*, 522.

McLuckie, A., Forbes, A. M., and Ilett, K. F. (1990). Role of repeated doses of oral activated charcoal in the treatment of acute intoxications, Anaesth. Intens. Care *18*, 375.

Mahutte, C. K., True, R. J., Michiels, T. M., Berman, J. M., and Light, R. W. (1983). Increased serum theophylline clearance with orally administered activated charcoal, Am. Rev. Respir. Dis. *128*, 820.

Mahutte, C. K., Berman, J. M., and Light, R. W. (1984). Increased serum theophylline clearance with orally administered activated charcoal (reply to letter), Am. Rev. Respir. Dis. *130*, 148.

Mauro, L. S., Mauro, V. F., Brown, D. L., and Somani, P. (1987). Enhancement of phenytoin elimination by multiple-dose activated charcoal, Ann. Emerg. Med. *16*, 1132.

Mofenson, H. C., Caraccio, T. R., Greensher, J., D'Agostino, R., and Rossi, A. (1985). Gastrointestinal dialysis with activated charcoal and cathartic in the treatment of adolescent intoxications, Clin. Pediatr. *24*, 678.

Mofenson, H. C. and Caraccio, T. R. (1992). Is activated charcoal useful for acetaminophen overdose beyond two hours after ingestion? (letter), Ann. Emerg. Med. *21*, 894.

Montgomery, B. R. (1978). Tricyclic overdose? Treat with activated charcoal, JAMA *240*, 424.

Moulin, M. A., Potier, J. C., Grollier, G., and Camsonne, R. (1982). [Digitoxin intoxication: Accelerated elimination by administration of active carbon] (letter), Nouvelle Presse Médicale *11*, 1079.

Neuvonen, P. J. and Elonen, E. (1980). Effect of activated charcoal on absorption and elimination of phenobarbitone, carbamazepine, and phenylbutazone in man, Eur. J. Clin. Pharmacol. *17*, 51.

Neuvonen, P. J., Elonen, E., and Mattila, M. J. (1980). Oral activated charcoal and dapsone elimination, Clin. Pharmacol. Ther. *27*, 823.

Neuvonen, P. J. and Kärkkäinen, S. (1983). Effects of charcoal, sodium bicarbonate, and ammonium chloride on chlorpropamide kinetics, Clin. Pharmacol. Ther. *33*, 386.

Neuvonen, P. J., Elonen, E., and Haapanen, E. J. (1983). Acute dapsone intoxication: Clinical findings and effect of oral charcoal and haemodialysis on dapsone elimination, Acta. Med. Scand. *214*, 215.

Neuvonen, P. J. and Olkkola, K. T. (1988). Oral activated charcoal in the treatment of intoxications: Role of single and repeated doses, Med. Toxicol *3*, 33.

Ohning, B. L., Reed, M. D., and Blumer, J. L. (1986). Continuous nasogastric administration of activated charcoal for the treatment of theophylline intoxication, Pediatr. Pharmacol. *5*, 241.

Park, G. D., et al. (1983). Effects of size and frequency of oral doses of charcoal on theophylline clearance, Clin. Pharmacol. Ther. *34*, 663.

Park, G. D., Spector, R. , Goldberg, M. J., Johnson, G. F., Feldman, R., and Quee, C. K. (1984). Effect of the surface area of activated charcoal on theophylline clearance, J. Clin. Pharmacol. *24*, 289.

Park, G. D., et al. (1985). The effects of activated charcoal on digoxin and digitoxin clearance, Drug Intell. Clin. Pharm. *19*, 937.

Payne, V. W., Sector, R. A., and Noback, R. K. (1982). Use of colestipol in a patient with digoxin intoxication, Drug Intell. Clin. Pharm. *15*, 902.

Pond, S., Jacobs, M., Marks, J., Garner, J. , Goldschlager, N., and Hansen, D. (1981). Treatment of digitoxin overdose with oral activated charcoal (letter), Lancet, November 21, p. 1177.

Pond, S., Olson, K. R., Osterloh, J. D., and Tong, T. G. (1984). Randomized study of the treatment of phenobarbital overdose with repeated doses of activated charcoal, JAMA *251*, 3104.

Pond, S., Osterloh, J. D., Olson, K. R., and Tong, T. G. (1985). Activated charcoal in phenobarbital overdose (reply to letter), JAMA *253*, 1121.

Pond, S. M. (1986a). A review of the pharmacokinetics and efficacy of emesis, gastric lavage and single and repeated doses of charcoal in overdose patients, in *New Concepts and Developments in Toxicology*, P. L. Chambers, P. Gehring, and F. Sakai, Eds., Elsevier, Amsterdam.

Pond, S. M. (1986b). Role of repeated oral doses of activated charcoal in clinical toxicology, Med. Toxicol. *1*, 3.

Prescott, L. F., Hamilton, A. R., and Heyworth, R. (1989). Treatment of quinine overdosage with repeated oral charcoal, Br. J. Clin. Pharmacol. *27*, 95.

Radomski, L., Park, G. D., Goldberg, M. J., Spector, R., Johnson, G. F., and Quee, C. K. (1984). Model for theophylline overdose treatment with oral activated charcoal, Clin. Pharmacol. Ther. *35*, 402.

Reigart, J. R., Trammel, H. L., Jr., and Lindsey, J. M. (1982-83). Repetitive doses of activated charcoal in dapsone poisoning in a child, J. Toxicol.-Clin. Toxicol. *19*, 1061.

Reissell, P. and Manninen, V. (1982). Effect of administration of activated charcoal and fibre on absorption, excretion, and steady state blood levels of digoxin and digitoxin: Evidence for intestinal secretion of the glycosides, Acta. Med. Scand. (Suppl.) *668*, 88.

Roberts, D., Honcharik, N., Sitar, D. S., and Tenenbein, M. (1991). Diltiazem overdose: Pharmacokinetics of diltiazem and its metabolites and effect of multiple-dose charcoal therapy, J. Toxicol.-Clin. Toxicol. *29*, 45.

Ros, S. P. and Black, L. E. (1989). Multiple-dose activated charcoal in management of phenytoin overdose, Pediatr. Emerg. Care *5*, 169.

Rowden, A. M., Spoor, J. E., and Bertino, J. S., Jr. (1990). The effect of activated charcoal on phenytoin pharmacokinetics, Ann. Emerg. Med. *19*, 1144.

Rygnestad, T., Walstad, R. A., and Dahl, K. (1986). Self poisoning with theophylline: The effect of repeated doses of oral charcoal on drug elimination, Acta. Med. Scand. *219*, 425.

Scheinen, M., Virtanen, R., and Iisalo, E. (1985). Effect of single and repeated doses of activated charcoal on the pharmacokinetics of doxepin, Int. J. Clin. Pharmacol. Ther. Toxicol. *23*, 38.

Sessler, C. N., Glauser, F. L., and Cooper, K. R. (1985). Treatment of theophylline toxicity with oral activated charcoal, Chest *87*, 325.

Sethna, M., Solomon, G., Cedarbaum, J., and Kutt, H. (1989). Successful treatment of massive carbamazepine overdose, Epilepsia *30*, 71.

Shannon, M., Amitai, Y., and Lovejoy, F. H., Jr. (1987). Multiple dose activated charcoal for theophylline poisoning in young infants, Pediatrics *80*, 368.

Shannon, M. W. and Woolf, A. (1992). The efficacy of elimination enhancement procedures after theophylline intoxication (abstract), Vet. Hum. Toxicol. *34*, 331.

Slaughter, R. L. and Dupuis, R. E. (1985). Severe theophylline toxicity (letter), Am. J. Med. *78*, A76, A78, A82.

Swartz, C. M. and Sherman, A. (1984). The treatment of tricyclic antidepressant overdose with repeated charcoal, J. Clin. Psychopharmacol. *4*, 336.

Tenenbein, M., Kirschenbaum, L. A., and Sitar, D. S. (1989). Multiple-dose charcoal therapy for salicylate poisoning (abstract), Ann. Emerg. Med. *18*, 444.

Tenenbein, M. (1991). Multiple doses of activated charcoal: Time for reappraisal?, Ann. Emerg. Med. *20*, 529.

Traeger, S. M. and Haug, M. T. (1986). Reduction of diazepam serum half life and reversal of coma by activated charcoal in a patient with severe liver disease, J. Toxicol.-Clin. Toxicol. *24*, 329.

True, R. J., Berman, J. M., and Mahutte, C. K. (1984). Treatment of theophylline toxicity with oral activated charcoal, Crit. Care Med. *12*, 113.

Turk, J., Aks, S. E., and Hryhorczuk, D. O. (1993). Constraints on the enhancement of elimination of drugs with activated charcoal, Vet. Hum. Toxicol. *35*, 489.

Vale, J. A., Ruddock, F. S., and Boldy, D. A. R. (1987). Multiple doses of activated charcoal in the treatment of phenobarbitone and carbamazepine poisoning (abstract), Vet. Hum. Toxicol. *29* (Suppl. 2), 152.

Vale, J. A. and Heath, A. (1992). Carbamazepine overdose (letter), J. Toxicol.-Clin. Toxicol. *30*, 481.

Veerman, M., Espejo, M. G., Christopher, M. A., and Knight, M. (1991). Use of activated charcoal to reduce elevated serum phenobarbital concentration in a neonate, J. Toxicol.-Clin. Toxicol. *29*, 53.

Venho, V. M. K., Salonen, R. O., and Mattila, M. J. (1978). Modification of the pharmacokinetics of doxycycline in man by ferrous sulphate or charcoal, Eur. J. Clin. Pharmacol. *14*, 277.

Vertrees, J. E., McWilliams, B. C., and Kelly, H. W. (1990). Repeated oral administration of activated charcoal for treating aspirin overdose in young children, Pediatrics *85*, 594.

Vicas, I. M. O. (1987). Digoxin overdose managed with multi-dose activated charcoal (abstract), Vet. Hum. Toxicol. *29*, 463.

Wason, S., Baker, R. C., Carolan, P., Seigel, R., and Druckenbrod, R. W. (1992). Carbamazepine overdose—The effects of multiple dose activated charcoal, J. Toxicol.-Clin. Toxicol. *30*, 39.

Watson, W. A., Jenkins, T. C., Velasquez, N., and Schentag, J. J. (1987). Repeated oral doses of activated charcoal and the clearance of tobramycin, a nonabsorbable drug, J. Toxicol.-Clin. Toxicol. *25*, 171.

Weichbrodt, G. D. and Elliott, D. P. (1987). Treatment of phenytoin toxicity with repeated doses of activated charcoal, Ann. Emerg. Med. *16*, 1387.

Weidle, P. J., Skiest, D. J., and Forrest, A. (1991). Multiple-dose activated charcoal as adjunct therapy after chronic phenytoin intoxication, Clin. Pharm. *10*, 711.

Wogan, J. M., Kulig, K., and Frommer, D. A. (1986). Multiple-dose activated charcoal in salicylate poisoning (abstract), Ann. Emerg. Med. *15*, 651.

Wogan, J., Frommer, D., Kulig, K., and Rumack, B. (1987). Multiple dose activated charcoal for intravenous salicylate intoxication in a dog model (abstract), Vet. Hum. Toxicol. *29* (Suppl. 2), 41.

Yeakel, D., Stemple, C., and Dougherty, J. (1988). A prospective human cross-over study on single versus multiple-dose charcoal in salicylate ingestion (abstract), Ann. Emerg. Med. *17*, 439.

15

Ipecac, Cathartics, and Charcoal: Interactions and Comparative Efficacies

In Chapter 7, we reviewed several methods for treating victims of drug overdoses or poisonings; however, we were primarily concerned with simply describing each of the various methods—oral dilution, administration of syrup of ipecac, gastric lavage, use of saline cathartics, use of sorbitol, whole-bowel irrigation, etc.—and some discussion of the efficacy of each individual treatment method was presented. In this chapter, we specifically compare the effectiveness of charcoal against the effectiveness of some of these other treatment modalities. In particular, we compare charcoal to ipecac, lavage, and whole-bowel irrigation. Additionally, we discuss the interactions which can or do occur when charcoal and other treatments are used together (e.g., charcoal and ipecac, charcoal and sorbitol, and charcoal and saline cathartics).

I. EFFECT OF CHARCOAL ON IPECAC

Several authors have mentioned that charcoal might interfere with the emetic effects of ipecac but without citation of any supporting scientific evidence. For example, Corby and Decker (1974) stated without any elaboration that: "since charcoal effectively adsorbs ipecac, syrup of ipecac should be given before the activated charcoal." Another example is a letter by Manoguerra (1975), referring to an article by another author, which recommended giving syrup of ipecac followed by a charcoal slurry and then repeating the ipecac in 20 min if vomiting had not yet occurred, stated that because of the interaction of charcoal and ipecac, charcoal should not be used until emesis has occurred.

Another example is an editorial by Levy (1982) in which he indicated his belief that: "orally administered medications (including ipecac) are not likely to be absorbed [i.e., absorbed from the GI tract] during the administration of charcoal ...." Llera (1983), commenting on the editorial, agreed and made reference to an article by Greensher et al. (1979) which stated: "If syrup of ipecac is to be used concomitantly [with charcoal], wait until it has induced vomiting before adding activated charcoal, as it may adsorb the emetic principles in ipecac and prevent vomiting."

Yet some scientific evidence does exist to support the idea that charcoal can suppress the emetic activity of ipecac. Perhaps the earliest reference to ipecac inactivation by charcoal is in an 1846 paper by Garrod (quoted by Holt and Holz, 1963). Garrod gave 10 grains (0.65 g) of ipecacuanha to an unspecified number of dogs, thereby inducing emesis. The emetic effect was negated when the ipecacuanha was administered with 0.5 oz (14.2 g) of animal charcoal. Pierce (1975), in a letter, pointed out that syrup of ipecac contains 70 g powdered ipecac per liter and thus even a 15 mL dose of the syrup contains about 1 g ipecac. Thus, Garrod's ipecac dose was not a large one. Additionally, dogs have been found to be less resistant to ipecac-induced emesis than humans. For these reasons, Garrod's study is not conclusive in suggesting that charcoal would inactivate ipecac in humans.

However, in support of the idea that charcoal could inactivate ipecac, there is an in vitro study by Cooney (1978) which determined quantitatively how strongly the alkaloid in ipecac which is primarily responsible for its emetic action (i.e., emetine) adsorbs to charcoal. Prior studies by others showed that alkaloids can adsorb well to charcoal; for example, studies by Anderson (1946, 1947, 1948), by Picchioni et al. (1966, 1974), and by Henschler (1970) on alkaloids such as strychnine, nicotine, atropine, morphine, yohimbine, veratrine, and aconitine clearly showed that charcoal generally adsorbs alkaloids well. However, the binding varies widely with pH. In particular, since alkaloids are basic and are thus ionized at low pH, the degree of adsorption in a gastric environment is much less than at higher pH.

Cooney used emetine hydrochloride in SGF and added different amounts of charcoal. The equilibrium adsorption data were fit over the concentration range of interest with a Freundlich isotherm equation. Cooney showed that for a usual dose of syrup of ipecac (2 tablespoonsful or 30 mL), which would contain about 42 mg alkaloids, one gram of charcoal would adsorb essentially 100% of the emetine. Thus, charcoal was predicted to nullify the effect of ipecac if given at the same time or before the ipecac had a chance to work.

The concurrent use of syrup of ipecac and charcoal was investigated by Krenzelok et al. (1986). The purpose of the study was to assess whether charcoal interferes with the emetic action of the ipecac. No mention of the subjects receiving any drug is made. Contrary to what might have been expected, emesis occurred in 8 out of 10 of the subjects in an average of 20 min when an aqueous slurry of 50 g charcoal was administered 5 min after 60 mL syrup of ipecac (note, however, that the ipecac dose was twice the normal amount). The total dose of charcoal was retained for a mean time of 6.75 min (range 0–17 min). In an earlier report on this same study (Freedman et al., 1985), the volunteers ingested 2.6 g aspirin as a marker drug; serum salicylate measurements 2 hr after the aspirin ingestion showed an average reduction of 57% from control values for the 8 subjects who vomited and an average reduction of 48% from control values for the two subjects who did not vomit. Thus, the authors stated that the charcoal was effective regardless of whether emesis occurred or not.

Based on this initial study, a follow-up investigation with overdose patients in an emergency department was done by the same researchers (Freedman et al., 1987). A dose of 60 mL syrup of ipecac (again, twice the usual dose) was given to 10 overdose patients by nasogastric tube and followed 10 min later with a slurry of 50 g charcoal in

500 mL water, also given via the nasogastric tube. Thirty minutes after emesis subsided, a second dose of 50 g charcoal (with sorbitol) was given orally. Emesis started an average of 13.8 min after the ipecac and ceased an average of 45.9 min after the ipecac. On average, 3.7 episodes of vomiting occurred per patient. Thus, it appears that charcoal, given after some delay, does not interfere with the induction of emesis by a 60 mL dose of ipecac.

The various results cited above are mixed, as some studies show that charcoal interferes with the emetic action of syrup of ipecac and other studies suggest that charcoal does not. Clearly, additional research is needed to clarify things, particularly when 30 mL doses of ipecac are employed.

However, it is not current practice to give syrup of ipecac and charcoal concomitantly as a treatment for overdose. Indeed, the use of syrup of ipecac in medical facilities is being phased out in favor of using charcoal instead. In this sense, a complete resolution of the questions surrounding the issue posed may be irrelevant.

II. COMPARATIVE EFFECTIVENESS OF IPECAC AND CHARCOAL

Neuvonen et al. (1983) compared charcoal and syrup of ipecac as separate therapies for preventing drug absorption. Three drugs (1000 mg paracetamol, 500 mg tetracycline, and 350 mg aminophylline) were given to healthy subjects and then charcoal (50 g) was administered either after 5 min or after 30 min. In other tests, syrup of ipecac (20 mL) was given after either 5 or 30 min. When charcoal was given at 5 min, the 0–24 hr AUC values for paracetamol, tetracycline, and aminophylline were reduced to 22, 3, and 25% of the control values, respectively. When charcoal was given at 30 min, the AUC values were reduced to 92, 36, and 53% of the control values. Syrup of ipecac at 5 min lowered the 0–24 hr AUC values for paracetamol, tetracycline, and aminophylline to 31, 21, and 48% of the control values, respectively. Finally, syrup of ipecac at 30 min lowered the AUC values to 107, 70, and 98% of the control values, respectively (the 107% figure means that the AUC was higher than in the control case). Figure 15.1 shows the results for the case of tetracycline. Syrup of ipecac caused emesis in all subjects, with a mean delay of 15 min. When given at the same time, charcoal was more effective than ipecac (ratios of AUC values for charcoal:ipecac were 0.71, 0.14, 0.52, 0.86, 0.51, and 0.54 when compared for the same times of administration). However, note that the ipecac dose was only two-thirds of the usual recommended amount for adults.

Neuvonen and Olkkola (1984) studied the effect of charcoal and syrup of ipecac on cimetidine and pindolol absorption in man after administration of metoclopramide as an antiemetic agent. Charcoal (50 g), given orally 5 min after 400 mg cimetidine plus 10 mg pindolol, reduced their absorption by 99% or more. Syrup of ipecac (again, only 20 mL), although producing emesis in every case, reduced the absorption of cimetidine and pindolol by 75 and 60%, respectively. Figure 15.2 presents the data on pindolol. Hence, the authors recommended that charcoal be given immediately, without preceding lavage or emesis.

The efficacy of ipecac versus charcoal/cathartic in a simulated aspirin overdose was studied by Curtis et al. (1984). They gave twenty-four 81-mg aspirin tablets to healthy volunteers, followed by 30 mL ipecac syrup (repeated if emesis did not occur within 30 min), or 60 g charcoal plus 15 g magnesium sulfate. Another group received the ipecac (again, repeated if necessary) and then the charcoal/MgSO$_4$ 90 min after the last vomiting

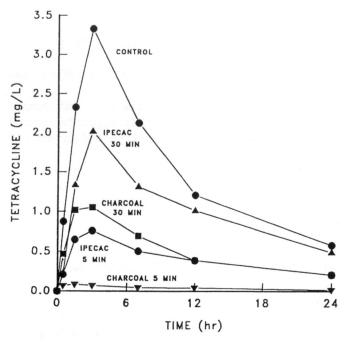

Figure 15.1 Effect of activated charcoal and syrup of ipecac on tetracycline levels in six volunteers. From Neuvonen et al. (1983). Figure 2. Copyright Springer-Verlag GmbH & Co. Used with permission.

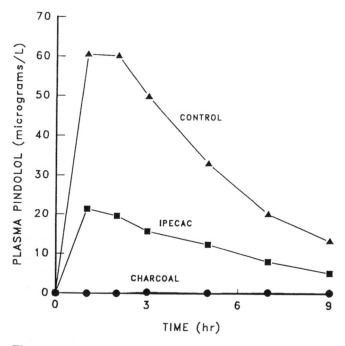

Figure 15.2 Plasma concentrations of pindolol for controls, charcoal-treated subjects, and ipecac-treated subjects. From Neuvonen and Olkkola (1984).

Table 15.1 Extent of Drug Absorption: Ipecac Versus Charcoal

| Drug | Time after drug (min) | Percent absorption | | Reference |
		Ipecac	Charcoal	
Acetaminophen	5	35	15	Neuvonen et al. (1983)
	30	98	60	
Tetracycline	5	28	3	Neuvonen et al. (1983)
	30	78	38	
Aminophylline	5	49	19	Neuvonen et al. (1983)
	30	85	25	
Cimetidine[a]	5	25	1	Neuvonen and Olkkola (1984)
Pindolol[a]	5	40	1	Neuvonen and Olkkola (1984)
Aspirin	0	70	57	Curtis et al. (1984)

[a] Subjects were given metoclopramide 1 hr before. From Pond (1986). Reprinted by permission of Elsevier Science Publishers BV.

episode. All treatments began 60 min after the aspirin ingestion. The mean recovery of salicylate from the urine over 0–48 hr was 96.3% in control tests, 70.3% when ipecac was used, 56.4% for the charcoal/$MgSO_4$ case, and 72.4% for the combined ipecac/charcoal/$MgSO_4$ case. Thus, activated charcoal/$MgSO_4$ given right away was the best of the strategies tried. Adding charcoal/$MgSO_4$ to the ipecac treatment 90 min after the last vomiting episode was not much different from using ipecac alone; thus, the implication is that the charcoal was delayed too long in this case to be of value. It should be mentioned that, even when the charcoal/$MgSO_4$ was delayed for 90 min, 8 of the 10 subjects in this group vomited the charcoal mixture immediately.

The results of these three studies are summarized in Table 15.1. It is clear that drug absorption was always less with charcoal treatment than with the use of ipecac.

In a four-phase cross-over study, Tenenbein et al. (1987) gave 5 g ampicillin to ten volunteers and then treated them with either gastric lavage, 50 g charcoal/30 g magnesium sulfate in a slurry taken orally, or a dose of 30 mL syrup of ipecac (repeated at 20 min if vomiting did not occur). Serum ampicillin levels for the four phases are shown in Figure 15.3. AUC values over 0–12 hr were reduced 32% by lavage, 38% by ipecac-induced emesis, and 57% by the charcoal/cathartic. Thus, charcoal was recommended over lavage or emesis.

Charcoal alone versus charcoal and ipecac combined was studied by Albertson et al. (1989; this paper also appears in abstract form in Foulke et al., 1988). Two hundred actual overdose patients presenting to an emergency department over a 24-month period were involved. One group received 30 mL syrup of ipecac orally (repeated in 30 min if there was no emesis) followed with enough of a slurry of 50 g charcoal/sorbitol in water (Actidose with Sorbitol) to give a charcoal dose of 1 g/kg. The other group received only the charcoal slurry (again, at 1 g/kg charcoal). The group receiving only charcoal spent an average of 6.0 hr in the ED, versus 6.8 hr for the charcoal/ipecac group. The percentage of patients requiring hospitalization was not significantly different for the two groups. A complication rate of 5.4% was found for the charcoal/ipecac group versus 0.9% for the charcoal group. Three episodes of aspiration pneumonitis occurred with the

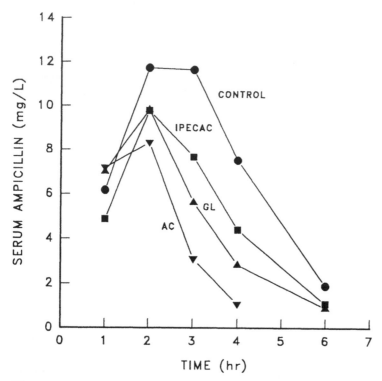

Figure 15.3 Ampicillin levels in subjects treated with ipecac, gastric lavage (GL), or activated charcoal (AC). From Tenenbein et al. (1987). Reprinted with permission of the American College of Emergency Physicians.

charcoal/ipecac group, versus none for the charcoal group. Thus, the use of ipecac in addition to charcoal does not appear to offer any advantage. A letter by Stiell (1990) questioned the design of the Albertson study with respect to randomness, criteria for patient inclusion/exclusion, assessment of outcomes, and other aspects. Albertsons's group (see Foulke et al., 1990) replied to these concerns.

McNamara et al. (1989) studied ipecac versus charcoal/cathartic as treatments for acetaminophen overdose. The drug (3.0 g) was given orally to fasting volunteers and followed 1 hr later by 30 mL syrup of ipecac or by 50 g charcoal in a 70% sorbitol solution (Charcoaid). Adequate emesis occurred in all ipecac-treated subjects, starting at a mean time of 25.5 min and with an average of four episodes per subject. AUC values over 0–8 hr for the ipecac group averaged 21.0% less than control values, while AUC values for the charcoal/sorbitol group were 25.5% less than controls. Thus, both treatments were moderately effective but were not significantly different from each other. Drug blood level data showed that by the time ipecac or charcoal/sorbitol were given at 1 hr, substantial drug absorption had already occurred. Hence, earlier intervention undoubtedly would have been much more effective (but the 1 hr delay is more representative of real-life conditions).

Kirk et al. (1991) carried out a study with 123 children aged 6 years or younger, referred to a regional poison center, who had ingested 150 mg/kg or more of acetaminophen. Groups were treated with : (1) ipecac at home within 1 hr, (2) charcoal in an ED within 2 hr, or (3) no gastrointestinal decontamination. Postingestion acetaminophen blood levels were obtained at 4.5 hr and for the three groups had mean values of 11, 20, and 39 mg/L, respectively. Hence, ipecac given within 1 hr was the best of the three protocols, while charcoal within 2 hr was also effective.

Kornberg and Dolgin (1991) described a two-year study carried out in a pediatric ED. The patients were 70 children, aged six years or less, all of whom were orally poisoned. One group received 15 mL syrup of ipecac (repeated 30 min later if no emesis occurred). After vomiting subsided, they received 1 g/kg charcoal in 40% sorbitol (Acti-dose with Sorbitol). The other group of patients was given 1 g/kg charcoal without any preceding gastric emptying. In all cases, the charcoal was offered orally and if refusal occurred it was given by nasogastric tube. It was concluded that for mild-to-moderate oral poisoning of young children, syrup of ipecac treatment delays the ultimate admin-istration of charcoal, interferes with its retention, and prolongs ED time. No advantage of syrup of ipecac prior to charcoal was seen. The data compared were those relating to time spent in the ED, retention of charcoal, patient improvement while in the ED, etc. No quantification of the effect of either treatment method on drug elimination was made. Thus, the study is somewhat qualitative. One interesting fact was that only 21 of the 70 patients (30%) were willing to take the charcoal orally. The mean age of the patients was about 2.4 yr; thus, it appears that children this young are quite resistant to accepting charcoal orally.

A review of the Kornberg/Dolgin study in *Emergency Medicine* (Anonymous, 1991) quotes Dr. Kornberg as stating: "We feel that if a child needs to be treated, ipecac is just going to get in the way. We no longer give it in the emergency department—it really has no benefit in that situation." He explains that "... in a busy emergency department, it can take hours for him [the child] to be seen, and giving ipecac that late may remove only 10% or 20%." Dr. Kornberg points out, however, that ipecac is "still an appropriate treatment for parents to use immediately after an ingestion" (i.e., in the home).

Danel et al. (1988) compared charcoal, emesis, and gastric lavage in aspirin overdose. Twelve healthy volunteers were given 1.5 g aspirin and were treated with nothing (controls), with 30 mL syrup of ipecac at 60 min (repeated if emesis did not occur in 30 min), gastric lavage started at 60 min, or 50 g oral charcoal (Carbomix) at 60 min. Salicylate levels in the urine were determined for 24 hr. The results are shown in Table 15.2. There were no significant differences in the three treatments in terms of the effects on aspirin absorption. However, it is clear that using a 60 min delay with a test drug like aspirin, which is quite rapidly absorbed, essentially precluded any differences from occurring, as the drug was probably almost totally absorbed by that time (note that the control results are barely different from the treatment results). Hence, the study provides

Table 15.2 Effects of Charcoal, Emesis, and Gastric Lavage

	Control	Emesis	Lavage	Charcoal
Mean salicylate in urine (%)	60.3	55.6	55.5	52.5
Fluid recovered (mL)	—	303	2755	—

From Danel et al. (1988). Reprinted with permission of the BMJ Publishing Group.

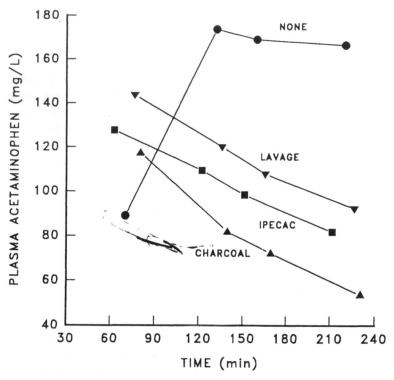

Figure 15.4 Mean plasma acetaminophen levels for patients treated with lavage, ipecac, or activated charcoal. From Underhill et al. (1990). Reproduced with permission of Blackwell Scientific Publications Ltd.

a very weak comparison of the three treatment protocols. The authors state that the emesis and lavage treatments both produced discomfort in the patients and that the lavage treatment was somewhat time-consuming (over 30 min) to perform. Thus, the authors concluded that charcoal should be used because of its equal effectiveness, its ease of use, and lack of adverse effects.

Underhill et al. (1990) compared gastric lavage, ipecac, and charcoal for reducing paracetamol (acetaminophen) absorption. Sixty patients 16 years of age or over who had ingested 5 g or more of paracetamol within 4 hr of admission to an ED were treated with lavage, ipecac (30 mL, repeated after 30 min if no response occurred), or activated charcoal (Carbomix, in a charcoal:drug ratio of 10:1). Blood samples taken prior to treatment and 60, 90, and 150 min later were analyzed for paracetamol. Figure 15.4 shows the results. The mean percentage decreases in the drug level between the first and last blood samples were: with lavage, 39.3%; with ipecac, 40.7%; and with charcoal, 52.2%. Thus, charcoal was the best treatment (the difference between lavage and ipecac was not significant).

In vivo tests of the effect of charcoal on tilidine HCl absorption were done by Cordonnier et al. (1987) and comparisons to the use of ipecac were made. The drug (50 mg) was given orally to healthy volunteers and then after 3 or 25 min either 20 g charcoal or 20 mL syrup of ipecac was given. With ipecac, emesis occurred in every subject, with first emesis at a mean time of 14.9 min (range 13.6–15.4 min). Urine was collected for 48 hr. Charcoal at 3 min reduced tilidine absorption by 89% and charcoal at 25 min reduced tilidine absorption by 66%. Syrup of ipecac at 3 min reduced tilidine absorption by 56% but when given at 25 min the ipecac had no significant effect (absorption was reduced by 1.1%). Thus, when given at the same time, charcoal was much more effective than ipecac.

The studies just reviewed make it clear that charcoal alone is consistently more effective than syrup of ipecac alone and that, in two studies, charcoal was also superior to gastric lavage. Moreover, syrup of ipecac is certainly uncomfortable for the patient and runs the risk of pulmonary aspiration of gastric contents and other complications (in one study, there were six times as many complications with ipecac use than with charcoal use). Thus, in a medical facility environment, syrup of ipecac should be shunned in favor of the immediate use of charcoal alone.

Notwithstanding this recommendation, if poisoning occurs in the home and syrup of ipecac is the only agent immediately available, it should be used (except when a physician or medical facility advises otherwise).

III. COMPARATIVE EFFECTIVENESS OF LAVAGE AND CHARCOAL

We have already compared lavage and charcoal above in two studies where lavage, emesis, and charcoal treatments were all employed. The studies involved were those of Tenenbein et al. (1987) and Danel et al. (1988). As the above discussion stated, lavage reduced ampicillin absorption by 32%, compared to 57% for charcoal/cathartic therapy, when both were carried out immediately. However, in the study involving aspirin, where the two treatments were delayed for 60 min, they produced small but identical beneficial effects (however, this study is of little significance, since drug absorption was probably almost complete prior to intervention).

Burton et al. (1984) compared charcoal and gastric lavage in the prevention of aspirin absorption in dogs. Thirty minutes after giving the dogs 500 mg/kg aspirin by gastric intubation, one group of dogs was treated with a slurry of charcoal in water (1.5 g/kg Norit A charcoal in 150 mL water). A second group was treated by lavage using 200 mL portions of water, repeated for a total of 3 L of water (each 200 mL took about 10 min and thus the 3 L lavage required about 2.5 hr), after which these dogs were given a 1.5 g/kg charcoal slurry. A third group of dogs received the charcoal slurry first then the lavage procedure followed by a second charcoal slurry (it was not stated how much time, if any, passed between administration of the first charcoal slurry and performing the first lavage; this seems to be a crucial piece of information). The effects of the different protocols are presented in Figure 15.5. Peak plasma drug concentrations were reduced (compared to control values) by an average of 17% with charcoal alone, by 37% for the lavage plus charcoal treatment, and by 48% for the charcoal plus lavage plus additional charcoal treatment. Although drug plasma concentrations were followed over 0–24 hr, no AUC values were computed. In any event, the authors concluded that lavage followed by charcoal is more effective than charcoal alone. However, it would appear that charcoal followed by lavage is even better. Whether similar results would occur in humans is quite uncertain, since studies with humans have shown lavage to be fairly inefficient for drug removal (see Chapter 7). Kulig (1984) makes this point, among others, in a commentary on the Burton study.

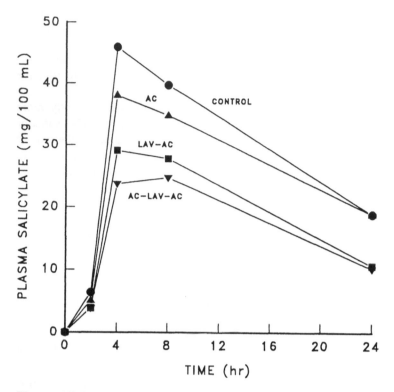

Figure 15.5 Salicylate levels in subjects treated with activated charcoal (AC), lavage plus charcoal (LAV-AC), and charcoal/lavage/charcoal (AC-LAV-AC). From Burton et al., 1984.

Comstock et al. (1982) assessed the efficacy of charcoal following gastric lavage in acute drug emergencies. Patients presenting to an ED were treated with gastric lavage; one group then received a slurry of 100 g Norit A charcoal via the gastric tube while the other group did not receive any charcoal. Efficacy was evaluated by subsequent blood drug concentration changes. The blood drug level changes in the two groups were similar and not statistically different. Comparisons using subgroups based on the type of drug, the treatment delay time, and entering functional decompensation showed a benefit of charcoal only for the less symptomatic patients. It was concluded that the use of charcoal following lavage may often be too late to be of benefit. Charcoal, if given, should be given in the home, in the emergency vehicle, or immediately upon admission.

These studies make it somewhat difficult to compare charcoal alone versus gastric lavage alone, since two of them (by the Burton and Comstock groups) mixed the two treatment methods. Thus, we must go back to the Danel study, which is not definitive, due to the use of a rapidly absorbed drug and a long delay time prior to intervention, or to the Tenenbein study. The Tenenbein study provides the best comparison and it shows that charcoal alone was quite superior to lavage alone. When one considers also that lavage is time-consuming, entails real risks, and is highly uncomfortable for the patient, it becomes clear that the immediate use of charcoal alone is much to be preferred.

IV. COMPARATIVE EFFECTIVENESS OF WHOLE BOWEL IRRIGATION AND CHARCOAL

Although whole-bowel irrigation (WBI) is a relatively new technique, quite a few studies comparing it to charcoal have already been reported. Much of this is due to the considerable activity of the Tenenbein group in Canada (Winnipeg, Manitoba), which has pioneered this approach. We review those studies now.

Brown et al. (1987) gave six healthy volunteers a simulated caffeine overdose (5 mg/kg) and then treated them with: (1) an irrigating fluid (Colyte) in an amount of 1 gallon over 2 hr, (2) the irrigating fluid followed by 1 g/kg oral charcoal, (3) 2 mL/kg of a suspension of charcoal in 70% sorbitol, or (4) nothing. Time to first stool, stool volume, and caffeine clearance and half-life were determined for all protocols. Times to first stool for the first three protocols were 74.2, 61.7, and 70.8 min, respectively (the time for the control case is not given). Caffeine clearance was greater in the two protocols involving charcoal than in the Colyte-alone case but were not different in the two charcoal treatments (this report, an abstract, does not cite the clearance values). We can conclude that charcoal was better than WBI in this case but can not tell by how much.

Tenenbein et al. (1988), in a brief abstract, have compared whole bowel irrigation and charcoal plus sorbitol in a randomized three-phase cross-over trial. Three grams of enteric-coated aspirin were given to ten volunteers. One phase was the control phase, one involved WBI 4 hr postingestion, and the third involved activated charcoal/sorbitol (ACS) 4 hr postingestion. Blood samples were collected over 48 hr. AUC values were: control, 51; ACS, 22; and WBI, 12 mg-hr/L-kg. Thus, ACS reduced absorption by 57%, and WBI reduced absorption by 76%. Peak aspirin levels were: control, 179; ACS, 128; and WBI, 84 mg/L. Clearances were: control, 14; ACS, 41; and WBI, 136 mg/hr-kg. Thus, in the case of enteric-coated aspirin, both treatments were effective but WBI was superior. The authors suggest that WBI may also be effective for other delayed-release drugs.

Rosenberg et al. (1988) presented the results of a study aimed at determining the effectiveness of WBI, alone or with charcoal, in reducing aspirin absorption in human subjects. The subjects were given 650 mg aspirin orally and then were treated with one of three methods: (1) immediate WBI using 4 L of fluid given at a rate of 100 mL/min for 40 min, (2) 50 g activated charcoal followed by WBI, or (3) 50 g activated charcoal alone. Aspirin absorption in the control group (no treatment) over 24 hr was 456 mg; with WBI, it was 354 mg and with charcoal plus WBI it was 321 mg. When only the charcoal was used, the aspirin absorption was 97.9 mg. Thus, charcoal alone was by far the best treatment.

Tenenbein (1989) commented on the Rosenberg study, stating in a letter that, in his view, whole-bowel irrigation had not been properly carried out. He stated that his recommended rate of administration of lavage fluid is 2 L/hr, continuing until the effluent is clear, as compared to the delivery of 4 L of lavage fluid over 40 min (a rate of 6 L/hr) used by Rosenberg et al.. Tenenbein contended that, because the lavage fluid was stopped after 40 min, a significant portion of the lavage solution and ingested aspirin was not expelled from the GI tract. Tenenbein mentioned "recent data" of his on the prevention of aspirin absorption by WBI and by charcoal/sorbitol in adult subjects, which showed WBI to be superior to the charcoal: a 73% reduction in aspirin absorption occurred with WBI and a 55% reduction occurred with charcoal/sorbitol (these may be Tenenbein's 1988 data even though the percentages implied there were 76 and 57%, respectively). Rosenberg and Livingstone's (1989) reply to these comments mentions that they were able to achieve dramatic bowel emptying despite the lower total volume of lavage fluid. They state that Tenenbein's aspirin data are not directly comparable to theirs, as he used a larger aspirin dose and did not indicate the amount or type of charcoal used. Thus, it is possible that WBI could be more effective than charcoal if the amount of lavage fluid is great and the amount of charcoal is small, since both treatments can be expected to be dose-related.

A paper by Kirschenbaum et al. (1989), of which Tenenbein is a coauthor, described a study comparing WBI with charcoal/sorbitol for reducing the absorption of enteric-coated aspirin (this is apparently the study published in abstract form in 1988 and referred to in Tenenbein's 1989 letter and in the reply of Rosenberg and Livingstone). Nine 325-mg aspirin tablets were ingested by 10 adult subjects in a three-phase randomized cross-over study. WBI was done using a polyethylene glycol electrolyte lavage solution given at a rate of 1.5–2 L/hr for 3–5 hr starting 4 hr after the drug ingestion. The activated charcoal/sorbitol (Charcodote, Pharmascience, Montreal) was given 4 hr after drug ingestion in the amount of 50 g charcoal in 70% sorbitol (total mixture volume 250 mL). The third phase was the control phase. There was no difference between the control, charcoal/sorbitol, and WBI interventions on the plasma salicylate concentration versus time curves before the eighth hour after drug ingestion; thereafter there was a difference. The WBI treatment over the time 0–∞ period decreased aspirin bioavailability by 73% compared to the controls and for charcoal/sorbitol the decrease was 57% (over the period of 4–14 hr, the decreases were 60 and 29%, respectively). WBI gave greater reductions in peak salicylate concentrations but the times of the occurrence of these peaks were not different. The time at which essentially a zero plasma salicylate concentration was reached was less for the WBI treatment than for the charcoal/sorbitol treatment. Based on these results, WBI was found to be superior to charcoal/sorbitol (both were, of course, far superior to doing nothing). The authors also state their conclusions that adverse effects (dehydration due to sorbitol) were greater with the charcoal/sorbitol treatment and that the subjects preferred WBI over charcoal/sorbitol.

Mayer et al. (1992) presented more results on WBI from the Tenenbein group in Winnipeg, Manitoba. A randomized three-phase study was done with nine humans who ingested aspirin. The subjects were given 2880 mg aspirin in suspension and then treated with PEG-ELS at 2 L/hr starting 4 hr after the drug administration for the next 3–5 hr, or were given 25 g charcoal (Charac-25, Laboratoires Lurabec, Montreal) 4, 6, 8, and 10 hr after the drug. This protocol is essentially the same as that of the Kirschenbaum et al. (1989) study from the same source (i.e., Tenenbein et al.) except that the earlier study employed a single 50-g dose of charcoal (different brand) in sorbitol, given at 4 hr, and 2925 mg of enteric-coated aspirin. Thus, the aspirin doses were very similar but this study had a multiple-dose charcoal protocol. There was no difference between the control, WBI, or charcoal group responses in terms of serum concentration versus time curves (the values of the 0–32 hr AUCs were 2320, 2040, and 2093 mg-hr/L for the control, charcoal, and WBI groups, respectively), for time to peak drug level, for peak serum level, or for urinary salicylates. Thus, because the aspirin was given in a very available form (suspension) and because the WBI and charcoal treatments were delayed for 4 hr, neither WBI nor charcoal had any effect—the great majority of the aspirin had been adsorbed by the time WBI or charcoal treatment was begun.

A study by Burkhart et al. (1992) involved a randomized cross-over study with dogs which had been given 75 mg/kg sustained-release theophylline. In Treatment 1, this was

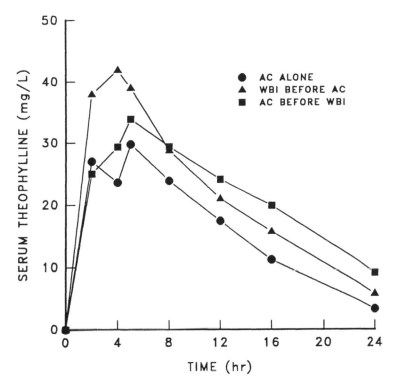

Figure 15.6 Serum theophylline in dogs for activated charcoal treatment, whole-bowel irrigation before activated charcoal, and activated charcoal before whole-bowel irrigation. From Burkhart et al. (1992). Reprinted with permission of the American College of Emergency Physicians.

followed by 1 g/kg charcoal administered via nasogastric tube 2 hr after the drug, followed by 0.5 g/kg doses of charcoal at 5 and 8 hr. In Treatment 2, the drug was followed by 25 mL/kg whole-bowel irrigation with PEG-ELS starting 2 hr after the drug, and repeated every 45 min for a total of four doses, followed by charcoal. In Treatment 3, the first dose of charcoal was given 10 min before beginning the whole-bowel irrigation protocol. Serum theophylline levels were determined at 0, 2, 4, 5, 8, 12, 16, and 24 hr. The serum drug values versus time are shown in Figure 15.6 and it clear that charcoal alone was the best of the three protocols. Thus, whole-bowel irrigation did not add to the effectiveness of charcoal.

However, Buckley and Dawson (1993) criticized the Burkhart study, stating that the theophylline doses used were fairly small and thus any effects of WBI would be small. Palatnick and Tenenbein (1993) offered other objections: (1) that no control [no treatment] existed and hence any benefit due to charcoal was unproved, and (2) relevance to human overdose cases was not established. Burkhart et al. (1993) replied that, among other things, the use of a control group would have been unethical.

Olsen et al. (1993) compared ipecac plus oral charcoal against low-volume WBI for reducing aspirin absorption in healthy subjects. Six adult men were given 3.25 g aspirin and were treated with 30 mL syrup of ipecac in 240 mL water 30 min later (the dose was repeated if no emesis occurred in 30 min). After initial emesis subsided, 50 g

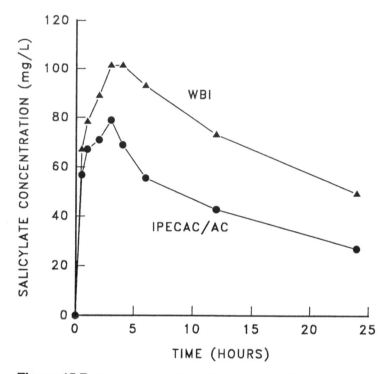

Figure 15.7 Serum salicylate in healthy volunteers after ipecac plus activated charcoal/sorbitol and after whole-bowel irrigation. From Olsen et al. (1993). Reproduced by permission of Pharmacotherapy Publications.

charcoal plus 96 g sorbitol (Actidose with Sorbitol) was administered. In a second treatment, 3000 mL WBI (30 mL/min for 100 min) was begun 30 min after the aspirin ingestion. Blood and urine samples were collected for 24 hr. Figure 15.7 shows the blood level data. The mean recovery of salicylates in the urine was 48.6% with WBI and 37.0% with charcoal/sorbitol. Peak serum drug levels averaged 112 mg/L with WBI and 74 mg/L for charcoal/sorbitol, while the AUC values were 1664 mg-hr/L for WBI and 952 mg-hr/L for charcoal/sorbitol. Thus, charcoal/sorbitol was superior to low-volume WBI.

When one looks carefully at these studies, it appears that whole-bowel irrigation was shown to be superior to charcoal only in the study where enteric-coated aspirin was used. In all other cases, charcoal was more effective. The implication is therefore that WBI may have a beneficial role in situations where the drug involved is slowly absorbed, either because it is poorly soluble or because it is in a delayed-release dosage form.

Before leaving the topic of whole-bowel irrigation, we consider two studies related to a concern which has arisen relative to irrigation fluids, namely, whether the constituents of such fluids affect the ability of charcoal to adsorb drugs. Why this is of concern is not entirely clear. The implication, perhaps, is that if charcoal therapy is followed by WBI, the irrigating fluid might cause desorption of a drug already bound to the charcoal.

Kirschenbaum et al. (1990) examined, in vitro, interactions between polyethylene glycol electrolyte lavage solution (PEG-ELS) and activated charcoal when PEG-ELS and charcoal powders were both added to water. Polyethylene glycol (PEG, mw 3,350 daltons) is present in PEG-ELS in the amount of 60 g/L. They found that PEG was adsorbed somewhat by the charcoal (16 to 38% adsorption over the wt:wt range of PEG:charcoal from 2.4:1 to 0.6:1). However, charcoal did not change the osmolarities of the PEG-ELS solutions (the osmolarities are determined almost totally by the electrolytes, which adsorb negligibly to charcoal). The presence of PEG in solution reduced the ability of charcoal to adsorb salicylic acid. A continual increase in the amount of added PEG steadily decreased the binding of salicylic acid to charcoal (at 30 g/L PEG, the salicylic acid adsorption was only 38% of what it was in the absence of PEG). This effect is clearly due to PEG binding competitively to the charcoal and reducing the adsorption sites available for salicylic acid adsorption. The authors concluded that combining WBI with charcoal would likely provide little added benefit. Exactly what is meant by "combining" the two therapies is unclear. Does this mean doing one and then the other successively or does it mean actually putting charcoal in the irrigation fluid?

Hoffman et al. (1991) studied the binding of theophylline to charcoal in vitro and determined the effect of added PEG-ELS (they claim to have showed "theophylline desorption from activated charcoal caused by whole bowel irrigation solution," but in fact they did not demonstrate *desorption* but only a decrease in *adsorption* when PEG-ELS was present). The theophylline was in the form of aminophylline, a compound of theophylline and 1,2-ethanediamine which dissociates upon dissolution to give free theophylline). Charcoal was added to solutions of aminophylline in an amount to give charcoal:theophylline ratios of 1:1, 3:1, and 10:1. PEG-ELS was added to some solutions. The PEG-ELS reduced theophylline adsorption from 67 to 37% for the 3:1 ratio and from 97 to 62% for the 10:1 ratio (for the 1:1 ratio, the adsorption was 16–17% with and without PEG-ELS). The Q_m value was decreased from 264 to either 218 or 203 mg/g (depending on the order of addition of charcoal, PEG-ELS, and aminophylline to the test batches) when PEG was present. Thus, it was concluded that PEG-ELS can interfere with the adsorption of drugs (in this case theophylline) by charcoal.

Although the two studies just discussed show that PEG does interfere somewhat with the ability of charcoal to bind test drugs, clinical studies are needed to determine what would actually happen in vivo if oral charcoal were followed by irrigation fluids.

V. EFFECTS OF SORBITOL ON THE ACTION OF CHARCOAL

The concept of using cathartics to reduce the gastrointestinal absorption of ingested drugs or toxins is based on the assumption that hastening the passage of such materials through the digestive tract will reduce their absorption (this is clearly not to be expected for rapidly absorbed drugs). However, the giving of cathartics causes a dilution effect due to the large volume of cathartic solution administered and an additional dilution effect due to the induced transfer of water into the GI tract. Thus, the net result of giving a cathartic solution may be the sum of at least three effects: (1) a dilutional effect which promotes gastric emptying and thus tends to increase drug absorption, (2) a dilution effect which lowers the drug concentration in the GI tract fluids, thus reducing the rate of absorption of the drug, and (3) a cathartic effect which reduces intestinal transit time and thus tends to decrease drug absorption.

A.   Some Preliminary Comments

We begin by discussing one study which dealt with the combined use of charcoal and sorbitol, and not either one alone, since it is different from the other studies which will be considered later. This study will also conveniently lead into a discussion of a common misconception concerning "70% sorbitol solution" which should be cleared up before proceeding further. Sorbitol is marketed as a 70% w/w solution because at concentrations of 70% w/w and higher, bacteria can not grow in the solution.

Lim et al. (1986) studied the effects of oral charcoal plus added sorbitol on the first 12 hr of absorption and elimination of sustained-release theophylline in 20 normal children, ages 8 to 18 years (no tests were done with charcoal alone, however, so the effects of adding sorbitol to the charcoal can not be assessed). The drug dose was 10 mg/kg. After receiving the drug, the children were given from one to four doses of 1 g/kg charcoal (InstaChar) as an aqueous slurry with 30 mL 70% sorbitol (presumably 70% w/w). Group 1 was given a single dose at 1 hr; Group 2 received four doses at 1, 4, 7, and 10 hr; Group 3 received four doses at 3, 6, 9, and 12 hr; and Group 4 was given three doses at 6, 9, and 12 hr. The reductions in 0–12 hr AUC values, relative to controls, were 61, 68, 37, and 18%, respectively, for the four groups. It was concluded that early administration of oral charcoal was important.

However, Krenzelok (1987b) stated, in a letter, that the effect of the sorbitol on the beneficial action of the charcoal was not acknowledged. He also criticized the use of sorbitol with every dose of charcoal. In reply, Lim and Nourtsis (1987) stated that no data have shown that sorbitol alone is superior to oral charcoal alone, and therefore they "continue to believe that OAC [oral activated charcoal] is the major determinant of the rate of elimination of theophylline from the GI tract and blood." This point of view, however, clearly does not allow for the possibility of synergistic effects between sorbitol and charcoal, and therefore may not be valid. With respect to using sorbitol with every charcoal dose, Lim and Nourtsis indicated that the total sorbitol dose used was relatively small compared to other studies and that the few instances of cramps and diarrhea encountered were readily resolved by bowel movements.

Weaver (1988) commented on the letters of Krenzelok and of Lim and Nourtsis and on an article by Krenzelok et al. (1985). Weaver stated that there is a great misconception of how to calculate sorbitol dosage. Sorbitol Solution USP is 70% w/w (i.e., 70 g sorbitol per 100 g of solution, not 70 g sorbitol per 100 mL of solution). One must use the fact that the density of a 70% w/w sorbitol solution at room temperature is 1.285 g/mL. Thus, for example, 250 mL of 70% sorbitol would weigh

$$250 \times 1.285 = 321.25 \text{ g}$$

and its sorbitol content would be

$$0.7 \times 321.25 = 224.88 \text{ g}$$

If one fails to use the 1.285 factor, one would calculate the sorbitol content of 250 mL of "70% sorbitol" to be

$$0.7 \times 250 = 175 \text{ g}$$

or 22% less.

Such confusion about "70% sorbitol" has been noted in many of the articles discussed in this and other chapters. Most investigators who use such solutions simply report using

"70% sorbitol" and, in such cases, it is probable (but not certain) that what they employed were indeed 70% w/w solutions. However, in several studies, researchers have stated that 70% w/v sorbitol was used. Since, as mentioned above, Sorbitol Solution USP is not 70% w/v, but rather 70% w/w, the implication is that the researchers were not aware of this fact and mistakenly assumed that "70% sorbitol" was to be prepared on a w/v basis.

With this matter clarified, we now review studies which have examined the effects of sorbitol and other cathartics on gastrointestinal transit times, and we consider two studies which were aimed at developing dosage recommendations for sorbitol. Then, we look at studies dealing with the effects of sorbitol on drug adsorption in vitro and, most important-ly, the effects of sorbitol on the in vivo efficacy of charcoal in reducing drug absorption.

B. Effects of Sorbitol on GI Transit Times

Minocha et al. (1984–85) gave 30 g charcoal in 150 mL of 70% sorbitol as a single dose to human subjects and monitored its effects on various serum chemistry values and hematological parameters. They measured the effects on serum osmolality, electrolytes, metabolic profile (SMAC), magnesium, hepatic enzymes, and complete blood count. No effects were observed that did not occur in control tests (sodium and phosphorus showed rises due to circadian rhythms). The mean transit time of the charcoal was 89 min, the mean duration of diarrhea was 12 hr, and the mean duration of black stools was 62 hr. All of the subjects described the diarrhea as severe and rated the abdominal discomfort (cramping, gurgling) associated with the diarrhea as mild to moderate. The authors concluded that, overall, charcoal/sorbitol was "an attractive combination for use as a gastrointestinal decontaminant."

Krenzelok (1985) studied human volunteers in which gastrointestinal transit times were determined after the administration of oral charcoal alone or in combination with several types of cathartics. No drugs were given. The charcoal (50 g in each case) was given in 240 mL water followed by 300 mL ginger ale, in 240 mL 70% sorbitol solution followed by 300 mL ginger ale, in 240 mL water followed by 300 mL magnesium citrate solution, or in 210 mL water followed by 15 g magnesium sulfate (30 mL of a 50% solution) and 300 mL ginger ale. The author stated that the 240 mL dose of 70% sorbitol solution contained 130 g sorbitol; however, as discussed previously, 70% sorbitol solution is 70% w/w and has a density of 1.285 g/mL, so the amount of sorbitol was

240 mL × 1.285 g solution/mL × 0.7 g sorbitol/g solution = 216 g sorbitol

or 66% more than the author's value. The mean bowel transit times were: with sorbitol, 77 min; with magnesium citrate, 244 min; and with magnesium sulfate, 1003 min. When no cathartic was used in combination with the charcoal, the transit time was 1491 min. Thus, sorbitol was the most effective cathartic for reducing transit time.

Some further data on transit times, in actual poison victims presenting to a Poison Center, have been given by Minocha et al. (1986). They gave charcoal and different cathartics (amounts unstated). The mean times to the first charcoal stool were: with no cathartic ($N = 23$), 33.3 hr; with magnesium sulfate ($N = 5$), 33.0 hr; with sodium sulfate ($N = 4$), 29.2 hr; with magnesium citrate ($N = 25$), 21.5 hr; and with sorbitol ($N = 91$), 15.5 hr. Thus, sorbitol gave the lowest transit time by far (again, the doses were not stated). One further point: vomiting followed sorbitol ingestion in 15.5% of the patients, versus 12.5% of the patients who took only charcoal (whether any patients vomited after taking any of the other cathartics is not stated).

Tenenbein (1987) commented on the 1985 Krenzelok transit time study, stating that the basic issue of whether any cathartic should be used was not addressed. He further questioned the endpoint (time to first charcoal-laden stool) as being subject to voluntary influences and the fact that diet and activity were not controlled after 2 hr. Finally, he stated that the doses of cathartic used were perhaps neither ideal nor comparative (e.g., a larger dose of any one cathartic would have caused a greater effect for that cathartic). Tenenbein's main point was, however, that "no studies confirming benefit from cathartic therapy for the overdose patient could be found." Krenzelok (1987a) replied that the differences between the various cathartic groups were so large that neurologic influences were highly doubtful and, he contended, the choices of the cathartic doses were reasonable. His main point, however, was that until research has proved that desorption of a drug from the drug/charcoal complex is insignificant, it is prudent to use cathartics to cause elimination of the drug/charcoal complex as quickly as possible. However, it would appear that the following opposite conclusion is more reasonable: until research has shown cathartics to be of significant benefit, they should not be given, as they are not harmless.

In another paper from Krenzelok's group, Harchelroad et al. (1989) analyzed the records on 276 patients presenting to an urban ED with acute drug or chemical ingestion. Of the 276 patients, 69 had received a charcoal/sorbitol slurry (Actidose with Sorbitol) containing 25 g charcoal and 96 g sorbitol as part of their treatment. The records were evaluated retrospectively with respect to gastrointestinal transit time. Of this group, 50.7% took less than 6 hr for their first charcoal stool and 26.1% had emesis of the charcoal/sorbitol slurry within 30 min of administration. Gastrointestinal transit time did not depend significantly on whether or not emesis occurred. Ingestion of drugs which decrease bowel motility (e.g., opiates, tricyclic antidepressants) correlated with prolonged time to stool despite treatment with the charcoal/sorbitol slurry.

C. Dosage Recommendations for Sorbitol

Minocha et al. (1985) developed dosage recommendations for charcoal/sorbitol treatment, based on the size of the patient, type of poison, and clinical status. In seriously ill adults, they recommend 1 g/kg charcoal in 4.3 mL/kg body weight 70% sorbitol every 4 hr, until the first stool containing charcoal appears. In children and ambulatory adults the same dose of charcoal may be given in 4.3 mL/kg body weight 35% sorbitol (it appears, but is not unequivocally stated, that "%" means % w/w). They state that patients requiring multiple doses of charcoal may be administered the charcoal every 2–6 hr with alternation between using aqueous slurries and sorbitol solutions. These patients should be monitored closely for any fluid or electrolyte imbalance, or depletion of essential vitamins.

Tominack and Spyker (1987) attempted to find an optimal dose of sorbitol which would produce catharsis reliably without excessive purgation or adverse effects. Thus, they gave different sorbitol amounts (30–120 g) and concentrations of sorbitol (10–37% sorbitol) plus charcoal to 15 healthy volunteers (the authors do not indicate if "%" means w/w or w/v). A relatively wide range of responses to sorbitol dose was found (e.g., one 50-kg female did not respond to a 135-g dose in a preliminary test and one 85-kg male had prolonged violent catharsis to doses greater than 45 g, with symptomatic orthostatic hypotension two hours after a 120-g dose). Based on their results, an initial adult dose of 60 g sorbitol (with charcoal) was recommended. If no bowel activity occurs in 2–3 hr, a different cathartic should be considered.

These two studies suggest quite different doses. For a 150-lb adult (68.0 kg), Minocha's recommended sorbitol dose would be

$$68 \text{ kg} \times 4.3 \text{ mL/kg} = 292.4 \text{ mL}$$

which, based on a density of 1.285 g/mL for 70% w/w sorbitol solution would be 292.4 mL × 1.285 g/mL = 375.7 g of sorbitol solution. This would deliver

$$0.70 \times 375.7 = 263 \text{ g}$$

of sorbitol. Thus, the recommended sorbitol dose for an adult according to Minocha et al. is more than 4 times that suggested by Tominack and Spyker and seems excessive.

As cited in Chapter 4, commercial charcoal/sorbitol formulations contain the following amounts of sorbitol: LiquiChar with Sorbitol, 27 and 54 g (in the 25- and 50-g charcoal sizes); Actidose with Sorbitol, 48 and 96 g (in the 25- and 50-g charcoal sizes); Activated Charcoal USP with Sorbitol, 27 g (in the 25-g charcoal size); and Charcoaid, 110 g (in the 30-g charcoal size).

Assuming one were to give an adult a charcoal dose of 100 g using these formulations (in the case of Charcoaid, 90 g delivered in three 30-g bottles), the patient would receive either 108, 192, 108, or 330 g of sorbitol.

Clearly, the 330 g of sorbitol delivered by three 30-g bottles of Charcoaid would be excessive. One could argue for using only two bottles of this formulation, since its charcoal (Norit B Supra) has a higher surface area (1500 m^2/g) than the charcoals in the other formulations (900–950 m^2/g). In this case, two 30 g bottles of Charcoaid would deliver 220 g of sorbitol. Even this seems high.

Based on a target sorbitol dose of around 100 g (much less than recommended by Minocha et al. but significantly more than recommended by Tominack and Spyker) and a target charcoal dose of 100 g, one could achieve this by using the following:

Two 50-g charcoal bottles of LiquiChar with Sorbitol (108 g sorbitol total)
Four 25-g charcoal bottles of Activated Charcoal USP with Sorbitol (108 g sorbitol total)
One 50-g charcoal bottle of Actidose with Sorbitol (96 g sorbitol) plus one 50-g bottle of Actidose Aqua (sorbitol-free)
One 30-g charcoal bottle of Charcoaid (110 g sorbitol) plus a sufficient amount of a sorbitol-free formulation to give the desired charcoal dose (one 50-g charcoal bottle of CharcoAid 2000 would be a good choice; although this would give a total of only 80 g charcoal, the charcoal in CharcoAid 2000 has a very high surface area (2000 m^2/g) and thus the 80 g total charcoal in this combination would be more effective than 100 g of 900–950 m^2/g charcoal).

Other target doses of charcoal and of sorbitol can be easily achieved by combining bottles having different charcoal contents, without sorbitol and with different amounts of sorbitol. Thus, EDs may wish to have more than one manufacturer's charcoal formulations available in order to provide flexibility in creating whatever charcoal/sorbitol doses are desired.

D. Effect of Sorbitol on the Efficacy of Charcoal

We turn now to studies dealing with the effects of sorbitol on the efficacy of charcoal. We begin with two in vitro studies and follow this with in vivo studies.

1. In Vitro Studies

Eyer and Sprenger (1991) studied the effects of sorbitol on the in vitro adsorption of acetaminophen, codeine, and diphenhydramine by Norit A charcoal in vitro. Their results on maximal binding capacities (Q_m values) are shown in Table 15.3. These results show that sorbitol reduced the Q_m values for diphenhydramine, codeine, and acetaminophen by 19, 72, and 15%, respectively, at neutral pHs (actually, the pH values in water and water/ sorbitol are not necessarily neutral), and by 24, 52, and 13%, respectively, in acidic media. The reason why sorbitol reduced drug adsorption, by as much as 72% in one case, is unclear. Sorbitol itself would not be expected to adsorb significantly to charcoal and thus would not offer competition for adsorption sites. This study (in German) needs further scrutiny.

An in vitro study of the effects of different cathartics on the adsorption of drugs to charcoal was given in a meeting abstract by Al-Shareef et al. (1992). Details are not included in the abstract, so the charcoal and cathartic amounts, and the pH levels are unknown. However, the results are presented anyway, in Table 15.4. The values shown are the maximal adsorption capacities (Q_m values) with added cathartic, expressed as a percentage of the Q_m values for charcoal alone. Lactulose and sorbitol appear to have had fairly small effects in reducing adsorption, while the magnesium sulfate gave mixed results. In two cases, $MgSO_4$ actually increased drug adsorption. A possible reason for this enhancement will be discussed in a following section of this chapter.

2. In Vivo Studies

An early study involving charcoal and sorbitol is one done by Mayersohn et al. (1977). They administered three 325 mg aspirin tablets plus 50 mL water to human subjects and followed this with: (1) 150 mL water, (2) 100 mL of an aqueous slurry of 20 g charcoal, (3) 100 mL of an aqueous 70% sorbitol solution, or (4) 100 mL of a slurry of 20 g charcoal in a 70% sorbitol solution (presumably 70% w/w). Salicylate excretion into the urine was followed for 48 hr. Table 15.5 shows the results obtained. The plain sorbitol solutions produced diarrhea in all subjects, and had little effect on salicylate recovery in the urine as compared to the controls. In all but one subject, the sorbitol/charcoal mixture also produced diarrhea and the data show a decidedly higher recovery compared to the charcoal alone. It is apparent that the sorbitol did interfere somewhat with the action of the charcoal.

Van de Graaff et al. (1982) studied the effects of sorbitol and other cathartics alone or in combination with charcoal on the absorption of acetaminophen in dogs. A dose of 0.6 g/kg acetaminophen was given by orogastric tube and then followed by either water, Norit A or Nuchar charcoal (3 g/kg), mannitol plus sorbitol (2 g/kg), castor oil (3 mL/kg), or both charcoal and either mannitol and sorbitol or castor oil. AUC values from 0–∞ were obtained by extrapolation of 0–11 hr AUC values. Cathartics alone decreased the drug AUC by 14% (castor oil) and by 31% (mannitol plus sorbitol). Both mannitol/sorbitol and castor oil produced diarrhea. The Nuchar and Norit A charcoals alone reduced the AUC by 92.7 and 94.5%, respectively. Each cathartic diminished the effect of charcoal: mannitol/sorbitol given with charcoal increased the AUC to 75% higher than for charcoal alone and castor oil given immediately after charcoal caused the AUC to be more than twice as large as with charcoal alone. In vitro tests were also done by Van de Graaff et al. using 16 different charcoals and resins. They found that acetaminophen adsorption was little changed by adding mannitol/sorbitol to the test mixtures.

Table 15.3 Maximal Binding Capacities (mg/g) With and Without Sorbitol

Medium	Diphenhydramine	Codeine	Acetaminophen
Water	120	173	193
50% Sorbitol	97	49	165
0.1 N HCl	106	86	196
50% Sorbitol/0.1 N HCl	81	41	171

From Eyer and Sprenger (1991). Table 1. Copyright Springer-Verlag GmbH & Co. Used with permission.

Table 15.4 Effect of Cathartics on In Vitro Drug Adsorption (Maximal Percentages Adsorbed Compared to Tests Without Cathartics)

Drug	With lactulose	With sorbitol	With MgSO$_4$
Antipyrine	97.0	90.0	109.5
Metoclopramide	75.8	83.7	70.7
Acetaminophen	89.3	91.4	110.3
Sodium warfarin	99.8	99.2	98.8

From Al-Shareef et al. (1992). Reprinted with permission from *Veterinary and Human Toxicology*.

Table 15.5 The Influence of Activated Charcoal and Sorbitol on Aspirin Absorption

Experiment	Dose recovered[a]	Maximal excretion rate[a]	Time of maximal excretion rate[a]
Control	100.0	100.0	100
Sorbitol	95.9	72.1	335
Charcoal slurry	38.7	34.2	201
Charcoal and sorbitol	51.3	55.7	315

[a]All values are means and are expressed relative to the control experiment. The results of the control experiment are as follows: dose recovered, 73.2%; maximal excretion rate, 67.3 mg/hr; t_{max}, 1.9 hr. From Mayersohn et al. (1977).

Table 15.6   AUC Values (% of Controls) for Drug Absorption in Rats

Drug	Charcoal	Sorbitol	Charcoal/Sorbitol
Chlorpheniramine	37.2	68.2	30.1
Chloroquine	52.3	69.0	27.4
Pentobarbital	70.7	40.6	27.0
Aspirin	74.9	109.4	73.3

From Picchioni et al. (1982).

Picchioni et al. (1982) studied the in vivo absorption of four drugs in rats and determined the effects of charcoal combined with sorbitol. The drugs were chlorpheniramine maleate (80 mg/kg), chloroquine diphosphate (100 mg/kg), aspirin (100 mg/kg), and sodium pentobarbital (50 mg/kg), and were given by stomach tube. Immediately after, the rats were treated with 20 mL/kg of one of the following: 70% w/v sorbitol solution, charcoal in water, or charcoal in sorbitol (the use of 70% w/v sorbitol is a bit strange, since Sorbitol Solution USP is 70% w/w, not 70% w/v). The charcoal:drug ratio was varied. The AUC values (as percentages of control values) were as shown in Table 15.6 (the time spans for the AUC values are not clearly stated). Charcoal alone was more effective than sorbitol alone, in general; however, charcoal plus sorbitol was better than charcoal alone.

Wieland et al. (1986) evaluated the effects of sorbitol and bentonite on the efficacy of charcoal in reducing aspirin absorption. They gave seven volunteers 20 mg/kg aspirin and then either: (1) water, (2) charcoal in water, (3) charcoal in 70% sorbitol (presumably 70% w/w), or (4) charcoal in a bentonite slurry (the exact amounts of each are not stated). Urine was collected and analyzed to determine aspirin absorption. The aspirin absorption percentages were 84, 70, 49, and 67%, respectively, for the four treatments. Thus, charcoal reduced aspirin absorption in all cases but the charcoal/sorbitol formulation was the best. The presence of bentonite had no significant effect on the action of charcoal.

Berg et al. (1987) studied the relative effects of charcoal and charcoal/sorbitol on the elimination of intravenous phenobarbital. They gave 200 mg/70 kg of the drug IV to healthy subjects over 1 hr. After the infusion had ended, the subjects were given either a no-charcoal control solution, multiple doses of a charcoal/water suspension, or multiple doses of a commercially available charcoal/sorbitol suspension (Charcoaid). The charcoal doses in the latter two cases were 30 g for the first dose and 15-g portions given at 6, 12, 18, 24, and 36 hr postinfusion. The charcoal and charcoal/sorbitol treatments decreased the $0-\infty$ AUC values by 36 and 38%, respectively. Half-life reductions were also very similar. All six subjects experienced diarrhea when they took the charcoal/sorbitol mixture; however, all subjects preferred this mixture. The data showed that the sorbitol did not interfere with the beneficial action of charcoal and, in fact, accelerated the onset of the charcoal's action (the 0–60 hr AUC values with charcoal and charcoal/sorbitol were 27 and 37% less than controls, respectively).

Goldberg et al. (1987) studied the effect on serum theophylline concentrations of adding sorbitol to an oral regimen of multiple doses of charcoal, after giving nine healthy subjects slow-release theophylline. At 6, 7, 8, 10, and 12 hr after ingestion of Theo-24 (1200 mg/70 kg), the subjects received 20 g charcoal in water or 20 g charcoal in water plus 75 mL 70% sorbitol (presumably 70% w/w) at 6 and 8 hr only. Figure 15.8 gives

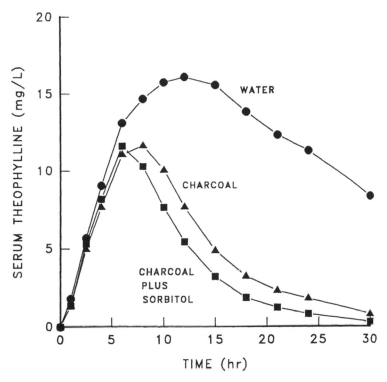

Figure 15.8 Mean serum theophylline concentrations with multiple doses of water, or activated charcoal, or activated charcoal plus sorbitol, beginning 6 hr after Theo-24 ingestion. From Goldberg et al. (1987). Reprinted by permission of Mosby-Year Book Inc.

the mean serum theophylline data determined in the study. Serum AUC values (6–30 hr) were 113 and 85 mg-hr/L for the charcoal and charcoal plus sorbitol, respectively, compared to a control value of 305 mg-hr/L. Thus, the addition of sorbitol significantly enhanced drug elimination.

Curd-Sneed et al. (1987) gave rats 40 mg/kg sodium pentobarbital by gavage, and then 5 min later gave them either 40 mg Darco G-60 charcoal, USP charcoal, or SuperChar charcoal. The charcoals were given in 1 mL plain water in one case and in 1 mL of 70% w/v sorbitol in another case (the use of 70% w/v sorbitol is anomolous). The 0–6 hr AUC values are given in Table 15.7. Clearly, all charcoals reduced the AUC values. When given in water, the SuperChar charcoal was best, as one would expect due to its very high surface area. When given in sorbitol, the efficacies of the Darco G-60 and the USP charcoals were enhanced but the efficacy of the SuperChar charcoal was decreased, relative to its performance in water. The reason for the mixed effects of sorbitol are not clear.

McNamara et al. (1988) studied the use of sorbitol along with charcoal in acetaminophen overdose. They gave eight healthy subjects 3 g acetaminophen orally followed either by 50 g charcoal at 1 hr or by 50 g charcoal/sorbitol at 1 hr. Serum drug concentrations were determined over the next 8 hr. Both interventions significantly reduced the drug AUC (29.5% for charcoal alone and 26.0% for charcoal/sorbitol) but they did not differ significantly from one another (Figure 15.9). Diarrhea, abdominal

Table 15.7 AUC Values (mg-hr/L) for Different Charcoals Plus Water or Sorbitol

	AUC	% of Control
Water alone (control)	93.4	100.0
Water plus Darco G-60 charcoal	82.2	88.0
Water plus USP charcoal	73.1	78.3
Water plus SuperChar charcoal	55.5	59.4
Sorbitol alone (control)	89.4	100.0
Sorbitol plus Darco G-60 charcoal	70.0	78.3
Sorbitol plus USP charcoal	57.0	63.8
Sorbitol plus SuperChar charcoal	64.6	72.3

From Curd-Sneed et al. (1987).

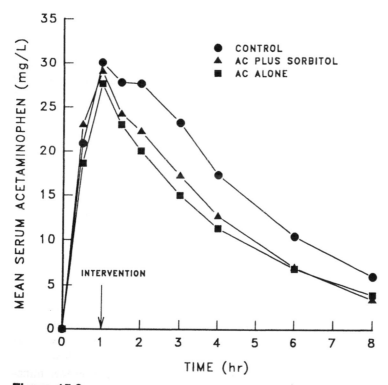

Figure 15.9 Mean serum acetaminophen levels in subjects treated with activated charcoal alone or with activated charcoal plus sorbitol. From McNamara et al. (1988). Reprinted with permission of the American College of Emergency Physicians.

cramping, and nausea occurred in all of the subjects given charcoal/sorbitol. In the charcoal group, two cases of nausea occurred (no diarrhea or cramping was noted). Neither group showed any instance of constipation. These data suggest that sorbitol does not provide any added benefit.

Krenzelok and Heller (1986) studied five adult volunteers who were given 2.6 g aspirin and then either 25 g aqueous charcoal or a slurry of 25 g charcoal/95 g sorbitol. Serum salicylate was determined over an 8 hr period. The charcoal alone reduced aspirin absorption by 51% (compared to controls) and the charcoal/sorbitol reduced aspirin absorption by 76%. Sorbitol-induced catharsis occurred in an average of 77 minutes. Thus, the addition of sorbitol to the charcoal was of significant benefit in this study.

In a similar study (Keller et al., 1990), ten healthy subjects were given 2.5 g aspirin orally followed by 25 g charcoal at 1 hr or 25 g charcoal plus 1.5 g/kg sorbitol at 1 hr. Urine was collected for 48 hr and analyzed for salicylate metabolites. Average aspirin absorption was 1.26 g with charcoal and 0.91 g with charcoal/sorbitol. Thus, sorbitol decreased aspirin absorption by 28%. These authors therefore recommended the combined use of charcoal and sorbitol for aspirin overdose.

Al-Shareef et al. (1990) investigated the effects of charcoal and sorbitol, alone and in combination, on the absorption of sustained-release theophylline in eight healthy female subjects. The subjects were given two 300-mg slow-release theophylline tablets, followed 2 hr later by either: (1) 400 mL water then 80 mL water every 6 hr up to 20 hr, (2) 20 g charcoal (Carbomix) in 400 mL water followed by 10 g of the charcoal in 80 mL water every 6 hr up to 20 hr, (3) 50 mL 70% sorbitol followed by 80 mL water every 6 hr up to 20 hr, or (4) a combination of charcoal and sorbitol at the same doses and times as the prior two regimens. The 0-24 hr AUC values for the four treatments were: 97.6, 10.5, 116.6, and 7.5 mg-hr/L, respectively. Thus, sorbitol increased the AUC relative to the "drug only" case (but not statistically significant). Both charcoal treatments dramatically lowered the AUC levels, with the charcoal alone and charcoal/sorbitol regimens being essentially equivalent (sorbitol helped but not to a statistically significant extent).

Figure 15.10 shows the plasma theophylline levels versus time for the four treatments involved. Clearly, the charcoal and charcoal/sorbitol regimens were effective. However, the use of sorbitol alone may be of more harm than good. The maximum plasma theophylline concentration with sorbitol was 30% higher than that which was seen after theophylline alone and occurred earlier, indicating a higher rate of absorption of theophylline in the presence of sorbitol. Al-Shareef et al. feel that this was due to changes in gastrointestinal motility.

Jessen and Barone (1992) warned that there are now so many ready-mix charcoal/sorbitol preparations on the market that it is easy to administer too much sorbitol during multiple-dose charcoal therapy. They state that many hospital pharmacies stock only charcoal/sorbitol mixtures and that the labels on these products do not always highlight warnings to avoid too much multiple use of them. Finally, the similarity of bottles of these charcoal/sorbitol products to those containing only charcoal/water increases the risk of confusion. They indicate that their policy is to give a cathartic only with the first dose of charcoal, since the benefits of cathartics are unproved, while the hazards of too much cathartic are indisputable.

In reply, Tenenbein (1992) stated that he concurs fully and cites a case where a physician ordered that a magnesium cathartic be given in conjunction with a charcoal/water slurry. What happened, through poor labeling of the product used, was that a charcoal/sorbitol slurry was given in addition to the magnesium cathartic.

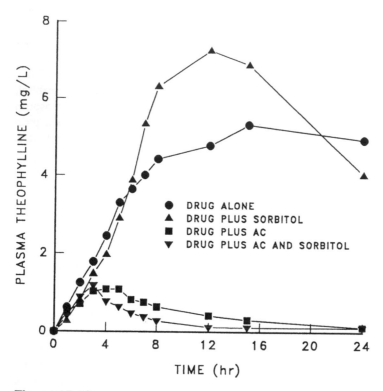

Figure 15.10 Plasma theophylline concentrations after the drug alone, drug plus sorbitol, drug plus activated charcoal, and drug plus activated charcoal and sorbitol. From Al-Shareef et al. (1990). Reproduced by permission of The Macmillan Press Ltd.

Eyer and Sprenger (1991) also did in vivo tests with the same three drugs that they used in their in vitro work. Volunteers were given 500 mg acetaminophen, 50 mg diphenhydramine HCl, or 50 mg codeine phosphate; then 30 g charcoal in 150 mL of 70% sorbitol was administered. The charcoal/sorbitol suspension resulted in drug absorption values (compared to controls) of 44% for acetaminophen, 28% for diphenhydramine, and 12% for codeine (i.e., reductions in absorption of 56, 72, and 88%, respectively). Unfortunately, no trials with charcoal alone were done, so the effects of sorbitol on the in vivo efficacy of charcoal were not determined. The authors mention that the charcoal/sorbitol mixture was generally well tolerated except for "marked flatulence."

E.   Conclusions and Recommendations

We have discussed six in vivo studies in which sorbitol appears to have had a definite positive effect, one in vivo study in which the effect was slightly positive, two in vivo studies in which there was no effect, and two in vivo studies in which a definite negative effect was found. Additionally, two in vitro studies indicated that sorbitol produced small to moderate reductions in the adsorption of various drugs to charcoal. Thus, on balance, it appears that sorbitol tends to enhance the in vivo action of charcoal in reducing drug absorption. However, this is not always the case and one must take into account the risks of over-zealous use of sorbitol (dehydration, hypernatremia, abdominal distention, etc., as discussed in Chapter 17).

A reasonable but tentative recommendation based on all evidence and considerations of risk would be to administer sorbitol (100 g or so for an adult and correspondingly less for children) with an initial charcoal dose (again, 100 g or so for an adult and correspondingly less for children). If multiple-dose charcoal therapy is to be employed, successive charcoal doses should be lower than the initial dose (e.g., 50 g for adults and less for children) and, likewise, additional sorbitol doses should be less than the initial one. Most importantly, sorbitol should not be given with every charcoal dose but, for example, given every third dose. Bowel sounds should be monitored closely to judge whether the gastrointestinal contents are moving along satisfactorily and fluid and electrolyte balances should be carefully scrutinized to see if the sorbitol has caused any problems.

VI. EFFECT OF SALINE CATHARTICS ON THE ACTION OF CHARCOAL

We have already discussed the study done by Krenzelok (1985) in which the effects of various cathartics on GI transit times was determined. This was discussed in the section dealing with sorbitol where some results on magnesium citrate and magnesium sulfate were given. To briefly reiterate those results, they showed GI transit times of 77 min for sorbitol, 244 min for magnesium citrate, 1003 min for magnesium sulfate, and 1491 min when no cathartic was used.

A. In Vitro Studies

A variety of in vitro studies showing the effects of saline cathartics on the adsorption of drugs to charcoal have been done. The effects of added magnesium citrate and other salts on the adsorption of sodium salicylate from aqueous solution to charcoal in vitro were studied by Ryan et al. (1980). This study was also reported in abstract form by Zeldes et al. (1979). The salicylate concentrations were 20, 60, and 100 mg/100 mL, or 0.2, 0.6, and 1.0 g/L. The initial pH values were all 4, but it is not clear that the pH was kept at a value of 4 after the addition of the various salts (the citrate salts, in particular, could have changed the solution pH). A fixed amount of charcoal (100 mg Norit A) was added to 30 mL of the test solution and the percent salicylate adsorbed at equilibrium was determined. The experiments were done with a 1:1 mixture containing magnesium citrate NF and in solutions containing 0.25 M magnesium sulfate, 0.2 M sodium citrate, 3.45% sodium sulfate, or 2% sodium chloride. The percentages of salicylate adsorbed in each case are shown in Table 15.8. All of the added salts enhanced the adsorption of salicylate, with the citrate salts being particularly effective. At pH 4, the carboxylic acid group on the salicylate molecule is nearly completely ionized and the salicylate molecule carries a charge of about −1 (see Figure 3.7). Thus, adjacently adsorbed salicylate anions on the charcoal surface tend to repel each other strongly. Adding any salt which can dissociate to produce cations, which can locate themselves between the salicylate anions and dampen out these electrical repulsions, will greatly enhance the adsorption of the salicylate. The effect of the citrate salts, however, is larger than one would expect. A likely explanation is that the added citrate salts lowered the

Table 15.8 Effect of Saline Cathartics on Salicylate Adsorption

| | Initial salicylate concentration | | |
| | 20 mg% | 60 mg% | 100 mg% |
Solution	Percent salicylate adsorbed		
Water	51	33.3	21.4
Magnesium citrate	100	82.5	60
Sodium citrate	100	83.8	60.4
Magnesium sulfate	93.5	58.2	42.2
Sodium chloride	85.3	45.5	32.9
Sodium sulfate	87.5	50	37

From Ryan et al. (1980).

pH of the test solutions, causing the salicylate to be in a more nonionized state, in which condition it can adsorb much better (due to the lessening of electrical repulsion effects).

LaPierre et al. (1981) also studied the effect of magnesium citrate on charcoal adsorption in vitro. The test drug was aspirin. Aspirin (40 mg) was mixed with 400 mg USP charcoal in 30 mL SGF (without pepsin) or SIF (without pancreatin) at 37°C for 20 min; 2.4 or 12 mL of a magnesium citrate solution was also added to some samples. Table 15.9 shows the results. The citrate solution lowered aspirin adsorption somewhat at gastric pH, although this could just be because the added citrate solution diluted the aspirin solution which, in turn, caused the charcoal to adsorb less. Additionally, at gastric pH, citrate forms undissociated citric acid, which can adsorb to charcoal and thus compete with the aspirin for adsorption sites. The added citrate increased aspirin adsorption somewhat at intestinal pH. The probable reason for this is that the acidic citrate solution decreased the solution pH and thus caused the aspirin to be less dissociated, in which state it adsorbed more strongly. Hence, failure to keep the solution pH constant and the dilutional effect mentioned above confound interpretation of the magnesium citrate effect.

Further in vitro work on the effect of magnesium citrate was done by Cooney and Wijaya (1986). They first determined that citrate itself adsorbs well to charcoal at low

Table 15.9 Effect of Magnesium Citrate on Adsorption of 40 mg Aspirin by 400 mg Activated Charcoal

| | Aspirin adsorbed (mg) | |
Magnesium citrate added (mL)	Gastric fluid	Intestinal fluid
0.0	40.0	31.5
2.4	39.5	35.3
12.0	36.7	34.2

From LaPierre et al. (1980).

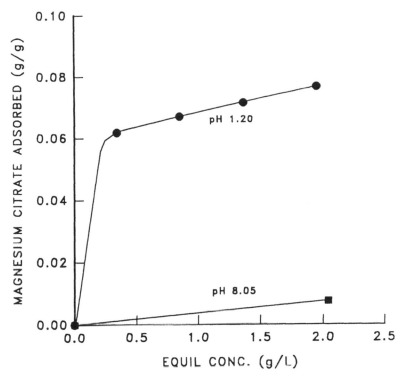

Figure 15.11 Adsorption of magnesium citrate to charcoal at pH 1.20 and pH 8.05. From Cooney and Wijaya (1986). Reprinted with permission from *Veterinary and Human Toxicology.*

pH (1.20) and poorly at higher pH (8.05), as shown in Figure 15.11. Then, they found that magnesium citrate added to sodium salicylate decreased salicylate adsorption slightly at pH 1.25 due to competition from the undissociated citric acid that is formed at low pH. At pH 9.0, magnesium citrate enhanced salicylate adsorption slightly, at low citrate concentrations, because the presence of Mg^{++} cations in solution helped to reduce repulsive forces between adjacently adsorbed salicylate anions, thus increasing salicylate adsorption. These results on the effects of magnesium citrate on salicylate adsorption to charcoal are given in Figure 15.12.

An in vitro study on the effect of magnesium citrate and other saline cathartics on the adsorption of aspirin was carried out by Czajka and Konrad (1986). They added either magnesium citrate, magnesium sulfate, or sodium citrate to aspirin solutions of initial pH 1.2. Different weights of charcoal were contacted with aliquots of these solutions and adsorption isotherms determined. Maximal adsorption capacities, Q_m, were in the range of 320–336 mg aspirin/g charcoal when adsorption took place from distilled water, tap water, magnesium sulfate, or sodium sulfate solutions. However, for a solution containing magnesium citrate, the Q_m was 283 mg/g. Thus, at low pH, the magnesium citrate reduced the adsorption of aspirin. This is consistent with the effect seen by Cooney and Wijaya that relates to the formation of undissociated citric acid at low pH, which competes with the test drug for adsorption sites on the charcoal.

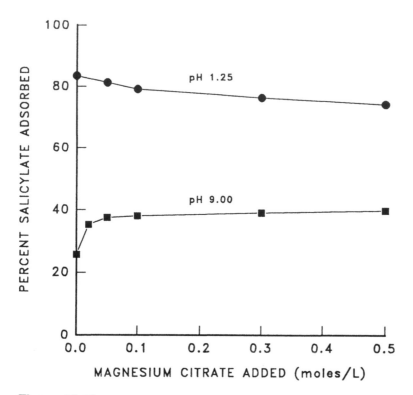

Figure 15.12 Effect of magnesium citrate on salicylate adsorption to charcoal at pH 1.25 and pH 9.00. From Cooney and Wijaya (1986). Reprinted with permission from *Veterinary and Human Toxicology*.

Akintonwa and Orisakwe (1988a) studied the effect of magnesium citrate on the in vitro adsorption of metronidazole (a protozoacide used for treating trichomoniasis, giardiasis, and amebiases) and tinidazole (also used for treating trichomoniasis and amebiasis). Various solutions of these drugs in distilled water were contacted with 50 or 200 mg of charcoal. The percentages of the drugs adsorbed in the various tests were determined; however, the data were not fit to isotherm equations and thus quantities such as maximal binding capacities, Q_m, were not computed. Solution pH values were not controlled (using buffers) and were not measured. Thus, pH variations could have occurred which would have confounded the results. Thus, it is difficult to summarize the data except to say that, in the proportions of solution and charcoal used, drug adsorption was fairly high (66–96%) when no $MgSO_4$ was added. The percentage of drug adsorbed seemed to be dose dependent in that the percentage changed when the amount of charcoal was kept constant and the drug concentration in the solution was varied; this is, of course, exactly as one would expect and is not the surprise that these authors imply.

When $MgSO_4$ was added in the amount of 7.5 mg/mL, the amount of charcoal required to adsorb 50% of the drug was reduced for metronidazole from 255 to 153 mg and was increased for tinidazole from 126 to 178 mg. Thus, $MgSO_4$ aided adsorption for metronidazole and inhibited adsorption for tinidazole. The reason for the difference was explained in terms of competition due to SO_4^{2-} adsorption but this seems unlikely.

Akintonwa and Orisakwe (1990a) did a similar study involving the adsorption of quinine and quinidine to charcoal, with and without added magnesium sulfate. Both drugs adsorbed to charcoal from 0 to 100%, depending on the charcoal:drug ratio. Magnesium sulfate enhanced quinine adsorption and inhibited quinidine adsorption. The reason for these opposite effects was not explained. The same comments apply here as apply to the previous study just discussed: there was no fitting of data to isotherm equations, no pH measurement or control, etc. Thus, the results are difficult to interpret.

A third study by Akintonwa and Orisakwe (1990b) involved the adsorption of sulphamethoxazole to charcoal, with and without added sodium sulfate and magnesium sulfate. Without added saline cathartics, the amount of drug adsorbed in vitro varied from 10.3 to 83.7%, depending on the charcoal:drug ratio. The addition of 7.5 mg/mL of sodium sulfate or magnesium sulfate to the solutions reduced the amount of charcoal needed for 50% adsorption from 472 mg (no added cathartic) to 274 mg with sodium sulfate and to 41 mg with magnesium sulfate. This latter figure for the magnesium sulfate seems surprisingly low. Thus, the saline cathartics greatly enhanced sulphamethoxazole adsorption to charcoal. Since pH was not controlled or measured, it is difficult to interpret these results. The pK_a of sulphamethoxazole is stated to be 5.6; the authors also state that the cathartics "have a pH range of 5.8–6.9." Not controlling solution pH with buffers when working near the pK_a of the adsorbing substance (where the degree of ionization of the substance, and hence its adsorbability, are most sensitive to slight pH changes) means that the results obtained with and without added cathartics will be at different pHs and hence can not be compared on an equal basis.

Yet a fourth study by Orisakwe and Akintonwa (1991) dealt with the effects of sodium sulfate on the adsorption of chloroquine and mefloquine to charcoal in vitro. Chloroquine phosphate at 125, 250, and 500 mg/L adsorbed to 50 mg charcoal to the extent of 10.0, 23.0, and 38.4%. Similar tests with 0.4 to 1.0 mg/L mefloquine showed 60.6 to 95.8% adsorption. When sodium sulfate was added to the test batches in the amount of 7.5 g/L, the amount of charcoal required for 50% adsorption increased slightly from 375 to 444 mg for chloroquine and had no effect on the amount of charcoal needed to adsorb 50% of mefloquine. Thus, added sodium sulfate had little or no effect with these two drugs, as one might expect, since sodium sulfate itself does not adsorb significantly to charcoal.

A charcoal formulation called AZU suspension, developed by Rademaker et al. (1987, 1989), was prepared using 40 g of either Carbo adsorbens charcoal (in 1987) or Norit A charcoal (in 1989), added to 10 g of 85% glycerol solution and sufficient purified water to make the final volume 200 mL. In vitro adsorption tests were done with phenazone, phenobarbital, and amitriptyline in a pH 6.5 buffer and in the same buffer with 10% w/v sodium sulfate. In the 1987 study, the maximal adsorption capacities for the three drugs were increased by about 14.5, 7.4, and 7.3%, respectively, when the sodium sulfate was added to the drug/buffer solutions. In the 1989 study, the same tests yielded 24.4, 6.5, and 10.2% increases in 10% w/v sodium sulfate solution.

B. In Vivo Studies

Chin et al. (1981) studied the effects of sodium sulfate (1.32 g/kg) on the effectiveness of charcoal in reducing aspirin (100 mg/kg), pentobarbital sodium (50 mg/kg), chlorpheniramine maleate (80 mg/kg), and chloroquine phosphate (100 mg/kg) absorption in rats. The charcoal:drug ratio was 2:1 for pentobarbital and chloroquine, and was 4:1 for aspirin and chlorpheniramine. They found that sodium sulfate, given alone, reduced the aspirin, pentobarbital, chlorpheniramine, and chloroquine bioavailabilities by about 29, 0, 9, and 5%, respectively. Charcoal alone was more effective than sodium sulfate alone (the respective AUC reductions were 39, 26, 64, and 54% for the four drugs) and charcoal plus sodium sulfate was even more effective than charcoal alone (the respective AUC reductions were 57, 39, 70, and 67% for the four drugs). Thus, at least in rats, the combined use of sodium sulfate and charcoal appears to be beneficial. The mechanism by which the sodium sulfate enhances drug adsorption by charcoal is unclear. One possibility is that sodium sulfate promotes mixing of the charcoal with the drug-containing intestinal contents. A preliminary report on this study (see Chin and Picchioni, 1980) appeared as one of several brief reports on the 1979 Annual Meeting of the American Academy of Clinical Toxicology, American Association of Poison Control Centers, American Board of Medical Toxicology, and the Canadian Association of Poison Control Centres.

Easom et al. (1982) studied eight healthy subjects who were given three 325-mg aspirin tablets plus: water alone, water and 10 g charcoal, water plus 10 g charcoal and 200 mL magnesium citrate solution, or water plus 10 g charcoal and 200 mL magnesium citrate solution delayed for 30 min. The percentages of the aspirin excreted in the urine over 0–36 hr after these treatments were 94.2, 73.8, 64.8, and 66.5%, respectively. Thus, the addition of magnesium citrate to the charcoal decreased aspirin absorption about 10–12% as compared to charcoal alone, even after a 30 min delay.

The effect of sodium sulfate on aspirin bioavailability with and without charcoal was examined in vivo by Sketris et al. (1982). They gave six subjects 975 mg aspirin followed 30 min later by either water alone, 15 g charcoal in water, or 15 g charcoal plus 20 g sodium sulfate in water. Blood and urine samples were taken up to 48 hr. Figure 15.13 shows the blood level versus time data for the three treatments. The calculated $0-\infty$ AUC values in mg-hr/L were: control, 847; charcoal alone, 428; and charcoal/sulfate, 619. Thus, charcoal reduced absorption by 49.5% and charcoal/sulfate reduced absorption by 26.9%. Therefore, sodium sulfate appears to have had a negative impact on the action of charcoal. However, the charcoal/sulfate result was not statistically different from either the control or charcoal alone results at the $p < 0.01$ level, and yet it seems that a real effect is evident here, so perhaps a less stringent p criterion should have been used. On the other hand, recoveries of salicylates in the urine were 94, 58, and 61% of the drug dose for the control, charcoal, and charcoal/sulfate cases, respectively. Based on these results, there was a slight and insignificant difference between charcoal alone and charcoal/sulfate.

The effect of sodium sulfate in conjunction with charcoal was also studied by Galinsky and Levy (1984). Eight normal adults were given 1 g acetaminophen alone or in combination with: (1) 18 g sodium sulfate, (2) 10 g charcoal, or (3) 18 g sodium sulfate plus 10 g charcoal. Urine was collected for 48 hr and assayed for acetaminophen and its major metabolites. The charcoal reduced the drug absorption by 37%, whereas sodium sulfate alone reduced the drug absorption by only 3%. The combination of charcoal and sodium sulfate reduced acetaminophen absorption by 35%; hence, sodium sulfate had no added effect. Concurrently, charcoal had no effect on the urinary recovery of sodium sulfate. It is of value to give sodium sulfate in acetaminophen poisoning because acetaminophen sulfate is a major metabolite of acetaminophen; thus, acetaminophen can deplete inorganic sulfate, decrease the rate of acetaminophen elimination, favor the formation of other more toxic metabolites, and affect the action of CNS

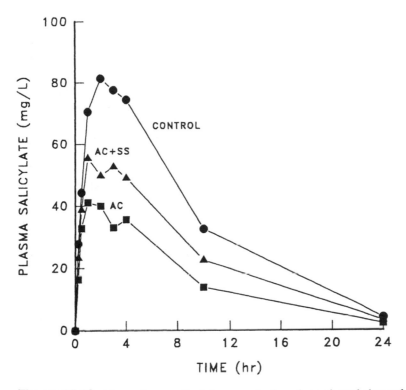

Figure 15.13 Mean plasma salicylate concentrations for activated charcoal treatment, and activated charcoal plus sodium sulfate treatment. From Sketris et al. (1982). Reprinted with permission of the J. B. Lippincott Company, Philadelphia, PA.

neurotransmitters. Thus, because charcoal and sodium sulfate do not negatively impact each other, both can be given.

Curtis et al. (1984) studied the effect of ipecac alone or of charcoal/MgSO$_4$ on aspirin absorption in human volunteers; this was discussed earlier in the section comparing ipecac and charcoal. These investigators found that charcoal/MgSO$_4$ reduced aspirin absorption by 39.9% (compared to controls) versus 26.0% for ipecac alone (both treatments were begun 60 min after the aspirin dose). Since charcoal alone was not compared to the charcoal/MgSO$_4$ treatment, the effect of MgSO$_4$ on the action of the charcoal can not be assessed in this study.

Another in vivo study in which magnesium citrate was combined with charcoal is that of Gaudreault et al. (1985). They first did some in vitro work in which charcoal was used to adsorb paraquat from aqueous solution, with and without added magnesium citrate. The citrate increased the percent paraquat bound at a charcoal:drug ratio of 10:1 from 37 to 53% and at a charcoal:drug ratio of 20:1 from 56 to 80%. This effect was probably caused by the magnesium citrate lowering the pH (pH values were not measured in these in vitro studies), decreasing the dissociation of paraquat (an acidic compound), and thus increasing its adsorption. The in vivo work involved mice given paraquat orally plus magnesium citrate alone, charcoal alone, or a mixture of magnesium citrate and charcoal. The survival percentages were 69, 63, and 94% for these treatments, respectively, as compared to a control value of 31%. Whether the favorable effect of the citrate on the charcoal treatment was due to its effect on paraquat binding to charcoal (the low pH results of Cooney and Wijaya would suggest that this was not the case) or due to its cathartic action can not be determined.

Neuvonen and Olkkola (1986) studied the effects of magnesium citrate and metoclopramide on the efficacy of oral charcoal in seven human volunteers. The subjects were given 1000 mg aspirin, 100 mg atenolol, and 50 mg phenylpropanolamine. In one phase, the subjects then received 25 g charcoal orally after 5 min, in a second phase they received 25 g charcoal orally and 20 mg metoclopramide rectally after 5 min plus 10 mg bisacodyl rectally 3 hr later, in a third phase they received 25 g charcoal with 250 mL magnesium citrate USP after 5 min, and in a fourth phase they received 25 g charcoal with metoclopramide after 60 min and bisacodyl after 3 hr. A fifth phase was the control phase. Plasma drug concentrations (0–24 hr) and urinary excretion (0–72 hr) were followed. Charcoal alone reduced the absorption of aspirin and phenylpropanolamine about 50% and that of atenolol by about 95%. Neither magnesium citrate or metoclopramide plus bisacodyl modified the efficacy of charcoal, although they did decrease the gastrointestinal transit time from 22 hr with charcoal to 7 hr for charcoal plus magnesium citrate, and 16 hr for charcoal plus the purgatives given after 5 min. When charcoal was delayed for 1 hr, its effectiveness was somewhat decreased. Thus study suggests that for the drugs used in this study, the use of magnesium citrate and the other purgatives was not of significant benefit.

Katona et al. (1989) studied the effect of superactive charcoal and magnesium citrate on ethanol pharmacokinetics in humans. The subjects were given 0.6 g/kg ethanol orally and then 60 g charcoal plus 300 mL of 5.8% magnesium citrate solution at 1 hr and again at 3 hr. Comparison to control tests showed no effect of the charcoal/citrate on the AUC value and the peak ethanol blood levels. These results are not surprising since ethanol is rapidly absorbed. By the time the first charcoal dose was given at 1 hr, very little ethanol would have been left in the gastrointestinal tract. In addition, ethanol adsorbs poorly to charcoal, so that even giving charcoal immediately would probably not have had a significant effect. The choice of ethanol and a charcoal/citrate delay time of 1 hr in this study made the insignificance of charcoal therapy a foregone conclusion.

Vuiginier et al. (1989) evaluated the effects of magnesium citrate on the excretion of charcoal given to healthy subjects after the subjects had been given clinidium bromide (an anticholinergic drug). Forty subjects were divided into four groups of 10 each. Group 1 received 5 mg clinidium bromide (CB) plus 15 g charcoal with magnesium citrate (amount unstated) 90 min later; Group 2 received CB and a placebo liquid 90 min later; Group 3 was given a placebo initially and then charcoal/citrate at 90 min; and Group 4 was given both the initial placebo and the placebo at 90 min. The times to first appearance of charcoal in the stool for the 4 groups were: 4.5, 17, 6.3, and 20.6 hr, respectively. Thus, magnesium citrate greatly reduced onset times, both with and without CB. A particularly interesting result is that the onset times for the groups that received CB were shorter than for those who received the placebo liquid, regardless of whether magnesium citrate or the placebo liquid was given. Thus, although CB is an anticholinergic drug, it did not slow gastrointestinal transit times in the dose used. Clearly, CB did not affect the cathartic efficacy of magnesium citrate. These results suggest that the dose of CB was too small to produce the lengthening of transit times that would be expected. This study was previously reported in abstract form by Insley et al. (1985).

C. Summary and Conclusions

We reviewed three in vivo studies involving sodium sulfate, one of which showed a positive effect, one of which showed essentially no effect, and one of which showed a negative effect. With magnesium citrate, one study showed a positive effect, and three others showed no significant effects. In vitro studies tend to support the notion that sodium or magnesium sulfates can slightly increase drug adsorption to charcoal, while citrates decrease drug adsorption (due to competitive adsorption by citric acid). Thus, citrates should be avoided for this reason. Sodium and magnesium sulfates also should be avoided because, while they might slightly enhance the effects of charcoal, they run the risk of hypernatremia or hypermagnesemia (see Chapter 17). Thus, there appears to be no compelling reason to use any saline cathartic in combination with oral charcoal.

VII. SUMMARY

The first topic addressed in this chapter concerned whether charcoal given after syrup of ipecac interferes with the emetic action of the ipecac. One study, by Garrod, with a lower than normal ipecac dose and dogs showed suppression of the ipecac's action by charcoal, while two studies by another research group showed no significant suppression of the emetic effects of 60 mL syrup of ipecac when charcoal was given 5 min later. The question still remains as to what would happen if a normal adult dose of 30 mL of syrup of ipecac were used. However, ipecac and charcoal are not normally given contiguously in time, so this question appears to be a moot one.

As far as the comparative efficacy of syrup of ipecac and charcoal, we reviewed several studies in which both were used. In two studies, syrup of ipecac given at 1 hr was as effective as charcoal given at 1 hr. However, in all studies where treatment was started immediately after drug ingestion, charcoal was superior. Thus, it appears that syrup of ipecac offers no advantage over oral charcoal. However, if poisoning occurs in the home and syrup of ipecac is the only agent immediately available, it should normally be used (except for certain substances, as discussed in Chapter 7, or when a physician or medical facility advises otherwise).

We next considered studies which compared gastric lavage against charcoal. One study showed them to be essentially equal in effectiveness, while another study showed charcoal to be superior. Considering this and the time/discomfort of lavage, charcoal is the better approach. Whole-bowel irrigation was then compared to charcoal. WBI was more effective in a study where enteric-coated aspirin was the drug but charcoal was the better treatment in the other studies we considered. Another factor to take into account is evidence that PEG-ELS solution can interfere with the adsorption capacity of charcoal. Thus, although WBI may have a role in poisonings involving substances which do not adsorb to charcoal (e.g., iron compounds) or enteric-coated drugs, charcoal is generally the better treatment.

The influence of sorbitol on the efficacy of charcoal was considered next. There were six in vivo studies in which sorbitol appears to have had a definite positive effect, one in vivo study in which the effect was slightly positive, two in vivo studies in which there was no effect, and two in vivo studies in which a definite negative effect was found. Thus, on balance, it appears that sorbitol tends to enhance the action of charcoal in reducing drug absorption. However, because excessive sorbitol can do harm (see Chapter 17), its use should be limited. Dosage recommendations for sorbitol were reviewed.

The effects of saline cathartics were summarized at the end of the previous section and the reader is referred to that discussion.

Taken as a whole, the studies reviewed in this chapter suggest that: (1) syrup of ipecac has a role only as an immediate aid in the home, (2) in a medical facility setting, charcoal is superior to ipecac, lavage, and WBI, (3) charcoal plus sorbitol is often more effective than charcoal alone but excessive sorbitol can cause problems (dehydration, hypernatremia, abdominal distention); thus, sorbitol probably should be included only in the first charcoal dose or perhaps in every third dose, and (4) there is no clear evidence that saline cathartics are of significant enough benefit in vivo to warrant the risks involved in using them.

VIII. REFERENCES

Akintonwa, A. and Orisakwe, O. E. (1988). The adsorption of metronidazole and tinidazole to activated charcoal and the effect of magnesium sulfate, Vet. Hum. Toxicol. *30*, 556.

Akintonwa, A. and Orisakwe, O. E. (1990a). The adsorption of quinine and quinidine to activated charcoal with and without magnesium sulfate, Vet. Hum. Toxicol. *32*, 567.

Akintonwa, A. and Orisakwe, O. E. (1990b). Effect of saline cathartics on the adsorption of sulphamethoxazole to activated charcoal, Arch. Int. Pharmacodyn. *304*, 290.

Al-Shareef, A. H., Buss, D. C., Allen, E. M., and Routledge, P. A. (1990). The effects of charcoal and sorbitol (alone and in combination) on plasma theophylline concentrations after a sustained-release formulation, Hum. Exp. Toxicol. *9*, 179.

Al-Shareef, A., Buss, D. C., Alldridge, G. L., and Routledge, P. A. (1992). Effects of cathartics on drug adsorption by activated charcoal in vitro (abstract), EAPCCT XV Congress, Istanbul, May 24-27, p.83.

Albertson, T. E., Derlet, R. W., Foulke, G. E., Minguillon, M. C., and Tharratt, S. R. (1989). Superiority of activated charcoal alone compared with ipecac and activated charcoal in the treatment of acute toxic ingestions, Ann. Emerg. Med. *18*, 56.

Andersen, A. H. (1946). Experimental studies on the pharmacology of activated charcoal. I. Adsorption power of charcoal in aqueous solutions, Acta Pharmacol. Toxicol. *2*, 69.

Andersen, A. H. (1947). Experimental studies on the pharmacology of activated charcoal. II. The effect of pH on the adsorption by charcoal from aqueous solutions, Acta Pharmacol. Toxicol. *3*, 199.

Andersen, A. H. (1948). Experimental studies on the pharmacology of activated charcoal. III. Adsorption from gastro-intestinal contents, Acta Pharmacol. Toxicol. *4*, 275.

Anonymous (1991). Pediatric ingestions: Hold the ipecac, Emerg. Med. *23*, 103.

Berg, M. J., Rose, J. Q., Wurster, D. E., Rahman, S., Fincham, R. W., and Schottelius, D. D. (1987). Effect of charcoal and sorbitol-charcoal suspension on the elimination of intravenous phenobarbital, Ther. Drug Monit. *9*, 41.

Brown, C. R., Becker, C. E., Osterloh, J. D., Olson, K. R., and Viadro, C. (1987). Whole gut lavage in a simulated drug overdose (abstract), Vet. Hum. Toxicol. *29*, 492.

Buckley, N. and Dawson, A. (1993). Whole-bowel irrigation for theophylline overdose (letter), Ann. Emerg. Med. *22*, 1774.

Burkhart, K. K., Wuerz, R. C., and Donovan, J. W. (1992). Whole-bowel irrigation as an adjunctive treatment for sustained-release theophylline overdose, Ann. Emerg. Med. *21*, 1316.

Burkhart, K. K., Wuerz, R. C., and Donovan, J. W. (1993). Whole-bowel irrigation for theophylline overdose (reply to letters), Ann. Emerg. Med. *22*, 1775.

Burton, B. T., Bayer, M. J., Barron, L., and Aitchison, J. P. (1984). Comparison of activated charcoal and gastric lavage in the prevention of aspirin absorption, J. Emerg. Med. *1*, 411.

Chin. L. and Picchioni, A. L. (1980). Charcoal and saline laxative for treatment of poison ingestion, Vet. Hum. Toxicol. 22, 211.

Chin, L., Picchioni, A. L., and Gillespie, T. (1981). Saline cathartics and saline cathartics plus activated charcoal as antidotal treatments, Clin. Toxicol. 18, 865.

Comstock, E. G., Boisaubin, E. V., Comstock, B. S., and Faulkner, T. P. (1982). Assessment of the efficacy of activated charcoal following gastric lavage in acute drug emergencies, J. Toxicol.-Clin. Toxicol. 19, 149.

Cooney, D. O. (1978). In vitro evidence for ipecac inactivation by activated charcoal, J. Pharm. Sci. 67, 426.

Cooney, D. O. and Wijaya, J. (1986). Effect of magnesium citrate on the adsorptive capacity of activated charcoal for sodium salicylate, Vet. Hum. Toxicol. 28, 521.

Corby, D. G., and Decker, W. J. (1974). Management of acute poisoning with activated charcoal, Pediatrics 54, 324.

Cordonnier, J., Van den Heede, M., and Heyndrickx, A. (1987). Activated charcoal and ipecac syrup in prevention of tilidine absorption in man, Vet. Hum. Toxicol. 29 (Suppl. 2), 105.

Curd-Sneed, C. D., Bordelon, J. G., Parks, K. S., and Stewart, J. J. (1987). Effects of activated charcoal and sorbitol on sodium pentobarbital absorption in the rat, J. Toxicol.-Clin. Toxicol. 25, 555.

Curtis, R. A., Barone, J., and Giacona, N. (1984). Efficacy of ipecac and activated charcoal/cathartic: Prevention of salicylate absorption in a simulated overdose, Arch. Intern. Med. 144, 48.

Czajka, P. A. and Konrad, J. D. (1986). Saline cathartics and the adsorptive capacity of activated charcoal for aspirin, Ann. Emerg. Med. 15, 548.

Danel, V., Henry, J. A., and Glucksman, E. (1988). Activated charcoal, emesis, and gastric lavage in aspirin overdose, Br. Med. J. 296, 1507.

Easom, J. M., Caraccio, T. R., and Lovejoy, F. H., Jr. (1982). Evaluation of activated charcoal and magnesium citrate in the prevention of aspirin absorption in humans, Clin. Pharm. 1, 154.

Eyer, P. and Sprenger, M. (1991). [Oral administration of a charcoal-sorbitol-suspension as a first-line treatment to counteract poison absorption?], Klin. Wochenschr. 69, 887.

Foulke, G. E., Albertson, T. E., and Derlet, R. W. (1988). Use of ipecac increases emergency department stays and patient complication rates (abstract), Ann. Emerg. Med. 17, 402.

Foulke, G. E., Albertson, T. E., Derlet, R., and Tharratt, R. S. (1990). Activated charcoal alone versus activated charcoal and ipecac (reply to letter), Ann. Emerg. Med. 19, 1203.

Freedman, G. E., Krenzelok, E. P., and Pasternack, S. (1985). Activated charcoal before syrup-of-ipecac-induced emesis (abstract), Ann. Emerg. Med. 14, 825.

Freedman, G. E., Pasternack, S., and Krenzelok, E. P. (1987). A clinical trial using syrup of ipecac and activated charcoal concurrently, Ann. Emerg. Med. 16, 164.

Galinsky, R. E., and Levy, G. (1984). Evaluation of activated charcoal-sodium sulfate combination for inhibition of acetaminophen absorption and repletion of inorganic sulfate, J. Toxicol.-Clin. Toxicol. 22, 21.

Gaudreault, P., Friedman, P. A., and Lovejoy, F. H., Jr. (1985). Efficacy of activated charcoal and magnesium citrate in the treatment of oral paraquat intoxication, Ann. Emerg. Med. 14, 123.

Goldberg, M. J., Spector, R., Park, G. D., Johnson, G. F., and Roberts, P. (1987). The effect of sorbitol and activated charcoal on serum theophylline concentrations after slow-release theophylline, Clin. Pharmacol. Ther. 41, 108.

Greensher, J., Mofenson, H. C., Picchioni, A. L., and Fallon, P. (1979). Activated charcoal updated, JACEP 8, 261.

Harchelroad, F., Cottington, E., and Krenzelok, E. P. (1989). Gastrointestinal transit times of a charcoal/sorbitol slurry in overdose patients, J. Toxicol.-Clin. Toxicol. 27, 91.

Henschler, D. (1970). Antidotal properties of activated charcoal, Arh. Hig. Rada. Toksikol. 21, 129.

Hoffman, R. S., Chiang, W. K., Howland, M. A., Weisman, R. S., and Goldfrank, L. R. (1991). Theophylline desorption from activated charcoal caused by whole bowel irrigation solution, J. Toxicol.-Clin. Toxicol. *29*, 191.

Holt, L. E., Jr., and Holz, P. H. (1963). The black bottle - a consideration of the role of charcoal in the treatment of poisoning in children, J. Pediatrics *63*, 306.

Insley, B. M., Oderda, G. M., Gorman, R. L., Klein-Schwartz, W., and Watson, W. A. (1985). The effects of a saline cathartic on the excretion of activated charcoal (abstract), Vet. Hum. Toxicol. *27*, 321.

Jessen, L. M. and Barone, J. A. (1992). Ready-mix charcoal/sorbitol (letter), Ann. Emerg. Med. *21*, 110.

Katona, B. G., Siegel, E. G., Roberts, J. R., Fant, W. K., and Hassan, M. (1989). The effect of superactive charcoal and magnesium citrate solution on blood ethanol concentrations and area under the curve in humans, J. Toxicol.-Clin. Toxicol. *27*, 129.

Keller, R. E., Schwab, R. A., and Krenzelok, E. P. (1990). Contribution of sorbitol combined with activated charcoal in prevention of salicylate absorption, Ann. Emerg. Med. *19*, 654.

Kirk, M. A., Peterson, J., Kulig, K., Lowenstein, S., and Rumack, B. H. (1991). Acetaminophen overdose in children: A comparison of ipecac versus activated charcoal versus no gastro-intestinal decontamination (abstract), Ann. Emerg. Med. *20*, 472.

Kirschenbaum, L. A., Mathews, S. C., Sitar, D. S., and Tenenbein, M. (1989). Whole-bowel irrigation versus activated charcoal in sorbitol for the ingestion of modified-release pharmaceuticals, Clin. Pharmacol. Ther. *46*, 264.

Kirschenbaum, L. A., Sitar, D. S., and Tenenbein, M. (1990). Interaction between whole-bowel irrigation solution and activated charcoal: Implications for the treatment of toxic ingestions, Ann. Emerg. Med. *19*, 1129.

Kornberg, A. E. and Dolgin, J. (1991). Pediatric ingestions: Charcoal alone versus ipecac and charcoal, Ann. Emerg. Med. *20*, 648.

Krenzelok, E. P. (1985). Gastrointestinal transit times of cathartics used with activated charcoal, Clin. Pharm. *4*, 446.

Krenzelok, E. P., Keller, R., and Stewart, R. D. (1985). Gastrointestinal transit times of cathartics combined with charcoal, Ann. Emerg. Med. *14*, 1152.

Krenzelok, E. P. and Heller, M. B. (1986). Comparison of activated charcoal and activated charcoal with sorbitol in human volunteers (abstract), Vet. Hum. Toxicol. *28*, 498.

Krenzelok, E. P., Freedman, G. E., and Pasternack, S. (1986). Preserving the emetic effect of syrup of ipecac with concurrent activated charcoal administration: A preliminary study, J. Toxicol.-Clin. Toxicol. *24*, 159.

Krenzelok, E. P. (1987a). Cathartics for drug overdose (reply to letter), Ann. Emerg. Med. *16*, 833.

Krenzelok, E. P. (1987b). Role of sorbitol in theophylline elimination (letter), Ann. Emerg. Med. *12*, 1409.

Kulig, K. (1984). Interpreting gastric emptying studies, J. Emerg. Med. *1*, 447.

LaPierre, G., Algozzine, G., and Doering, P. L. (1981). Effect of magnesium citrate on the in vitro adsorption of aspirin by activated charcoal, Clin. Toxicol. *18*, 793.

Levy, G. (1982). Gastrointestinal clearance of drugs with activated charcoal (editorial), New Engl. J. Med. *307*, 676.

Lim, D. T., Singh, P., Nourtsis, S., and Dela Cruz, R. (1986). Absorption inhibition and enhancement of elimination of sustained-release theophylline tablets by oral activated charcoal, Ann. Emerg. Med. *15*, 1303.

Lim, D. T. and Nourtsis, S. (1987). Role of sorbitol in theophylline elimination (reply to letter), Ann. Emerg. Med. *16*, 1410.

Llera, J. L. (1983). Charcoal for gastrointestinal clearance of drugs (letter), New Engl. J. Med. *308*, 157.

Manoguerra, A. S. (1975). The use of charcoal (letter), Pediatrics *55*, 445.

Mayer, A. L., Sitar, D. S., and Tenenbein, M. (1992). Multiple-dose charcoal and whole-bowel irrigation do not increase clearance of absorbed salicylate, Arch. Intern. Med. *152*, 393.

Mayersohn, M., Perrier, D., and Picchioni, A. L. (1977). Evaluation of a charcoal-sorbitol mixture as an antidote for oral aspirin overdose, Clin. Toxicol. *11*, 561.

McNamara, R. M., Aaron, C. K., Gemborys, M., and Davidheiser, S. (1988). Sorbitol catharsis does not enhance efficacy of charcoal in a simulated acetaminophen overdose, Ann. Emerg. Med. *17*, 243.

McNamara, R. M., Aaron, C. K., Gemborys, M., and Davidheiser, S. (1989). Efficacy of charcoal cathartic versus ipecac in reducing serum acetaminophen in a simulated overdose, Ann. Emerg. Med. *18*, 934.

Minocha, A., Herold, D. A., Bruns, D. E., and Spyker, D. A. (1984-85). Effect of activated charcoal in 70% sorbitol in healthy individuals, J. Toxicol.-Clin. Toxicol. *22*, 529.

Minocha, A., Krenzelok, E. P., and Spyker, D. A. (1985). Dosage recommendations for activated charcoal-sorbitol treatment, J. Toxicol.-Clin. Toxicol. *23*, 579.

Minocha, A., Wiley, S. H., Chabbra, D. R., Harper, C. R., and Spyker, D. A. (1986). Superior efficacy of sorbitol cathartics in poisoned patients (abstract), Vet. Hum. Toxicol. *28*, 494.

Neuvonen, P. J., Vartiainen, M., and Tokola, O. (1983). Comparison of activated charcoal and ipecac syrup in prevention of drug absorption, Eur. J. Clin. Pharmacol. *24*, 557.

Neuvonen, P. J. and Olkkola, K. T. (1984). Activated charcoal and syrup of ipecac in prevention of cimetidine and pindolol absorption in man after administration of metoclopramide as an antiemetic agent, J. Toxicol.-Clin. Toxicol. *22*, 103.

Neuvonen, P. J. and Olkkola, K. T. (1986). Effect of purgatives on antidotal efficacy of oral activated charcoal, Hum. Toxicol. *5*, 255.

Olsen, K. M., Ma, F. H., Ackerman, B. H., and Stull, R. E. (1993). Low-volume whole bowel irrigation and salicylate absorption: A comparison with ipecac-charcoal, Pharmacother. *13*, 229.

Orisakwe, O. E. and Akintonwa, A. (1991). Effect of sodium sulfate on the adsorption of chloroquine and mefloquine to activated charcoal, East Afr. Med. J. *68*, 420.

Palatnick, W. and Tenenbein, M. (1993). Whole-bowel irrigation for theophylline overdose (letter), Ann. Emerg. Med. *22*, 1774.

Picchioni, A. L., Chin, L., Verhulst, H. L., and Dieterle, B. (1966). Activated charcoal versus universal antidote as an antidote for poisons, Toxicol. Appl. Pharmacol. *8*, 447.

Picchioni, A. L. (1974). Research in the treatment of poisoning, in *Toxicology Annual—1974*, pp. 27-51.

Picchioni, A. L., Chin, L., and Gillespie, T. (1982). Evaluation of activated charcoal-sorbitol suspension as an antidote, J. Toxicol.-Clin. Toxicol. *19*, 433.

Pierce, A. W., Jr. (1975). The use of charcoal (reply to letter), Pediatrics *55*, 445.

Pond, S. M. (1986). A review of the pharmacokinetics and efficacy of emesis, gastric lavage, and single and repeated doses of charcoal in overdose patients, in *New Concepts and Developments in Toxicology*, P. L. Chambers, P. Gehring, and F. Sakai, Eds., Elsevier, Amsterdam.

Rademaker, C. M. A., van Dijk, A., Glerum, J. H., and van Heijst, A. N. P. (1987). A practical and effective formulation of activated charcoal, Vet. Hum. Toxicol. *29* (Suppl. 2), 42.

Rademaker, C. M. A., van Dijk, A., de Vries, M. H., Kadir, F., and Glerum, J. H. (1989). A ready-to-use activated charcoal mixture: Adsorption studies in vitro and in dogs: Its influence on the intestinal secretion of theophylline in a rat model, Pharm. Weekbl. [Sci] *11*, 56.

Rosenberg, P. J., Livingstone, D.J., and McLellan, B. A. (1988). Effect of whole-bowel irrigation on the antidotal efficacy of oral activated charcoal, Ann. Emerg. Med. *17*, 681.

Rosenberg, P. and Livingstone, D. (1989). Whole bowel irrigation and activated charcoal (reply to letter), Ann. Emerg. Med. *18*, 708.

Ryan, C. F., Spigiel, R. W., and Zeldes, G. (1980). Enhanced adsorptive capacity of activated charcoal in the presence of magnesium citrate, N.F., Clin. Toxicol. *17*, 457.

Sketris, I. S., Mowry, J. B., Czajka, P. A., Anderson, W. H., and Stafford, D. T. (1982). Saline catharsis: Effect on aspirin bioavailability in combination with activated charcoal, J. Clin. Pharmacol. *22*, 59.

Stiell, I. G. (1990). Activated charcoal alone versus activated charcoal and ipecac (letter), Ann. Emerg. Med. *19*, 1202.

Tenenbein, M. (1987). Cathartics for drug overdose (letter), Ann. Emerg. Med. *16*, 832.

Tenenbein, M., Cohen, S., and Sitar, D. S. (1987). Efficacy of ipecac-induced emesis, orogastric lavage, and activated charcoal for acute drug overdose, Ann. Emerg. Med. *16*, 838.

Tenenbein, M., Kirschenbaum, L. A., Mathews, S. C., and Sitar, D. S. (1988). Whole bowel irrigation versus activated charcoal/sorbitol for the ingestion of delayed release pharmaceuticals (abstract), Vet. Hum. Toxicol. *30*, 353.

Tenenbein, M. (1989). Whole bowel irrigation and activated charcoal (letter), Ann. Emerg. Med. *18*, 707.

Tenenbein, M. (1992). Ready-mix charcoal/sorbitol (reply to letter), Ann. Emerg. Med. *21*, 111.

Tominack, R. L. and Spyker, D. A. (1987). Cathartic response to sorbitol with activated charcoal (abstract), Vet. Hum. Toxicol. *29*, 491.

Underhill, T. J., Greene, M. K., and Dove, A. F. (1990). A comparison of the efficacy of gastric lavage, ipecacuanha and activated charcoal in the emergency management of paracetamol overdose, Arch. Emerg. Med. *7*, 148.

Van de Graaff, W. B., Thompson, W. L., Sunshine, I., Fretthold, D., Leickly, F., and Dayton, H. (1982). Adsorbent and cathartic inhibition of enteral drug absorption, J. Pharmacol. Exp. Ther. *221*, 656.

Vuignier, B. I., Oderda, G. M., Gorman, R. L., Klein-Schwartz, W., and Watson, W. A. (1989). Effects of magnesium citrate and clinidium bromide on the excretion of activated charcoal in normal subjects, Drug Intell. Clin. Pharm. *23*, 26.

Weaver, W. R. (1988). Calculating sorbitol dosage (letter), Ann. Emerg. Med. *17*, 661.

Wieland, M. J., Ling, L. J., and Thompson, J. D. (1986). In vivo effects of excipient agents on the adsorptivity of activated charcoal (abstract), Vet. Hum. Toxicol. *28*, 495.

Zeldes, G., Ryan, C. F., and Spigiel, R. W. (1979). The effect of the cathartic, magnesium citrate, on the ability of activated charcoal to bind sodium salicylate (abstract), Clin. Res. *27*, 723A.

16

The Development Of Palatable Formulations

The difficulties of administering sufficient amounts of activated charcoals to victims of poisoning, especially children, has been frequently mentioned in the literature (e.g., Holt and Holz, 1963). Dr. J. M. Arena (1970), for example, has stated the following view:

> To get a two-year-old child to swallow powdered charcoal is a rare accomplishment indeed. The drawback to this compound is that it is black and many children will refuse to drink it and if spewed, it spots uniforms, clothes, walls, and personnel.

Another report which describes the difficulty in getting children to accept charcoal is that by Grbcich et al. (1987). Their study selected children, ages 1–5, whose parents had called a poison center and who were judged to not need a trip to a medical facility. Trained persons were sent to the children's homes with charcoal and the charcoal was given to a parent to administer. Twenty minutes were allowed. Of the six children involved, none took a dose equivalent to 1 g/kg and only one took 50% of this amount. The report further states that: "All parents had considerable difficulty getting the child to drink the charcoal and most indicated that they would not choose this method of oral decontamination in the event of a future poisoning."

In contrast to this, Calvert et al. (1971) described how 50 children presenting consecutively to an ED for treatment of accidental ingestion of drugs or household products were given paper cups containing slurries of 10 g activated charcoal in water. There was prior use of ipecac in 35 of the children. The charcoal dose was presented "in a firm but kindly manner" and the children were told that it would not taste bad, would not make them sick, and would make them feel better. Eighty-six percent of the children readily drank the slurry and 76% of these consumed 95–100% of the dose.

The problems in swallowing significant amounts of a water/charcoal slurry are clear to anyone who has attempted such. The charcoal immediately sticks in the throat because it adheres strongly to the mucosal surfaces and begins to "cake." Also, a pronounced gritty texture is apparent, unless the charcoal is very finely divided. The obvious solution to the grittiness problem is to use only very finely powdered grades of charcoal.

Another aspect of attempting to develop ready-made formulations of charcoal is to try to overcome the reluctance of medical personnel in administering such a black, messy substance. Pilcher (1980) mentioned that he has used charcoal less than he should because of the mess. Greensher et al. (1980) replied to Pilcher's letter with a suggestion to premix 30 g charcoal with 150 mL of a 70% sorbitol solution (a few years later, such mixtures did become available commercially; see Chapter 4). Krenzelok (1980) recommended a simple method of filling an empty one-pound ointment jar half-full with charcoal (40–60 g), adding 30 mL tap water per each estimated 10 g charcoal, placing the cover on, and shaking the jar for 30 seconds. He also mentioned making palatable mixtures with bentonite and chocolate syrup (this was later described in the paper by Navarro et al., 1980). Jaeger and de Castro (1977) described a method they developed for packaging mixtures of charcoal and 5% w/v carboxymethylcellulose gel solution, using heat-sealable plastic bags. They report successful use of these packages in three participating ED's.

An interesting attempt to make powdered charcoal more palatable has been described by Nakano et al. (1984). They prepared 48–250 mesh spherical beads of powdered activated charcoal dispersed in a porous agarose matrix. The dried beads contained 67% charcoal, and had "good flow and handling properties." The in vitro adsorption powers of the encapsulated charcoal were tested with salicylic acid and acetaminophen, and were found to be unaffected by the agarose matrix. Moreover, the rate of approach to adsorption equilibrium was reduced only slightly by the charcoal being dispersed in the agarose matrix (the "percentage approach to equilibrium" values for salicylic acid at 5 and 15 min in one medium were 98.5 and 99.3% for charcoal powder and 90.7 and 96.5% for the beads; for a second medium, the values were 96.7 and 99.4% for the powder and 94.2 and 97.1% for the beads). In vivo tests with rats showed that the beads reduced plasma salicylate levels as well as did an equivalent amount of charcoal in the form of powder. It would have been interesting if palatability studies had been done with human volunteers, as this novel form of preparation might solve at least three problems associated with pure charcoal powder: a gritty texture, a tendency to stick to the mucosal surfaces of the mouth, and an extremely black appearance.

I. PROVIDING LUBRICITY

Manes and Mann (1974) attempted to solve the stickiness problem by testing a number of high molecular weight thickening agents which, when mixed with charcoal and water, would provide lubricity to such mixtures (see also the patent by Manes, 1975). They investigated sodium carboxymethylcellulose (NaCMC), carrageenan, sodium alginate, gelatin, and bentonite; bentonite and NaCMC were both greatly effective in enhancing the ease with which the charcoal mixture could be swallowed. Moreover, tests on the equilibrium adsorption of aspirin and sodium benzoate from a 0.1 M HCl solution showed that neither of these additives interfered with the adsorptive capacity of the charcoal (clearly because they themselves do not adsorb). Experiments on the effect of these thickeners on the rate of aspirin adsorption from a well-shaken 0.1 M HCl solution suggested that very little reduction in the adsorption rate was caused. However, these rate tests involved 1 g charcoal (plus thickener, if any) and 100 mL solution, so, in

Table 16.1 The Influence of Different Activated Charcoal Preparations on the Gastrointestinal Absorption of Aspirin

	Relative absorption of aspirin		
Subject	Charcoal slurry	Charcoal-bentonite	Charcoal-carboxy-methylcellulose
1	69.0	63.3	61.8
2	65.7	73.7	75.4
3	77.0	73.5	—
4	81.8	59.8	—
Mean	73.4	67.6	68.6

From Gwilt and Perrier (1976).

effect, the thickeners were rapidly diluted via the shaking which was imposed. It is therefore not surprising that little effect on the adsorption rate was found.

Gwilt and Perrier (1976) also confirmed, by in vivo tests with four male volunteers, that NaCMC and bentonite do not interfere with the gastrointestinal adsorption of aspirin by charcoal. In these tests, three 325-mg aspirin tablets were given orally with a mixture of 200 mL water and 10 g Norit powdered charcoal plus: (1) no additive, (2) 50 g bentonite magma USP, or (3) 25 g of a 2% NaCMC solution (whether this was 2% w/w or 2% w/v is not stated). Their results are shown in Table 16.1 (the relative absorption is the percent recovery of salicylate in the urine compared with the recovery of salicylate when aspirin was administered without activated charcoal). The bentonite and sodium carboxymethylcellulose actually decreased aspirin absorption but to a statistically insignificant extent (however, the number of data points was relatively small in this study). It was further noted that the time of maximum salicylate excretion was also unaltered by the addition of bentonite or NaCMC. Hence, both the rate and extent of adsorption were not significantly affected.

In contrast to the results discussed thus far, Mathur et al. (1976a) reported that carboxymethylcellulose (CMC) significantly reduced the efficacy of charcoal in vivo. In their study, three 325-mg aspirin tablets were given orally to six male volunteers, followed immediately by 180 mL of water plus: (1) nothing else, (2) 20 g charcoal and 45 mL water as a slurry, (3) 65 g of an aqueous charcoal/CMC mixture, or (4) 65 g of the aqueous charcoal/CMC mixture with 10 mL Hershey's chocolate syrup mixed in. The effects of each treatment on salicylate excretion in the urine are shown in Figure 16.1. It is clear that CMC does reduce the efficacy of charcoal and that the chocolate syrup had an additional, but slight, negative effect.

Manes (1976), in trying to explain the differences in the effects of CMC in the three studies just discussed, pointed out that "Mathur et al. used about twice the highest CMC concentration of Manes and Mann. Moreover, a sample of the 'gel' of Mathur et al. shows it to be hard and rubbery in consistency, not readily suspended in water" and that, after 2 hr of shaking such a gel in water, "more than half the gel remained in the original rubbery state." This resistance to dissolution would certainly account for the interference observed by Mathur et al. Indeed, the recipe of Manes and Mann consisted of 25 wt% charcoal and 1.5 wt% NaCMC in water; that of Mathur et al. consisted of

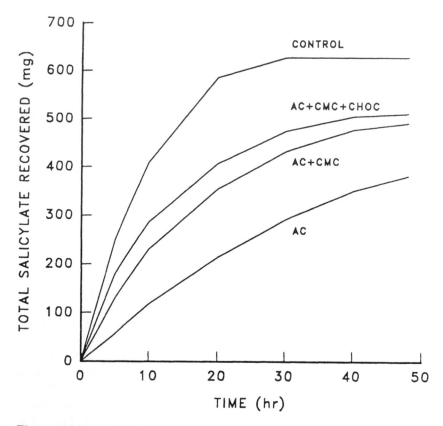

Figure 16.1 Mean cumulative urinary excretion of total salicylate after oral administration of three aspirin tablets alone (control) or with various combinations of activated charcoal (AC), CMC gel, and chocolate syrup (CHOC). Originally published in Mathur et al. (1976a), American Society of Hospital Pharmacists, Inc. All rights reserved. Reprinted with permission.

Table 16.2 Mean Percent Aspirin Dose Recovered in the Urine of Six Subjects in 48 hr

Treatment	CMC gel 3.46 wt%	CMC gel 1.5 wt%
Aspirin alone	84.7	80.6
Aspirin + charcoal	49.5	47.5
Aspirin + charcoal/CMC gel	66.4	64.0
Aspirin + charcoal/CMC gel + chocolate syrup	70.4	65.5

Originally published in Mathur et al. (1976b), American Society of Hospital Pharmacists. All rights reserved. Reprinted with permission.

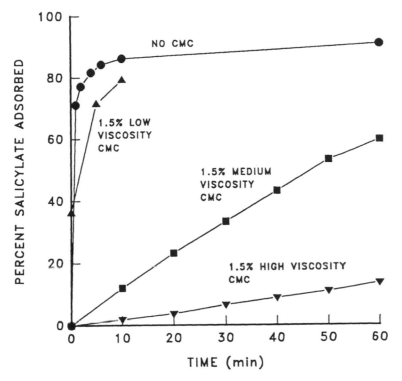

Figure 16.2 Rate of salicylate adsorption by 1.5 wt% CMC-30 wt% activated charcoal mixtures in horizontal shaken vessels. From Cooney, 1982.

31 wt% charcoal and 3.5 wt% CMC in water. The difference in CMC contents in the two studies would easily explain the different results.

In rebuttal, Mathur et al. (1976b) replied that they repeated their studies exactly using a 1.5 wt% CMC level and obtained the results shown in Table 16.2, results which show that the reduction in charcoal efficacy occurs to almost the same degree with 1.5 wt% CMC as with 3.5 wt% CMC. However, a likely explanation for this, apart from noting that Mathur employed CMC whereas Manes and Mann employed NaCMC, is that Mathur's CMC was a "High Viscosity Grade." Neither Manes and Mann nor Gwilt and Perrier specify the source of their CMC, but it is possible that it was a type of much lower viscosity. Since CMC is ordinarily available in a variety of molecular weights (e.g., 80,000, 250,000, 700,000 daltons) and since polymer viscosities increase substantially with molecular weight, large differences in the viscosities of different CMCs exist.

The effect of the type (low-viscosity, medium-viscosity, high-viscosity) and amount of CMC, added to charcoal slurries as a lubricant, on the in vitro adsorption of sodium salicylate was studied by Cooney (1982). He made mixtures of 30 wt% charcoal, 1.5 wt% CMC, and 68.5 wt% water, using the three grades of CMC. He then took 5 g portions of such mixtures and put them into a horizontal vessel (plastic bottle on its side) mounted on a shaker platform which oscillated at 60 cycles per minute. The vessel contained 400 mL of a 1 g/L pH 1.2 sodium salicylate solution. Samples were withdrawn at various times and analyzed for salicylate. Figure 16.2 shows the results. The mixture made with the high-viscosity CMC broke up very slowly; thus, adsorption of the test drug by the charcoal was slow. The low-viscosity CMC mixture broke up reasonably quickly. These results show that, when using CMC to add lubricity to an antidotal charcoal mixture, enough water should be used, regardless of what type of CMC is employed, to make the mixture pourable and not gel-like. If the mixture is too gel-like, it will not break up and allow the charcoal to adsorb the drug in a reasonable time.

It is clear from all of these CMC/bentonite studies that lubricity can be imparted to antidotal charcoal mixtures without adsorptive interference, if bentonite or CMC is used in the proper amount. If CMC is employed, a low-viscosity grade is obviously preferable.

Before leaving the subject of using CMC, an interesting study by Levy and Jusko (1965) needs mention. They administered two test drugs (ethanol and salicylic acid) to rats in solutions having different added amounts of a low viscosity CMC and found that the rate of drug absorption into the rats' systems decreased as the solution viscosity increased. They were able to show that the increased viscosity decreased the rate of drug absorption by retarding the rate of movement of the drug molecules to the absorbing GI tract membranes and by slowing the rate of gastrointestinal transit of the drug. This suggests that a charcoal/CMC formulation, if administered promptly, could have a benefit other than increasing the charcoal's palatability, namely, that the ingestion of CMC would, after it mixes with the stomach's contents, reduce the rate of drug absorption by increasing the viscosity of the stomach's contents.

II. PROVIDING FLAVOR

A second aspect of making charcoal mixtures more appealing concerns flavor. Levy et al. (1975) attempted to provide flavor as well as lubricity to activated charcoal formulations by mixing charcoal with ice cream or sherbet. In vitro studies were done by mixing 0.25 g aspirin in 100 mL 0.01 N HCl with a suspension of 5 g charcoal in either 25 mL water or in 12.5 mL water plus 12.5 g ice cream or sherbet. After 5 min shaking, the residual aspirin was determined. With no added ice cream, the unadsorbed aspirin was present at a concentration of 17.4 mg/100 mL (average of seven trials). With regular and diet vanilla ice creams, the final concentrations averaged 54.2 and 53.5 mg/100 mL, respectively. The use of lemon sherbet gave a final concentration of 56.1 mg/100 mL. Considering that the samples initially contained about 200 mg/100 mL aspirin, it can be said that the ice cream and sherbet reduced aspirin adsorption from 91.3 to roughly 73%.

Levy et al. (1975) also carried out an in vivo study with five male volunteers. Three 325-mg aspirin tablets were given to each subject along with 50 mL water; 20 g Norit charcoal was given either in 100 mL water or in 50 mL water plus 50 g ice cream. Data on salicylate excretion in the urine over a period of 48 hr showed an average of 36% recovery of aspirin into the urine with charcoal alone and 60% recovery when charcoal plus ice cream was used. In other words, aspirin absorption was reduced by 64% with charcoal alone and by 40% with charcoal/ice cream. While Levy et al. concluded that ice cream should not be used as a vehicle for administering charcoal, Oderda (1975) subsequently commented, "Why could not the dose of charcoal be raised by half again ... and thus negate the loss of charcoal (activity) by the ice cream. Perhaps children would accept 30 grams of charcoal mixed with ice cream more readily than 20 grams of charcoal mixed with water." This would seem to be a logical possibility; thus, the use of ice cream as a vehicle for charcoal perhaps should not be discounted. In reply to

Oderda's letter, Levy (1975) reiterated his view that the use of ice cream should be avoided or at least minimized, especially if other suspending agents are found which would not interfere with the adsorptive action of charcoal.

Subsequently, the idea of mixing charcoal with ice cream was studied by Cheng and Robertson (1989). They used a 3:1 ice cream:charcoal mixture and solutions of salicylate between 0.5 and 2.0 g/L in simulated gastric fluid, pH 3 (rather than the more traditional pH 1.2). These were combined to give an overall charcoal:salicylate ratio of 10:1 in the test samples. In control (ice cream-free) samples, 74% of the salicylate was adsorbed, while in the samples containing ice cream an average of 44% of the salicylate was adsorbed. Although the ratio 44/74 = 0.60 and thus a 40% reduction in salicylate adsorption occurred with ice cream, a child who will consume charcoal-laced ice cream is better off than a child who refuses to take any, or very little, charcoal.

De Neve (1976) pursued the notion of adding flavors to charcoal with an in vitro study of the effects of jams, marmalades, starches, cocoa powder, sugar, and milk on aspirin adsorption to charcoal. To 100 mL solution of 250 mg/100 mL aspirin in 0.01 M HCl, De Neve added 1 g charcoal in 25 mL water plus various additives (e.g., 3 g of a jam). After 5 min shaking, the percentage of unadsorbed aspirin was determined. Table 16.3 shows the results on the jams, marmalade, and jellies. At the dose levels used, the reductions in aspirin adsorption are not large (the maximal reduction is from 98.4 to 91.8% adsorbed). Differences between the last two jams listed and the others are probably related to the fact that they were manufactured by a different company. Figure 16.3 shows the effects of increasing the jam or jelly dose; it can be seen that even when one uses 12 g jam or jelly, more than 80% aspirin adsorption is still achieved.

De Neve also obtained similar data using starches, cocoa powder, sugar, and different types of milk. As shown in Table 16.4, only the milk and simple syrup offered significant interference. The question that De Neve provides little answer to is that of whether these additives give lubricity and/or flavor to the charcoal. He stated that "the preparations do not look very palatable because of their black color; also the taste of the jams is lost to a considerable degree."

The effects of adding sorbitol, sucrose, cocoa powder, milk, and starch on the in vitro adsorption of tilidine hydrochloride, a narcotic analgesic, to Norit A charcoal was studied by Cordonnier et al. (1986–87). At pH 1.2 (in HCl solution), with 100 mg charcoal added to 10 mL of a solution containing 10 mg of the drug, tilidine adsorption was 72.9% and decreased to the values shown in Table 16.5 with the addition of different materials. When the amount of charcoal was doubled to 200 mg, tilidine adsorption was 95.0% and decreased to the values shown in the same table with the addition of the same materials. Thus, when the amount of charcoal relative to the amount of added ingredient was increased, cocoa powder, potato starch, corn starch, and milk had no deleterious effect, and the effects of the other ingredients was made much less. At the lower charcoal:drug ratio, sucrose and sorbitol showed surprising interference, considering that neither adsorbs significantly to charcoal.

Cooney (1977) described a study in which the jam formulations of De Neve (1 g charcoal, 3 g jams, 25 mL water) were prepared (using Norit A powdered charcoal) and were taste-tested. It was found that essentially no residual flavor remained. The charcoal/chocolate syrup recipe of Mathur et al. (1976a) was also taste-tested with the same result. Cooney also tried mixing moderate amounts of flavor extracts like orange oil and peppermint oil with charcoal slurries and found that, even after a minute or two, no taste or odor remained. Cooney (1977) then investigated the use of saccharin sodium as a flavor,

Table 16.3 Effect of Jams and Jellies on the Adsorption of Acetylsalicylic Acid by Activated Charcoal

Additive	Amount (g)	Acetylsalicylic acid in filtrate (%)
None	—	1.6
Orange marmalade	3	4.9
Pear jelly	3	4.2
Red berry jelly	3	4.2
Peach jam	3	4.2
Strawberry jam	3	4.7
Black cherry jam[a]	3	8.2
Bilberry jam[a]	3	8.0

[a]These two jams are of a different brand than the other products. Originally published in De Neve (1976), American Society of Hospital Pharmacists. All rights reserved. Reprinted with permission.

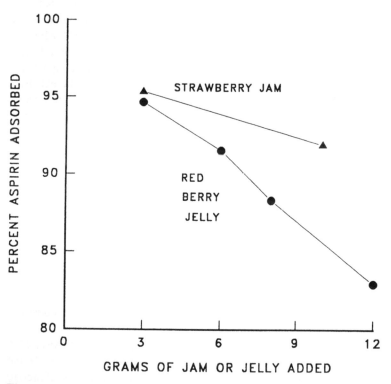

Figure 16.3 Percentage of aspirin adsorbed from a dilute HCl solution by activated charcoal as a function of the quantity of jam or jelly added. Originally published in De Neve (1976), American Society of Hospital Pharmacists, Inc. All rights reserved. Reprinted with permission.

Table 16.4 Effect of Starches, Milk, Cocoa Powder, and Sugar on the Adsorption of Acetylsalicylic Acid by Activated Charcoal

Additive	Amount	Acetylsalicylic acid in filtrate (%)
None	—	1.6
Starch (potato)	1 g	1.5
Starch (corn)	1 g	1.5
Starch (wheat)	1 g	1.4
Starch (rice)	1 g	1.5
Cocoa powder	1 g	7.4
Sugar	1 g	3.2
Simple syrup	25 mL	21.1
Milk (brand A)	25 mL	8.3
Milk (brand B)	25 mL	19.4
Skim milk	25 mL	22.5

Originally published in De Neve (1976), American Society of Hospital Pharmacists. All rights reserved. Reprinted with permission.

Table 16.5 Effect of Various Additives on the Adsorption of Tilidine HCl by Norit A Activated Charcoal (AC)

Additive	Amount	Percentage of drug adsorbed (AC:drug) 10:1	20:1
None	—	72.9	95.0
Cocoa powder	300 mg	34.0	94.7
Sucrose	300 mg	49.0	81.0
Starch (potato)	300 mg	60.0	95.4
Starch (corn)	300 mg	69.1	95.5
Milk	10 mL	84.2	96.6
Sorbitol	300 mg	59.8	74.0
Sorbitol	600 mg	56.5	72.7
Sorbitol	900 mg	58.0	78.4
Sorbitol	1200 mg	58.1	79.5

From Cordonnier et al. (1986–87).

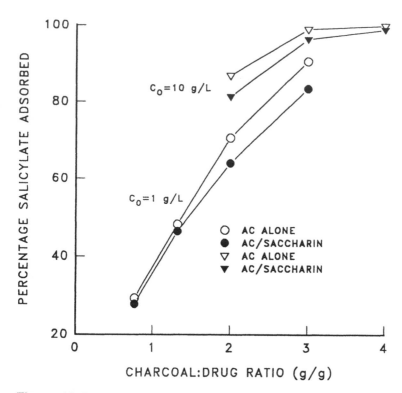

reasoning that because saccharin is roughly 500 times sweeter than sugar, only a small amount would be needed. Certainly the saccharin itself would adsorb to the charcoal (as with all of the other flavors mentioned above). This might cause loss of the sweet flavor, as well as taking up some of the adsorption capacity of the charcoal. However, the addition of 1 part saccharin per 10 parts charcoal provided enough residual unadsorbed saccharin to give a pleasant sweetness to an antidotal mixture of the Manes and Mann type (25 wt% charcoal plus 1.5 wt% CMC in pure water). Moreover, this amount of saccharin did not interfere significantly with the adsorption of a test drug (sodium salicylate) from SGF USP (pepsin omitted). Figure 16.4 shows the results of the interference tests. These results show that at a charcoal:salicylate ratio of 2:1, for example, the saccharin reduced the salicylate adsorption from 70 to 64% from a fluid of 1 g/L initial salicylate concentration. At a 10 g/L initial concentration, the reduction was from 87 to 82% adsorbed. Considering that sweetness might significantly enhance patient acceptance of the antidotal mixture, such a small loss could be easily counteracted by administration of a slightly larger dose of the charcoal mixture.

A nearly identical study was then done by Cooney and Roach (1979) using sucrose (household sugar) as a sweetener. They found that one part of sucrose per part of charcoal was needed to impart a pleasantly sweet taste. Again using sodium salicylate as a test

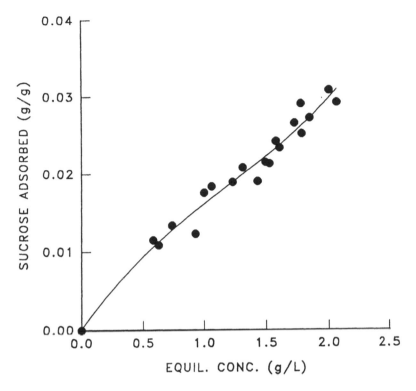

Figure 16.5 Adsorption of sucrose to activated charcoal. Originally published in Cooney and Roach (1979), American Society of Hospital Pharmacists, Inc. All rights reserved. Reprinted with permission.

drug and a charcoal:salicylate ratio of 2:1, the presence of sucrose (one part per part of charcoal) reduced salicylate adsorption, from a fluid of 1 g/L initial salicylate concentration, from 87% absorbed to 84% adsorbed. Hence, sucrose is an acceptable (and familiar) flavoring agent. Figure 16.5 shows the isotherm for sucrose adsorption from simulated gastric fluid, and it can be seen that sucrose adsorbs poorly (at a fluid-phase concentration of 1 g/L, the amount adsorbed is less than 0.02 g per gram of charcoal). Figure 16.6 shows the results of the tests conducted to determine the extent of interference of sucrose with salicylate adsorption.

Another test of a flavoring agent was reported by Yancy et al. (1977). They added cherry flavoring (imitation wild cherry extract) to charcoal and conducted both in vitro and in vivo tests. In the in vitro tests, 100 mL of a 320 mg/100 mL solution of sodium salicylate in SGF was contacted for 20 min with either 5 g charcoal in 50 mL water or 5 g charcoal in 48 mL water and 2 mL cherry extract. Residual salicylate was nearly the same in both cases (0.133 and 0.144 mg/100 mL, respectively). In vivo studies with rats showed reductions in plasma salicylate levels at 1 hr of 55 and 49%, compared to control animals, for rats treated with 125 mg charcoal alone and 125 mg charcoal plus 0.2 mL cherry extract, respectively. Further in vivo tests were done with five human volunteers who received ten 325-mg aspirin tablets followed 30 min later by either 15 g charcoal in 150 mL water or this same slurry with 20 mL cherry extract added. The

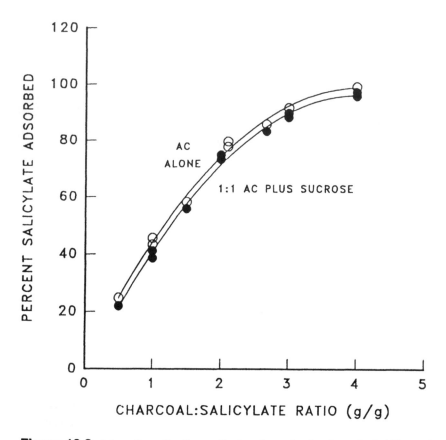

Figure 16.6 Adsorption of sodium salicylate from simulated gastric fluid by activated charcoal alone and by a 1:1 mixture of charcoal:sucrose. From Cooney and Roach, 1979.

Table 16.6 The Influence of Activated Charcoal and Sorbitol on Aspirin Absorption

Experiment	Dose recovered (%)[a]	Maximal excretion rate (mg/hr)[a]	Time of maximal excretion rate (hr)[a]
Control	100.0	100.0	100
Sorbitol	95.9	72.1	335
Charcoal slurry	38.7	34.2	201
Charcoal and sorbitol	51.3	55.7	315

[a]All values are means and are expressed relative to controls. The results of the control experiment were: dose recovered, 73.2%; maximal excretion rate, 67.3 mg/hr; t_{max}, 1.9 hr. From Mayersohn et al. (1977).

plasma salicylate concentrations were essentially identical for both types of treatments during the 0–2 hr period. Thereafter, from about 4 to 8 hr, the plasma aspirin levels were about 10–15% higher for the group treated with the cherry-flavored charcoal. The amount of cherry extract used in these human trials was apparently enough to impart a pleasant taste to the charcoal without interfering very much with the charcoal's efficacy. However, the subjects still complained that the preparation was black and had a chalk-like consistency.

Another study along these lines is one done by Mayersohn et al. (1977) using sorbitol as the flavoring. They administered three 325-mg aspirin tablets plus 50 mL water to human subjects and followed this immediately with: (1) 150 mL water, (2) 100 mL of a slurry of 20 g charcoal/100 mL, (3) 100 mL of a 70% sorbitol solution, or (4) 100 mL of a slurry of 20 g charcoal in a 70% sorbitol solution. Salicylate excretion into the urine was followed for 48 hr. Table 16.6 shows the results obtained.

The plain sorbitol solutions produced diarrhea in all subjects and led to a decrease in the salicylate recovery in the urine as compared to the controls and, in all but one subject, the sorbitol/charcoal mixtures also produced diarrhea. While this might tend to decrease the salicylate recovery (during diarrhea the aspirin is moved down the gastrointestinal tract to regions where it is not as rapidly absorbed), the data show a decidedly higher recovery compared to the charcoal alone. It is apparent that the sorbitol does interfere somewhat with the action of the charcoal. Here again, however, if the sweetness imparted by the sorbitol increases the amount of charcoal the patient will accept, its use may be of benefit. In addition, since diarrhea reduces drug absorption, the possibility arises that for drugs which are much more slowly absorbed than aspirin, a charcoal/sorbitol mixture might be more effective than charcoal alone.

Scholtz et al. (1978) presented results on taste tests of several activated charcoal formulations and on in vivo aspirin absorption tests using them. The formulations consisted of 20 g activated charcoal plus: (1) 60 g water; (2) 43 g of 70% sorbitol solution USP; (3) 77 g of 70% sorbitol solution USP; (4) 10 mL chocolate syrup, 35 g water, and 0.3 g sodium CMC; (5) 10 mL chocolate syrup plus 43 g of a 4% corn starch solution; and (6) 31 g of 70% sorbitol solution USP plus 31 g of a mineral oil emulsion (45.9 wt% mineral oil, 49.5 wt% 70% sorbitol solution USP, and 4.6 wt% of Span 80 and Tween 80 surfactants). The taste tests, done with the first five mixtures only, revealed that the flavored mixtures were much preferred over the nonflavored formulation and that the high-sorbitol mixture (number 3) was liked best. However, the sorbitol formulations were often judged to be somewhat too sweet. Additionally, the chocolate flavor in mixtures 4 and 5 was found to disappear quickly, which agrees with an observation made by Mathur et al. (1976a) two years earlier.

When all six mixtures were given to adults immediately after a 972-mg dose of aspirin, the mean reductions in aspirin absorption, relative to controls, were 32, 40, 33, 29, 27, and 8%, respectively. Note that the idea, mentioned above, that charcoal/sorbitol mixtures might be more effective than charcoal alone (because of the diarrhea effect) is supported slightly by these results. Clearly the mineral oil formulation had little effect, probably because the mineral oil itself adsorbed to the charcoal and used up much of its capacity, or perhaps because it coated the charcoal, thereby preventing good contact of gastrointestinal contents with the charcoal. The plain slurry, the chocolate formulations, and the sorbitol formulations were essentially statistically equivalent. The main conclusion of this study was that sorbitol was the best flavoring agent of those tested.

An article in the *Drug and Therapeutics Bulletin* (Anonymous, 1979) discussed the charcoal formulation that appeared in the late 1970s called Medicoal (described earlier in Chapter 4). Medicoal is not currently available. This formulation consisted of 5 g portions of charcoal in an effervescent medium, all packaged in sealed sachets. It was said to be more palatable than charcoal in water alone and "less likely to deteriorate with time than earlier preparations" (although any sealed, sterile charcoal preparation will last indefinitely). The article mentioned the study of Crome et al. (1977) in which Medicoal was given to volunteers 30 min after a 75-mg dose of nortriptyline and caused a 60% reduction in the drug's absorption. While Dawling et al. (1978) stated that, after mixing with water, the resultant slurry is "not unpleasant to drink", no further information on its palatability has been reported.

Cooney (1980) carried out taste tests on activated charcoal formulations containing saccharin sodium, sorbitol, and sucrose in the proportions of 0.05, 1, and 1 parts of these per part of charcoal, respectively. All three flavored formulations were strongly preferred over unflavored charcoal by 16 adult subjects but were not rated statistically different from one another. He determined that the relative volumes of the mixtures that one would need to administer to a person in order to deliver a given amount of charcoal would be 1.02, 1.22, 1.20, and 1.00 for the saccharin, sorbitol, sucrose, and flavorless formulations, respectively. Thus, one advantage of using saccharin as a flavor is that it increases the formulation volume, per unit weight of charcoal, by only 2%.

Boehm et al. (1978) described a study in which 20 g strawberry flavor (a powdered flavor prepared by spray drying with acacia) and 40 mg fructose per each 100 g Norit Medicinal activated charcoal were used. When dogs were given 20 mg/kg pheniramine maleate orally and were treated with flavored or unflavored charcoal, the drug plasma levels (followed for 24 hr) were lowered dramatically by both charcoals (e.g., the peak drug level was 2.73 mg/L without charcoal, 0.15 mg/L with unflavored charcoal, and 0.17 mg/L with flavored charcoal). Figure 16.7 shows their results in graphical form. Thus, the flavoring did not significantly interfere with the efficacy of the charcoal.

Boehm et al. also made some interesting comments but without any details or supporting data: "Saccharin and cyclamate solutions were tested as sweetening agents, but lost their flavour when charcoal was added. Sucrose and fructose produced a sweetness which was still present after addition of activated charcoal, but sucrose interfered slightly with adsorption whereas the effect of fructose was negligible." However, as mentioned earlier, Cooney (1977) found that saccharin, in a proper amount, did not lead to a loss of flavor. The statement that sucrose interferes only "slightly" with adsorption agrees with the observations of Cooney and Roach (1979), mentioned earlier.

A short report on strawberry-flavored charcoal was given by Oppenheim (1980). He developed a recipe consisting of 20 g Norit V charcoal, 8 g fructose, and 4 g imitation strawberry flavor. He stated that, upon addition of water, an acceptable sweetened flavor is generated which lasts for at least 7 min, which, he concluded, "should be long enough for administration to a child."

As a sequel to the two studies just discussed, Oppenheim and Miles (1981) evaluated the in vitro effectiveness of a dry powder mixture of activated charcoal, strawberry flavor, and fructose for adsorbing mianserin (mianserin is a quadricyclic antidepressant). Tests were done in SGF (without pepsin) and in SIF (without pancreatin). They used either charcoal alone or formulations that were 62.5 wt% charcoal, 25 wt % fructose, and 12.5 wt% of strawberry flavor. At a charcoal:drug ratio of 10:1 in the SGF, Norit charcoal alone adsorbed 100% of the drug, whereas the flavored/sweetened formulation

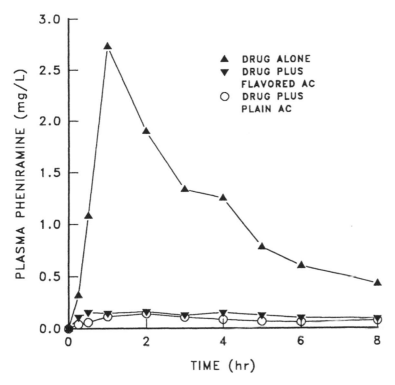

Figure 16.7 Plasma levels of pheniramine in whole blood for the drug alone, the drug and activated charcoal, and the drug plus flavored activated charcoal. From Boehm et al. (1978). Reproduced by permission of the Australian Pharmaceutical Publishing Company.

adsorbed 80% of the drug. In the SIF, Norit alone again adsorbed 100% of the drug, compared to 98% for the flavored/sweetened formulation. Hence, the flavor and/or fructose interfered somewhat with drug adsorption at low pH but not at the higher pH. The reason for this difference is not clear.

The relative efficacy and palatability of three charcoal mixtures was studied by Navarro et al. (1980). One mixture was simply 10 g charcoal in 45 mL water. A second was 10 g charcoal and 2.5 g bentonite clay in 45 mL water. The third was 10 g charcoal, 2.5 g bentonite, and 10 mL Hershey's chocolate syrup in 45 mL water. These were given to healthy volunteers in two ways: first, in a palatability test, and second, in an in vivo aspirin absorption test. All four adults in the study agreed that the charcoal/bentonite/chocolate mixture was the most palatable, and the plain charcoal/water mixture the least palatable. With ten children, seven picked the charcoal/bentonite/chocolate mixture as best, and nine picked the charcoal/water mixture as worst. Thus, the mixture containing chocolate was clearly favored.

In the in vivo tests, four adult males took aspirin followed by one of the three mixtures. Salicylate excretion in the urine was monitored for 48 hr. Figure 16.8 shows the cumulative salicylate excreted in the urine for the different treatments. The percentage decreases in aspirin absorption, relative to control tests, were: for the charcoal/water mixture, 44.6%; for the charcoal/bentonite/water mixture, 35.2%; and for the charcoal/

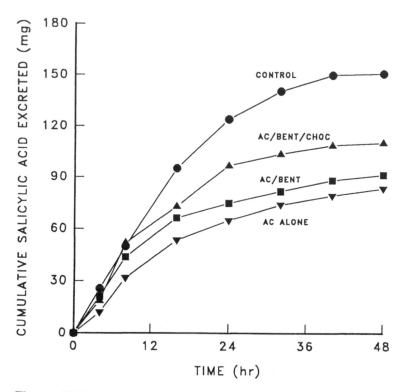

Figure 16.8 Mean urinary excretion of salicylates following administration of three 5-grain aspirin tablets with various mixtures of activated charcoal, bentonite, and chocolate syrup. From Navarro et al. (1980). Reprinted with permission from *Veterinary and Human Toxicology*.

bentonite/chocolate syrup/water mixture, 24.8%. Thus, the results showed that bentonite reduces the effectiveness of charcoal to a modest extent and that the addition of chocolate syrup had a comparable and additional negative effect. Nevertheless, as in the case of adding ice cream to charcoal, the negative effect of chocolate syrup might be tolerable if the end result is that a patient would accept a significantly greater amount of the mixture, such that the benefits of the greater dose of charcoal outweigh the effects of the chocolate.

Chung et al. (1982) compared the effectiveness of two kinds of charcoal with added fructose on aspirin absorption in seven healthy subjects. Fructose was chosen as the sweetener because it is the sweetest of all of the naturally occurring sugars and thus a relatively dilute solution (10%) could be used to provide good palatability. The subjects were given 975 mg aspirin with 240 mL water in one phase, the same aspirin dose followed immediately with 20 g Norit A charcoal in 120 mL 10% fructose solution in a second phase, and, in a third phase, 20 g superactive charcoal was used with 120 mL 10% fructose. Urine samples were collected for 48 hr. The mean salicylates recovered in the urine were 773 mg (control phase), 480 mg (Norit/fructose phase), and 265 mg (superactive charcoal/fructose phase). The charcoals were not tried without the fructose. In any case, both charcoal formulations were effective in reducing aspirin absorption, with the superactive charcoal clearly superior.

Krenzelok and Heller (1987) studied the effectiveness of five commercially available charcoal formulations (ActaChar, Actidose Aqua, InstaChar, LiquiChar, and Super-Char) on the absorption of aspirin in seven healthy adult subjects. The subjects were given 2,592 mg aspirin (eight intact 324-mg tablets) followed by 25 g charcoal in the form of the five different products. The study was a double-blind cross-over study. Each subject took each formulation, as well as serving as a control in one phase. The 0–8 hr reductions in aspirin absorption compared to control values, as determined from blood samples, for each product were: SuperChar, 57.8%; Actidose Aqua, 50.4%; InstaChar, 39.6%; LiquiChar, 33.4%; and ActiChar, 27.5%. The higher surface area products, SuperChar (3,150 m^2/g) and Actidose Aqua (1,500 m^2/g), prevented the absorption of aspirin more effectively than the other three products, whose charcoal surface areas were about 950 m^2/g.

A charcoal formulation called AZU suspension was developed by Rademaker et al. (1987, 1989) in The Netherlands (AZU stands for the hospital where this was developed, the Academisch Ziekenhuis Utrecht). It was prepared using 40 g charcoal added to 10 g of 85% glycerol solution and sufficient purified water to make the final volume 200 mL. The mixture was then autoclaved for 20–30 min to sterilize it. In the first study (1987), the formulation was made with Carbo adsorbens charcoal and was tested against a simple aqueous suspension of Carbo adsorbens charcoal in water. The second study (1989) was essentially the same except that Norit A Supra charcoal was used instead of Carbo adsorbens charcoal. In vitro adsorption tests were done with phenazone, pheno-barbital, and amitriptyline in a pH 6.5 medium, and in vivo tests were done to determine the efficacy of the AZU mixtures and the charcoal/water suspensions for treating acet-aminophen poisoning in dogs. Neither the in vitro nor the in vivo tests showed any significant difference between the AZU formulations and the glycerol-free charcoal/water suspensions. Since glycerol would not be expected to adsorb to charcoal to any significant extent (it is a simple, very hydrophilic molecule), then the presence of glycerol in the AZU mixtures would not be expected to cause any interference effects.

Eisen et al. (1991) evaluated the effectiveness of a milk chocolate/charcoal mixture in reducing aspirin absorption in six healthy subjects in a randomized cross-over study. They gave the subjects 975 mg of crushed aspirin followed by either water (control), 10 g of a milk chocolate/charcoal mixture, 10 g of a SuperChar charcoal suspension, or 10 g of Actidose Aqua suspension. The chocolate/charcoal mixture was made by melting a 1.65 oz milk chocolate bar in a microwave oven for 1 min and then mixing it manually with 10 g of a SuperChar charcoal suspension just before administration. Serum salicylate concentrations were determined over 0–24 hr. Total aspirin absorption was reduced as follows: SuperChar, 67%; milk chocolate/charcoal mixture, 50%; and Actidose Aqua, 2%. Although the addition of chocolate to the SuperChar charcoal suspension affected its ability to reduce aspirin absorption somewhat, it made the mixture much more palat-able. And the chocolate/charcoal mixture was still much more effective than the Actidose Aqua product, which contained Darco KB-B charcoal having a surface area of about 1,500 m^2/g (versus 3,150 m^2/g for SuperChar charcoal) at the time of the study. Still, the poor effectiveness of the Actidose Aqua product in this study is somewhat surprising.

Prior to this in vivo study, Eisen et al. (1988) evaluated the palatability of a their new milk chocolate-charcoal mixture (MCCM). Thirty-six children, ages 3–7, were asked to rank three mixtures on a scale of 1 (least palatable) to 5 (most palatable). The children used their most favorite, most disliked, and most tolerated foods as a reference for the scale. The charcoal slurry received a mean rating of 1.67, the MCCM received a 3.58

score, and chocolate alone was rated 4.77. Thus, it was concluded that the MCCM was quite palatable and certainly far better than charcoal/water.

III. OTHER APPROACHES TO PALATABILITY

One interesting approach to the charcoal palatability problem is a device marketed by the Starling Bay Products Group of San Jose, California. Their "First Responder Oral Tube" makes use of an ordinary white plastic straw having an accordion type bend joint about one-third from the upper end. Just below the bend is a disc, roughly 4 inches in diameter, which is made of a white opaque plastic material. Surrounding the upper part of the straw, above the bend, is a spongy material roughly 1 inch in diameter, which is to be dipped into a packet of peppermint-flavored syrup which comes with the tube. The lower end of the straw is placed in an aqueous activated charcoal suspension of one's choice and the flavored spongy end of the tube is inserted well into the patient's mouth. The charcoal suspension is then sipped through the straw, "bypassing the taste buds," while "the unappealing sight of charcoal is shielded from [the] patient's view by [the] opaque disc." N-acetylcysteine and barium solutions are other formulations recommended for Oral Tube use. This device appears to have been marketed to date mainly along the West Coast of the USA.

Still another approach to the palatability problem is a product being developed by Dr. Michael Stang of De Novo, Inc. (Pikesville, MD). The item is similar to an "Oreo" cookie and consists of two chocolate-flavored cookie type wafers, with a sugary filling in between. The wafers contain a substantial fraction by weight of powdered activated charcoal. Preliminary in vitro testing has shown that these charcoal cookies are able to adsorb a significant amount of a test drug. The drug uptake is reasonably rapid, if the cookies are first well-chewed (as they would be in actual application).

Another, similar attempt to provide charcoal in the form of a pleasant-tasting solid is one described to the writer by Mr. Paul Tower (1994), who was the developer. He created a "mint candy" using activated charcoal, starch, gelatin, and mint flavor. The mint contained 60 wt% charcoal. This formulation was never marketed, however.

As a final example of attempts to deal with the palatability problem, a Poison Control Center Director told this writer that, when his group is trying to get a child to consume charcoal, they run to a McDonald's which is adjacent to their hospital (Denver General Hospital), buy a milk shake, run back, mix in the charcoal, and then give the mixture to the child (Dart, 1993). It apparently works quite well.

IV. SUMMARY

In summary, it appears that by using finely powdered charcoal and bentonite or CMC in the proper amounts, an easily swallowed charcoal formulation can be prepared. As far as adding flavoring, many flavoring agents have been found to disappear when mixed well with charcoal (i.e., the agents, or at least the chemical compounds in them which provide the flavor taste, are adsorbed so well to the charcoal that not enough flavor remains in solution to provide any pleasant taste). However, studies have shown that saccharin, sorbitol, sucrose, chocolate syrup, cherry extract, powdered strawberry/acacia, and several other materials do not totally disappear. These flavoring agents, in the proper proportions, can provide a pleasant taste to charcoal/water mixtures without impairing the adsorptive action of the charcoal very seriously.

Studies with ice creams and sherbets have indicated that, while these agents cause a more serious interference with the adsorption capacity of charcoal, their use could, on balance, be positive if the extra charcoal accepted due to their presence is enough to counteract the loss of adsorption capacity per unit weight of charcoal (mixing charcoal into a milk shake would probably yield similar results). With respect to the unattractive appearance of charcoal mixtures, there seems to be no way in which to avoid a very black appearance.

Attempts to develop formulations that would be particularly attractive to children, such as "Oreo" charcoal cookies and charcoal-laden mint candies, appear to offer significant promise, and should be pursued further. The Oral Tube device also appears to be effective and it will be interesting to see if it becomes more widely adopted.

V. REFERENCES

Anonymous (1979). Medicoal (effervescent activated charcoal) in the treatment of acute poisoning, Drug Ther. Bull. *17*, January 19, pp. 7-8.

Arena, J. M. (1970). Aspirin poisoning - gastric lavage, ipecac, or activated charcoal?, JAMA *212*, 327.

Boehm, J. J., Brown, T. C. K., and Oppenheim, R. C. (1978). Flavoured activated charcoal as an antidote, Aust. J. Pharm. Sci. *7*, 119 .

Calvert, W. E., Corby, D. G., Herbertson, L. M., and Decker, W. J. (1971). Orally administered activated charcoal: Acceptance by children, JAMA *215*, 641.

Cheng, M. and Robertson, W. O. (1989). Charcoal "flavored" ice cream (abstract), Vet. Hum. Toxicol. *31*, 332.

Chung, D. C., Murphy, J. E., and Taylor, T. W. (1982). In vivo comparison of the adsorption capacity of "superactive charcoal" and fructose with activated charcoal and fructose, J. Toxicol.-Clin. Toxicol. *19*, 219.

Cooney, D. O. (1977). Saccharin sodium as a potential sweetener for antidotal charcoal, Am. J. Hosp. Pharm. *34*, 1342.

Cooney, D. O. and Roach, M. (1979). Sucrose as a sweetener for activated charcoal, Am. J. Hosp. Pharm. *36*, 797.

Cooney, D. O. (1980). Palatability of sucrose-, sorbitol-, and saccharin-sweetened activated charcoal formulations, Am. J. Hosp. Pharm. *37*, 237.

Cooney, D. O. (1982). Effect of type and amount of carboxymethylcellulose on in vitro salicylate adsorption by activated charcoal, J. Toxicol.-Clin. Toxicol. *19*, 367.

Cordonnier, J. A., Van den Heede, M. A., and Heyndrickx, A. M. (1986-87). In vitro adsorption of tilidine HCl by activated charcoal, J. Toxicol.-Clin. Toxicol. *24*, 503.

Crome, P., Dawling, S., Braithwaite, R. A., Masters, J., and Walkey, R. (1977). Effect of activated charcoal on absorption of nortriptyline, Lancet, December 10, p. 1203.

Dart, R. C., Director, Rocky Mountain Poison Control Center, Denver (1993). Personal communication.

Dawling, S., Crome, P., and Braithwaite, R. (1978). Effect of delayed administration of activated charcoal on nortriptyline absorption, Eur. J. Clin. Pharmacol. *14*, 445.

De Neve, R. (1976). Antidotal efficacy of activated charcoal in presence of jam, starch, and milk, Am. J. Hosp. Pharm. *33*, 965.

Eisen, T. F., Lacouture, P. G., and Woolf, A. (1988). The palatability of a new milk chocolate-charcoal mixture in children (abstract), Vet. Hum. Toxicol. *30*, 351.

Eisen, T. F., Grbcich, P. A., Lacouture, P. G., Shannon, M. W., and Woolf, A. (1991). The adsorption of salicylates by a milk chocolate-charcoal mixture, Ann. Emerg. Med. *20*, 143.

Grbcich, P. A., Lacouture, P. G., and Woolf, A. D. (1987). Administration of charcoal in the home (abstract), Vet. Hum. Toxicol. *29*, 458.

Greensher, J., Mofenson, H. C., and Picchioni, A. L. (1980). Activated charcoal (reply to letter), Ann. Emerg. Med. *9*, 111.

Gwilt, P. R. and Perrier, D. (1976). Influence of "thickening" agents on the antidotal efficacy of activated charcoal, Clin. Toxicol. *9*, 89.

Holt, L. E., Jr. and Holz, P. H. (1963). The black bottle - a consideration of the role of charcoal in the treatment of poisoning in children, J. Pediatrics *63*, 306.

Jaeger, R. W. and de Castro, F. J. (1977). Packaging method for charcoal gels (letter), Am. J. Hosp. Pharm. *34*, 562.

Krenzelok, E. P. (1980). Activated charcoal (letter), Ann. Emerg. Med. *9*, 111.

Krenzelok, E. P. and Heller, M. B. (1987). Effectiveness of commercially available aqueous activated charcoal products, Ann. Emerg. Med. *16*, 1340.

Levy, G. and Jusko, W. J. (1965). Effect of viscosity on drug absorption, J. Pharm. Sci. *54*, 219.

Levy, G. (1975). Activated charcoal and ice cream (reply to letter), Am. J. Hosp. Pharm. *32*, 562.

Levy, G., Soda, D. M., and Lampman, T. A. (1975). Inhibition by ice cream of the antidotal efficacy of activated charcoal, Am. J. Hosp. Pharm. *32*, 289.

Manes, M. and Mann, J. P., Jr. (1974). Easily swallowed formulations of antidote charcoals, Clin. Toxicol. *7*, 355.

Manes, M. (1975) Palatable activated carbon, U. S. Patent 3,917,821 issued November 4, 1975.

Manes, M. (1976). Effect of carboxymethylcellulose on the adsorptive capacity of charcoal (letter), Am. J. Hosp. Pharm. *33*, 1120, 1122, 1127.

Mathur, L. K., Jaffe, J. M., Colaizzi, J. L., and Moriarty, R. W. (1976a). Activated charcoal-carboxymethylcellulose gel formulation as an antidotal agent for orally ingested aspirin, Am. J. Hosp. Pharm. *33*, 717.

Mathur, L. K., Jaffe, J. M., and Colaizzi, J. L. (1976b). Effect of carboxymethylcellulose on the adsorptive capacity of charcoal. (reply to letter), Am. J. Hosp. Pharm. *33*, 1122.

Mayersohn, M., Perrier, D., and Picchioni, A. L. (1977). Evaluation of a charcoal-sorbitol mixture as an antidote for oral aspirin overdose, Clin. Toxicol. *11*, 561.

Nakano, N. I., Shimamori, Y., Umehashi, M., and Nakano, M. (1984). Preparation and drug adsorption characteristics of activated carbon beads suitable for oral administration, Chem. Pharm. Bull. *32*, 699.

Navarro, R. P., Navarro, K. R., and Krenzelok, E. P. (1980). Relative efficacy and palatability of three activated charcoal mixtures, Vet. Hum. Toxicol. *22*, 6.

Oderda, G. M. (1975). Activated charcoal and ice cream (letter), Am. J. Hosp. Pharm. *32*, 562.

Oppenheim, R. C. (1980). Strawberry flavoured activated charcoal (letter), Med. J. Aust. *1*, 39.

Oppenheim, R. C. and Miles, A. P. (1981). Adsorption of pheniramine and mianserin onto formulated activated charcoal, Aust. J. Pharm. Sci. *10*, 74.

Pilcher, C. A. (1980). Activated charcoal (letter), Ann. Emerg. Med. *9*, 111.

Rademaker, C. M. A., van Dijk, A., Glerum, J. H., and van Heijst, A. N. P. (1987). A practical and effective formulation of activated charcoal, Vet. Hum. Toxicol. *29* (Suppl. 2), 42.

Rademaker, C. M. A., van Dijk, A., de Vries, M. H., Kadir, F., and Glerum, J. H. (1989). A ready-to-use activated charcoal mixture: Adsorption studies in vitro and in dogs: Its influence on the intestinal secretion of theophylline in a rat model, Pharm. Weekbl. [Sci] *11*, 56.

Scholtz, E. C., Jaffe, J. M., and Colaizzi, J. L. (1978). Evaluation of five activated charcoal formulations for inhibition of aspirin absorption and palatability in man, Am. J. Hosp. Pharm. *35*, 1355.

Tower, P. M., Filtercorp, Woodinville, WA (1994). Personal communication.

Yancy, R. E., O'Barr, T. P., and Corby, D. G. (1977). In vitro and in vivo evaluation of the effect of cherry flavoring on the adsorptive capacity of activated charcoal for salicylic acid, Vet. Hum. Toxicol. *19*, 163.

17

Hazards Associated with Antidotal Charcoal Use

In this chapter we review studies dealing with the intrinsic toxicity of powdered activated charcoal itself, various problems which have occurred affecting the lungs when charcoal or charcoal/gastric contents have entered the lungs, problems with perforations associated with intubation, and various types of intestinal obstructions which have occurred when multiple doses containing rather large amounts of charcoal have been administered. Problems with hypernatremia, hypermagnesemia, and abdominal distention due to the concurrent administration of cathartics with charcoal are also discussed in some detail.

I. GENERAL TOXICITY OF CHARCOAL

Powdered carbon has been studied for toxicity due to ingestion, skin contact, and inhalation; all studies show it to be harmless when properly handled. Nau et al. (1958a) found that feeding carbon black (essentially identical to powdered activated charcoal, except for the absence of a fine internal pore structure) mixed with dog chow in the amount of 10 wt% to mice for extended periods of time (12–18 months) produced no adverse effects on any organs or tissues.

Nau et al. (1958b) also reported that applications of carbon black in water or oil suspensions to the skins of monkeys, mice, and rabbits for periods of up to 30 months produced no significant changes from normal. Indeed, the effect of activated charcoal on skin and, in particular, on wounds has been observed to be quite positive. As discussed in Chapter 23, the use of charcoal cloth dressings and the use of charcoal poultices for treating wounds and other skin disturbances has shown remarkable results. The reason for this is that activated charcoal powerfully adsorbs bacteria (as discussed in Chapter 24), thus rendering them inactive.

Nau et al. (1962) similarly found that mice and monkeys exposed by inhalation to carbon black for periods of up to 13,000 hr showed no significant changes in lung tissue; however, physical deposition of the carbon black in all parts of the lungs occurred but with minimal or no fibrous tissue proliferation (little of this carbon was "cleared" spontaneously after termination of the exposure).

In clinical experiments, Yatzidis and Oreopoulos (1976) relate that uremic patients were given 20–50 g activated charcoal per day for up to 4 months without any side effects. In fact, it was noted that "the patients had a marked subjective improvement of gastrointestinal symptoms and signs, such as anorexia, nausea, vomiting, and uremic odor. The constipating effect of charcoal was overcome easily by either sorbitol or paraffin oil." In a similar study, Friedman et al. (1978) fed 35 g charcoal per day to six adult patients for up to 2 months and found that "all patients accepted charcoal therapy without difficulty or adverse reaction. There was no apparent interference in appetite, sleep pattern, or general well-being that could be attributed to charcoal ingestion."

Wehr et al. (1975) carefully studied the lung conditions of workers in activated charcoal manufacturing plants and found definite radiographic evidence of pneumoconiosis (chronic reaction to dust collection) in 9.6% of the subjects. However, even with extensive bronchial and interstitial dust accumulation, minimal fibrosis was noted. Also, the incidence of respiratory symptoms in the population was remarkably low and dust accumulation was not a significant determinant of any pulmonary functional parameter, nor was it an important factor in the production of respiratory tract symptoms. Thus, the authors concluded that even chronic exposure to activated charcoal dust is relatively innocuous.

II. ASPIRATION

We first discuss the general topic of the aspiration of gastric contents and then consider reports of the aspiration of gastric contents containing activated charcoal. Bynum and Pierce (1976) provided an excellent review of the dangers of aspirating gastric contents. Their analysis of 50 cases showed that the patients invariably had impaired consciousness, most commonly due to sedative drug overdose. The onset of clinical signs was prompt and similar in all patients, regardless of the subsequent course and outcome. These signs usually included fever, tachypnea, diffuse rales, and serious hypoxemia. In about one-third of the cases, cough, cyanosis, wheezing, and apnea were seen. Initial roentgenograms showed diffuse or localized alveolar infiltrates which progressed during the following 24–36 hr. Treatment, even from the onset, with adrenocortical steroids or antimicrobial agents had no demonstrable effect on the outcome. The outcomes fell into three classes: 12% of the patients died relatively shortly, 62% showed early and progressive clearing with resolution of infiltrates over 2–16 days (mean time 4.5 days), and 26% showed early clearing but then developed new or extending infiltrates (8 of the 13 patients in this group ultimately died).

Aspiration of activated charcoal and gastric contents appears to have been reported first by Pollack et al. (1981). An eight-month-old girl was given 8 mL syrup of ipecac by mouth, producing three episodes of vomiting. Ninety minutes later, 9 g activated charcoal in 35 mL water was given through a nasogastric tube. The child vomited again, causing her trachea to plug up. An endotracheal tube was inserted, which twice occluded and was replaced each time. Prolonged respiratory insufficiency with severe bronchospasm then developed. Mechanical ventilation and tracheal suction were provided and, after about a week, the patient's condition returned more or less to normal.

The problems experienced here were due to several factors: (1) the end of the nasogastric tube may not have been in the stomach when the charcoal slurry was given,

(2) the charcoal slurry was thick and may have come out into the esophagus when the nasogastric tube was being removed, and (3) residual ipecac may have prompted the vomiting which occurred after the charcoal was given. Thus, when administering charcoal via a nasogastric tube, it is very important that the tube be firmly anchored to assure that the tube end is in the stomach, and that the charcoal slurry is thin enough to flow easily through the tube (and not stick somewhere in the tube). Many nasogastric tubes have a white strip of material along their length which will show up in an x-ray. Thus, the position of the tube can be confirmed radiographically. A less accurate but easier and far cheaper technique is to introduce air into the tube with a syringe. As the air bubbles out through the holes at the end of the tube, a stethoscope placed over the stomach will pick up the sound of the bubbles and confirm the general location of the tip. Aspiration of gastric contents up through the tube can also be used to confirm that the tip is in the proper location. Additionally, observation of the patient's respiration status and chest movements can indicate if inadvertent tracheal intubation has occurred. To make the charcoal slurry thinner and more comfortable for the patient, it can be warmed by placing the slurry bottle in a tub of warm water while the patient is being intubated. Then, by the time the tube has been placed and checked for position, the charcoal slurry will be warm and will flow more easily through the tube.

The severe bronchospasm which is observed after aspiration could possibly be caused just as much by the effects of gastric acid as by charcoal itself, since wheezing is common after acid aspiration. However, Dunbar et al. (1981) showed that charcoal aspiration can cause severe bronchospasm even without acid aspiration. They report a case of severe bronchospasm in a child due to aspiration of charcoal during treatment for poisoning, and describe a study they did with dogs as a result of this observation. In their study, 13 anesthetized, intubated, paralyzed, and ventilated dogs were given (via the tracheal tube) one of four aspirates: (1) 50 mL of pH 6 water, (2) 50 mL of pH 3 water, (3) 50 mL of a pH 6 charcoal slurry containing 8.7 g charcoal, or (4) 50 mL of a pH 3 charcoal slurry containing the same amount of charcoal. Various cardiorespiratory parameters were measured over 1–30 min. Charcoal aspiration caused a rapid and sustained 15–20% increase in cardiac output, wheras charcoal-free aspirates decreased cardiac output 10–30%. More significantly, airway resistance nearly doubled due to the pH 6 charcoal slurry and more than tripled after the pH 3 charcoal slurry. Shunt fraction increased rapidly with all aspirates, in the order: pH 3 charcoal slurry > pH 6 charcoal slurry > pH 3 water > pH 6 water. Their results show that charcoal aspiration, with or without acid aspiration, can contribute significantly to bronchospasm.

Harsch (1986) described the death of a drug overdose victim due to charcoal aspiration. The victim, a 29-year-old male, was treated by gastric lavage followed by nasogastric administration of 25 g activated charcoal every two to four hours (the total number of times this was done was not stated). After 24 hr, the patient seemed stable and was removed from ventilator assistance. He then vomited a black mixture and soon became febrile and tachypneic. He vomited again, went into asystole, and could not be resuscitated. Large quantities of charcoal were aspirated from his lungs during the resuscitation effort. This report suggests that the use of multiple-dose charcoal therapy should be done carefully. The total accumulation of charcoal in the patient after many large doses of charcoal may be so great as to present serious problems if regurgitation occurs.

Also, in connection with multiple-dose charcoal therapy, after the last dose has been given and time is allowed for it to empty from the stomach, the nasogastric or orogastric tube should be suctioned to remove residual charcoal from the stomach prior to removal of the tube. Then, when the tube is removed, the chances of charcoal being pulled up from the stomach with the tube, or charcoal hung up in the tube falling out as the tube is drawn upwards, will be minimized, and potential aspiration should be avoided.

Menzies et al. (1988) described a fatality due to aspiration. A 58-year-old male overdose victim was treated with gastric lavage and then was given 50 g charcoal (Medicoal) in 200 mL water. Further doses of 12.5 g charcoal were given hourly after lavage of the stomach contents. After 12 hr, the patient vomited, with pulmonary aspiration of gastric contents. Despite ventilation, suctioning, and IV antibiotics, fever and tachycardia developed, and lung function deteriorated progressively. The patient died 15 days after admission. One contributing factor was that the Medicoal formulation (no longer available) contained povidone, a suspending agent which can cause chronic inflammatory reactions and which has been implicated in pneumonitis after inhalation of hairsprays. Obviously, this agent should not be used in charcoal formulations.

Rau et al. (1988) commented on Menzies' report, stating that three patients of theirs died from pulmonary aspiration after charcoal administration. They subsequently started using a cuffed endotracheal tube prior to giving the charcoal, keeping the tube in for 48 hr, and have had no problems with the six patients they have treated in this way since.

Power (1988) also remarked, relative to pulmonary aspiration, that the airway should be protected by a cuffed endotracheal tube prior to gastric lavage "when the protective laryngeal reflexes are likely to be impaired." He stated that many cases of postoperative chest infections are now thought to be due to low grade aspiration. However, it should be repeated that endotracheal intubation is not without risks, and that these risks must be weighed against any potential benefits.

Danel (1988), in commenting on the Menzies' case report, remarked that in his experience nausea and vomiting after charcoal administration are not rare. He stated that intubation for ventilatory support is carried out in his unit prior to lavage or charcoal delivery by orogastric tube and that this has prevented aspiration. Despite this letter, it should be emphasized that tracheal intubation can involve significant complications— especially in alert, struggling patients—and may require sedation to accomplish (which is undesirable in poisoned patients). Thus, it is not current clinical practice to endotracheally intubate patients prior to gastric lavage unless such is needed for airway control or assisted ventilation due to CNS or respiratory depression.

Watson et al. (1986a) reported a study in EDs that showed that 13.3% of patients given charcoal after gastric lavage vomited, 43.3% given charcoal after syrup of ipecac vomited, and one patient who had been treated only with charcoal vomited. This suggests that the residual emetic action of ipecac was a factor in the greater incidence of charcoal vomiting after syrup of ipecac use. Indeed, several literature articles have recommended that an antiemetic such as ranitidine or metoclopramide be given after the effects of syrup of ipecac have waned, if charcoal is to be administered subsequent to ipecac use. Based on the results of Watson's group, such a recommendation seems reasonable but further studies are needed on this point. Present clinical practice is generally to not give an antiemetic between syrup of ipecac and charcoal. However, the most enlightened current approach is to not bother with syrup of ipecac in the first place, since it is of doubtful benefit, and proceed directly to using charcoal, which is of proven benefit.

In a study similar to that of Watson et al., Minocha et al. (1986) reported that 12.5% of ED patients vomited when given charcoal/water and 15.5% vomited when given charcoal/sorbitol formulations.

Hoffman (1983) also mentioned that, in his experience, vomiting frequently occurs after oral charcoal administration. However, in a reply, Levy (1983) indicated that vomiting in normal adult volunteers (not patients) did not occur in any of several studies he carried out. He suggested that vomiting by actual patients may be due to stress or other influences of the clinical setting.

Donovan (1987) mentioned problems with vomiting due to prior ipecac use. He stated that persistent vomiting has been controlled using the antiemetic ranitidine or by use of slow nasogastric administration as opposed to bolus therapy. Nelli and Rau (1988), commenting on Donovan's paper, stated that they tried to treat the vomiting problem by injecting metoclopramide each time charcoal is given or by emulsifying the charcoal with sorbitol and instilling it through an orogastric Ryles tube (a thin rubber tube with an olive-shaped end, often used to give test meals). They use a loading dose of 50 g charcoal, followed by 25 g charcoal every 4 hr to a maximum of 8 doses. They found that the emulsified charcoal blocked the tube. Donovan's (1988) response to these comments was that he found that repeated administration of 50 g or greater boluses of charcoal are poorly tolerated and that slow nasogastric administration of smaller amounts (0.25–0.50 g/kg) of charcoal over a one-hour period every three hours is preferable, with antiemetics used as necessary. Brief nasogastric suction just prior to charcoal administration is also very helpful. Reed (1988) indicated that his group gives charcoal nasogastrically continuously at a rate of 0.25–0.50 g/kg-hr successfully and that this has avoided nausea/vomiting problems without the need for antiemetics.

Another case report of charcoal aspiration was given by Dammann et al. (1988). A 2-year-old girl was given charcoal via nasogastric tube for an ibuprofen overdose. When the tube was removed, she became apneic and cyanotic, and required mouth-to-mouth resuscitation and endotracheal intubation. Charcoal was suctioned from the endotracheal tube during the entire period of intubation. That night, she developed pneumomediastinum and subcutaneous emphysema followed by bilateral pneumothoraces. After lengthy and complicated treatments of various kinds, the child finally recovered. The exact time and cause of the charcoal aspiration is not clear from this report.

Silberman et al. (1990) described the case of a 20-year-old male who had overdosed on tetracycline. A charcoal suspension was given via nasogastric tube during some pre-hospital care, following which some gagging without any observed emesis occurred. When he was in the ED, he was given some sorbitol orally but no more charcoal, and was discharged a while later. He developed abdominal cramps, headache, shortness of breath, and had multiple bowel movements during the night. The patient returned to the ED the next morning, with acute respiratory distress. Lung auscultation revealed little air movement in any lobe and he was given 100% O_2 by breathing mask with little improvement. Endotracheal aspiration was performed and charcoal-containing mucus was obtained repeatedly. The patient was discharged on the 5th day. This report indicates that care must be taken when EMT's give charcoal in the field, where conditions are much less orderly than in an ED.

Aspiration of charcoal by a 30-year-old male intoxicated with amitriptyline was described by Harris and Filandrinos (1993). The patient was treated with gastric lavage using a 16 French nasogastric tube (clearly, a tube this small is a poor choice for lavage). Charcoal (50 g) was then administered through the nasogastric tube. Then, because of persistent hypercarbia, the patient was tracheally intubated. The patient pulled out the nasogastric tube and subsequently it was reinserted. Following this, more charcoal/sorbitol suspension (15 mL) was given through the nasogastric tube. A sudden fall in oxygen saturation was noted and a radiograph of the chest showed that the nasogastric tube was in the right mainstem bronchus. The nasogastric tube was repositioned into the stomach and more charcoal was administered. The patient subsequently developed respiratory distress syndrome and became febrile. Repeated suctioning of the bronchus was performed (even on the third day, "moderate" to "copious" amounts of charcoal-laden sputum were still being recovered), accompanied by administration of antibiotics and nebulized beta antagonist therapy. The patient recovered and was discharged 14 days later. The authors emphasize the need to confirm the position of the nasogastric tube by auscultation or radiography prior to charcoal administration. They also indicate that, in this case, it is likely that the endotracheal tube cuff was not properly inflated, as it is clear that the nasogastric tube was able to pass around the cuffed area of the endotracheal tube.

Still another case of charcoal aspiration was reported by Givens et al. (1992). A 14-year-old girl who attempted suicide with thioridazine and imipramine was given gastric lavage and then 50 g charcoal via a nasogastric tube. She later received another 25-g dose of charcoal through the nasogastric tube. Through an error, the patient was given 50 g charcoal with sorbitol every hour for the next 6 hr (thus, she received a total of 375 g of charcoal). The patient improved steadily and the tube was removed. However, she then vomited substantial amounts of black material "several times in rapid succession and displayed clinical evidence of aspiration with decreasing oxygen saturations, wheezing, cyanosis, and increased agitation." Over the next several days, she showed evidence of pneumonitis. Large amounts of charcoal were suctioned from an emplaced endotracheal tube. The patient was extubated on the tenth day and tests indicated that the aspiration episode had caused some hypoxic or ischemic injury. The authors indicate, among other things, that adequate airway protection be insured and that repeated doses of charcoal be given only if bowel sounds indicate that it is reasonable to do so.

A death apparently caused by charcoal aspiration was reported by Benson et al. (1989). A 66-year-old male with a Theo-Dur overdose was treated with ipecac and then with multiple-dose charcoal therapy (50 g each 2 hr, given orally). The patient had an abrupt onset of seizures and aspirated charcoal-laden vomitus 9 hr post-admission. Attempts to resuscitate the patient were hampered by charcoal blocking the trachea and the patient died of cardiorespiratory arrest. Examination showed charcoal in the trachea and right main bronchus, extending deeply into the lung.

A case involving a 60-year-old man with chronic obstructive pulmonary disease who had an elevated theophylline level and who was treated with 75 g charcoal/sorbitol plus 30 g charcoal/water 4 hr later was described by Geller and Ekins (1993). About 14 hr after admission, a sudden junctional rhythm was seen on telemetry. A nurse rushed to his room and found him with charcoal covering his face and airway. The patient was in respiratory arrest and died. Autopsy revealed the cause of death as "asphyxiation secondary to aspiration of activated charcoal."

III. PERFORATIONS

Several reports in the literature have described perforations of the esophagus or stomach by lavage tubes, with subsequent introduction of charcoal into the abdominal space, when charcoal therapy was performed. Such perforations have sometimes led to death.

Geller and Ekins (1993), in addition to the case of the asphyxiation/aspiration death of a patient just discussed, described the case of an 86-year-old woman who was treated for camphor intoxication by lavage, followed by charcoal. Within 2 hr she developed subcutaneous emphysema of the neck, dyspnea, pneumomediastinum, and a pleural effusion. She died after 24 days. Autopsy revealed "mediastinitis, pleuritis, and pneumonitis causing death secondary to perforation of esophageal diverticulum, with charcoal hemothorax."

Mariani and Pook (1993) also described a case of perforation. A 35-year-old female was being treated for tricyclic intoxication with lavage and charcoal, and then developed an acute abdomen. She underwent laparotomy and charcoal was found throughout the peritoneum. Her hospital course was long and characterized by "tenacious peritoneal charcoal deposition, persistent peritonitis, and adhesion and abscess formation." Various procedures (abscess drainage, small-bowel resection, parenteral nutrition, etc.) resulted in the ultimate recovery of the patient. This paper reviews a few reports of perforations resulting from gastric lavage procedures and offers several recommendations for minimizing the risk of such.

IV. BRONCHIOLITUS OBLITERANS

Elliott et al. (1989) described the case of a 16-year-old female who had taken a nortriptyline overdose. A nasogastric tube was put in place, gastric lavage was performed, and then 75 g activated charcoal with sorbitol (Actidose with Sorbitol) was given through the tube. Ten minutes later, the patient had a grand mal seizure and cardiac arrest. She was revived, survived, and was finally discharged 19 days after admission. During her stay, extensive examinations revealed that considerable aspiration of charcoal and gastric contents had occurred at some point, probably during the seizure. Six days after discharge, she returned with complaints of fever, cough, headache, and dyspnea. She was readmitted and after a long bout of progressive respiratory failure, she died ten weeks after readmission. An autopsy disclosed bronchiolitis obliterans (cirrhosis of the bronchiole walls due to hardening caused by contact with the charcoal) with fibrous obliteration and stenosis of most of the small airways. Embedded in the bronchiolar scar tissue were massive amounts of black material. The exact pathogenic role of the charcoal remains uncertain.

V. EMPYEMA

Justiniani et al. (1985) reported one case of a man who was treated with gastric lavage and activated charcoal via an Ewald tube for alcohol and methaqualone overdose. In the next few hours he developed pneumothorax, which was successfully treated using a chest tube. He was discharged four days later but soon experienced pleuritic pain, chills, fever, sweating, and a cough. He was found to have a charcoal-containing empyema (accumulation of pus) in the chest cavity. This was successfully treated with antibiotics. The authors admit that the original pneumothorax was likely due to perforation of the esophagus when the charcoal slurry was administered via the Ewald tube and that this allowed gastric contents to leak into the chest cavity, causing the empyema. Thus, activated charcoal was not the cause of these problems. This report does, however, underline the need for care in using tubes to administer charcoal slurries.

VI. INFECTIONS IN THE RESPIRATORY TRACT FROM NONSTERILE CHARCOAL

Contaminated commercial charcoal preparations have been found by George et al. (1991) to be a source of fungi leading to respiratory tract infection. A single patient (a 71-year-old man) with chronic obstructive lung disease and an elevated theophylline blood level was given activated charcoal (50 g) in 40% sorbitol (Actidose with Sorbitol) at 4 hr intervals for five doses. The patient aspirated the charcoal suspension, and thereafter developed respiratory tract infection with Aspergillus niger, Paecilomyces variotti, and Penicillium spp. The patient died on the 11th day (it should be noted that the patient was severely immunocompromised from corticosteroids used to treat his lung disease).

Cultures of the charcoal product obtained from the hospital pharmacy grew moderate to heavy amounts of *Paecilomyces* spp. and *Penicillium* spp. Cultures of the same product from a separate lot obtained from a different hospital grew moderate to heavy amounts of *Paecilomyces* and *Aspergillus* spp. According to the product marketer (Paddock Laboratories, Minneapolis), the charcoal is shipped from the original maker (undoubtedly the American Norit Company) to a second company, which then prepares the single-dose containers. At this second company, the conditions are not sterile; however, a mixture of antimicrobial preservatives (propyl paraben and butyl paraben) is added. Samples of the product before and after preparation are cultured for *Escherichia coli* and *Salmonella*, and lots which fail these tests are not sold.

This report indicates that charcoal products are not totally sterile and that sorbitol can be used as a nutrient by *Aspergillus niger*. Thus, the added sorbitol itself could be a source of contamination. It would be important to know when the paraben compounds are added during the preparation of the suspensions, and how. Paraben compounds (p-hydroxybenzoates), possessing aromatic rings, would be expected to be adsorbed strongly by activated charcoal and thus very little residual of the parabens would remain in the free solution of the charcoal suspensions.

What we have described here is not inconsistent with what was stated at the outset of this chapter regarding the innocuousness of inhaled carbon black (Nau et al., 1962) and inhaled activated charcoal (Wehr et al., 1975), for in those cases dry particles were involved. In the present section, we have been dealing with the aspiration of liquid suspensions of charcoal with added sorbitol, which is clearly quite a different situation.

VII. ABDOMINAL DISTENSION FROM SORBITOL

Anker and Smilkstein (1993) described a death due to sorbitol-induced massive abdominal distension in a 49-year-old female who was treated for a triple-drug overdose. She was treated with lavage, charcoal with sorbitol, tracheal intubation, and IV sodium bicarbonate. In all, five charcoal/sorbitol doses were given (total charcoal 250 g, total sorbitol 270 g). After extubation, belching, vomiting, abdominal pain, and abdominal distension occurred. Two hours later, bradycardia and apnea developed rapidly. The abdominal distention became progressively worse and at 20 hr post-admission fatal ventricular fibrillation occurred. Autopsy revealed compressed lungs (no evidence of aspiration), massive small bowel and colon gas, intramucosal gas blebs, impacted stool in the rectum, but no other obstructions and no perforations. It was concluded that "death resulted from respiratory failure due to massive gas accumulation consistent with fermentation of sorbitol by gut flora."

Hyams (1982) described a case of chronic abdominal pain caused by sorbitol mal-absorption in a 15-year-old girl who had been chewing sugar-free sorbitol-containing gum several times per week. Hyams states that: "Malabsorbed sorbitol is fermented by colonic bacteria with the subsequent generation of hydrogen gas. As little as 5 g of ingested sorbitol has been shown to generate a significant rise in the hydrogen excreted in the breath ... and to cause abdominal discomfort."

In two cases described by Longdon and Henderson (1992), which will be discussed in some detail under "Obstructions Caused by Charcoal," it was reported that due to the use of sorbitol in combination with charcoal, the "patients retained 600 mL of 70% sorbitol which was then a substrate for intestinal bacteria, leading to the production of copious volumes of gas and gut distention."

Eyer and Sprenger (1991), describing in vivo studies of reductions in drug absorption (diphenhydramine, codeine, acetaminophen) in volunteers given single doses of 30 g charcoal in 150 mL 70% sorbitol, state that the charcoal/sorbitol suspension caused "marked flatulence."

VIII. HYPERNATREMIA AND HYPERMAGNESEMIA

Many reports in the literature indicate that sorbitol can cause vomiting, as well as diarrhea. Recognizing the incidence of adverse effects due to sorbitol use in multiple-dose charcoal regimens, Wax et al. (1993) carried out a survey of all of the EDs in the area served by their New York City Poison Control Center. Sixty-seven EDs responded. They found that administration of charcoal/sorbitol formulations during multiple-dose charcoal therapy was always done in 30% of the EDs, was sometimes done in 19% of the EDs, and was never done in 51% of the EDs. The inconsistency of charcoal formulations stocked by the hospital pharmacies suggests the lack of any clear consensus regarding sorbitol use: 16% stocked only formulations containing sorbitol, 52% stocked only sorbitol-free formulations, and 31% stocked both kinds of formulations (these figures add only to 99% due to rounding-off). In the 21 hospitals stocking both types, 38% of those used the formulations containing sorbitol during all doses of multiple-dose therapy.

Mann (1988) also conducted a survey of charcoal products stocked and used in 54 hospitals served by the Delaware Valley Regional PCC. Two hospitals did not stock any charcoal; 34 stocked only one formulation (22 stocked sorbitol-containing products and 12 stocked sorbitol-free products). Eighteen hospitals stocked products both with and without sorbitol. Of 29 hospitals using multiple-dose charcoal therapy, 22 used sorbitol in all doses and 6 employed another cathartic in addition to the sorbitol. Of the 18 hospitals stocking products both with and without sorbitol, none alternated these during multiple-dose charcoal therapy. Mann concluded that the potential for overuse of sorbitol during multiple-dose charcoal therapy existed.

Fish et al. (1989) did a randomized double-blind study in which the incidence of vomiting and/or diarrhea associated with sorbitol-containing charcoal formulations was evaluated. They gave adult patients 50 g charcoal and children 1 g/kg charcoal, along

Table 17.1 Vomiting and Diarrhea Due to Sorbitol/Charcoal

Formulation	62 g sorbitol/ 50 g charcoal	96 g sorbitol/ 50 g charcoal	196 g sorbitol/ 50 g charcoal
Vomiting	10/32 (31%)	11/25 (44%)	20/35 (57%)
Diarrhea	7/29 (24%)	6/23 (26%)	16/37 (43%)

From Fish et al. (1989). Reprinted with permission from *Veterinary and Human Toxicology*.

with different amounts of sorbitol: (1) 62 g sorbitol per 50 g charcoal, (2) 96 g sorbitol per 50 g charcoal, or (3) 192 g sorbitol per 50 g charcoal. Table 17.1 shows the results. The numbers of patients experiencing vomiting or diarrhea and the total numbers of patients in each group are shown. It is clear that the incidence of both vomiting and diarrhea increased with the amount of sorbitol given and that the incidence rates were significant. Thus, the use of sorbitol is not without adverse effects.

Severe dehydration and hypernatremia after use of an activated charcoal/sorbitol suspension was first reported in depth by Farley (1986). A 3-month-old boy experiencing theophylline toxicity was given 220 mL of a suspension of activated charcoal in 70% sorbitol over 3–4 hr, of which he retained 150 mL, containing 30 g charcoal and 110 g sorbitol. After about 10 days, a black stool was passed. The infant was ashen and a blood pressure was unobtainable. Fluid support was initiated and finally the child stabilized after several days. After the period of hypotension, the child developed an obstruction of the intestines that lasted 24 hr. The sorbitol in the charcoal mixture is known to act osmotically to draw water into the gut lumen, usually causing diarrhea. In this case, with a small infant, the effect was powerful enough to cause sudden hypotension, dehydration, and a rise in serum osmolarity (causing hypernatremia). However, some sodium bicarbonate was used during treatment, which may explain part of the hypernatremia.

McCord and Okun (1987) responded to this case report by Farley with a letter reporting a similar incident with a 2-year-old child who was treated for imipramine overdose. After gastric lavage, a charcoal/sorbitol formulation (Actidose with Sorbitol) containing 15 g charcoal and 25 g sorbitol in 70 mL total volume was administered (this size of Actidose with Sorbitol is no longer marketed but the 120 mL and 240 mL sizes presently available have sorbitol:charcoal ratios of 48 g:25 g and 96 g:50 g, or 1.92, versus the 25 g:15 g or 1.67 ratio in the 70 mL product used by McCord and Okun). Three more doses were given over the next 14 hr and sodium bicarbonate was given IV to maximize serum protein binding of the drug. After 14 hr, moderate dehydration had developed. The charcoal/sorbitol was stopped, oral rehydration was begun, and 24 hr later all signs had normalized. It was stated that the large volume of stool fluid loss produced by the sorbitol was believed to be the reason for the dehydration. The amount of sodium bicarbonate given was not enough to explain the profound hypernatremia noted.

In response to McCord and Okun, Farley (1987) remarked: "No compelling data exist showing that cathartics augment the effect of charcoal on toxin elimination, and constipation does not seem to be a problem in children. Therefore, I believe that ... cathartics should either be dispensed with entirely or limited to the first charcoal dose."

Another follow-up letter on Farley's report by Massanari (1987) describes the treatment of human volunteers for nontoxic levels of theophylline ingestion using water alone,

70% sorbitol alone, charcoal plus water, and charcoal plus 70% sorbitol. Severe vomiting occurred with the sorbitol alone in five of the eight subjects. Although sorbitol, alone or with charcoal, decreased GI transit times, no effect on the absorption of theophylline was found. Thus, at least for theophylline poisoning, Massanari does not recommend sorbitol or sorbitol/charcoal mixtures.

Moore (1988), also in response to Farley, mentioned a case brought up by Klein et al. (1985) in which the use of a charcoal/70% sorbitol mixture given nasogastrically at the rate of 0.5 g charcoal/kg-hr for the treatment of chlorpromazine overdose in two children was described. That report stated that dehydration, hypernatremia, and neurologic deterioration developed after 8 hr and 20 hr, respectively. Although Klein et al. believed that the chlorpromazine caused these, Moore suggested that sorbitol was the more likely cause.

Caldwell et al. (1987) reported three cases of severe hypernatremia associated with three different multidose charcoal/cathartic regimens. In the first, a 22-year-old woman with an amitriptyline/naproxen overdose was given 50 g charcoal and 300 mL magnesium citrate solution every 4 hr for six doses. The 50 g charcoal was in the form of a preparation containing 70% sorbitol (apparently called Actidose, rather than Actidose with Sorbitol, at that time, since the authors state that "the sorbitol vehicle was not specified on the front label of the Actidose preparation"). A solution of 5% dextrose and 0.45% saline was administered at the rate of 150 mL/hr. The patient's serum sodium level rose from 138 mmol/L initially to as high as 170 mmol/L. The IV solution was changed to one without saline at 26 hr postadmission and, by 50 hr, the patient's serum sodium had returned to normal.

In the second case described by Caldwell et al., a 34-year-old woman with an amitriptyline/barbiturate overdose was given 50 g charcoal plus 150 mL magnesium citrate every 4 hr for a total of 4 doses. Dextrose/saline solutions were given IV. Her serum sodium rose from an initial 138 mmol/L to as high as 175 mmol/L. By 48 hr, however, her serum sodium had returned to normal. In the third case, a 37-year-old man with alcohol/PCP toxicity was given 50 g charcoal and 300 mL magnesium citrate solution initially and then 50 g charcoal in 70% sorbitol every 4 hr for three doses. His serum sodium rose to as high as 165 mmol/L. Continued water replacement efforts returned his sodium levels to normal by 30 hr postadmission. Caldwell et al. concluded that "acute colonic water loss due to the osmotic or saline cathartics administered [was] the most likely cause of acute hypernatremia. In all three cases, nursing notes specifically document excessively loose and watery stools." The use of multiple doses of saline or osmotic cathartics was concluded to be dangerous.

Sullivan and Krenzelok (1988), commenting on a study of repetitive-dose charcoal by Jones et al. (1987), advised of caution in using charcoal/sorbitol mixtures for multiple-dose treatment regimens. They mentioned the case of an adult who experienced hypernatremia and hypokalemia secondary to multiple doses of a standard charcoal/sorbitol mixture. They stated that more emphatic warnings need to be given on labels and in package inserts. Jones et al. (1988) replied by endorsing the call for caution in sorbitol use but further supporting its use.

Gorchein et al. (1988) mentioned that the highly effective formulation available in the UK named Medicoal contained undisclosed excipients and had 18 millimole sodium per 5 g packet. They warned that it be used in infants with caution because of the possible danger of hypernatremia. The manufacturers subsequently clarified their data sheets to disclose the sodium content of Medicoal (which is no longer available).

Brent et al. (1989) described an iatrogenic death due to sorbitol and magnesium sulfate during treatment of a patient for salicylism. A 55-year-old female was admitted with a 60 mg/100 mL acetylsalicylic acid (ASA) level and was treated by hemodialysis, after which her ASA level was 17 mg/100 mL. She was then given multiple-dose charcoal and cathartic therapy over 24 hr. The $MgSO_4$ given was four 30-g doses and the sorbitol given was two 120 mL doses of 70% sorbitol plus four doses (volume unstated) of a charcoal preparation containing 70% sorbitol over the 24 hr period. She became hypotensive and bradycardic and was found to have a serum Mg^{++} of 17.8 mEq/L. She was again dialyzed and stabilized, but the next day she developed an acute abdomen and died. Postmortem examination revealed a profoundly dilated bowel containing fluid and charcoal, and a perforation at the hepatic flexure. The cause of death was judged to be peritonitis secondary to cathartic-induced bowel perforation.

Another case report on hypernatremia in a 55-year-old man following treatment for theophylline toxicity with a charcoal/70% sorbitol mixture was given by Gazda-Smith and Synhavsky (1990). Actidose with Sorbitol was given by nasogastric tube in an amount equivalent to 10 g charcoal, each hour for 12 hr. Overnight, profuse diarrhea developed (2700 mL total stool output) and hypernatremia occurred. Serum theophylline was effectively lowered in this patient but the profuse diarrhea was an indication that the sorbitol dose was excessive.

Allerton and Strom (1991) mentioned hypernatremic dehydration due to the use of the osmotic cathartic lactulose in treating hepatic encephalopathy. The article then described severe hypernatremic dehydration due to administration of charcoal/sorbitol and saline/dextrose solutions for phenobarbital intoxication in a 23-year-old woman. Charcoal/sorbitol, 30 g in 150 mL 70% sorbitol (Charcoaid), was given by nasogastric tube along with additional 30 mL doses of 70% sorbitol and 200 mL doses of magnesium citrate. Dosing was repeated approximately every 3 hr for 27 hr, for a total of nine doses of charcoal/sorbitol, seven 30 mL doses of 70% sorbitol, and three doses of 200 mL magnesium citrate. Saline in dextrose was infused to counteract fluid loss. At 18 hr after initiation of therapy, the patient's serum sodium was 183 mmol/L. Infusion of saline-free fluids ultimately corrected the hypernatremia. Allerton and Strom offered recommendations concerning the dose of charcoal/sorbitol, the monitoring of fluid/electrolyte balance, and the choice of replacement fluids.

Related to these cases of hypernatremia just discussed are several cases of hypermagnesemia which have occurred when magnesium-containing cathartics have been used in conjunction with charcoal. Hypermagnesemia can cause refractory hypotension, bradyardia, CNS depression, muscle weakness and paralysis with secondary respiratory failure, bowel hypomotility, and hypocalcemia. We first describe a study by Smilkstein et al. (1987), who studied the elevation of serum magnesium due to single and repeated doses of magnesium sulfate in the absence of any charcoal. When a single 30-g dose of magnesium sulfate was given to ten patients, the mean serum magnesium rose only to 1.76 mEq/L at 1 hr and was 1.73 mEq/L at 4 hr, compared to a baseline value of 1.68 mEq/L. However, when 30-g doses of magnesium sulfate were given to 14 patients at 0, 4, and 8 hr, serum magnesium was 2.55 mEq/L 1 hr after the final dose, and 2.51 mEq/L 4 hr after the final dose (the baseline value was 1.69 mEq/L). Thus, the authors concluded that "significant hypermagnesemia can occur rapidly after the use of multiple doses of Mg cathartics even at 'standard' doses in patients with normal renal function."

Woodard et al. (1990) performed a similar prospective study in which 102 patients were given an average of 960 mL magnesium citrate (9.22 g Mg) as part of normal treatment for overdoses. The final mean serum magnesium level in the group was 2.5 mEq/L, with 12 patients having levels of greater than 3.0 mEq/L. No correlation was found between the final serum magnesium levels and the amounts of magnesium citrate given. The authors concluded that "with close monitoring, repetitive magnesium citrate can be administered without inducing severe hypermagnesemia (serum magnesium concentration > 5.0 mEq/L)." While Smilkstein et al. were concerned with *significant* hypermagnesemia, this group was oriented towards *severe* hypermagnesemia.

Three additional cases of hypermagnesemia have occurred when magnesium citrate cathartics were used in conjunction with oral charcoal to treat theophylline toxicity in a 61-year-old woman, in a 77-year-old woman, and in an 87-year-old man. Weber and Santiago (1989) reported on the first two patients. One received 55 g magnesium over an 8 hr period and had a serum magnesium level of 6.9 mg/100 mL (the normal range is 1.8–2.4 mg/100 mL). This patient suffered permanent morbidity. The other patient received enough magnesium to produce a serum magnesium level as high as 10.3 mg/100 mL. During the course of treatment, charcoal aspiration occurred. The patient died 48 hr after theophylline toxicity. The authors state that the magnesium citrate caused, or contributed to, refractory hypotension and contributed significantly to the morbidity and mortality of the two patients. Weber and Santiago concluded that sorbitol should be used instead of magnesium citrate if a cathartic is to be used, since sorbitol appears to increase GI transit times better. The report on the third patient (Garrelts et al., 1989) involved an 87-year-old man having theophylline toxicity who was given 40 g charcoal (Liqui-Char) along with 4 oz of magnesium citrate. More charcoal (25 g) was given every 2 hr, plus 4 oz of magnesium citrate with every other charcoal dose. A total of nine charcoal doses and five magnesium citrate doses were given over a 16 hr period. The patient's serum magnesium rose to as high as 5.3 mg/100 mL; he became disoriented, lethargic, confused, and had diminished deep tendon reflexes. His serum magnesium gradually fell after cessation of the charcoal plus magnesium citrate therapy and was 1.1 mg/100 mL on day 7. While the charcoal/citrate therapy did reduce the patient's serum theophylline rapidly (from an initial 75 mg/100 mL to 10 mg/100 mL after only 24 hr), the use of too much magnesium citrate caused marked magnesium toxicity.

Two other reports of hypermagnesemia as a result of treatment with oral charcoal and magnesium-containing cathartics are those of Jones et al. (1986) and Fassler et al. (1985). Both cases resulted in either respiratory failure or prolongation of ventilatory support. In one of the cases, diminished bowel function due to theophylline overdose may have contributed to magnesium toxicity. In the case described by Jones et al., a 39-year-old woman was treated for a tricyclic antidepressant overdose using repeated doses of 50 g charcoal and 300 mL magnesium citrate solution for 72 hr. Her serum magnesium rose from 1.9 mEq/L upon admission to a peak of 11.4 mEq/L at 72 hr. She exhibited acute neuromuscular deterioration and respiratory depression. Hemodialysis was started at 76 hr to reduce serum magnesium and by 80 hr (at which time dialysis was stopped) it had fallen to 5.1 mEq/L. It fell steadily to 3.0 mEq/L over the next 24 hr. In the case described by Fassler et al., a 50-year-old man with an overdose of an unknown toxin was given gastric lavage, then 50 g charcoal and one bottle of magnesium citrate (containing 18 g magnesium). He was then admitted to the ICU and given 30 g charcoal and 100 g magnesium sulfate every 4 hr. Profound catharsis and neuromuscular rigidity occurred, and the charcoal/magnesium therapy was stopped after 18 hr. He had a respiratory arrest at 24 hr post-admission and developed bradycardia. His serum magnesium level was 13.2 mEq/L at this time. IV calcium gluconate and oral sodium and potassium phosphates gradually brought this down to 2.7 mEq/L 24 hr later, and the patient recovered fully.

Another report of severe hypermagnesemia was given by Smilkstein et al. (1986). A 56-year-old female with multiple-drug overdose was treated with lavage and 100 g charcoal plus 15 g magnesium sulfate. Five hours later, she was started on 50 g charcoal each 2 hr and 75 g magnesium sulfate each hour. After six $MgSO_4$ doses, cardiopulmonary arrest occurred; resuscitation was successful. Her serum Mg^{++} post-arrest was 21.3 mEq/L and, with supportive care alone, it fell gradually to normal levels over the next 72 hr, although during that time hypotension, coma, hyperglycemia, and hypokalemia all occurred. Since the patient's renal function was always normal, this case shows that excessive doses of magnesium cathartics in patients with normal renal function can occur.

Yet another report on hypermagnesemia is that of Woolf and Gren (1988), who treated a 16-year-old girl for acute salicylate intoxication (her peak salicylate level was 114 mg/100 mL) with multiple doses of charcoal/magnesium citrate. Her serum Mg^{++} (baseline level 1.8 mg/100 mL) rose to a peak of 9.8 mg/100 mL after two doses of the charcoal/magnesium citrate. It fell back to normal levels after sorbitol was substituted and the patient was hemodialyzed. The authors feel that the patient's severe underlying condition of anorexia nervosa with chronic laxative abuse, combined with the acute salicylism, caused disordered magnesium metabolism in this case and that repeated magnesium citrate may still be of benefit in most cases—as long as careful monitoring of serum magnesium is done.

IX. CONSTIPATING EFFECT OF CHARCOAL

Bauer (1928) reported that the administration of charcoal doubles the period required for material to pass through the intestinal canal of rats. Charcoal likewise inhibited the increased peristalsis induced by castor oil.

Many of the studies discussed in Chapter 19 on the use of oral charcoal to reduce blood uremic metabolites or blood lipids have involved feeding up to 35 g/day of charcoal to human volunteers for up to several months. No particular problems with constipation have been encountered. In one such study, Yatzidis and Oreopoulos (1976) reported that the constipating effect of charcoal given to uremic patients "was overcome easily with either sorbitol or paraffin oil."

Also in Chapter 19, studies on the use of oral charcoal to treat erythropoietic porphyria in humans using large daily doses of oral charcoal have generally failed to show any serious problems with constipation. Pimstone et al. (1987), for example, gave a patient 60 g charcoal (in a water slurry) three times/day for nine months! The patient "had no side effects due to charcoal and no constipation." The only metabolic abnormalities observed were decreases in a few vitamins, which returned to normal levels after vitamin supplementation. In some contrast to this was another study of theirs on the treatment of porphyria in which 75 g/day of "superactivated" charcoal was given to one patient for 30 days. Constipation did occur in this patient, but the patient tolerated it for a full 25 days, so it could not have been unduly severe.

These reports, involving very large amounts of oral charcoal given in water suspension, show a striking absence of constipation as a serious side effect. It does occur but not often. One can conclude that in the absence of factors which would significantly suppress gastric motility (e.g., the ingestion of anticholinergic drugs), the occurrence of constipation from charcoal/water formulations is generally low.

Boldy and Vale (1986) described two cases of salicylate poisoning treated with repeated oral charcoal (Carbomix). Longer elimination half-lives were noted in their two cases, as compared to a previous report in the literature in which Medicoal was employed, and they conjectured that "the influence of a different type of charcoal cannot be discounted ... Carbomix may not be as effective as Medicoal." The interesting thing about this letter is their statement that, in their experience, "Carbomix produces marked constipation, whereas repeated doses of Medicoal may cause diarrhea. Therefore when prolonged treatment is required ... we now alternate 50 g Carbomix and Medicoal." If it were known what added ingredients exist in these two formulations, the explanation for the differences observed by Boldy and Vale might be evident. Their observations suggest that the Carbomix formulation has no added cathartic, while the Medicoal formulation does. Since Medicoal has been reported to have a significant sodium content (see Gorchein et al., 1988), it is possible that it contains sodium sulfate (a cathartic). However, the fact that this formulation is effervescent suggests that the sodium is in the form of sodium bicarbonate.

X.   OBSTRUCTIONS CAUSED BY CHARCOAL

Watson et al. (1986b) described a case of gastrointestinal obstruction associated with multiple-dose activated charcoal. A 30-year-old man, overdosed on carbamazepine, was given four 30-g doses of a charcoal formulation (Arm-A-Char) via a nasogastric tube during the first day, along with two 10 oz portions of magnesium citrate solution. On the morning of the second day, the abdomen was distended; x-rays showed gastrointestinal obstruction due to ileus. After more citrate solution and three enemas, more charcoal was given (60 g every 4 hr). Thirty-six hours after admission, emesis with a large volume of expelled charcoal occurred. A graph presented in the Watson paper implies that seven 30-g doses and four 60-g doses of charcoal were given, for a total of 450 g. Thus, the total charcoal dose was quite substantial. It was concluded that the anticholinergic properties of carbamazepine, which caused cessation of intestinal peristalsis, was a major cause of the obstruction, along with the large cumulative amount of charcoal.

Another report of obstruction was given by Anderson and Ware (1987). They treated a 32-year-old man for a suicide attempt involving amitriptyline and chlorpromazine by giving the patient gastric lavage followed by 50 g Carbomix charcoal. He was then given a further 300 g charcoal over the next 36 hr. He recovered but became severely constipated 10 days later. By the 12th day, he was in great discomfort, with spurious diarrhea secondary to fecal impaction. The obstruction, a mass consisting mainly of charcoal and resembling a charcoal briquette, was removed surgically from his anal region. The authors concluded that "the combination of the anticholinergic action of both drugs taken in overdose together with 3/4 lb of high residue charcoal led to this result." Note that, in this case, as in the case of constipation described by Boldy and Vale, the formulation involved was Carbomix. The involvement of Carbomix might not be related to its specific formulation, but may reflect the fact that large amounts of charcoal of any type, if given without any sort of cathartic, can potentially result in constipation or obstruction.

Flores and Battle (1987) described the case of a 49-year-old woman who was given charcoal every 3 hr for an overdose of haloperidol and maprotiline. No cathartic was administered. The patient developed obstructions due to many large briquette-like lumps of dehydrated charcoal in her jejunum and colon.

A case report by Ray et al. (1988) described a small-bowel bezoar (hard gastrointestinal mass of material) that developed in a patient overdosed with amitriptyline (which suppresses peristalsis due to its anticholinergic properties). The patient, a 21-year-old male, was treated with gastric lavage and then with 30–60 g activated charcoal administered through a nasogastric tube every 4–6 hr for five days. On the third day, the patient's abdomen became distended and tympanitic. Early small-bowel obstruction was found by radiography. Five days post-admission, complete small-bowel obstruction was confirmed and a laparotomy revealed a 3 × 3 cm (the third dimension is not stated) hard black mass in the distal ileum consisting of densely packed charcoal, which was presumably removed. The patient eventually recovered. The total charcoal administered can be inferred from the stated figures of 30–60 g per dose at intervals of 4–6 hr for 5 days as being anywhere between 600 and 1800 g total. Assuming the midpoint of this range (i.e., 1200 g), it can be seen that the total quantity of charcoal was very large. This, combined with the lack of peristalsis which accompanies amitriptyline overdoses, makes the development of a charcoal bezoar not unsurprising.

A case of a "pharmacobezoar" is interesting and will be mentioned here. Although the bezoar apparently did not contain charcoal, it is important to realize that drugs themselves can form bezoars which might be make the drugs resistant to the action of charcoal, due to the resulting poor dispersal of the drug. The case, described by Bernstein et al. (1992), involved an instance of fatal theophylline overdose in a 54-year-old woman. Her initial serum theophylline level was 31 mg/L and she showed signs of only mild toxicity, which disappeared after treatment in the ED with gastric lavage, charcoal, and a cathartic. The patient was discharged but arrested 8 hr later. At autopsy, her serum drug level was 190 mg/L and a white waxy mass weighing 319 g and containing 29 g theophylline (the residue of many sustained-release tablets) was found in her stomach.

A case report of bowel obstruction by charcoal requiring surgery was given by Atkinson et al. (1992). A 24-year-old unconscious male who had overdosed on barbiturates and benzodiazepines was given 25 g Medicoal with 200 mL water via a nasogastric tube every 4 hr for five doses. About two days later, the patient's abdomen became distended and fever developed. Radiography confirmed a small bowel obstruction. A laparotomy revealed a large bolus of inspissated charcoal in the cecum. It could not be fragmented, so a limited right hemicolectomy was undertaken. Bowel sounds returned after three days and the patient ultimately recovered completely. The authors recommended that charcoal be given with an osmotically-active cathartic and that charcoal administration be stopped if it does not appear in the stools within 12 hr.

Two cases of intestinal pseudo-obstruction due to charcoal, one fatal, have been described by Longdon and Henderson (1992). In the first, a 57-year-old male with combined theophylline, aspirin, and acetaminophen overdose was treated with 4 doses of 50 g charcoal in 70% sorbitol (150 mL). At 48 hr, he had not passed any charcoal and developed a distended tympanic abdomen with no bowel sounds. At laparotomy there was "massive distension of the large and small intestines without evidence of mechanical obstruction or other pathology." The patient died of progressive sepsis at 77 hr. In the second case, a 35-year-old male with a theophylline overdose was given the same charcoal treatment. At 30 hr, he also had not passed any charcoal and developed a distended and tympanic abdomen. The large and small intestines were grossly distended. Over the next 24 hr the pseudo-obstruction subsided and the patient recovered.

Both cases involved theophylline, which reduces intestinal motility. Also, in both cases, opiate (papaveretum) infusions were performed in order to facilitate mechanical ventilation. Opiates reduce the propulsive contractility of the large and small intestines. The authors state that in many other cases in which theophylline was not involved, the use of papaveretum did not cause the problems described here; thus, it appears that the combination of theophylline overdose and opiate infusion was the key factor in the pseudo-obstructions observed. The authors state that the sorbitol used was probably not a factor in the gut hypomotility, but "it almost certainly contributed to the morbidity. Both patients retained 600 mL of 70% sorbitol which was then a substrate for intestinal bacteria, leading to the production of copious volumes of gas and gut distension."

Goulbourne and Cisek (1994) reported a case involving a 64-year-old woman patient having theophylline toxicity who was given 50 g charcoal with 96 g sorbitol (Actidose with Sorbitol) followed by six subsequent 50-g doses of plain charcoal every 2 hr. At discharge, she had active bowel sounds and had passed one charcoal stool. However, 5 days later, she returned with vomiting, periumbilical pain, abdominal distension, and dehydration. She was found to have a small bowel obstruction and, after IV rehydration, the patient underwent laparatomy with lysis of low-grade adhesions at the ileocecal region, for which an ileotransverse colostomy was performed. Several pieces of charcoal, totalling $4.5 \times 5 \times 3$ cm overall, were removed from the bowel.

Gomez et al. (1994) described a case of a 39-year-old female who developed an intestinal perforation secondary to a charcoal stercolith (fecal concretion). The patient received two 50-g doses of charcoal given 4 hr apart (she refused further doses) for an amitriptyline ingestion. Peritoneal signs developed several days after admission and an exploratory laparotomy was performed. A 4-cm diameter perforation in the posterior wall of the sigmoid colon was discovered. At this site, there was a 120 g obstructing charcoal mass. This report, and reports of similar obstructions, suggests that multiple-dose charcoal be used with care in the case of drugs having antiperistaltic properties.

XI. RECTAL ULCER WITH HEMORRHAGE

Mizutani et al. (1991) described the case of a 42-year-old woman who was treated for organophosphate poisoning with gastric lavage, mechanical ventilation, IV atropine/pralidoxime, and repeated nasogastric administration of a charcoal/magnesium sulfate suspension. The suspension (50 g charcoal plus 50 g $MgSO_4$ in 1 liter of water) was given every 4–6 hr for 50 hr. The patient developed massive rectal bleeding after passing several hard masses of charcoal resembling "barbecue briquettes" on the 10th day. Surgical hemostasis was required to control the bleeding. Although cathartics were administered after completion of the charcoal treatment, the amounts were probably too little, as the amount of stool passed after the charcoal treatment was rather small. The early and vigorous treatment of constipation is essential when large amounts of charcoal (on the order of 500 g total) are given.

XII. CORNEAL ABRASIONS

Two cases of corneal abrasions secondary to charcoal therapy have been described by McKinney et al. (1993). Both patients were combative during lavage and subsequent charcoal administration, and both had some of the charcoal spilled into their eyes. Corneal abrasions, confirmed by fluorescein staining and magnified Wood's light examination, occurred. The patients' eyes were irrigated, treated with antibiotics, and patched. Both patients recovered without ocular complications. This report does point out one problem associated with trying to give charcoal suspensions to combative patients.

XIII. "BLACK SMOKE SYNDROME"

Before leaving this discussion of the hazards of charcoal, we mention a rather humorous "hazard" described by Raikhlin (1988). A young woman who went to a carefully planned rendezvous took two capsules of a charcoal/simethicone formulation, which had been prescribed for intestinal gas. She then sprayed her mouth with a breath deodorant containing ethanol, glycerine, saccharin, and menthol. As the story goes, "when she opened her mouth to speak, black smoke flowed out of her throat and onto the surprised listener."

XIV. SUMMARY

Based on the reports we have reviewed in this chapter, several recommendations can be made concerning how to avoid complications when using oral charcoal.

1. If the patient is combative, enough sedation (e.g., with diazepam) should be given to insure that proper intubations (endotracheal, and orogastric or nasogastric) can be performed.
2. Orogastric or nasogastric intubation should not be carried out if there is a significant risk of emesis (e.g., if ipecac had been given not long before). An anti-emetic should be given if the risk of emesis is judged to be significant.
3. A cuffed endotracheal tube should be inserted to protect against aspiration of charcoal and gastric contents into the lungs if regurgitation or vomiting were to occur.
4. Intubation for charcoal instillation should be done carefully with either an orogastric tube or a nasogastric tube, with confirmation of the tube's position by auscultation or radiography.
5. The charcoal slurry used should be thin enough to flow readily through the tube.
6. If multiple doses of charcoal are given, the existence of sufficient bowel motility should be confirmed by listening for bowel sounds and passage of charcoal rectally after a certain time period (e.g., 12 hr) should be confirmed if charcoal is to be continued beyond that time.
7. Consideration should be given to using continuous slow nasogastric administration of charcoal as opposed to large single-bolus delivery.
8. If formulations using sorbitol are used or if magnesium cathartics are used, careful monitoring of the patient's fluid balance and electrolyte levels must be done. Fluid replacement and electrolyte adjustment should be carried out as necessary. In general, sorbitol or other cathartics need not be given with each charcoal dose—one should at least alternate between doses having a cathartic and doses having no cathartic.

XV. REFERENCES

Allerton, J. P., and Strom, J. A. (1991). Hypernatremia due to repeated doses of charcoal-sorbitol, Am. J. Kidney Dis. *XVII*, 581.

Anderson, I. M. and Ware, C. (1987). Syrup of ipecacuanha (letter), Br. Med. J. *294*, 578.

Anker, A. and Smilkstein, M. (1993). Fatal sorbitol-induced abdominal distention (abstract), Vet. Hum. Toxicol. *35*, 334.

Atkinson, S. W., Young, Y., and Trotter, G. A. (1992). Treatment with activated charcoal complicated by gastrointestinal obstruction requiring surgery, Br. Med. J., *305*, Sept. 5, p. 563.

Bauer, H. (1928). Constipating effect of charcoal, Arch. Exp. Pathol. Pharmakol. *134*, 185.

Benson, B., Van Antwerp, M., and Hergott, T. (1989). A fatality resulting from multiple dose activated charcoal therapy (abstract), Vet. Hum. Toxicol. *31*, 335.

Bernstein, G., Jehle, D., Bernaski, E., and Braen, G. R. (1992). Failure of gastric emptying and charcoal administration in fatal sustained-release theophylline overdose: Pharmacobezoar formation, Ann. Emerg. Med. *21*, 1388.

Boldy, D. and Vale, J. A. (1986). Treatment of salicylate poisoning with repeated oral charcoal (letter), Br. Med. J. *292*, 136.

Brent, J., Kulig, K., and Rumack, B. H. (1989). Iatrogenic death from sorbitol and magnesium sulfate during treatment for salicylism (abstract), Vet. Hum. Toxicol. *31*, 334.

Bynum, L. J. and Pierce, A. K. (1976). Pulmonary aspiration of gastric contents, Am. Rev. Respir. Dis. *114*, 1129.

Caldwell, J. W., Nava, A. J., and De Haas, D. D. (1987). Hypernatremia associated with cathartics in overdose management, West. J. Med. *147*, 593.

Dammann, K. Z., Wiley, S. H., and Tominack, R. L. (1988). Aspiration pneumonia following activated charcoal—A case report (abstract), Vet. Hum. Toxicol. *30*, 353.

Danel, V. (1988). Fatal pulmonary aspiration of oral activated charcoal (letter), Br. Med. J. *297*, 684.

Donovan, J. W. (1987). Activated charcoal in management of poisoning: A revitalized antidote, Postgrad. Med. *82*, 52.

Donovan, J. W. (1988). Activated charcoal therapy (letter), Postgrad. Med. *83*, 38.

Dunbar, B. S., Pollack, M. M., and Shahvari, M. B. G. (1981). Cardiorespiratory changes after charcoal aspiration (abstract), Crit. Care Med. *9*, 221.

Elliott, C. G., Colby, T. V., Kelly, T. M., and Hicks, H. G. (1989). Charcoal lung: Bronchiolitis obliterans after aspiration of activated charcoal, Chest *96*, 672.

Eyer, P. and Sprenger, M. (1991). [Oral administration of a charcoal-sorbitol-suspension as a first-line treatment to counteract poison absorption?], Klin. Wochenschr. *69*, 887.

Farley, T. A. (1986). Severe hypernatremic dehydration after use of an activated charcoal-sorbitol suspension, J. Pediatr. *109*, 719.

Farley, T. A. (1987). Toxicity of sorbitol-charcoal suspension (reply to letter), J. Pediatr. *111*, 308.

Fassler, C. A., Rodriguez, R. M., Badesch, D. B., Stone, W. J., and Marini, J. J. (1985). Magnesium toxicity as a cause of hypotension and hypoventilation: Occurrence in patients with normal renal function, Arch. Intern. Med. *145*, 1604.

Fish, S., Munier-Sham, J., and Blansfield, J. (1989). Activated charcoal/sorbitol preparations: Incidence of adverse effects (abstract), Vet. Hum. Toxicol. *31*, 350.

Flores, F. and Battle, W. S. (1987). Intestinal obstruction secondary to activated charcoal, Contemp. Surg. *30*, 57.

Friedman, E. A., Feinstein, E. I., Beyer, M. M., Galonsky, R. S., and Hirsch, S. R. (1978). Charcoal-induced lipid reduction in uremia, Kidney Int. *13* (Suppl. 8) S-170.

Garrelts, J. C., Watson, W. A., Holloway, K. D., and Sweet, D. E. (1989). Magnesium toxicity secondary to catharsis during management of theophylline poisoning, Am. J. Emerg. Med. *7*, 34.

Gazda-Smith, E. and Synhavsky, A. (1990). Hypernatremia following treatment of theophylline toxicity with activated charcoal and sorbitol (letter), Arch. Intern. Med. *150*, 689.

Geller, R. J. and Ekins, B. R. (1993). Death complicating gastrointestinal decontamination — Time to rethink "routine therapy"? (abstract), Vet. Hum. Toxicol. 35, 335.

George, D. L., McLeod, R., and Weinstein, R. A. (1991). Contaminated commercial charcoal as a source of fungi in the respiratory tract, Infect. Contr. Hosp. Epidemiol. 12, 732.

Givens, T., Holloway, M., and Wason, S. (1992). Pulmonary aspiration of activated charcoal: A complication of its misuse in overdose management, Pediatr. Emerg. Care 8, 137.

Gomez, H. F., et al. (1994). Charcoal stercolith with intestinal perforation in a patient treated for amitriptyline ingestion, J. Emerg. Med. 12, 57.

Gorchein, A., Chong, S. K. F., and Mowat, A. P. (1988). Hazards of oral charcoal (letter), Lancet, May 28, p. 1220.

Goulbourne, K. B. and Cisek, J. E. (1994). Small-bowel obstruction secondary to activated charcoal and adhesions, Ann. Emerg. Med. 24, 108.

Harris, C. R. and Filandrinos, D. (1993). Accidental administration of activated charcoal into the lung: Aspiration by proxy, Ann. Emerg. Med. 22, 1470.

Harsch, H. H. (1986). Aspiration of activated charcoal (letter), New Engl. J. Med. 314, 318.

Hoffman, J. R. (1983). Charcoal for gastrointestinal clearance of drugs (letter), New Engl. J. Med. 308, 157.

Hyams, J. S. (1982). Chronic abdominal pain caused by sorbitol malabsorption, J. Pediatr. 100, 772.

Jones, J., Heiselman, D., Dougherty, J., and Eddy, A. (1986). Cathartic-induced magnesium toxicity during overdose management, Ann. Emerg. Med. 15, 1214.

Jones, J., McMullen, M. J., Dougherty, J., and Cannon, L. (1987). Repetitive doses of activated charcoal in the treatment of poisoning, Am. J. Emerg. Med. 5, 305.

Jones, J., McMullen, M. J., and Dougherty, J. (1988). Repetitive doses of the activated charcoal-sorbitol combination: A word of caution (reply to letter), Am. J. Emerg. Med. 6, 202.

Justiniani, F. R., Hippalgaonkar, R., and Martinez, L. O. (1985). Charcoal-containing empyema complicating treatment for overdose, Chest 87, 404.

Klein, S. K., Levinsohn, M. W., and Blumer, J. L. (1985). Accidental chlorpromazine ingestion as a cause of neuroleptic malignant syndrome in children, J. Pediatr. 107, 970.

Levy, G. (1983). Charcoal for gastrointestinal clearance of drugs (reply to letters), New Engl. J. Med. 308, 157.

Longdon, P. and Hendersen, A. (1992). Intestinal pseudo-obstruction following the use of enteral charcoal and sorbitol and mechanical ventilation with papaveretum sedation for theophylline poisoning, Drug Safety 7, 74.

Mann, K. V. (1988). Activated charcoal products stocked and dispensed by hospital pharmacists for multiple dose activated charcoal therapy (abstract), Vet. Hum. Toxicol. 30, 351.

Mariani, P. J. and Pook, N. (1993). Gastrointestinal tract perforation with charcoal peritoneum complicating orogastric intubation and lavage, Ann. Emerg. Med. 22, 606.

Massanari, M. J. (1987). Toxicity of sorbitol-charcoal suspension (letter), J. Pediatr. 111, 308.

McCord, M. M. and Okun, A. L. (1987). Toxicity of sorbitol-charcoal suspension (letter), J. Pediatr. 111, 307.

McKinney, P., Phillips, S., Gomez, H. F., and Brent, J. (1993). Corneal abrasions secondary to activated charcoal (letter), Am. J. Emerg. Med. 11, 562.

Menzies, D. G., Busuttil, A., and Prescott, L. F. (1988). Fatal pulmonary aspiration of oral activated charcoal, Br. Med. J. 297, 459.

Minocha, A., Wiley, S. H., Chabbra, D. R., Harper, C. R., and Spyker, D. A. (1986). Superior efficacy of sorbitol cathartics in poisoned patients (abstract), Vet. Hum. Toxicol. 28, 494.

Mizutani, T., Naito, H., and Oohashi, N. (1991). Rectal ulcer with massive haemorrhage due to activated charcoal treatment in oral organophosphate poisoning, Hum. Exp. Toxicol. 10, 385.

Moore, C. M. (1988). Hypernatremia after the use of an activated charcoal-sorbitol suspension (letter), J. Pediatr. 112, 333.

Nau, C. A., Neal, J., and Stembridge, V. (1958a). A study of the physiological effects of carbon black. I. Ingestion, Arch. Ind. Health 17, 21.

Nau, C. A., Neal, J., and Stembridge, V. (1958b). A study of the physiological effects of carbon black. II. Skin contact, Arch. Ind. Health *18*, 511.

Nau, C. A., Neal, J., Stembridge, V. A., and Cooley, R. N. (1962). Physiological effects of carbon black. IV. Inhalation, Arch. Environ. Health *4*, 45.

Nelli, P. and Rau, N. R. (1988). Activated charcoal therapy (letter), Postgrad. Med. *83*, 38.

Pimstone, N. R., Gandhi, S. N., and Mukerji, S. K. (1987). Therapeutic efficacy of oral charcoal in congenital erythropoietic porphyria, New Engl. J. Med. *316*, 390.

Pollack, M. M., Dunbar, B. S., Holbrook, P. R., and Fields, A. I. (1981). Aspiration of activated charcoal and gastric contents, Ann. Emerg. Med. *10*, 528.

Power, K. J. (1988). Fatal pulmonary aspiration of oral activated charcoal (letter), Br. Med. J. *297*, 919.

Raikhlin, B. (1988). Black smoke syndrome, Vet. Hum. Toxicol. *30*, 485.

Rau, N. R., Nagaraj, M. V., Prakash, P. S., and Nelli, P. (1988). Fatal pulmonary aspiration of oral activated charcoal (letter), Br. Med. J. *297*, 918.

Ray, M. J., Padin, D. R., Condie, J. D., and Halls, J. M. (1988). Charcoal bezoar: Small-bowel obstruction secondary to amitriptyline overdose therapy, Dig. Dis. Sci. *33*, 106.

Reed, M. D. (1988). Oral activated charcoal therapy (letter), Am. J. Emerg. Med. *6*, 318.

Silberman, H., Davis, S. M., and Lee, A. (1990). Activated charcoal aspiration, N. Car. Med. J. *51*, 79.

Smilkstein, M. J., Smolinske, S. C., Kulig, K. W., and Rumack, B. H. (1986). Severe hyper-magnesemia due to multiple-dose charcoal therapy (abstract), Vet. Hum. Toxicol. *28*, 494.

Smilkstein, M. J., Steedle, D., Kulig, K. W., Marx, J. A., and Rumack, B. H. (1987). Magnesium levels after magnesium-containing cathartics (abstract), Vet. Hum. Toxicol. *29*, 458.

Sullivan, J. B., Jr. and Krenzelok, E. P. (1988). Repetitive doses of the activated charcoal-sorbitol combination: A word of caution (letter), Am. J. Emerg. Med. *6*, 201.

Watson, W. A., Guy, J. C., and Leighton, J. (1986a). The incidence of emesis after activated charcoal in emergency room patients (abstract), Vet. Hum. Toxicol. *28*, 498.

Watson, W. A., Cremer, K. F., and Chapman, J. A. (1986b). Gastrointestinal obstruction associated with multiple-dose activated charcoal, J. Emerg. Med. *4*, 401.

Wax, P. M., Wang, R., Hoffman, R. S., Mercurio, M., Howland, M. A., and Goldfrank, L. R. (1993). Prevalence of sorbitol in multiple-dose activated charcoal regimens in emergency departments, Ann. Emerg. Med. *22*, 1807.

Weber, C. A. and Santiago, R. M. (1989). Hypermagnesemia: A potential complication during treatment of theophylline intoxication with oral activated charcoal and magnesium-containing cathartics, Chest *95*, 56.

Wehr, K. L., Johanson, W. G., Jr., Chapman, J. S., and Pierce, A. K. (1975). Pneumoconiosis among activated-carbon workers, Arch. Environ. Health *30*, 578.

Woodard, J. A., Shannon, M., Lacouture, P. G., and Woolf, A. (1990). Serum magnesium concentrations after repetitive magnesium cathartic administration, Am. J. Emerg. Med. *8*, 297.

Woolf, A. D. and Gren, J. (1988). Hypermagnesemia associated with catharsis in a salicylate intoxicated patient with anorexia nervosa (abstract), Vet. Hum. Toxicol. *30*, 352.

Yatzidis, H., and Oreopoulos, D. (1976). Early clinical trials with sorbents, Kidney Int. *10* (Suppl. 7), S-215.

18

Effect of Charcoal on Various Inorganic Substances

As stated in Chapter 3, inorganic compounds display a wide range of adsorbability to charcoal. Strongly dissociated salts (NaCl, KNO_3, etc.) essentially do not adsorb to charcoal. On the other hand, when pH conditions are such that an inorganic species is undissociated (e.g., I_2, $HgCl_2$), fairly strong adsorption often occurs. Indeed, we have pointed out in Chapter 2 that I_2 and $HgCl_2$ have been widely used to test the general adsorption ability of medicinal charcoals. Even today, the "iodine number" is one standard measure of the adsorption power of charcoals.

In Chapter 10, we quoted a maximal adsorption capacity of a charcoal for $HgCl_2$ of 1,800 mg/g, an extremely high capacity (Andersen, 1946). $HgCl_2$ in solution dissociates very little and it has an unusual ability to form chemical complexes with chemical groups that occur on the surface of charcoals. In Chapter 3, we discussed the ability of charcoals to adsorb HCl and NaOH; Figure 3.4 indicates that charcoals can adsorb up to about 0.024 g NaOH/g and up to about 0.015 g HCl/g. However, from a clinical standpoint, these NaOH and HCl adsorption capacities are insignificant and thus charcoal is not recommended for treating cases of acid or base ingestion. Besides the possibility that charcoal might provoke vomiting, which would expose the esophagus a second time to injury, charcoal can obscure the evaluation of esophageal damage by esophagoscopy.

We have already described a study by Decker et al. (1968) (Chapter 11) in which 5 g Norit A charcoal was added to 50 mL of simulated gastric fluid at 37°C for 20 min (with shaking). The solutions had different amounts of organic and inorganic compounds. They found that, of the inorganic compounds, mineral acids, sodium and potassium hydroxides, and sodium metasilicate (an ingredient in several cleansing preparations) were "not adsorbed to any measurable extent." Highly ionic substances such as cupric copper, ferrous iron, and boric acid were "very poorly adsorbed." Table 18.1 shows the extent of adsorption of three inorganic substances for which quantitative results were given. Clearly, ferrous sulfate and boric acid are adsorbed poorly, while iodine is ad-

Table 18.1   Adsorption of Inorganic Substances in Vitro

Substance	Percent adsorbed by 5 g Norit A charcoal		
Ferrous sulfate	1 tablet	7 tablets	20 tablets
	9%	4%	1%
Boric acid	10 mL	25 mL	100 mL
(3% solution)	11%	6%	4%
Iodine	1 g	3 g	
(tincture)	97%	91%	

Based on figures of Decker et al. (1968). Used with permission of Academic Press, Inc.

sorbed well. Since the weight of the ferrous sulfate tablets was not stated, it is difficult to fully assess the ferrous sulfate results.

Mitchell et al. (1989) mentioned in vitro studies of theirs which confirmed that arsenic, boric acid, and potassium bromide do not adsorb significantly to charcoal, whereas mercuric chloride, iodine, and silver do adsorb (but, no quantitative results are given). It is not clear what the "silver" was: elemental silver, or a silver salt? The fact that mercuric chloride and iodine adsorb well to charcoal was established long ago (see Chapter 2).

We now review the fairly small number of other studies in the literature dealing with the adsorption of various inorganic compounds or ions to charcoal. The discussion will not include any of the older literature dealing with iodine or mercury salts, as these have been considered in earlier chapters.

I.   ARSENIC

Veenendaal (1951) found that a preparation of charcoal and magnesium sulfate was relatively ineffective as an antidote for arsenic poisoning in animals.

Al-Mahasneh et al. (1990) investigated the use of charcoal for poisoning by arsenic salts (sodium arsenite, sodium arsenate) in rats. The animals were dosed at the LD_{90} levels of 72 mg/kg (sodium arsenite) and 874 mg/kg (sodium arsenate), which give As^{3+} and As^{5+}, respectively, upon dissociation in the body fluids. Some rats received, additionally, a single dose of charcoal (10 times the poison weight) 5–10 min later; other rats received three doses of charcoal (again, 10 times the poison weight) at 0, 60, and 120 min. No protective effect of charcoal was noted; all rats died within 6–24 hr. The authors refer to previous in vitro data which also showed no significant binding of arsenic to charcoal. Since arsenic in solution exists as inorganic cations, the whole class of which tend to bind poorly to charcoal, then these results are not surprising.

II.   BORIC ACID

The results of Decker et al. (1968) on boric acid adsorption have been presented above in Table 18.1. Oderda et al. (1987) also studied the adsorption of boric acid (dissolved in distilled water) to activated charcoal. Three samples were prepared of 100 mL size, each containing 1 g boric acid, and 7.5, 15, and 30 g charcoal. The percentages of boric acid adsorbed in these three samples were 5.7, 17.6, and 38.6%, respectively. Thus, the amount of charcoal needed to adsorb 39% of the boric acid is 30 times the weight of boric acid. This is not surprising, as boric acid, H3BO3 dissociates in solution to give the borate ion, BO_3^{3-}, which—being an ion—would be expected to adsorb poorly to charcoal. Considering the toxic and potentially fatal doses of boric acid in children (5 g) and in adults (20 g), activated charcoal administration of greater than 150 g in children or 600 g in adults would be required. These amounts are clearly impractical in clinical situations.

III. CESIUM

The binding of radioactively-labeled cesium, ^{137}Cs, to two types of Prussian blue (potassium ferrihexacyanoferrate), to sodium polystyrene sulfonate (Resonium-A), and to a Norit charcoal (exact type not stated) was studied at pH 1.0, 6.5, and 7.5 by Verzijl et al. (1992). Cesium was used as the chloride salt, CsCl, which in solution totally dissociates to give Cs$^+$ and Cl$^-$ ions. Binding of the cesium ions to the charcoal was negligible at all three pH values. The resin binding capacities (Q_m values) were low (< 10 mg/g) at all pH levels and the Prussian blue binding capacities ranged from 48–238 mg/g. Since the resin employed was a cation exchanger, one might expect it to have bound Cs$^+$ to a greater degree. However, the three pH levels were achieved using HCl and phosphate buffer salts (KH_2PO_4 and NaH_2PO_4), and it is likely that competition for cation exchange by H$^+$, K$^+$, and Na$^+$ was responsible for the poor binding of Cs$^+$. This brings up an important point: when studying the binding of ions from aqueous solution, it can make an enormous difference whether the medium is simply distilled water or contains other ionic species, used to establish certain pH levels. Clearly, solutions in distilled water are not relevant to in vivo conditions.

IV. IRON

Iron in the form of ions (e.g., ferrous or ferric ions) adsorb poorly to charcoal. However, Gomez et al. (1993) found that adding a chelating agent, deferoxamine (DFO), to pH 2.5 test solutions gave substantial Fe^{2+} adsorption. Three mixtures of different amounts of Fe^{2+} were prepared, with Fe^{2+} amounts of 120, 600, and 947 μg. Addition of charcoal alone or DFO alone produced no significant changes in unbound Fe^{2+}. However, addition of a charcoal/DFO slurry (the exact amount is unclear) gave unbound Fe^{2+} of 93, 480, and 720 μg, respectively. Thus, 22.5, 20.0, and 24.0% of the Fe^{2+} were taken up from solution in the three cases. In similar tests at pH 7.5, insoluble precipitates of iron compounds occurred, which apparently adsorbed to the charcoal, and both the charcoal alone and charcoal/DFO tests gave zero residual Fe^{2+} in the solutions.

Despite these encouraging in vitro results with deferoxamine, there is some concern that oral deferoxamine may actually increase iron absorption; additionally, the amounts of deferoxamine needed are very expensive. Thus, at this point, the clinical use of deferoxamine is not recommended.

V. LITHIUM

The adsorption of lithium to activated charcoal was studied in vitro by Favin et al. (1988). Lithium carbonate was added to water (6 g/L) and then 50 mL aliquots were mixed with 50 mL of more water or 50 mL of SGF. Then 1.5, 3, or 9 g SuperChar charcoal was added. After adsorption equilibrium was reached, residual Li_2CO_3 was determined. In water, 14.7, 26.5, and 40.4% of the lithium compound was adsorbed at the three increasing charcoal dose levels. In SGF, no significant adsorption occurred. Inorganic compounds such as Li_2CO_3 generally adsorb well to charcoal only when undissociated. At neutral pH, lithium carbonate is about 72% dissociated; presumably the 28% undissociated was the major part of what was found to adsorb to the charcoal. In SGF at pH 1.2, the Li_2CO_3 reacts with the HCl to form CO_2, most of which then escapes from the solution as a gas. Then, the Li exists as LiCl, which is essentially totally dissociated. Thus, lithium in this form adsorbs very poorly. Since SGF mimics in vivo conditions, it appears that charcoal would not be effective in lithium carbonate overdoses.

Linakis et al. (1989a,b) tested activated charcoal and sodium polystyrene sulfonate (SPS) for their potential to bind lithium in mice. The mice were given orogastric doses of LiCl (250 mg/kg) followed immediately by oral doses of 10 g/kg SPS or 6.7 g/kg charcoal (the 1989a paper quotes the 6.7 g/kg figure but the 1989b abstract quotes 8.68 g/kg as the charcoal dose—otherwise the two reports appear to be identical). The mice were killed at 1, 2, 4, and 8 hr after treatment and their serum analyzed for Li concentration. There were no statistical differences between the charcoal-treated mice and the control mice. However, the SPS group had significantly lower Li concentrations than either of these two groups (Li^+ serum levels were 60.6, 59.4, 41.9, and 60.3% of control levels at 1, 2, 4, and 8 hr, respectively, when SPS was used). Thus, it appears that in order to effectively bind an inorganic cation such as Li^+, a cation exchange resin is more effective than charcoal.

VI. PHOSPHORUS

Snodgrass and Doull (1982) report the case of an 18-month-old boy who ingested about 65 mg of elemental yellow phosphorus contained in a rodenticide paste (doses in adults of as little as 10 mg, and in children of as little as 3 mg, have been reported to be lethal). Yellow phosphorus is a general protoplasmic poison and is toxic to the liver, intestines, kidney, heart, and blood vessels. Ipecac-induced emesis, lavage, and initial charcoal administration were all done within 90 min of the estimated ingestion time. The initial charcoal dose was 30 g and subsequent 10-g doses were given every 4 hr for 24 hr. The patient recovered without undue complications. Whether the successful outcome in this one case was due to effective removal of the phosphorus by emesis and/or lavage, and to what extent it was due to the charcoal therapy, is totally speculative. Elemental phosphorus would not be expected to adsorb to charcoal very well but it is highly reactive chemically and presumably could react with chemical groups on the charcoal surface. Further research would be needed to explain what really occurred in this isolated anecdotal report.

Table 18.2 Effect of Different Surface Area Charcoals on KCN-Dosed Rats

	Control	SC	NS	D	USP
Onset of symptoms (min)	1.7	8.9	9.9	7.9	8.6
Percent with symptoms	100	92	80	70	70
Time of death (min)	8.0	25.2	21.5	26.9	96.8
Percent mortality	100	67	30	30	30
Drug level (mg/L)[a]	8.0	2.0	2.5	3.1	4.0

[a] Time of determination not stated. From Goetz et al. (1990). Reprinted with permission of the American College of Emergency Physicians.

VII. POTASSIUM

The binding of potassium ions in vitro to charcoal and to sodium polystyrene sulfonate (SPS) resin from solutions of KCl in 0.1 N HCl at 37°C was studied by Welch et al. (1986). KCl in aqueous solution dissociates completely to K^+ and Cl^-. This report, an abstract, does not state the concentrations and volumes of the KCl solutions used. Not surprisingly, the charcoal did not bind any K^+; however, the SPS—acting as an ion exchange resin—did take up "significant" K^+. Apparently, competition for binding from the H^+ ions was not severe, perhaps because the ratio of K^+ to H^+ was high enough.

VIII. CYANIDE

Lambert et al. (1988) tried using oral superactive charcoal (SuperChar) to treat rats overdosed with 35 or 40 mg/kg potassium cyanide (KCN). The KCN was given orally in a gelatin capsule, followed immediately with 4 g/kg charcoal in a 20% suspension in water or with a similar volume of just water. Of course, KCN dissociates in aqueous media to give K^+ (harmless, unless present in excessive amounts) and the cyanide ion CN^-, which is highly toxic. Signs of cyanide toxicity occurred rapidly (in about 3 min). In the control group, 25 of 26 rats died within 19 min, while in the charcoal-treated group only 12 of 26 rats showed signs of cyanide toxicity and only 8 of those rats died. Thus, it was concluded that oral charcoal is effective for treating cyanide poisoning in rats. This is encouraging, as the in vitro work of Anderson (1946) suggested that CN^- binds poorly to activated charcoal. However, by using enough charcoal and one of very high surface area, this tendency for low binding can be overcome to a significant extent.

Goetz et al. (1990) also studied the effects of charcoal on cyanide poisoning in rats, using four different charcoals. The rats were given oral doses of 10 mg/kg KCN, followed by 0.5 g charcoal. The development of symptoms, mortality, and time of death were monitored. The charcoals used were: SC (SuperChar, 3150 m^2/g), NS (Norit A Supra, 2000 m^2/g), D (Darco, 1500 m^2/g), and USP (USP XXII, 950 m^2/g). Table 18.2 shows the results. Clearly, all four charcoals were effective in reducing morbidity and mortality. However, the differences between the charcoals were generally not significant and did not relate in the expected fashion to surface area (e.g., the SuperChar charcoal, which one would expect to work the best, was best in only one category: drug level).

IX. THALLIUM

Activated charcoal and Prussian Blue dye are effective in acute thallium (Tl^+) intoxication in rats. They bind the metal present in the gut or that is secreted into the gut by the GI tract epithelium. Pedro et al. (1984) did in vitro studies to determine thallium adsorption by these agents from distilled water solution. The thallium was used in the form of the sulfate salt, Tl_2SO_4. The equilibrium adsorption data were fit well by the Langmuir equation. The Q_m values obtained were: for charcoal, 124 mg/g, and for Prussian Blue, 72 mg/g. Thus, charcoal was the more effective adsorbent. A Q_m of 124 mg/g is rather good for metal ion adsorption by charcoal. It is likely that the thallium bound well because it is a heavy metal and, like mercury, existed significantly in the form of an undissociated salt in solution (Tl_2SO_4 in this case).

X. SUMMARY

Studies on the binding of inorganic chemical species are quite consistent even though they are few in number. When the sorbent is activated charcoal, the binding of cations (H^+, Na^+, K^+, Li^+, As^{3+}, As^{5+}, Fe^{2+}, Fe^{3+}, Cs^+) is poor, except when the cation exists to a significant extent in the form of an undissociated salt ($HgCl_2$, Tl_2SO_4) or is complexed with a chelating agent (as in the case of Fe^{2+}). On the other hand, the CN^- anion appears to bind well to charcoal, based on in vivo results in rats (however, no in vitro work has been reported to quantify the degree of binding), while other anions like OH^-, Cl^-, and BO_3^{3-} have been found to bind poorly. Iodine in undissociated form, I_2, binds quite strongly to charcoal.

XI. REFERENCES

Al-Mahasneh, Q. M., Rodgers, G. C., and Benz, F. W. (1990). Activated charcoal (AC) as an adsorbent for inorganic arsenic: Study in rats (abstract), Vet. Hum. Toxicol. *32*, 351.

Andersen, A. H. (1946). Experimental studies on the pharmacology of activated charcoal. I. Adsorption power of charcoal in aqueous solutions, Acta Pharmacol. Toxicol. 2, 69.

Decker, W. J., Combs, H. F., and Corby, D. G. (1968). Adsorption of drugs and poisons by activated charcoal, Toxicol. Appl. Pharmacol. *13*, 454.

Favin, F. D., Klein-Schwartz, W., Oderda, G. M., and Rose, S. R. (1988). In vitro study of lithium carbonate adsorption by activated charcoal, J. Toxicol.-Clin. Toxicol. *26*, 443.

Goetz, C. M., Dricker, J. D., and Krenzelok, E. P. (1990). Comparative efficacy of different activated charcoal surface areas in adsorbing potassium cyanide (abstract), Ann. Emerg. Med. *19*, 453.

Gomez, H. F., McClafferty, H., Horowitz, R., Brent, J., Dart, R. C., and Flory, D. (1993). Adsorption of Fe^{++} to a charcoal (AC) deferoxamine (DFO) slurry (abstract), Vet. Hum. Toxicol. *35*, 366.

Lambert, R. J., Kindler, B. L., and Schaeffer, D. J. (1988). The efficacy of superactivated charcoal in treating rats exposed to a lethal oral dose of potassium cyanide, Ann. Emerg. Med. *17*, 595.

Linakis, J. G. et al. (1989a). Administration of activated charcoal or sodium polystyrene sulfonate (Kayexalate) as gastric decontamination for lithium intoxication: An animal model, Pharmacol. Toxicol. *65*, 387.

Linakis, J. G. et al. (1989b). Activated charcoal and sodium polystyrene sulfonate (Kayexalate) in gastric decontamination for lithium intoxication: An animal model (abstract), Ann. Emerg. Med. *18*, 445.

Mitchell, R. D., Walberg, C. B., and Gupta, R. C. (1989). In vitro adsorption properties of activated charcoal with selected inorganic compounds (abstract), Ann. Emerg. Med. *18*, 444.

Oderda, G. M., Klein-Schwartz, W., and Insley, B. M. (1987). In vitro study of boric acid and activated charcoal, J. Toxicol.-Clin. Toxicol. *25*, 13.

Pedro, A., Lehmann, F. and Favari, L. (1984). Parameters for the adsorption of thallium ions by activated charcoal and Prussian Blue, J. Toxicol.-Clin. Toxicol. *22*, 331.

Snodgrass, W. R. and Doull, J. (1982). Early aggressive activated charcoal treatment of elemental yellow phosphorus poisoning, Vet. Hum. Toxicol. *24* (Suppl.), 96.

Veenendaal, E. M. (1951). [Arsenic antidote versus adsorbent charcoal], Nederl. Tijdschr. Geneesk. *95*, 3481.

Verzijl, J. M., et al. (1992). In vitro binding characteristics for cesium of two qualities of Prussian blue, activated charcoal and Resonium-A, J. Toxicol.-Clin. Toxicol. *30*, 215.

Welch, D. W., Johnson, P. N., Driscoll, J. L., and Lewander, W. J. (1986). In vitro potassium binding: A comparison of activated charcoal and sodium polystyrene sulfonate (abstract), Vet. Hum. Toxicol. *28*, 495.

19

Effect of Charcoal on Endogenous Biochemicals

Activated charcoal, taken orally, has an astonishing ability to counteract the abnormal buildup of a variety of exogenous toxins. For example, bile acids, which in excessive amounts can cause pruritus, and bilirubin, which in excessive amounts causes jaundice, are well adsorbed by charcoal and, therefore, have shown considerable benefit in treating such conditions. Likewise, charcoal can adsorb the excess porphyrins which cause erythropoietic porphyria; further, some of the toxins which accumulate in patients who have deficient kidney or liver function bind well to charcoal in vivo. When charcoal was used to try to adsorb uremic toxins in patients having kidney failure, it frequently reduced the levels of cholesterol and triglycerides in such patients (who typically have elevated lipid levels). In this chapter, we review the effects of charcoal on the wide variety of endogenous toxins just mentioned. In addition, we review some reports from Russia which have shown that charcoal has prolonged the lifespans of experimental animals, presumably by adsorbing harmful endogenous biochemicals.

I. BILE SALTS

Krasopoulos et al. (1980) studied the in vitro binding of five bile salts (sodium cholate, etc.) from pH 7–9 buffer solutions (containing NaHCO3, NaCl, and NaOH) by two types of powdered activated charcoals. The charcoals adsorbed the bile salts rapidly and strongly. Maximal adsorption capacities (Q_m values) were on the order of 400 mg/g charcoal (i.e., the charcoals bound the bile salts up to around 40% of their own weights; this is a large adsorption capacity). Tests with cholestyramine showed that the charcoals and cholestyramine were equally effective when 5 mM solutions were used but that the charcoals were somewhat less effective than cholestyramine when 10 mM solutions were used. The potential for charcoal to bind bile salts in the gut would explain its effectiveness in reducing serum cholesterol and triglycerides in vivo, as discussed in Section VI of this chapter.

II. BILIRUBIN

Neonatal jaundice caused by excess bilirubin in the blood occurs relatively frequently in premature infants. Its cause is thought to be the inefficient hepatic removal of bilirubin. Unconjugated bilirubin is excreted into the lumen of the gut (in the bile) and from there it undergoes reabsorption. Several investigators have considered whether orally administered charcoal would be effective in adsorbing unconjugated bilirubin, thereby preventing its reabsorption.

Künzer et al. (1963) found that charcoal adsorbs bilirubin well in vitro from duodenal fluid. They added 1.0 g Carbo medicinalis to 10 mL samples of duodenal fluids taken from eight premature infants and found that the charcoal reduced the total bilirubin from an average of 10.3 mg/dL (range 3.0–21.5) to an average of less than 0.1 mg/dL (range 0–0.3). However, it should be noted that the charcoal:bilirubin ratio was approximately 1000:1, so it is not surprising that the extent of adsorption was so high.

Künzer et al. (1964) did in vitro studies of bilirubin adsorption from alkaline solutions and found that activated charcoal (Carbo medicinalis) was a much more effective adsorbent than several other adsorbents (e.g., talc, a silicate, and a Dowex resin). They also varied the charcoal:bilirubin ratio and concluded that 0.1 g charcoal can bind at least 1 mg bilirubin (this is not a particularly large amount, however). They also did an in vivo study involving two groups of premature infants (25 per group). One group was given 1 g/day of charcoal from the second day of life onwards. There was a slight but statistically insignificant reduction in serum bilirubin levels during days 3–6; however, it should be noted that the charcoal dose was not very large.

In a study by Lücking and Künzer (1966a) the amounts of free bilirubin in the stools of premature infants were compared for a group given 1 g/day charcoal during days 2–7 and a group given 4.5 g/day charcoal during days 1–3. In the 4.5 g/day group, the amount of free bilirubin was significantly smaller than the amount of free bilirubin in the stools of the 1 g/day group. For example, the average milligrams of bilirubin in the stools for days 2, 3, and 4 were 5.8, 1.9, and 4.6, respectively, for the 1 g/day group, and 2.7, 0.5, and 0.6, respectively, for the 4.5 g/day group. In a second report, Lücking and Künzer (1966b) gave charcoal orally to premature infants at a dosage of 4.5 g/day for three days, starting at 4 hr after delivery. The mean serum bilirubin values over the first few days of life were lower in the charcoal-treated group but there was no difference in the maximal bilirubin levels in the treated and untreated groups. These two reports show that, while charcoal can indeed bind bilirubin in vivo, the effect on serum bilirubin levels in jaundiced infants is not significant at the charcoal dose levels employed.

Ulstrom and Eisenklam (1964) found that if charcoal feeding was started at 12 hr of age, no difference in bilirubinemia occurred between test and control infants. However, when the first dose of charcoal was given at age 4 hr, the charcoal-fed infants had significantly less bilirubinemia than the controls. Figure 19.1 shows the results for charcoal given at 4 hr. This suggests that enterohepatic circulation of bilirubin may play a more critical role in determining the size of the bilirubin pool during the first few hours of life than it does a short time later.

Canby (1965) reported on a study involving large numbers of babies. Exchange transfusions were given whenever it appeared that the serum bilirubin level would exceed 20 mg/dL by the fifth day of life. During one trial period, no charcoal was given and 53 exchange transfusions were performed out of 3,009 live births. During a second trial per-

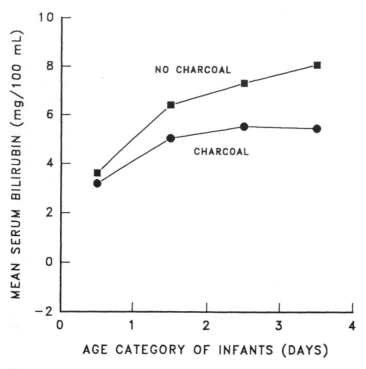

Figure 19.1 Mean values of serum bilirubin in 15 normal newborn infants treated with activated charcoal beginning at 4 hr and in 16 untreated normal infants. Values are plotted for infants aged 0–24 hr at 0.5 day, for infants aged 24–48 hr at 1.5 days, for infants aged 48–72 hr at 2.5 days, and for infants aged 72–96 hr at 3.5 days. From Ulstrom and Eisenklam (1964). Reprinted by permission of Mosby-Year Book, Inc.

iod, infants showing signs of jaundice were given Darco G-60 charcoal, 0.5 g in water every two hours for periods of 120–168 hr. These infants were selected from a group of 1,562 live births and yet only 12 exchange transfusions were performed. Thus, the use of charcoal reduced exchange transfusions from 17.6 to 7.7 per thousand live births.

Davis et al. (1983a) studied the effects of feeding a 5 wt% and a 10 wt% activated charcoal diet to jaundiced female Gunn rats. The feeding of charcoal reduced plasma bilirubin levels by as much as 40%. Figure 19.2 shows the bilirubin levels for both charcoal diets versus time. Females were bred for 21 days and it was found at necropsy that 58% of the charcoal-treated females had live embryos, versus 0% in the control group. Continuously-mated females produced live litters as follows: charcoal group, 48%; control group, 7%. These findings indicate that activated charcoal can reduce the adverse effect of hyperbilirubinemia on reproduction in rats and also suggest its use for treating hyperbilirubinemia in humans.

Davis et al. (1983b) performed another study with hyperbilirubinemic adult Gunn rats, comparing the effects of phototherapy and charcoal. Charcoal at 5 wt% in the diet (giving roughly 3–4 g/kg ingested per 24 hr) was just as effective as continuous phototherapy in reducing plasma bilirubin levels, and in combination their effects were addi-

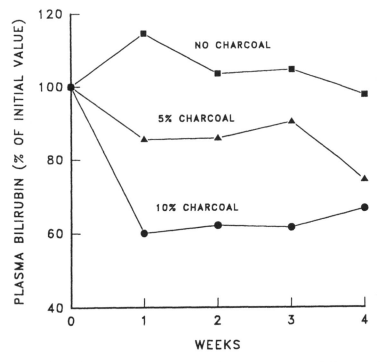

Figure 19.2 Changes in plasma bilirubin levels in jaundiced Gunn rats fed different levels (w/w) of activated charcoal. From Davis et al. (1983a). Reproduced by permission of S. Karger AG, Basel.

tive. In separate tests, on weanling rats, a diet of 5 wt% charcoal for 8 wk was found to have no effect on growth rate.

Davis and Yeary (1987) did a further study along the same lines as the one just described. In adult jaundiced rats, charcoal and phototherapy were again comparably effective in reducing bilirubin levels, and their effects were additive. However, in suckling jaundiced rats, phototherapy was ineffective. Charcoal alone reduced plasma bilirubin levels by 2.2 mg/dL and charcoal plus phototherapy reduced bilirubin by 4.5 mg/dL. Thus, phototherapy strongly enhanced the effect of charcoal even though when used alone it had no effect.

Amitai et al. (1993) gave a brief report on the effects of oral charcoal for treating neonatal hyperbilirubinemia. Thirty newborns with jaundice receiving phototherapy were randomly assigned to either a control group or a study group. The study group was given 0.98-g doses (infants weighing 2.4–3.0 kg) or 1.3-g doses (infants weighing more than 3.0 kg) of SuperChar charcoal before meals. Bilirubin levels in the control group at 0, 24, and 48 hr were 253, 240, and 232 μmol/L, respectively. In the charcoal-treated group the values were 265, 235, and 209 μmol/L. Thus, charcoal gave a significantly ($p < 0.02$) greater decrease in bilirubin over 48 hr (56 versus 21 μmol/L). It should be mentioned that charcoal was stopped in one patient due to vomiting.

These studies, although producing mixed results, suggest that charcoal can lower serum bilirubin levels moderately, if given in sufficient amount. Charcoal also appears to enhance the effects of phototherapy. However, much more research will be needed before more definitive conclusions can be reached.

III. EFFECT ON PRURITUS

Relief of generalized itching (pruritus) by oral activated charcoal was reported by Pederson et al. (1980). Eleven artificial kidney patients, who were experiencing pruritus as a side effect of their kidney disease, were given 6 g per day activated charcoal by mouth for 8 weeks. The pruritus was relieved in all but one patient. Additionally, scratch-induced skin sores were greatly cleared up. A possible explanation of the charcoal's effect is that it adsorbs compounds producing the pruritus in the gastrointestinal tract. No adverse effects of the charcoal were noted; hence, it is highly recommended for relieving itching conditions.

Svensson and Cali (1981) commented on the paper by Pederson et al. with the warning that the use of activated charcoal for treating pruritus might interfere with concurrent drug therapies; thus, administration of charcoal and drugs should be separated by as large a time interval as possible to decrease the chance of significant interaction taking place. This is an important point: when charcoal is used for changing the concentration of any endogenous chemical, it can obviously have an effect on any exogenous substance (e.g., drug) which is being given to the patient concurrently.

Fusaro (1981) also warned that the use of charcoal in large amounts for the relief of pruritus in uremic patients could interfere with other drugs which the patient may be taking. In a reply, Thiers (1981) stated that the ability of charcoal to bind the kinds of drugs which are routinely used in treating uremia has not been established.

Lauterberg et al. (1980), citing the implication of bile acid accumulation in the pathogenesis of pruritus, tried plasma perfusion on 8 patients with chronic cholestasis and intractable pruritus. The perfusion columns contained glass beads coated with USP charcoal. During 22 perfusions, 975 to 6100 mL of plasma were passed through the charcoal columns, resulting in the removal of 126–795 µmole of bile acids. The mean extraction of the bile acids was 81%. After the perfusions, all patients experienced relief of their pruritus, lasting from 24 hr to 5 months. Although the removal of other unidentified pruritogens other than bile acids is possible, the fact remains that—whatever the true pruritogens might be—they appear to be effectively adsorbed by charcoal.

Pruritus is the dominant symptom in intrahepatic cholestasis of pregnancy. It is believed that increased bile acids in the blood are the causative factor. Kaaja et al. (1994) gave charcoal, 50 g three times per day, to nine pregnant women suffering from cholestasis and monitored their serum total bile acids, bilirubin, and other blood chemicals. Ten other women served as controls. By the eighth day, serum total bile acids were 33.9 µmol/L in the charcoal-treated women versus 79.1 µmol/L in the controls. However, no other blood chemicals showed any significant effect of charcoal therapy. Four of the nine charcoal-treated women experienced relief of their pruritus. Perhaps charcoal treatment for longer than 8 days would have produced more dramatic results.

These studies, although few in number, show that charcoal therapy has been generally successful in treating pruritus, presumably by adsorbing bile acids and other pruritogens effectively in the gut. However, as in the case of hyperbilirubinemia, much more research is needed to better define the effects of charcoal.

IV. EFFECT ON ERYTHROPOIETIC PORPHYRIA

Guenther's disease (congenital erythropoietic porphyria) is a rare autosomal recessive disorder of heme biosynthesis. It results from a deficiency in uroporphyrinogen III cosynthetase activity. The disease is characterized by the presence of large amounts of series I porphyrin isomers. Disfiguring photomutilation begins in childhood and avoidance of sunlight has been the only certain way of controlling or preventing these effects. Orally administered cholestyramine, which can bind porphyrins in the gut and prevent their enteral absorption, has been tried with some success. We review here studies which have employed activated charcoal for the same purpose.

Mukerji et al. (1985) gave oral activated charcoal and cholestyramine to a 53-year-old white male who had a rare form of familial porphyria cutanea tarda with bone marrow rather than hepatic expression of the disease. The charcoal dose (ActaChar) was 30 g given every 3 hr and the cholestyramine dose was 4 g given every 6 hr. In less than one day, the charcoal caused a reduction of total plasma porphyrins from over 30 µg/dL to less than 0.5 µg/dL (a normal level); also within one day, the cholestyramine reduced plasma porphyrins to less than 12 µg/dL (a 60+% decrease). Within two days, the charcoal therapy reduced the total skin porphyrin content from 1–5 µg/g wet weight to about 0.01–0.04 µg/g wet weight (the ranges of values were due to the different sites from which skin samples were taken). Clearly, the charcoal therapy was dramatically effective in lowering blood and tissue porphyrins, presumably by trapping endogenous porphyrins in the gut lumen and preventing their enteral absorption. This study is very similar to one reported two years earlier by Gandhi and Pimstone (1983). The same two adsorbents and the same doses were used in the same type of patient. The charcoal lowered plasma porphyrins by 25% in 3 hr, by 77% in 4 hr, by more than 95% in 6 hr, and to barely detectable levels at 13 hr. The cholestyramine was less effective: plasma porphyrins fluctuated between 25 and 50% of pretreatment values after cholestyramine was given.

In another study, Pimstone et al., (1986) gave oral charcoal to a patient suffering from protoporphyric cirrhosis. ActaChar charcoal was given every 6 hr in the amount of 60 g/dose. One week later, skin and liver protoporphyrin levels were 0.11 and 37 µg/g wet weight versus levels of 12.1 and 92.2 µg/g wet weight prior to charcoal therapy. Plasma protoporphyrin was reduced 40% by the charcoal. While the charcoal therapy greatly reduced blood and tissue protoporphyrin, it was not successful in preventing further deterioration of liver function. It was stated that "removal of this pigment [protoporphyrin] in a cirrhotic, jaundiced patient is too slow to reverse the disease process."

Tishler and Winston (1985a) did an in vitro study in which the maximal adsorption capacities, Q_m s, of cholestyramine and eight charcoals were determined for uroporphyrin, protoporphyrin, and coproporphyrin, which accumulate within tissues or vasculature in certain porphyrias. These species were dissolved in a pH 8.2 solution of 0.5% desoxycholate. For uroporphyrin, the Q_m values were 26.5 mg/g for Amoco PX-21 charcoal, 17.0 mg/g for cholestyramine, and from 1.6 to 5.7 mg/g for the other seven charcoals. For protoporphyrin, the Q_m values were 32.4 mg/g for cholestyramine, 30.9 mg/g for the PX-21 charcoal, and 3.7 to 29.4 for the other seven charcoals. For coproporphyrin, the cholestyramine and PX-21 Q_m values were 39.2 and 35.1 mg/g, respectively. It was concluded that both cholestyramine and Amoco PX-21 superactive charcoal would be good candidates for trying to interrupt the enterohepatic circulation of porphyrins.

Winston and Tishler (1986) did another in vitro study in which the capacity of several activated charcoals and of cholestyramine for binding the porphyrin precursors aminolevulinic acid (ALA) and porphobilinogen (PBG) were evaluated. These two pre-cursors were dissolved in a pH 8.2, 0.1% desoxycholate solution. High Q_m values were obtained for all of the charcoals, but those for cholestyramine were much lower. For ALA, the charcoal Q_m values ranged from 72 to 110 mg/g and the cholestyramine Q_m was only 4 mg/g. For PBG, the charcoal Q_m values were 27 to 68 mg/g, whereas the cholestyramine Q_m was only 0.03 mg/g. The best of the charcoals studied was SuperChar charcoal (this is the same as Amoco PX-21 charcoal). It appears that this charcoal has a reasonable affinity for ALA and PBG (Q_m values of around 200 mg/g are typical of substances that adsorb "well" to charcoal) and that cholestyramine does not. This same study was reported in abstract form a year earlier (Tishler and Winston, 1985b).

Tishler and Winston (1988) described an in vivo study in which a 17-year-old male with severe congenital erythropoietic porphyria was given SuperChar charcoal in doses of 25 g, three times per day, for 31 days (constipation halted the study at this point). Excretion of porphyrin in the urine increased from 81 mg/day to 126 mg/day, and plasma uroporphyrin I decreased from 1.3 to 0.58 mg/L. Erythrocyte uroporphyrin I fell from 16.0 to 5.6 mg/L. However, no objective clinical improvement was noted in this brief trial.

Tishler and Winston (1990) described another in vivo study in which SuperChar charcoal was given orally at a dose of 25 g, three times daily, to another patient (a 16-year-old male) suffering from probable congenital erythropoietic porphyria. Erythrocyte porphyrin fell from 21.4 to 7.4 µmole/L and plasma porphyrin fell from 1.56 to 0.70 µmole/L over 49 days. Urinary porphyrin rose from 103 to 160 µmole/dL but this rise was not statistically significant. Constipation was somewhat of a problem in the study.

Pimstone et al. (1985) described a case in which a patient having photomutilating porphyria was given 60 g charcoal, three times per day, for 12 months. Plasma porphyrins decreased from 25–30 to 0.5–2.0 µg/dL, daily urinary porphyrin excretion dropped by approximately 50%, fecal porphyrin concentrations roughly doubled, skin porphyrin levels fell from 1.5 to 0.4 µg/g wet weight, and clinical activity of the disease was significantly diminished, with minimal new photocutaneous lesions.

Pimstone et al. (1987) reported a case study of a male in his mid-50s who had progressive photomutilation. This report appears to be a longer, follow-up study on the 53-year-old patient involved in the Mukerji et al. (1985) study. The patient was given 60 g ActaChar charcoal orally (as a slurry with cold water) three times a day for nine months. His plasma porphyrin returned to normal, and his skin porphyrin content fell to 1% of the level prior to charcoal therapy. The patient had no photocutaneous activity, felt well, and had no constipation or other side effects. He maintained a steady weight and had normal liver, renal, and hemopoietic function at all times. Some vitamin defi-ciencies were found, which were corrected by vitamin therapy. Cholestyramine was also tried orally in another test period on the same subject but was found to be less effective than charcoal. Figure 19.3 shows the results obtained with charcoal and with cholestyram-ine. It appears that charcoal lowers plasma porphyrins by either interfering with the enterohepatic circulation of them or by preventing the enteral absorption of porphyrins secreted by the mucosal cells in the gut lumen. Tishler (1988) commented on the Pimstone study, recommending that superactive charcoal be evaluated.

Hift et al. (1993) described a case in which a patient suffering from congenital erythropoietic porphyria was given oral charcoal, 60 g every 8 hr for the first 12 days, 20 g every 8 hr after the twelfth day, and then 10 g twice daily after one month had passed. Within hours of starting the charcoal therapy, urine porphyrin levels decreased,

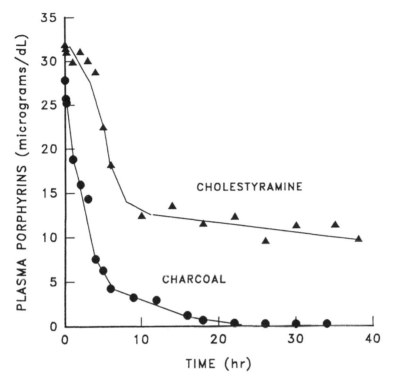

Figure 19.3 Changes in total plasma porphyrins after administration of activated charcoal or cholestyramine resin. From Pimstone et al. (1987). Reprinted by permission of the New England Journal of Medicine.

and reached 8% of pretreatment values by the tenth day. Plasma porphyrins declined to 20% of pretreatment levels within 5 days. However, after about 3 months, both urinary and plasma porphyrin levels mysteriously began to rise—ultimately the plasma porphyrin levels were higher than before the charcoal treatment had begun—and the disease symptoms reappeared. Three years after this first trial, the same patient was readmitted for a second course of charcoal therapy. This time, the urine and plasma porphyrin levels showed no initial decrease but rather increased significantly over two weeks, before gradually decreasing back to pre-treatment levels. No clear explanation for the responses shown in either treatment course is evident.

Gorchein et al. (1989) studied the effect of charcoal given over 1 year to a 5-year-old patient having protoporphyria and cirrhosis. Three different charcoals (Medicoal, Carbomix, and SuperChar) were evaluated. Plasma protoporphyrin was reduced from pretreatment levels of 1200–1750 nmol/L to 400 nmol/L by the use of Medicoal (10–15 g per day) for 9 days. Later administration of Carbomix (up to 30 g per day) for 4 months reduced the plasma protoporphyrin level to only 800 nmol/L. Reintroduction of the Medicoal lowered the protoporphyrin level to 450 nmol/L initially but a rise to about 800 nmol/L occurred over the next 2 months. Then the use of SuperChar charcoal (up to 60 g per day) was associated with an increase in protoporphyrin to about 2000 nmol/L over the next 11 weeks. Reintroduction of Medicoal did not reduce this level. The tendency for porphyrin levels to rise with time, despite the continued use of charcoal, is consistent with the observations of Hift et al., discussed in the previous paragraph.

The studies just discussed are mysterious, as in some cases charcoal has provided lasting reductions in serum porphyrins, while in other cases a rapid decline in serum porphyrins has been followed by steady rises back to pretreatment levels. Hopefully, further research will unravel this mystery.

V.   EFFECT ON UREMIC TOXINS

The issue of whether orally administered sorbents can be of use in uremia is one that has been around for a long time. As early as 1972, Yatzidis (1972) wrote that he had used activated charcoal in uremic patients, and stated that:

> We had ... found that charcoal effectively adsorbed certain uraemic substances directly from plasma or gastrointestinal contents (creatinine, uric acid, urea, phenols, guanidines, organics acids). A dose of 20–50 g of activated charcoal daily often results in impressive clinical improvement, mainly on the gastro-intestinal disturbances of the patient (oral fetor, anorexia, nausea, vomiting). No side effects were noted during continuous treatment for 4–20 months.

Friedman (1976) provided a review of the uremic toxin adsorption issue. Of course, a build-up of toxic metabolites—such as urea, uric acid, and creatinine—occurs in uremic patients. Also, electrolyte abnormalities occur (e.g., phosphate increases). In the early 1930s it was shown that urea diffuses from plasma into the intestinal lumen, where it is ultimately degraded by bacteria to ammonia. The use of charcoal to attempt to bind urea in the gut was studied by Yatzidis (1964). He gave patients with end-stage renal disease 20–50 g/day charcoal orally and was able to maintain them without dialysis for 4–20 months. However, these patients all had relatively good renal function (creatinine clearances of 10–15 mL/min) and were on a restricted protein diet (20 g protein/day), so it is difficult to say how much effect the charcoal had. Denti et al. (1972) gave 50 g/day of charcoal to healthy volunteers to see how well it would be tolerated but problems with nausea, vomiting, and constipation were severe enough to interrupt the study.

Maxwell et al. (1972) gave various forms of charcoal to uremic dogs in amounts up to 90 g/day and found no "significant effect on the lowering of any of the metabolites associated with the uremic state." Sparks (1975) explained this lack of effect by saying that "... carbon adsorbs a wide variety of molecules in the intestinal milieu in preference to creatinine and uric acid. For example, the removal capacity of activated carbon for creatinine from pig gut fluid is 25-fold less than from buffer." Thus, he continues, "... a major problem with gastrosorbents [in uremia] is obtaining selectivity of removal in intestinal fluid." Spark's research group has clearly demonstrated the much lower adsorption of creatinine in pig small-intestine fluid (they also gave data for creatinine adsorption from pig large-intestine fluid, human duodenal fluid, and for dog/pig bile) as compared to buffer solutions (see Goldenhersh et al., 1976).

Yatzidis and Oreopoulos (1976) described some specific results obtained by giving 20–50 g charcoal per day to uremic patients. They gave the charcoal for two 8-week periods, with a "no charcoal" period in between (as well as before and after, of course). While no effect of charcoal on urea and electrolytes was found (as one would expect), serum phenol and guanidines fell significantly—about 25–30%, judging from a figure of theirs—and uric acid fell about 4%. In contrast to the Yatzidis' 1972 report, this later article stated that no significant decreases in creatinine and organic acids occurred. Yatzidis and Oreopoulos also presented some data on uremic patients treated by hemoperfusion over 200 g of granular charcoal at a blood flow rate of 250 mL/min. Over a 2 hr period, the following decreases were found: creatinine, 29.7%; uric acid, 22.2%; phenols, 28.6%; guanidines, 35.3%; organic acids, 20.0%; and urea, 6.7%. Thus, these data suggest that creatinine and organic acids can adsorb from blood to charcoal to a fair degree.

Later in vitro studies have shown that urea, in particular, adsorbs very poorly to charcoal (it is a simple molecule, $CO(NH_2)_2$, and has such little organic character that it is not attracted to charcoal surfaces to a significant extent). Thus, the adsorption of urea by activated charcoal—even the adsorption of urea from hemodialyzer dialysate solutions, in an attempt to regenerate such solutions—has never been very successful. Many attempts have been made over the years to get around this problem. We only mention one: the use of urease enzyme to degrade urea in dialysate solutions to ammonium ions which are then captured by ion exchange using cation exchange resins. This has turned out to be an expensive way to rid such solutions of urea and has created the additional problem of an electrolyte imbalance in the dialysate due to the exchanged cations. Further accounts of the urea/uric acid/creatinine removal story, both in vivo and from dialysate solutions, could be written and the full the story is very long; space does not permit the pursuit of this interesting subject any further here.

Sinclair et al. (1979) administered large amounts of charcoal directly into the intestines of aneuric rats and goats. With charcoal doses of up to 47 g/kg per day, urea and potassium levels in the animals' serum were stabilized. However, no effect on creatinine occurred. The rise in serum creatinine was the same as in animals which received no charcoal.

In contrast to charcoal, oxystarch has been found, in vitro, to adsorb urea somewhat well but it does not adsorb creatinine or uric acid. Nevertheless, several clinical studies involving the feeding of oxystarch to uremic patients have been carried out. For example, Giordano et al. (1973) found that giving 20 g/day oxystarch for two months to uremic patients having small renal function (creatinine clearances of 0.4 to 3.2 mL/min) caused a significant fall in the blood urea nitrogen levels. Friedman et al. (1974) gave 29 g/day of oxystarch in four equal doses to seven uremic patients (creatinine clearances of 6 to 30 mL/min) in a double-blind study and found that BUN levels decreased 33%, from a mean of 93 mg/dL to 62 mg/dL. No significant changes in serum creatinine, uric acid, or amino acid levels occurred. The oxystarch significantly increased fecal nitrogen (from 1.4 g to 2.5 g/day) and fecal potassium (from 5 to 22 mEq/day). These were counterbalanced by decreases in urinary excretions of nitrogen and potassium, however.

Two interesting studies which do not relate directly to the reduction of uremic toxins per se but rather to other ramifications of uremia are discussed here. Hoffman and Levy (1990), noting that uremic rats show greater CNS sensitivity to the hypnotic action of drugs like phenobarbital and to the neurotoxic (convulsive) action of drugs like theophylline, wondered if oral charcoal might cause a reduction in the concentration of the circulating endogenous substances which cause such altered drug effects. Thus, they created acute renal failure in rats by bilateral ligation of the ureters and then gave some of the rats oral charcoal (1 g/kg every 8 hr for 6 doses). About 2 hr after the last charcoal dose, the rats were infused IV with phenobarbital until the onset of the loss of the righting reflex, or with theophylline until maximal seizure onset. The CSF drug levels required to produce these effects were higher in charcoal-treated rats as compared to control rats. For example, the CSF theophylline concentrations needed to produce seizures in the controls averaged 137 mg/L versus 189 mg/L in the charcoal-treated rats. For both drugs, charcoal treatment caused a reversal of the hyperalgesia (sensitivity to pain) associated with renal failure, as determined before drug infusion by tail flick latency.

In a follow-up study with normal rats, however, Hoffman (1992) found no effect of pre-administered oral charcoal on the neurotoxic/convulsive effect of IV theophylline (given in the form of aminophylline). Although the charcoal-treated rats required larger theophylline doses to induce convulsions, the theophylline concentrations in the serum and brain at the onset of convulsions were not affected by the charcoal pretreatment. It was concluded that the gastrointestinal dialysis produced by charcoal had no effect on theophylline-induced neurotoxicity in normal rats.

To summarize the effects of sorbent therapy on uremic toxins, we can state that the disruption of normal blood levels of different substances (urea, uric acid, creatinine, electrolytes) in uremia is so broad in scope that little hope for treatment with oral adsorbents alone seems possible. Charcoal will not correct electrolyte abnormalities in vivo and does not adsorb urea. Resins can affect electrolytes but not urea, uric acid, and creatinine; oxystarch can affect urea somewhat but not creatinine or uric acid. Finally, no sorbents can rid the body of the water which inexorably builds up in aneuric patients. The use of sorbents for regenerating dialysate solutions has been more promising, but dialysate is so inexpensive that regeneration makes little economic sense. Regeneration of dialysate would be important in trying to create a portable dialysis system, where light weight is more important than economics. However, attempts to design effective and truly potable dialysis systems have not been successful; all systems are simply too cumbersome.

VI. HYPOLIPIDEMIC EFFECTS OF CHARCOAL

In 1976, Friedman (1976) and Friedman et al. (1976) reported a striking finding: four uremic patients who had been given large amounts of oral charcoal and oxidized starch daily (35 g of each) in an attempt to bind urea and other uremic metabolites (uric acid, creatinine) showed a decrease in serum cholesterol from an average of 200 mg/dL to an average of 166 mg/dL within one week. All patients tolerated the charcoal and oxystarch without complaint. In a subsequent study done in greater depth, Friedman (1977) gave three uremic patients 35 g/day oral activated charcoal alone (i.e., without oxystarch) for two weeks. All serum lipids fell. Figure 19.4 shows the results. The charcoal reduced cholesterol from an average of 282 mg/dL to as low as 230 mg/dL, a fall of 18.4%.

Another study by Friedman et al. (1978a) involved six uremic patients who were given 8.5 g oral activated charcoal (without oxystarch) four times per day (34 g/day total) for a total of 24 weeks. The phases of the study were: 4 wk control (no charcoal)/4 wk charcoal/4 wk control/8 wk charcoal/and 4 wk control. All patients tolerated the charcoal well. There was no apparent interference in appetite, sleep pattern, or general well-being. Although no reductions in serum urea or creatinine concentrations were noted, significant reductions in serum cholesterol and serum triglycerides were observed in three of the six patients. Maximal decreases in these three subjects, compared to their control period values, were 43, 23.4, and 40.4% for cholesterol and 76, 60.3, and 64.3% for triglycerides. It was concluded that charcoal binds these lipids or their metabolic

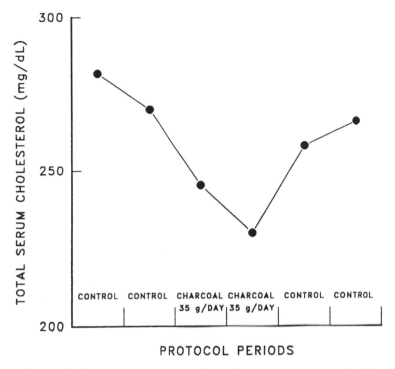

Figure 19.4   Effect of activated charcoal on mean serum total cholesterol in three patients. From Friedman (1977). Reproduced by permission of the Cahners Publishing Company.

precursors in the gut. Charcoal was thus recommended for the treatment of the hyperlipidemia which occurs in most uremic patients.

Friedman et al. (1978b) published another report in which 35 g/day charcoal plus 35 g/day oxidized starch was given orally to four uremic patients. No adverse reactions of any kind were noted. Mean serum cholesterol fell from control values of 200 mg/dL to 140 mg/dL (a 30% reduction) during the four weeks of sorbent therapy. Triglycerides fell from a range of 181–543 mg/dL to less than 150 mg/dL.

Manis et al. (1979) reported studies with normal rats and rats suffering from streptozotocin-induced diabetes. Groups of each were given diets containing 5, 10, or 15 wt% Norit charcoal for 15 days. Normal rats receiving 25 wt% charcoal had serum cholesterol reduced from an average of 53.4 mg/dL down to an average of 32.8 mg/dL (a 38.6% fall) and serum triglycerides reduced from an average of 117.6 mg/dL down to an average of 48.6 mg/dL (a 58.7% fall). Normal rats on the 10 wt% charcoal diet showed only a fall in triglycerides and normal rats on the 5 wt% charcoal diet showed no reductions in any lipids. The diabetic rats had their cholesterol and triglyceride levels restored essentially to normal levels. The major part of the response occurred in the first 12 hr. Charcoal diets of 1 and 3 wt% were tried with the rats and were found to be equally as effective as the higher wt% charcoal diets.

Friedman et al. (1979) described another study dealing with the effects of low dose oral charcoal in diabetic rats. Diabetes was again induced using streptozotocin and the rats became hyperlipidemic as expected. They were then given diets with 1, 3, and 5 wt% charcoal. After 12 hr, cholesterol was unchanged but the triglycerides had fallen dramatically. Over the test period of 33 days, cholesterol fell by almost half and triglycerides fell by about 80%. However, no significant differences were found among the 1, 3, and 5 wt% charcoal diets.

A year later, Manis et al. (1980) described additional experiments with rats having hyperlipidemia caused by streptozotocin-induced diabetes or by azotemia created by subtotal nephrectomy. The diabetic rats had a 10-fold rise in triglycerides and a doubling of cholesterol compared to normal values. Ingestion of 5 wt% charcoal normalized these values. Additionally, the HDL pattern on polyacrylamide gel electrophoresis was restored towards normal. In the azotemic rats, the charcoal significantly lowered cholesterol and triglycerides, again to essentially normal levels.

These studies show clearly that both diabetic and azotemic animals respond to oral charcoal with normalization of cholesterol and triglyceride levels. However, the diabetic rats respond to lower doses of charcoal and respond more quickly. The reasons for the different responses of diabetic rats versus azotemic rats are unknown.

The Finnish group of Neuvonen, Kuusisto, Vapaatalo, Manninen, and associates has also studied hypercholesterolemia. In their first report, Kuusisto et al. (1986) treated seven patients with 8 g charcoal (Carbomix) three times a day for four weeks. Plasma total cholesterol and LDL-cholesterol decreased by 25 and 41%, respectively, whereas HDL-cholesterol increased by 8%. Figure 19.5 shows the results for total cholesterol. No side effects were noted.

In a comment on this work, Hoekstra and Erkelens (1987) mentioned a double-blind study they did (published the next year in greater length; see Hoekstra and Erkelens, 1988) with 12 patients who received 15 g/day Norit charcoal ($N = 6$) or a nonactivated charcoal placebo ($N = 6$) in one 12-week test period, and 30 g/day Norit charcoal ($N = 6$) or a nonactivated charcoal placebo ($N = 6$) in another 12-week test period. Mean serum cholesterol levels did not change significantly (10.2 to 9.7 mmol/L with the 15 g charcoal doses and 10.5 to 9.7 mmole/L with the 30 g charcoal doses). In three hypertriglyceridemic patients, no decrease in serum triglycerides was observed. The authors felt that the nonplacebo design of the Kuusisto study may explain the differences in the results. However, there have been so many studies showing that charcoal does lower cholesterol levels that it is difficult to accept this conclusion. A reason may be that the type of charcoal used by Hoekstra and Erkelens may have had a pore structure which was not large enough to permit adequate penetration of the cholesterol, or whatever compounds are involved in the effect, into the charcoal. The authors state that the charcoal used was "Norit," but in fact the Norit company makes a wide variety of charcoals, so the simple designation "Norit" is not adequate to describe the exact type of charcoal used. Since the charcoal also was stated to be in granulated form, rather than the powdered form typical of antidotal charcoals, this further suggests that the charcoal employed may have been different from usual antidotal types.

The Finnish group published a second study (Neuvonen et al., 1989a), involving seven patients, again treated with 8 g activated charcoal three times per day for four weeks. This time, Norit A Supra charcoal, having a much higher surface area than the Carbomix charcoal used before, was employed. Serum cholesterol levels decreased by 27%. This was accompanied by a significant rise in serum squalene (55%) and demosterol (30%), and by a marked increase in lathosterol (175%) and delta-8-lathosterol (420%). These four compounds are all cholesterol precursors. Cholesterol decrease also was associated with decreases in serum cholestanol (39%), campesterol (24%), and β-sitos-

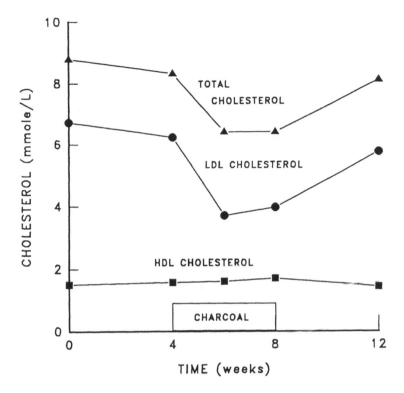

Figure 19.5 Effect of 4 weeks' intake of activated charcoal (8 g three times a day) on serum total, LDL, and HDL cholesterol concentrations in seven hypercholesterolemic patients. From Kuusisto et al. (1986). Copyright 1986 by The Lancet Ltd. Reproduced by permission.

terol (18%) for the first two weeks, but by four weeks time the concentrations of these three compounds had returned to their control levels. The authors concluded that charcoal causes increased cholesterol synthesis by interfering with the enterohepatic circulation of bile acids. Since cholesterol malabsorption is known to reduce campesterol and β-sitosterol and since these were essentially unchanged at four weeks, it is believed that cholesterol absorption was not strongly affected by the charcoal. Hence, the reduction in cholesterol must have been due to the adsorption of bile acids by charcoal (Krasopoulos et al., 1980, have shown that charcoal adsorbs bile acids well, at least in vitro, as mentioned earlier in this chapter). An interruption of the enterohepatic circulation of bile acids causes a feedback stimulation of bile acid synthesis, depletes hepatocyte cholesterol, and reduces serum LDL-cholesterol levels.

A third report from the Finnish group (Neuvonen et al., 1989b), described a study with seven patients who were given 4, 8, 16, or 32 g/day charcoal in three-week phases. Average total and LDL-cholesterol decreased by as much as 29 and 41%, respectively, and the HDL/LDL cholesterol ratio increased by as much as 121%. Ten more patients were treated in different ways for three-week phases as follows: 16 g/day charcoal, 16 g/day cholestyramine, and 8 g/day charcoal plus 8 g/day cholestyramine. These treatments reduced total and LDL-cholesterol as follows: 23 and 29%, 31 and 39%, and 30 and 38%, respectively. The HDL/LDL ratio changed from 0.13 to values of 0.23, 0.29, and 0.25, respectively. Thus, the cholestyramine and charcoal/cholestyramine treatments were somewhat better than charcoal alone. Triglycerides, however, were increased by cholestyramine but not by charcoal.

Park et al. (1988) also compared charcoal and cholestyramine for lowering cholesterol. Six patients with elevated cholesterol were treated with 20 g/day superactive charcoal or 16 g/day of cholestyramine (8 g twice/day) in a randomized cross-over trial. Each phase lasted three weeks, with a one week control period in between. The charcoal and cholestyramine reduced total cholesterol by 21.8 and 16.2%, respectively. Side effects were mild and similar for both treatments.

A final study on the use of charcoal and cholestyramine in lowering cholesterol is that of Tishler et al. (1987). This study was reported in abstract form by Bell et al. (1986) a year earlier. In vitro adsorption studies showed maximal adsorption capacities (Q_m values, determined by use of the Langmuir adsorption isotherm equation) for cholesterol adsorption from acidic solution by various adsorbents as follows: SuperChar charcoal, 277 mg/g; Norit USP charcoal, 33 mg/g; ActaChar charcoal, 26 mg/g; Mallinckrodt USP charcoal, 26 mg/g; Norit A charcoal, 22 mg/g; and cholestyramine, 0 mg/g. The Q_m values for the adsorption of the bile salt sodium desoxycholate from a pH 8.2 solution were: cholestyramine, 4161 mg/g; and SuperChar, 2814 mg/g. The effect of cholestyramine and SuperChar added to the diets of rabbits was then studied. The reductions in cholesterol were: 61% in one rabbit fed 1 wt% cholestyramine, 61 and 67% in two rabbits fed 1 wt% SuperChar, and 90% in one rabbit fed 2 wt% SuperChar. In WHHL homozygous rabbits (which lack cellular receptors for LDL's) the reductions from pre-treatment and post-treatment levels were: 52 and 38% for two rabbits fed 2 wt% cholestyramine, 70 and 43% for two rabbits fed 2 wt% SuperChar, and 70 and 63% for three rabbits fed 4 wt% SuperChar. Thus, they concluded that SuperChar charcoal appears to be an effective hypocholesterolemic agent in rabbits, warranting tests in humans.

Except in the study reported by Hoekstra and Erkelens, oral charcoal has been moderately effective in lowering many blood lipids, particularly total cholesterol, LDL-cholesterol, and triglycerides. However, it does not appear to be a viable replacement for the many drugs which are currently available for the same purpose, as these drugs are easier to take and are more effective, in general (cholestyramine resin is also effective for lowering blood lipids, as will be discussed in Chapter 22). It would be interesting if it could be determined why the results of the Dutch study are at odds with most of the other studies.

VII. EFFECTS IN LIVER DISEASE

Takahama et al. (1983) described the use of a spherical carbonaceous adsorbent called AST-120 (0.2–0.4 mm diameter), given orally, in the treatment of liver disease. Preliminary in vitro experiments showed that AST-120 was superior to "conventional charcoal" (type unstated) in the adsorption of toxic substances such as octopamine, dimethylamine, methylguanidine, etc., in the presence of bile acids (the details of these in vitro tests are not given). In vivo tests with dogs which had received portacaval end-to-side anastomoses were then carried out. One group of dogs was given 10–20 g/day AST-120 in their diet for 1–2 wk starting 1 wk after the operation, then meat for 1 wk, then AST-120 again thereafter. Dogs not treated with AST-120 progressively lost body weight and died within

2–4 wk, but those receiving AST-120 experienced no significant weight loss during the time that AST was given. Moreover, their plasma bile acids, immunoreactive insulin, and amino acid levels decreased to almost normal levels when AST was given. In further experiments, rats underwent end-to-end porta-renal anastomoses and then were given a diet with 5 wt% AST. Ten days later the rats were sacrificed. Total bile acids markedly increased after the porta-renal shunt to 171 μmole/L versus 22.4 μmole/L in normal rats. In rats receiving the AST, the total bile acids rose only to 116 μmole/L. In the untreated and AST-treated rats, the following were noted as percentages of values in normal rats: liver DNA synthesis, 53 and 89%, respectively, and liver protein synthesis, 62 and 88%, respectively.

In clinical studies, AST was given over 3 months to 15 liver cirrhosis patients, four of whom had hepatic encephalopathy and nine of whom had itching, at doses of 5–10 g/day. Hepatic encephalopathy occurred less frequently or was diminished by the AST therapy (specific data are not given). In most patients, itching diminished after 4 wk of AST therapy. Abnormally high plasma bile acid levels became almost normal after 4–12 wk of AST. The AST therapy thus led to adsorption of hepatic metabolites in the gastrointestinal tract. Further, it is believed that during the AST therapy liver function itself improved, as blood NH_3 (which can not be adsorbed by AST) also declined.

Studies with powdered charcoal in this area would be extremely interesting, as would further work with AST. The one study we have discussed is certainly intriguing, but requires confirmation by another research group.

VIII. PROLONGATION OF ANIMAL LIFESPAN

We review four reports on the effects of charcoal in prolonging the lifespans of rats. One pair of reports deals with nephrectomized rats and the second pair deals with normal rats. Friedman et al. (1975) fed oxystarch (200 mg/day), charcoal (200 mg/day), or a charcoal/oxystarch mixture (200 mg/day of each) to nephrectomized rats. Rats given no sorbents lived 2.6–2.8 days after nephrectomy (depending on whether they were fed or starved), while rats receiving oxystarch lived an average of 3.3 days. Rats given charcoal alone lived 2.9 days (not significantly different from the untreated rats). Rats receiving both charcoal and oxystarch lived 4.2 days. The rise in blood urea nitrogen (BUN) which occurred after nephrectomy did not differ among the treatment groups. The mechanism of life prolongation due to oxystarch and charcoal/oxystarch was not determined in this study.

However, a follow-up study (Saltzman et al., 1976) determined the mechanism. In this study, the same protocol was used with rats but the doses of charcoal and oxystarch were increased five-fold to 1 g/day. In addition, another group of rats was fed sodium polystyrene sulfonate (Kayexalate) and sorbitol and still another group was given Kayexalate, sorbitol, charcoal, and oxystarch. Mean survival times were: 3.1 days (control), 4.1 days (charcoal), 4.8 days (oxystarch), 5.0 days (charcoal/oxystarch), 5.0 days (Kayexalate/sorbitol), and 5.7 days (Kayexalate, sorbitol, charcoal, oxystarch). It was observed that feeding oxystarch increased the fecal excretion of potassium and both charcoal and oxystarch mitigated the rise in blood urea nitrogen to some degree. The Kayexalate/sorbitol therapy decreased blood potassium levels. The prolongation of life was thus assumed to be due to a combination of lower BUN levels and lower potassium levels; however, in all likelihood, the reasons are not that simple.

Dramatic increases in the lifespans of normal rats fed activated charcoal have been reported by Frolkis et al. (1984). They hypothesized that, since toxic metabolites are believed to play a role in aging, the purification of digestive juices in the intestinal tract using activated charcoal ("enterosorption") could potentially remove such toxic substances from the animals' systems. They first noted, in a preliminary study, that rats fed charcoal (amount not stated) for a significant length of time (length not stated) developed arterial hypertension 4–6 weeks after the cessation of charcoal. The mean arterial pressures rose to 118 mm Hg versus 85 mm Hg prior to the study and the hypertension lasted for several months. These observations suggest that the feeding of charcoal "is linked not merely with the sorption of some physiologically active substance, but rather with the slowly evolving regulatory renal and hypothalamic rearrangements."

In the primary study, one group of 28-month-old rats was given charcoal in their diets for 10 days, followed by no charcoal for one month. This cycle was repeated until each animal died. Charcoal increased the mean lifespans at 50% mortality from 937 to 977 days, at 80% mortality from 972 to 1023 days, and at 100% mortality from 993 to 1055 days. The mean increases in the lifespans were 47.3, 41.4, and 43.7%, respectively. The maximal lifespan was increased by 34.4%. The animals demonstrated less marked age-related structural and ultrastructural changes in the liver, kidneys, myocardium, intestines, and pancreas, as compared to controls. For example, the charcoal-fed rats had no sclerosed renal glomeruli and heart myofibrosis was less marked. Other things noted in the charcoal-fed group were decreases in liver cytochrome P-450, blood triglycerides and cholesterol, cardiac and cerebral tissue cholesterol, total lipids, and liver cholesterol and lipids. The charcoal increased RNA and protein biosynthesis in the kidneys, livers, and adrenal glands. In one group that was fed charcoal in cycles of one month of charcoal plus 10 days of no charcoal, the lifespan increases were much lower than those noted above; hence, there appears to be an optimal schedule of feeding the charcoal.

A second paper by Frolkis et al. (1989), although coming five years after publication of their first report, presented essentially the same results. Thus, it must be assumed that this paper, although shorter than the first paper, is merely a condensation and rewrite of the first paper.

Further research is needed to determine if the dramatic effects of enteric activated charcoal are due to the adsorption of known or supposed metabolites, or due to changes in the amounts of physiologically active substances and subsequent regulatory transformations, or both.

IX. SUMMARY

In this chapter, we considered the in vivo effectiveness of charcoal in binding (and thereby normalizing) the excess amounts of various endogenous biochemicals (bile salts, bilirubin, nitrogenous metabolites, lipids) which occur in disease states. Bile acids and their salts bind well to charcoal in vitro and this may explain the generally good in vivo effects of charcoal in treating pruritus (two separate studies showed 91% success in uremic patients and 44% success in pregnant women suffering from cholestasis).

On the other hand, success in lowering elevated serum bilirubin levels in jaundiced patients using oral charcoal has been quite mixed in nature. Charcoal appears to help when given in a sufficient amount or when combined with phototherapy, but the benefits have generally been modest. The effectiveness of charcoal in treating erythropoietic porphyria has also been quite variable: in some patients, lasting benefits have been achieved, while in other patients a rapid decline in serum porphyrins has been followed by a steady rise back to pretreatment levels.

The efficacy of charcoal in lowering serum levels of nitrogenous metabolites in uremic patients has been poor, especially with respect to urea (which simply does not adsorb to charcoal significantly), and of course charcoal has no effect on the large electrolyte imbalances which occur in uremia.

Despite these partial or outright failures, charcoal has had some notable successes. There is little doubt that charcoal can significantly lower total cholesterol, LDL-cholesterol, and triglycerides in hyperlipidemic patients. Also, in dogs and humans suffering from liver disease, a carbonaceous adsorbent which is similar to charcoal has shown promise in alleviating many of the symptoms of the disease. Reports from Russia have shown a striking prolongation of lifespan in rats given a steady diet of charcoal, presumably due to the binding of a variety of endogenous toxins which are produced as byproducts of normal metabolic processes.

Clearly, before more definitive conclusions can be drawn, much more research must be done on the use of charcoal in all of the areas mentioned. Well-designed clinical trials are urgently needed.

X.　REFERENCES

Amitai, Y., Regev, M., Arad, I., Peleg, O., and Boehnert, M. (1993). Treatment of neonatal hyperbilirubinemia with repetitive oral activated charcoal as an adjunct to phototherapy, J. Perinat. Med. 21, 189.

Bell, S. M., Tishler, P. V., and Winston, S. H. (1986). Treatment of casein-induced hypercholesterolemia in rabbits with highly activated charcoal (abstract), Clin. Res. 34, 282A.

Canby, J. P. (1965). Charcoal therapy of neonatal jaundice, Clin. Pediatr. 4, 178.

Davis, D. R., Yeary, R. A., and Lee, K. (1983a). Improved embryonic survival in the jaundiced female rat fed activated charcoal, Pediatr. Pharmacol. 3, 79.

Davis, D. R., Yeary, R. A., and Lee, K. (1983b). Activated charcoal decreases plasma bilirubin levels in the hyperbilirubinemic rat, Pediatr. Res. 17, 208.

Davis, D. R. and Yeary, R. A. (1987). Activated charcoal as an adjunct to phototherapy for neonatal jaundice, Dev. Pharmacol. Ther. 10, 12.

Denti, E., Gonella, M., Moriconi, L., Barsott, G., and Giovannetti, S. (1972). Attempts to remove uremic toxins through the gastrointestinal tract, in Uremia, R. Kluthe, G. Berlyne, and B. Burton, Eds., Georg Thieme Verlag, Stuttgart, p. 239.

Friedman, E. A., Fastook, J., Beyer, M. M., Rattazzi, T., and Josephson, A. S. (1974). Potassium and nitrogen binding in the human gut by ingested oxidized starch, Trans. Am. Soc. Artif. Int. Organs XX, 161.

Friedman, E. A., Laungani, G. B., and Beyer, M. M. (1975). Life prolongation in nephrectomized rats fed oxidized starch and charcoal, Kidney Int. 7 (Suppl.), S-377.

Friedman, E. A. (1976). Sorbents in the management of uremia, Am. J. Med. 60, 614.

Friedman, E. A., Saltzman, M. J., Beyer, M. M., and Josephson, A. S. (1976). Combined oxystarch-charcoal trial in uremia: Sorbent-induced reduction in serum cholesterol, Kidney Int. 10 (Suppl.), S-273.

Friedman, E. A. (1977). Oral sorbents in uremia: Charcoal-induced reduction in plasma lipids, Am. J. Med. 62, 541.

Friedman, E. A., Feinstein, E. I., Beyer, M. M., Galonsky, R. S., and Hirsch, S. R. (1978a). Charcoal-induced lipid reduction in uremia, Kidney Int. 13 (Suppl. 8), S-170.

Friedman, E. A., Saltzman, M. J., Delano, B. G., and Beyer, M. M. (1978b). Reduction in hyperlipidemia in hemodialysis patients treated with charcoal and oxidized starch (oxystarch), Am. J. Clin. Nutrition *31*, 1903.

Friedman, E. A., Manis, T., Zeig, S., and Lum, G. Y. (1979). Hypolipidemic effect of low-dose oral charcoal in diabetic rats, Clin. Nephrol. *11*, 79.

Frolkis, V. V., et al. (1984). Enterosorption in prolonging old animal lifespan, Exp. Gerontol. *19*, 217.

Frolkis, V. V. et al. (1989). Effect of enterosorption on animal lifespan, Biomater., Art. Cells, Art. Organs *17*, 341.

Fusaro, R. M. (1981). A contraindication for the use of charcoal in uremic patients (letter), J. Am. Acad. Dermatol. *5*, 219.

Gandhi, S. N. and Pimstone, N. R. (1983). Charcoal far superior to cholestyramine in blocking enterohepatic circulation of porphyrins: A major therapeutic tool in porphyria (abstract), Gastroenterol. *84*, 1372.

Giordano, C., Esposito, R., and Pluvio, M. (1973). The effects of oxidized starch on blood and fecal nitrogen in uremia, Proc. Eur. Dial. Transpl. Assoc. *10*, 136.

Goldenhersh, K. K., Huang, W. D., Mason, N., and Sparks, R. E. (1976). Effect of micro-encapsulation on competitive adsorption in intestinal fluids, Kidney Int. *10* (Suppl.), S-251.

Gorchein, A., Chong, S. K. F., and Mowat, A. P. (1989). Oral activated charcoal in protoporphyria with liver damage, Br. J. Clin. Pharmacol. *27*, 703P.

Hift, R. J., Meissner, P. N., and Kirsch, R. E. (1993). The effect of oral activated charcoal on the course of congenital erythropoietic porphyria, Br. J. Dermatol. *129*, 14.

Hoekstra, J. B. L. and Erkelens, D. W. (1987). Effect of activated charcoal on hypercholesterolae-mia, Lancet, August 22, p. 455.

Hoekstra, J. B. L. and Erkelens, D. W. (1988). No effect of activated charcoal on hyperlipidaemia: A double-blind prospective trial, Neth. J. Med. *33*, 209.

Hoffman, A. and Levy, G. (1990). Kinetics of drug action in disease states. XXXIX. Effect of orally administered activated charcoal on the hypnotic activity of phenobarbital and the neurotoxicity of theophylline administered intravenously to rats with renal failure, Pharm. Res. *7*, 242.

Hoffman, A. (1992). Potential pharmacodynamic effect of charcoal on theophylline neurotoxicity in normal rats, Pharmacol. Biochem. Behav. *43*, 621.

Kaaja, R. J., Kontula, K. K., Räihä, A, and Laatikainen (1994). Treatment of cholestasis of pregnancy with peroral activated charcoal: A preliminary study, Scand. J. Gastroenterol. *29*, 178.

Krasopoulos, J. C., De Bari, V. A., and Needle, M. A. (1980). The adsorption of bile salts on activated carbon, Lipids *15*, 365.

Künzer, W., Schenck, W., and Vahlenkamp, H. (1963). [Bilirubin adsorption from duodenal fluid by charcoal], Klin. Wochenschr. *41*, 1108.

Künzer, W., Vahlenkamp, H., Jarre, W., and Fuss, W. (1964). [On the treatment of jaundice in the newborn with charcoal], Ann. Paediat. *203*, 247.

Kuusisto, P., Vapaatalo, H., Manninen, V., Huttunen, J. K., and Neuvonen, P. J. (1986). Effect of activated charcoal on hypercholesterolaemia, Lancet, August 16, p. 366.

Lauterberg, B. H., Taswell, H. F., Pineda, A. A., Dickson, E. R., Burgstaler, E. A., and Carlson, G. L. (1980). Treatment of pruritus of cholestasis by plasma perfusion through USP-charcoal-coated glass beads, Lancet, July 12, p. 53.

Lücking, T. and Künzer, W. (1966a). Adsorption von Darmbilirubin an Kohle, Klin. Wochenschr. *44*, 469.

Lücking, T. and Künzer, W. (1966b). [Charcoal treatment of neonatal jaundice. 2nd communica-tion], Ann. Paediat. *206*, 258.

Manis, T., Zeig, S., Feinstein, E. I., Lum, G., and Friedman, E. A. (1979). Oral sorbents in uremia and diabetes: Charcoal-induced hypolipidemia, Trans. Am. Soc. Artif. Intern. Organs *XXV*, 19.

Manis, T., Deutsch, J., Feinstein, E. I., Lum, G. Y., and Friedman, E. A. (1980). Charcoal sorbent-induced hypolipidemia in uremia and diabetes, Am. J. Clin. Nutr. *33*, 1485.

Maxwell, M. H., Gordon, A., and Greenbaum, M. (1972). Oral sorbents in medicine, in *Uremia*, R. Kluthe, G. Berlyne, and B. Burton, Eds., Georg Thieme Verlag, Stuttgart, p. 220.

Mukerji, S. K., Pimstone, N. R., Gandhi, S. N., and Tan, K. T. (1985). Biochemical diagnosis and monitoring therapeutic modulation of disease activity in an unusual case of congenital erythropoietic porphyria, Clin. Chem. *31*, 1946.

Neuvonen, P. J., Kuusisto, P., Manninen, V., Vapaatalo, H., and Miettinen, T. A. (1989a). The mechanism of the hypercholesterolaemic effect of activated charcoal, Eur. J. Clin. Invest. *19*, 251.

Neuvonen, P. J., Kuusisto, P., Vapaatalo, H., and Manninen, V. (1989b). Activated charcoal in the treatment of hypercholesterolaemia: Dose-response relationships and comparison with cholestyramine, Eur. J. Clin. Pharmacol. *37*, 225.

Park, G. D., Spector, R., and Kitt, T. M. (1988). Superactivated charcoal versus cholestyramine for cholesterol lowering: A randomized cross-over trial, J. Clin. Pharmacol. *28*, 416.

Pederson, J. A., Matter, B. J., Czerwinski, A. W., and Llach, F. (1980). Relief of idiopathic generalized pruritus in dialysis patients treated with oral activated charcoal, Annals Intern. Med. *93*, 446.

Pimstone, N. R., Gandhi, S., and Mukerji, S. K. (1985). Therapeutic efficacy of chronic charcoal therapy in photomutilating porphyria (abstract), Gastroenterol. *88*, 1684.

Pimstone, N. R., Mukerji, S., Chow, H., and Goldstein, L. (1986). Oral charcoal therapy for protoporphyric cirrhosis (abstract), Hepatology *6*, 1176.

Pimstone, N. R., Gandhi, S. N., and Mukerji, S. K. (1987). Therapeutic efficacy of oral charcoal in congenital erythropoietic porphyria, New Engl. J. Med. *316*, 390.

Saltzman, M. J., Beyer, M. M., and Friedman, E. A. (1976). Mechanism of life prolongation in nephrectomized rats treated with oxidized starch and charcoal, Kidney Int. *10* (Suppl.), S-343.

Sinclair, A., Griffin, D. D., Voreis, J. D., and Ash, S. R. (1979). Sorbent binding of urea and creatinine in a Roux-Y intestinal segment, Clin. Nephrol. *11*, 97.

Sparks, R. E. (1975). Gastrosorbents in the therapy of uremia: Inferences from intestinal loop dialysis, Kidney Int. (Suppl.) *7*, S-373.

Svensson, C. K. and Cali, T. J. (1981). Drug adsorption by charcoal (letter), Annals Intern. Med. *94*, 281.

Takahama, T. et al. (1983). Effect of oral adsorbent on chronic hepatic disturbance, Trans. Am. Soc. Artif. Intern. Organs *XXIX*, 704.

Thiers, B. H. (1981). A contraindication for the use of charcoal in uremic patients (reply to letter), J. Am. Acad. Dermatol. *5*, 219.

Tishler, P. V. and Winston, S. H. (1985a). Sorbent therapy of the porphyrias. IV. Adsorption of porphyrins by sorbents in vitro, Meth. Find. Exp. Clin. Pharmacol. *7*, 485.

Tishler, P. V. and Winston, S. H. (1985b). Sorbent therapy of the porphyrias. Oral sorbents for the acute porphyrias (abstract), Am. J. Hum. Gen. *37* (Suppl.), A19.

Tishler, P. V., Winston, S. H., and Bell, S. M. (1987). Correlative studies of the hypercholesterolemic effect of a highly activated charcoal, Meth. Find. Exp. Clin. Pharmacol. *9*, 799.

Tishler, P. V. and Winston, S. H. (1988). Congenital erythropoietic porphyria (CEP): Porphyrin economy during treatment with superactivated charcoal (abstract), Clin. Res. *36*, 379A.

Tishler, P. V. (1988). Oral charcoal therapy of congenital erythropoietic porphyria, Hepatology *8*, 183.

Tishler, P. V. and Winston, S. H. (1990). Rapid improvement in the chemical pathology of congenital erythropoietic porphyria with treatment with superactivated charcoal, Meth. Find. Exp. Clin. Pharmacol. *12*, 645.

Ulstrom, R. A. and Eisenklam, E. (1964). The enterohepatic shunting of bilirubin in the newborn infant: I. Use of oral activated charcoal to reduce normal serum bilirubin values, J. Pediatrics *65*, 27.

Winston, S. H. and Tishler, P. V. (1986). Sorbent therapy of the porphyrias. V. Adsorption of the porphyrin precursors delta-aminolevulinic acid and porphobilinogen by sorbents in vitro, Meth. Find. Exp. Clin. Pharmacol. *8*, 233.

Yatzidis, H. (1964). Researches sur l'épuration extra rénale à l'aide de charbon actif, Nephron *1*, 310.

Yatzidis, H. (1972). Activated charcoal rediscovered (letter), Br. Med. J., October 7, p. 51.

Yatzidis, H. and Oreopoulos, D. (1976). Early clinical trials with sorbents, Kidney Int. *10* (Suppl.), S-215.

20

Use of Charcoal to Treat Poisoning in Animals

Up to this point, we have considered the effects of activated charcoal in treating drug overdoses and poisonings in humans, and in altering the effects of abnormal accumulations of endogenous toxins (as in uremia, liver disease, pruritus, porphyria, etc.). in humans. However, as many veterinarians are aware, charcoal has a considerable history of use for treating animals and it is widely used for such purposes today. In this chapter, we review many studies concerning the use of charcoal for treating animals poisoned by plants, herbicides, pesticides, insecticides, fungal and algal toxins, and miscellaneous chemicals.

I. INCIDENCE OF ANIMAL POISONING AND GENERAL GUIDELINES FOR CHARCOAL USE

Buck and Bratich (1986) recommended the use of activated charcoal to treat animals exposed to a variety of natural toxins and man-made chemicals. They mention that, in 1984, the National Animal Poison Control Center (NAPCC; located in the College of Veterinary Medicine at the University of Illinois, Urbana) received nearly 10,000 inquiries. Approximately 100,000 animals were at risk, with half being cattle and one-quarter being swine. More than 6,000 dogs and more than 2,000 each of cats, horses, and sheep were also involved. Among the toxins and chemicals for which Buck and Bratich recommended charcoal are: antihelmintics (used to treat parasitic intestinal worms), avicides, herbicides, insecticides, fungicides, parasiticides, molluscacides, rodenticides, antimicrobials, feed additives (monensin and lasalocid), bacterial toxins, mycotoxins, zootoxins, rumen and other gastrointestinal toxins, and plant alkaloids. They also include antifreeze (ethylene glycol) and "gases of all kinds" in their list, but charcoal is not very effective for these.

Information from Dr. Buck (Director, NAPCC) indicates that the NAPCC receives about 100 calls per day, or about 30,000 calls per year, most of which involve pets (in terms of the numbers of animals involved, cattle and swine represent a majority, as suggested by the figures given above; however, in terms of the numbers of calls, pets predominate). The NAPCC recommends the use of activated charcoal in approximately half of the cases. Their recommendations also include using sorbitol, especially with the first charcoal dose, in a sorbitol:charcoal ratio of 2:1, since many animals (particularly pigs, for some reason) tend to "plug up" with charcoal if this is not done. Saline cathartics had been recommended in the past but were found to produce undesirably high magnesium/sodium levels in some cases. The NAPCC usually recommends the charcoal dose be 1 g/lb (2–2.5 g/kg) of body weight and that the charcoal be repeated every 6–8 hr or every 8–12 hr until a time equivalent to the half-life of the ingested substance (assuming the substance involved is known) has passed (Buck, 1994).

A detailed compilation of information on animal poisoning is contained in a report by Hornfeldt and Murphy (1992), which summarizes data compiled by the American Association of Poison Control Centers (AAPCC) relative to the incidence of poisonings in animals reported to the AAPCC in 1990. This report cites a total of 41,854 cases, of which 90.9% were accidental. Acute poisonings accounted for 90.2% of the cases. Ingestion was the most common route of exposure (76.7%). There were only 419 deaths (1.0%) and, of these, 91 were due to insecticides, 82 involved human prescription and OTC medications, 60 were due to rodenticides, 41 involved plants, 41 resulted from ethylene glycol, 12 were due to herbicides, and 10 involved fertilizers (the remaining 82 deaths were due to 16 other known substances and 66 unknown substances). It is clear that substances typically kept in storage sheds and garages (e.g., antifreeze, insecticides, herbicides, fertilizers) account for many of the cases (whether the rodenticides were consumed in a storage location or where they were placed to kill the rodents is unknown). Interestingly, in only 2.8% of the cases was activated charcoal used to treat the animals, despite the fact that 56.9% of the animals received some sort of therapy (the most common therapies being dilution, 24.1%, or an emetic, 10.1%).

With respect to charcoal, the 2.8% figure might seem unreasonably low until one recalls that a certain degree of voluntary cooperation from the "patient" is needed for its administration and that animals are unlikely to offer such cooperation. With respect to the 10.1% receiving an emetic, ipecac use was 0.9%, with the remainder of 9.2% being listed as "other." One might wonder what the "other" emetics were, since many historical emetics (salts, mustard, etc.) have long since been disreputed. Perhaps some of them were detergents.

II. VETERINARY ACTIVATED CHARCOAL PRODUCTS

A company called Vet-A-Mix, Shenandoah, IA, headed by Dr. W. Eugene Lloyd, markets three activated charcoal products for veterinary use in poisonings. The products are:

1. *ToxiBan Suspension*, containing 104 mg/mL MedChar charcoal and 62.5 mg/mL kaolin in an aqueous base (water plus "suspending agents" and preservatives).
2. *ToxiBan Suspension with Sorbitol*, which is essentially the same as Toxiban Suspension but with 100 mg/mL of Sorbitol Solution USP added (the charcoal content is listed as 100 mg/mL rather than the 104 mg/mL in the ToxiBan Suspension—why there is a slight difference is unclear).

3. *ToxiBan Granules*, which contain 47.5 wt% MedChar charcoal, 10 wt% kaolin, and 42.5 wt% wetting and dispersing agents (plus a small amount of sorbitol). Whatever binder is used to granulate the charcoal would be part of the 42.5 wt% also (it may be that the sorbitol performs this role). The granule form is easier to handle and less messy than powdered charcoal, and yet dissolves to give a finely powdered charcoal when added to water. It is recommended that 1 part of the granules (by volume) be mixed with roughly 5–7 parts of water (by volume) and the mixture shaken well for 10–30 seconds, prior to administering it to animals. Alternatively, the dry granules can be mixed in with the animals' feed when subacute or chronic toxicosis is involved.

MedChar is the name Vet-A-Mix has given to the charcoal involved, which is a Westvaco Nuchar powdered charcoal having a surface area in the range of 1400–1800 m^2/g (Lloyd, 1994). The recommended doses of the Vet-A-Mix products are:

1. For ToxiBan or ToxiBan with Sorbitol, 10–20 mL per kg of body weight for small animals, and 4–12 mL per kg of body weight for large animals. This would deliver 1–2 g/kg or 0.4–1.2 g/kg of charcoal (dry basis), respectively.
2. For ToxiBan Granules, 2–4 g/kg for small animals and 0.75–2 g/kg for large animals. This would deliver about 0.95–1.9 g/kg and 0.36–0.95 g/kg charcoal (dry basis), respectively.

Thus, a 100 lb (45.35 kg) animal (which would be regarded as being "small") would be given the equivalent of about 45–90 g of charcoal and a 1000 lb (453.5 kg) animal would be given the equivalent of about 160–540 g charcoal.

The inclusion of kaolin is based on the concept that it is "an intestinal protectant for inflamed GI tract mucosa" and is also effective for the adsorption of bacteria and endotoxins. The few studies reviewed in Sections IV and V of Chapter 24 do not suggest that kaolin is particularly effective for these purposes but it may be that a thorough examination of the veterinary literature would uncover some support for this notion.

Sorbitol is included in a significant amount in one of the ToxiBan products for its cathartic action (Vet-A-Mix does warn, appropriately, that sorbitol "should only be used intermittently during multiple-dose activated charcoal use"). Regarding the suspending/dispersing agents, it is believed that one of them is the surfactant Tween. As for the preservatives, propylene glycol was used at one time but was found to give a false positive in blood tests for ethylene glycol poisoning and was subsequently replaced by other preservatives (Lloyd, 1994).

III. COMMON SOURCES OF ANIMAL POISONS AND TOXINS

Poisoning in animals most frequently occurs with grazing animals (cattle, sheep) which eat plants containing toxins. Such poisoning can be frequent and a source of substantial economic loss (either through outright death of the animal or through very poor weight gain due to toxicity) in many parts of the world. It is particularly prevalent in tropical and subtropical parts of the world where a greater variety and profusion of toxic plants grow.

A second source of poisoning of grazing animals is due to their eating of grasses which have been treated with herbicides, pesticides, or insecticides. Still another source of poisoning to grazing types of animals is from toxic algae which grow in ponds (which often are their water supply).

In addition to poisoning which actually kills the animal involved, there is a serious problem with the buildup of toxic levels of pesticides, herbicides, or insecticides in animals such as chickens—a buildup which produces undesirable levels of the organic toxin in the meat and eggs, and causes poor weight gain. If the toxins build up to a high enough level, mortality can occur.

Stored grains, which are often used as feed for cattle and poultry, can produce a variety of toxins (aflatoxin B_1 being one of the more common and more potent ones). Such grains, when fed to cattle or poultry, can cause severe toxic reactions and death.

Another class of poisoning in animals is that due to the ingestion of poisons by household pets (dogs, cats). Examples of such cases are the ingestion of chemicals which are stored in sheds or garages—such as antifreeze (ethylene glycol)—and the ingestion of poisons that have been placed inside or outside of a home in order to kill off troublesome pests (rodents).

There have been several studies aimed at evaluating the effectiveness of activated charcoal in the treatment of animal poisonings from the causes just mentioned. We review a variety of such cases here.

IV. POISONING BY PLANTS

In 1982, Joubert and Schultz (1982a,b,c) did a series of studies of the activated charcoal treatment of poisoning of sheep and cattle by various substances found in plants which are indigenous to South Africa. In one study (1982a), sheep were given lethal doses of *Urgenea sanguinea* plant bulbs, which contain a bufadienolide cardiac glycoside called transvaalin. The sheep were given 2.25 g/kg of the ground-up plant bulbs; then, at the first clinical signs of poisoning, they were given 5 g/kg activated charcoal plus 1 g/kg potassium chloride in water via a stomach tube and a second identical charcoal-KCl dose 24 hr later. Eight out of ten treated sheep survived versus only two of ten control sheep.

Their next study (1982b) concerned poisoning of sheep and cattle by *Morea polystachya* plants, which also contain toxic bufadienolide cardiac glycosides. Three out of four sheep survived a plant dose of 1.25 g/kg when treated with one dose of 5 g/kg activated charcoal plus 1 g/kg KCl 4–6 hr after ingestion of the plant. None of four control sheep survived. Steers were given 0.75 g/kg, 1 g/kg, or 1.25 g/kg of the plant and no further treatment. All died. Six steers were given 1.25 g/kg of the plant and 5 g/kg, 10 g/kg, or 20 g/kg (two steers at each dose level) of barbecue charcoal 8–12 hr later; they all died. A third group of six steers were given 1.25 g/kg of the plant and 5 g/kg activated charcoal only (no KCl) 8–12 hr after the plant ingestion; all survived. This study shows that: (1) barbecue charcoal, which has almost no internal surface area because it is not activated, is ineffective, (2) activated charcoal was effective without added potassium chloride, and (3) activated charcoal is effective both in sheep and in cattle.

The third paper by Joubert and Schultz (1982c) was aimed at seeing whether activated charcoal doses of 5 g/kg (the amounts used in their previous studies) could be made smaller and still be effective against *Morea polystachya* poisoning. The approximate minimal effective activated charcoal dose, given to sheep 12 hr after the administration of a lethal dose of the toxic plant, was found to be approximately 2 g/kg. No potassium chloride was used.

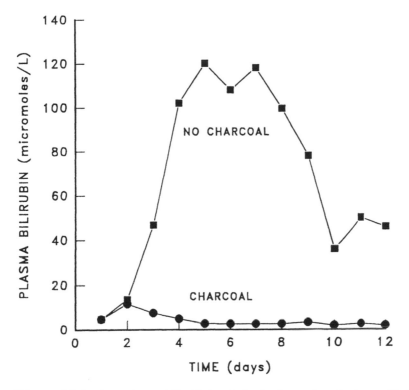

Figure 20.1 Mean plasma bilirubin levels in six sheep given lantana, with and without 500 g activated charcoal. From Pass and Stewart (1984). Copyright 1984, John Wiley & Sons, Ltd. Reproduced by permission.

Pass and Stewart (1984) studied the effect of charcoal in preventing poisoning in sheep by the plant *Lantana camara*, which grows in tropical and subtropical climates. Coastal regions of New South Wales and Queensland in Australia are areas where such poisoning is common, both in sheep and in cattle. The toxins of this plant produce rumen stasis, with most of the toxins being retained in the rumen, from which they are absorbed continuously and produce liver damage. Sheep were given 2 g/kg ground-up dried lantana, with or without 500 g Norit A activated charcoal. Figure 20.1 shows the results from this part of the study. The sheep that did not receive the charcoal all developed anorexia, jaundice, and elevated plasma bilirubin levels. Those that did receive the charcoal showed no such effects. Other test sheep were given charcoal six days after lantana doses. All of the animals recovered, whereas six of eight control sheep that received no charcoal died. Similar but less clear-cut results were observed in the same types of tests with cattle.

McLennan and Amos (1989) therefore decided to study the poisoning of cattle in Australia by *Lantana camara*, to clarify the results of Pass and Stewart. Four cattle with naturally occurring lantana poisoning were treated with 5 g/kg charcoal in water and two others were left untreated to serve as controls. The four which were treated right away recovered in 9–12 days and the two untreated animals recovered on days 20 and 22. Thus, charcoal significantly shortened the recovery period (however, some animals were also given electrolyte solutions, antibiotics, or antihistamines, so the effects due solely to the charcoal are not totally clear).

McKenzie (1991) has compared activated charcoal to bentonite as a treatment for *Lantana camara* poisoning in cattle. Each was given to calves, dosed 5 days previously with 5 g/kg of the plant material (as a slurry delivered through a stomach tube). Five of six calves in each group survived while five of six calves in an untreated control group did not. Calves given the bentonite took, on average, three days longer to recover clinically than did those given the charcoal (13 days versus 10 days). However, the plasma total bilirubin levels in the two groups were not statistically different. Thus, bentonite has promise as a cheaper alternative in this type of poisoning.

McKenzie and Dunster (1987) gave charcoal to calves after giving them 20 g/kg each (a lethal dose) of minced flower heads of the plant *Bryophyllum tubiflorum* ("mother-of-millions"). The flowers of this plant contain various bufadienolide cardiac glycosides. The charcoal was given as a slurry in electrolyte replacement fluid. When charcoal in the amount of 5 g/kg was given within 7 to 24 hr after the flowers, 9 of 11 calves survived.

Dollahite et al. (1973) reported that feed containing activated charcoal was used to prevent poisoning in sheep due to bitterweed ingestion. However, they concluded that the use of charcoal was not practical on ranches and would not make a significant difference.

Wright and Cordes (1971) found that the mycotoxin sporidesmin was adsorbed by activated charcoal in vitro and that in experiments with sheep the use of charcoal did reduce the toxic effects of sporidesmin. The extent of the protection was related to how much charcoal was used and the time interval between administration of the poison and the charcoal, as one would expect.

V. POISONING DUE TO HERBICIDES, PESTICIDES, AND INSECTICIDES

Cook and Wilson (1971) described the general problem of contamination of cattle and other ruminants (sheep, goats) with chlorinated hydrocarbon pesticides such as dieldrin, heptachlor, hexachlorobenzene, and DDT. Since these pesticides are highly soluble in lipids, they are stored in the adipose tissue of the animal following ingestion and are extremely difficult to remove. Urinary excretion is insignificant and fecal elimination is slow. The primary route of excretion is in the milk of a lactating animal.

Animals can come in contact with pesticides in various ways. Feed crops can be contaminated by direct application of pesticides or by drift from aerial or ground rig spraying. Feed crops can also be tainted by residual pesticides in the soil. To counteract pesticide contamination, one can use: (1) a chemical antidote such as phenobarbital which stimulates the body to metabolize the toxin at a faster rate (it was discovered in the 1950s that phenobarbital markedly induces the activity of toxin-metabolizing enzymes in liver microsomes, especially mixed-function oxidases), (2) a physiological antidote such as atropine, which counteracts poisons such as parathion which interfere with cholinesterases, or (3) mechanical antidotes which physically adsorb such toxins (charcoal being the primary such antidote).

Wilson et al. (1968) did experiments involving dieldrin given to sheep and goats. Dieldrin was introduced at 11 mg/kg into the rumens of the animals, some of whom had 3 g/kg charcoal placed into their rumens 1 hr prior to the dieldrin. Blood dieldrin levels were 5 times higher in the untreated group. Over three days, 2 and 22% of the dieldrin

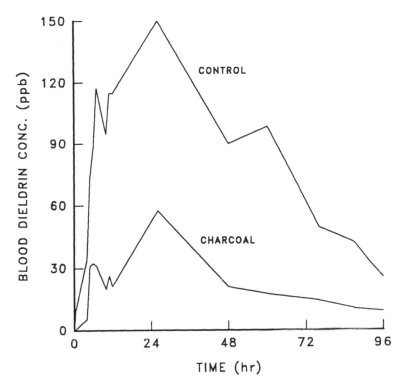

Figure 20.2 Effect of activated charcoal on blood dieldrin concentrations in goats. From Wilson and Cook (1970). Copyright 1970 American Chemical Society. Reprinted with permission.

doses were excreted in the feces of the animals in the control and charcoal-treated groups, respectively. Wilson and Cook (1970) studied goats, sheep, and heifers which were given dieldrin 1 hr after charcoal had been added to their rumens (the dieldrin doses were in the range of 5–15 mg/kg and the charcoal doses were in the range of 2.3–4.1 g/kg). Figures 20.2 and 20.3 show typical results on dieldrin concentrations in the blood and in the feces of goats, both with and without charcoal treatment. Both studies by Wilson et al. clearly showed that charcoal decreased the absorption of dieldrin and increased its appearance in the feces.

Several investigators have studied the effects of including both charcoal and phenobarbital in the diets of animals in order to try to remove pesticides already present in the tissues of such animals. We consider four studies here; one study involved hexachlorobenzene while the others all involved dieldrin. These studies showed that a regimen of charcoal/phenobarbital can accelerate the removal of pesticides from the tissues of animals. However, by giving charcoal alone and phenobarbital alone, it has been demonstrated fairly clearly that charcoal given after the animals tissues have been loaded with the pesticide has no significant ability to accelerate the removal of the pesticide. In contrast, phenobarbital given after loading of the animals tissues with the pesticide was able to increase the pesticide clearance.

Braund et al. (1970, 1971) mentioned that the sale of 4,000 lb/day of milk from a 105–cow dairy herd was prohibited because of dieldrin residues in the milk (the source

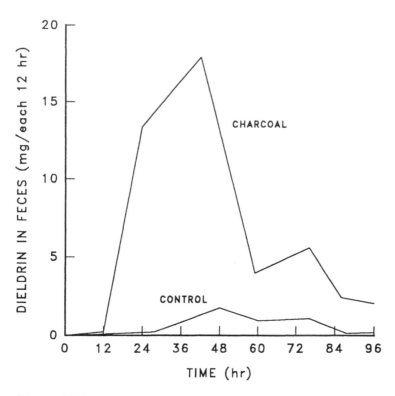

Figure 20.3 Effect of activated charcoal on dieldrin excretion in the feces of goats. From Wilson and Cook (1970). Copyright 1970 American Chemical Society. Reprinted with permission.

of the dieldrin was contaminated oats). The cows were divided into two groups and each animal in one group was given 10 mg/kg phenobarbital (about 5 g/day) for 24 days and 0.91 kg/day (2.0 lb/day) powdered charcoal (in the cows' corn silage diet) for 67 days. After one week, the dieldrin in the milk of the treated group had declined 64%, as compared to 36% for the control group. The dieldrin in the milk of the treated group fell to 50% of initial levels after 22 days. The phenobarbital/charcoal regimen shortened the time that the milk had to be withheld from market by at least one month.

Harr et al. (1974) described the case of 1013 steers which had become contaminated with dieldrin due to being fed with cull seed potatoes containing 0.1–0.2 ppm dieldrin. The animals' fat contained 0.76 ppm dieldrin, a level in excess of federal guidelines (0.30 ppm maximum). The animals were given phenobarbital (0.8–1.2 g/1000 lb) and charcoal (0.1–0.25 lb/1000 lb) daily. The rate of dieldrin loss from the animals was increased six-fold by this treatment and the residual dieldrin half-life was reduced from 150–250 days to 25–80 days.

Unglaub (1979) performed field tests on the acceleration of the excretion of the hexachlorobenzene (HCB) in dairy cows. He administered 500 g/day charcoal plus some thyroxine and phenobarbital for 30 days after the animals were dosed with the HCB. He found that this reduced the amount of HCB in the milk over the first 10 days by 16.7%, whereas in untreated cows the amount of HCB in the milk fell by only 2.7% over the same 10 day period.

Dieldrin was given by Engebretson and Davison (1971) to rats maintained for 6 wk on diets having 2 ppm dieldrin. After the 6 wk dieldrin loading period, the dieldrin was stopped and the rats were then given 0 or 1 g/day charcoal plus 0, 40, or 80 mg/kg per day phenobarbital in a 2x3 factorial experiment for an additional 2 wk. Urine and fecal samples were analyzed regularly. Carcass samples were analyzed at the end of the 2 wk charcoal period. In other experiments, the dieldrin and phenobarbital/charcoal were all given simultaneously for 6 wk. Charcoal effectively reduced dieldrin storage in the tissues when it was fed simultaneously with the dieldrin; however, when the dieldrin had already been loaded into the rats, charcoal was ineffective in reducing tissue dieldrin levels. The phenobarbital reduced dieldrin storage dramatically whether fed simultaneously with dieldrin or after dieldrin accumulation.

Hershberger et al. (1971) added 5 wt% charcoal to a ration of 43% apple pomace being fed to lambs; the apple pomace was contaminated with 2.3 ppm DDT. No significant effect on weight gain or feed efficiency (weight gain per weight of feed ingested) occurred compared to lambs not receiving charcoal. No effects on hematocrit or serum copper or iron were noted. The percent nitrogen, percent volatile fatty acids, and pH of the rumen fluid were unaffected. However, the protozoa count in the rumen fluid decreased from 2.2×10^6/mL to 1.19×10^6/mL when charcoal was used. DDT in the body fat of the lambs was apparently not measured.

Wilson et al. (1971) determined the effects of charcoal on growth rates and adipose concentrations of steers fed apple waste contaminated with dieldrin, DDT, DDE, and TDE (0.14, 0.40, 0.57, and 0.64 ppm, respectively). In the charcoal-treated group, charcoal (0.91 kg/day per animal) was mixed in with the feed, and so it was consumed at the same time as the contaminated feed. The charcoal had no significant effect on growth rates (i.e., weight gain) over 75–80 day periods. Adipose levels of dieldrin, DDE, and TDE were significantly reduced by including charcoal in the feed. DDT was not affected. In a follow-up experiment, steers which had been loaded with these pesticides were fed charcoal to try to remove them. Charcoal had no effect on the rate of decontamination versus controls. This study again confirms other evidence that charcoal is effective in reducing pesticide uptake when fed simultaneously with the pesticide but that, once the pesticide has been taken up into the fatty tissues, the use of charcoal has small or slight effects.

Smalley et al. (1971) found that the addition of 5 wt% activated charcoal to feed containing 1000 ppm of the organophosphorus compound ronnel (a systemic grub insecticide and larvicide) over an 84 day period decreased the ronnel residue levels in sheep to 10% of those in untreated animals. Later, Crookshank et al. (1972) dealt with assessing changes over 84 days in the well-being of sheep fed 5 wt% charcoal in their diet, 1000 ppm ronnel in their diet, or 5 wt% charcoal plus 1000 ppm ronnel in their diet, as determined by changes in blood serum composition, urine excretion, weight, feed consumption, and feed efficiency. None of the blood or urine results varied significantly from control results. Feed consumption decreased with added charcoal alone or with added ronnel alone but did not decrease further when they were combined in the diet. Feed efficiency expressed as kg weight gain/kg feed ingested was 0.068 for controls, 0.049 for added ronnel, and 0.064 for added charcoal plus ronnel. Thus, inclusion of charcoal with the ronnel brought the feed efficiency essentially back to the control level.

Rumsey et al. (1975) studied steers which had been fed ronnel. Ronnel (4.5 mg/kg) and charcoal (5 wt%) were added to their feed. Charcoal decreased the accumulation of ronnel in fat tissue by 31%, as compared to animals not fed any charcoal.

Crookshank and Smalley (1978) added 1 wt% charcoal to a diet containing 2.5 ppm dieldrin or 10 ppm each of benzene hexachloride, chlorophenothane, and heptachlor epoxide. When fed to sheep, the incorporation of the charcoal markedly reduced omental fat levels of dieldrin but did not significantly lower the fat levels of the other three compounds.

Kobel and Sumner (1985) studied cattle poisoned with the herbicide atrazine. Six heifers, about 2 years old, were given lethal doses of atrazine (500 mg/kg, by gavage). Three were then given 1 lb each of Nuchar S-A powdered charcoal. The three animals that did not receive any charcoal died at 48, 56, and 72 hr; the charcoal-treated animals all survived. The charcoal-treated animals had slightly increased body temperatures, higher pulse rates, and higher respiration rates for about 6 days after receiving atrazine, and exhibited salivation, ataxia, nervousness, diarrhea, and black feces between day 1 and day 7. Otherwise, they were normal.

A number of studies involved hens as the subject animals. Foster et al. (1972) studied the effects of charcoal in reducing residues of different pesticides in the eggs of hens. Chlordane (0.1 ppm), DDT (0.1 ppm), lindane (0.1 ppm), ethion (0.5 ppm), atrazine (0.5 ppm), and linuron (0.5 ppm) were added to the feed of laying hens. Chlordane, DDT, and lindane were subsequently detected in the yolks of the eggs, rising to maximal levels after 6 wk. Three weeks after resuming normal rations, residues either decreased to undetectable or to control levels. No residues of atrazine, linuron, or ethion were detected in the hens' eggs, abdominal fat, or minced tissues. Adding 1 wt% charcoal to the diets did not significantly affect the uptake of, or hasten the elimination of, the low levels of the pesticides fed in the diets. Neither charcoal nor the pesticides significantly affected mortality, weight gain, feed consumption, egg production, egg weight, or eggshell quality. Thus, it may be concluded that the pesticides involved, in the low quantities employed, had no significant deleterious effects. Since the pesticide levels were so low, it is not surprising that the addition of charcoal to the feed would have no observable effect.

Waibel et al. (1972) studied the effects of 300 ppm DDT in the diets of white leghorn hens, with and without 3 wt% charcoal, over a 30–day period. The charcoal reduced deposition of DDT in the hens' abdominal fat and reduced yolk pigmentation (by binding colored pigments and preventing their uptake into the yolk). However, hen fertility, egg weight, egg shell thickness, embryo development, hatchability, and progeny growth were not significantly affected by either the DDT or the charcoal. Had greater doses of DDT and charcoal been used and a longer test period been employed, perhaps significant effects would have occurred.

The contamination of chickens with dieldrin has occurred with some frequency. Mick et al. (1973) have studied the effects of adding phenobarbital and charcoal to the diets of laying hens given dieldrin. The chickens were administered capsules of 3 mg dieldrin orally every other day until six doses had been given (11 days). Some of these chickens received phenobarbital (100 mg) along with each dieldrin dose, while others had charcoal mixed with their feed over the first 11 days, in the amount of 1 g charcoal (Norit A) per pound of feed. Liver, fat, and brain tissue samples were analyzed from sacrificed chickens on days 19, 29, 83, and 105. Eggs were analyzed periodically. Dieldrin levels in the eggs peaked nine days after the end of dieldrin administration. Compared to controls, no statistical differences were noted in dieldrin residues in the eggs with the various treatments. Dieldrin in the liver tissue was less with the charcoal-fed chickens but amounts in the fatty tissue were unaffected by charcoal (no dieldrin was found in brain tissue in any samples). One problem with this study is the fact that 1 g charcoal per lb of feed represents only 0.22 wt%, a rather small amount.

Several studies have employed rodents such as rats, mice, and guinea pigs as test animals (we have already described above one study with rats, by Engebretson and Davison, 1971). Fries et al. (1970) gave rats 5 mg/kg DDT for 14 days. Other rats were given the same DDT but with 5 wt% charcoal in their feed. In this latter group, the retention of DDT in the rats' bodies after the 14 days was only about 27% versus about 68% for the untreated rats. Thus, DDT retention was reduced significantly. In another experiment, rats were given 5 mg/kg DDT for 14 days and a DDT-free diet for 14 more days. A second group was also given DDT for 14 days but then received a 5 wt% charcoal diet for the next 14 days. After the 28 day period, both groups of rats had retained about 62–63% of the DDT they had received. Thus, charcoal given after DDT loading had no effect in removing the DDT. These authors also studied cows which had been loaded with dieldrin, DDE, and DDT. The cows were then given 1 kg/day charcoal for 26 days. No significant reductions in the pesticide levels in the milk fat or body fat were noted. Milk production fell (from 20.4 to 18.5 kg/day) and milk fat percent fell slightly (from 4.0 to 3.9%) when charcoal was given. Charcoal had no effect on feed consumption, however.

Coccia et al. (1980) studied mice given a single oral dose (7.6 µg/kg) of tetrachloro-dibenzo-p-dioxin (TCDD) followed by feeding with standard chow or chow with 5 wt% charcoal. The mice were killed at 14 days and their livers were analyzed for TCDD. With charcoal-free chow, 17.3% of the TCDD dose was found in the livers, whereas with charcoal-laden chow only 6.3% of the TCDD dose appeared in the livers. The use of 4 wt% cholestyramine in the chow resulted in 13.1% of the TCDD dose in the livers. Thus, charcoal was more effective than cholestyramine in eliminating TCDD from the mice.

Four years later, Manara et al. (1984) described a study in which mice, rats, and guinea pigs were given single lethal parenteral doses of TCDD. Mortalities were greatly reduced by including 2.5 or 5 wt% activated charcoal in their diets immediately after the TCDD doses had been given. Mortalities during 60 days fell from 93 to 53% for the mice, from 80 to 50% for the rats, and from 64 to 25% for the guinea pigs.

An interesting study related to fish has been reported by Engstrom-Heg (1974). Rotenone is used in lakes and streams to kill unwanted fish. Once used, the problem is to remove or otherwise detoxify the rotenone so that desired fish species can be introduced. Potassium permanganate can counteract rotenone but has the disadvantages of coloring the water reddish-purple and, in high enough concentrations, is itself toxic to game fish species. Thus, Engstrom-Heg evaluated the use of charcoal added to streams as a way of removing rotenone. Preliminary in vitro tests showed that charcoal, in sufficient amounts, adsorbed rotenone well in batch tests if a contact time of about 15 min was allowed. Then, a 200–foot long section of a small stream was treated for 90 min with 0.8 ppm rotenone (as a 5% solution). Charcoal was introduced as a 10% slurry at the downstream end of the 200–ft section via a siphon drum and mixing chamber at a rate of 38 ppm for 150 min. The stream flow rate was 1.4 ft^3/sec and its mean velocity was 0.46 ft/sec. Fingerling rainbow trout were placed in liveboxes at points 300, 400, and 500 feet downstream of the charcoal introduction point. A complete kill of trout placed in the 200–foot section upstream of the charcoal introduction point occurred. Live fish fry were observed starting at about 180 feet downstream of the charcoal introduction point and all of the fish in the liveboxes at 300, 400, and 500 feet downstream of the charcoal introduction point survived. The stream bottom color varied from black at the charcoal introduction point to a dull gray 1,000 feet downstream; however, this color dissipated over a period of a few days. The discolored zone appeared to support normal insect fauna. Thus, the use of charcoal to clear a stream of toxic rotenone levels was successful in this test case.

Buck and Bratich (1986) compared the effects of three different charcoals for reducing mortality due to carbaryl (a carbamate insecticide), as well as by strychnine and T-2 toxin. Sixty female rats were given supralethal doses of each of these, followed by 9 mL/kg of charcoal suspensions (each having 104 mg charcoal per mL). The three charcoals were: Calgon, SuperChar Vet (a superactive charcoal, no longer available), and Toxiban (a suspension of 104 mg/mL charcoal made by Vet-A-Mix, Inc., Shenandoah, IA). The carbaryl, strychnine, and T-2 doses were 300 mg/kg, 65 mg/kg, and 25 mg/kg, respectively. Control tests gave 0/10 survival in all cases (mineral and castor oils were tried as treatments in ancillary tests, and also gave zero survival). For carbaryl, strychnine, and T-2 toxin, the Calgon charcoal gave survivals of 0/10, 4/10, and 0/10, while Toxiban gave survivals of 0/10, 3/10, and 0/10. These results are essentially identical. The survivals with SuperChar Vet were 7/10, 9/10, and 7/10, respectively. Hence, the superactive charcoal was quite effective.

VI. FUNGAL TOXINS FROM GRAINS

Aflatoxin B_1 is a deadly toxin produced by the fungus *Aspergillus flavus* which is commonly found growing as mold on corn, peanuts, and seeds such as cottonseed. Animals, and occasionally humans, have died as a result of eating such contaminated plants. Decker and Corby (1980) showed that activated charcoal can adsorb aflatoxin effectively in vitro. At pH 7.0, 1 mg of aflatoxin B_1 in 10 mL of solution was combined with 100 mg of Norit A activated charcoal. After adsorption equilibrium was achieved, no free aflatoxin B_1 could be detected in the solution.

Dalvi and McGowan (1984) mixed pure aflatoxin B_1 (AFB$_1$) into feed given to young chicks for 8 wk, at levels of 0, 2.5, 5, and 10 ppm. Reductions in weight gain and feed consumption were AFB$_1$-dose-dependent. Addition of 0.1 wt% activated charcoal into the feed of the chicks receiving 10 ppm AFB$_1$ caused a 10% reversal in feed consumption decline and a 28% reversal in weight gain suppression. Additionally, activated charcoal provided moderate protection against AFB$_1$-induced liver injury (as manifested by an elevation of serum glutamic oxaloacetic transaminase—SGOT).

Ademoyero and Dalvi (1983) studied the effects of charcoal in reducing the hepatotoxic effects of aflatoxin B_1 in chickens. Charcoal (200 mg/kg) was given orally with aflatoxin B_1 (6 mg/kg). After 72 hr the chickens were killed and blood and liver samples were analyzed. SGOT activity was 38% less, hepatic cytochrome P-450 content was 56% greater, and the rate of benzphetamine metabolism was 29% greater when the charcoal was given than when the charcoal was not given. All of these results demonstrate that charcoal can considerably reduce toxic liver injury by aflatoxin B_1.

Dalvi and Ademoyero (1984) observed improved chick performance when 200 ppm charcoal was added to diets contaminated with 10 ppm aflatoxin B_1, and hepatic microsomal activity was restored. They concluded that charcoal adsorbs aflatoxin B_1 and prevents its absorption from the GI tract.

Hatch et al. (1982) injected lethal doses (3 mg/kg) of aflatoxin B_1 intraruminally into 5 groups of male goats and then treated them with a slurry of activated charcoal in a phosphate buffer, with or without added stanozolol or oxytetracycline. The group receiving the activated charcoal slurry (8 hr after injection) all survived but showed signs of illness for 133 hr, on average. Those receiving the slurry plus stanozolol (8 hr after injection) all survived and were ill for only 19 hr, on average. Those receiving the slurry plus oxytetracycline (8 hr after injection) all survived and showed signs of illness for 44 hr, on average. The control group all died. Therefore, it was concluded that a charcoal slurry plus either stanozolol or oxytetracycline (but not both since they are antagonistic) is the best treatment. The amount of charcoal employed was 6.7 g/kg of Norit A activated charcoal as a 30% w/v slurry in a pH 7 phosphate buffer.

Bonna et al. (1991) described the use of hydrated sodium calcium aluminosilicate (HSCAS) and activated charcoal in treating mink exposed to aflatoxins in their diets. Mink were fed diets having 0 or 34 ppb aflatoxins, or 102 ppb aflatoxins with or without 0.5 wt% HSCAS and/or 1 wt% charcoal for 77 days (the HSCAS and charcoal were added only to the 102 ppb aflatoxin diets). The 34 ppb aflatoxin diet killed 20% of the mink and the 102 ppb diet killed 100% of the mink, within 53 days. Adding HSCAS to the diet entirely prevented mortality, whereas adding charcoal to the diet decreased mortality to 50% and increased the mean time to death from 35 days to 61 days. Thus, HSCAS was more effective than charcoal in this study.

Six reports dealing with T-2 toxin adsorption by activated charcoal have appeared. T-2 toxin, produced by *Fusarium* fungal molds, can contaminate grains and other foods, and is highly toxic to animals and humans. Galey et al. (1987) gave oral lethal doses (8 mg/kg) to rats along with superactive charcoal. The median effective charcoal dose in preventing death was 0.175 g/kg. Concurrent use of cathartics (sorbitol, magnesium sulfate, sodium sulfate) had no added effect. Doses of 1 g/kg charcoal increased survival times and rates as late as 3 hr after T-2 toxin administration. Figure 20.4 shows the results with the control and charcoal-treated rats.

Poppenga et al. (1987) gave 3.6 mg/kg T-2 toxin intravenously to swine (the IV LD_{50} is about 1.2 mg/kg) and then gave the swine a mixture of substances, including metoclopramide, charcoal, dexamethasone sodium phosphate, sodium bicarbonate, and normal saline. This regimen increased the mean survival time from 8.6 hr to 35.2 hr. They then looked at survival rates when each one of the ingredients was removed from the regimen separately. Removal of the charcoal had the largest effect. When charcoal was removed, the mean survival time was 18.0 hr. However, there were only three swine in each test group and lifetimes varied significantly. Thus, the differences observed, although in the expected directions, must be viewed with some caution.

Fricke and Jorge (1990) also gave T-2 toxin to rodents, in this case to mice, along with superactive charcoal. After giving 5 mg/kg T-2 toxin alone, only 6% of the mice survived after 72 hr. But, with the concurrent administration of 7 g/kg oral charcoal, 100% of the mice survived. With the charcoal given after 1 hr, the survival rate was 75%. Following parenteral exposure to T-2 toxin (2.8 mg/kg subcutaneously), untreated and charcoal-treated survival rates were 50 and 90%, respectively. The LD_{50} for T-2 toxin, determined at 96 hr, rose from 2 mg/kg to 4.5 mg/kg when charcoal was used. In vitro equilibrium adsorption tests, with application of the Langmuir equation to the data, showed a maximal binding capacity of 480 mg T-2/g charcoal (an excellent value).

Biehl et al. (1989) studied the effect of superactive charcoal paste in the prevention of T-2 toxin-induced local cutaneous effects in topically exposed swine. T-2 aliquots (6 mg dissolved in 90% dimethylsulfoxide) were applied to different skin sites 9 cm^2 in area. The charcoal paste was then applied at times ranging from 5 to 65 min. Skin lesions were examined at 1, 3, and 6 days. The charcoal paste significantly reduced lesion

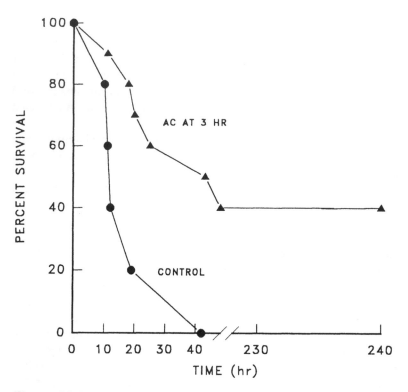

Figure 20.4 Proportion surviving versus time for rats given 8 mg/kg of T-2 toxin, with and without 1 g/kg of superactive charcoal at 3 hr. From Galey et al. (1987). Copyright 1987 Elsevier Science Ltd. Reprinted with kind permission of Elsevier Science Ltd, Kidlington, UK.

severity. However, charcoal preceded by a soap-and-water wash was even more effective. It seems reasonable to assume that the charcoal was able to bind the T-2 toxin effectively and reduce its absorption into the skin.

Bratich et al. (1990) gave T-2 toxin orally to female rats at 25 mg/kg (roughly six times the LD_{50}). Half of the rats received 0.936 mg/kg charcoal (SuperChar) in water, while the other half received the equivalent amount of water. A charcoal-treated rat was killed at the same time that a control rat died. Severe necrosis was seen in the spleen, thymus, stomach, small intestine, liver, and adrenal glands of the controls. Lesions were absent or minimal in the charcoal-treated rats. A related paper is that of Buck et al. (1985) in which rats were again given 25 mg/kg T-2 toxin and then three different charcoals. The 7-day survival rate for the control group was 0/10 and for the three charcoals were 0/10, 0/10, and 7/10. The 7/10 result was obtained with SuperChar charcoal.

Rotter et al. (1989) studied the ability of activated charcoal to adsorb ochratoxin A (OA), a potent nephrotoxin produced in stored grains by certain *Aspergillus* and *Penicillium* spp. of fungi, in vitro, and to reduce the toxic effects of OA in vivo when added to the diet of growing Leghorn chicks. In vitro, 50 mg charcoal adsorbed 90% of a 150 μg dose of OA from pH 7.0 buffer. In vivo, adding charcoal up to 1 wt% to diets containing 4 ppm OA did not reduce OA toxicity. However, the chick diets contained significant amounts of tallow, which may have absorbed the OA and prevented the charcoal from adsorbing it. Charcoal might have been effective if tallow had been omitted from the diet, if OA had been given as a single dose rather than being always present in the diet, or if a greater ratio of charcoal:OA had been used.

A report dealing with the use of adsorbents other than charcoal for preventing T-2 toxicosis in rats has been given by Carson and Smith (1983). The rats were fed diets containing 5 wt% of one of the following: bentonite, an anion exchange resin, a cation exchange resin, or a vermiculite/hydrobiotite mixture. Each diet was fed with or without 3 μg T-2 toxin per gram of feed for 2 wk. In other experiments, the bentonite and anion exchange resin were tested at 2.5, 7.5, and 10 wt% in the feed. Growth depression and feed refusal were monitored in all tests. Bentonite at 10 wt% was the most effective.

VII.   TOXINS FROM ALGAE

Kerr et al. (1987) studied toxicosis in five Holstein cows caused by ingestion of water from a pond containing the blue-green algae *Microcystis aeruginosa*. Toxins from these algae produce ataxia, muscle tremors, recumbency, and bloody diarrhea. The cows were treated with activated charcoal, procaine, penicillin, glucose, and calcium/magnesium gluconate. All cows were clinically normal ten days later.

The important toxin in blue-green algae is microcystin-LR, a cyclic heptapeptide. This toxin was studied by Mereish and Solow (1989) to determine how well SuperChar charcoal adsorbs this toxin in vitro and in vivo. In vitro, adsorption from a distilled water solution having a trace (< 1%) of ethanol showed a maximal binding capacity of 0.692 mmole toxin/g charcoal. This is an excellent Q_m value. In vivo experiments were done with mice using 75 μg/kg intraperitoneal doses or 5 mg/kg oral doses of the toxin. Administration of an oral charcoal dose of 10 mg per mouse did not prevent 100% mortality of the mice in all cases within 1–2 hr, due to massive intrahepatic hemorrhage. Mice who were given an oral dose of microcystin-LR mixed with SuperChar prior to its administration had 100% survival, although their liver weights did become significantly greater than those of control animals (indicating some degree of bleeding in the liver). One might wonder, in the in vivo tests in which all of the mice died, what the results might have been if a charcoal dose larger than 10 mg (which, even for a mouse, is rather small) had been employed.

Comments on algae poisoning received by the writer from Dr. William Buck indicate that algae poisoning is somewhat like cyanide poisoning in that death often occurs very rapidly. Quite commonly, the animals simply "die right on the banks of the pond." Thus, for activated charcoal to be of value in counteracting algal toxins, it would have to be administered before or very soon after ingestion of the toxins (Buck, 1994).

VIII.   OTHER TOXINS

Mice have been used in two other tests dealing with the use of charcoal to counteract toxins. Rivera et al. (1993) dealt with potent neurotoxins produced in shellfish, usually caused by the shellfish feeding on certain species of marine dinoflagellates (human intoxications due to contaminated shellfish harvested during "red tide" periods have been reported from time to time in the popular press). The toxins used in this study included a toxic clam extract (CE), saxitoxin (STX), tetrotoxin (TTX), and brevetoxin (PbTx-3). First, the minimum lethal dose of each was determined in the mice for each toxin. In "prophylaxis" charcoal studies, the mice were given 10 mg Norit A charcoal in water orally and then were given lethal doses of various sizes of each toxin 15–30 min later. In "therapy" charcoal studies, the mice were given the lethal toxin doses of various sizes first, followed by the 10 mg charcoal in water orally within 5 min. The animals were monitored for 7 days. Control animals died in all cases. For CE and TTX, charcoal given either before the toxin (in the prophylaxis tests) or after the toxin (in the therapy tests) resulted in zero deaths. For STX, 1 of 5 mice died with either protocol. For PbTx-3, no mice died in the prophylaxis tests but all died in the therapy tests. The results show that prior charcoal administration is extremely effective and that administration of charcoal within 1 min is effective unless the toxin is very fast-acting (as in the case of PbTx-3).

The second toxin study dealing with mice is that of Gomez et al. (1992). Food contaminated with various doses of toxin from *Clostridium botulinum* (type A toxin) was homogenized in a pH 6.2 phosphate buffer. One group of mice was injected intraperitoneally with 0.5 mL of the food/toxin mixtures. In other tests, charcoal was added to the food/toxin mixtures in the amount of "20% concentration" (whether this means 20 g charcoal/100 mL or 20 g charcoal per 100 g is not clear). Then, the mixtures were centrifuged and 0.5 mL of the supernates were injected IP into the mice. The food/toxin mixtures which had been treated with charcoal (to adsorb the toxin) produced no deaths and no reactions (piloerection, labored breathing, paralysis). In contrast, the mice given the untreated food/toxin mixtures showed 100% piloerection at all toxin levels and 100% labored breathing at all toxin levels. At toxin dilutions of 1:10, 1:50, and 1:100, the incidences of paralysis were 10/10, 9/10, and 5/10, respectively. Incidences of death were almost the same, being 10/10, 9/10, and 4/10, respectively. Clearly, charcoal adsorbed the botulin toxin extremely effectively from the food/toxin mixtures.

IX. POISONING OF HOUSEHOLD PETS

In the introduction to this chapter, we mentioned information obtained from Dr. William Buck which indicated that the majority of the roughly 30,000 calls per year currently received by the NAPCC involve pets (Buck, 1994).

Two specific studies of poisoning in dogs will be described here. In the first (Waters et al., 1992), a Doberman Pinscher who had ingested mole bait containing thallium was successfully treated with activated charcoal as well as with antibiotics, warm-water enemas, and fluids. The second study (Hansen et al., 1992) described a dog who had ingested 850 levothyroxine sodium tablets. Fortunately, the dog subsequently vomited at least three times. The dog was then given a mixture of 23 g charcoal and magnesium sulfate (125 mg/kg) diluted in water. Serum T4 levels were followed for 36 days. The initial T4 level was 4,900 nmol/L and decreased progressively but was not in the normal range (5 to 27 nmol/L) until day 36. Serum T3 levels normalized by the sixth day. In general, the dog showed few outward signs of thyroid hormone toxicosis. One mitigating factor is that dogs are known to absorb thyroxine only to the extent of 10–50%.

X. SUMMARY

The studies we have reviewed in this chapter have shown that charcoal given after the poisoning of animals by plant toxins has been quite effective in reducing morbidity and mortality. However, with more potent toxins—such as shellfish toxins, botulin toxins, toxins from grains (aflatoxin B_1, T-2 toxin), and algal toxins—charcoal must be given either before or simultaneously with the toxin. When the charcoal is delayed, these powerful toxins cause death before the charcoal can be delivered. However, in those cases when charcoal has been given in sufficient amount before or with such toxins, the charcoal has generally been very effective in reducing mortality. We may thus conclude that most toxins adsorb quite tenaciously to charcoal.

With respect to chronic poisoning by pesticides, herbicides, and insecticides, it is clear that once these chemicals become lodged in the tissues of an animal (particularly the adipose tissues), they are essentially inaccessible to removal by orally-fed charcoal. However, phenobarbital usually can induce their removal by accelerating metabolic enzyme activity.

XI. REFERENCES

Ademoyero, A. A. and Dalvi, R. R. (1983). Efficacy of activated charcoal and other agents in the reduction of hepatotoxic effects of a single dose of aflatoxin B_1 in chickens, Toxicol. Letters *16*, 153.

Biehl, M. L., Lambert, R. J., Haschek, W. M., Buck, W. B., and Schaeffer, D. J. (1989). Evaluation of a superactivated charcoal paste and detergent and water in prevention of T-2 toxin-induced local cutaneous effects in topically exposed swine, Fundam. Appl. Toxicol. *13*, 523.

Bonna, R. J., Aulerich, R. J., Bursian, S. J., Poppenga, R. H., Braselton, W. E., and Watson, G. L. (1991). Efficacy of hydrated sodium calcium aluminosilicate and activated charcoal in reducing the toxicity of dietary aflatoxin to mink, Arch. Environ. Contam. Toxicol. *20*, 441.

Bratich, P. M., Buck, W. B., and Haschek, W. M. (1990). Prevention of T-2 toxin-induced morphologic effects in the rat by highly activated charcoal, Arch. Toxicol. *64*, 251.

Braund, D. G., Langlois, B. E., Connor, D. J., and Moore, E. E. (1970). Acceleration of dieldrin residue removal in a contaminated dairy herd by feeding phenobarbital and activated carbon (abstract), J. Dairy Sci. *53*, 379.

Braund, D. G., Langlois, B. E., Conner, D. J., and Moore, E. E. (1971). Feeding phenobarbital and activated carbon to accelerate dieldrin removal in a contaminated dairy herd, J. Dairy Sci. *54*, 435.

Buck, W. B., Bratich, P. M., and Ernst, M. (1985). Experimental studies with activated charcoals and oils in preventing toxicoses (abstract), Vet. Hum. Toxicol. *27*, 311.

Buck, W. B. and Bratich, P. M. (1986). Activated charcoal: Preventing unnecessary death by poisoning, Vet. Med. *81*, 73.

Buck, W. B. (1994). Personal communication.

Carson, M. S. and Smith, T. K. (1983). Role of bentonite in prevention of T-2 toxicosis in rats, J. Animal Sci. *57*, 1498.

Coccia, P., Croci, T., and Manara, L. (1980). Less TCDD persists in liver 2 weeks after a single dose to mice fed chow with added charcoal or cholic acid, Br. J. Pharmacol. *72*, 181P.

Cook, R. M. and Wilson, K. A. (1971). Removal of pesticide residues from dairy cattle, J. Dairy Sci. *54*, 712.

Crookshank, H. R., Smalley, H. E., and Radeleff, R. D. (1972). Effect of prolonged oral administration of activated charcoal and of ronnel on the well-being of sheep, J. Animal Sci. *34*, 322.

Crookshank, H. R. and Smalley, H. E. (1978). Activated charcoal and the omental fat residues of BHC, DDT, dieldrin and heptachlor epoxide in sheep, Southwest. Vet. *31*, EN.

Dalvi, R. R. and Ademoyero, A. A. (1984). Toxic effects of aflatoxin B1 in chickens given feed contaminated with *Aspergillus flavus* and reduction of toxicity by activated charcoal and some chemical agents, Avian Dis. *28*, 61.

Dalvi, R. R. and McGowan, C. (1984). Experimental induction of chronic aflatoxicosis in chickens by purified aflatoxin B1 and its reversal by activated charcoal, phenobarbital, and reduced glutathione, Poultry Sci. *63*, 485.

Decker, W. J. and Corby, D. G. (1980). Activated charcoal adsorbs aflatoxin B1, Vet. Hum. Toxicol. *22*, 388.

Dollahite, J. W., Rowe, L. D., Kim, H. L., and Camp, B. J. (1973). Control of bitterweed (Hymenoxys odorata) poisoning in sheep, Tex. Agr. Exp. Sta. Progr. Rep. PR-3149.

Engebretson, K. A. and Davison, K. L. (1971). Dieldrin accumulation and excretion by rats fed phenobarbital and carbon, Bull. Environ. Contam. Toxicol. *6*, 391.

Engstrom-Heg, R. (1974). Use of powdered activated carbon to eliminate rotenone toxicity in streams, N. Y. Fish and Game J. *21*, 153.

Foster, T. S., Morley, H. V., Purkayastha, R., Greenhalgh, R., and Hunt, J. R. (1972). Residues in eggs and tissues of hens fed a ration containing low levels of pesticides with and without charcoal, J. Econ. Entomol. *65*, 982.

Fricke, R. F. and Jorge, J. (1990). Assessment of efficacy of activated charcoal for treatment of acute T-2 toxin poisoning, J. Toxicol.-Clin. Toxicol. *28*, 421.

Fries, G. F., Marrow, G. S., Jr., Gordon, C. H., Dryden, L. P., and Hartman, A. M. (1970). Effect of activated carbon on elimination of organochlorine pesticides from rats and cows, J. Dairy Sci. 53, 1632.

Galey, F. D., Lambert, R. J., Busse, M., and Buck, W. B. (1987). Therapeutic efficacy of super-active charcoal in rats exposed to oral lethal doses of T-2 toxin, Toxicon. *25*, 493.

Gomez, H. F., et al.(1992). Adsorption of botulinum toxin to activated charcoal using a mouse bioassay (abstract), Vet. Hum. Toxicol. *34*, 361.

Hansen, S. R., Timmons, S. P., and Dorman, D. C. (1992). Acute overdose of levothyroxine in a dog, J. Am. Vet. Med. Assoc. *200*, 1512.

Harr, J. R., Gillett, J. W., Exon, J. H., and Clark, D. E. (1974). A therapeutic regimen for purging dieldrin from adulterated individuals, Bull. Environ. Contam. Toxicol. *12*, 433.

Hatch, R. C., Clark, J. D., Jain, A. V., and Weiss, R. (1982). Induced acute aflatoxicosis in goats: Treatment with activated charcoal or dual combinations of oxytetracycline, stanozolol, and activated charcoal, Am. J. Vet. Res. *43*, 644.

Hershberger, T. V., Wilson, L. L., Chase, L. E., Rugh, M. C., and Varela-Alvarez, H. (1971). Effect of activated carbon on lamb performance and rumen parameters, J. Dairy Sci. *54*, 693.

Hornfeldt, C. S. and Murphy, M. J. (1992). Poisonings in animals: A 1990 report of the American Association of Poison Control Centers, Vet. Hum. Toxicol. *34*, 248.

Joubert, J. P. J. and Schultz, R. A. (1982a). The treatment of *Urginea sanguinea* Schinz poisoning in sheep with activated charcoal and potassium chloride, J. S. Afr. Vet. Assoc. *53*, 25.

Joubert, J. P. J. and Schultz, R. A. (1982b). The treatment of *Moraea polystachya* (Thunb) Ker-Gawl (cardiac glycoside) poisoning in sheep and cattle with activated charcoal and potassium chloride, J. S. Afr. Vet. Assoc. *53*, 249.

Joubert, J. P. J. and Schultz, R. A. (1982c). The minimal effective dose of activated charcoal in the treatment of sheep poisoned with the cardiac glycoside containing plant *Moraea polystachya* (Thunb) Ker-Gawl, J. S. Afr. Vet. Assoc. *53*, 265.

Kerr, L. A., McCoy, C. P., and Eaves, D. (1987). Blue-green algae toxicosis in five dairy cows, J. Am. Vet. Med. Assoc. *191*, 829.

Kobel, W., Sumner, D. D., Campbell, J. B., Hudson, D. B., and Johnson, J. L. (1985). Protective effect of activated charcoal in cattle poisoned with atrazine, Vet. Hum. Toxicol. *27*, 185.

Lloyd, W. E. (1994). Personal communication.

Manara, L., Coccia, P., and Croci, T. (1984). Prevention of TCDD toxicity in laboratory rodents by addition of charcoal or cholic acids to chow, Food Chem. Toxicol. *22*, 815.

McKenzie, R. A. and Dunster, P. J. (1987). Curing experimental *Bryophyllum tubiflorum* poisoning of cattle with activated carbon, electrolyte replacement solution, and antiarrhythmic drugs, Aust. Vet. J. *64*, 211.

McKenzie, R. A. (1991). Bentonite as therapy for *Lantana camara* poisoning of cattle, Aust. Vet. J. *68*, 146.

McLennan, M. W. and Amos, M. L. (1989). Treatment of lantana poisoning in cattle, Aust. Vet. J. *66*, 93.

Mereish, K. A. and Solow, R. (1989). Interaction of microcystin-LR with Superchar: Water decontamination and therapy, J. Toxicol.-Clin. Toxicol. *27*, 271.

Mick, D. L., Long, K. R., and Aldinger, S. M. (1973). The effects of dietary dieldrin on residues in eggs and tissues of laying hens and the effects of phenobarbital and charcoal on these residues, Bull. Environ. Contam. Toxicol. *9*, 197.

Pass, M. A. and Stewart, C. (1984). Administration of activated charcoal for the treatment of lantana poisoning of sheep and cattle, J. Appl. Toxicol. *4*, 267.

Poppenga, R. H., Lundeen, G. R., Beasley, V. R., and Buck, W. (1987). Assessment of a general therapeutic protocol for the treatment of acute T-2 toxicosis in swine, Vet. Hum. Toxicol. *29*, 237.

Rivera, V. R., Poli, M. A., and Hewetson, J. F. (1993). The effect of charcoal or specific antibodies in oral intoxication by marine toxins in mice, J. Nat. Toxins *1*, 47.

Rotter, R. G., Frohlich, A. A., and Marquardt, R. R. (1989). Influence of dietary charcoal on ochratoxin A toxicity in leghorn chicks, Can. J. Vet. Res. *53*, 449.

Rumsey, T. S., Williams, E. E., and Evans, A. D. (1975). Tissue residues, performance and ruminal and blood characteristics of steers fed ronnel and activated carbon, J. Animal Sci. *40*, 743.

Smalley, H. E., Crookshank, H. R., and Radeleff, R. D. (1971). Use of activated charcoal in preventing residues of ronnel in sheep, J. Agr. Food Chem. *19*, 331.

Unglaub, W. (1979). [Field test on the acceleration of pesticide excretion (hexachlorobenzene) in dairy cows], Arch. Lebensmittelhyg. *30*, 48.

Waibel, G. P., Speers, G. M., and Waibel, P. E. (1972). Effects of DDT and charcoal on performance of white leghorn hens, Poultry Sci. *51*, 1963.

Waters, C. B., Hawkins, E. C., and Knapp, D. W. (1992). Acute thallium toxicosis in a dog, J. Am. Vet. Med. Assoc. *201*, 883.

Wilson, K. A., Cook, R. M., and Emery, R. S. (1968). Effects of charcoal feeding on dieldrin excretion in ruminants (abstract), Fed. Proc. *27*, 558.

Wilson, K. A. and Cook, R. M. (1970). Metabolism of xenobiotics in ruminants: Use of activated carbon as an antidote for pesticide poisoning in ruminants, J. Agr. Food Chem. *18*, 437.

Wilson, L. L., et al. (1971). Effects of feeding activated carbon on growth rate and pesticide concentrations in adipose tissues of steers fed apple waste, J. Animal Sci. *33*, 1361.

Wright, D. E. and Cordes, D. O. (1971). Adsorption of sporidesmin by activated carbon and lignin, N. Z. J. Agr. Res. *15*, 321.

21

Reports from the Soviet Union

I. INTRODUCTION

There have been a number of reports from research groups in the former Soviet Union (primarily from Russia) in which peroral administration of carbonaceous adsorbents (a technique which is called enterosorption) was used to treat a remarkably wide range of diseases and disorders. Unfortunately, these reports are in Russian journals, in Russian, and are difficult to access. Thus, much of what will be reviewed here is based on English abstracts as obtained from Med-Line, a data base compiled by the U.S. National Library of Medicine. These abstracts provide few real details and thus these reports, when evaluated from such abstracts alone, must be viewed with some reservation. Persons who can access and read these papers may find them to be of substantial interest, although an evaluation of their validity may pose a problem.

However, there is a review paper in English by Nikolaev (1990) that covers many uses of enterosorption. He developed the "enterosorption" technique in about the late 1970s. The technique does not use ordinary powdered activated charcoal but relies on carbonaceous adsorbents made by the pyrolysis of synthetic resins. These resins are in the form of hard, black, spherical beads of 0.2–1.0 mm diameter. Carbonaceous adsorbents of this general type have been made in the USA by the Rohm and Haas Company (Philadelphia). Rohm and Haas developed these new adsorbents (called Ambersorb adsorbents) by pyrolyzing a variety of their own synthetic polystyrene resins (their Amberlite line of resins) and began marketing them in about 1977. The development, properties, and uses of these carbonaceous adsorbents have been described in detail by Neely and Isacoff (1982). They have properties which are intermediate between those of activated charcoal and those of unpyrolyzed polymeric adsorbents. Typical properties are shown in Table 21.1.

The U.S. mesh sizes of 20 and 50 correspond to diameters of 0.833 and 0.295 mm, respectively. Thus, the Ambersorb diameter range of about 0.3–0.8 mm is comparable to the size range of Nikolaev's enterosorbents (0.2–1.0 mm).

The largest difference in these three types of Ambersorb adsorbents is in their pore-size distributions. The XE-340 type is designed to remove nonpolar organics or halogenated organic molecules from aqueous or gas streams. The XE-347 type is like a

Table 21.1 Typical Properties of Ambersorb Adsorbents

	Ambersorb XE-340	Ambersorb XE-347	Ambersorb XE-348
Surface area (m^2/g)	400	350	500
Pore volume (cm^3/g)	0.34	0.41	0.58
Ash content (wt%)	< 0.5	< 0.5	< 0.5
Particle size (U.S. mesh series)	20–50	20–50	20–50
Bulk density (lb/ft^3)	37	43	37
Pore size distribution (%)			
< 6 Å	0	50	16
6–40 Å	18	0	21
40–100 Å	13	0	9
100–300 Å	69	50	51
> 300 Å	0	0	3

From *Technical Notes: Ambersorb Carbonaceous Adsorbents*, Rohm and Haas Company, Philadelphia, 1992. Used with permission.

macroporous molecular sieve, and is most often used to remove organic substances from air streams. The XE-348 material has the broadest adsorption spectrum of the three, and is the most similar to activated charcoal. It is designed to remove more polar organic compounds from either water or air.

The carbonaceous adsorbents used by Nikolaev are undoubtedly similar to these Ambersorb materials. He often used a type called SKN but has also used types called SUGS and SUA (these latter types have cation exchange capacities of 1–2 mEq/g). The doses involved with these resinous adsorbents are large, usually 100–150 g/day. Nikolaev stated that even with these large doses there are fewer side effects than with powdered charcoals. One benefit of the large doses is that they increase the undigested stool matter by a factor of 1.5–2.5, which enhances GI motility and moves toxins out of the body faster. Nikolaev also used "fibrous" types of adsorbents (called "Vaulen" and "Gastrosorb") at dosages of 3–6 g/day. What is meant by fibrous is unclear.

Nikolaev stated that the basic mechanisms of action of his enterosorbents are the adsorption of toxins (toxins ingested, toxins which diffuse or are secreted into the GI tract, and toxins formed in the GI tract), the modification of the GI tract chemical environment created by pathogenic microorganisms, and the aforementioned enhancement of the movement of toxins through the GI tract by virtue of the greater bulk of undigestible matter.

II. ENTEROSORPTION STUDIES DESCRIBED BY NIKOLAEV

Nikolaev reviewed his work on the use of enterosorption in treating viral hepatitis. In hepatitis B patients, enterosorption lowered total and direct bilirubin (to 56 and 34 mole/L, respectively, versus 99 and 63 mole/L in controls) after only 2 wk. (These concentrations are enormous; for example, 56 mole/L bilirubin would be 32,700 g/L! One possible explanation is that the concentration units should be mg/L or µmole/L).

Aspartate aminotransferase activity was reduced from 3.5 (controls) to 1.5. With hepatitis A patients, the results were also dramatic. For example, total and direct bilirubin were 29 and 16 mole/L versus 38 and 26 mole/L in controls (again, the concentration units can not possibly be mole/L). Nikolaev mentioned other Russian studies on the enterosorption treatment of hepatitis. In one, enterosorption relieved pruritus, reduced hyperbilirubinemia, and reduced transaminasemia in patients with severe hepatitis B.

Nikolaev also mentioned the use of enterosorption in the following diseases and disorders but gives no indication of the degree of success: neonatal hyperbilirubinemia, liver cirrhosis, portal cirrhosis, obstructive jaundice, cholecystoangiocholitis, cholecystitis, leptospirosis aggravated with hepato-renal syndrome, diabetes mellitus, alcohol abstinence syndrome, burn disease, and ischemic heart disease.

Other uses mentioned, again without any description of the enterosorption regimen involved and what success was achieved, include: acute pancreatitis, wound infection and sepsis, chronic osteomyelitis, the treatment of intestinal infections (salmonelloses, escherichioses, shigelloses), viral respiratory diseases, and immunodependent diseases (urticaria, food and drug allergies, and allergic dermatoses such as eczema, lupus erythematosus, psoriasis, and drug dermatitis). Studies involving patients with gout, hyperuricemia, atherosclerosis, and patients undergoing chemotherapy are mentioned (without elaboration). Various uses in conjunction with kidney diseases are mentioned: for supplementation of basic hemodialysis or hemoperfusion therapy, for acute uremic pruritus, and for lengthening the interdialysis period.

With respect to metabolic disorders, Nikolaev briefly described a study involving one month of enterosorption with SCN adsorbent (30 mL, three times/day) given to elderly patients having ischemic heart disease. The treatment lowered total lipids by 26%, triglycerides by 41%, and total cholesterol by 46% (this hypolipidemic effect is well known and was discussed in detail in Chapter 19). Nikolaev also mentions the work by Frolkis and others dealing with the prolongation of the lifespans of rats via enterosorption (this work is also described in Chapter 19).

Unfortunately, this review article by Nikolaev gives almost no details on the vast number of enterosorption studies referenced and, in particular, makes virtually no comments on whether any degree of success was achieved. The interested reader will have to check the references cited by Nikolaev and try to determine what results were obtained. In the interests of conserving space, these references will be omitted from the References section at the end of this chapter.

III.   SOME OTHER STUDIES

A review paper by Mamyrbaev and Takhtaev (1990) summarized the use of enterosorption as a general method of detoxification and may be consulted by those readers who can read Russian.

A short selection of some studies abstracted by Med-Line is briefly reviewed here. Dzhugostran et al. (1991) used enterosorption (30–60 g/day of SKNP-2 enterosorbent) combined with a fasting diet to treat patients with bronchial asthma and chronic obstructive bronchitis. The authors state that "the combination of the two treatment methods appeared effective." No further details are offered in the English abstract of this paper.

Barbas et al. (1991) used enterosorption for treating multiple sclerosis. Fifty-nine patients were give Vaulen three times per day in the amount of 50–60 mg/kg for 20 days. The relief of neurological symptoms began after 3–4 days, with continued decreases during the remainder of the treatment. There was "recovery of normal immunological spectra of serum parameters."

Myasthenia gravis was treated in 17 patients with the enterosorbent Vaulen, given for 20 days, by Klimova et al. (1991). Twelve of the patients showed appreciable improvement, as evidenced by the enhancement of the power of oculomotor, mimic, bulbar, respiratory, and bodily musculature. As a result, the doses of anticholinesterase drugs could be reduced by half in nine patients and totally discontinued in three.

Cerebral spinal fluid (CSF) hypertension in 40 patients was treated with enterosorption by Skoromets et al. (1991). The adsorbent was "a crumbled up filamentous carbonic adsorbent" at a dose of 50–60 mg/kg, three times per day for 20 days. Regression of neurological symptoms began after 5–6 days and regression continued throughout the 20-day period. CSF hypertension disappeared and there was "normalization of the immunological spectrum of the blood serum."

A vast number of reports from the USSR have dealt with the effects of charcoal hemoperfusion, as opposed to oral charcoal therapy, in the treatment of various conditions. We only describe one such hemoperfusion study here. Pinchuk et al. (1991), from the Ukrainian SSR Academy of Sciences in Kiev, investigated the use of charcoal for treating radiation disease. They cited a mortality rate in one group of lethally irradiated dogs ($N = 31$) of 97%, with an average lifespan of 13.4 days. However, in a second group of irradiated dogs who were given one charcoal hemoperfusion treatment 2 hr after irradiation, the mortality rate was 32% ($N = 19$) and the average lifespan was 17.0 days. Along with a higher survival rate in charcoal perfused dogs, "an impressive positive modification of their hematological indices was observed."

This completes our brief survey of enterosorption and hemoperfusion studies from the former USSR. There have been so many studies that a whole volume could be spent on them. The variety of diseases and disorders involved is astonishing, as are many of the results. The mechanisms by which such dramatic results have occurred are generally unclear, however. It would be interesting to see if similar studies done in other countries would yield comparable results. In the absence of confirmation of such work by research groups in other countries, the bulk of these Russian studies must be viewed with extreme caution. That is not to say that the writer questions the legitimacy of these reports, but only that the results are so striking that independent confirmations are greatly needed.

IV. REFERENCES

Barbas, I. M., et al. (1991). [Enterosorption in the combined treatment of patients with multiple sclerosis], Klin. Med. Mosk. *69*, 88.

Dzhugostran, V. I., Niamtsu, E. T., Zlepka, V. D., and Marchenko, I. G. (1991). [Enterosorption and therapeutic fasting in the treatment of patients with bronchial asthma], Klin. Med. Mosk. *69*, 54.

Klimova, T. T., Sanadze, A. G., Skoromets, A. A., and Khliustova, O. V. (1991). [Possibilities of using enterosorption in the treatment of myasthenia gravis], Zh. Nevropatol. Psikhiatr. *91*, 25.

Mamyrbaev, A. M. and Takhtaev, F. K. (1990). [Enterosorption as a method of detoxification (review of the literature)], Gig. Tr. Prof. Zabol. *3*, 40.

Neely, J. W. and Isacoff, E. G. (1982). *Carbonaceous Adsorbents for the Treatment of Ground and Surface Waters*, Marcel Dekker, New York.

Nikolaev, V. G. (1990). Peroral application of synthetic activated charcoal in USSR, Biomater., Art. Cells, Art. Org. *18*, 555.

Pinchuk, L. B., et al. (1991). Early experimental studies of sorption therapy of acute radiation disease (abstract), Artif. Organs *15*, 339.

Skoromets, A. A., et al. (1991). [Efferent methods of the treatment of cerebrospinal fluid hypertension in exacerbation of multiple sclerosis], Zh. Nevropatol. Psikhiatr. *91*, 23.

Technical Notes: Ambersorb Carbonaceous Adsorbents, Rohm and Haas Company, Philadelphia, 1992.

22

Resins and Clays as Sorbents

Resins and clays have potential for binding certain drugs and poisons but only under relatively specific conditions. Clays are cation-exchange materials; thus, they can bind species which are in cation form (e.g., basic drugs at pH levels low enough—as in the stomach-such that they are cations). The resins in common use (cholestyramine, colestipol) are anion-exchangers; thus, they can bind drugs or poisons which are in anion form (e.g., acidic drugs at pH levels which are high enough to ionize them). The pK_a values given in Chapter 5 give an indication of what pH levels are needed to make basic and acidic drugs ionize (a pH level 2 units away from a compound's pK_a is sufficient to make the compound exist either 99.01% undissociated or 99.01% dissociated).

However, in any real drug overdose or poisoning situation, one often does not know what kind of substance was ingested. Therefore, in general, activated charcoal should be used because its spectrum of the kinds of substances it can bind is much wider than that of clays or ion exchange resins. With this preamble, we will give a brief survey of studies of drug/poison binding involving these two specific types of sorbents. We will use the term "sorbent" in this chapter to refer generally to charcoals, resins, clays, and other materials (e.g., talc) which have been evaluated for drug/poison binding. Traditionally the term "sorbent" has been used to denote any material which binds or otherwise takes up a distinct chemical species by any kind of mechanism (ion exchange, adsorption, absorption).

I.   BASIC TYPES AND PROPERTIES OF RESINS USED AS ANTIDOTES

Cholestyramine is a strongly basic anion-exchange resin in the chloride form. It consists of a styrene-divinylbenzene copolymer with quaternary ammonium functional groups. The resin is made by emulsion polymerization and ends up as hard beads having a color roughly white to buff. The beads are sieved to a fine powder. Each gram exchanges 1.8–2.2 g of sodium glycolate, calculated on a dry resin basis.

Cholestyramine has been used widely to treat elevated blood lipids, especially cholesterol and lipoproteins. Cholestyramine has an affinity for bile salt anions and lowers cholesterol by binding them. The mechanism is as follows: bile acids (e.g., cholic acid) are synthesized in the liver from cholesterol and exist in their salt forms (e.g., sodium cholate) which are dissociated to give anions (e.g., the cholate anion). These anions can exchange with the Cl⁻ ions on cholestyramine, thus binding them by electrostatic attraction (a powerful binding mechanism). During normal digestion, bile acids are secreted into the intestines, from which major portions are absorbed and returned to the liver via the enterohepatic cycle. Cholestyramine binds the bile acid anions in the intestines to form an insoluble complex which is excreted in the feces. The increased fecal loss of bile acids causes an increased oxidation of cholesterol to bile acids (cholesterol is essentially the sole precursor of bile acids), a decrease in beta lipoprotein (low density lipoprotein) plasma levels, and a decrease in serum cholesterol levels.

In its use to lower elevated blood lipids, cholestyramine is often combined with drugs such as clofibrate, fenofibrate, bezafibrate, lovastatin, simvastatin, probucol, compactin, and gemfibrozil. Synergistic effects often occur with such combinations.

Cholestyramine is also used to relieve the pruritus associated with bile stasis which occurs in biliary cirrhosis and various forms of partial obstructive jaundice. In patients with such conditions, the reduction in serum bile acid levels reduces excess bile acids deposited in dermal tissues and thus decreases pruritus.

Cholestyramine may bind drugs given concurrently for therapeutic purposes (e.g., anticoagulants, antibiotics). In particular, it can bind acidic drugs in the small intestine region, because such drugs are ionized to the anion form at the higher pH levels which exist in the small intestine. Binding of drugs in the stomach is generally small, since at low pH acidic drugs are neutral and basic drugs are in cation form.

Other medical uses of cholestyramine besides its already-mentioned uses in patients with elevated blood lipids and pruritus include the treatment of erythropoietic porphyria, hyperbilirubinemia, diarrhea (acute bacterial diarrhea in infants, diarrhea caused by intestinal resection, etc.), gastritis (bile reflux-induced gastritis, postgastrectomy gastritis, etc.), colitis (especially that associated with antibiotics) and gastric or intestinal lesions/ulcers. Thompson (1971) provided an excellent review of many of these medical uses.

Among adverse effects due to cholestyramine therapy which have been reported include intestinal obstruction (see Lloyd-Still, 1977, or Cohen et al., 1969, for example), bleeding/hemorrhage associated with hypoprothrombinemia, hypernatremia, eosinophilia, hyperchloremic metabolic acidosis (especially in patients with insufficient renal function), and urethral calculi. However, the incidence of adverse sequelae is quite low.

Colestipol hydrochloride is also a bile acid sequestrant antilipemic agent, similar to cholestyramine. It is also an anion exchange resin, light yellow to orange in color. In this case the copolymer is one of diethylenetriamine and 1-chloro-2,3-epoxypropane. The polymer contains secondary and tertiary amines, with about 1 of 5 of the amine nitrogens protonated with chloride ions. A comparison of the binding capacities of cholestyramine and colestipol in terms of micromoles of taurocholate and glycocholate is shown in Table 22.1.

Thus, cholestyramine has about 14–20% higher capacity than colestipol. These values represent capacities in the range of about 0.38–0.57 g bile acid per gram of resin, which are substantial amounts.

Another resin, often used to treat hyperkalemia but useful also in treating excesses of other cations, is sodium polystyrene sulfonate (SPS). Two brand names for this resin are Kayexalate and Resonium-A. The resin is made by sulfonating a conventional polystyrene resin and exchanging the H⁺ cations of the sulfonic acid with Na⁺ ions, using strong NaCl solution. The resin is a golden brown, fine powder which is odorless and

Table 22.1 In Vitro Binding Capacities of Cholestyramine and Colestipol (μmoles/g dry resin)

	Cholestyramine	Colestipol
Taurocholate	1100	938
Glycocholate	913	825

From *AHFS Drug Information'92* (1992), p. 945.

tasteless. Each gram of the resin contains approximately 4.1 mEq of sodium. At the end of this chapter, we discuss some studies in which SPS was used to treat lithium intoxication.

II. BASIC TYPES AND PROPERTIES OF CLAYS

Clays are minerals which are predominantly hydrated silicates of aluminum, iron, or magnesium. Some are crystalline, while others are amorphous. They exist as very fine particles of colloidal size. We will discuss only crystalline types, as they are the ones which have been used as potential binding agents for toxins.

One group of important clays is the kaolins. This class includes kaolinite, dickite, and nacrite (all with the formula $Al_2O_3 \cdot 2SiO_2 \cdot H_2O$) , halloysite ($Al_2O_3 \cdot 2SiO_2 \cdot 2H_2O$), and other types. Another class of clays is the smectites. An important member of this class is montmorillonite, which has the formula $[Al_{1.76}Mg_{0.33}(Na_{0.33})]Si_4O_{10}(OH)_2$. The parentheses around $Na_{0.33}$ indicate that montmorillonite can exist with sodium ions or with magnesium ions. Fuller's earth, a widely used clay, is calcium montmorillonite. Most sorbent clays and bleaching clays are smectites, although some are of the attapulgite type (see later).

Bentonite is a rock rich in montmorillonite that has usually resulted from the alteration of intermediate siliceous types of clays. Attapulgite is a clay with a combination chain-sheet type structure and is capable of binding cations and neutral molecules. Its cation exchange capacity is about 20 mEq/100 g dry clay. A typical analysis shows it to be about 55.0% SiO_2, 10.2% Al_2O_3, 3.5% Fe_2O_3, 10.5% MgO, and 0.5% K_2O.

Another material which has sometimes been evaluated as a sorbent is talc, a mineral consisting of hydrated magnesium silicate (often with a small amount of aluminum silicate). It is a fine whitish powder and, in fact, when it is perfumed or medicated it is sold as the familiar item called talcum powder.

Crystalline clays are capable of cation exchange and thus are of potential use in binding drugs which are in the cation form. This means that under low pH conditions (as in the stomach), basic drugs, which are in cationic form, can be bound. However, when basic drugs pass into the more neutral pH or even mildly alkaline regions of the intestines, they become uncharged and can no longer bind to clays by an ion exchange mechanism; any such drug previously bound to clay in the stomach is likely to "desorb." Acidic drugs, of course, are never in cationic form (they are uncharged at low pH and are anionic at high pH) and thus would not be expected to be bound significantly by clays at any pH. Since a large fraction of drugs are acids and since basic drugs can desorb from clays under intestinal conditions, clays have limited potential for antidotal applications. Nevertheless, a variety of studies (mostly old) been carried out with clays, particularly in vitro type studies.

III. STUDIES INVOLVING CLAYS

Evcim and Barr (1955) studied the binding of three alkaloids (strychnine, atropine, and quinine) by different clays and found that attapulgite and halloysite were better binding agents than kaolinite. The clay dickite showed very poor binding.

Wai and Banker (1966) investigated the binding of four compounds in vitro by montmorillonite. Brucine was found to be bound both by adsorption and by ion exchange, methapyrilene and Et_3N were bound by ion exchange, and niacinamide was neither adsorbed nor ion exchanged.

Sorby and Plein (1961) examined the in vitro binding of 15 different phenothiazines from water solutions by kaolin, talc, and Norit A charcoal. The average Q_m values were 42 times higher for charcoal than for kaolin and 18 times higher for charcoal than for talc. Sorby et al. (1966) later showed that the binding of these drugs to kaolin and talc was fairly sensitive to pH, whereas adsorption to charcoal was much less affected by pH. Sorby (1965) also investigated the in vivo binding of promazine by both activated charcoal and by attapulgite. Very little decrease in the availability of promazine was found when promazine was mixed with attapulgite before administration; however, when activated charcoal was used, the bioavailability was significantly less than control (i.e., promazine alone) cases. In vitro studies, which were also done, indicated that charcoal has an affinity for promazine which is roughly four times that of attapulgite, per unit weight. We should point out that there is one significant incentive for the use of clays instead of charcoal, namely the fact that clay has a greater aesthetic appeal than charcoal. When mixed with water it is fluffy, tasteless, and well wetted; thus, it doesn't have a gritty texture nor does it tend to stick to the mucosal surfaces of the mouth and esophagus.

Weber et al. (1965) found that the herbicides diquat and paraquat were rapidly taken up and bound as organic cations by montmorillonite at pH 6.0, up to the exchange capacity of the clay (0.85 mEq/g). Neither temperature (10 and 55°C) or concentration in the fluid phase (up to 0.001 M) affected the uptake. X-ray analysis indicated that the herbicide cations were bound in the clay lattice with the planes of their rings parallel to the silicate sheets. On kaolinite, the two chemicals were bound to a much lower extent, about 0.05 mEq/g. This study also examined the binding of prometone and 2,4-D to the clays as well as the binding of all four herbicides to an anion exchange resin (Amberlite IR-120) and to activated charcoal.

Attapulgite was also studied by Atkinson and Azarnoff (1971). In vivo work was done with pentobarbital-dosed rats and it was shown that giving 250 mg charcoal 1 min later resulted in plasma and brain tissue drug levels, at 60 min time, which were only 27 and 13% of control values, respectively. However, when 250 mg attapulgite was given, the drug levels in the plasma and brain at 60 min were 84 and 61% of control values, respectively. Clearly, the charcoal was superior. Using rats dosed with Neguvon (a pesticide), charcoal and attapulgite were about equally effective in preventing depression of cholinesterase activity. However, when dogs were dosed with sodium salicylate and were then given either charcoal or attapulgite, the charcoal was much superior.

Whereas the charcoal decreased the peak plasma salicylate concentration by 40%, an equivalent amount of attapulgite had no effect.

Chin et al. (1969) compared Arizona montmorillonite against charcoal and evaporated milk for their in vivo effects in rats given kerosene, aspirin, or strychnine phosphate. The montmorillonite was nearly as good as charcoal in lowering plasma kerosene levels, equally as good in lowering plasma strychnine levels, and was nearly ineffective in the aspirin case. The good binding of strychnine is probably due to its being in a cationic form in solution. Since most clays are excellent cation exchangers, it is not surprising that aspirin, which exists as an anion in solution, was essentially not bound at all. The binding of kerosene, which is probably an undissociated compound in solution, must, however, occur by some adsorption mechanism, since ion exchange would not be possible.

These results also agree with those of Smith et al. (1967) who reported on the in vitro binding of several drugs from SGF and SIF to either activated charcoal or Alaskan montmorillonite. In gastric fluid, the charcoal bound 3 and 13 times more sodium salicylate and sodium pentobarbital, respectively, than did the montmorillonite. This is because both drugs exist as undissociated acids at the prevailing pH, and thus would not be ion-exchanged by the clay. On the other hand, it was found that two other drugs (tripelennamine and D-amphetamine) which are fully ionized as cations at gastric pH, were bound 89 and 18% better by the montmorillonite, as compared to an equal amount of charcoal. Finally, at pH 7.4, in SIF, the charcoal bound sodium salicylate nearly 70 times better than did the montmorillonite. At this pH, the salicylate exists mainly as anions and clearly the clay would have very little affinity for these. It is clear, therefore, that cationic clays have a special affinity only for drugs which would exist as cations under gastric conditions (such as alkaloids and amines). For nonionized drugs their affinity is generally less than that of charcoal (but, sometimes, not greatly so), and for drugs in anionic form their affinity is nearly nil.

Armstrong and Johns (1974) investigated the binding of morphine HCl by kaolin at 25°C and found that their data fit the Langmuir equation. At 0.1% morphine HCl in solution, 88% was bound (it was probably largely in cationic form, and thus was bound well by ion exchange). However, when 4.5% $NaHCO_3$ was added to the solution, only 18% of the morphine HCl was bound, presumably because the Na^+ cations competed strongly against it.

Brown and Juhl (1976) studied the effectiveness of kaolin plus pectin (Kaopectate, a mixture of 20% kaolin and 1% pectin in an aqueous medium) and of several antacids on digoxin pharmacokinetics in healthy volunteers. The subjects were given 0.75 mg of digoxin and then 60 mL of one of the following solutions: Kayopectate, 4% aluminum hydroxide gel suspension, 8% magnesium hydroxide gel, or magnesium trisilicate suspension. Blood samples were taken at various times up to 8 hr and urine was collected for 6 days. The 0–8 hr blood level AUC values in ng-min/mL were 559 in control tests, 414 with aluminum hydroxide, 418 with magnesium hydroxide, 349 with magnesium trisilicate, and 329 with Kaopectate. The percent of the drug recovered in the urine was 40.1% in controls, 30.7% with aluminum hydroxide, 27.1% with magnesium hydroxide, 29.1% with magnesium trisilicate, and 23.4% with Kaopectate. The reductions in bioavailability of digoxin with these different materials can not be explained solely by the physical adsorption of the drug, since Khalil (1974) has shown that digoxin adsorbs poorly to aluminum and magnesium hydroxides.

Albert et al. (1978) also studied the influence of kaolin-pectin on digoxin bioavailability. Healthy subjects were given two 0.25 mg digoxin tablets plus 90 mL of Kaopectate. In other tests, the Kaopectate was administered either 2 hr before or 2 hr after the digoxin. Blood AUC values for 0–48 hr showed that giving the drug and Kaopectate together reduced the drug bioavailability by 62%. Giving the Kaopectate 2 hr before the drug caused a 20% reduction and giving the Kaopectate 2 hr after the drug had no effect.

Juhl (1979) investigated the effects of a kaolin/pectin mixture versus activated charcoal for reducing aspirin absorption in ten human volunteers. The subjects were given three 325 mg aspirin tablets and then 30, 60, or 90 mL of Kaopectate, or 10 g of Norit A charcoal. Salicylate excretion in the urine was monitored for 48 hr. Whereas the charcoal reduced aspirin absorption from 98.6% (control) to 69.5%, the 30, 60, and 90 mL kaolin/pectin doses reduced the mean aspirin absorption to only 90.6, 94.6, and 95.3% (note that, in contrast to what might have been expected, the larger kaolin/pectin doses were less effective). Thus, kaolin/pectin had an insignificant effect on aspirin absorption.

The in vitro binding of acetohexamide, tolazamide, and tolbutamide at pH 7.4 by kaolin as well as by charcoal was studied by Said and Al-Shora (1980). By fitting the data to the Langmuir equation, maximal binding capacities, Q_m, were determined. The Q_m values for these drugs were (in the order mentioned): 692, 502, and 501 mg/g for charcoal, and 653, 481, and 456 mg/g for kaolin. Thus, the kaolin was very nearly (91–96%) as effective as charcoal.

Okonek et al. (1982a) studied the binding of paraquat to charcoal, bentonite clay, and Fuller's earth in vitro using SGF as the medium. Paraquat is a herbicide which is extremely toxic. Paraquat in the amount of 10, 20, 30, 40, 50, 60, 70, or 80 mg in 5 mL of SGF was mixed with 1 g of sorbent in a flask, and the mixture agitated until binding equilibrium was reached. At a paraquat dose of 80 mg, for example, the amount of unbound paraquat was determined to be: for charcoal, 0.75 mg; for one bentonite, 18 mg; for a second bentonite, 23 mg; for a third bentonite, 27 mg; and for Fuller's earth, 23 mg. Thus charcoal was superior, in general, and the bentonites and the Fuller's earth were about the same. In vivo tests were done with rats, who were given 40 mg paraquat alone or 40 mg paraquat which had previously been incubated with 1 g of sorbent. Mortality over 14 days was 6/6 for controls, and 0/6 when any of the sorbents were used (charcoal, bentonite, Fuller's earth). Additional in vivo tests were done in which rats were given 200 or 300 mg/kg paraquat and 1-g doses of sorbent at 0.5, 1, 2, and 3 hr. Charcoal was the most effective sorbent in these tests, with bentonite being the next most effective, and Fuller's earth the least effective. These same results, plus additional data on the effect of hemoperfusion in paraquat poisoning, were given in another paper by Okonek et al. (1982b).

The potential of montmorillonite (Wyoming bentonite) for binding atrazine was studied by Browne et al. (1980). As mentioned earlier, clays such as montmorillonite have a high binding capacity for cations but do not interact strongly with neutral molecules. Thus, weak bases, which constitute a large chemical class of potential poisons, are well bound by montmorillonite if the pK_a is high so that the protonated form of the weak base is present in the pH range encountered in the gastrointestinal tract. It was hypothesized that montmorillonite which was treated to replace the naturally occurring inorganic cations with an organic exchangeable cation (in this study, 3-hydroxypropyl-ammonium cations) would have a desirable effect on the acid-base equilibrium of weak bases (in this study, atrazine). Indeed, this study showed that this was the case. The ability of bentonite USP, sodium-saturated bentonite, and 3-hydroxypropylammonium-saturated (3-HPA) bentonite to bind atrazine as a function of pH showed that having

3-HPA cations on the bentonite caused atrazine to exist in nonionized form over a much wider range of pH values than otherwise. In this nonionized form, it bound much more strongly to the clay.

The binding of various drugs to montmorillonite in vitro was studied by McGinity and Lach (1976). Cationic drugs (chlorpheniramine maleate, amphetamine sulfate, and propoxyphene HCl) generally bound quite strongly to the clay. Amphoteric substances (caffeine, theophylline) bound moderately well and anionic drugs (sodium salicylate, sulfanilamide) bound poorly. The proposed mechanism for binding involves a two-step process: a cation-exchange reaction followed by strong surface chemisorption.

The in vitro binding of phenethylamines to bentonite clay and to a resin (Lewatite resin SP, Merck, Germany) was examined by Stul et al. (1984). The studies were done in pH 4 and pH 5 solutions. The binding was essentially by ion exchange up to a solid phase loading of about 0.8 mEq/g. The organic cations bound more strongly on the bentonite than on the lewatite.

IV. STUDIES INVOLVING RESINS

We will not be concerned with the use of resins alone to lower blood lipids (cholesterol, lipoproteins) and to treat conditions related to hepatic disturbances (pruritus, jaundice, porphyria, cholestasis), since a separate volume could be written on such (the literature by now concerning these applications, particularly in hyperlipidemia, is very large). The reader may recall that in Chapter 19 we did give consideration to such disorders; however, all of the studies discussed in that chapter involved activated charcoal as one of the sorbents (if not the only one) evaluated.

We restrict our discussion here to general investigations of the binding of drugs by resins and to the use of cholestyramine/colestipol in two types of studies: (1) those in which cholestyramine/colestipol have been used as oral antidotes to counteract toxic levels of drugs/poisons and (2) those aimed at determining whether cholestyramine/colestipol interfere with, or reduce, the beneficial effects of drugs taken concomitantly in therapeutic amounts. We also review a few studies on the effects of cholestyramine/colestipol on bacterial toxins and on endogenous chemicals such as amino acids, vitamins, and hormones.

A. General Studies of Ion Exchange Resins

Edwards and McCredie (1967) studied the in vitro binding of 17 different drugs, from plain water, to nine kinds of ion exchange resins and one type of activated charcoal. The acidic drugs were in the sodium salt form (e.g., sodium phenobarbital, NaBr, NaI, etc.). Table 22.2 shows results for one of the anion exchange resins (cholestyramine), one of the cation exchange resins (Katonium), and for the charcoal. The results in Table 22.2, as well as all of the other results, clearly show that the cation exchange resins were much superior to the charcoal in binding the cationic drugs and that the anion exchange resins were much superior to the charcoal in binding the anionic drugs. However, for the neutral drugs, the charcoal was much superior to any resin (the binding of these occurs by ordinary adsorption).

Edwards and McCredie also studied the binding of several drugs from a 0.1 M sodium bicarbonate solution, and showed that the presence of the added ions (H^+,

Table 22.2 Percentage Removal of Drugs from Six mL of 0.025 M Aqueous Solutions by 0.2 Grams of Resin or Charcoal

	Cholestyramine	Katonium	Charcoal
Acidic drugs			
Phenobarbital	90	0	61
Secobarbital	86	0	69
Pentobarbital	85	0	69
Barbital	73	0	38
Salicylate	96	3	65
Bromide	73	3	6
Iodide	77	0	8
Basic drugs			
Quinidine	24	53	38
Quinine	0	39	15
Mecamylamine	9	92	30
Strychnine	0	97	98
Strychnine	0	97	98
Morphine	0	90	28
Chlorpromazine	0	42	63
Neutral drugs			
Gluthethimide	51	40	100
Carbromal	2	0	46
Meprobamate	36	26	100

From Edwards and McCredie (1967). Copyright *The Medical Journal of Australia* 1967. Reprinted with permission.

HCO_3^-) interfered more with the ability of the resins to bind the drugs than it did with the charcoal. Thus, one might expect that in the environment of an actual digestive tract, where a whole range of ionic species may be present, that the resins might not perform as well as tests involving the binding of drugs from pure water solutions would indicate.

Khouw et al. (1978) studied the binding of 13 frequently abused drugs from SGF to two resins, Amberlite XAD-2 and Amberlite XAD-4. Although these investigators refer to the resins as "ion exchange" resins, they are in fact just plain polystyrene type resins with no exchangeable ionic groups. They found that these resins had much lower binding rates and capacities than activated charcoal.

The in vitro binding of oxalic acid and glyoxylic acid to charcoal, resins, and zirconium oxide was studied by Scholtens et al. (1982). The charcoal was Norit RBX-1 and the resins were Amberlite XAD-4, Imac S5-50, and Lewatit MP-7080. The pH was 7.4 in all tests. The tests involved adding 0.5 g of each sorbent to 50 mL of solutions containing 100 μmole/L of either oxalic acid or glyoxylic acid, allowing sorption to proceed for 6 or 24 hr at 25 or 37°C, and measuring the residual solute in solution. Only the hydrous zirconium oxide bound significant quantities of the two compounds (charcoal would not be expected to bind either very well, as they are simple molecules without

Table 22.3 Q_m Values for Paraquat Binding by Different Adsorbents

Adsorbent	Distilled water	SGF	SIF
Kayexalate	526	357	419
Kalimate	400	305	409
Activated charcoal	54	20	70
Bentonite	60	44	61
Adsorbin	88	68	81

From Tagaki et al. (1983). Reprinted with permission from *Veterinary and Human Toxicology*.

any ring structures and are highly water-soluble). The zirconium oxide bound up to 5.5 μmole oxalic acid and 8 μmole glyoxylic acid per gram.

Two cation exchange resins (Kayexalate and Kalimate), bentonite, activated charcoal, and a material called Adsorbin were evaluated by Tagaki et al. (1983) for their capacities to bind paraquat in vitro from distilled water, SGF (pH 1.2), and SIF (pH 6.8). Q_m values in mg/g were determined by fitting the data to the Langmuir equation. Table 22.3 shows the values. It is clear that the resins were superior to the other adsorbents. Tagaki et al. also did in vivo studies with paraquat-dosed rats and determined LD_{50} values when rats were given 1 g/kg of three of the adsorbents along with paraquat. For Adsorbin, Kayexalate, and Kalimate, the values were 213, 296, and 269 mg/kg, respectively (bentonite and charcoal were not tested), compared to a control value of 144 mg/kg. Thus, all three adsorbents evaluated increased the LD_{50} value substantially, with the resins being better than the Adsorbin.

Al-Shareef et al. (1990) studied the binding of four drugs (warfarin, paracetamol, antipyrine, and metoclopramide) to two anionic resins (cholestyramine, colestipol) as well as to two charcoal formulations (Carbomix, Medicoal) in vitro in a pH 7.4 phosphate buffer at 37°C. Q_m values were obtained by fitting the data to the Langmuir equation. The Q_m values are shown in Table 22.4. There is no consistent relationship between the

Table 22.4 Langmuir Q_m Values (mg/g) for Charcoals and Resins

	Warfarin (acid)	Paracetamol (acid)	Antipyrine (base)	Metoclopramide (base)
pK_a	5.0	9.5	1.4	9.2
% ionized, pH 7.4	91.3	10.9	0.2	85.8
Molecular weight	330	151	188	354
Carbomix Q_m	637	523	964	1154
Medicoal Q_m	536	558	891	999
Cholestyramine Q_m	1234	231	787	488
Colestipol Q_m	1435	19	237	1706

From Al-Shareef et al. (1990). Reprinted by permission of The Macmillan Press Ltd.

Q_m values and the drug properties (acid/base, pK_a, molecular weight). The ability of the charcoals to bind drugs is more consistent in that their Q_m values do not vary so widely as do those of the resins. This is generally well known as one of the strengths of activated charcoal. The resins are both clearly excellent for binding warfarin but both bound paracetamol much less well than did the charcoals. However, the resins showed quite variable behavior. For example, colestipol was surprisingly good for binding metoclopramide and surprisingly poor for binding paracetamol. Since the behavior of the resins varied so widely, charcoal is a more conservative choice of sorbent, especially if the toxin involved in an overdose situation is not known.

Arimori et al. (1992) studied the in vitro binding of imipramine to charcoal and a cation exchange resin (sodium polystyrene sulfonate) in a macrogol (polyethylene glycol) electrolyte solution (PEG-ELS) and in a solution called JP XII second medium, at 37°C. The Q_m values in mg/g were: for charcoal and PEG-ELS, 610; for charcoal and JP XII, 372; for the resin and PEG-ELS, 272; and for the resin and JP XII, 667. The pH of PEG-ELS is 8.5, and that of JP XII is 6.8. Thus, imipramine, being basic, is more nonionized in PEG-ELS solution and would be expected to bind better to charcoal than in the JP XII second medium solution. This is exactly what the charcoal results show. With the cation exchange resin, binding is better at lower pH (in the JP XII second medium) because imipramine has a greater positive charge at this lower pH, and therefore exchanges better with the resin.

By omitting constituents from the PEG-ELS solution one by one, it was found that eliminating the polyethylene glycol gave a substantially greater adsorption of imipramine to charcoal, while omitting Na_2SO_4, KCl, or $NaHCO_3$ gave significantly less drug adsorption. However, the reasons for these effects were not examined further. The authors concluded that using PEG-ELS solution to accelerate the passage of previously-administered charcoal through the GI tract would be an effective strategy for imipramine overdose, since this drug binds well to charcoal in the presence of PEG-ELS.

Yamashita et al. (1987) found that the cation exchange resin sodium polystyrene sulfonate (Kayexalate) has a binding capacity for paraquat which is 15 times greater than that of an activated charcoal called "Adsorbin" (origin unknown but possibly Japanese). The oral LD_{50} of paraquat was increased by a factor of 2.1 when Kayexalate was given intragastrically. The survival rates over 7 days for groups of rats given 200 mg/kg paraquat orally was 0/8 for control rats, and 3/8 for rats given either 500, 1000, or 2000 mg/kg charcoal (i.e., the charcoal dose caused increased survival but the rates did not vary with the charcoal dose). Rats given 500, 1000, or 2000 mg/kg had survival rates of 5/8, 6/8, and 7/8, respectively. Thus, Kayexalate led to better survival than did charcoal, and the survival rates increased with the Kayexalate dose.

B. General Studies of Cholestyramine and Colestipol

In this section we focus on studies that have confined themselves to cholestyramine and/or colestipol exclusively, rather than ion exchange resins (and, sometimes, charcoal) in general.

Edwards (1965) studied the binding of the sodium salts of several barbiturates to cholestyramine in vitro. He prepared 0.025 M solutions of each drug and contacted 6 mL with 0.2 g of the resin. The percentages of three barbiturates bound ranged approximately from 72–90%. With the addition of 0.05 M sodium bicarbonate, the percentages fell but were still in the range of 46–87% (for eight barbiturates). The addition of 0.05 M sodium taurocholate caused greater reductions in binding and the addition of both sodium bicarbonate and sodium taurocholate at 0.05 M each reduced binding further (to a range of about 15–66% for eight barbiturates). In vivo studies were done with rats given 125 mg/kg sodium secobarbital, with or without 0.4 g of the resin. Drug plasma levels were greatly reduced by the resin. Additionally, 14 of 30 rats died within 30 min when no resin was given, as compared to zero deaths in 18 rats which did receive the resin.

Gallo et al. (1965) performed a general study of the in vitro and in vivo effects of cholestyramine on a variety of drugs. The in vitro tests involved equilibrating 100 mg (434 µEq) of the resin with solutions of 4.34 µEq of each drug in 10 mL of a 0.15 M phosphate buffer (pH 7.5). The in vivo studies involved giving rats the anionic drugs sodium cholate, tetracycline, nicotinic acid, aspirin, phenobarbital, chlorothiazide (dogs were used for this drug), phenylbutazone, or warfarin, and following plasma drug levels with time. Different amounts of cholestyramine (e.g., 14.3, 71.5, or 357.5 mg/kg) were given orally along with the drugs. In vitro, cationic and neutral drugs were bound only weakly, if at all. For example, dextromethorphan, dihydrocodeinone, and chlorpheniramine were not appreciably bound, while quinidine was only 28% bound. The neutral drug digoxin was 33% bound. Anionic drugs were generally well bound. Binding percentages were: sodium cholate, 83%; warfarin, 97%; phenylbutazone, 98%; chlorothiazide, 88%; and hydrochlorothiazide, 86%. However, the anionic drugs phenobarbital, tetracycline, and aspirin were bound less well, at 58, 60, and 32%, respectively. In vivo, cholestyramine decreased plasma drugs levels of sodium cholate well but had only modest effects with the other drugs. However, only a single resin dose was given and the possibility that repeated resin doses would have significant effects remains.

Johns and Bates (1969, 1970a,b) produced a series of three reports on the "Quantification of the Binding Tendencies of Cholestyramine." The first paper deals with the effects of the structure of the adsorbing molecule, added electrolytes, and temperature on the binding of conjugated and unconjugated bile salt anions to cholestyramine. However, with respect to the in vivo use of cholestyramine, only one temperature (37°C) is relevant, so we will not review temperature effects here. With regard to structure, it was found that dihydroxy anions bound more strongly than trihydroxy derivatives. Whereas conjugation with glycine had little effect on the degree of binding of bile salt anions, conjugation with taurine markedly increased the binding. Added chloride and bicarbonate ions significantly reduced bile salt anion binding.

In their second paper, Johns and Bates studied the interactions between conjugated bile salt anions and fatty acid anions with cholestyramine; in their third report they discuss studies on the rates of adsorption of conjugated bile salt anions to cholestyramine as a function of added inorganic electrolyte concentration, temperature, and agitation intensity. We will not discuss these results here, as the effect of fatty acid anions is just one of many possible interactions and by itself is not determinant of in vivo performance. Further, rates of adsorption to cholestyramine are not usually limiting (equilibrium capacity aspects are more important).

Johns and Bates (1972a,b) studied the in vitro binding of sodium fusidate, an antibiotic used to treat staphylococcus infections, to cholestyramine. Both equilibrium and rate tests were done. Sodium fusidate bound well, independent of temperature, and the binding was decreased significantly by added sodium chloride or sodium bicarbonate. Detailed rate-of-adsorption studies were done but will not be discussed since rate effects are of secondary importance in vivo. In their second paper, Johns and Bates present in vivo results on sodium fusidate absorption in rats. The rats were given 500 mg/kg of the drug orally along with 214.5 mg/kg cholestyramine. In other tests, the resin dose of 214.5 mg/kg was delayed for either 1 or 2 hr, and in other tests the resin dose given with the drug was changed to 71.5 or 357.5 mg/kg. With simultaneous resin and drug, mean serum drug levels at 1, 2, 3, and 4 hr were reduced by 69, 32, 49, and 55%, respectively, compared to control values. The 3-hr drug levels in mg/L for simultaneous resin doses of 0, 71.5, 214.5, and 357.5 mg/kg were 3.71, 2.50, 1.89, and 0.84, respectively. Thus, giving more resin had the expected effect. When the 214.5 mg/kg resin dose was given at 0, 1, or 2 hr, the 3-hr drug levels in mg/L were 1.89, 2.96, and 3.32, respectively (a value of 3.71 mg/L was predicted statistically for an infinite delay). Thus, early administration of the resin was significantly better.

Ko and Royer (1974) studied the in vitro binding of various drugs to colestipol hydrochloride, as a function of ionic strength, pH, and the type of competing ions present. The drug and resin quantities employed were stated to be in the ratios one would have if both were administered orally in therapeutically effective amounts (however, note from the discussion in Chapter 3 that the fraction of a drug adsorbed is not a unique function of the drug:adsorbent ratio but rather depends also on the amounts of each). Chlorpropamide, niacin, ascorbic acid, aspirin, salicylic acid, phenobarbital, sulfadiazine, penicillin G, and lincomycin HCl were less than 30% bound in a pH 7.5 buffer, whereas warfarin and tetracycline HCl were bound 59 and 30%, respectively. The degrees of binding of the drugs were considered to be upper limits on the percentages that would be bound in vivo. The effects of added monoolein, oleic acid, and taurocholate were determined. Also, cholestyramine was used in some of the tests and was found to bind most of the drugs better than did the colestipol.

The absorption of amitriptyline, desipramine, doxepin, imipramine and nortriptyline to cholestyramine from 1.2 N HCl solutions at 37°C was studied by Bailey et al. (1992). For the resin:drug ratios employed, about 80% of these drugs were adsorbed by the cholestyramine. Amberlite XAD-2 resin bound these drugs to about the same degree. In contrast, five other drugs (acetaminophen, chlordiazepoxide, procainamide, quinidine, and theophylline) were found to bind poorly to cholestyramine. Thus, they concluded that cholestyramine may be useful for the treatment of tricyclic overdoses and, in normal patients receiving tricyclic therapy for depression, should not be used (e.g., to lower blood lipids). However, a direct extrapolation of in vitro data to recommendations for human clinical therapy is extremely tenuous and, at best, such data suggest a need for animal studies.

Bailey et al. also found that activated charcoal adsorbed 100% of all of the drugs tested. Sea sand (SiO_2) was tried as an adsorbent and it was, as one would expect, generally quite poor. The ratios of adsorbent:drug employed were either 10:1 or 20:1, except for desipramine and nortriptyline (for which the ratios were about 133:1 to 222:1). Since the adsorbent:drug ratio affects the percentage of drug adsorbed and since variation of this ratio can produce almost any "percent adsorbed" result (assuming that the adsorbent is not an unreasonable one, like sea sand), results such as these are not very meaningful. What should have been determined and reported are maximal adsorption capacities (Q_m values).

Bailey (1992) also studied the effects of pH changes and ethanol on the binding of tricyclic antidepressants in SGF. The percent binding of amitriptyline, desipramine, doxepin, imipramine, and nortriptyline decreased from 79–90% at pH 1.0 to 36–48% at pH 4.0. Above pH 4.0, the amounts bound increased up to 62–76% at pH 6.5. This behavior can be explained in terms of the effect of pH on the states of ionization of the resin and the drugs. The addition of ethanol in increasing amounts decreased the adsorption of the drugs steadily to the point where, in pure ethanol, adsorption ranged from 0% (nortriptyline) to 32% (doxepin). These results suggest that cholestyramine would be less effective in TCA overdoses when ethanol is ingested with the drugs. However, animal studies, at the very least, are needed to show whether this would be true in vivo.

C. Effects of Cholestyramine and Colestipol in Cardiac Glycoside Overdoses

Since cholestyramine and colestipol have been widely used to treat overdoses of cardiac glycosides (mainly digoxin and digitoxin), we devote a separate section to reports of those uses. Because most cardiac glycosides undergo enterohepatic circulation, resin therapy is suitable not only for oral but also for intravenous glycoside intoxication.

Saral and Spratt (1967) conducted in vitro tests with four glycosides and in vivo experiments with digitoxin using cholestyramine. The in vitro tests showed that the resin bound the glycosides well but in decreasing order according to the sequence digitoxin, digitoxigenin, digoxin, and digoxigenin. In vivo, mice were given digitoxin doses of 2.5, 5, 10, 20, or 40 μmol/kg along with no resin or with 5 g/kg of the resin. The LD_{50} values in μmol/kg determined with these digitoxin doses were: 9.93 and 12.9 in male and female mice, respectively, when no resin was given, and 17.1 and 22.5 in male and female mice, respectively, when the resin was included. The increase factors in the LD_{50} values for male and female mice due to the resin were almost identical, 1.72 and 1.74, respectively. Tests were also done with IP-administered digitoxin, in which case the use of cholestyramine made no significant difference.

Bazzano et al. (1970) examined the use of colestipol in treating patients with digitalis intoxication. The resin was given in the amount of 10 g initially and then 5 g every 5–8 hr to patients having digitoxin or digoxin plasma levels well above normal. Plasma drug levels fell to normal within 15 hr with this therapy. Bazzano and Bazzano (1972) showed that both cholestyramine and colestipol effectively bind digoxin and digitoxin in vitro. Aqueous solutions of each glycoside were passed through columns containing 1 g of resin and the amounts of unbound glycoside in the eluates were determined (the flow rates used are not stated). Cholestyramine bound 6.4 mg/g digoxin and 14.4 mg/g digitoxin, while colestipol bound 4.8 mg/g digoxin and 7.4 mg/g digitoxin. The tests were repeated with duodenal juice added to the solutions (the amounts of duodenal juice added are not stated). In these tests, cholestyramine bound 2.8 mg/g digoxin and 4.6 mg/g digitoxin, while colestipol bound 4.7 mg/g digoxin and 5.5 mg digitoxin. One can see that cholestyramine bound both glycosides from water solutions better than did colestipol; however, with the addition of duodenal juice, binding to cholestyramine was decreased far more than binding to colestipol, to the extent that colestipol was then the better resin. They then did an in vivo study in which colestipol was given (10 g at first, then 5 g every 6–8 hr) to patients who had received a nontoxic dose of digitoxin. The half-life was decreased from a control value of 9.3 days (in one control patient) to an average of 2.75 days in the four subjects tested. Using digoxin as the drug, a half-life reduction from 1.8–2.0 days (two control patients) to 16 hr in one colestipol-treated patient was seen.

Caldwell and Greenberger (1971a) found that cholestyramine bound tritiated digoxin and digitoxin well in vitro and that the binding was only modestly inhibited by the presence of bile. In vivo experiments were then done with rats injected with digitoxin.

One group of rats received an 80-mg dose of resin 2 hr after a subcutaneous injection of a LD_{100} dose of digitoxin (10 mg/kg), three more 80 mg resin doses at 2 hr intervals on the first day, and further 80-mg doses three times daily for 5 days or until they died. A second group of rats were given the first resin dose 2 hr before the drug and resin doses thereafter according to the same schedule as the first group. Of 20 control rats, all died. In the group given the first resin dose 2 hr after the drug injection, 4/18 (22%) survived, whereas in the group given the first resin dose 2 hr prior to the drug, 7/10 (70%) survived. Similarly, guinea pigs given 4, 7.5, and 10 mg/kg digitoxin subcutaneously had survival rates of 70, 50, and 25%, respectively, when given cholestyramine beforehand, compared to survival rates of 30, 20, and 0% in control animals.

Caldwell and Greenberger (1971b) described studies with humans in which the subjects were given 4-g doses of cholestyramine 8, 12, and 16 hr after receiving 1.2 mg digitoxin orally, and additional 4 g resin doses four times daily for the next 5 days. The cholestyramine reduced the serum half-life from 11.5 to 6.6 days.

Haacke et al. (1973) showed that ^{3}H-labeled digitoxin and its metabolites are bound well in vitro from guinea pig bile, by both activated charcoal and cholestyramine. Further data showed that the extent of binding of unmetabolized digitoxin versus that of a mixture of digitoxin and its metabolites was about equal (percentage-wise) with the charcoal but that the cholestyramine showed only about half as much binding of the digitoxin as compared to the mixture of digitoxin and its metabolites. Haacke et al. estimated that a toxic dose of from 2 to 4 mg digitoxin could be satisfactorily counteracted by an oral dose of 30 g activated charcoal or 40 g cholestyramine.

The effect of cholestyramine on the absorption of tritiated digoxin in rats was studied by Thompson (1973). The rats were started on an 8 wt% resin diet and then given the drug orally. A significant decrease in stool radioactivity and an increase in urine radioactivity was observed, as compared to rats not given any resin. In other experiments, rats which had undergone bile duct ligation also showed decreased stool radioactivity and increased urine radioactivity; the addition of cholestyramine did not alter this. Thus, it was concluded that the apparent enhancement of digoxin elimination by cholestyramine is due to a reduction in bile flow caused by interruption of the enterohepatic circulation of bile salts. This decreases the biliary excretion of digoxin.

The effect of colestipol on digitoxin plasma levels in 11 patients having excessive drug plasma concentrations was examined by van Bever et al. (1976). The resin dose was 5 g four times daily. No effect on digitoxin half-life was found between resin-treated patients ($t_{1/2}$ = 6.3 day) and nonresin-treated patients ($t_{1/2}$ = 6.8 day). Why these results differ so markedly from those of Bazzano and Bazzano is unclear.

Hall et al. (1977) gave 0.5 mg of tritiated digoxin to six subjects receiving digoxin therapy for heart disease. In one case, cholestyramine was also given (4 g each 6 hr) before as well as after the tritiated drug, while in another case the tritiated drug was given 1 month after cholestyramine had been initiated. Digoxin serum levels and excretion in the urine and in the stools was essentially unaffected by the cholestyramine, as compared to control results. Thus, the authors concluded that the resin did not bind digoxin significantly in vivo.

Gilfrich et al. (1978) used charcoal hemoperfusion and cholestyramine therapy in a 47-year-old female overdosed on digitoxin. After two periods of successful hemoperfusion, her plasma digitoxin level had been lowered from 125 ng/mL to 70 ng/mL. Cholestyramine was then started at 4 g three times daily for the next 5 days. The drug elimination half-life during the resin therapy period was 62 hr. Three weeks later, 0.75

mg digitoxin was given to the patient IV and the subsequent drug half-life was determined to be 145 hr.

Frésard et al. (1979) measured plasma levels in a 57-year-old patient treated with multiple doses of cholestyramine after a massive digoxin overdose. They determined a $t_{1/2}$ of 27.6 hr during the resin therapy period. By reference to $t_{1/2}$ values from literature cases in which only supportive therapy was provided, they concluded that cholestyramine was probably of significant benefit to their patient.

Cholestyramine in three 4-g doses was given by Cady et al. (1979) to two patients intoxicated with digitoxin. Both patients showed a rapid decline in serum digitoxin concentrations, with decreased signs and symptoms of toxicity. Similarly, Payne et al. (1982) reported a digoxin half-life decrease in one patient given colestipol (10 g initially, followed by 5 g every six hours for 48 hr) from an estimated 85 hr down to 55 hr. In a case study by Kilgore and Lehmann (1982), a patient with a digoxin overdose had a drug half-life of 2.0 days when receiving colestipol (10 g every 8 hr) versus a value of 3.2 days without colestipol.

Carruthers and Dujoven (1980) studied the effects of cholestyramine (4 g eight times daily) and spironolactone (300 mg daily) on digitoxin elimination, alone and in combination, in six healthy subjects. Cholestyramine decreased the drug half-life from 141.6 to 84.4 hr. Spironolactone increased the $t_{1/2}$ to 192.2 hr and the combined treatment gave a $t_{1/2}$ of 102.9 hr. Thus, cholestyramine alone was superior.

Neuvonen et al. (1988) studied the effects of resins on the absorption of digoxin, carbamazepine, and furosemide. The absorption of a 0.25-mg dose of digoxin was 98% prevented by 8 g of charcoal, 40% reduced by 8 g of cholestyramine, and not significantly (3%) reduced by 10 g of colestipol (all sorbents were given simultaneously with the drug). In another phase of this same study, 40 mg furosemide was used and these same sorbent doses reduced the furosemide absorption by 99% (charcoal), 94% (cholestyramine), and 79% (colestipol). Figure 22.1 shows the furosemide results. In a third phase, a 400-mg dose of carbamazepine was employed and the reductions in absorption were: charcoal, 92%; cholestyramine, none; and colestipol, 10%. All percent reductions in absorption were based on 0–72 hr AUC values. Thus, charcoal is to be preferred over the two resins.

The use of cholestyramine resin in digoxin intoxication was examined by Henderson and Solomon (1988). A 94-year-old man was treated for digoxin overdose with 4 g cholestyramine every 6 hr, starting at about 86 hr post-admission. In all, eight cholestyramine doses were given. The resin reduced the digoxin half-life from 75.5 to 19.9 hr. All signs and symptoms of toxic reaction subsided during the period of cholestyramine therapy.

In a comment on the study by Henderson and Solomon, Neuvonen and Kivistö (1989) brought up the point that the half-life of 19.9 hr was determined using only two data points and thus is uncertain. Neuvonen and Kivistö then mentioned the results of one of their own studies (Neuvonen et al., 1988). In reply, Henderson and Solomon (1989) stated that the comparison of sorbent effects when the sorbent is given simultaneously with the drug (in Neuvonen and Kivistö's study) to sorbent effects when the sorbent is given well after the drug has been absorbed into the body (in Henderson and Solomon's study) "may not be fair." In any event, the 40% reduction in digoxin absorption by cholestyramine which was observed by Neuvonen's group is not necessarily inconsistent with the effects of cholestyramine seen by Henderson and Solomon. It appears that cholestyramine can be moderately effective for digoxin overdose, as both

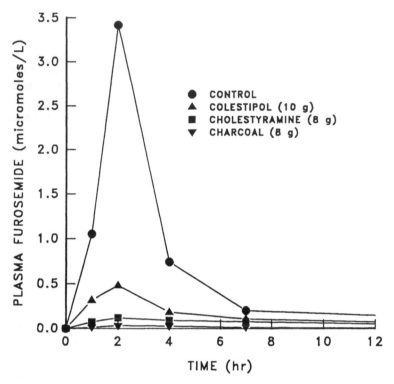

Figure 22.1 Effect of activated charcoal, cholestyramine, and colestipol on the absorption of simultaneously ingested furosemide. From Neuvonen et al. (1988). Reprinted by permission of Blackwell Scientific Publications Ltd.

studies have shown. The point is simply that charcoal is, weight for weight, perhaps even better.

Pieroni and Fisher (1981) reported the case of a 72-year-old woman having a digitoxin overdose who was given cholestyramine doses of 4 g every 6 hr. This resulted in a drug half-life of 3.7 days, considerably shorter than a reported digitoxin half-life of 11.5 days in normal volunteers (Caldwell and Greenberger, 1971b). Four patients with digitoxin intoxication were studied by Demers et al. (1982). They found that oral chole-styramine (8 g every 6 hr) reduced the total amount of drug in the body by one-half in two days. No change in the percentage protein-binding of the drug was observed. Baciewicz et al. (1983) reported on the use of cholestyramine to enhance digitoxin elimination in a 72-year-old female overdose patient; they mentioned that initiation of resin therapy significantly enhanced the decline in the serum drug level. However, no quantification in terms of the reduction in the $t_{1/2}$ value was performed.

Overdose cases involving two other glycosides in addition to digitoxin were presented by Kuhlmann (1984). He gave cholestyramine (8 g every 6 hr) to three different patients overdosed with β-acetyldigoxin, β-methyldigoxin, and digitoxin, respectively. He found that during the resin treatment the plasma concentrations of β-acetyldigoxin and β-methyldigoxin declined with half-lives of 20.4 and 30.0 hr, respectively (values much shorter than therapeutic half-life values in the literature). The digitoxin half-life was 74.5 hr during the first 2 days of resin therapy; however, when the plasma digitoxin level fell below 40 ng/mL, the half-life increased and became similar to that without cholestyramine administration.

D. Effects of Cholestyramine and Colestipol on Other Drugs

We turn now to a discussion of studies in which cholestyramine and/or colestipol have been used to enhance the elimination of drugs other than cardiac glycosides.

Armstrong and Edwards (1967) studied the effects of using cholestyramine to treat aspirin poisoning in rats. Thirty-two rats were given an LD_{50} dose of aspirin, with or without different resins, activated charcoal, or sodium bicarbonate. A multifactorial design was used in which all combinations of six factors at two different levels were tested. Total plasma salicylate levels at 20 and 40 min were the criteria by which treatment effectiveness was judged. Average salicylate levels at 20 and 40 min were 78 and 86 mg/dL with cholestyramine, and 88 and 102 mg/dL without cholestyramine. It is stated that the differences "were probably significant ($p < 0.05$)." Higher cholestyramine doses undoubtedly would have had a larger effect but the authors state that the resin could only be given in one-third the maximal dose because it had to be combined with other substances due to the multifactorial design of the study.

In Chapter 11, we mentioned the use of cholestyramine by Dordoni et al. (1973) for reducing acetaminophen absorption. Cholestyramine or charcoal were given right after acetaminophen to human volunteers and it was found that the drug absorption was reduced to 38 and 37% of control values, for the resin and the charcoal, respectively. When given after a 60 min delay, the charcoal and cholestyramine reduced acetaminophen absorption to 77 and 84% of control values, respectively. Thus, cholestyramine was just as effective as charcoal.

Siegers and Möller-Hartmann (1989) evaluated cholestyramine as an antidote against acetaminophen-induced hepatotoxicity and nephrotoxicity in rats. The drug dose was 1.5 g/kg and the resin doses, given 4 and 24 hr after the drug, were 1 g/kg. The resin markedly reduced both hepatotoxicity and nephrotoxicity, as evidenced by a variety of tests (enzyme activity, creatinine retention, etc.). The authors concluded that cholestyramine almost completely interrupts the enterohepatic circulation of acetaminophen and conjugates undergoing biliary excretion (other studies have shown that up to 30% of an acetaminophen dose is excreted in the bile).

Rosenberg and Bates (1974) studied the inhibitory effect of cholestyramine on the absorption of flufenamic and mefenamic acids in rats. Flufenamic acid (50 mg/kg) or mefenamic acid (100 mg/kg) were given orally to rats along with cholestyramine (500 mg/kg). Plasma drug levels were determined at 15, 30, 60, 90, 150, 210, and 270 min. The resin lowered the flufenamic acid plasma levels at these times by 48, 72, 64, 63, 62, 61, and 35%, respectively. For mefenamic acid, 0–∞ AUC values in mg-hr/L averaged 145.0 for control rats and 43.4 for resin-dosed rats. Thus, cholestyramine significantly inhibited the absorption of both drugs.

Audétat and Bircher (1976) studied two patients being treated with prednisolone. The drug was given (75 mg to one patient and 37.5 mg to the other), both with and without 8 g cholestyramine, and plasma drug levels were determined over 6 hr. The resin had no effect.

Hunninghake and Pollack (1977) studied the effects of cholestyramine and colestipol on aspirin, tolbutamide, and warfarin absorption in normal adult subjects. Colestipol (10

g) or cholestyramine (8 g) were administered 2 min before and 6 and 12 hr after administration of 650 mg aspirin, 500 mg tolbutamide, or 40 mg warfarin. No significant effect of either resin on aspirin or tolbutamide absorption was found, but with warfarin the AUC was reduced 32% by cholestyramine and 5% (not significant) by colestipol.

Calvo and Dominguez-Gil (1984) studied the effects of cholestyramine on naproxen pharmacokinetics. In vitro, naproxen bound to the resin up to a maximal level of about 2.2 mmol/g. In vivo studies with healthy volunteers given concurrent single doses of 250 mg naproxen and 4 g cholestryramine showed that the resin delayed the absorption of the drug but had no effect on any other pharmacokinetic parameter.

Erttmann and Landbeck (1985) found that cholestyramine binds methotrexate (MTX) effectively in vitro. Compared to an activated charcoal (Merck), its binding capacity was 5.4 times higher. In two patients, 8 g per day (2 g every 6 hr) of the resin dramatically increased the rate of elimination of MTX following high-dose MTX therapy.

Activated charcoal and cholestyramine were evaluated by Guentert et al. (1986) for their ability to bind tenoxicam (a nonsteroidal anti-inflammatory agent) in vitro. Q_m values determined in pH 7.5 buffer solutions were 1.63 g/g for cholestyramine and 0.18 g/g for the charcoal. In vivo studies with IV drug administration to dogs followed by repeated oral cholestyramine gave a much increased total body clearance (0.65 L/hr versus a control value of 0.13 L/hr) and reduced the elimination half-life from 31 to 5.4 hr. In vivo, charcoal had little effect. Thus, cholestyramine was far superior to charcoal in both the in vitro and the in vivo tests.

Guentert et al. (1988) studied the effects of multiple doses of cholestyramine on the single-dose pharmacokinetics of tenoxicam and piroxicam in eight healthy subjects. IV injections of 20 mg tenoxicam or oral doses of 20 mg piroxicam were employed. Doses of cholestyramine (4 g) were given 2 hr before the drug, in the case of tenoxicam, or 3.5 hr after the drug, in the case of piroxicam. Subsequent doses were taken 0.5 hr before the morning and evening meals. The multiple-dose resin therapy reduced the half-life of tenoxicam from 67.4 to 31.9 hr, while for piroxicam the reduction was from 46.8 to 28.1 hr.

Ferry et al. (1990), working with healthy adults given piroxicam, found that repeated 4-g doses of cholestyramine reduced the drug half-life from 53.1 hr (control) to 29.6 hr versus a reduction with repeated 5-g doses of charcoal to 40.0 hr. Thus, in this case the resin was more effective.

Lyddane et al. (1988) tried using cholestyramine to reduce the absorption of phencyclidine given to rats but it had no significant effect (it was given 45 min after the drug, however). Charcoal, in contrast, reduced low, medium, and high drug dose absorption by 30, 4, and 0%, respectively.

Kivistö and Neuvonen (1990) used single 8-g doses of cholestyramine to reduce glipizide absorption in human subjects. It was somewhat effective (29% reduction in AUC) but not as effective as single 8-g doses of charcoal (81% reduction in AUC).

Gendrel et al. (1990) gave 10 mg/kg chloroquine to five children, with or without 4 g cholestyramine. Plasma drug levels at 6 hr were significantly lower with cholestyramine than without cholestyramine. However, the study design makes more definitive conclusions impossible, since initial plasma chloroquine levels ranged widely (from 0 to 160 mg/L) and since some of the children were receiving the drug prior to the study period.

In vitro tests by Herman and Chaudhary (1991) with lorazepam and lorazepam gluconuride showed the following percentages bound: by cholestyramine, 23.7 and 74.3%; for colestipol, 11.3 and 20.8%, and for charcoal, 100 and 100%, respectively, for the two compounds.

E. Effects of Cholestyramine and Colestipol on Concomitant Drugs Taken Therapeutically

Cholestyramine and/or colestipol are often given to patients who are taking other drugs. For example, patients with heart disease may be taking one of these resins along with a lipid-lowering agent such as clofibrate, or an anticoagulant such as warfarin, phenprocoumon, dicumarol, or tromexan. Other drugs frequently given to persons with heart disease are diuretics such as hydrochlorothiazide, beta-blockers such as propranolol, antiarrhythmics such as amiodarone and phenytoin, antidiarrheal agents such as loperamide, antidepressants, and various antibiotics. In these situations, one must be concerned with whether the resin will adsorb or otherwise interfere with the bioavailability of the other drugs. We consider studies of this kind in the present section.

Sedaghat and Ahrens (1975) studied whether cholestyramine had any effect on the pharmacokinetics of clofibrate in man. Fifteen patients taking 1 g of clofibrate twice daily were given 16 g/day cholestyramine. The resin had no effect.

Robinson et al. (1971) gave 40 mg warfarin to six normal subjects, along with three 4-g doses of cholestyramine. Two of the doses were given 3 and 6 hr after the drug. The other dose was given 3 hr before the drug in one test and simultaneous with the drug in another test. Plasma drug levels determined at 6, 10, 24, and 48 hr showed that both protocols lowered the drug levels modestly, with the lowering being larger when the first resin dose was simultaneous with drug administration. For example, at 6 hr, plasma drug levels were lowered about 11% (compared to controls) when the first resin dose was given 3 hr before the drug, and by about 28% when the first resin dose was simultaneous with the drug. Since cholestyramine and warfarin are sometimes used together in patients with heart disease, they should taken at different times to avoid a decrease in warfarin effectiveness due to binding to the cholestyramine.

Jänchen et al. (1978) showed that cholestyramine decreased plasma warfarin levels by 30% and decreased the drug elimination half-life by a similar amount in five healthy volunteers given a single IV dose of warfarin (1.0–1.2 mg/kg) followed by 4 g cholestyramine three times a day. Presumably, the resin interrupted enterohepatic circulation of the drug. An overdose case with warfarin was reported by Renowden et al. (1985). A 25- year-old man who had overdosed on warfarin was given cholestyramine (4 g four times daily), starting on about the seventh day, and a decrease in the drug elimination half-life of 38% was observed (note that the resin dose was higher than in the Jänchen study).

Hahn et al. (1972) studied the effect of 8 g of cholestyramine in reducing the absorption of a 15-mg dose of phenprocoumon given simultaneously to human volunteers. The resin significantly lowered serum phenprocoumon levels over the 0–96 hr period that they were determined (AUC values are not given but a figure in their paper suggests a reduction of about 40% due to the resin). Similar tests using 500 mg of acetylsalicylic acid showed no significant effect of cholestyramine.

Meinertz et al. (1977a) described the case of a 35-year-old man, on phenprocoumon therapy, who took a suicidal dose of the drug. He was given 4 g cholestyramine three times daily, starting on day seven. This lowered the drug elimination half-life from a value of 6.8 days prior to resin therapy to 3.5 days during the resin therapy. Meinertz et al. (1977b) found that cholestyramine binds phenprocoumon strongly in vitro. They gave a single IV dose (30 mg) of phenprocoumon to six normal subjects, followed every day by three doses of 4 g cholestyramine each. Total drug clearance increased by 1.5

to 2.0-fold. Thus, both studies by Meinertz et al. showed that cholestyramine interrupts the enterohepatic cycling of phenprocoumon.

Tembo and Bates (1974) found that 250 mg/kg cholestyramine given to rats along with 100 mg/kg of either dicumarol or tromexan caused peak plasma levels of the two drugs to decrease by 66 and 80%, respectively. The 0–∞ AUC values for the two drugs were decreased by 45 and 71%, respectively.

In another study, 5 g cholestyramine and 10 g colestipol were given to healthy subjects 2 min before and 6 and 12 hr after a 75-mg dose of hydrochlorothiazide (Hunninghake et al., 1982). Urinary excretion of the drug over 24 hr was decreased 43% by colestipol and 85% by cholestyramine, and peak plasma drug levels were decreased 14% by colestipol and 69% by cholestyramine (AUC values were not computed).

Hunninghake and Hibbard (1986) then studied the effect of time intervals for cholestyramine dosing on the absorption of hydrochlorothiazide. The oral drug dose was 75 mg. Dosing of cholestyramine (8 g per dose) followed several schedules: (1) 2 hr before the drug, (2) 2 hr after the drug, (3) 4 hr after the drug, and (4) at 24 and 12 hr before the drug and 4 hr after the drug. These schedules reduced the amount of the drug excreted unchanged in the urine over 0–24 hr by 65, 26, 4, and 35%, respectively. AUC values (0–∞) in mg-hr/L for some of the regimens were: control, 2549; single-dose resin at 4 hr, 2334; and the multiple-dose resin schedule, 1737. Thus, to minimize the effect of the resin on the drug, the best dosing schedule for the resin is a single dose at 4 hr.

Hibbard et al. (1984) gave 10 g colestipol or 8 g cholestyramine to 12 healthy volunteers 2 min prior to an oral dose of 120 mg propranolol. In a second test, resin doses were given 12 hr before and 2 min before the drug. Blood samples were taken for up to 24 hr. AUC values (0–∞) for propranolol in mg-hr/L in the one-dose resin and two-dose resin tests were: controls, 732 and 752; with colestipol, 870 and 523; and with cholestyramine, 638 and 432, respectively. Thus, the two-dose resin regimen reduced the AUC by 30% (colestipol) and by 43% (cholestyramine). Peak plasma concentrations in mg/L in the one-dose resin and two-dose resin tests were: controls, 103 and 104; with colestipol, 133 and 67; and with cholestyramine, 77 and 46. Again, reductions with the two-dose resin regimen were 36% (colestipol) and 56% (cholestyramine).

Nitsch and Lüderitz (1986) gave single oral doses (400 mg) of the antiarrhythmic drug amiodarone to 11 healthy volunteers, with and without cholestyramine. The resin doses (4 g each) were given at 1.5, 2.5, 3.5, and 4.5 hr. Mean serum levels of the drug at 7.5 hr were 0.42 mg/L without the resin and 0.21 mg/L with the resin. The elimination half-lives in three subjects were 23.5, 29, and 32 days, compared to half-lives in eight controls ranging from 35–58 days. Thus, cholestyramine significantly interrupted the enterohepatic circulation of the drug.

Callaghan et al. (1983) found that 5 g cholestyramine and 10 g colestipol given to healthy subjects 2 min before and 6 and 12 hr after the administration of 500 mg phenytoin had no effect on the drug AUC values. The 0–∞ AUC values in g-min/L were 16.4 in control tests, 17.9 with colestipol, and 14.8 with cholestyramine. Elimination half-lives were 26.5, 27.2, and 23.8 hr, respectively.

Barzaghi et al. (1988), like Callaghan et al., also found that cholestyramine had no effect on phenytoin bioavailability. The drug (400 mg) was given orally to normal subjects, followed by cholestyramine (4 g four times a day for five days). AUC values with and without resin treatment were virtually identical.

Ti et al. (1978) reported the case of a 55-year-old man with an ileostomy who was given loperamide HCl, an antidiarrheal agent, both with and without cholestyramine (2 g every 4 hr). The output of fluid from the ileostomy was significantly higher ($p < 0.05$) when cholestyramine was given with the drug. However, this difference was significant only for fluid intake values between 1 and 3 L. Below 1 L, no data for the combined treatment existed, and above 3 L there was no difference due to including cholestyramine.

Johansson et al. (1978) gave 50 mg of hydrocortisone to healthy subjects along with 8 g cholestyramine (hydrocortisone is frequently combined with cholestyramine, as in recurrent Crohn's disease after ileal resections). The resin reduced the drug 0–200 min AUC from 103.5 to 58.9 μmol-min/L, a decrease of 35%. The time to the peak plasma concentration of the drug was 50 min later when the resin was used. Hence, cholestyramine both delayed and reduced the absorption of hydrocortisone.

Allgayer et al. (1982) studied D-penicillamine binding to cholestyramine as a function of the resin amount, pH, and the presence/absence of other compounds such as bile salts. For a drug:resin ratio of 150 mg drug per 4–8 g of resin, about 10% of the drug bound to the resin. Bile salts (10 mmole/L) reduced this small degree of adsorption by 87%. Thus, the use of cholestyramine in patients receiving D-penicillamine would likely not cause much interference with the effectiveness of the penicillamine.

Coltman et al. (1990) studied the effect of cholestyramine on the in vitro activity of gentamicin (these agents are used in combination to treat severe diarrhea in infants). In vitro binding tests were done and cultures of *E. coli* were used to determine the antibiotic activity of gentamicin in the presence and absence of the resin. It was found that cholestyramine did not inhibit the growth of *E. coli*, did not bind gentamicin, and did not affect the antibacterial activity of the antibiotic. Thus both may be given simultaneously to patients.

Geeze et al. (1988) presented a case report of a 69-year-old man who was taking 300 mg doxepin each day at bedtime for depression and who was started on cholestyramine (6 g twice a day) for diarrhea problems. Within 2 wk, his affective symptoms recurred. His blood levels of doxepin and its active metabolite *n*-desmethyldoxepin decreased whenever the time interval between taking the drug and taking the resin decreased. These results suggest that cholestyramine impairs the absorption of doxepin.

F. Effects of Resins on Various Endogenous Biochemicals

In Chapter 19 we reviewed the effects of charcoal and other sorbents on a variety of endogenous biochemicals. We just briefly mention here those cases which involved resins. The reader is referred to Chapter 19 for details.

Krasopoulos (1980) studied the in vitro binding of bile acid salts and found that cholestyramine was as effective as charcoal, in some cases more effective than charcoal.

In vitro tests with charcoal, cholestyramine, and various porphyrins and porphyrin precursors (Tishler and Winston, 1985; Winston and Tishler, 1986) showed that cholestyramine was generally much better than various charcoals for binding the porphyrins (except for superactive charcoal) but much worse than the charcoals in the cases of the porphyrin precursors. Related to this was a study by Pimstone et al. (1987) which showed that cholestyramine was inferior to ActaChar charcoal for treating a patient with erythropoietic porphyria.

The use of Kayexalate resin to try to bind uremic toxins in rats (Saltzman et al., 1976) was mentioned in Chapter 19. This resin increased survival times by lowering blood potassium levels significantly.

Cholestyramine for lowering blood lipids was shown in several studies to be quite effective. Neuvonen et al. (1989), Park et al. (1988), and Tishler et al. (1987) all showed very significant cholesterol decreases in humans and in rabbits. Cholestyramine appeared to be less effective than superactive charcoal but a bit better than ordinary charcoals. Triglycerides were not decreased by cholestyramine.

Finally, the use of a carbonaceous sorbent (a pyrolyzed resin) in liver disease was studied by Takahama et al. (1983), who showed that repeated oral administration to dogs (in which liver failure was surgically created) had remarkable positive effects. The same resin was given to human patients with cirrhosis and essentially alleviated their symptoms after 4 wk.

These reports, taken as a whole, show that resins of various kinds can be very effective in binding endogenous substances in bile disorders, porphyria, uremia, hypercholesterolemia, and liver failure. Compared to charcoal they are generally not quite as effective but in some cases they may be equivalent or potentially superior. Some additional reports, not discussed previously in Chapter 19, will now be considered.

Longenecker and Basu (1965) showed in human subjects that 4- and 8-g doses of cholestyramine given with test meals had no effect on plasma amino acid levels. However, when vitamin A was added to the test meals, the 8-g dose significantly lowered plasma vitamin A levels (this abstract does not state how much).

Bergman et al. (1966) showed that cholestyramine given to hamsters who had been fed diets containing radioactively-labeled L-thyroxine had less thyroxine absorption and more fecal excretion of thyroxine than control animals.

Leonard et al. (1979) studied the ability of cholestyramine and colestipol to bind vitamin B_{12}, vitamin B_{12} intrinsic factor complex, folic acid, and iron citrate as functions of pH in solutions of 0.01 M NaCl, plus added HCl or NaOH to adjust the pH. Cholestyramine adsorbed vitamin B_{12} intrinsic factor, folic acid, and iron citrate strongly (80% or greater removal) at all pH levels, while colestipol adsorbed them well but not quite as strongly. Vitamin B_{12} was 14–19% bound over the entire pH range by cholestyramine and 1–14% bound by colestipol. Studies were also done in simulated gastric and duodenal juices. In these media, the folic acid was strongly bound by both resins. Significant binding (50% or greater) occurred otherwise only for iron citrate with cholestyramine, in both media. The authors suggest that regular monitoring of levels of such biochemicals be carried out during therapy with resins such as cholestyramine and colestipol.

Lindenbaum and Higuchi (1975) studied the binding of three bile acids (glycocholic acid, taurocholic acid, and taurodeoxycholic acid) to cholestyramine resin at pH 1.0, 3.0, and 4.2. The pH levels were established using HCl. The ionic strength was kept constant at 0.1 molar (except in one case) by adding NaCl as necessary. Cholestyramine binds the dissociated anion forms of bile acids. The anion A^- is formed by the dissociation

$$HA \rightleftarrows H^+ + A^-$$

and is then bound by the resin according to

$$RCl + A^- \rightleftarrows RA + Cl^-$$

where RCl is the chloride form of the resin R. Table 22.5 shows the results for 25°C.

At pH 1, glycocholic acid ($pK_a = 3.95$) is only slightly ionized and the glycocholic acid anion has to compete with Cl^- ions for binding sites on the resin. Thus, the resin bound the glycocholic acid poorly. At higher pH values (which generate more bile acid anions), the glycocholate bound much better. Note that when the ionic strength was

Table 22.5 Binding of Bile Acids by Cholestyramine[a]

Bile acid	pH	Ionic strength	Mmoles bound/ gram resin	Grams bound/ gram resin
Glycocholic acid	1.0	0.1	0.08	0.039
	3.0	0.001	2.1	1.00
	3.0	0.1	0.1	0.048
	4.2	0.1	0.26	0.125
Taurocholic acid	1.0	0.1	0.50	0.269
Taurodeoxycholic acid	1.0	0.1	3.63	1.89

[a]Values are those corresponding to equilibrium concentrations of the bile acids in solution of 0.0004 moles/L. From Lindenbaum and Higuchi (1975). Reproduced by permission of the American Pharmaceutical Association.

reduced to 0.001 molar, the amount of glycocholic acid bound by the resin was dramatically increased, since competition from Cl⁻ ions was much less.

Because taurine-conjugated acids are much stronger acids and dissociate well even at pH 1, taurocholate acid and taurodeoxycholate acid (actually, the anions of the acids) bound well even at pH 1, despite competition from Cl⁻ ions. It appears from this study that cholestyramine has the potential to bind bile acids in vivo, at least those that are reasonably strong acids.

Kos et al. (1991) studied the binding of the anions of bile salts by cholestyramine and the effect of competing anions. They first determined that pH had only a small effect on the binding of cholate ions to cholestyramine (0.36 mmol bound at pH 2–4 and 0.30 mmol bound at pH 7–8, in their experiments). They next determined Q_m values for different bile salt anions binding to the resin at pH 7.5 and 30°C. The values in mmol/g were: cholate anion, 3.49; glycocholate anion, 3.79; and taurocholate anion, 3.59. Under the same conditions, citrate anion had a Q_m of 1.48 mmol/g. When the competing citrate anion was added to solutions of sodium cholate in amounts of 0, 0.5, 1, 2, and 5 parts per part of cholate, the mmoles of cholate bound (pH 7.5, 37°C) were 0.95, 0.87, 0.76, 0.69, and 0.60, respectively. Thus, citrate did interfere with cholate binding but not to a large extent.

Zhu et al. (1992) compared cholestyramine and colestipol as adsorbents for bile acid anions (cholate, glycocholate) in vitro, in pH 7.4 tris and phosphate buffers. The results suggest that binding is mainly electrostatic (i.e., ionic), but that hydrophobic interactions are also important. Cholestyramine had a higher capacity for the bile acids than did colestipol. An increase in the ionic strength of the buffers decreased adsorption, as one would expect since it would increase competition for binding from the competing buffer anions.

Henning et al. (1982) did in vitro studies on the binding of bilirubin by cholestyramine from a pH 7.8 buffer at different temperatures. Q_m values in molecules of bilirubin per gram of resin at 10, 20, and 25°C were: 6.4×10^{19}, 7.9×10^{19} and 9.4×10^{19}, respectively. Binding increased with temperature, which is the opposite of what is expected for physical adsorption. This suggests that some activated chemical adsorption process must be involved. On the basis of the manufacturer's claim of 4×10^{-3} equivalents of quaternary ammonium ions per gram of resin, it was concluded that the resin binds one bilirubin molecule for every 25 ammonium ions, at 25°C.

G.   Use of Cholestyramine for Treating Hydrocarbon Toxicity

In Chapter 12, the use of sorbents for treating intoxications due to various hydrocarbons was discussed. Here we only mention those cases in which resins were involved. The reader is referred back to Chapter 12 for details.

Cholestyramine was used by McConnell et al. (1980) to try to induce polybrominated biphenyl (PBB) to desorb from the fatty tissues of PBB-loaded rats. This resin did not lower tissue PBB levels but it did seem to prevent progressive nephropathy.

Boylen et al. (1978) showed in vitro that cholestyramine can bind the organic pesticide chlordecone (Kepone); in experiments with chlordecone-loaded rats they found that cholestyramine bound chlordecone in the rats' intestines, increased its elimination in the feces, and decreased its content in the tissues. Based on these results in rats, Cohn et al. (1978) did trials in industrial workers who had been exposed to chlordecone and had developed symptoms of toxicity. They gave 16 g/day cholestyramine to 22 patients for five months; the half-life of chlordecone in the blood was decreased from 165 to 80 days and the half-life in the fatty tissues was decreased from 125 to 64 days. Concomitantly, neurological abnormalities in the patients decreased. Additional tests, by Guzelian (1981), showed that the important mechanism for chlordecone excretion into the intestines is not via the bile but by some other mechanism.

Kassner et al. (1993) used cholestyramine and charcoal to treat toxicity in mice due to lindane (a chlorinated hydrocarbon pesticide). Mice were given lindane either orally or intraperitoneally, followed by oral cholestyramine or charcoal (the dose amounts of these are not stated but presumably they were equal). The lindane dose at which 50% of the mice had convulsions, CD_{50}, or died, LD_{50}, were determined. The CD_{50} values for oral lindane were: control, 76 mg/kg; charcoal, 119 mg/kg; and cholestyramine, 193 mg/kg. The LD_{50} values for oral lindane were: control, 186 mg/kg; charcoal, 242 mg/kg; and cholestyramine, 375 mg/kg. Thus, both sorbents were effective but cholestyramine was the better of the two. In the intraperitoneally-administered lindane tests, the CD_{50} values were: control, 25 mg/kg; charcoal, 30 mg/kg; and cholestyramine, 34 mg/kg. The LD_{50} values were: control, 101 mg/kg; charcoal, 122 mg/kg; and cholestyramine, 82 mg/kg. These intraperitoneal results were not statistically different ($p > 0.05$). Note that the CD_{50} and LD_{50} values were lower than for orally administered lindane, which means that IP lindane is more toxic than oral lindane.

Some studies not discussed previously in Chapter 12 are discussed at this point. Bioulac et al. (1981) fed rats a 4 wt% cholestyramine diet and then, 5 days later, they were given IP injections of 5 mL/kg of a 20% (v/v) carbon tetrachloride solution in corn oil or 2.1 mL/kg of a 20% (v/v) solution of bromobenzene in corn oil. The rats were killed at different times thereafter and their livers were examined for signs of hepatotoxicity. The rats which had been pretreated with cholestyramine exhibited significantly less liver damage (less inflammation, less necrosis, etc.).

Rozman et al. (1982) examined the effect of cholestyramine on the disposition of pentachlorophenol in rhesus monkeys. Three monkeys were dosed orally with 50 mg/kg pentachlorophenol. Starting 24 hr later, 4 wt% cholestyramine was added to their diets for 6 days. As compared to control animals, the resin reduced the cumulative urinary excretion of the chemical from 35 to 5% and increased its cumulative fecal excretion from 3 to 54% of the dose administered. Total excretion was increased by 40%. Other tests showed that the chemical is strongly excreted into the bile (30% of the dose appeared in the bile in the first day). Cholestyramine interrupts the enterohepatic circulation of the chemical, as well as increasing its elimination directly across the intestinal wall.

H. Effect of Cholestyramine and Colestipol on Bacterial Toxins and Antibiotics

Nolan and Vilayat (1972) studied the toxicity of intraperitoneal injections of a cholestyramine-endotoxin suspension and the absorption of the toxin through the intestinal wall, in rats. The toxin was derived from *E. coli*. Cholestyramine significantly reduced the toxicity of the suspension placed in the peritoneal cavity and effectively inhibited the passage of the toxin through the intestinal wall.

Mullan et al. (1979) studied the binding of *E. coli* enterotoxins by cholestyramine and other adsorbents in vitro and in vivo (in mice and piglets). Kaolin and attapulgite showed no in vitro binding activity. Cholestyramine and one type of bentonite adsorbed 98% of the enterotoxin and the other adsorbents (other bentonite and resin types) showed intermediate results (18–39% binding). The two most effective adsorbents were selected for in vivo studies. In vivo, reduction of toxic effects by bentonite given with the toxin was zero, while the reduction was 89% with cholestyramine. Cholestyramine also reduced endotoxin-induced diarrhea in piglets significantly when given with the toxin. However, when the toxin was given beforehand to establish diarrhea first, cholestyramine had no beneficial effect. They hypothesized that the presence of sow's milk in the intestinal tracts of the piglets caused this poor result, but in fact it could be due to interference by several GI tract chemicals.

Taylor and Bartlett (1980) studied the binding of *Clostridium difficile* cytotoxin and vancomycin by cholestyramine and colestipol. Both resins bound the toxin very strongly in the amounts used. Also, vancomycin was bound well by the resins, with cholestyramine being somewhat more effective. The effect of the resins and of vancomycin in treating clindamycin-induced cecitis in hamsters showed that all were effective, with vancomycin being the best. The results suggest that the therapeutic benefit of cholestyramine in some patients with antibiotic-associated pseudomembranous colitis is due to its binding of the *C. difficile* cytotoxin.

King and Barriere (1981) studied the binding of vancomycin by cholestyramine in vitro using saline solutions buffered with sodium phosphate (pH 7.0). Solutions of 2 g/L vancomycin and 12 g/L cholestyramine were contacted for different periods of time. Vancomycin activity in the supernatants was determined by an agar diffusion technique involving *Bacillus subtilis* as the test organism (one wonders why a chemical analysis of unbound vancomycin would not have been simpler and more accurate). When the contact time was 15 min, 71% of the vancomycin activity disappeared. However, with a 72 hr contact time, only 48% of the vancomycin activity was lost. This strange result (one would expect the longer contact time to give an activity loss of at least 71%) may be explained by the methods used to separate the supernatant—centrifugation for the 15 min sample and mere settling of the resin for the 72 hr sample. Other researchers (Taylor and Bartlett, 1980) showed vancomycin binding to cholestyramine of roughly 90% when low concentrations of vancomycin (0.125 g/L) were mixed with high concentrations of cholestyramine (100 g/L). However, this is not inconsistent with the King/Barriere results, as a higher adsorbent:drug ratio always increases the percent bound of the drug.

Pantosti et al. (1985) studied the in vitro activity of the antimicrobial agents teicoplanin and vancomycin against fecal isolates of *Clostridium difficile* in the presence and absence of cholestyramine. Teicoplanin was found to be four times more potent than vancomycin when no resin was used. However, when the resin was added it bound the teicoplanin almost completely and reduced its activity by 99.7%. The resin reduced the vancomycin activity by 81.1%.

I. Effect of Sodium Polystyrene Sulfonate on Inorganic Species

In Chapter 18, we reviewed some studies in which both charcoal and sodium polystyrene sulfonate (SPS) were used to treat intoxications due to cesium, lithium, and potassium. We will not review those studies again but will consider a few other studies in which only SPS, and not charcoal, was used for treating lithium intoxication.

The binding of lithium ions (Li^+) in vitro by sodium polystyrene sulfonate was described by Gehrke and Gehrke (1986). Different amounts of the resin (0.25, 0.50, and 2.00 g) were exposed to varying concentrations of Li_2CO_3 solutions and the amounts of Li^+ bound were determined by flame photometry analysis. Three pH levels were used: 2.11, 7.21, and 10.8 (this report, an abstract, does not state what chemicals were used to establish the three pH levels). Additionally, competing K^+ ions were introduced in some cases by adding K_2CO_3 to the solutions. The resin was found to bind 0.452 mEq/g of Li^+. Variation of the pH and the presence of K^+ had little effect on this value.

Welch et al. (1987) also studied lithium binding by sodium polystyrene sulfonate. Using solutions of 100 mEq/L Li_2CO_3 in 0.1 N HCl (pH 7.0) and 0.2 N HCl (pH 1.0) at 37°C, the resin bound 0.7 mEq/g of Li^+ at both pH levels. Why this value is higher than that observed by Gehrke and Gehrke (0.452 mEq/g) is not clear but may be the result of the use of different buffer salts in the Gehrke and Gehrke study, salts which offered greater competition for binding.

As mentioned in Chapter 18, Linakis et al. (1989a,b) tested activated charcoal and sodium polystyrene sulfonate for their potential to bind lithium in mice. The mice were given orogastric doses of LiCl (250 mg/kg) followed immediately by oral doses of 10 g/kg SPS or 6.7 g/kg charcoal (the 1989a paper quotes the 6.7 g/kg figure but the 1989b abstract quotes 8.68 g/kg as the charcoal dose; otherwise the two reports appear to be identical). The mice were killed at 1, 2, 4, and 8 hr after treatment and their serum analyzed for Li concentration. There were no statistical differences between the charcoal-treated mice and the control mice. However, the SPS group had significantly lower Li concentrations than either of these two groups (Li^+ serum levels were 60.6, 59.4, 41.9, and 60.3% of control levels at 1, 2, 4, and 8 hr, respectively, when SPS was used). Thus, it appears that in order to effectively bind an inorganic cation such as Li^+, a cation exchange resin is more effective than charcoal.

In a second study, Linakis et al. (1989c, 1990a) did the same type of investigation except that multiple-dose SPS was used. The LiCl was again given as 250 mg/kg orally but then was followed with either water or 5 g/kg SPS at 0, 30, 90, 180, and 360 min. A third group received 2.5 g/kg SPS at these times. A fourth group was given water at 0 and 30 min, and 5 g/kg SPS at 90, 180, and 360 min. The data showed that the SPS significantly lowered serum Li^+ levels and that the effect was related to the SPS dose.

Linakis et al. (1990b) followed up their prior two studies to determine if multiple oral doses of SPS would enhance the elimination of intravenously administered Li. Mice were give LiCl (125 mg/kg) IV. Then, half were given water at 20, 40, 90, 150, and 210 min, while the other half received oral SPS (5 g/kg per dose) at the same times. Serum Li levels at 1, 2, 4, and 6 hr were determined for both groups. At these times, the serum levels of the SPS-treated groups averaged 87.4, 61.5, 36.5, and 29.7%, respectively, of the levels of the control group. Thus, SPS was effective in eliminating IV-administered lithium.

Tomaszewski et al. (1990) also showed that SPS reduces lithium absorption, this time in humans. Subjects were given 18.5 mg/kg (0.5 mEq/kg) lithium carbonate followed 1 hr later by 60 g/70 kg SPS in water or just the equivalent amount of water. Serum lithium levels were determined over 0–24 hr. The SPS reduced the AUC from a mean of 18.1 to 13.1 mEq-hr/L and peak lithium concentrations from a mean of 1.05 to 0.85 mEq/L. There was no significant effect on Li^+ excretion in the urine. Thus, SPS appears to bind Li^+ effectively in vivo in humans.

Roberge et al. (1993) reported a single case of lithium intoxication in which a 23-year-old woman was given 30 g of SPS in sorbitol orally every six hours for a total of five doses. During the 33 hr following initiation of SPS, her serum lithium fell from 4.20 to 0.68 mEq/L. The half-life value observed (12 hr) was significantly shorter than noted in other similar overdoses. The SPS therapy caused no adverse effects.

V. SUMMARY

The many studies on clays discussed in this chapter demonstrate several points: (1) since clays are cation exchangers, they can bind only molecules which are in cation form, (2) in vivo, cations which are normally present reduce the ability of clays to bind cationic drugs, by competing with them, (3) when the effectiveness of clays has been compared to that of charcoal, charcoal has usually been superior. While most in vivo studies with clays have shown poor effectiveness, montmorillonite has demonstrated significant benefit in rats dosed with kerosene and strychnine, and kaolin/pectin has shown an ability to reduce digoxin absorption in humans.

With respect to resins, it is also true that anion-exchange resins generally bind only anions well and that cation-exchange resins generally bind only cations well. However, the anion exchange resins cholestyramine and colestipol have often shown a remarkable ability to bind a wide variety of substances which are neutral in form.

A large number of studies (on the order of 16) showed that cholestyramine and colestipol significantly reduce the absorption of digitoxin, digoxin, and digoxin derivatives. One study concluded that the effect is not due to the binding of the glycosides directly but rather to a reduction in bile flow caused by interruption of the enterohepatic circulation of bile salts. This decreases the biliary excretion of the glycosides and increases their elimination by metabolization and excretion in the urine. Whether this is true deserves further scrutiny. Two studies (van Bever et al., 1976; Hall et al., 1977), however, have shown no effect of the resins on glycoside elimination.

Cholestyramine and colestipol have shown clear in vivo effectiveness in binding acetaminophen, sodium fusidate, warfarin, phenprocoumon, flufenamic and mefenamic acids, dicumarol, tromexan, hydrocortisone, hydrochlorothiazide, propranolol, tenoxicam, amiodarone, doxepin, piroxicam, chloroquine, glipizide, and methotrexate. On the other hand, no significant effects have been found with aspirin, clofibrate, prednisolone, phenytoin, phencyclidine, tolbutamide, and gentamicin.

With respect to endogenous chemicals, these resins have been found to bind bile acid anions, porphyrins (but not porphyrin precursors), lipids, folic acid, vitamin A, and iron citrate, but not triglycerides or amino acids. Bacterial toxins have been shown to bind well to cholestyramine, as does the antibiotic vancomycin.

Cholestyramine has effectively reduced the toxicity of several kinds of hydrocarbons: chlordecone, lindane, carbon tetrachloride, bromobenzene, and pentachlorophenol. However, the resins must be administered at the same time as the hydrocarbon. Once a hydrocarbon becomes lodged in the adipose tissues, the use of resins does not accelerate its removal (the same is true of charcoal, as we have seen in Chapter 20).

We have seen that the binding of ions by resins can be relatively strong. Studies performed to date with inorganic ions and resins have dealt only with cations (Cs^+, K^+, Li^+); the cation-exchange resin sodium polystyrene sulfonate has been shown to be effective in vivo for hyperkalemia and for lithium intoxication.

Overall, the picture that emerges is not clearly defined. More research needs to be done to determine why cholestyramine and colestipol are effective for some types of drugs/chemicals/toxins and are ineffective for others. The key may relate to the states of ionization of both the resins and the binding molecules at the prevailing pH levels in the GI tract, as well as to the chemical nature of the binding molecule species (hydrophobic/hydrophilic character, etc.).

VI. REFERENCES

AHFS Drug Information '92, American Hospital Formulary Service, American Society of Hospital Pharmacists, Bethesda, Maryland, 1992.

Albert, K. S., et al. (1978). Influence of kaolin-pectin suspension on digoxin bioavailability, J. Pharm. Sci. *67*, 1582.

Allgayer, H., Kruis, W., and Paumgartner, G. (1982). Studies on the in vitro binding of D-penicillamine to cholestyramine, Experientia *38*, 482.

Al-Shareef, A. H., Buss, D. C., and Routledge, P. A. (1990). Drug adsorption to charcoals and anionic binding resins, Hum. Exp. Toxicol. *9*, 95.

Arimori, K., Furukawa, E., and Nakano, M. (1992). Adsorption of imipramine onto activated charcoal and a cation exchange resin in macrogol-electrolyte solution, Chem. Pharm. Bull. *40*, 3105.

Armstrong, C. and Edwards, K. D. G. (1967). Multifactorial design for testing oral ion exchange resins, charcoal, and other factors in the treatment of aspirin poisoning in the rat: Efficacy of cholestyramine, Med. J. Aust. *2*, 301.

Armstrong, N. and Johns, A. (1974). Adsorption of morphine by kaolin, J. Hosp. Pharm. *32*, 185.

Atkinson, J. P. and Azarnoff, D. L. (1971). Comparison of charcoal and attapulgite as gastrointestinal sequestrants in acute drug ingestions, Clin. Toxicol. *4*, 31.

Audétat, V. and Bircher, J. (1976). Bioavailability of prednisolone during simultaneous treatment with cholestyramine (letter), Gastroenterol. *71*, 1110.

Baciewicz, A. M., Isaacson, M. L., and Lipscomb, G. L. (1983). Cholestyramine resin in the treatment of digitoxin toxicity, Drug Intell. Clin. Pharm. *17*, 57.

Bailey, D. N. (1992). Effect of pH changes and ethanol on the binding of tricyclic antidepressants to cholestyramine in simulated gastric fluid, Ther. Drug Monit. *14*, 343.

Bailey, D. N., Coffee, J. J., Anderson, B., and Manoguerra, A. S. (1992). Interaction of tricyclic antidepressants with cholestyramine in vitro, Ther. Drug Monit. *14*, 339.

Barzaghi, N., Monteleone, M., Amione, C., Lecchini, E., Perucca, E., and Frigo, G. M. (1988). Lack of effect of cholestyramine on phenytoin bioavailability, J. Clin. Pharmacol. *28*, 1112.

Bazzano, G., Gray, M., and Sansone-Bazzano, G. (1970). Treatment of digitalis intoxication with a new steroid-binding resin (abstract), Clin. Res. *18*, 592.

Bazzano, G. and Bazzano, G. S. (1972). Digitalis intoxication: Treatment with a new steroid-binding resin, JAMA *220*, 828.

Bergman, F., Heedman, P. A., and van der Linden, W. (1966). Influence of cholestyramine on absorption and excretion of thyroxine in Syrian hamster, Acta Endocrinol. *53*, 256.

Bioulac, P., Despuyoos, L., Bedin, C., Iron, A., Saric, J., and Balabaud, C. (1981). Decreased acute hepatotoxicity of carbon tetrachloride and bromobenzene by cholestyramine in the rat, Gastroenterol. *81*, 520.

Boylan, J. J., Egle, J. L., and Guzelian, P. S. (1978). Cholestyramine: Use as a new therapeutic approach for chlordecone (Kepone) poisoning, Science *199*, 893.

Brown, D. D. and Juhl, R. P. (1976). Decreased bioavailability of digoxin due to antacids and kaolin-pectin, New Engl. J. Med. *295*, 1034.

Browne, J. E., Feldkamp, J. R., White, J. L., and Hem, S. L. (1980). Potential of organic cation-saturated montmorillonite as treatment for poisoning by weak bases, J. Pharm. Sci. *69*, 1393.

Cady, W. J., Rheder, T. L., and Campbell, J. (1979). Use of cholestyramine resin in the treatment of digitoxin toxicity, Am. J. Hosp. Pharm. *36*, 92.

Caldwell, J. H. and Greenberger, N. J. (1971a). Interruption of the enterohepatic circulation of digitoxin by cholestyramine. I. Protection against lethal digitoxin intoxication, J. Clin. Invest. *50*, 2626.

Caldwell, J. H. and Greenberger, N. J. (1971b). Interruption of the enterohepatic circulation of digitoxin by cholestyramine. II. Effect on metabolic disposition of tritium-labeled digitoxin and cardiac systolic intervals in man, J. Clin. Invest. *50*, 2638.

Callaghan, J. T., Tsuru, M., Holtzman, J. L., and Hunninghake, D. B. (1983). Effect of cholestyramine and colestipol on the absorption of phenytoin, Eur. J. Clin. Pharmacol. *24*, 675.

Calvo, M. V. and Dominguez-Gil, A. (1984). Interaction of naproxen with cholestyramine, Biopharm. Drug Dispos. *5*, 33.

Carruthers, S. G. and Dujovne, C. A. (1980). Cholestyramine and spironolactone and their combination in digitoxin elimination, Clin. Pharmacol. Ther. *27*, 184.

Chin, L., Picchioni, A. L., and Duplisse, B. R. (1969). Comparative antidotal effectiveness of activated charcoal, Arizona montmorillonite, and evaporated milk, J. Pharm. Sci. *58*, 1353.

Cohen, M. I., Winslow, P. R., and Boley, S. J. (1969). Intestinal obstruction associated with cholestyramine therapy, New Engl. J. Med. *280*, 1285.

Cohn, W. J., et al. (1978). Treatment of chlordecone (Kepone) toxicity with cholestyramine: Results of a controlled clinical trial, N. Engl. J. Med. *298*, 243.

Coltman, D., Mann, M. D., and Bowie, M. D. (1990). Effect of cholestyramine on activity of gentamicin in vitro (letter), Pediatrics *85*, 390.

Demers, H. G., Pabst, J., and Piper, C. (1982). [Pharmacokinetics in cholestyramine treatment of digitoxin intoxication], Deutsch. Med. Wochenschr. *107*, 1476.

Dordoni, B., Willson, R. A., Thompson, R. P. H., and Williams, R. (1973). Reduction of absorption of paracetamol by activated charcoal and cholestyramine: A possible therapeutic measure, Br. Med. J. *3*, 86.

Edwards, K. D. G. (1965). The use of oral ion exchange resins in drug poisoning: General hypothesis and some studies on cholestyramine versus barbiturates, Med. J. Aust. *2*, 925.

Edwards, K. D. G. and McCredie, M. (1967). Studies on the binding properties of acidic, basic, and neutral drugs to anion and cation exchange resins and charcoal in vitro, Med. J. Aust. *1*, 534.

Erttmann, R. and Landbeck, G. (1985). Effect of oral cholestyramine on the elimination of high-dose methotrexate, J. Cancer Res. Clin. Oncol. *110*, 48.

Evcim, N. and Barr, M. (1955). Adsorption of some alkaloids by different clays, J. Am. Pharm. Assoc. *44*, 570.

Ferry, D. G., Gazeley, L. R., Busby, W. J., Beasley, D. M. G., Edwards, I. R., and Campbell, A. J. (1990). Enhanced elimination of piroxicam by administration of activated charcoal or cholestyramine, Eur. J. Clin. Pharmacol. *39*, 599.

Frésard, F., Balant, L., Noble, J., Garcia, B., and Muller, A. F. (1979). Cholestyramine intoxication à la digoxine: Efficacité thérapeutique?, Schweiz. Med. Wschr. *109*, 431.

Gallo, D. G., Bailey, K. R., and Sheffner, A. L. (1965). The interaction between cholestyramine and drugs, Proc. Soc. Exp. Biol. Med. *120*, 60.

Geeze, D. S., Wise, M. G., and Stigelman, W. H., Jr. (1988). Doxepin-cholestyramine interaction, Psychosomatics *29*, 233.

Gehrke, J. and Gehrke, C. W. (1986). In vitro binding of Li^+ ion using a clinical cation exchange resin (Kayexalate) (abstract), Ann. Emerg. Med. *15*, 651.

Gendrel, D., Verdier, F., Richard-Lenoble, D., and Nardou, M. (1990). Interaction enter cholestyramine et chloroquine, Arch. Fr. Pediatr. *47*, 387.

Gilfrich, H. J., Kasper, W., Meinertz, T., Okonek, S., and Bork, R. (1978). Treatment of massive digitoxin overdose by charcoal hemoperfusion and cholestyramine, Lancet, March 4, p. 505.

Guentert, T. W., Schmitt, M., and Defoin, R. (1986). Acceleration of the elimination of tenoxicam by cholestyramine in the dog, J. Pharmacol. Exp. Ther. *238*, 295.

Guentert, T. W., Defoin, R., and Mosberg, H. (1988). The influence of cholestyramine on the elimination of tenoxicam and piroxicam, Eur. J. Clin. Pharmacol. *34*, 283.

Guzelian, P. S. (1981). Therapeutic approaches for chlordecone poisoning in humans, J. Toxicol. Environ. Health *8*, 757.

Haacke, H., Johnsen, K., and Kolenda, K. D. (1973). [On the therapy of digitalis intoxication: Another experimental indication of the efficacy of adsorbents], Med. Welt. *24*, 1374.

Hahn, K. J., Eiden, W., Schettle, M., Hahn, M., Walter, E., and Weber, E. (1972). Effect of cholestyramine on the gastrointestinal absorption of phenprocoumon and acetylsalicylic acid in man, Eur. J. Clin. Pharmacol. *4*, 142.

Hall, W. H., Shappell, S. D., and Doherty, J. E. (1977). Effect of cholestyramine on digoxin absorption and excretion in man, Am. J. Cardiol. *39*, 213.

Henderson, R. P. and Solomon, C. P. (1988). Use of cholestyramine in the treatment of digoxin intoxication, Arch. Intern. Med. *148*, 745.

Henderson, R. P. and Solomon, C. P. (1989). Activated charcoal should replace the resins in the treatment of digoxin intoxication (reply to letter), Arch. Intern. Med. *149*, 2603.

Henning, D. S., Brown, G. R., and St-Pierre, L. E. (1982). The adsorption of bilirubin from aqueous solution onto solid cholestyramine and polyvinylpyrrolidone, Int. J. Artif. Organs *5*, 373.

Herman, R. J. and Chaudhary, A. (1991). In vitro binding of lorazepam and lorazepam gluconuride to cholestyramine, colestipol, and activated charcoal, Pharm. Res. *8*, 538.

Hibbard, D. M., Peters, J. R., and Hunninghake, D. B. (1984). Effects of cholestyramine and colestipol on the plasma concentrations of propranolol, Br. J. Clin. Pharmacol. *18*, 337.

Hunninghake, D. B. and Pollack, E. (1977). Effect of bile acid sequestering agents on the absorption of aspirin, tolbutamide, and warfarin (abstract), Fed. Proc. *36*, 966.

Hunninghake, D. B., King, S., and LaCroix, K. (1982). The effect of cholestyramine and colestipol on the absorption of hydrochlorothiazide, Int. J. Clin. Pharmacol. Ther. Toxicol. *20*, 151.

Hunninghake, D. B. and Hibbard, D. M. (1986). Influence of time intervals for cholestyramine dosing on the absorption of hydrochlorothiazide, Clin. Pharmacol Ther. *39*, 329.

Jänchen, E., Meinertz, H. J., Gilfrich, F., Kersting, F., and Groth, U. (1978). Enhanced elimination of warfarin during treatment with cholestyramine, Br. J. Clin. Pharmacol. *5*, 437.

Johansson, C., Adamsson, U., Stierner, U., and Lindsten, T. (1978). Interaction by cholestyramine on the uptake of hydrocortisone in the gastrointestinal tract, Acta Med. Scand. *204*, 509.

Johns, W. H. and Bates, T. R. (1969). Quantification of the binding tendencies of cholestyramine. I. Effect of structure and added electrolytes on the binding of unconjugated and conjugated bile-salt anions, J. Pharm. Sci. *58*, 181.

Johns, W. H. and Bates, T. R. (1970a). Quantification of the binding tendencies of cholestyramine. II. Mechanism of interaction with bile salt and fatty acid anions, J. Pharm. Sci. *59*, 329.

Johns, W. H. and Bates, T. R. (1970b). Quantification of the binding tendencies of cholestyramine. III. Rates of adsorption of conjugated bile salt anions onto cholestyramine as a function of added inorganic electrolyte concentration, temperature, and agitation intensity, J. Pharm. Sci. *59*, 788.

Johns, W. H. and Bates, T. R. (1972a). Drug-cholestyramine interactions. I. Physicochemical factors affecting in vitro binding of sodium fusidate to cholestyramine, J. Pharm. Sci. *61*, 730.

Johns, W. H. and Bates, T. R. (1972b). Drug-cholestyramine interactions. II. Influence of cholestyramine on GI absorption of sodium fusidate, J. Pharm. Sci. *61*, 735.

Juhl, R. P. (1979). Comparison of kaolin-pectin and activated charcoal for inhibition of aspirin absorption, Am. J. Hosp. Pharm. *36*, 1097.

Kassner, J. T., Maher, T. J., Hull, K. M. and Woolf, A. D. (1993). Cholestyramine as an adsorbent in acute lindane poisoning: A murine model, Ann. Emerg. Med. *22*, 1392.

Khalil, S. A. H. (1974). The uptake of digoxin and digitoxin by some antacids, J. Pharm. Pharmacol. *26*, 961.

Khouw, V., Giles, H.G., and Sellers, E. M. (1978). Binding of drugs to ion-exchange resins in simulated gastric fluid, J. Pharm. Sci. *67*, 1329.

Kilgore, T. L. and Lehmann, C. R. (1982) Treatment of digoxin intoxication with colestipol, South. Med. J. *75*, 1259.

King, C. Y. and Barriere, S. L. (1981). Analysis of the in vitro interaction between vancomycin and cholestyramine, Antimicrob. Agents Chemother. *19*, 326.

Kivistö, K. T. and Neuvonen, P. J. (1990). The effect of cholestyramine and activated charcoal on glipizide absorption, Br. J. Clin. Pharmacol. *30*, 733.

Ko, H. and Royer, M. E. (1974). In vitro binding of drugs to colestipol hydrochloride, J. Pharm. Sci. *63*, 1914.

Kos, R., White, J. L, Hem, S. L., and Borin, M. T. (1991). Effect of competing anions on binding of bile salts by cholestyramine, Pharm. Res. *8*, 238.

Krasopoulos, J. C., De Bari, V. A., and Needle, M. A. (1980). The adsorption of bile salts on activated carbon, Lipids *15*, 365.

Kuhlmann, J. (1984). Use of cholestyramine in three patients with β-acetyldigoxin, β-methyldigoxin and digitoxin intoxication, Int. J. Clin. Pharmacol. Ther. Toxicol. *22*, 543.

Leonard, J. P., Desager, J. P., Beckers, C., and Harvengt, C. (1979). In vitro binding of various biological substances by two hypercholesterolaemic resins: Cholestyramine and colestipol, Arzneim.-Forsch./Drug. Res. *29*, 979.

Linakis, J. G. et al. (1989a). Administration of activated charcoal or sodium polystyrene sulfonate (Kayexalate) as gastric decontamination for lithium intoxication: An animal model, Pharmacol. Toxicol. *65*, 387.

Linakis, J. G. et al. (1989b). Activated charcoal and sodium polystyrene sulfonate (Kayexalate) in gastric decontamination for lithium intoxication: An animal model (abstract), Ann. Emerg. Med. *18*, 445.

Linakis, J. G. et al. (1989c). Multiple-dose sodium polystyrene sulfonate (SPS) in lithium (Li) intoxication: An animal model (abstract), Vet. Hum. Toxicol. *31*, 364.

Linakis, J. G. et al. (1990a). Role of repetitive Kayexalate in lowering serum lithium concentrations in the mouse (abstract), Ann. Emerg. Med. *19*, 465.

Linakis, J. G., Hull, K. M., Lacouture, P. G., Maher, T. J., and Lewander, W. J. (1990b). Enhancement of lithium (Li) elimination by multiple dose sodium polystyrene sulfonate (abstract), Vet. Hum. Toxicol. *32*, 351.

Lindenbaum, S. and Higuchi, T. (1975). Binding of bile acids to cholestyramine at gastric pH conditions, J. Pharm. Sci. *64*, 1887.

Lloyd-Still, J. D. (1977). Cholestyramine therapy and intestinal obstruction in infants, Pediatrics *59*, 626.

Longenecker, J. B. and Basu, S. G. (1965). Effect of cholestyramine on absorption of amino acids and vitamin A in man (abstract), Fed. Proc. *24*, 375.

Lyddane, J. E., Thomas, B. F., Compton, D. R., and Martin, B. R. (1988). Modification of phencyclidine intoxication and biodisposition by charcoal and other treatments, Pharmacol. Biochem. Behav. *30*, 371.

McConnell, E. E., Harris, M. W., and Moore, J. A. (1980). Studies on the use of activated charcoal and cholestyramine for reducing the body burden of polybrominated biphenyls, Drug. Chem. Toxicol. *3*, 277.

McGinity, J. W. and Lach, J. L. (1976). In vitro adsorption of various pharmaceuticals to montmorillonite, J. Pharm. Sci. *65*, 896.

Meinertz, T., Gilfrich, H. J., Bork, R., and Jänchen, E. (1977a). Treatment of phenprocoumon intoxication with cholestyramine, Br. Med. J., August 13, p. 439.

Meinertz, T., Gilfrich, H. J., Groth, U., Jonen, H. G., and Jänchen, E. (1977b). Interruption of the enterohepatic circulation of phenprocoumon by cholestyramine, Clin. Pharmacol. Ther. *21*, 731.

Mullan, N. A., Burgess, M. N., Bywater, R. J., and Newsome, P. M. (1979). The ability of cholestyramine resin and other adsorbents to bind *Escherichia coli* enterotoxins, J. Med. Microbiol. *12*, 487.

Neuvonen, P. J., Kivistö, K., and Hirvisalo, E. L. (1988). Effects of resins and activated charcoal on the absorption of digoxin,

carbamazepine, and frusemide, Br. J. Clin. Pharmacol. *25*, 229.

Neuvonen, P. J. and Kivistö, K. T. (1989). Activated charcoal should replace the resins in the treatment of digoxin intoxication (letter), Arch. Intern. Med. *149*, 2603.

Neuvonen, P. J., Kuusisto, P., Manninen, V., Vapaatalo, H., and Miettinen, T. A. (1989). The mechanism of the hypercholesterolaemic effect of activated charcoal, Eur. J. Clin. Invest. *19*, 251.

Nitsch, J. and Lüderitz, B. (1986). [Enhanced elimination of amiodarone by cholestyramine], Deutsch Med. Wochenschr. *111*, 1241.

Nolan, J. P. and Vilayat, M. (1972). Effect of cholestyramine on endotoxin toxicity and absorption, Dig. Dis. *17*, 161.

Okonek, S., Setyadharma, H., Borchert, A., and Krienke, E. G. (1982a). Activated charcoal is as effective as Fuller's earth or bentonite in paraquat poisoning, Klin. Wochenschr. *60*, 207.

Okonek, S., et al. (1982b). Successful treatment of paraquat poisoning: Activated charcoal per os and "continuous hemoperfusion", J. Toxicol.-Clin. Toxicol. *19*, 807.

Pantosti, A., Luzzi, I., Cardines, R., and Gianfrilli, P. (1985). Comparison of the in vitro activities of teicoplanin and vancomycin against *Clostridium difficile* and their interactions with cholestyramine, Antimicrob. Agents Chemother. *28*, 847.

Park, G. D., Spector, R., and Kitt, T. M. (1988). Superactivated charcoal versus cholestyramine for cholesterol lowering: A randomized cross-over trial, J. Clin. Pharmacol. *28*, 416.

Payne, V. W., Sector, R. A., and Noback, R. K. (1982). Use of colestipol in a patient with digoxin intoxication, Drug Intell. Clin. Pharm. *15*, 902.

Pieroni, R. E. and Fisher, J. G. (1981). Use of cholestyramine resin in digitoxin toxicity, JAMA *245*, 1939.

Pimstone, N. R., Gandhi, S. N., and Mukerji, S. K. (1987). Therapeutic efficacy of oral charcoal in congenital erythropoietic porphyria, New Engl. J. Med. *316*, 390.

Renowden, S., Westmoreland, D., White, J. P., and Routledge, P. A. (1985). Oral cholestyramine increases elimination of warfarin after overdose, Br. Med. J. *291*, 513.

Roberge, R. J., Martin, T. G., and Schneider, S. M. (1993). Use of sodium polystyrene sulfonate in a lithium overdose, Ann. Emerg. Med. *22*, 1911.

Robinson, D. S., Benjamin, D. M., and McCormack, J. J. (1971). Interaction of warfarin and nonsystemic gastrointestinal drugs, Clin. Pharmacol. Ther. *12*, 491.

Rosenberg, H. A. and Bates, T. R. (1974). Inhibitory effect of cholestyramine on the absorption of flufenamic and mefenamic acids in rats, Proc. Soc. Exp. Biol. Med. *145*, 93.

Rozman, T., Ballhorn, L., Rozman, K., Klaassen, C., and Greim, H. (1982). Effect of cholestyramine on the disposition of pentachlorophenol in rhesus monkeys, J. Toxicol. Environ. Health *10*, 277.

Said, S. and Al-Shora, H. (1980). Adsorption of certain oral hypoglycaemics on kaolin and charcoal and its relationship to hypoglycaemic effects of drugs, Int. J. Pharmaceut. *5*, 223.

Saltzman, M. J., Beyer, M. M., and Friedman, E. A. (1976). Mechanism of life prolongation in nephrectomized rats treated with oxidized starch and charcoal, Kidney Int. *10* (Suppl.), S-343.

Saral, R. and Spratt, J. L. (1967). Alteration of oral digitoxin toxicity and its in vitro binding by cholestyramine, Arch. Int. Pharmacodyn. *167*, 10.

Scholtens, R., et al. (1982). In vitro adsorption of oxalic acid and glyoxylic acid onto activated charcoal, resins, and hydrous zirconium oxide, Int. J. Artif. Organs *5*, 33.

Sedaghat, A. and Ahrens, E. H. (1975). Lack of effect of cholestyramine on the pharmacokinetics of clofibrate in man, Eur. J. Clin. Invest. *5*, 177.

Siegers, C. P and Möller-Hartmann, W. (1989). Cholestyramine as an antidote against paracetamol-induced hepato- and nephrotoxicity in the rat, Toxicol. Lett. *47*, 179.

Smith, R. P., Gosselin, R. E., Henderson, J. A., and Anderson, D. M. (1967). Comparison of the adsorptive properties of activated charcoal and Alaskan montmorillonite for some common poisons, Toxicol. Appl. Pharmacol. *10*, 95.

Sorby, D. L. and Plein, E. M. (1961). Adsorption of phenothiazine derivatives by kaolin, talc, and Norit, J. Pharm. Sci. *50*, 355.

Sorby, D. L. (1965). Effect of adsorbents on drug absorption: I. Modification of promazine absorption by activated attapulgite and activated charcoal, J. Pharm. Sci. *54*, 677.

Sorby, D. L., Plein, E. M., and Benmaman, J. D. (1966). Adsorption of phenothiazine derivatives by solid adsorbents, J. Pharm. Sci. *55*, 785.

Stul, M. S., Vliers, D. P., and Uytterhoven, J. B. (1984). In vitro adsorption-desorption of phenethylamines and phenylimidazoles by a bentonite and a resin, J. Pharm. Sci. *73*, 1372.

Tagaki, S., Yamashita, M., Suga, H., and Naito, H. (1983). The effectiveness of cation exchange resin as an adsorbent of paraquat both in vitro and in vivo, Vet. Hum. Toxicol. *25* (Suppl. 1), 34.

Takahama, T., et al. (1983). Effect of oral adsorbent on chronic hepatic disturbance, Trans. Am. Soc. Artif. Intern. Organs *XXIX*, 704.

Taylor, N. S. and Bartlett, J. G. (1980). Binding of *Clostridium difficile* cytotoxin and vancomycin by anion-exchange resins, J. Infect. Dis. *141*, 92.

Tembo, A. V. and Bates, T. R. (1974). Impairment by cholestyramine of dicumarol and tromexan absorption in rats: A potential drug interaction, J. Pharmacol. Exp. Ther. *191*, 53.

Thompson, W. G. (1971). Cholestyramine, Can. Med. Assoc. J. *104*, 305.

Thompson, W. G. (1973). Effect of cholestyramine on absorption of ^{3}H digoxin in rats, Am. J. Dig. Dis. *18*, 851.

Ti, T. Y., Giles, H. G., and Sellers, E. M. (1978). Probable interaction of loperamide and cholestyramine, Can. Med. Assoc. J. *119*, 607.

Tishler, P. V. and Winston, S. H. (1985). Sorbent therapy of the porphyrias. IV. Adsorption of porphyrins by sorbents in vitro, Meth. Find. Exptl. Clin. Pharmacol. *7*, 485.

Tishler, P. V., Winston, S. H., and Bell, S. M. (1987). Correlative studies of the hypercholesterolemic effect of a highly activated charcoal, Methods Find. Exp. Clin. Pharmacol. *9*, 799.

Tomaszewski, C., Musso, C., Pearson, J. R., Kulig, K., and Rumack, B. (1990). Prevention of lithium absorption by sodium polystyrene sulfonate in volunteers (abstract), Vet. Hum. Toxicol. *32*, 351.

van Bever, R. J., Duchateau, A. M. J. A., Pluym, B. F. M., and Merkus, F. W. H. M. (1976). The effect of colestipol on digitoxin plasma levels, Drug Res. *26*, 1891.

Wai, K. N. and Banker, G. S. (1966). Some physicochemical properties of the montmorillonites, J. Pharm. Sci. *55*, 1215.

Weber, J. B., Perry, P. W., and Upchurch, R. P. (1965). Influence of temperature and time on the adsorption of paraquat, diquat, 2,4-D, and prometone by clays, charcoal, and an anion-exchange resin, Soil Sci. Soc. Am. Proc. *29*, 678.

Welch, D. W., Driscoll, J. L., Lewander, W. J., and Johnson, P. N. (1987). In vitro lithium binding with sodium polystyrene sulfonate (abstract), Vet. Hum. Toxicol. *29*, 472.

Winston, S. H. and Tishler, P. V. (1986). Sorbent therapy of the porphyrias. V. Adsorption of the porphyrin precursors delta-aminolevulinic acid and porphobilinogen by sorbents in vitro, Meth. Find. Exp. Clin. Pharmacol. *8*, 233.

Yamashita, M., Naito, H., and Takagi, S. (1987). The effectiveness of a cation resin (Kayexalate) as an adsorbent of paraquat: Experimental and clinical studies, Hum. Toxicol. *6*, 89.

Zhu, X. X., Brown, G. R., and St-Pierre, L. E. (1992). Polymeric adsorbents for bile acids: I. Comparison between cholestyramine and colestipol, J. Pharm. Sci. *81*, 65.

23

Other Medicinal Uses of Charcoal in Humans

Activated charcoal has a long history of medical applications other than as an oral antidote. Charcoal has been used to treat body surface wounds and ulcers, insect bites, and other skin conditions. It has also been used for a wide variety of gastrointestinal problems (including excessive gas, diarrhea, peptic ulcers, and the deodorizing of ostomies). One interesting study recommended it for the treatment of alcoholism. A dramatic but poorly appreciated application of charcoal is its enteral use for the treatment of endotoxin shock due to abdominal sepsis in patients having acute renal failure.

Additionally, drugs adsorbed to charcoal have been used as prolonged-release medications. In the older literature, there are several reports of the use of intravenously administered slurries of charcoal in water for treating various diseases (an incredibly dangerous idea which, fortunately, was soon abandoned).

For the purpose of bringing these other in vivo medical applications to the attention of present-day audiences, we now review some of the studies in these various areas of application.

I. EFFECT OF CHARCOAL ON SURFACE WOUNDS

In addition to Kehls (1793), mentioned under Section II in Chapter 2, in which the external application of charcoal to gangrenous ulcers for the removal of bad odors was cited, several other works recommend the use of charcoal for treating surface wounds. Schobesch (1938) indicated that aqueous solutions or pastes of yperite (dichlorodiethyl sulfide) readily irritated the skin of rabbits. Activated charcoal, applied to such injured areas, was the most effective treatment found. Even after a 10 min delay, good results were obtained. Peyer (1940) reported that "coffee chars" are efficacious on various surface wounds; however, Riedel (1940) suggested that the therapeutic value of charcoals derived from coffee could not be due to their adsorptive powers, since he found that they were not very adsorptive (since these charcoals are not of the activated type, it is

525

probably true that they have little adsorptive power). However, he did agree that the coffee charcoals were for some reason therapeutic.

One might be tempted to dismiss these old studies because they are so out-of-date; certainly by their very nature they were very qualitative. However, from more modern studies which have shown that charcoal can adsorb bacteria effectively, we may have significant confidence in much of this older work.

The most scientific study carried out on the effect of activated charcoal for treating wounds is by Beckett et al. (1980). They used activated charcoal cloth (made by pyrolyzing a rayon fabric and then activating the resulting charred cloth by exposure to an oxidizing agent). Twenty-six patients with chronic leg ulcers and 13 patients with suppurating post-operative wounds had the charcoal cloth applied as a dressing. All wounds were malodorous. Wound odor was reduced noticeably in 95% of the patients and self-cleansing of the wounds occurred in 80% of the patients. No adverse reactions to the charcoal cloth were observed, nor did the cloth adhere to the wounds and cause any difficulties in removal of the dressings. In vitro experiments with patches of the cloth dropped into solutions of bacteria showed that bacterial counts in the solutions decreased by 1000- to 100,000-fold due to binding of the bacteria by the cloth. Hence, it appears that bacterial adsorption by the cloth accounts for its effectiveness as a wound dressing.

By the early 1980's, an activated charcoal cloth (derived from rayon, as mentioned above) was available commercially in the United Kingdom. The brand name was Actisorb (Johnson & Johnson, Ltd.). Mulligan et al. (1986) indicated that by 1986 this cloth had already received extensive use in hospitals, owing to its ability to improve wounds and reduce offensive odors. Mulligan's group treated 97 patients for ulcerous wounds with Actisorb and found significant improvement in the condition of the ulcerous wounds. Increased epithelialization occurred and there were greatly reduced levels of exudate, odor, and edema (Figure 23.1). An average reduction of 0.8% per day in the size of the wounds was noted, compared to 0.3% per day in untreated patients.

Another study of charcoal cloth, reported in the Russian literature by Ustinova et al. (1987) mentioned that portions of charcoal cloth were implanted subcutaneously in experimental animals and that there was an insignificant soft tissue response over the time period of two months involved. They recommend such cloth material as a wound dressing.

From Martindale's *The Extra Pharmacopeia*, Twenty-Ninth Edition (1989), one may note other wound dressing materials containing charcoal which are available in the United Kingdom. Besides the Actisorb cloth, there is Actisorb Plus (also made by Johnson & Johnson), which contains silver impregnated into the activated charcoal cloth (silver is bactericidal). A charcoal cloth dressing called Carbonet (Smith & Nephew) is also available. This has absorbent and nonadherent layers of material in addition to the charcoal cloth. Lyofoam C (Ultra), having a charcoal cloth layer plus hydrophilic, hydrophobic, and polyurethane foam layers is a third charcoal dressing product available in the United Kingdom.

A study by Wunderlich and Orfanos (1991) involved the use of a dry wound dressing (SIAX) composed of silver-impregnated activated charcoal "xerodressing." This was applied to 19 patients with venous leg ulcers. A comparison group of 19 more patients was treated with conventional ointments (e.g., zinc paste). The ulcers of 6 of the patients in the SIAX group completely healed in the 6 wk study, compared to those of only 2 patients in the other group. The SIAX group showed significantly increased epithelialization and reduced ulcer size.

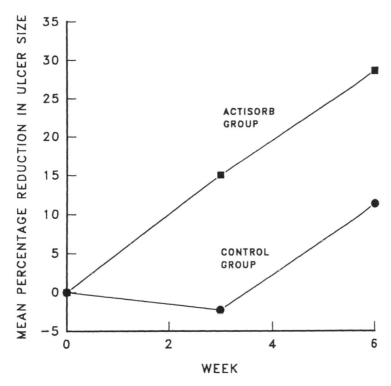

Figure 23.1 Mean reduction in ulcerous wound size with and without the use of Actisorb activated charcoal cloth. From Mulligan et al. (1986). Reproduced by permission of Medicom UK Ltd.

II. EFFECT OF CHARCOAL ON VARIOUS SKIN PROBLEMS

Rosenvold (1968) mentioned the use of charcoal as a "remedy in ophthalmology and otolaryngology," namely, that charcoal poultices (pastes of charcoal made by mixing powdered charcoal with a liquid—usually water—and placed on the skin, often with a cloth covering) had been used much in the era prior to the advent of antibiotics. They were used to treat gangrenous and other foul wounds. Uses that Rosenvold found for such poultices included the treatment of inflamed tissues of the face, eyelids, and ears. His method was, in general, to place a few charcoal tablets on a piece of thin cotton cloth, and then fold the cloth over the tablets from all four sides, making four layers of cloth on the top side and leaving a single layer of cloth on the side that was applied to the skin. This bundle was then dipped into hot water and the imbibed water disintegrated the tablets into a paste. The compress was then gently wrung out, to eliminate excess dripping water, and applied to the affected area for 10–15 minutes several times daily.

Rosenvold applied this type of compress to infected areas of skin (anywhere on the body) and, more specifically, to swollen eyes. He used charcoal pastes (without the cloth) placed in the ear canal to treat inflammations of the external ear and used charcoal tablets placed against oral ulcers to treat such ulcers. He claimed that when the charcoal poultice is in direct contact with skin infections, it adsorbs the toxins and promotes rapid healing. The mechanism of action whereby charcoal compresses can effect relief of deeper swellings and indurations is not clear.

In a book on home remedies, Thrash and Thrash (1981) described the preparation of various charcoal poultices. For treating large areas, they recommended a poultice made with ground flaxseed (which is a thickener) and boiling water. The paste, after cooling, is spread 1/4 inch deep on a piece of cloth of appropriate size, and covered with another piece of cloth. After placing this "patch" on the affected area, it is covered with a layer of plastic wrap (to keep it from drying out), held in place with a roller bandage, and is left in place for 6–10 hr. Small portions of poultice, for treating bee stings, spider bites, etc., are more easily made by just mixing powdered charcoal or crushed charcoal tablets with sufficient water and putting the paste on a piece of facial tissue or paper towel. Thrash and Thrash used these poultices for all varieties of insect bites (bees, wasps, ants, mosquitos, chiggers) and for skin rashes such as those caused by poison ivy. They had particular success in treating bites from the very toxic brown recluse spider.

The uses of activated charcoal for the home remedy purposes just described need to be viewed with a certain amount of reservation, as the cases reported are anecdotal and are not the result of well-designed scientific studies. Despite this caution, there is ample reason to believe that charcoal poultices, especially when applied in more or less direct contact with infected areas of skin, can be effective, as it is well known that activated charcoal can effectively adsorb bacteria and toxins. The reported efficacy of poultices for treating swollen tissues and similar subcutaneous problems is more in doubt, however, as the mechanisms by which benefits might occur are not clear.

Finally, we have mentioned in Chapter 20 that Biehl et al. (1989) studied the effect of superactive charcoal paste in the prevention of T-2 toxin-induced local cutaneous effects in topically exposed swine. T-2 aliquots (6 mg dissolved in 90% dimethylsulfoxide) were applied to different skin sites 9 cm^2 in area. The charcoal paste was then applied at times ranging from 5 to 65 min. Skin lesions were examined at 1, 3, and 6 days. The charcoal paste significantly reduced lesion severity. It seems reasonable to assume that the charcoal was able to bind the T-2 toxin effectively and reduce its absorption into the skin.

III. EFFECT ON GI TRACT BACTERIAL SEPSIS

Whereas the ability of activated charcoal to adsorb bacterial toxins was well known and fairly extensively studied in the period from 1910 to the mid-1930s, its value in treating infections of the digestive tract seems thereafter to have been unappreciated, if not forgotten. It is therefore very interesting to note that a paper by Kopp (1978) appeared having the title "The Unexpected Success of Enteral Activated Carbon in Acute Renal Failure Patients in the Prevention and Therapy of Abdominal Sepsis."

Kopp stated that the sequence of paralytic ileus, septic peritonitis, and endotoxin shock has been one of the major unconquered and frustrating problems in the treatment of patients with acute renal failure. Despite heroic attempts, these complications invariably have led to death. The use of charcoal in treating drug intoxications gave Kopp the idea of trying it for abdominal sepsis:

The results are indeed impressive. Toxic degradation products within the intestinal lumen are efficiently adsorbed. Paralytic small and large bowel distension disappears, toxic damage to the liver and ... the "spill over" phenomenon of bacterial toxins into the systemic circulation are obviously prevented. Other conventional attempts, e.g., whole gut sterilization, etc., were nonsuccessful in our experience. Survival of patients has been achieved in a number of cases even in the presence of incipient abdominal catastrophe, using enteral carbon. The results have also important implications in the management of surgical and trauma patients in general.

IV. EFFECT ON INTESTINAL GAS

Riese and Damrau (1964) gave results of a study in which charcoal was used to treat flatulence, as well as other intestinal disorders (diarrhea, foul-smelling stools). We discuss the flatulence results here and the other results in later sections. Forty-three patients having flatulence and distention were given 2 capsules of vegetable charcoal, four times daily. Each capsule had 4 grains of charcoal (0.26 g). Thus, the daily total charcoal dose was 2.08 g. Good results (disappearance of flatus and distention, with minimal rectal expulsion of gas) were obtained in 28 cases (65%).

Some additional impressive evidence that activated charcoal can be effective in treating various intestinal disorders was reported by Chevrel (1978). Using a mixture called Carbomucil (30 g activated charcoal powder, 14 g magnesium bicarbonate, 33 g sterculia gum, and 100 g excipient), Chevrel studied its effects, when given a minimum of twice per day, on a variety of intestinal disorders, such as diarrhea, constipation, cramps, and flatulence. In 60 cases, excellent results (all troubles gone in 2–4 wk) were achieved in 70% of them and very good results (notable amelioration but requiring more than 4 wk treatment) were obtained in another 15%. Only 15% of the cases showed little or no improvement. With respect to flatulence, in particular, there were eight patients, four of which had very good results, one who had good results, and three who had no significant benefit. While the Carbomucil ingredients other than the charcoal might have been of real importance, there are substantial reasons for believing that the activated charcoal was the crucial substance.

Hall et al. (1981) gave 13 human adult subjects various types of meals, with or without activated charcoal, over a period of several weeks. The meals were: (1) a normal meal, (2) a high gas-producing bean meal followed by three capsules of activated charcoal (194 mg in each capsule) immediately, plus three more capsules after 2 hr, and (3) a high gas-producing bean meal followed by three starch-filled placebo capsules immediately, plus three more placebo capsules after 2 hr. The study was done in a double-blind manner. Breath hydrogen was measured periodically in all subjects. Hydrogen is produced in the body almost entirely by bacterial action in the GI tract (carbon dioxide and methane are the other major gases produced in this way); it is absorbed from the GI tract into the blood and carried to the lungs for excretion in the breath. Hydrogen in the breath has been proved to be proportional to the amount of hydrogen produced in the colon. Over the period from 0 to 7 hr, the mean number of flatus events resulting from the three meals cited above were found to be 3.0, 2.7, and 14.5, respectively. Figure 23.2 shows the breath hydrogen results. Breath hydrogen levels were elevated after 4

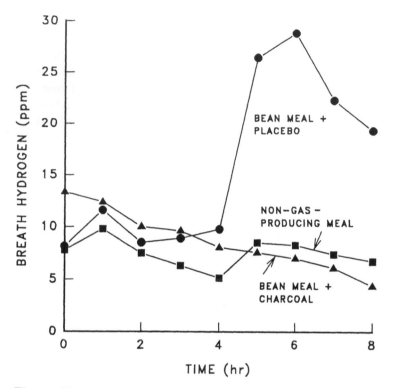

Figure 23.2 Breath hydrogen levels in subjects following test meals, with and without activated charcoal. From Hall et al. (1981). Copyright 1981 by Williams & Wilkins. Reprinted by permission.

hours in the "bean meal plus placebo" case and were essentially normal for the other two meals. It was concluded that the use of activated charcoal can keep gas production at normal levels, even when a high gas-producing type of meal is consumed.

In stark contrast to this study, Potter et al. (1985) found no effect of activated charcoal. In vitro studies were performed with human fecal homogenates, incubated with or without additional carbohydrate, and with or without activated charcoal. In these in vitro studies, hydrogen and carbon dioxide production rates were found to be unaffected by the addition of activated charcoal. In vivo tests on human subjects were done in much the same manner as the studies by Hall et al. (i.e., bean meals, double-blind protocol, placebo capsules versus charcoal capsules). The charcoal (16 capsules containing 250 mg each) was given as follows: 4 at the start, 4 at 30 min, 4 at 60 min, and 4 at 90 min. Figure 23.3 shows the breath hydrogen data. No significant differences in breath hydrogen or in the number of flatus events were noted between the charcoal-treated and the untreated groups.

There is no clear explanation for the differences between these two studies. Potter's group used the same brand of activated charcoal (Norit USP) as Hall's group; in fact, their dosage was slightly greater. Also, the dosage of beans in the two studies was similar, although the brands were different. Clearly, additional research is needed to explain the lack of agreement in these two studies.

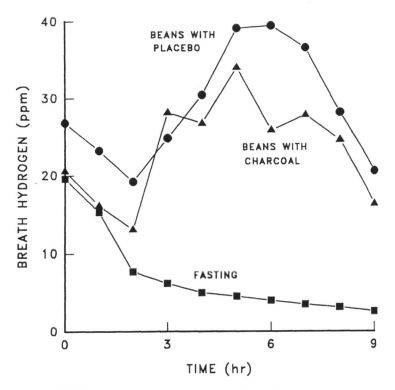

Figure 23.3 Breath hydrogen levels in fasting subjects and in subjects given a bean meal, with and without activated charcoal. From Potter et al. (1985). Reproduced by permission of the W. B. Saunders Company.

Two studies concerning the effect of charcoal on intestinal gas have been reported by Jain et al. (1986a,b). In the first study, nine adults were treated with three types of regimens: activated charcoal capsules containing 260 mg charcoal each, capsules containing 20 mg simethicone each, and placebo capsules. All capsules looked identical. The study was a randomized double-blind one. After collection of breath samples for analysis, the subjects ingested four capsules. Thirty minutes later they ate a meal of 8 oz baked beans. Thirty minutes after the meal, four more capsules were ingested. Breath samples were collected each 30 min for 7 hr and were analyzed for hydrogen content. Figure 23.4 shows the breath hydrogen results. Compared to the placebo group, the charcoal group's peak hydrogen breath levels were reduced from an average of 24 to 8 ppm. Areas under the curves of hydrogen level versus time over 0–7 hr were reduced by charcoal from 618 to 178 ppm/hr. The effects were most pronounced between 2.5–3 hr and 4.5–6.5 hr post-ingestion. Simethicone had no significant effect. Charcoal also significantly reduced episodes of abdominal discomfort from seven to one (again, simethicone had no effect). The reason why these results are different from those of Potter et al. (1985), who found no effect using the same brands of beans and charcoal, was conjectured to be that, in this study, some charcoal was given 30 min before the bean meal rather than with the beans.

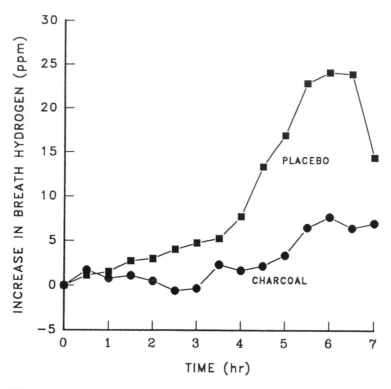

Figure 23.4 Comparison of efficacy of a placebo and activated charcoal in reducing breath H_2 levels after a bean meal. From Jain et al. (1986a). Reproduced with permission from the American College of Physicians.

In their second study, Jain et al. (1987b) conducted a randomized double-blind study on U.S. subjects ($N = 30$) and subjects from India ($N = 69$). It is known that these groups differ in their dietary habits and in their gut flora. Lactulose was used as the substrate. The procedure was the same as in the first study except for the use of 15 mL lactulose rather than beans. Tests were also tried in which the charcoal was given either with the lactulose or after it, but no significant effect of the charcoal was seen. However, when the charcoal was given 30 min before the lactulose and 30 min after, breath hydrogen levels were reduced, compared to the placebo tests, from 28 to 11 ppm (U.S. subjects) and from 40 to 7 ppm (Indian subjects). Numerical hydrogen AUC data were not stated but the figures in the article show a clear and dramatic decrease in the AUC values for the two groups when charcoal was given. Subjects experiencing abdominal discomfort fell from 29 (19 with bloating, 10 with cramps) to 3 (3 with bloating, 0 with cramps) when charcoal was used. The charcoal was, overall, roughly equally effective in the two populations.

Bigard et al. (1989) studied the effect of charcoal on gas volume and gas composition by collecting gas passed from subjects after a meal of beans and onions, both with charcoal and with placebo administration. The gas was collected between 6 and 10 hr after a meal of 250 g beans and 200 g onions, with 50 g tomatoes and 10 g butter. A group of 40 subjects was given charcoal capsules (Carbolevure) containing 100 mg charcoal per capsule in one trial and placebo capsules in a repeat trial. The capsules were given as follows: ten per day for five days preceding the day of the test meal and six capsules right after the test meal. The volume of gas collected averaged 217 mL with charcoal and 233 mL with the placebo, an insignificant difference. When the authors discarded the data of six subjects who had passed less than 100 mL gas, the average volumes were 253 mL with charcoal and 320 mL with placebos (the difference was insignificant at the $p < 0.05$ level). Slight changes in the gas composition with and without charcoal were noted (less hydrogen and more nitrogen when charcoal was used) but the differences were not large. The lack of effect of charcoal could easily be ascribed to the very low doses of charcoal employed, 1 g/day for each of 5 days and 0.6 g after the test meal. The authors admit that this is a likely explanation.

The mechanisms by which charcoal may reduce intestinal gas are unknown. Charcoal probably does not adsorb the gases themselves to any significant degree. Simple gases, especially hydrogen, do not adsorb in large quantity to charcoal. The charcoal could perhaps adsorb the substrate itself. However, whether it could do this before bacteria would have a chance to interact with the substrate is unclear. A more likely hypothesis is that charcoal, by the adsorption of various key chemicals in the intestinal milieu, changes the conditions for effective bacterial fermentation of the substrate. An even more plausible cause would be the binding of bacteria by the charcoal in such a way as to render them less effective in digesting the substrate (it is well known that bacteria generally adsorb very strongly to activated charcoal).

V.   EFFECT ON DIARRHEA

We mentioned earlier a study by Riese and Damrau (1964) and discussed results dealing with flatulence. They also treated six patients with nervous diarrhea due to irritable or spastic colon with the same charcoal doses they used for treating flatulence (2 capsules, four times daily, having a total of 2.08 g charcoal). In a group of six patients, four showed good results, one fair, and one poor. In a group of five patients having ulcerative colitis with severe diarrhea, results were fair in two cases and poor in three.

We also mentioned the study of Chevrel (1978); with respect to the 60 cases of intestinal problems, 18 involved diarrhea. The Carbomucil regimen described earlier produced very good results in 14 of these patients, good results in two, and no significant benefit in the remaining two. We also summarize here Chevrel's results on the opposite problem of constipation: in 34 patients, the Carbomucil regimen gave very good results in 24 cases, good results in 6 cases, and no significant benefit in 4 cases.

Sebedo et al. (1982) studied the effect of activated charcoal (Norit brand) on diarrhea in 39 children in Indonesia. Group I (23 children) received only oral glucose electrolyte solution (this combats the severe dehydration resulting from diarrhea). Group II (16 children) received the oral glucose solution plus a dose of activated charcoal (166 mg to 750 mg, depending on body weight, three times per day). The charcoal lowered the duration of diarrhea from 3.0 days to 2.1 days. The difference was not significantly different at $p > 0.05$; however, it may be that a much larger activated charcoal dose would have had a bigger effect. Unfortunately, only fairly low charcoal doses were employed.

Alestig et al. (1979) studied the effect of activated charcoal, kaolin (a clay), and diphenoxylate (a drug with antiperistaltic properties) on acute nonspecific diarrhea. The charcoal dose used was 5 g, three times per day. No statistical difference between control subjects and patients treated with any of the three regimens was noted. However, acute nonspecific diarrhea is a disorder of short duration (2–4 days) and tends to clear up fairly quickly regardless of the treatment used.

Gaussen (1963) described the treatment of various diarrhetic syndromes with nitro 5-hydroxy-8-quinoline (a compound which has antibacterial action, especially on pathogenic staphylococci) adsorbed to activated charcoal. At an average dose of 0.05 g/day (the paper does not say whether the 0.05 g is the quinoline, the charcoal, or the combination) for 15 days, this substance gave excellent results in 23 of 25 patients having parasitic or bacterial intestinal disorders, and good/excellent results in 29 of 36 patients having functional colopathies.

VI. EFFECT ON PEPTIC ULCERS

Ellis et al. (1970) created peptic ulcers by surgical techniques in 70 dogs. The hypothesis behind this study was that peptic ulceration requires pepsin, the major acid protease in gastric juice, in addition to low pH. The study was thus aimed at determining whether compounds which have the ability to inhibit protease activity in vitro without changing the pH of the gastric juice would affect the incidence and severity of chronic peptic ulceration. After the surgery, the dogs were divided into 7 groups (ten per group) and each group was treated with a different regimen. All dogs underwent necropsy at death or were sacrificed at 60 days, and the number of ulcers, their locations and sizes, and whether they were perforated were noted. The different groups received a standard kennel diet, water ad libitum, and gelatin capsules. The capsules contained: nothing, 1 g antacids (aluminum hydroxide and magnesium trisilicate), 3 g of a cation exchange resin, three different types of sulfated polysaccharides (3 g cellulose sulfate, 1 g carrageenan, or 0.5 g sulfated amylopectin), or 3 g activated charcoal.

The most effective substance for reducing ulceration was the sulfated amylopectin. Carrageenan was also effective but at a larger dose. The mode of action of these is unclear. The cation exchange resin was by far the worst treatment, while the antacids had little or no effect. The activated charcoal was moderately effective: the charcoal-treated group had a 40% incidence of ulceration versus 80% in the controls, 20% perforation versus 60% in the controls, a mean ulcer size of 2.0 cm versus 3.25 cm in the controls, and a mean number of 1 ulcer/dog versus 2 ulcers/dog in the control group.

VII. USE IN DEODORIZING OSTOMIES

Kappeler et al. (1984) tested the effect of an activated charcoal preparation consisting of 15 g activated charcoal plus 12 g sorbitol sweetener and some excipient (an inert material added to produce a thicker consistency to the mixture) in 100 mL of water on stool consistency, flatulence, and odor in 93 patients having intestinal stomas (an opening created surgically in the abdominal wall, through which intestinal matter is excreted rather than having it pass through the lower intestines and rectum). More than two-thirds of the patients had improved stool consistency, reduced flatulence, and reduced odor. A granular charcoal preparation (granular charcoal plus bentonite clay) was also tried. Patients with colostomies (in which the opening is connected to the colon, or large intestine) had good results, as before, but patients with ileostomies (in which the opening is connected to the ileum, or small intestine) did not show significant improvement.

Johnson (1977) stated that the use of activated charcoal capsules three times daily reduces stool odor and intestinal gas in patients having colostomies. Sparberg (1977) replied that in his experience charcoal indeed is of some benefit in controlling odor (although he mentioned that agents such as bismuth subgallate are better) but that "no agent can absorb intestinal gas in quantity."

An older paper which deals with this same general topic is by Riese and Damrau (1964), mentioned previously in connection with flatulence and diarrhea. They also treated seven patients complaining of fetid stools. Capsules of 4 grains (0.26 g) vegetable charcoal were again used. Two patients responded well to six capsules daily, while one patient required 8 capsules per day for good results. Two patients obtained fair results with 1 capsule four times per day, and one person was not helped by a dosage as high as 12 capsules daily.

VIII. USE IN THE TREATMENT OF ALCOHOLISM

An extremely interesting study of the use of "animal carbon" for treating alcoholism was described by Royer et al. (1961). Based on observations in 1917 by Moench, they carried out a study with 18 patients being treated for alcoholism. The subjects were given three 1 g sachets of charcoal, three times per day (the total charcoal dose was thus 9 g/day). The patients (fasting) were then asked to drink, at intervals of 10–20 min, a glass of wine, or a can of beer, or a small glass of a liquor such as brandy or rum.

Within 5–10 minutes, various reactions occurred. However, the report indicates that they did not usually become severe until the second drink had been consumed. The patients reacted with redness of the skin (first the face, then the neck, chest, and arms), an increased pulse rate (to 110 beats/min), hypotension, palor, accelerated respiration (to 25 breaths/min), watering of the eyes, headache, vertigo, and psychic disturbances (some patients became excited and verbose, while others sank into lassitude). However, such reactions did not occur unless the patients had taken the charcoal for at least 3 days.

Although the patients were recovering alcoholics, by the time they took the second drink, they were not very motivated to drink it; by the time they were to consume the third drink, they had to be prompted to drink it for the sake of the study. In many cases, just the odor of the second or third drink caused the subjects to become nauseous.

The researchers conjectured that perhaps an impurity in the charcoal was responsible for the observed effects. They mention a chemical called disulfiram which is used to treat alcoholism by causing adverse reactions when alcohol is consumed. The authors recommend charcoal over disulfiram because the adverse effects do not set in if a relatively small amount of alcohol is consumed, only when the amount starts to become excessive. Thus, total abstinence is not required.

IX. CHARCOAL AS A VEHICLE FOR PROLONGED-RELEASE MEDICATIONS

We previously mentioned, in discussing enzyme adsorption to charcoal, that Falk and Sticker (1910) used trypsin adsorbed to charcoal as a type of prolonged-release medication. They found that the subcutaneous injection of the trypsin/charcoal complex (called Carbenzyme) into dogs suffering from carcinomas was more beneficial than plain trypsin injections. Goiffon (1930) found that eserine adsorbed to charcoal is useful in treating atonia of the intestine. Likewise, 0.05 g hydroxyaminophenylarsenic acid on a teaspoonful of charcoal (called Carbarsene) was effective against amebic, parasitic, and certain bacterial infections. A combination of 0.15 g iodine on a similar amount of charcoal was also recommended as a "general antiseptic of the intestine." While these two rather old reports might be viewed with suspicion, because of their age, more modern studies suggest that they are probably valid.

Two reports by Sul'din et al. (1976a,b) in the Russian literature showed that drugs pre-adsorbed onto activated charcoal have prolonged action in vivo. Phenobarbital and chloracon had prolonged and stable antispasmodic action when adsorbed on charcoal; similarly dibazole administered intragastrically to rats was absorbed more slowly and gave lower peak plasma concentrations when it was adsorbed to charcoal (the peak levels were 4.38 mg/L at 0.5 hr for the drug alone and 0.96 mg/L at 2 hr for the drug/charcoal combination).

Kawa et al. (1987) studied in vitro the rate of desorption of adriamycin (ADM) previously adsorbed to charcoal and found that the drug "was released so slowly and continuously that the concentration of ADM around the charcoal particles remained high for a long time." They suggested that ADM/charcoal might be effective for the treatment of peritonitis carcinomatosa and lymph node metastases.

Bonhomme et al. (1992) used a dosage form comprised of 5-fluorouracil (5-FU) adsorbed on activated charcoal to treat mammary carcinomas in mice. They combined 25 mg/mL 5-FU and 100 mg/mL charcoal and injected this directly into the tumors. Desorption of the drug provided an effective method for achieving tumor regression without causing undue toxicity.

Hagiwara et al. (1987) created a prolonged release form of an anticancer drug, mitomycin C, by adsorbing it onto activated charcoal particles roughly 40 µm or less in diameter. This combination has an affinity for tumor surfaces and lymph nodes. Experimental tests in rats showed that the charcoal/drug dosage form had therapeutic effects 3.1 and 2.9 times higher for treating peritoneal and lymph node cancers, respectively. In human trials, the charcoal/drug complex was injected into the peritoneal or pleural cavities of 81 patients having cancers. Fifty-one responded well to the therapy, with a marked improvement in subjective symptoms.

Hagiwara et al. (1992a), in a subsequent study, adsorbed the anticancer drug etopside to charcoal and injected the drug/charcoal complex into the lymph nodes of mice to test for its therapeutic effect on distal lymph node metastases. Metastases had been established in the lymph nodes by prior injection, 8 days before, of leukemia cells into the left hind foot pads of the mice. A second group received injections of the drug alone (same drug dose). After two days, the number of leukemia cells in the secondary draining nodes was determined; the group receiving the drug/charcoal complex had significantly fewer cells. It would have been interesting if cell counts at other times post-injection had been determined.

Hagiwara et al. (1992b) similarly studied a delayed-release charcoal formulation for treating cancer. Fifty patients with gastric cancer who had gastrectomies were treated with the drug mitomycin adsorbed to charcoal (M-CH). This dosage form was placed into the peritoneal cavity. Controls received no M-CH. Survival was tracked at 2, 2.5, and 3 years, and it was found that the M-CH group had significantly higher survival rates than the control group, presumably because the M-CH inhibited a recurrence of

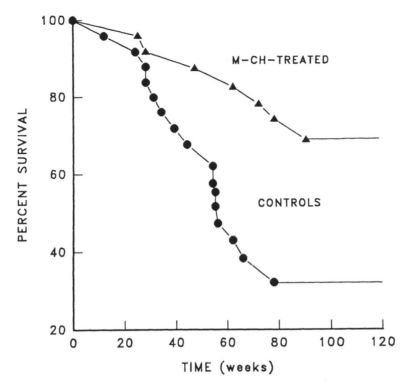

Figure 23.5 Survival of control and M-CH treated patients. From Hagiwara et al. (1992b). Copyright 1992 by The Lancet Ltd. Reproduced by permission.

the cancer. Figure 23.5 shows the survival data. Differences in the long-term survival rates were about 36%.

Sakakura et al. (1992) adsorbed an anticancer agent (aclarubicin) to activated charcoal and injected the charcoal subcutaneously into the left hind paws of mice which had received injections of leukemia cells seven days before. Survival and lymph node metastases were evaluated for those mice and compared to results for mice treated with injections of aclarubicin solution alone. The charcoal/aclarubicin recipients showed statistically better survival.

An interesting paper by Sklow (1943) concerned the prolongation of sex hormone effects by adsorption to charcoal. The charcoal/hormone combinations were prepared as follows: estrone, estradiol benzoate, and diethylstilbestrol, 0.01 mg each, were dissolved in 2 mL 96% ethyl alcohol, 4 mg charcoal was added, and the mixtures were evaporated to dryness.

Roivas and Neuvonen (1992) evaluated the use of activated charcoal as a suitable matrix for the sustained-release delivery of drugs by studying the reversible adsorption and desorption of nicotinic acid (NA); NA/charcoal was not intended as a desired formulation but only an example for studying such. Unfortunately, NA did not adsorb very well to the charcoal in acidic or neutral media. Desorption was carried out in a Sartorius apparatus into which fluid could be continuously added and withdrawn. The desorption process showed two phases: an initial rapid release in which equilibrium conditions were closely approached and a subsequent slower release as fresh fluid entering the apparatus permitted further desorption. The release during continuous flow conditions provided a sustained-release type of behavior.

Although we have been discussing the use of charcoal as a vehicle for prolonged-release medications, research has also evaluated the use of ion exchange resins for the same purpose. Two such studies are those of Becker and Swift (1959) and Schlichting (1962). More modern developments in making better prolonged-release dosage forms, by microencapsulation techniques and the development of improved polymeric matrices, have made this older work obsolete.

X. CHARCOAL INJECTED INTRAVENOUSLY

As irrational and dangerous as the notion may seem today, in the period up to about 1938 several investigators recommended the direct intravenous injection of powdered charcoal slurries. Izar and Patane (1913) seem to be the first to have experimented with intravenous slurries. In a paper entitled "Physiological Action of Colloidal Carbon," they indicate that so-called mellogen, produced by the disintegration of a carbon anode by galvanic current, was injected intravenously in large quantities into rabbits, white rats, and pigeons. This produced dyspnea (labored breathing), undoubtedly because of blockage of capillaries in the lungs, but few of the animals died. Conklin (1927) reported much later that suspensions of colloidal carbon had been used very successfully in 100 cases of septicemia, metritis, mastitis, lymphangitis, and external pyrogenic wounds in humans and animals. The dose used was very small, 0.001 g/kg body weight. He indicates that such treatment increased the number of polymorphonuclear leukocytes. Body temperatures were elevated initially but returned to normal within an hour. External signs of improvement were noted on the fourth day in pyrogenic infections. Whether the "carbon" referred to in this work was a true activated charcoal is not clear; certainly in the study of Izar and Patane it was not an activated charcoal.

Saint-Jacques (1934) recommended using a dose of 3–4 mL of a suspension of 2% finely ground animal charcoal in water, up to six times, as a general treatment for infection. He claims reports of successful results. Touraine and Menetrel (1934) also declared that "intravenous charcoal treatment is harmless and of decided value in infections." Selvaggi (1935) reported that charcoal given intravenously deposited in the following sequence: spleen, liver, lungs, and some in the lymph glands and in the bone marrow. He states that foci of infection with *Staphylococcus* are surrounded by a "filter" of charcoal particles. He further claims that "elimination" (implying removal from the body) of the charcoal occurs in the liver but it is difficult to imagine the process by which such could occur.

Lumiere and Meyer (1936) later reported that injections of 1 mL/kg of a 2% suspension of animal charcoal, "polishing rouge" (Fe_2O_3), or Troyes whiting into rabbits caused a 20–160% increase in blood sugar. The effect reached its maximum 2–4 hr after the injection. Lémetayer and Uhry (1938) also studied rabbits and found that intravenous injections of suspensions of animal or vegetable charcoals alone had no protective action against tetanus toxin but that charcoal or kaolin suspensions injected with antiserum reinforced the protective action of antiserum.

With these studies, the notion of using charcoal suspensions for intravenous injection appears to have faded. Certainly one might accept that a minor amount of injected charcoal may not be harmful but knowing that the charcoal must certainly end up blocking capillaries and other small blood vessels surely suggests that its use is fraught with potential hazard. Nevertheless, it is an interesting concept and does appear to have worked beneficially in several cases.

XI. SUMMARY

Many of the uses of charcoal described in this chapter have involved rather qualitative estimates of its efficacy. For example, the reduction of the severity of diarrhea, the abatement of odor from ostomies, the alleviation of the pain of insect bites, the reduction of swelling/inflammation of the skin, and a decrease in flatus events are all very subjective and qualitative. Thus, many of the studies discussed were not very scientific in nature. Part of this is because the effects to be observed were indeed qualitative in nature but much of the lack of rigor resulted from the failure to develop more quantitative measurement indices. Another problem with many of the studies is that many are old and, in many areas, few in number (e.g., the effect of charcoal on peptic ulcers involved only one investigation).

Despite these drawbacks, several conclusions appear to be valid. First, the use of charcoal cloth in treating skin wounds is clearly effective. Charcoal is known to adsorb bacteria and bacterial toxins well. For the same reason, there is reason to believe that the use of poultices for treating insect bites does have a sound basis (i.e., the adsorption of toxins). The benefits of poultices in reducing swelling are much less certain.

Six studies dealing with the effects of charcoal on intestinal gas production have shown clear reductions in gas, while two have shown no effect (however, one of these studies, by Bigard, employed very low charcoal doses). Given the variability of intestinal flora and normal gas production from person to person, it is not surprising that charcoal failed in some cases.

With respect to diarrhea and similar intestinal problems, most studies have shown a spectrum of results, with decreasing numbers of cases falling into the categories of "excellent", "very good", "good", "fair", and "poor." That is, on balance most of the results were favorable. One study which showed no effect (Sebedo et al.) was flawed by the use of very small doses of charcoal.

Charcoal has been shown, in isolated reports, to be effective in treating GI tract bacterial sepsis, in diminishing ostomy odor, and in reducing peptic ulceration. However, these reports are so few in number that much more research is called for.

The French study dealing with the use of "animal carbon" for treating alcoholism is certainly intriguing and one wonders why further work in this area appears not to have been forthcoming.

Finally, charcoal has been demonstrated to be a good vehicle for the prolonged-release of drugs of various kinds. The desorption of a drug previously adsorbed on charcoal is generally slow when the drug/charcoal complex is implanted or injected into the body. However, the drug release rate is probably far from constant (most likely, it is highest at the start and tapers off substantially with time). Thus, more modern formulations which provide a more constant release would probably be better.

XII. REFERENCES

Alestig, K., Trollfors, B., and Stenqvist, K. (1979). Acute nonspecific diarrhoea: Studies on the use of charcoal, kaolin-pectin, and diphenoxylate, Practitioner 222, 859.

Becker, B. A. and Swift, J. G. (1959). Effective reduction of the acute toxicity of certain pharmacologic agents by the use of synthetic ion exchange resins, Toxicol. Appl. Pharmacol. 1, 42.

Beckett, R., Coombs, T. J., Frost, M. R., McLeish, J. and Thompson, K. (1980). Charcoal cloth and malodorous wounds (letter), Lancet, September 13, p. 594.

Biehl, M. L., Lambert, R. J., Haschek, W. M., Buck, W. B., and Schaeffer, D. J. (1989). Evaluation of a superactivated charcoal paste and detergent and water in prevention of T-2 toxin-induced local cutaneous effects in topically exposed swine, Fundam. Appl. Toxicol. 13, 523.

Bigard, M. A., Gilbert, C., and Bloom, M. (1989). La mesure du débit et de la composition des gaz rectaux après repas-test "haricots-oignons": Description de la méthode et étude en double aveugle du Carbolevure, Gastroenterol. Clin. Biol. 13, 312.

Bonhomme, L., et al. (1992). Intratumor treatment of C3H mouse mammary carcinoma with 5-fluorouracil adsorbed on activated charcoal particles, Anticancer Drugs 3, 261.

Chevrel, B. (1978). Traitment des troubles functionnels intestinaux par le Carbomucil, Med. et Chir. Digest (Paris) 7, 443.

Conklin, R. L. (1927). Intravenous injections of colloidal carbon in human and veterinary medicine, Sci. Agr. 8, 112.

Ellis, C. M., Lunseth, J. B., and Nicoloff, D. M. (1970). Effect of protease inhibitors on experimental peptic ulceration, Am. J. Surgery 119, 213.

Falk, E. and Sticker, A. (1910). Carbenzyme, Munch. Med. Wochenschr. 57, 4.

Gaussen, L. (1963). Traitment des syndromes diarrhéiques par une association charbon-nitro 5-hydroxy 8-quinoléine, La Presse Medicale 71, 2301.

Goiffon, R. (1930). Medicaments adsorbed on charcoal in the therapeutics of the digestive organs (eserine, iodine, arsenic), Semana Med. (Buenos Aires) 37, 1133.

Hagiwara, A., Takahishi, T., Lee, R., Ueda, T., Takeda, M., and Itoh, T. (1987). Chemotherapy for carcinomatous peritonitis and pleuritis with MMC-CH, mitomycin C adsorbed on activated carbon particles, Cancer 59, 245.

Hagiwara, A. et al. (1992a). Enhanced therapeutic efficacy of intralymph-nodal etopside on distal lymph mode metastases using a new dosage format — activated carbon particles adsorbing etopside, Anticancer Drug Des. 7, 163.

Hagiwara, A. et al. (1992b). Prophylaxis with carbon-adsorbed mitomycin against peritoneal recurrence of gastric cancer, Lancet 339, 629.

Hall, R. G., Jr., Thompson, H., and Strother, A. (1981). Effects of orally administered activated charcoal on intestinal gas, Am. J. Gastroenterol. 75, 192.

Izar, G. and Patané, C. (1913). Physiological action of colloidal carbon, Biochem. Z. 56, 307.

Jain, N. K., Patel, V. P., and Pitchumoni, S. (1986a). Activated charcoal, simethicone, and intestinal gas: A double-blind study, Annals Intern. Med. 105, 61.

Jain, N. K., Patel, V. P., and Pitchumoni, C. S. (1986b). Efficacy of activated charcoal in reducing intestinal gas: A double-blind clinical trial, Am. J. Gastroenterol. 81, 532.

Johnson, J. B. (1977). Try activated charcoal for ostomates? (letter), Patient Care, October 30, p. 152.

Kappeler, M., Rüfenacht, R., Müller, S., and Halter, F. (1984). [Effect of a charcoal preparation on fecal consistency and odor of patients with intestinal stomas], Schweiz. Rundschau Med. Prax. 73, 351.

Kawa, J., Taniguchi, H., Maeda, T., Hagiwara, A., Yamaguchi, T., and Takahashi, T. (1987). [Fundamental studies on the properties of a new adriamycin delivery system involving adsorption to activated charcoal], Gan. To. Kagaku. Ryoho. 14, 606.

Kehls, D. M. (1793). Memoire sur le charbon végétal, observations et journal sur la physique, de chemie et l'histoire naturelle et des arts, Paris, Tome XLII, 250.

Kopp, K. F. (1978). The unexpected success of enteral activated carbon in acute renal failure patients in the prevention and therapy of abdominal sepsis, Abstr. Am. Soc. Artif. Intern. Organs *VII*, 31.

Lemétayer, E. and Uhry, P. (1938). Prevention and treatment of tetanus intoxication in rabbits by intravenous injection of charcoal suspension and specific antiserum, Compt. Rend. Soc. Biol. *125*, 823.

Lumiere, A. and Meyer, P. (1936). Effects of intravenous injections of suspensions of granular solids on blood sugar, Compt. Rend. Soc. Biol. *123*, 606.

Martindale: The Extra Pharmacopeia, Twenty-ninth edition, J. E. F. Reynolds, Ed., The Pharmaceutical Press, London, 1989.

Mulligan, C. M., Bragg, A. J. D., and O'Toole, C. B. (1986). A controlled comparative trial of Actisorb activated charcoal cloth dressings in the community, Br. J. Clin. Pract., *40*, 145.

Peyer, W. (1940). Coffee char and its possibilities, Pharm. Zentralhalle *81*, 1.

Potter, T., Ellis, C., and Levitt, M. (1985). Activated charcoal: In vivo and in vitro studies of effect on gas formation, Gastroenterol. *88*, 620.

Riedel, H. (1940). The adsorptive capacity of coffee carbon, Klin. Wochenschr. *19*, 1064.

Riese, J. A. and Damrau, F. (1964). Use of activated charcoal in gastroenterology: Value for flatulence and nervous diarrhea, J. Am. Geriatr. Soc. *12*, 500.

Roivas, L. and Neuvonen, P. J. (1992). Reversible adsorption of nicotinic acid onto charcoal in vitro, J. Pharm. Sci. *81*, 917.

Rosenvold, L. (1968). Charcoal as a remedy in ophthalmology and otolaryngology, The Eye, Ear, Nose, and Throat Monthly *47*, 55.

Royer, P., Poiré, R., Colmart, C., and Thouvenot, P. (1961). Prophylaxie de la récidive alcoolique par le charbon animal, Rev. Med. Nancy *86*, 1173.

Saint-Jacques, S. (1934). Treatment of infection by intravenous injections of animal charcoal, Bull. Acad. Med. *88*, 169.

Sakakura, C., et al. (1992). Enhancement of therapeutic efficacy of aclarubicin against lymph node metastases using a new dosage form: Aclarubicin adsorbed on activated carbon, Anticancer Drugs *3*, 233.

Schlichting, D. A. (1962). Ion exchange resin salts for oral therapy. I. Carbinoxamine, J. Pharm. Sci. *51*, 134.

Schobesch, O. (1938). Prophylactic treatment of yperite injuries, Antigaz (Bucharest) *12*, 436.

Sebedo, T., Iman, S., Sobiran, H., Setiyono, Suminta, and Ismangoen (1982). Carbo-adsorbent (Norit) in the treatment of children with diarrhoea, S. E. Asian J. Trop. Med. Pub. Health *13*, 424.

Selvaggi, G. (1935). Intravenous animal charcoal in physiological and some pathological conditions, Sperimentale *89*, 386.

Sklow, J. (1943). Prolongation of sex hormone effects by adsorption on powdered carbon, Endocrinology *32*, 109.

Sparberg, M. (1977). Try activated charcoal for ostomates? (reply to letter), Patient Care, October 30, p. 152.

Sul'din, A. V., Zalesov, V. S., Starkova, S. M., and Fedorov, A. A. (1976a). Prolongation of the effect of dibazole by adsorption on activated carbon, Deposited Doc. VINITI 2176-76.

Sul'din, A. V., Zalesov, V. S., Fedorov, A. A., and Kalugin, Y. P. (1976b). Study of the prolongation of the action of antispasmodic preparations by their adsorption on activated charcoal, Nauchn. Tr.-Permsk. Gos. Farm. Inst. *10*, 35.

Thrash, A. M. and Thrash, C. L., Jr. (1981). *Home Remedies: Hydrotherapy, Massage, Charcoal, and Other Simple Treatments*, Thrash Publications, Seale, Alabama, pp. 143-152.

Touraine, A. and Menetral, B. (1934). Intravenous charcoal treatment, Presse Méd. *42*, 1997.

Ustinova, T. S., Kaem, R. I., Ianshevskii, A. V., and Lopatto, I. (1987). [Possibility of using activated carbon tissue in combination bandages], Biull. Eksp. Biol. Med. *104*, 244.

Wunderlich, U. and Orfanos, C. E. (1991). [Treatment of venous ulcera cruris with dry wound dressings. Phase overlapping use of silver impregnated activated charcoal xerodressing], Hautartz. *42*, 446.

24

Other Biochemical and Biological Uses of Charcoal

Activated charcoal has an impressively wide ability to adsorb substances other than just drugs and various common chemicals. For example, it has been shown that activated charcoal can adsorb bacteria, viruses, bacterial and fungal toxins, snake venoms, enzymes, hormones, vitamins, and various other biochemicals, both in vitro and in vivo.

Because of charcoal's ability to adsorb almost any organic substance, it has been used in a host of biochemical and biological applications other than for the in vivo treatment of poisonings and overdoses. Charcoal has been used as a way of separating and concentrating organic substances in various assay procedures for biochemicals, and in the detection of drugs in blood, urine, and other body fluids. Charcoal also has been widely used in culture media, to enhance microorganism and plant embryo growth. It has also been used to protect newly-planted crop seedlings from damage by herbicides subsequently applied to fields to suppress weed growth.

Some interesting studies involving charcoal in such applications will be reviewed in this chapter, so as to round-out our treatment of charcoal, and to demonstrate the exceptional breadth of applications of activated charcoal. One note of caution should be sounded: many of the studies are of the in vitro type and, therefore, their relevance to the clinical treatment of patients is not clear and certainly not proven. Moreover, many of the studies were carried out a long time ago (often several decades) and require confirmation using more modern techniques.

I. SNAKE VENOM ADSORPTION

Houssay (1921) was apparently the first person to report on the adsorption of snake venom components by charcoal. She found that the hemolytic substance of snake venoms is adsorbed by animal charcoal. Antitoxic serum is not adsorbed. After contact with charcoal there remains in the venom a residue which can combine antitoxin but which is not hemolytic. The remaining substance and that which combines with the charcoal,

542

neither of which alone has any hemolytic action, are therefore the two constituents of the hemolysin.

Boquet (1928) studied the adsorption of cobra venom by charcoal. Fifty milligrams of desiccated cobra venom were dissolved in 50 mL physiological salt solution, pH 6.8, and 1 g of sterilized activated charcoal was added and agitated. After 2 hr, the suspension was filtered; the suspension, filtrate, and charcoal were equally nontoxic. Injections equivalent to 100 fatal doses had no venomous action. This inactivation of venom by charcoal was independent of temperature between 12 and 38°C, it required only 8–10 sec of contact, and it acted both in the serum of the horse and in physiological salt solution. Neither heat to 70°C for 30 min nor acids, 0.2–0.5 mL of 0.1 N HCl per 10 mL, liberated the venom fixed on the charcoal. Tests with diphtheria toxin gave similar results.

II. VIRUS ADSORPTION

Poppe and Busch (1930) showed that three strains of foot-and-mouth disease viruses having isoelectric points at pH 7.6, 7.8, and 8.0 were capable of adsorbing to charcoal so strongly that neither the charcoal nor the supernatant were infectious upon injection into guinea pigs. Although adsorption occurred over a pH range of 6.5–8.4, it tended to be weakest at the isoelectric point of the virus (this is opposite to the behavior of molecular species, which tend to adsorb most strongly at their isoelectric points; however, with moieties as large as viruses, adsorption may involve electrical attractive forces, in which case binding would indeed be weakest when the species is, overall, electrically neutral). Kaolin did not adsorb the virus.

In contrast to these results, Pyl (1931) found that in buffer solutions charcoal adsorbs these viruses independently of pH. Also, kaolin and $Al(OH)_3$ were found to adsorb such viruses at neutral and alkaline pHs, respectively.

Cordier (1939) later reported that foot-and-mouth disease virus in a 1% suspension in a physiological solution can be entirely adsorbed on bone charcoal if used in the amount of 10 g charcoal per 100 mL fluid. The virus is not destroyed but its activity is much reduced. Wood and coconut charcoal adsorb the virus less well. Injections of the virus/charcoal complex were found to create immunity to the virus.

Cordier (1940) also discussed the adsorption of foot-and-mouth disease virus and the practice of immunization by means of injecting the adsorbed virus. He further indicates that sheep pox virus is also adsorbed by charcoal. This was confirmed by Stamatin (1937), who studied sheep pox virus adsorption on bone charcoal and on kaolin. He found that bone charcoal adsorbs this virus, especially under acidic conditions, but the kaolin had no affinity for the virus.

Galvez (1964) reviewed some studies in which charcoal was used, not to adsorb viruses, but to purify some virus solutions by adsorbing normal plant components (mainly brown pigments) from the solutions. He then presented results on experiments in which solutions of three plant viruses (tobacco mosaic virus, southern soybean mosaic virus, and brome mosaic virus) were treated with charcoal and then either filtered or centrifuged. He showed that the removal of the brown pigments was the same with either treatment but that when filtration was employed the layer of charcoal which built up on the filter removed much of the viruses from solution also. Virus removal did not occur when centrifugation was used. This suggests that these plant viruses are removed by charcoal via physical entrapment in the layer of charcoal which builds up during filtration, rather than by adsorption per se.

III. BIOCHEMICAL FACTOR ADSORPTION

Sladek and Kyer (1938) indicated that an active hemopoietic (i.e., antianemic) factor can be adsorbed from purified liver extract on Norit charcoal at pH 5.0. Cook and Walter (1939) similarly reported that respiratory stimulating factors can be decolorized and deodorized effectively by charcoal. They identified both yeast-active and skin-active factors, the former being more readily adsorbed by charcoal.

Cook et al. (1941) found that charcoal can effectively adsorb the factors in aqueous/alcohol extracts of baker's yeast which stimulate the respiration of rat liver slices. The factors were readily desorbed by alkali.

IV. BACTERIA ADSORPTION

Wiechowski (1914) was one of the first to report that "animal charcoal is capable, not only in the test tube but in the human body, of adsorbing ... bacteria ... thereby preventing ... their poisonous action." Salus (1916) wrote of experiments in which contaminated water was shaken with animal charcoal and then filtered through paper. The filtrate was generally free of bacteria, although when milk or blood was used as the fluid such was not the case. He also observed that cocci were better adsorbed than typhoid bacteria and the latter better than *Bacillus coli*.

Oksent'yan (1940) reported that up to 90–100% of *Thermobacterium helveticum*, *Streptococcus lactis*, and *Saccharomyces ellipsoideus* are adsorbed by animal and wood charcoals, kaolin, and talc. The difference in the degree of adsorption was related to the values of the negative charges of the adsorbents. Kaolin and talc had the highest charges and the lowest powers of adsorption; animal charcoal had the lowest charge and was the best adsorbent. Adsorption was not found to affect the reproduction of the bacteria but it did sharply lower their physiological functions.

Gunnison and Marshall (1937) found that *Escherichia coli*, *Clostridium welchii*, and *Lactobacillus acidophilus* adsorption by particulate kaolin, $CaCO_3$, $Al(OH)_3$, and $BaSO_4$ was not marked. Charcoal was found to remove *L. acidophilus* and *L. staphylococci*, but not *E. coli* or *C. welchii*, from suspension. Kaolin adsorbed *Staphylococcus aurea*, *Sarcina lutea*, and *Bacillus subtilis*. Gram-positive organisms did not generally adsorb any better than gram-negative ones. It was concluded that clinical improvements observed following the administration of adsorbents are probably due to the adsorption of bacterial toxins or enzymes rather than the bacterial cells themselves.

Lasseur et al. (1934) observed that cells of *Pseudomonas chlororaphis* adsorb on activated charcoal at one end and rarely parallel to the surface. *Bacillus megatherium* were found to exhibit rapid adsorption on charcoal, again in a polar fashion. These cells often came to rest a short distance from the charcoal granules.

Sands et al. (1976) attempted to remove patulin (the epithet for *Penicillium patulum*) from apple juice and cider. They found that a patulin concentration of 30 µg/mL was reduced to undetectable levels by adding 5 mg/mL activated charcoal and that 20 mg/mL completely removed ^{14}C-labeled patulin.

V. BACTERIAL TOXIN ADSORPTION

Zunz (1911) seems to have been the first to study the binding of bacterial toxins to adsorbents. Table 24.1 indicates the qualitative results reported. It appears that, in general, the adsorption of toxin or antitoxin is possible but that the combination of toxin and antitoxin is not adsorbable.

Kraus and Barbara (1915) demonstrated the excellent binding of diphtheria toxin to Merck animal charcoal by contacting a 10% solution of the toxin with the charcoal for 1 hr and then injecting the supernate subcutaneously into a guinea pig. The animal survived without even showing a local edema, whereas a control animal died within 48 hr from 1% as much toxin. Similar results were obtained with tetanus and dysentery toxins. A study of the same nature on diphtheria toxin by Boquet (1928) has already been mentioned in the discussion of snake venom adsorption.

Signorelli (1933) reported that animal charcoal is capable of diminishing the toxic properties of tuberculin toxin as well as of tetanus toxin. In contrast to Zunz and Boquet, however, no reduction of diphtheria toxin activity was found. This may be related to either the use of a different kind of charcoal or to the use of different experimental conditions.

Seibert (1935) found that a purified tuberculin protein derivative was not itself able to stimulate antibody production. However, when injected in particulate form after adsorption to $Al(OH)_3$ or charcoal, it became truly antigenic (i.e., stimulated antibody production).

Hemolysins for sheep erythrocytes were found to quantitatively adsorb from rabbit serum to animal charcoal, according to a study by Roffo and Barbara (1925).

An interesting paper by Emery (1937) reviewed several reports on the use of kaolin in the treatment of Asiatic cholera. Emery mentioned that "studies in vitro suggest that the beneficial action of kaolin depends on the adsorption of the cholera toxin and inclusion of the bacteria." He described a report by Braafladt (1923) which indicated that kaolin, together with hypertonic salt solutions, reduced the mortality from cholera among soldiers in the Balkan War of 1910 from 60% to 3%, and a report by McRobert (1934) which advocated the use of kaolin for treating acute bacterial food poisonings which were encountered in the British Army in India.

Table 24.1 Zunz's Results on the Adsorption of Bacterial Toxins by Various Materials

Material	Diphtheria toxin	Diphtheria antitoxin	Diphtheria toxin-antitoxin	Tetanus toxin	Tetanus antitoxin	Tetanus toxin-antitoxin
Animal charcoal	Yes	Yes	No	Yes	Yes	No
Wood charcoal	No	No	No	—	—	—
Kaolin	No	No	No	Yes	Yes	No
Talc	No	No	No	Yes	No	No
Kiesel-guhr	—	—	—	No	Yes	No
Clay	—	—	—	Yes	Yes	No
BaSO4	No	No	No	No	Yes	No

Adapted from Zunz (1911).

Drucker et al. (1977) discussed very clear evidence that activated charcoal and activated attapulgite (a heated magnesium aluminum silicate) both prevent the toxic effects of endotoxins produced by *Vibrio cholerae* and *Escherichia coli*. When the adsorbents and isolated toxins were either preincubated together or injected simultaneously into the intestinal loops of rabbits, no toxic effects occurred. However, when the whole bacterial cells were used, rather than the isolated toxins, toxic reactions did occur in most cases. This is in conflict with the report described in Part III of Chapter 23 in which charcoal was found to be extremely effective in combating intestinal bacterial infections in humans.

Stoll et al. (1980) discovered that the enterotoxin produced by *Vibrio cholerae* bacteria could be bound by GM1 ganglioside adsorbed onto activated charcoal. They found that in patients with severe cholera, oral administration of GM1 attached to charcoal completely bound the free luminal enterotoxin. Charcoal without any attached receptor ganglioside had no such effect. Patients given GM1/charcoal had a significantly lower rate of stool output in the early stages of the disease but not later on. This suggests that free toxin produced in the gut lumen affects fluid loss early but that, later in the disease, toxin which is less accessible to luminal binding agents is the major cause of purging.

Ebata et al. (1980) found that endotoxin (lipopolysaccharide B from *Escherichia coli*) placed in rabbit blood which was then passed through a granular charcoal hemoperfusion column was completely removed after circulating the blood for 10–45 min through the column (the time required depended on how much endotoxin was initially added to the blood).

Ditter et al. (1983) studied the binding of endotoxins derived from *Escherichia coli* in vitro and in vivo in mice, using kaolin, bentonite, and activated charcoal as adsorbents. In vitro, the amounts of each adsorbent required to bind 50% of endotoxin from a solution containing 10 μg of endotoxin in 20 μL volume were: at pH 3.0, 20 μg bentonite, 75 μg charcoal, and 1100 μg kaolin. At pH 7.4, the amounts were 30, 100, and 900 μg, respectively. Thus, bentonite was superior to charcoal. In the tests with mice, an endotoxin dose was given, followed 30 min later by the administration of different amounts of adsorbent by stomach tube. At 4 hr, blood samples were collected and analyzed to determine whether endotoxemia existed. In reducing the percentage of mice with endotoxemia, the bentonite was again more effective than charcoal, with kaolin the least effective of the three.

Du Xiang-Nan et al. (1987) studied the adsorption of an endotoxin produced from the cell walls of gram-negative bacteria by granular activated charcoal coated with cross-linked agarose in both batch systems and in hemoperfusion column systems. This coated charcoal was proposed for use in hemoperfusion columns to remove endotoxins associated with liver failure. Two batch tests gave 39 and 50% removal of the endotoxin from the solution at 60 min, and one column experiment gave 63 and 69% removal from the circulating solution after 60 min and 120 min, respectively. It thus appears that such a coated charcoal, when used in a hemoperfusion system, might be effective in removing the endotoxins associated with liver failure.

VI. FUNGAL TOXIN ADSORPTION

The production of fungal toxins such as aflatoxin B_1 and T-2 toxin was discussed in some detail in Chapter 20. The reason for this is that the most common instances of such fungal toxin poisonings arise out of the production of these toxins in stored grains, which are commonly fed to cattle and poultry.

In humans, the most frequent instances of fungal toxin poisonings by far are those due to the eating of toxic mushrooms. Little research on such poisonings in humans has been done. The most common type of mushroom poisoning in humans is due to the deadly mushroom *Amanita phalloides*. Its toxin can be effectively counteracted by hemoperfusion over granular activated charcoal. No research using orally administered powdered charcoal has been reported to date but the hemoperfusion results strongly suggest that oral charcoal might also be effective in mushroom poisoning, if given early enough.

VII. ADSORPTION OF ENZYMES AND OTHER PROTEINS

As early as 1906, Hedin (1906) reported that charcoal neutralizes trypsin by adsorption. He stated that if sufficient charcoal is used, the trypsin solution has no ability to split casein. Four years later, Falk and Sticker (1910) described similar studies in which 10 g wood charcoal, shaken for 1 hr with 10 mL of a 0.2% trypsin solution, was found to significantly adsorb the trypsin. The charcoal/trypsin complex, named Carbenzyme by these authors, was injected subcutaneously into dogs and rabbits. Experiments with dogs suffering from carcinomas showed improved beneficial action as compared to plain trypsin injections. Presumably the Carbenzyme complex acted in a prolonged-release fashion (this may be one of the earliest examples of a method for prolonging the release of a therapeutic substance).

Keefer (1920), Kikawa (1926), Husa and Littlejohn (1952), and Daly and Cooney (1978) all reported on the adsorption of pepsin by charcoals. Kikawa indicates that a pH of 1.0 or 2.0 is optimum. Piper and Fenton (1961) studied the adsorption of pepsin on various adsorbents at pH 1.5. For the amounts of pepsin and adsorbent used in each test (same values in every case), they found the following percentages of adsorption of the pepsin: charcoal, 90%; aluminum hydroxide gel, 73%; Katonium, 46%; Fuller's earth, 31%; dihydroxyaluminum sodium carbonate, 30%; magnesium trisilicate, 28%; aluminum hydroxide tablets, 20%; kaolin, 16%; silicic acid, 14%; bismuth carbonate, 13%; and aluminum hydroxide powder, 7%. These results have implications for the treatment of peptic ulcers.

Daly and Cooney (1978) investigated whether the presence or absence of pepsin had any effect on the in vitro adsorption of a test drug (sodium salicylate) to charcoal. Daly and Cooney showed first that pepsin itself does adsorb strongly to charcoal (a Q_m value on the order of 0.25 g/g was determined), but that it did not significantly reduce the adsorption of sodium salicylate. This study was discussed in greater detail in Chapter 9, Section VI.

Przylecki et al. (1927) studied the system urea/urease/animal charcoal and found that urea was only slightly adsorbed, while urease was 85–95% adsorbed. They also studied the system amylase/animal charcoal/polysaccharides (dextrins or glycogens). In both systems the velocities of the enzyme reactions were not affected by the presence of the charcoal. This would seem to suggest that the enzymatically active sites are different from (and spatially distant from) the sites at which adsorption to these enzymes occurs.

In contrast to this, Unna (1926) found that diastase enzyme adsorbed to charcoal had little effect in hydrolyzing starch and was entirely ineffective in splitting glycogen.

Piper and Fenton (1961) also showed that mixing charcoal with pepsin largely inactivates the activity of the pepsin.

Mixed results regarding the effect of adsorption on enzyme activity were obtained by Sabalitschka and Weidlich (1929). They found that the activity of malt amylase adsorbed on kaolin varied with pH. When the pH was set in the range for optimal adsorption (4.5–6.2) the enzymatic activity of the amylase toward dextrin was not affected. However, at lower pH values (3.0, 4.0) where adsorption was less, the enzymatic activity was also found to be correspondingly lower. Blood charcoal was found to adsorb amylase also (with an optimum at pH 3.2) but less effectively than did kaolin.

Miller and Bandemer (1930) found that commercial (impure) invertase was 75–100% inactivated by adsorption to blood charcoal, whereas purified invertase retained higher activity. They suggested that the presence of acidic or alkaline impurities in charcoals may be primarily responsible for altering the activity of adsorbed enzymes.

Nonenzymatic proteins also have been found to adsorb well to charcoal. Michaelis and Rona (1910) showed, for example, that serum albumin adsorbs to both kaolin and charcoal in acid (0.01–0.1 N) media. Bleyer (1922) also reported that both serum proteins and agglutinins adsorb to animal charcoal.

VIII. HORMONE ADSORPTION

Indications that various types of hormones can adsorb to charcoal have appeared in the literature. Both Moloney and Findlay (1924) and Dingemanse and Laqueur (1927) described techniques for the purification of insulin via its adsorption to charcoal and its subsequent desorption by various solvents. Zondek and Bansi (1928) found that adrenalin adsorbs to charcoal in conformity to a Freundlich type of isotherm.

Sato (1928) reported that the active principles of the posterior lobe of the hypophysis, both a uterus-stimulating agent and a melanophore-expanding principle, are adsorbed by animal charcoal. Freeman et al. (1935) described methods of purifying the oxytocic hormone of the posterior lobe of the pituitary gland by adsorption on Fuller's earth and Norit charcoal.

Heimer and Englund (1986) showed that oral charcoal can interrupt the enterohepatic cycling of the hormone oestriol (estriol), an estrogen. Twelve milligrams of oestriol were given to women volunteers, without charcoal, or with 20 g charcoal at 3 hr. When no charcoal was given, plasma oestriol rose and remained high for about 12 hr after the drug intake and then slowly declined over the next half day. When charcoal was given, plasma oestriol levels declined soon after and reached pretreatment levels after 6–7 hr. The charcoal clearly interrupted enterohepatic cycling of the oestriol.

Bergman et al. (1967) found that feeding charcoal to Syrian hamsters increased the fecal elimination of thyroxine. In their first test, hamsters were fed radioactive L-thyroxine (3 µg/100 g food) both with and without 3 wt% charcoal in their diet. At sacrifice, there was significantly less thyroxine in the animals that had received the charcoal. In their second test, hamsters were fed radioactive sodium iodide ($Na^{131}I$, 5 µCi/100 g food) both with and without 3 wt% charcoal. In this case, the charcoal had no effect on the sodium iodide amounts in the animals. In a third test, the hamsters received the sodium iodide with or without 3 wt% cholestyramine. In this case, the cholestyramine significantly reduced the amount of sodium iodide in the animals' bodies. In a fourth test, L-thyroxine given once intravenously was reduced by feeding 3 wt% charcoal. Since earlier tests by this research group showed that cholestyramine can reduce L-thyroxine, the overall conclusions were: (1) both 3 wt% charcoal and 3 wt% cholestyramine enhance thyroxine elimination, and (2) 3 wt% cholestyramine, but not 3 wt% charcoal, enhances NaI elimination.

IX. VITAMIN ADSORPTION

Guha and Chakravorty (1934) found that Fuller's earth, silica gel, $BaSO_4$, and charcoal are all capable of binding vitamin B_2 from ox and buffalo liver and kidney extracts. Vitamin B_6 was found to adsorb well to charcoal at pH 1.0 and only slightly at pH 7.0 by Schultz and Mattill (1937). Cheldelin and Williams (1942) showed that various vitamins adsorb reasonably well to charcoal in vitro.

Almquist and Zander (1940) found that charcoals in chick diets "interfered more or less with the absorption of the gizzard factor and vitamins A, G, and K." Matet and Matet (1945) determined that inclusion of 2 wt% charcoal in the diets of rats readily produced deficiencies of vitamin A.

X. ADSORPTION OF WHISKEY CONGENERS

It is well known that activated charcoal, especially granular charcoal, is widely used to "filter" whiskeys. One example is the popular "charcoal filtered" Jack Daniels whiskey. The reason for this is to remove substances called congeners. These are the substances which cause a hangover (spirits such as vodka have fewer congeners and generally produce less of a hangover when consumed in equivalent amounts). Typical hangover symptoms include headache, halitosis, thirst, gastric irritation, fatigue, and dizziness.

Damrau and Goldberg (1971) carried out a study dealing with the in vitro adsorption of congeners by charcoal. They took a standard brand 86-proof whiskey and first determined its total congener content as 2.94% (presumably volume percent). The primary congeners were fusel oil (a mixture containing about 95% of amyl, butyl, and propyl alcohols), acetaldehyde, furfural, esters such as ethyl acetate, and "probably several hundred unidentified constituents." The amounts of each major congener were: esters (as ethyl acetate), 2.667%; fusel oil, 0.25%; aldehydes (as acetaldehyde), 0.02%; and furfural, 0.0016%.

Into each of two 1-L flasks was poured 240 mL of the whiskey plus 400 mL of 0.1 N HCl. One gram of activated charcoal was added to one flask. Both flasks were equilibrated at 37°C for 1 hr with moderate shaking (to simulate the conditions in the human stomach). The solutions were then assayed for fusel oil, furfural, acetaldehyde, and ethyl acetate. The results are shown in Table 24.2.

The results show significant adsorption of these congeners and the values for mg adsorbed/g seem quite believable. Clinical studies have shown that various distilled spirits having higher congener contents undergo alcohol metabolization more slowly, which prolongs the effects of the alcohol and the attendant hangover symptoms. Damrau and Goldberg also showed that administration of a whiskey congener solution containing no alcohol (the ethyl alcohol was distilled off) to human volunteers elicited many unpleasant responses: bad taste, gastric irritation, flushing, heartburn, anorexia, sweating, nystagmus (rapid eyeball movement), impaired perceptual motor performance, etc. Thus, if one drinks

Table 24.2 Whiskey Congener Adsorption[a]

Type of congener	Fusel oil	Furfural	Acetaldehyde	Ethyl acetate
Control	0.25	1.56	20.02	2.66
AC-treated	0.16	0.10	3.48	1.83
Difference	0.09	1.46	16.53	0.83
Percent removed	36.0	93.8	82.6	31.3
mg adsorbed/g AC	0.54	0.88	9.9	0.50

[a]All values are in mL per 100 mL. From Damrau and Goldberg (1971).

distilled spirits and wishes to avoid/minimize a hangover, one should choose a type with small amounts of congeners (e.g., vodka, gin) or choose a brand that has been charcoal filtered.

XI. CATALYSIS OF REACTIONS BY ACTIVATED CHARCOAL

Several investigators have reported on the tendency of charcoal to act as a catalyst in the deamination of amino acids. Fürth and Kaunitz (1930) found that higher molecular weight amino acids were more readily deaminated than lower molecular weight kinds. Amides, some dipeptides, proline, and histidine were not affected by charcoal, whereas phenolic amino acids were rapidly destroyed. Gradwohl (1930) found that increasing alkalinity generally favors the oxidation process. Bergel and Bolz (1933) pointed out that oxygen must be present, that the final products are CO_2, NH_3, and aldehydes, and that the first step in the reaction is probably a dehydrogenation to amino acids. A comparison of alanine and leucine showed that the longer chain reacts much faster. Wunderly (1933a) showed, by use of other nitrogen-containing compounds (urea, biuret, barbituric acid, etc.), in addition to amino acids, that the deaminating effect of charcoal is specific for the true amino acid grouping. Wunderly (1933b) also showed that the amino acid DL-aspartic acid can be hydrolyzed (not oxidized) by charcoal and that increasing alkalinity decreases the rate.

Two related studies are those by Fox and Levy (1936), who showed that ascorbic acid was rapidly (10–15 min) oxidized entirely to dehydroascorbic acid when contacted with Norit charcoal, and by Gompel (1929), who showed that blood charcoal oxidizes oxalic acid.

Fürth and Kaunitz (1929) reported that, in addition to deaminating certain amino acids, charcoal is capable of attacking phenol, lactic acid, dihydroxyacetone, acetone, and β-hydrobutyric acid. Glucose, acetamide, benzamide, and asparagine were not decomposed by charcoal. Ito (1936) also reported that certain amino acids (lysine, histidine, and arginine) were "markedly adsorbed by animal charcoal and the maximum adsorption was at their isoelectric points." Apparently, no decompositions were observed.

Cheldelin and Williams (1942) studied the adsorption of 15 amino acids to Darco G-60 in vitro and found that those with aromatic nuclei adsorbed particularly well, as one would expect. They did not observe any oxidation because they separated the charcoal from the amino acid solutions after only 30 min of contact (and they were probably not checking for any decomposition either).

Chamuleau et al. (1981), in studying hemoperfusion as a means of temporary artificial liver support, recirculated a solution of amino acids and other compounds through a bed of granular activated charcoal. They noted a steady rise in ammonia in the fluid and determined that it came from the breakdown of L-methionine, L-tyrosine, and L-cysteine. Under sterile conditions, the same ammonia production was noted; hence, it came from catalysis of the breakdown of the amino acids by the charcoal and not from bacterial action. When oxygen was kept out of the system, no ammonia was produced. Thus, the reactions involved were oxidation reactions. Tijssen (1980) also showed that, in the presence of oxygen, activated charcoal causes creatinine and uric acid to convert to ammonia, urea, and other products.

Maeda et al. (1979) also mentioned the problem, when using granular activated charcoal in hemoperfusion columns, of the production of methylguanidine from creatinine and the production of lipoperoxide from fatty acids. The charcoal catalyzes both types of reactions.

Druart and De Wulf (1993) found that charcoal catalyzes the hydrolysis of sucrose (a disaccharide) to fructose and glucose (monosaccharides) during the autoclaving of culture media at 120°C for 20 min. With 0.1 wt% charcoal, 60% of the sucrose was hydrolyzed, while 1 wt% charcoal caused more than 96% hydrolysis.

Cooney and Xi (1994) presented data which show that many aromatic chemicals (aniline, phenolics) undergo charcoal-catalyzed reactions after adsorption. The reaction rates were increased under alkaline conditions and in the presence of molecular oxygen (i.e., O_2). However, the reactions (most of which are oxidative coupling reactions) were found to be quite slow, and would occur to an insignificant extent in the time frame involved in antidotal charcoal applications.

XII. MISCELLANEOUS COMPOUNDS ADSORBED BY CHARCOAL

Bancroft and Fry (1933) showed that glycogen adsorbs to charcoal and in that state resists hydrolysis to glucose. Cholesterol and saponin adsorption by charcoal was mentioned by Eisler (1926). Fantus and Dyniewicz (1936) indicated that phenolphthalein adsorbs well to charcoal, especially under less alkaline conditions. Loch (1938) reported that thiocyanates, which are excellent bactericides and sporicides, adsorb to charcoal in appreciable amounts and De (1937) indicated that carotene adsorbs to charcoal.

Kawano (1949) reported that penicillin is effectively adsorbed by charcoal from fermentation broths. Use of 1–1.5 wt% charcoal removed 99% of the penicillin and the adsorption followed the Freundlich equation. Similarly, De Duve and De Somer (1947) also found that penicillin adsorbs well to charcoal.

Streptomycin adsorption to charcoal was used by Carter et al. (1945) and by Silcox (1946) to recover and concentrate streptomycin from fermentation broths.

XIII. CHARCOAL USE IN CONCENTRATING DRUGS FROM BIOLOGICAL FLUIDS

We have discussed the topic of columns containing granular charcoal in the section concerning hemoperfusion (Chapter 7, Section XI). However, somewhat less elaborate cartridges containing granular charcoal have also been employed for other medically related purposes, namely, for the concentration of drugs from urine (or other biological fluids) prior to chemical analysis. If a drug is present in very low concentration in the fluid, one can often adsorb the drug onto granular charcoal by passing the fluid through a cartridge containing charcoal. Then, if one elutes the cartridge with a solvent (e.g., ethanol, chloroform) in which the drug is highly soluble, one can recover nearly all of the drug in only a small volume of the solvent. The net result is that the drug ends up in a much more concentrated form. The solution can then be treated by thin-layer chromatography, gas-liquid chromatography, or similar analytical techniques. The overall purpose of the analysis is usually to identify the drugs present in the body of a person or animal. This procedure has many forensic applications (e.g., determining drugs in murder victims, racehorses, athletes, etc.).

Bastos and Hoffman (1976) provided a review of the use of solid adsorbents (charcoal, ion exchange resins, nonionic resins) for the extraction of drugs of abuse from biologic materials. Basic principles and techniques are discussed.

Meola and Venko (1974) related their experience in isolating barbiturates, glutethimide, ethchlorvynol, amphetamines, phenothiazines, quinine, morphine, cocaine, diazepam, and chlordiazepoxide from urine using cartridges containing granular activated charcoal. They eluted these drugs from the charcoal cartridges using organic solvents and then employed thin-layer chromatography to separate and identify them. The average detection limit for all of the drugs was 1 μg/mL urine. They concluded that the procedure was fast, economical, and adaptable to any urine screening program.

Hindmarsh et al. (1975) described the isolation of the basic drugs chlordiazepoxide, cocaine, diazepam, diphenhydramine, and methaqualone from urine using charcoal cartridges. They found that the method is extremely sensitive, with microgram quantities being easily isolated. Hindmarsh and Hamon (1977) discussed the general procedures in using charcoal cartridges for these purposes. They state that the method is useful for acidic, basic, and neutral drugs, without the need for prior fractionation of the fluid involved.

A comparison of activated charcoal, XAD-4 resin (an uncharged polystyrene resin cross-linked with 4% divinylbenzene), and a Dowex resin in adsorbing hypnotic drugs from human serum was reported by Harstick et al. (1979). The adsorbents were also tried in an agarose-encapsulated bead-like form (5–10 mm diameter beads). The charcoal was generally better than the two resins; of the two resins, XAD-4 was the better. Agarose encapsulation reduced drug adsorption somewhat but not enough to make the adsorbents ineffective. This study was not aimed at detecting drugs but rather at producing biocompatible adsorbents for removing drugs from the blood of overdose victims by hemoperfusion. However, we discuss this study here, as it relates strongly to the present topic.

Charcoal encapsulated within a spherical, porous, polypropylene capsule, 1 cm in diameter, was used by Elahi (1979) to adsorb barbiturates from blood. After desorption with diethyl ether, the drugs were quantified by gas chromatography. The lower limit of detection in the blood was found to be 0.05–0.10 μg/L.

XIV. USE OF CHARCOAL IN THE ASSAY OF BIOCHEMICALS

Charcoal has been used for almost four decades as a separating agent in ligand assays. Frequently, the charcoal is mixed with dextran (a high molecular weight polysaccharide) or a protein (albumin, hemoglobin, etc.) which coats the charcoal surface and provides a sieving effect. Small molecules can pass through this sieve and be adsorbed inside the charcoal, while the large molecules, which are generally the ones of interest, are left outside. In most applications, the small molecules are ligands, and the large molecules are protein-bound ligands. Thus, the separation which occurs is between free ligands and protein-bound ligands. The ligands involved are frequently radioactively-labeled. Thus, their concentration in the fluid external to the charcoal is easily quantified. This method allows one to determine the extent of protein binding of various ligands. This basic technique has been used widely to assay for hormones and other immunological biochemicals.

This method was developed by Miller (1957) for vitamin B_{12}, who determined the difference between free vitamin and protein-bound vitamin B_{12}. The method was further developed by Herbert et al. (1966, 1968), who successfully applied it to a number of radioassay methods. The concept of "instant dialysis" (i.e., the separation of small and large molecules via the sieving action of the macromolecule coating layer) was demonstrated by them. The "coated charcoal" method has now become quite a standard procedure.

However, Mortensen (1974) pointed out that charcoal will adsorb not only the ligands but also some of the proteins and some of the protein/ligand complexes. Also, the free ligands will not all be adsorbed, as the degree of adsorption depends on the intrinsic binding affinity between the ligand and the charcoal, the amount of charcoal used, and the presence of other molecules which may compete with the ligands for adsorption sites. Mortensen discussed how to minimize errors arising from these effects.

Binoux and Odell (1973) studied the adsorption of ^{125}I-HTSH (human thyrotropin) and ^{125}I-HTSH-antibody complex at high and low serum protein concentrations, by uncoated charcoal and by charcoal coated with D10, D40, or D50 dextran. Normal rabbit serum and dog serum were added to the hormone solutions in varying amounts, so as to give samples with varying protein levels. Both the bound and free hormone forms adsorbed to charcoal, but the bound hormone adsorbed far less than the free hormone. Regardless of the protein concentration, uncoated charcoal was able to efficiently separate the bound and free hormone forms. At low protein concentrations, coated charcoal slightly improved the separation, but at high protein concentrations, uncoated charcoal was more efficient. The data showed that the proteins competed for adsorption with the free and bound hormone forms, since increased protein concentrations decreased the adsorption of both free and bound hormone. Dextran coating diminished the adsorption of both hormone forms and thus it was of no value in enhancing the separation. Contrary to opinion prevailing at the time, the authors concluded that dextran does not act as a molecular sieve but instead merely reduces the number of available adsorption sites for both free and bound hormone.

De Hertogh et al. (1975) studied estrogen binding to dextran-coated charcoal and determined the effects of incubation time, incubation temperature, and the amount of charcoal used. The affinity of charcoal for estrogens was high; for example, at 40 mg/100 mL charcoal in solution, more than 98% of free estradiol was adsorbed.

Boxen and Tevaarwerk (1982) presented results of a study in which various molecular weight dextrans, serum bovine albumin, immunoglobulin G (IgG), or insulin (a polypeptide) were used to coat charcoal. These charcoals were then used to adsorb various ligands (radiolabeled hormones such as triiodothyronine). They found that varying the amount or molecular weight of the dextrans had no appreciable effect on the adsorption of the ligands. However, there was a definite effect with the proteins (albumin, IgG, insulin). Under some conditions, in fact, ligand adsorption was enhanced. They concluded that the use of dextran to give a sieve effect has no basis in fact and that proteins should be used instead.

It should be mentioned that one area in which coated charcoal has been used very extensively in assaying ligands is in the area of determining concentrations of thyroid hormones. The quantitative measurement of the amounts of such hormones present allows one to assess whether deviations from normal amounts exist and therefore determine if hyperthyroidism or hypothyroidism exists.

Polak (1974) developed a quick test for detecting hepatitis-B antigen (HBAg) in serum using charcoal. The charcoal powder is suspended in a buffer and coated with gamma-globulin. A drop of this suspension is combined with a drop of the blood serum to be tested on a glass plate. After 5 min, if agglutination of the charcoal particles occurs, the test is positive for the HBAg. Otherwise, the mixture remains homogenous.

XV. PROTECTION OF CROP PLANT SEEDLINGS FROM HERBICIDE DAMAGE

Activated charcoal has been used to protect seeds and seedlings of desired plants from the effects of herbicides applied to soils to inhibit weed growth. There have been many reports in this subject area and here we summarize some of the major studies.

Lucas and Hamner (1947) were one of the first researchers to demonstrate that activated charcoal can inactivate a potent herbicide, in this case 2,4-dichlorophenoxy-acetic acid (2,4-D). They mixed a solution of 1,000 ppm of the sodium salt of 2,4-D with 1 wt% Norit A activated charcoal and sprayed the solution on bean plants shortly after emergence. The plants showed little effect. Higher amounts of 2,4-D and of charcoal (e.g., 10,000 ppm 2,4-D and 10 wt% charcoal) were tested with similar results.

Arle et al. (1948) tested whether charcoal could protect a sensitive field crop such as sweet-potato seedlings from injury due to the herbicide 2,4-D. They first treated the soil in various test plots with different amounts of 2,4-D. Half of each test plot was then planted with untreated sweet-potato sprouts and the other half with sprouts whose roots had been moistened and dusted with about 0.1 lb/100 lb Norit A charcoal. The results are shown in Table 24.3. Clearly the use of charcoal on the seedling roots greatly reduced injury to the sprouts by the 2,4-D.

Table 24.3   Effect of Charcoal on Sweet-Potato Seedling Survival

Amount of 2,4-D used (lb/acre)	Seedling survival (%)	
	Charcoal treatment	Untreated
0	100.0	93.5
1.3	95.0	2.5
5.2	32.0	0.0

Ahrens (1965a) mentioned experiments which showed that amounts of activated charcoal 100 to 400 times that of the triazine herbicides simazine and atrazine protected oats and tomato plants from injury, by adsorbing residues of these herbicides from the soil. Further work with other plants established that the sensitivity of the other plants to these herbicides increased in the order: cabbage, tobacco, beans, and beets. In soil treated with 1 lb/acre of simazine or atrazine, 200 lb/acre of charcoal provided good protection to beans, tobacco, and cabbage (plant yields decreased only 10–15%). When the herbicides were applied at 2 lb/acre, 800 lb/acre of charcoal was needed to prevent plant injury (this charcoal:herbicide ratio also protected beets well). Applying the charcoal dry was more effective than applying it as a suspension in water, for beans and beets. However, for tobacco and cabbage, no consistent difference was noted.

Ahrens (1965b) also reported that a pre-planting root dip in activated charcoal (the roots were moistened with water and dipped into powdered charcoal) gave variable protection from simazine, depending on the plant type and the simazine level. Undipped strawberry plants were severely injured or killed by 2–3 lb simazine/acre, while charcoal-dipped plants showed no injury. Undipped cabbage growth was reduced 28% by simazine, while charcoal-dipped cabbage growth was only 12% less than control plants.

Ahrens (1967) gave additional results on root-dipping studies two years later. Five field experiments were conducted during the 1966 growing season, using eight different herbicides. Injury to tomatoes and peppers from diphenamid and injury to peppers from trifluralin was slightly decreased by charcoal root-dipping. In strawberries, root-dipping prevented injury from simazine/DCPA and from diphenamid. Growth and survival of sod-planted white spruce trees was improved by the charcoal root-dipping treatment when the herbicides simazine, dichlobenil, or bromacil were involved. Finally, in soils treated with simazine or dichlobenil, growth depression was greatly reduced or eliminated in 1–2 year old forsythia, euonymus, and weigela plants.

Robinson (1965) dipped the runners of strawberry plants into several types of adsorbents (charcoal, kieselguhr, vermiculite) prior to planting them. The planted field was then sprayed a few days later with a solution of 1 lb/acre simazine (a broad spectrum herbicide). No significant damage to the new plants occurred when charcoal was used; dipping the roots into the dry powdered charcoal was more effective and easier than placing a similar quantity of charcoal around the roots at the time of planting.

Linscott and Hagin (1967) reported that alfalfa seedlings can be protected from the herbicide triazine using activated charcoal, by applying bands 2.5-cm wide (in greenhouses) or 3.2-cm wide (in fields) of powdered charcoal directly over the seeds. The charcoal was suspended in water and was sprayed at rates of 0, 50, 100, and 150 lb/acre directly above the seed rows. The use of 25 lb/acre charcoal or above gave adequate protection from 1.5 lb/acre of triazine, and 50 lb/acre charcoal or above gave good protection against 3.0 lb/acre triazine. However, when 1.5 lb/acre of a different herbicide (atrazine) was used, none of the charcoal applications provided significant protection.

Brenchley (1968) used the same technique to protect sugar beet seedlings from damage by herbicides. A 2.5-cm wide band of charcoal was applied in the furrow over the seeds at a rate of 190 lb charcoal/acre. This gave excellent protection against a herbicide known as R-11913 and significantly reduced damage from the herbicides pyrazon and propachlor. Another herbicide (bromacil) was only partly deactivated by the charcoal, however. Brenchley also did tests with oat fields sprayed with atrazine (applied at 3 or 6 lb/acre) and found that different charcoals (applied at rates of 200,

400, and 600 lb/acre) showed different degrees of protection. Aqua Nuchar was slightly more effective than Darco S-51, which was twice as effective as Darco DM, which, in turn, was three times more effective than a nonactivated charcoal. More specifically, 200 lb/acre of Aqua Nuchar provided excellent protection against 1.6 lb/acre atrazine (essentially zero injury versus 45% injury when no charcoal was used). From 10 to 60% injury was found on oats seeded in plots containing 4.8 lb/acre atrazine plus 200 lb/acre charcoal (versus 97% injury with no charcoal). From 10–80% injury was found when bromacil was used at 1.6 lb/acre with 200 lb/acre charcoal (versus 100% injury with no charcoal), and about 20% injury was noted with 4.8 lb/acre diuron plus 200 lb/acre charcoal (versus 72% injury with no charcoal).

Lee (1973) also used a 2.5-cm wide band method to protect various grass seedlings from diuron, atrazine, and simazine. The amount of charcoal required for protection depended on the type of herbicide, the amount of herbicide, and the type of grass involved. The effectiveness of this technique has also been found to depend on the soil type, and the sowing depth.

Kennedy and Talbert (1973) discussed the use of activated charcoal as a herbicide antidote, and discussed studies they carried out with a charcoal product called Gro-Safe. They determined the amount of Gro-Safe needed to inactivate 1 lb/acre of nitralin and 1.5 lb/acre of fluometuron, using sorghum and squash as the test plants. Charcoal was applied at rates of 0, 63, 125, 250, and 500 lb/acre. At 125 lb/acre or higher, the charcoal totally protected squash from growth depression by fluometuron. With sorghum, and nitralin as the herbicide, the charcoal was less effective and showed a marked dose relationship; charcoal applied at 63, 125, 250, and 500 lb/acre gave growth losses of 99, 89, 60, and 18%, respectively, when the nitralin was applied at the time of planting.

Toth and Milham (1975) compared activated charcoal and carbon ash (made by burning a carbonaceous material, without activating it) for their relative effectiveness in preventing injury to barley by diuron. The activated charcoal protected barley seedlings from death by 32 kg/ha diuron when used in a charcoal:diuron ratio of 5:1 or greater. However, carbon ashes prepared in different ways were much less effective and only one such ash gave any significant protection against 16 kg/ha diuron.

Nangju et al. (1976) used the 2.5-cm wide band technique to try to protect upland rice from chloramben, butachlor, oxadiazon, 2,4-D, and atrazine. In this study, the application of a band of charcoal slurry was sometimes ineffective but the application of a 2.5-cm wide and 2.5-cm deep layer of a charcoal and vermiculite mixture (1:1) was effective (note that the charcoal slurry probably delivered the charcoal only to the surface region, whereas the charcoal/vermiculite method delivered the charcoal 2.5-cm deep).

Andersen (1968; apparently the same Andersen who carried out the classic studies on antidotal charcoal that are described in Chapter 10) reported that the herbicides simazine and linuron can be inactivated in soils by charcoal. Andersen mixed these herbicides into boxes of sandy loam soil in various amounts, mixed in various amounts of charcoal, and then planted oat seeds in the boxes. There was a marked reduction in injury to the new plants when charcoal was added to the soils. For simazine, reasonably complete deactivation required a charcoal:herbicide weight ratio of about 200:1 to 400:1, whereas for linuron the required ratio was about 100:1.

Coffey and Warren (1969) carried out a root bioassay to compare the adsorption of herbicides by charcoal, muck soil, bentonite clay, a cation exchange resin, and an anion exchange resin. The effectiveness of the adsorbents was determined by comparing the concentrations of herbicide required to give 50% root inhibition of the test plant. Of eight herbicides tested, charcoal adsorbed six better than any of the other adsorbents.

Bovey and Miller (1969) studied various herbicides, various herbicide dosages, different plant species, and different amounts of charcoal in order to assess the effect of charcoal on reducing the phytotoxicity of herbicides in a tropical type of soil. The charcoal protected oats well but did not completely protect cucumbers or beans even in amounts up to 600 lb/acre. In other tests, with charcoal but without herbicides, it was found that charcoal alone was not harmful to the growth of beans, oats, or cucumbers in the silty clay soil used.

Jordan and Smith (1971) found that charcoal can deactivate 0.3 ppm atrazine and 1.0 ppm diuron in soils when the charcoal amounts were 30 and 50 ppm, respectively. Oats were used as the test plant. Kratky and Warren (1971) used a mixture of charcoal and vermiculite added to soil to increase the tolerance of direct-seeded cucumbers and direct-seeded tomatoes to simazine. When the mixture was placed in a hole 1-inch in diameter by 3/4-inch deep, with the seed at the bottom of the hole, the crop plant was not injured by the herbicide.

Burr et al. (1972) applied a slurry of charcoal to rye grass seedlings in a greenhouse. Then, a diuron solution was sprayed over the planted area. It was found that when a sandy soil was used, at least three times as much charcoal was required to protect the seedlings than when a clay loam soil was used. Seedling depth was a factor: seeds growing at a depth of 1.3 cm were less protected than seeds growing at depths of 0.6, 1.9, or 2.5 cm depth. Wetting agents added to the charcoal slurry reduced the degree of protection when used in more than a 0.3% v/v amount. The soil moisture level at the time of planting and the amount of subsequent irrigation did not influence the protective action of the charcoal.

Chandler et al. (1978) investigated the protection of seedling cotton from the effects of diuron. Diuron was applied at 1.78 and 3.55 kg/ha under field conditions for 5 yr. Charcoal did not provide adequate protection when applied through the seed hopper box or sprayed in the seed furrow over each hill. However, charcoal applied at 83 and 167 kg/ha as a spot over each hill of cotton gave adequate protection. If the charcoal was blended shallowly into the soil, the degree of protection was significantly reduced.

Rolston et al. (1979a) used the 2.5-cm wide charcoal banding method to control volunteer legume growth in fields which were planted with rows of white clover and then sprayed with various herbicides. The charcoal:herbicide ratios needed to protect the white clover seedlings with 0.5 kg/ha diuron, atrazine, and simazine were 50:1, 200:1, and 200:1, respectively. Rolston et al. (1979b) also studied volunteer legume control in fields planted with red clover and alfalfa, using the same 2.5-cm wide charcoal banding technique. Atrazine or diuron at 1 and 3 kg/ha, respectively, controlled 85–99% of red clover volunteers, while providing good protection to the seeded rows.

Lee (1978) applied the 2.5-cm wide charcoal banding method to the control of volunteer Kentucky bluegrass while Kentucky bluegrass was established for seed production. Volunteer control of 80–99+% was achieved with terbacil, diuron, atrazine, and simazine. The use of charcoal bands prevented adverse effects on crop establishment and seed yields.

Ogg (1978) studied the effect of applying powdered activated charcoal to protect direct-seeded asparagus from the effects of various herbicides (chloramben, chlorbromuron, linuron, nitralin). The herbicides were applied before emergence of the asparagus in order to reduce a variety of weeds (barnyard grass, redroot pigweed, hairy nightshade, common lambsquarter, etc.). In general, these herbicides control weeds well but have the disadvantage of injuring the asparagus and reducing the amount of asparagus. Activated charcoal applied at a rate of 56 and 112 kg/ha in 3-cm wide bands over seeded rows of asparagus protected the asparagus from these harmful effects and only reduced weed control about 10%. Weed control was still 85 to 90%.

Ogg (1982) also studied the effects of terbacil (a broad spectrum soil and foliage-active herbicide) on alfalfa, corn, spring wheat, and sugar beets. With 0.10 ppm in the soil, all of these plants were severely injured by terbacil. However, charcoal at 150 kg/ha prevented any significant weight loss of beans and corn in soil containing 0.10 ppm terbacil, and charcoal at 300 kg/ha fully protected sugar beets and significantly protected alfalfa in soil containing 0.19 ppm terbacil. Charcoal at 600 kg/ha did not protect the wheat from 0.19 ppm terbacil, however.

Rydrych (1985) investigated the protection of winter wheat from metribuzin (the only postemergence herbicide then registered for downy brome control in winter wheat in the Pacific Northwest) using charcoal. Whereas charcoal applied in 5-cm wide bands over seeded rows at 84, 167, and 336 kg/ha protected the wheat from metribuzin on a silt loam soil, only the 336 kg/ha application rate was protective on a sandy loam soil. Downy brome control was not reduced by charcoal applied over the wheat rows. The best treatment was found to be charcoal at 336 kg/ha applied pre-emergence over the rows and metribuzin applied at 0.6–1.1 kg/ha ten weeks later. No chemical injury occurred with this protocol.

Majek (1986) also studied the use of charcoal to offer protection against metribuzin, using peppers as the test plants. Charcoal applied in the amount of 1 g/plant protected bare-root transplants and plug transplants grown in cells having 10–20 g of greenhouse soil mix. The charcoal was either added to the transplant water of bare-root transplanted peppers or incorporated into the growing medium of plug-transplanted peppers. The soil in which the transplants were placed contained up to 0.56 kg/ha metribuzin. When charcoal was not used, soil having half as much metribuzin killed bare-root and plug-transplanted peppers.

XVI. EFFECT OF CHARCOAL ON PLANT TISSUE CULTURES

Many studies have shown that activated charcoal often improves plant tissue cultures by adsorbing growth inhibitors, preventing unwanted callus growth, promoting morphogenesis (particularly embryogenesis), and enhancing root growth. Charcoal exerts its beneficial action in various ways: by adsorbing phenols produced by wounded tissues (Weatherhead et al., 1978), by adsorbing agar impurities (Kohlenbach and Wernicke, 1978), by adsorbing components such as ethylene from the gas phases of the cultures (Horner et al., 1977), and by adsorbing cytokinins and auxins (Fridborg et al., 1978). Activated charcoal also adsorbs medium components such as ascorbic acid and other vitamins (Takayama and Misawa, 1980; Heberle-Bors, 1980; Johansson et al., 1982; Misson et al., 1983).

Vantis and Bond (1950) added charcoal to the rooting medium of peas and found marked increases in the weight of the pea plants and in nitrogen fixation by the plants, as compared to controls. They suggested that the favorable growth resulted from adsorption of harmful excretions of the roots, or harmful microorganisms, and from the maintenance of a favorable pH in the rooting medium.

Constantin et al. (1977) described how activated charcoal in culture media reportedly either stimulates growth, organogenesis, and embryogenesis, in a relatively wide range of plant species, differentiation in lower plants, growth of dikaryons and heterokaryon formation, or completely inhibits growth of in vitro-cultured plants. Their results with cultured anthers of *Nicotiana tabacum* showed that charcoal in the medium increased the frequency of plantlet development. They report that charcoal in a so-called MS culture medium inhibits callus growth and shoot development in Wisconsin-38 tobacco by the adsorption of hormones required for growth.

Proskauer and Berman (1970) reported favorable effects of adding activated charcoal to a nutrient medium for the culture of algae and mosses. Klein and Bopp (1971) also observed that charcoal enhances the growth of mosses. Ernst (1974) found that a certain variety of orchid showed dramatic increases in shoot and root growth when charcoal was added to seedling cultures. Anagnostakis (1974) found that charcoal stimulates haploid plants from anthers of tobacco. Fridborg and Eriksson (1975) found that charcoal added to culture media favored the growth of roots and plantlets in cultures of *Allium cepa* and *Daucus carota*. Without charcoal, no roots or plantlets formed, but when 1 wt% charcoal was added the *Allium* colonies developed roots and the *Daucus* colonies produced embryos (nearly 100% in both cases). The authors state that: "It seems that the effects of activated charcoal are due to the removal of substances from the medium which promote unorganized growth, inhibit embryogenesis, root formation and elongation."

Wang and Huang (1976) cultured several plants (palm embryos, ginger shoot tips, two kinds of plant seedlings, and one plant tissue culture) in agar media. When charcoal was added at concentrations of 0.5 and 3 g/L, improved development of all plants, often very dramatic, occurred. The effects of activated charcoal were attributed to the adsorption of toxic metabolites released by the plant tissues. The release of growth inhibitors by plants is well known; indeed, this phenomenon is known as allelopathy.

Ui et al. (1990) reported that cell division often failed to occur in protoplast cultures of grapes (*Vitus vinifera*) on agar media but that success was achieved when Gellan gum and activated charcoal were used. The frequency of cell division (the number of cells that divided more than once, divided by the initial number of cells, and multiplied by 100 to give a percentage) was determined for various media. The values were: for 0.6% agar, 0; for 0.6% purified agar, 0; for 0.6% agarose, 1.2; for 0.8% sodium alginate, 7.7; and for 0.4% Gellan gum, 24%. When 0.2% w/v activated charcoal was added to each medium, the frequency values rose to 1.7, 1.8, 12, 17, and 69%, respectively. Thus, charcoal had a large positive effect.

XVII. EFFECT OF CHARCOAL ON BACTERIAL CULTURES

A variety of papers have appeared over the years which deal with the effects of activated charcoal added to culture media. In some cases the charcoal accelerates growth; less frequently it suppresses growth. A sampling of these studies is given here to show the wide range of effects that charcoal can have on microorganisms.

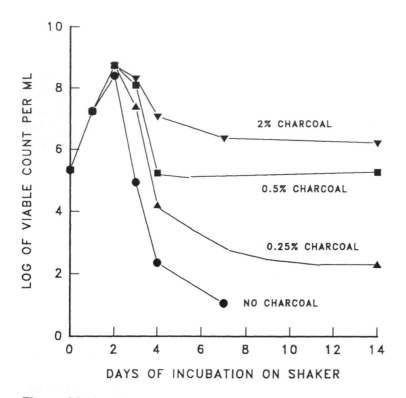

Figure 24.1 Effect of different levels of activated charcoal upon the growth curve of *Bacillus popilliae*. From Haynes and Rhodes (1966). Reproduced by permission of the American Society for Microbiology.

Roberts and Baldwin (1942) found that a greater percentage of *Bacillus subtilis* cells sporulated in peptone treated with charcoal than in untreated peptone broth. The charcoal effect was greatest at a pH of 3.0–5.0

Feeley et al. (1979) compared a charcoal-yeast extract (CYE) agar against F-G agar for supporting the growth of *Legionella pneumophila*. The CYE agar had 0.20% w/v of Norit A or Norit SG activated charcoal. The CYE agar gave 90 times more colony-forming units (4.35×10^6 versus 4.85×10^4) than did the F-G agar.

Haynes and Rhodes (1966) discussed the fact that it had not been previously possible to evoke sporulation of *Bacillus popilliae* in liquid culture media, but that they had discovered a way to cause this. They discovered that sporulation will occur in shaken cultures of tryptone-glucose-yeast extract broth if activated charcoal is added. Figure 24.1 shows the effects of different amounts of charcoal in the culture medium on the growth of this organism. One explanation for the positive effect of charcoal is its ability to combine with toxic fatty acids liberated into the medium by the cells themselves as they grow. This paper mentioned a group (Hardwick, Foster, and Guirard) which had previously found that activated charcoal adsorbed factors from complex organic media which inhibited the growth and sporulation of *B. larvae*, and also mentioned that this group subsequently discovered that the factors were largely fatty acids.

Smith and Alwen (1966) described problems encountered with the "swarming" of species of *Proteus* on solid culture media, a phenomenon which makes it difficult or impossible to detect the presence of other organisms and causes considerable delay in their isolation and subsequent identification. They discovered that adding 1% charcoal to the agar media prevented the proliferation of swarmers and gave a discrete central *Proteus* colony in every case (they studied five different strains of *Proteus*). They hypothesized that swarming results from "the synthesis of some extracellular factor, as yet unidentified. If this factor is adsorbed by charcoal, then swarmers are not produced and a discrete colony results." In a follow-up paper, Alwen and Smith (1967) showed that charcoal again inhibited the proliferation of swarming *Proteus* spp. but did not interfere with the growth of any other organisms. Thus, the isolation of organisms contaminated with *Proteus* was facilitated. Flagella were normal. Whereas some agents used to suppress swarming interfere with flagella production, charcoal was observed not to do so.

El-Mishad et al. (1975) described a study in which charcoal (50–200 mesh) was mixed and injected with D9 lymphoma cells into hamsters, in order to see if "coinoculated nonviable particulate matter could potentiate tumor incidence," presumably by some mechanism involving macrophages. Charcoal doses of 10–100 µg significantly increased the tumor incidence but higher doses tended to decrease the incidence of tumors. Charcoal had no effect when injected along with CILT/2 carcinoma cells. The reason for this lack of effect with carcinoma cells may be related to their longer latent period. Much more research needs to be done in this area to determine what mechanisms are operative.

Schuster et al. (1977) investigated the use of charcoal to remove ligands bound to cytochrome P-450, without any detectable destruction of the cytochrome and its catalytic properties. A charcoal suspension was combined with either microsomes (from rat livers) or partially-purified cytochrome P-450 (from rabbit livers), with stirring at 0°C for 20 min. Next, the charcoal was removed by centrifugation at 11,000 G for 10 min. A radioactively-labeled binder then was used to determine the degree of removal of the previously bound ligands.

Hoffman et al. (1983) discussed the use of charcoal in yeast extract culture media as a way of preventing the photochemical oxidation of media constituents by fluorescent lights. Such oxidation creates hydrogen peroxide and superoxide radicals which can kill off sensitive organisms such as *Legionella*. The use of charcoal prevented photochemical oxidation, catalytically decomposed hydrogen peroxide and superoxide radicals, and prevented the light-accelerated autoxidation of cysteine (a medium component).

Barker and Farrell (1986) described techniques used to determine the minimum inhibitory concentrations (MIC's) of antimicrobial agents used for preventing the growth of *Legionella pneumophila* (the microbe that causes Legionnaire's Disease). Strains of this organism grown in a lysed red blood cell agar medium and in a buffered yeast agar medium gave similar MIC results. However, when charcoal was added to the buffered yeast agar medium, the MIC values were at least 100 times higher. It was suggested that the charcoal may have adsorbed and bound a large fraction of the antimicrobial agents and thus reduced their effectiveness.

Torre et al. (1988) also studied the effect of incubation media which contain charcoal on the MIC values of an antibiotic, in this case the drug ciprofloxacin (which is used to treat Legionnaire's disease). However, they used the bacterium *Escherichia coli* for their tests. When an agar medium (CYE, Oxoid) containing charcoal (amount unstated) was used, the MIC for ciprofloxacin inhibition of *E. coli* growth increased by a factor of 4 (i.e., four times as high a concentration of the ciprofloxacin was required for inhibition).

In a reply to a paper dealing with the in vitro activity of ciprofloxacin, Torre (1988) explains that the four-fold effect "is probably due to the aspecific absorption (sic) properties of charcoal" and that the effect "depends on the total available surface area of charcoal granules and is, therefore, dose dependent." He then stated that, "in vivo, the concomitant administration of charcoal (1 g) and ciprofloxacin (500 mg) does not affect the kinetics of ciprofloxacin." This latter statement, according to Torre, is based on his unpublished observations.

Murate et al. (1988) used fetal bovine sera (FBS) to culture erythroid progenitor cells and granulocyte-macrophage progenitor cells. They found that when the sera had been treated by passage through a column containing granular activated charcoal, erythroid colony formation was markedly more efficient. This confirms earlier reports that certain sera contain erythropoietic-inhibiting factors. Lindquist and De-Alarcon (1987) reported the removal of inhibitors from fetal bovine serum by dextran-coated charcoal treatment but have not identified the molecular structures of them. Their method originated with the somewhat classic paper by Herbert et al. (1965), in which albumin- or dextran-coated charcoal was used to separate free hormones such as insulin from protein-bound ones (some of Herbert's work was described in Section XIV).

Padilla et al. (1988) cultured two-cell mouse embryos in media supplemented with 15% charcoal-extracted serum or with 15% nonextracted serum. Charcoal extraction significantly improved embryo development. Thus, it was concluded that charcoal extraction "significantly reversed the 'embryotoxic' effects of some sera in the two-cell mouse embryo model."

Since the discovery of the family *Legionellaceae*, buffered charcoal-yeast extract (BCYE) agar media have become the standard medium for isolating these organisms from cultures. In addition, Raad et al. (1990) used BCYE media for isolating *Brucellae* organisms, while Vickers et al. (1992) and Boquest and Tosolini (1993) used BCYE agar to isolate *Nocardia asteroides* organisms.

XVIII.   EFFECTS OF CHARCOAL ON INSECTS

There is an intriguing report by Majumder et al. (1959) concerning the insecticidal effects of activated charcoal and clays. They tested the ability of a wide variety of activated charcoals and clays in killing adults of the insect species *Tribolium castenum*. The insects were placed on powdered charcoal or clay which had been spread on glazed porcelain test plates. Each batch of insects was contained within glass rings. The mortalities at room conditions of 26°C and 68% relative humidity were noted at 4, 8, 16, and 24 hr. These were related to the adsorptive properties of the charcoals and clays, as indicated by tests such as CCl_4 adsorption, methylene blue solution decolorizing power, or bleaching qualities. Some sample results are given in Table 24.4.

One can conclude that: (1) the charcoals are more potent insecticides than the clays, and (2) the insecticidal power of each is generally related to its degree of activation, that is, its adsorbing/bleaching power. The hypothesis as to how these powdered materials function as insecticides involves desiccation (i.e., removal of water) of the insects by close contact with a material which can adsorb, among other things, water vapor. This theory remains to be proved (no references beyond this one from 1959 which follow up on this interesting topic have been found).

Table 24.4 Effect of Charcoal on Insects

	Mortality (%) at different times				CCl_4 adsorption (g/g)	MB decolorized (%)
	4 hr	8 hr	16 hr	24 hr		
Charcoal 1	100	100	100	100	1.26	100
Charcoal 2	46	68	100	100	0.95	100
Charcoal 3	28	79	100	100	0.79	100
Charcoal 4	30	75	100	100	0.74	96
Charcoal 5	30	54	83	100	0.62	61
Charcoal 6	0	12	23	66	0.50	20
Charcoal 7	0	0	0	28	0.05	20
						Bleaching power (%)
Clay 1	0	43	92	100	0.24	32
Clay 2	0	12	89	100	0.21	40
Clay 3	0	0	96	100	0.63	31
Clay 4	0	0	0	10	0.02	24

From Majumdar et al. (1959). Reprinted with permission from *Nature*. Copyright 1959 Macmillan Magazines Ltd.

XIX. SUMMARY

The "smorgasbord" of studies just discussed has shown that the following are generally well adsorbed by charcoal: snake venom components, viruses, biochemical factors, hormones, vitamins, fungal toxins (as mentioned in greater detail in Chapter 20), and antibiotics such as penicillin and streptomycin. Whisky congeners also adsorb well, in general (they are relatively small nonaromatic organic chemicals and thus their chemical structures strongly influence how well they adsorb).

With respect to bacteria, mixed results have been noted. Some bacteria adsorb well, while others (*E. coli*, *C. welchii*) do not. Likewise, some bacterial toxins adsorb in significant amounts, while others do not. The combination of toxin and antitoxin appears not to adsorb.

Proteins adsorb well but those which are enzymes sometimes lose part or all of their enzymatic activity upon adsorption. Some enzymes remain fully active, however. This suggests that enzymatic activity is impaired only if the site of adsorption of the protein and the site of enzyme activity are identical or close to each other.

Activated charcoal has been shown to catalyze many decomposition and oxidation reactions, some of which require the presence of dissolved oxygen in solution. Examples of chemicals which are susceptible to catalysis are amino acids (especially aromatic ones), ascorbic acid, phenolic compounds, lactic acid, creatinine, and uric acid.

Many biological and analytical uses of charcoal have evolved over the years. Among these are: the use of charcoal (and resins) to concentrate drugs for subsequent analysis, the use of dextran-coated charcoal to separate free and protein-bound ligands, and the use of charcoal in plant tissue growth media and bacterial culture media. In the case of plant tissue cultures, plant growth is usually accelerated due to the binding of growth inhibitors by the charcoal. With respect to bacterial cultures, the growth of bacteria can be made faster and more selective with added charcoal. However, if the bacteria are cultured with an added antibiotic, the purpose being to determine the minimum inhibitory concentration (MIC) of the antibiotic, the use of charcoal usually increases the MIC greatly, since the charcoal takes up much of the antibiotic from the culture medium.

The use of charcoal to protect seedling crop plants from damage by herbicides applied to fields to control weed growth was an active area of study in the past and many investigations have shown what charcoal doses are needed and what methods of application are best. However, no recent work in this subject area seems to have been carried out.

By now, the reader should be convinced that ever-new uses for the remarkable substance known as activated charcoal will continue to evolve.

XX. REFERENCES

Ahrens, J. F. (1965a). Improving herbicide selectivity in horticultural crops with activated carbon (abstract), Proc. North East Weed Contr. Conf. *19*, 366.

Ahrens, J. F. (1965b). Detoxification of simazine- and atrazine-treated soil with activated carbon (abstract), Proc. North East Weed Contr. Conf. *19*, 364.

Ahrens, J. F. (1967). Improving herbicide selectivity in transplanted crops with root dips of activated carbon, Proc. North East Weed Contr. Conf. *21*, 64.

Almquist, H. J. and Zander, D. (1940). Adsorbing charcoals in chick diets, Proc. Soc. Exp. Biol. Med. *45*, 303.

Alwen, J. and Smith, D. G. (1967). A medium to suppress the swarming of *Proteus* species, J. Appl. Bact. *30*, 389.

Anagnostakis, S. L. (1974). Haploid plants from anthers of tobacco—enhancement with charcoal, Planta *115*, 281.

Andersen, A. H. (1968). The inactivation of simazine and linuron in soil by charcoal, Weed Res. *8*, 58.

Arle, H. F., Leonard, O. A., and Harris, V. C. (1948). Inactivation of 2,4-D on sweet-potato slips with activated carbon, Science *107*, 247.

Bancroft, G. and Fry, E. G. (1933). Adsorption and hydrolysis of glycogen, J. Biol. Chem. *100*, 255.

Barker, J. and Farrell, I. D. (1986). The effect of charcoal on MICs for *Legionella* (letter), J. Antimicrob. Chemother. *17*, 127.

Bastos, M. L. and Hoffman, D. B. (1976). Liquid-solid extraction techniques for the isolation of drugs of abuse from biological material, Acta. Pharm. Jugoslav. *26*, 87.

Bergel, F. and Bolz, K. (1933). The autoxidation of amino acid derivatives with animal charcoal and hemin. I., Z. Physiol. Chem. *215*, 25.

Bergman, F., Halvorsen, P., and van der Linden, W. (1967). Increased excretion of thyroxine by feeding activated charcoal to Syrian hamsters, Acta Endocrin. (Kobenhaven), *56*, 521.

Binoux, M. A. and Odell, W. D. (1973). Use of dextran-coated charcoal to separate antibody-bound from free hormone: A critique, J. Clin. Endocrinol. Metab. *36*, 303.

Bleyer, L. (1922). The adsorption of bacteria and agglutinins by suspensions and colloids, Z. Immunitats. Abt. I, Orig. *33*, 478.

Boquest, A. L. and Tosolini, F. A. (1993). Isolation of *Nocardia asteroides* on buffered charcoal-yeast extract agar (letter), J. Clin. Microbiol. *31*, 1400.

Boquet, A. (1928). Adsorption of cobra venom and diphtheria toxin by carbon, Compt. Rend. *187*, 959.

Bovey, R. W. and Miller, F. R. (1969). Effect of activated carbon on the phytotoxicity of herbicides in a tropical soil, Weed Sci. *17*, 189.

Boxen, I. and Tevaarwerk, G. J. M. (1982). Charcoal as a phase separating agent in ligand assays: Mechanism of action and the effect of dextran and various proteins on the adsorption of small molecules, J. Immunoassay *3*, 53.

Braafladt, L. H. (1923). The effect of kaolin on the intestinal flora in normal and pathologic conditions, J. Infect. Dis. *33*, 434.

Brenchley, R. G. (1968). Charcoal, a means of protecting crops in Oregon, Proc. West. Soc. Weed Sci. *22*, 10.

Burr, R. J., Lee, W. O., and Appleby, A. P. (1972). Factors affecting use of activated carbon to improve herbicide selectivity, Weed Sci. *20*, 180.

Carter, H. E., Clark, R. K., Jr., Dickman, S. R., Loo, Y. H., Skell, P. S., and Strong, W. A. (1945). Isolation and purification of streptomycin, J. Biol. Chem. *160*, 337.

Chamuleau, R. A. F. M., Dupont, A., Brugman, A. M., and De Koning, H. W. M. (1981). Activated charcoal and ammonium production (letter), Lancet, September 19, p. 633.

Chandler, J. M., Wooten, O. B., and Fulgham, F. E. (1978). Influence of placement of charcoal on protection of cotton (*Gossypium hirsutum*) from diuron, Weed Sci. *26*, 239.

Cheldelin, V. H. and Williams, R. J. (1942). Adsorption of organic compounds. I. Adsorption of ampholytes on an activated charcoal, J. Am. Chem. Soc. *64*, 1513.

Coffey, D. L. and Warren, G. F. (1969). Inactivation of herbicides by activated carbon and other adsorbents, Weed Sci. *17*, 16.

Constantin, M. J., Henke, R. R., and Mansur, M. A. (1977). Effect of activated charcoal on callus growth and shoot organogenesis in tobacco, In Vitro *13*, 293.

Cook, E. S. and Walter, E. M. (1939). Charcoal as an adsorbent for respiratory factors, Studies Inst. Divi Thomae *2*, 189.

Cook, E. S., Walter, E. M., Waring, M. G., Eilert, M. R., and Rack, F. J. (1941). Effect of pH on adsorption by charcoal of factors increasing the respiration of yeast and liver, Studies Inst. Divi Thomae *3*, 139.

Cooney, D. O. and Z. Xi (1994). Activated carbon catalyzes reactions of phenolics during liquid phase adsorption, Am. Inst. Chem. Eng. J. *40*, 361.

Cordier, G. (1939). Adsorption of the virus of foot and mouth disease by bone carbon and tricalcium phosphate. Applications to the immunization of the cavy, Rec. Med. Vet. *115*, 599.

Cordier, G. (1940). Symbiosis "in vivo" of the virus of sheep pox and of the virus of foot and mouth disease. Simultaneous adsorption of the two viruses by animal carbon, Rec. Med. Vet. *116*, 254.

Daly, J. S. and Cooney, D. O. (1978). Omission of pepsin from simulated gastric fluid in evaluating activated charcoals as antidotes, J. Pharm. Sci. *67*, 1181.

Damrau, F. and Goldberg, A. H. (1971). Adsorption of whisky congeners by activated charcoal, Southwest. Med. *52*, 179.

De, N. K. (1937). The adsorption of vitamin A and carotene and the isolation of vitamin A from associated pigments, Indian J. Med. Res. *25*, 17.

De Duve, C. and De Somer, P. (1947). [Treatment of penicillin solutions with adsorbent charcoals], Bull. Soc. Chim. Biol. *29*, 367.

De Hertogh, R., Van Der Heyden, I., and Ekka, E. (1975). Unbound ligand adsorption on dextran-coated charcoal. Practical considerations, J. Steroid Biochem. *6*, 1333.

Dingemanse, E. and Laqueur, E. (1927). The purification of insulin, Arch. Neerland. Physiol. *12*, 259.

Ditter, B., Urbaschek, R., and Urbaschek, B. (1983). Ability of various adsorbents to bind endotoxins in vitro and to prevent orally induced endotoxemia in mice, Gastroenterol. *84*, 1547.

Druart, P. and De Wulf, O. (1993). Activated charcoal catalyses sucrose hydrolysis during autoclaving, Plant Cell, Tissue and Organ Culture *32*, 97.

Drucker, M. M., Goldhar, J., Ogra, P. L., and Neter, E. (1977). The effect of attapulgite and charcoal on enterotoxicity of *Vibrio cholera* and *Escherichia coli* enterotoxins in rabbits, Infection *5*, 211.

Du Xiang-nan, Niu Zhen, Zhou Guo-zheng, and Li Zong-ming (1987). Effect of activated charcoal on endotoxin adsorption. Part. I. An in vitro study, Biomater. Art. Cells Art. Organs *15*, 229.

Ebata, T., Kobayashi, K., Denno, R., Totsuka, M., and Hayasaka, H. (1980). Removal of endotoxin by activated charcoal in the endotoxin shock, Igaku no Ayumi *113*, 173.

Eisler, M. V. (1926). Further experiments on the influence of adsorption by charcoal on poisoning and detoxification, Biochem. Z. *172*, 154.

Elahi, N. (1979). Encapsulated charcoal extraction technique for a rapid quantitation of barbiturates in whole blood, J. Anal. Toxicol. *3*, 35.

El-Mishad, A. M., McCormick, K. J., McCormick, N. K., and Trentin, J. J. (1975). Potentiation of hamster tumors by normal cells or charcoal, Cancer Res. *35*, 2098.

Emery, E. S., Jr. (1937). The use of adsorbents in gastro-intestinal diseases, JAMA *108*, 203.

Ernst, R. (1974). The use of activated charcoal in a symbiotic seedling culture of *Paphiopedilum*, Am. Orchid Soc. Bull. *43*, 35.

Falk, E. and Sticker, A. (1910). Carbenzyme, Munch. Med. Wochenschr. *57*, 4.

Fantus, B. and Dyniewicz, J. M. (1936). Phenolphthalein studies: III. Phenolphthalein and activated carbon, Am. J. Dig. Dis. Nutr. *3*, 337.

Feeley, J. C., et al. (1979). Charcoal-yeast extract agar: Primary isolation medium for *Legionella pneumophila*, J. Clin. Microbiol. *10*, 437.

Fox, F. W. and Levy, L. F. (1936). Reversible oxidation of ascorbic acid by Norit, Biochem. J. *30*, 208.

Freeman, M., Gulland, J. M., and Randall, S. S. (1935). The oxytocic hormone of the posterior lobe of the pituitary gland: VII. Adsorption and electrodialysis, Biochem. J. *29*, 2211.

Fridborg, G. and Eriksson, T. (1975). Effects of activated charcoal on growth and morphogenesis in cell cultures, Physiol. Plant *34*, 306.

Fridborg, G., Pedersen, M., Landström, L., and Eriksson, T. (1978). The effect of activated charcoal on tissue culture: Absorption of metabolites inhibiting morphogenesis, Physiol. Plant *43*, 104.

Fürth, O. and Kaunitz, H. (1929). The oxidation of some physiological substances through animal charcoal, Monatsh. *53-54*, 127.

Fürth, O. and Kaunitz, H. (1930). Oxidation of some constituents of the body by activated charcoal, Bull. Soc. Chim. Biol. *12*, 411.

Galvez, G. E. (1964). Loss of virus by filtration through charcoal, Virology *23*, 307.

Gompel, M. (1929). Action of anaesthetics upon the oxidation of oxalic acid in the presence of charcoal, Ann. Physiol. Physiochem. Biol *5*, 761.

Gradwohl, M. (1930). Influence of reaction on the oxidation of amino acids by charcoal, Biochem. Z. *219*, 136.

Guha, B. C. and Chakravorty, P. N. (1934). Chemical behavior of vitamin B_2, J. Indian Chem. Soc. *11*, 295.

Gunnison, J. P. and Marshall, M. S. (1937). Adsorption of bacteria by inert particulate reagents, J. Bacteriol. *33*, 401.

Harstick, K., Holloway, C. J., Brunner, G., Kuelpmann, W. R., and Petry, K. (1979). The removal of hypnotic drugs from human serum. A comparative investigation of the adsorptive properties of native and agarose-encapsulated resins and charcoal, Int. J. Artif. Organs 2, 87.

Haynes, W. C. and Rhodes, L. J. (1966). Spore formation by *Bacillus popilliae* in liquid medium containing activated carbon, J. Bacteriol. *91*, 2270.

Hedin, S. G. (1906). An antitryptic effect of charcoal and a comparison between the action of charcoal and that of the tryptic antibody in the serum, Biochem. J. *1*, 484.

Heimer, G. M. and Englund, D. E. (1986). Enterohepatic recirculation of oestriol: Inhibition by activated charcoal, Acta Endocrinol. *113*, 93.

Heberle-Bors, E. (1980). Interaction of activated charcoal and iron chelate in anther culture of *Nicotiana* and *Atropa belladona*, Z. Pflanzenphysiol. *99-S*, 339.

Herbert, V., Lau, K. S., Gottlieb, C. W., and Bleicher, S. J. (1965). Coated charcoal immuno-assay of insulin, J. Clin. Endocrinol. *25*, 1375.

Herbert, V., Gottlieb, C. W., and Lau, K. S. (1966). Hemoglobin coated charcoal assay for serum vitamin B_{12}, Blood *28*, 130.

Herbert, V., Lau, K. S., and Gottlieb, C. W. (1968). Coated charcoal assay of vitamins, minerals, hormones, and their binders, Advan. Tracer Methodol. *4*, 273.

Hindmarsh, K. W., Hamon, N. W., and LeGatt, D. F. (1975). Use of a charcoal cartridge in isolating basic drugs from urine, Clin. Chem. *21*, 1852, 1854.

Hindmarsh, K. W. and Hamon, N. W. (1977). The charcoal cartridge its use in the isolation of drugs from biological fluids, J. Can. Soc. Forensic Sci. *10*, 1.

Hoffman, P. S., Pine, L., and Bell, S. (1983). Production of superoxide and hydrogen peroxide in medium used to culture *Legionella pneumophila*: Catalytic decomposition by charcoal, Appl. Environ. Microbiol. *45*, 784.

Horner, M., McComb, J. A., McComb, A. J., and Street, H. E. (1977). Ethylene production and plantlet formation by *Nicotiana* anthers cultured in the presence and absence of charcoal, J. Exp. Bot. *28*, 1365.

Houssay, M. A. (1921). Adsorption of snake venom by charcoal, Rev. Inst. Bacteriol. Dep. Nac. Hig. (Buenos Aires) 2, 197.

Husa, W. J. and Littlejohn, O. M. (1952). Some pharmaceutical uses of activated charcoal, J. Am. Pharm. Assoc (Pract. Pharm. Ed) *13*, 412.

Ito, T. (1936). The surface activity and the adsorption of amino acids, J. Agr. Chem. Soc. Japan *12*, 204.

Johansson, L., Andersson, B., and Eriksson, T. (1982). Improvement of anther culture technique: Activated charcoal bound in agar medium in combination with liquid medium and elevated CO_2 concentration, Physiol. Plant *54*, 24.

Jordan, P. D. and Smith, L. W. (1971). Adsorption and deactivation of atrazine and diuron by charcoals, Weed Sci. *19*, 541.

Kawano, Y. (1949). Adsorption of penicillin with activated carbon, J. Ferm. Technol. (Japan) *27*, 177.

Keefer, N. D. (1920). Pepsin adsorption by charcoal, Am. J. Pharm. *92*, 160.

Kennedy, J. M. and Talbert, R. E. (1973). Activated charcoal as a herbicide antidote, Arkansas Farm Res. 22(No. 1), 12.

Kikawa, K. (1926). Adsorption of pepsin, Biochem. J. (Japan) *6*, 275.

Klein, B. and Bopp, M. (1971). Effect of activated charcoal in agar on the culture of lower plants, Nature *230*, 474.

Kohlenbach, H. W. and Wernicke, W. (1978). Investigations on the inhibitory effect of agar and the function of active carbon in anther culture, Z. Pflanzenphysiol. *86-S*, 463.

Kratky, B. A. and Warren, G. F. (1971). Activated carbon-vermiculite mixture for increasing herbicide selectivity, Weed Sci. *19*, 79.

Kraus, R. and Barbara, B. (1915). Adsorption of toxins by animal charcoal, Deutsch. Med. Wochenschr. *41*, 393.

Lasseur, P., Dombray, P., and Palgen, W. (1934). [Adsorption of bacteria by strips of filter paper, paper pulp, and activated carbon], Trav. Lab. Microbiol. Faculté Pharm. Nancy 7, 117.

Lee, W. O. (1973). Clean grass seed crops established with activated carbon bands and herbicides, Weed Sci. *21*, 537.

Lee, W. O. (1978). Volunteer Kentucky bluegrass (*Poa pratensis*) control in Kentucky bluegrass seed fields, Weed Sci. *26*, 675.

Lindquist, D. L. and De Alarcon, P. A. (1987). Charcoal-dextran treatment of fetal bovine serum removes an inhibitor of human CFU-Megakaryocytes, Exp. Hematol. *15*, 234.

Linscott, D. L. and Hagin, R. D. (1967). Protecting alfalfa seedlings from a triazine with activated charcoal, Weeds *15*, 304.

Loch, P. (1938). The adsorption of thiocyanates, Vorratspflege u. Legensmittelforsch *1*, 469.

Lucas, E. H. and Hamner, C. L. (1947). Inactivation of 2,4-D by adsorption on charcoal, Science *105*, 340.

Maeda, K., et al. (1979). Problems with activated charcoal and alumina as sorbents for medical use, Artif. Organs *3*, 336.

Majek, B. A. (1986). Activated carbon for safening peppers (*Capsicum annuum* L.) in soils treated with metribuzin, Weed Sci. *34*, 467.

Majumder, S. K., Narasimhan, K. S., and Subrahmanyan, V. (1959). Insecticidal effects of activated charcoal and clays, Nature *184*, 1165.

Matet, A. and Matet, J. (1945). [Diets containing active charcoal for rapid production of avitaminosis A. Preparation of vitamin A-free diets], Bull. Soc. Chim. Biol. *27*, 513.

McRobert, G. R. (1934). The treatment of bacterial food poisoning, Br. Med. J. *2*, August 18, p. 304.

Meola, J. M. and Vanko, M. (1974). Use of charcoal to concentrate drugs from urine before drug analysis, Clin. Chem. *20*, 184.

Michaelis, L. and Rona, P. (1910). Influencing adsorption by reaction of the medium, Biochem. Z. *25*, 359.

Miller, E. J. and Bandemer, S. L. (1930). Adsorption from solution by ash-free adsorbent charcoal. VI. Adsorption of invertase, J. Phys. Chem. *34*, 2666.

Miller, O. N. (1957). Determination of bound vitamin B_{12}, Arch. Biochem. Biophys. *68*, 255.

Misson, J. P., Boxus, P., Coumans, M., Giot-Wirgot, P., and Gaspar, T. (1983). Rôle du charbon de bois dans les milieux de culture de tissus végétaux, Med. Fac. Landbouw, Rijksuniversitat, Gent, Belgie, *48* (4), 7.

Moloney, P. J. and Findlay, D. M. (1924). The purification of insulin and similar substances by sorption on charcoal and subsequent recovery, J. Phys. Chem. *28*, 402.

Mortensen, E. (1974). Separation of free and protein-bound ligand molecules by means of protein-coated charcoal, Clin. Chem. *20*, 1146.

Murate, T., Hotta, T., Goto, S., Ichikawa, A., Inoue, C., and Kaneda, T. (1988). Activated charcoal diminishes the lot difference of fetal bovine sera in erythroid colony formation of human bone marrow cells (42781), Proc. Soc. Exp. Biol. Med. *189*, 66.

Nangju, D., Plucknett, D. L., and Obien, S. R. (1976). Some factors affecting herbicide selectivity in upland rice, Weed Sci. *24*, 63.

Ogg, A. G., Jr. (1978). Herbicides and activated carbon for weed control in direct-seeded asparagus (*Asparagus officinalis*), Weed Sci. *26*, 284.

Ogg, A. G., Jr. (1982). Effect of activated carbon on phytotoxicity of terbacil to several crops, Weed Sci. *30*, 683.

Oksent'yan, U. G. (1940). [Activity of lactic acid bacteria in connection with adsorption], Microbiology (USSR) *9*, 3.

Padilla, S. L., Howe, A. M., and Boldt, J. P. (1988). Effects of charcoal-extracted serum as a growth medium supplement on in vitro development of mouse embryos, J. In Vitro Fert. Embryo Trans. *5*, 286.

Piper, D. W. and Fenton, B. (1961). The adsorption of pepsin, Am. J. Dig. Dis. *6*, 134.

Polak, S. (1974). Quick test for HBAg using charcoal, Lancet, March 9, p. 406.

Poppe, K. and Busch, G. (1930). [Physical and chemical studies of the virus of foot and mouth disease. I. Isoelectric point and adsorption], Z. Immunitats. *68*, 510.

Proskauer, J. and Berman, R. (1970). Agar culture medium modified to approximate soil conditions, Nature *227*, 1161.

Przylecki, S. J., Niedzwiedzka, H., and Majewski, T. (1927). Enzymic reactions in a medium macroscopically heterogeneous, Compt. Rend. Soc. Biol. *97*, 937.

Pyl, G. (1931). [Adsorption experiments with the virus of foot and mouth disease in buffer solutions], Zentr. Bakt. Parasitenk. Infek. *121*, 10.

Raad, I., Rand, K., and Gaskins, D. (1990). Buffered charcoal-yeast extract medium for the isolation of *Brucellae*, J. Clin. Microbiol. *28*, 1671.

Roberts, J. L. and Baldwin, I. L. (1942). Spore formation by *Bacillus subtilis* in peptone solutions altered by treatment with activated charcoal, J. Bact. *44*, 653.

Robinson, D. W. (1965). The use of adsorbents and simazine on newly planted strawberries, Weed Res. *5*, 43.

Roffo, A. H. and Barbara, B. (1925). [The adsorption of hemolysins], Bol. Inst. Med. Exptl. Estud. Cancer *1*, 280.

Rolston, M.P., Lee, W. O., and Appleby, A. P. (1979a). Volunteer legume control in legume seed crops with carbon bands and herbicides. I. White clover, Agron. J. *71*, 665.

Rolston, M. P., Lee, W. O., and Appleby, A. P. (1979b). Volunteer legume control in legume seed crops with carbon bands and herbicides. II. Red clover and alfalfa, Agron. J. *71*, 671.

Rydrych, D. J. (1985). Inactivation of metribuzin in winter wheat by activated carbon, Weed Sci. *33*, 229.

Sabalitschka, T., and Weidlich, R. (1929). [Malt amylase: VII. Adsorption of amylase from malt extract on kaolin and its elution], Biochem. Z. *211*, 229.

Salus, G. (1916). [Blood charcoal as a disinfectant of small quantities of water], Wein. Klin. Wochenschr. *29*, 846.

Sands, D. C., McIntyre, J. L., and Walton, G. S. (1976). Use of activated charcoal for the removal of patulin from cider, Appl. Environ. Microbiol. *32*, 388.

Sato, G. (1928). Adsorption of the active principle of the posterior lobe of the hypophysis by animal charcoal, Arch. Exptl. Path. Pharmakol. *130*, 323.

Schultz, H. W. and Mattill, H. A. (1937). The vitamin B complex, J. Biol. Chem. *122*, 183.

Schuster, I., Helm, I., and Fleschurz, C. (1977). The effect of charcoal treatment on microsomal cytochrome P-450, FEBS Letters *74*, 107.

Seibert, F. B. (1935). The chemical composition of the active principle of tuberculin: XIX. Difference in the antigenic properties of various tuberculin fractions; adsorption to aluminum hydroxide and charcoal, J. Immunol. *28*, 425.

Silcox, H. (1946). Production of streptomycin, Chem. Eng. News *24*, 2762.

Signorelli, S. (1933). [Adsorption power of carbon on complex toxins and tuberculosis of miners], Bull. Soc. Ital. Biol. Sper. *8*, 116.

Sladek, J. and Kyer, J. (1938). Preparation and properties of the charcoal adsorbate antianemic principle of liver extract, Proc. Soc. Exptl. Biol. Med. *39*, 227.

Smith, D. G. and Alwen, J. (1966). Effect of activated charcoal on the swarming of *Proteus*, Nature *212*, 941.

Stamatin, N. (1937). Adsorption of the sheep-pox virus on kaolin and animal charcoal, Compt. Rend. Soc. Biol. *124*, 984.

Stoll, B. J., et al. (1980). Binding of intraluminal toxin in cholera: Trial of GM1 ganglioside charcoal, Lancet, October 25, p. 888.

Takayama, S. and Misawa, M. (1980). Differentiation in *Lilium* bulb scales grown in vitro. Effects of activated charcoal, physiological age of bulbs, and sucrose concentration on differentiation and scale leaf formation in vitro, Physiol. Plant *48*, 121.

Tijssen, J. A. (1980). A haemoperfusion column based on coated activated carbon, Ph. D. dissertation, Twente University.

Torre, D. (1988). Influence of charcoal on ciprofloxacin activity (letter), Rev. Infect. Dis. *10*, 1231.

Torre, D., Sampietro, C., Quadrelli, C., Bianchi, W., Maggioli, F., and Ohnmeiss, H. (1988). In vitro influence of charcoal on ciprofloxacin activity, Drugs Exp. Clin. Res. *XIV*, 333.

Toth, J. and Milham, P. J. (1975). Activated-carbon and ash-carbon effects on the adsorption and phytotoxicity of diuron, Weed Res. *15*, 171.

Ui, S., et al. (1990). Cooperative effect of activated charcoal and Gellan gum on grape protoplast culture, Agric. Biol. Chem. *54*, 207.

Unna, Z. (1926). Diastase adsorption, Biochem. Z. *172*, 392.

Vantis, J. T. and Bond, G. (1950). The effect of charcoal on the growth of leguminous plants in sand culture, Ann. Appl. Biol. *37*, 159.

Vickers, R. M., Rihs, J. D., and Yu, V. L. (1992). Clinical demonstration of isolation of *Nocardia asteroides* on buffered charcoal-yeast extract media, J. Clin. Microbiol. *30*, 227.

Wang, P. J. and Huang, L. C. (1976). Beneficial effects of activated charcoal on plant tissue and organ cultures, In Vitro *12*, 260.

Weatherhead, M. A., Burdon, L., and Henshaw, G. G. (1978). Some effects of activated charcoal as an additive to plant tissue culture media, Z. Pflanzenphysiol. *89*, 141.

Wiechowski, W. (1914). [Pharmacological basis for a therapeutic use of charcoal], Z. Kinderheilk. *8*, 285.

Wunderly, K. (1933a). Behavior of several urea derivatives, amino acids, and peptides with animal black, Helv. Chim. Acta *16*, 1009.

Wunderly, K. (1933b). Arresting charcoal contact aminolysis, Helv. Chim. Acta *16*, 515.

Zondek, H. and Bansi, H. W. (1928). Hormones and adsorption, Biochem. Z. *195*, 376.

Zunz, E. (1911). Role of surface tension in the adsorption of toxins and antitoxins, Bull. Acad. Roy. Med. Belg. *1911*, Oct. 29.

25

Summary

We have come a long way in the course of this volume and have considered a wide range of topics and issues. To summarize in any simple fashion all that has been covered is difficult but several things tend to stand out more than others. In this summary, we consider those things which are remarkable about activated charcoal in a positive sense and then we discuss the few negative aspects of activated charcoal. Next, we identify some of the issues related to antidotal charcoal which need further resolution. Finally, we make some recommendations.

I. POSITIVE ASPECTS OF ACTIVATED CHARCOAL

We have seen that the recognition of the remarkable adsorptive powers of charcoals (which, since about 1900, have been increased through activation) goes back more than three millennia. Research over the past several decades, especially over the past 20 years or so, has shown repeatedly and conclusively that just a single charcoal dose of reasonable size can reduce the absorption of a wide variety of drugs and poisons tremendously. It is not uncommon for a charcoal dose of 30–50 g, administered without significant delay, to lower drug/poison absorption by 90% or more.

Moreover, for drugs which delay gastric emptying and/or reduce gastric motility, charcoal administered even after several hours has been shown to be surprisingly effective. Whether single doses of charcoal are sufficient or whether multiple doses are needed for effective inhibition of drug absorption has been seen to depend strongly on the drug involved, how quickly it is absorbed and/or eliminated, and particularly whether it undergoes significant enterohepatic and/or enteroenteric cycling. While resins and some clays have also been shown to be of benefit in reducing the absorption of various drugs and poisons, they are generally of benefit only for specific types of substances. Thus, charcoal is to be preferred, especially if the identity of the toxic substance is unknown or if multiple drugs are involved.

Studies have shown that lavage and emesis carried out in a medical facility (as opposed to the home) are time-consuming, uncomfortable for the patient, and generally produce relatively small recoveries of drugs. Thus, charcoal alone, along with appropriate supportive therapies, appears to be the treatment of choice. Charcoal therapy is inexpensive, easy to carry out, and is reasonably well tolerated by patients. Charcoal itself is nontoxic and indeed several studies in which patients have been given doses of 30–50 per day for several months have shown no adverse effects.

Presently, charcoals with quite high surface areas (e.g., the 1400 and 2000 m^2/g Norit B Supra and Norit A Supra charcoals) are available, as are ready-mixed formulations of several types of charcoals (with and without sorbitol). Average wholesale prices for ready-mixed formulations reported in 1993 by McFarland and Chyka (1993) are in the range of $5.40–$7.65 per unit for 25–50 g formulations and are about $1.50–$3.45 per unit for pediatric size units (12.5–15 g charcoal). Compared to other costs of treating an overdose/poisoning victim in a medical facility, the cost of the ready-mixed charcoal units themselves is obviously trivial. Thus, expense is not an issue with charcoal and indeed the use of charcoal offers a tremendous cost advantage over treatments such as lavage, hemoperfusion, and hemodialysis. It is also much simpler and faster.

Charcoal has been shown to be effective in treating various conditions caused by excessive amounts of endogenous chemicals, (e.g., pruritus and erythropoietic porphyria). Charcoal incorporated into wound dressings has shown remarkable healing and antiseptic properties.

II. SOME NEGATIVE ASPECTS OF ACTIVATED CHARCOAL

Charcoal is messy and can be unpleasant to work with if the patient resists taking it or vomits it after administration. However, the reluctance of medical personnel to employ charcoal on such grounds alone is indefensible and not valid. Yet, this reluctance should at least be acknowledged.

Charcoal is not generally effective in reducing the absorption of inorganic substances or simple organic substances such as alcohols (methanol, ethanol). Indeed, any simple organic substance which is highly hydrophilic (i.e., readily miscible with water in all proportions) tends to adsorb poorly to charcoal. Substances which have many hydroxyl groups are usually hydrophilic (the more hydroxyl groups, the more hydrophilic) and thus prefer to stay in the aqueous phase rather than attach to the charcoal surface. This property can be taken advantage of, however, in the case where one wishes to flavor a charcoal suspension with a substance like sucrose. In this case, sucrose, having many hydroxyl groups, adsorbs poorly to the charcoal and thus causes little interference with drug adsorption by the charcoal.

One may regard all of the hazards discussed in Chapter 17 as negative features of charcoal use. However, the various hazards (perforation, aspiration, obstructions, dehydration, hypernatremia, hypermagnesemia) may be readily avoided, as explained in the recommendations at the end of Chapter 17. There is no reason why a patient who is properly intubated, protected by a cuffed endotracheal tube, and given reasonable amounts of charcoal with minimal cathartic, should experience any adverse effects. Charcoal per se is not to be blamed for the hazards which have been encountered in its use.

III. UNSETTLED ISSUES

In the course of this volume, several issues have arisen which require further resolution. One concerns the use of osmotic cathartics (e.g., sorbitol) or saline cathartics (magnesium and sodium sulfates, magnesium citrate). Because in vivo studies have shown no consistent benefits from combining cathartics with charcoal in terms of patient clinical outcome, and because cathartics themselves can, in excessive amounts, cause dehydration and/or electrolyte imbalances, the use of cathartics along with charcoal needs further evaluation. It seems apparent that, if cathartics are to be used, they should be used in small amounts and not with every dose of charcoal.

Another topic area which needs further exploration is that of formulations containing charcoal, a lubricating agent, and flavors. While much research in this subject area has been done, the results appear to have been largely ignored. Many studies have shown that appropriate amounts of carboxymethylcellulose or bentonite make charcoal suspensions much more easily swallowed, without compromising the charcoal's adsorption capacity. Yet, no person or medical facility known to this writer uses such lubricating agents and many, many studies have shown that sucrose, fructose, chocolate syrup, and other flavoring agents make charcoal suspensions much more agreeable, again without producing much loss in adsorptive power. Ice cream, sherbet, and milk do cause some loss of adsorption capacity when mixed with charcoal but make the charcoal so much more acceptable to pediatric patients that the extra amount of charcoal taken in this form more than counteracts the loss of adsorption capacity. Again, the use of these vehicles appears to be negligible if nonexistent.

Of course, one can argue that formulations containing sorbitol are flavored but such formulations contain sorbitol more for its role as a cathartic than as a flavor. Notwithstanding this, sorbitol is an acceptable flavor, in that: (1) it is sweet, (2) it does not adsorb significantly to charcoal itself, and (3) it therefore does not significantly reduce the adsorptive capacity of the charcoal. So, charcoal formulations with sorbitol may be logical as flavored formulations but present-day sorbitol/charcoal suspensions have far too much sorbitol for just flavoring purposes, and are sickeningly sweet. For simple flavoring purposes, the amount of sorbitol needs to be reduced (but this may cause potential problems with bacterial growth).

Indeed, one problem with charcoal in general and with formulations flavored with sugars and similar agents, is the potential for bacterial contamination. As George et al. (1991) showed (see the discussion in Chapter 17), even dry powder USP grade charcoals have been found to be bacterially contaminated. More research and development needs to be done on the bacterial contamination problem, both with dry charcoals and with ready-mixed flavored charcoals.

The effects of charcoal on general intestinal problems have been inadequately studied. This is extremely surprising considering the popularity of charcoal formulations in some parts of the world (especially Europe) for treating intestinal maladies. Indeed, in the course of writing this volume, only 5 studies (with an average publication date of 1977) dealing with the general effects of charcoal on gastrointestinal maladies were found. With respect to the effects of charcoal on intestinal gas after gas-producing meals, only 8 studies were found and these have produced conflicting results. Only one study, now more than 30 years old, was found concerning the effect of charcoal on peptic ulcers. Thus, much more research needs to be done related to the effects of charcoal on various gastrointestinal problems.

An astonishing point which was brought up in Chapter 13 is that, of the hundreds of in vivo studies dealing with the effects of oral charcoal on drug absorption, only four have attempted to determine the effects of food. Considering that many drug overdose victims have substantial food in their stomachs and upper intestines, it is almost unbelievable that so few studies have tried to evaluate the effects of food on the efficacy of charcoal. Further, only one study attempted to modify gastric pH to see what effects would occur and that study used insufficient amounts of antacids. Thus, the research topic of modifying gastric pH remains open.

IV. RECOMMENDATIONS

Several recommendations were stated explicitly or implicitly in the last section. These included a call for further studies on the effect of food on the efficacy of charcoal, on the effects of charcoal on various intestinal problems (gas, ulcers), and more work on the effects of combining different cathartics with charcoal, with the aim being to determine which cathartic (perhaps none or perhaps more than one kind) to give, when to give it, and how much to give. In the interim, a reduction in the sometimes over-zealous use of cathartics is recommended.

We also mentioned that further work on, and acceptance of, formulations containing a lubricating agent and a flavoring agent is needed. Of particular concern here, as was also mentioned, will be potential bacterial contamination problems.

However, by far the most important recommendation by this writer is that charcoal suspensions of some sort be made readily available to the general public at local pharmacies and that the general public be educated in the administration of such suspensions in the home. There is no excuse for the almost complete absence of charcoal formulations in the home environment, particularly in the USA. What are the barriers? One, certainly, is that there is little or no profit in a private company doing enough advertising to sell formulations to the general public. For people to respond strongly enough to such advertising to actually go to their local pharmacy and buy a bottle of a charcoal suspension would require fairly extensive and persistent advertising (the advertising would also have to educate the public on how to use the charcoal, namely, it should emphasize that the formulation be shaken very well before use). Such advertising would be very expensive. Suppose such advertising were successful and that many people did in fact buy a bottle of a charcoal suspension. Most, thankfully, would never need to use it. Thus, after an initial groundswell of sales, additional sales would be quite small and would make the continuation of further advertising uneconomic. One could, of course, put an expiration date on the product and thereby periodically induce people to throw out their old bottle and buy a new one. But, this would be unethical, as we know that bottled charcoal suspensions, if unopened, remain viable essentially indefinitely (it might be mentioned in this regard that syrup of ipecac bottles carry expiration dates but a large number of studies have shown that "expired" ipecac is every bit as effective as newly-stocked bottles).

Another side to the economics (and politics) of encouraging the general public to stock charcoal formulations at home is that it would take at least the initial administration of charcoal out of the hands of medical personnel. Many medical personnel still feel that patients should not treat themselves in any significant way but should simply come to them as soon as possible. Then, there is the economic loss to the medical personnel and medical facilities which would occur if one were to substitute home treatment for treatment in a medical facility.

The answer then is that federal governments should step in and bear enough of the costs (e.g., of educating the general public through advertising) to make it possible for private companies to market their products to community pharmacies, to encourage such pharmacies to stock them, and to induce the general public to buy and correctly use them. Unless some profit motive exists for manufacturers and local pharmacies, the use of charcoal in the home will never come about.

V. REFERENCES

George, D. L., McLeod, R., and Weinstein, R. A. (1991). Contaminated commercial charcoal as a source of fungi in the respiratory tract, Infect. Contr. Hosp. Epidemiol. *12*, 732.

McFarland, A. K., III and Chyka, P. A. (1993). Selection of activated charcoal products for the treatment of poisoning, DICP—Ann. Pharmacother. *27*, 358.

INDEX

Milton Keynes UK
Ingram Content Group UK Ltd.
UKHW052028071024
449327UK00027B/2480